Handbook of
Experimental Pharmacology

Volume 152

Springer-Verlag Berlin Heidelberg GmbH

Endothelin and Its Inhibitors

Contributors

B. Battistini, N. Berthiaume, G. Bkaily, R. Corder,
G. Cournoyer, P.K. Cusick, A.P. Davenport, P. D'Orléans-Juste,
S.A. Douglas, M. Duval, J.D. Elliott, R.G. Goldie, P.J. Henry,
A.Y. Jeng, H. Kurihara, Y. Kurihara, J. Labonté, S.J. Leslie,
T. Masaki, J. McCulloch, S.J. Morgan, E.H. Ohlstein,
T.J. Opgenorth, M. Plante, D.M. Pollock, F.D. Russell,
Z. Shraga-Levine, M.S. Simonson, M. Sokolovsky, D.J. Stewart,
O. Touzani, B.A. Trela, D.J. Webb, J.R. Wu-Wong, J.-N. Xiang,
Y. Yazaki, Y.D. Zhao

Editor
Timothy D. Warner

 Springer

Timothy D. Warner, Ph.D
Department of Cardiac, Vascular
and Inflammation Research
The William Harvey Research Institute
St. Bartholomew's and
the Royal London School of Medicine and Dentistry
Charterhouse Square
London EC1M 6BQ
United Kingdom
e-mail: t.d. warner@mds.qmw.ac.uk

With 73 Figures and 17 Tables

ISBN 978-3-642-63237-2

Library of Congress Cataloging-in-Publication Data

Endothelin and its inhibitors / contributors, B. Battistini . . . [et al.]; editor: Timothy D. Warner.
    p. cm. – (Handbook of experimental pharmacology; v. 152)
    Includes bibliographical references and index.
    ISBN 978-3-642-63237-2   ISBN 978-3-642-56899-2 (eBook)
    DOI 10.1007/978-3-642-56899-2
    1. Endothelins. 2. Endothelins–Antagonists. I. Battistini, B. (Bruno) II. Warner, Timothy D.,
1963– III. Series.
    [DNLM: 1. Endothelins. 2. Receptors, Endothelin–antagonists & inhibitors. 3. Receptors,
Endothelin–physiology. QV 150 E567 2001]
    QP552.E54 E533 2001
    612′.015756 – dc21   00-067928

http://www.springer.de

© Springer-Verlag Berlin Heidelberg 2001
Originally published by Springer-Verlag Berlin Heidelberg in 2001
Softcover reprint of the hardcover 1st edition 2001

Cover design: design & production GmbH, Heidelberg
Typesetting: Best-set Typesetter Ltd., Hong Kong

SPIN: 10695556   27/3020xv-5 4 3 2 1 0 – printed on acid-free paper

Preface

Since the demonstration by Furchgott and Zawadzki in 1980 that the vascular endothelium formed a potent vasodilator (now known to be nitric oxide), tens of thousands of researchers have been drawn to the study of endothelial cells and thousands of papers have been published. Of all these publications, however, that by Yanagisawa and coworkers in 1988 reporting the discovery of endothelin must rank as one of the very best. It demonstrated the gene and peptide sequences, the synthetic pathway, and the biological effects of endothelin-1. Many researchers realized the significance of the findings and there was an explosion of endothelin research. Almost immediately it was demonstrated, again by Dr. Masashi Yanagisawa and colleagues in the research team of Prof. Tomoh Masaki, that there are in fact three members of the endothelin family. Others used the few pharmacological tools available to characterize endothelin receptors; but as always in this field, molecular biology quickly supplied definitive results with, in 1990, the isolation and cloning of two endothelin receptors. The pharmacological effort picked up speed as endothelin was implicated in diseases as varied as hypertension, asthma, inflammatory bowel disease, and stroke. The pharmaceutical industry produced first peptide and then non-peptide antagonists of the endothelin receptors, while efforts continued to understand the synthetic enzymes underlying endothelin production. Once again molecular biology supplied many interesting answers as well as provocative results: who would have imagined that endothelin-1 was so important for brachial arch development, or that endothelin-3 and ET_B receptors were involved in Hirschsprung's disease? Then, as we reached the end of the 1990s, endothelin antagonists and inhibitors became available for testing in disease states, and their efficacy was demonstrated, for instance, in congestive heart failure.

Now seems a good time to review this field of research. Molecular biological tools have supplied a wealth of information about the roles of the endothelins in development and in the healthy animal. Pharmacology has exhaustively studied endothelin receptors in every imaginable tissue. Endothelin expression and levels have been characterized in dozens of disease states. Potent, orally active drugs are available for clinical testing. A little over 10 years of research has produced the most dramatic developments.

What of the researchers involved in this field? Not all, but many, were young post-doctoral research fellows, junior academic staff, or low-ranking employees in pharmaceutical companies. One of the pleasures of research in this friendly field has been watching the progression of these individuals in parallel with our gaining of knowledge about endothelin. Many of those young researchers are now professors or directors of large academic or commercial research teams. However, the same friendliness remains and our regular international conferences on endothelin are always a great pleasure. The present volume collects reviews from representatives of this research community. They cover, systematically, every area of our knowledge of endothelins. Hopefully they will also convey at least some of the pleasure that we as scientists have derived from exploring this most interesting of peptides.

London, Great Britain   T.D. WARNER

List of Contributors

BATTISTINI, B., Centre de recherche, Hôpital Laval, Institut de cardiologie et de pneumologie, Department of Medicine, Laval University, 2725 Chemin Ste-Foy, Ste-Foy, QC G1V 4G5, Canada
e-mail: bruno.battistini@med.ulaval.ca

BERTHIAUME, N., Department of Pharmacology, Medical School, Université de Sherbrooke, Sherbrooke, QC J1H 5N4, Canada

BKAILY, G., Department of Anatomy and Cell Biology, Medical School, Université de Sherbrooke, Sherbrooke, QC J1H 5N4, Canada

CORDER, R., The William Harvey Research Institute, St. Bartholomew's and the Royal London School of Medicine and Dentistry, Queen Mary and Westfield College, Charterhouse Square Campus, London, EC1 M6BQ, UK
e-mail: R. Corder@mds.qmw.ac.uk

COURNOYER, G., Department of Pharmacology, Medical School, Université de Sherbrooke, Sherbrooke, QC J1H 5N4, Canada

CUSICK, P.K., Drug Safety Evaluation, Dept. 469, Bldg. AP-13A, 100 Abbott Park Road, Abbott Laboratories, Abbott Park, IL 60064-3500, USA

DAVENPORT, A.P., Clinical Pharmacology Unit, University of Cambridge, Addenbrooke's Hospital, Cambridge, CB2 2QQ, UK
e-mail: apd10@medschl.cam.ac.uk

D'ORLÉANS-JUSTE, P., Department of Pharmacology, Medical School, Université de Sherbrooke, Sherbrooke, QC J1H 5N4, Canada
e-mail: labpdj@courrier.usherb.ca

DOUGLAS, S.A., Department of Cardiovascular Pharmacology, GlaxoSmith Kline Pharmaceuticals, 709 Swedeland Road, King of Prussia, PA 19406-0939, USA

DUVAL, M., Department of Pharmacology, Medical School, Université de
Sherbrooke, Sherbrooke, QC J1H 5N4, Canada

ELLIOTT, J.D., Department of Medicinal Chemistry UW2430,
GlaxoSmith Kline Pharmaceuticals, 709 Swedeland Road,
PO Box 1539, King of Prussia, PA 19406-0939, USA
e-mail: John_D_Elliott@sbphrd.com

GOLDIE, R.G., Department of Pharmacology, University of Western
Australia, 35 Stirling Highway, Crawley, 6009, Western Australia, Australia
e-mail: rgoldie@receptor.pharm.uwa.edu.au

HENRY, P.J., Department of Pharmacology, University of Western Australia,
35 Stirling Highway, Crawley, 6009, Western Australia, Australia
e-mail: phenry@receptor.pharm.uwa.edu.au

JENG, A.Y., Metabolic and Cardiovascular Diseases, Novartis Institute for
Biomedical Research, Summit, NJ, 07901, USA

KURIHARA, H., Division of Integrative Cell Biology, Department of
Embryogenesis, Institute of Molecular Embryology and Genetics,
Kumamoto University, 2-2-1 Honjo, Kumamoto 860-0811, Japan
e-mail: kurihara@kaiju.medic.kumamoto-u.ac.jp

KURIHARA, Y., Department of Cardiovascular Medicine, Graduate
School of Medicine, University of Tokyo, 7-3-1 Hongo,
Bunkyo-ku, Tokyo 113, Japan
e-mail: yuki-tky@umin.ac.jp

LABONTÉ, J., Department of Pharmacology, Medical School, Université de
Sherbrooke, Sherbrooke, QC J1H 5N4, Canada

LESLIE, S.J., Clinical Pharmacology Unit & Research Centre, The University
of Edinburgh, Western General Hospital, Edinburgh EH4 2XU,
Scotland, UK
e-mail: s.j.leslie@ed.ac.uk

MASAKI, T., National Cardiovascular Center Research Institute,
5-7-1 Fujishirodai, Suita, Osaka 565-8565, Japan
e-mail: masaki@ri.ncvc.go.jp

McCULLOCH, J., Wellcome Surgical Institute & Hugh Fraser Neuroscience
Labs., University of Glasgow, Glasgow, UK

MORGAN, S.J., Drug Safety Evaluation, Dept. 469, Bldg. AP-13A, 100 Abbott
Park Road, Abbott Laboratories, Abbott Park, IL 60064-3500, USA
e-mail: sherry.j.morgan@abbott.com

OHLSTEIN, E.H., Department of Cardiovascular Pharmacology, GlaxoSmith Kline Pharmaceuticals, 709 Swedeland Road, King of Prussia, PA 19406-0939, USA
e-mail: eliot_h_ohlstein@sbphrd.com

OPGENORTH, T.J., Pharmaceutical Products Division, Abbott Laboratories, 100 Abbott Park Road, Abbott Park, IL 60064-3500

PLANTE, M., Department of Pharmacology, Medical School, Université de Sherbrooke, Sherbrooke, QC J1H 5N4, Canada

POLLOCK, D.M., Vascular Biology Center, Departments of Surgery, Physiology & Endocrinology, and Pharmacology & Toxicology, Medical College of Georgia, Augusta, GA 30912-2500, USA
e-mail: dpollock@mail.mcg.edu

RUSSELL, F.D., Clinical Pharmacology Unit, University of Cambridge, Addenbrooke's Hospital, Cambridge, CB2 2QQ, UK

SHRAGA-LEVINE, Z., Dept. of Neurobiochemistry, Tel Aviv University, Tel Aviv 69978, Israel

SIMONSON, M.S., Department of Medicine, Division of Nephrology, School of Medicine, Case Western Reserve University, Cleveland, Ohio 44106, USA
e-mail: mss5@po.cwru.edu

SOKOLOVSKY, M., Dept. of Neurobiochemistry, Tel Aviv University, Tel Aviv 69978, Israel
e-mail: sokol@post.tau.ac.il

STEWART, D.J., St. Michael's Hospital, 30 Bond Street, Room 7-081 Queen Wing, Toronto, ON M5B 1W8, Canada
e-mail: stewartd@smh.toronto.on.ca

TOUZANI, O., Centre Cyceron, CNRS UMR 6551, Boulevard H. Becquerel, BP 5229, 14074 Caen Cedex, France
e-mail: omar@cyceron.fr

TRELA, B.A., Drug Safety Evaluation, Dept. 468, Bldg. AP-13A, 100 Abbott Park Road, Abbott Laboratories, Abbott Park, IL 60064-3500, USA

WEBB, D.J., Clinical Pharmacology Unit & Research Centre, The University of Edinburgh, Western General Hospital, Edinburgh EH4 2XU, Scotland, UK

Wu-Wong, J.R., Pharmaceutical Products Division, Abbott Laboratories, 100 Abbott Park Road, Abbott Park, IL 60064, USA
e-mail: ruth.r.wuwong@abbott.com

Xiang, J.-N., Department of Medicinal Chemistry UW2430, GlaxoSmith Kline Pharmaceuticals, 709 Swedeland Road, PO Box 1539, King of Prussia, PA 19406-0939, USA

Yazaki, Y., International Medical Center of Japan, Shinjuku-ku Tokyo, Japan

Zhao, Y.D., St. Michael's Hospital, 30 Bond Street, Room 7-081 Queen Wing, Toronto, ON M5B 1W8, Canada
e-mail: yidan.zhao@utoronto.ca

Contents

CHAPTER 4

Endothelin Receptors

CHAPTER 5

Cell Signaling by Endothelin Peptides
M.S. SIMONSON. With 3 Figures 115

CHAPTER 6

Lessons from Gene Deletion of Endothelin Systems

CHAPTER 7

Endothelin-Converting Enzyme Inhibitors and their Effects

CHAPTER 8

**Endothelin Converting Enzymes and Endothelin Receptor
Localisation in Human Tissues**
A.P. DAVENPORT and F.D. RUSSELL. With 6 Figures.................   209

CHAPTER 12

**The Roles of Endothelins in Proliferation, Apoptosis,
and Angiogenesis**
J.R. WU-WONG and T.J. OPGENORTH. With 5 Figures 299

CHAPTER 13

**The Involvement of Endothelins in
Cerebral Vasospasm and Stroke**
O. TOUZANI and J. McCULLOCH. With 7 Figures 323

CHAPTER 14

Involvement of the Endothelins in Airway Reactivity and Disease
R.G. GOLDIE and P.J. HENRY. With 1 Figure ....

CHAPTER 15

Endothelin-1 and Pulmonary Hypertension
D.J. STEWART and Y.D. ZHAO. With 3 Figures 389

Contents

CHAPTER 16

Vascular and Cardiac Effects of Endothelin

CHAPTER 17

Endothelins Within the Liver and the Gastrointestinal System

CHAPTER 1

Historical Perspective and Introduction to the Endothelin Family of Peptides

T. Masaki

It is generally accepted that endothelial cells play a critical role in the regulation of vascular tone. This concept initially emerged after the discovery of endothelium-derived relaxing factor (EDRF) (Furchgott and Zawadzki 1980). EDRF was later proved to be a simple substance, nitric oxide (Palmer et al. 1987). At that time, numerous studies provided evidence of a role for arachidonic acid and its metabolites in the endothelium-dependent responses. Prostacyclin, a potent relaxing factor, was shown to be produced and released from endothelial cells in response to endothelium-dependent agonists. However, EDRF was not prostacyclin. Obviously, the regulatory mechanism of vascular tone involved several chemical substances, which are produced and released from endothelial cells in response to various physical and chemical stimuli and act on the underlying smooth muscle cells to elicit relaxation.

Numerous papers regarding the EDRF appeared in the early 1980s, and the number of publications increased thereafter. However, less attention was paid to endothelium-derived constricting factors until the mid-1980s. Endothelium was known to be a source of constricting factors as from the late 1970s into the 1980s papers were published describing endothelium-derived constricting factors. Since, in a variety of blood vessels, endothelium-dependent contraction was found to be blocked by inhibitors of cyclooxygenase, an arachidonic metabolite was suggested to be an endothelium-derived constrictor. For example, thromboxane A_2, a potent vasoconstrictor, was such a candidate. In the meantime, in 1985, Highsmith and his colleagues published a paper that described the existence of a constricting factor in the conditioned medium of cultured bovine endothelial cells (Hickey et al. 1985). They suggested that this factor was a peptide because it was sensitive to trypsin. However, they did not succeed in purifying it. This report evoked a worldwide competition for the identification of the peptide with several laboratories trying to isolate it quietly. In May 1987 we also began our attempts to identify the peptide. We collected from cultured bovine endothelial cells a large amount of conditioned medium that contained the peptide. We had an excellent team for this project at the University of Tsukuba, and quickly succeeded in purifying and identifying the peptide. Fortunately, in the early 1980s, techniques for the culture of endothelial cells and high performance liquid chromatography had become available. Also new methods of molecular biology were rapidly being developed at that time. Such advances expedited our success. We had finished the purification of the substance by the end of July, and started to identify the amino acid sequence of the peptide using a high performance auto-peptide sequencer, which had also just become available. Fortunately the substance that we had isolated was a peptide. We identified the whole amino acid sequence very quickly then conducted cloning and sequence analysis of the cDNA for the peptide. Results of the analyses of peptide and cDNA accorded exactly. Then we asked Dr. Fujino of the Takeda pharmaceutical company to synthesize this peptide. The contractile activity of the synthetic and native peptide on the rat aortic strip matched exactly. We named this new peptide endothelin (ET). Our first report was published in Nature in early spring 1988 (YANAGISAWA et al. 1988).

The first report stimulated worldwide interest particularly among cardiologists and endocrinologists, because the peptide had a unique sequence and interesting pharmacological properties. An amino acid sequence like this was unknown in higher animals. Furthermore, ET proved a most potent vasoconstrictive peptide. Following intravenous administration of ET into animals, blood pressure was elevated and remained so for some time. This finding suggested a role for the peptide in maintenance of blood pressure, or in the generation of hypertension. Thereafter, an enormous number of reports regarding this peptide appeared within a short period. The first conference on ET was held in London in December 1988, the year of its discovery. Ever since, similar international meetings have been held every other year.

ET is a 21 amino acid residue peptide with free amino and carboxy termini including four cysteine residues, which form two intramolecular disulfide bonds (Fig. 1) (YANAGISAWA et al. 1988). The calculated molecular weight of porcine ET-1 is 2492. Interestingly, the two disulfide bonds between the amino acids in position 1–15 and 3–11 are important for the biological activity of ET. For example, native ET shows the most potent vasoconstrictive activity whereas ETs with two different disulfide bonds have low biological activity.

Perhaps the most remarkable progress to be made in ET research, following the initial discovery, was the identification of three isopeptides of ET (INOUE et al. 1989). Analysis of the ET gene revealed two other ET-like peptide genes with the existence of these gene products in vivo being demonstrated soon after. Accordingly, we named these related peptides ET-1 (endothelin-1), ET-2, and ET-3. They are produced by both endothelial cells and non-endothelial cells, with endothelial cells producing predominantly ET-1. In general terms, however, these endogenous ETs are expressed with different patterns in a wide variety of cell types.

Shortly after the discovery of endothelin, it was demonstrated that the structure of rare cardiotoxic snake venom sarafotoxins was very similar to that of ETs. Both consist of 21 amino acid residues with four cysteine residues and two intrachain disulfide bridges. To date, four sarafotoxins have been identified which are similar to ETs (LANDAN et al. 1991).

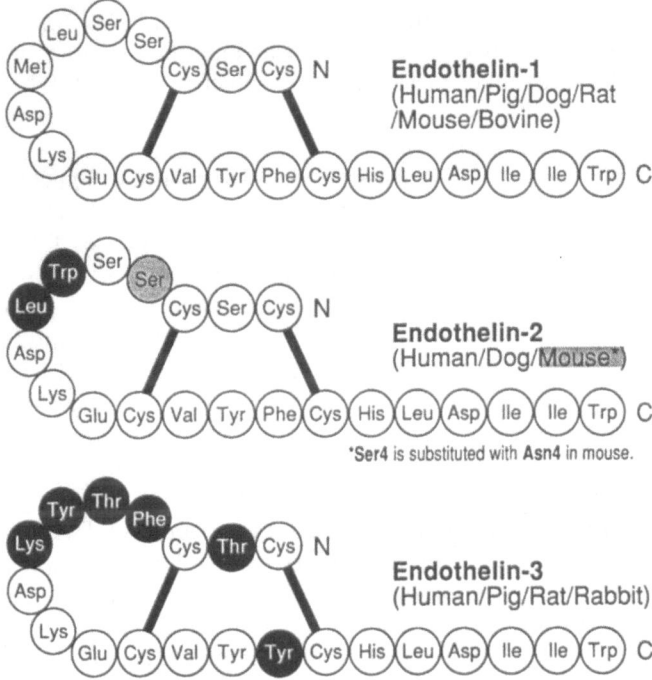

Fig. 1. Structures of endothelins. In mouse endothelin-2, Ser-4 of endothelin-1 is replaced by Asn-4

Sequence analysis of the cDNA for the ET-1 precursor revealed that ET-1 is processed from a long 212 (human) or 203 (porcine) amino acid peptide by several enzymes leading to the production of a 38 amino acid residue (human) or 39 amino acid residue (porcine) intermediate peptide, called big ET-1 (YANAGISAWA et al. 1988). ET-2 and ET-3 are also processed from their own precursors, i.e., big ET-2 or big ET-3. The workings of this processing pathway were made clearer later. In the case of porcine ET-1, after the signal peptide is cleaved, the resulting propeptide is further cleaved by a furin-like enzyme at the C-terminus of two Arg Ser Lys Arg motifs flanking the big ET-1 within proET-1 on either side (KIDO et al. 1997). The four carboxy terminal amino acid residues of the resulting peptide are then cleaved by carboxy-peptidase to produce big ET-1. The cleavage of the precursor by furin-like protease is an essential step before the cleavage of big ET-1. Big ET-1 must be further processed by an enzyme named endothelin converting enzyme (ECE) (YANAGISAWA et al. 1988) to produce ET-1. Later ECE was identified and cloned (SHIMADA et al. 1994; XU et al. 1994). It has a similar structure to neutral endopeptidase. Two isoforms of the ECE were reported in 1995, ECE-1 α/a and ECE-1 β/b, which differ only in the N-terminal amino acid sequence, and are active at neutral pH (SHIMADA et al. 1995). Another type of ECE named ECE-2, active at acidic pH, was identified (EMOTO and YANAGI-SAWA 1995). However, all of these ECEs are selective for big ET-1 rather than big ET-2 or big ET-3. Another type of ECE specific for ET-3 was purified recently (HASEGAWA et al. 1998).

In the early stages of ET research, detailed pharmacological studies on the three isoforms of ET revealed that responses to the isoforms can be classified into two types (MASAKI et al. 1994). In type I responses, ET-1 and ET-2 are more potent than ET-3, while in type II responses, all of the three isotypes are equally potent. In addition, experiments on the replacement of iodine-labeled ET-1 bound to the ET-receptor by non-labeled isopeptides demonstrated the existence of several distinct binding sites for ET, suggesting multiple ET receptors. Indeed, two types of ET-receptor were cloned in 1990 (ARAI et al. 1990; SAKURAI et al. 1990). We named them ET_A and ET_B. These receptors belong to a family of heptahelical G-protein coupled receptors, and exhibit a high polypeptide sequence identity with each other. ET_A and ET_B are distinct in their ligand binding affinity for endogenous ETs. ET_A has a high affinity for ET-1 and ET-2, but low affinity for ET-3, while ET_B has an equal affinity for all of the three endogenous ETs. ET_A and ET_B were shown to be distributed in a variety of tissues and cells in different proportions. In vascular tissues, ET_A and ET_B are present on smooth muscle cells and mediate vasoconstriction, whereas ET_B exists on endothelial cells and mediates the production and release of relaxing factors such as prostacyclin and nitric oxide (DOUGLAS et al. 1995). Soon after the discovery of ET-receptor subtypes, several reports demonstrated the existence of two ET_B subtypes, which could be discriminated by ET-receptor antagonists, such as PD142893 (DOUGLAS et al. 1994). These were named ET_{B1} and ET_{B2}. The former is present on endothelium mediating the release of relaxing factor, and the latter on smooth muscle such as rabbit saphenous vein mediating contraction. In addition, another type of ET receptor, named ET_C, was cloned from the dermal melanophores of *Xenopus laevis*, and found to be selective for ET-3 over ET-1 (KARNE et al. 1993). The existence of this type of receptor in rabbit saphenous vein was suggested by pharmacological experiments (DOUGLAS et al. 1995), but this has yet to be confirmed in mammalian tissue by conventional protein purification/molecular cloning. Additionally, several papers reported ET_B splice variants that were similar to conventional ET_B in ligand binding but not in functional features (ELSHOURBAGY et al. 1996; TANAKA et al. 1998).

ET receptors are good models for investigations into structure-function relationships of G-protein coupled receptors, because ligand binding to the receptor is rarely affected by structural modification of the intracellular domain. Therefore, studies on mutated ET_A or ET_B receptors have provided important information. Both ET_A and ET_B are able to couple to $G\alpha s$, $G\alpha i$, and $G\alpha q$ proteins (TAKAGI et al. 1995), depending on the cell types, and probably also depending on the intracellular environment. Each of the three $G\alpha$-proteins couples to different domains of the ET-receptor (TAKAGI et al. 1995).

Therefore, modification of the intracellular loop greatly affects the specificity of ET-receptor for the coupling of different G-proteins. Palmitoylation of the C-terminal cysteine residues of ET_B is particularly important (OKAMOTO et al. 1998).

As a result of the coupling of G-proteins to ET-receptors, multiple pathways of intracellular signal transduction are activated (MIYAUCHI and MASAKI 1999). At first, ET_A-mediated smooth muscle contraction was reported to be mediated by activation of phospholipase C (PLC) and opening of voltage operated calcium channels. However, those effects were observed at unphysiologically high concentrations of ET-1. At physiologically low concentrations of ET-1, ET-1-induced contractions via ET_A are mediated by activation of non-selective cation channels (IWAMURO et al. 1999; ZHANG et al. 1998). Interestingly, in guinea pig tracheal smooth muscle, ET_B-mediated but not ET_A-mediated contraction is inhibited by an inhibitor of voltage operated calcium channels, nifedipine (INUI et al. 1999), suggesting mediation by $G\alpha i$ activation. The latter ET_A-mediated contraction is neither inhibited by voltage dependent calcium channel inhibitor nor by a PLC-inhibitor. Another important function of ET is promoting growth of cells such as smooth muscle or endothelial cells. This activity is mostly mediated by ET_A via activation of MAP kinase. The MAP kinase activation is mediated in two ways, i.e., classical PKC-dependent and independent ways (SUGAWARA et al. 1996). Cell growth activity was also shown by activation of ET_B in some cells such as glia cells. However, in other cells ET_B receptor activation appears linked to differentiation and apoptosis together with suppressed cell growth in a cell cycle dependent manner (OKAZAWA et al. 1998).

Soon after the discovery of ET, the pharmaceutical industry started to search for ET-receptor antagonists (RUFFOLO 1995). The first important finding in this field was the discovery by at least two laboratories of an antagonistic action on ET of BE-18257B, one of the fermentation products of *Streptomyces misakiensis*. Modification of BE-18257B led to the identification of several active ET-receptor antagonists such as BQ-123 and FR139317 (IHARA et al. 1991; ARAMORI et al. 1993). These peptide antagonists had higher affinities for ET_A than ET_B receptors. Since its discovery, BQ-123 has been used widely for experiments. Following a series of studies on peptide antagonists, ET_B-selective antagonists such as BQ-788, and RES-701–1 were also discovered. Subsequently many selective and non-selective antagonists for ET_A and ET_B emerged (MIYAUCHI and MASAKI 1999). The first non-peptide ET receptor antagonist for oral administration was Ro46–2005, a non-selective (ET_A/ET_B) antagonist. A modified form of Ro46–2005, Ro47–0203 or bosentan, has been widely tested for preclinical and clinical uses.

ET-1 is not a circulating hormone because the plasma ET-1 concentration is very low. The production of ET-1 in endothelium is regulated by a variety of local and systemic hormones and physical stimulation such as blood flow. ET-1 released from endothelium acts back on the neighboring endothelial cells to release relaxing factor(s). The released relaxing factor(s) acts on the underlying smooth muscle cells to induce relaxation. The released ET-1 also acts directly on the underlying smooth muscle cells to elicit contraction. Therefore ET-1 may act either as a relaxing or constricting factor, probably depending on the concentration of ET-1 and ET receptor in local areas of the vascular bed (MASAKI 1995).

ET interacts with other endothelium-derived vasoactive factors such as nitric oxide or prostanoid(s), but the exact mechanism of the interaction is not clear. For example, several studies have revealed regional differences in the degree of involvement in the endothelium-dependent regulatory mechanism of vascular tone. Therefore, elucidation of the role of endogenous ET in systemic blood pressure is difficult. Here, ET receptor antagonists and ET or ET receptor knockout mice are helpful. Recent studies using ET-antagonists have demonstrated that endogenous ET is responsible for basal constrictor tone in peripheral vascular beds and plays a fundamental physiological role in the maintenance of blood pressure in humans (HAYNES et al. 1996). In contrast, in knockout mice experiments, heterozygous ET-1-deficient mice showed elevated systemic blood pressure despite a low level of plasma ET-1 (KURIHARA et al. 1994). This result may be partly explained by high sympathetic nerve activity in these mice due to hypoxia (LING et al. 1998). Indeed, ET has been shown to be involved in the mechanism of chemoreception of CO_2. An additional explanation is that ET-1 acts as a vascular relaxing factor in the normal state by stimulating the release of relaxing factors from the endothelium via ET_B activation, and that ET-1 deficient mice lose this function (OHUCHI et al. 1999). The result that the arterial blood pressure of ET_B-deficient mice was significantly higher than that of normal mice supports this hypothesis. The plasma concentration of ET-1 did not differ between ET_B-deficient and normal mice. Furthermore, a selective ET_B antagonist BQ-788 increased the arterial pressure of normal mice but not ET_B-deficient mice. These results suggest that in mice endogenous ET-1 is hypotensive. In addition, ET_B may be involved in the natriuretic receptive mechanism in kidney, because hypertension in ET_B-deficient mice is salt-sensitive (WEBB et al. 1998). From this we can conclude that the roles of ET in the mechanisms maintaining blood pressure are complicated.

Many reports have suggested that ET plays important roles in the pathological processes within the vascular system, such as endothelial dysfunction. As above, the pathophysiological significance of ET has also been elucidated by use of ET-receptor antagonists and/or observation of ET and ET-receptor knockout mice. It has been demonstrated that ET plays critical roles in several pathological conditions including chronic heart failure, pulmonary hypertension, cerebrovascular spasm after subarachnoid hemorrhage, acute renal failure, and essential hypertension (MIYAUCHI and MASAKI 1999; LEWIN 1995). ET antagonists have demonstrated significant beneficial effects under pathological conditions in which plasma ET-1 levels are elevated, and probably where expression of ET_A is enhanced. For example, in chronic heart failure, the expression of ET-1 and its receptor in cardiomyocytes was enhanced.

Treatment with ET receptor antagonists markedly improved the survival rate and various parameters of cardiac functions. Similarly, ET is implicated in vascular neointimal formation after balloon angioplasty. In this case, smooth muscle cells were found to proliferate and express ET-1 and ET receptors, as is the case for cardiac muscle in chronic heart failure. ET acts as a growth factor for myocardiac and smooth muscle cells. ET also promotes extracellular matrix synthesis.

At the very early stages of ET research, autoradiographic experiments using radioligand ET-1 revealed that ET-1 binding sites are distributed not only in the vascular system but also in other tissues and organs including the brain, adrenal gland, lung, kidney, intestine, etc., suggesting both the existence of ET receptors in these organs and multiple functions of ET (KOSEKI et al. 1989). Particularly interesting was that ET-3 and ET_B distributed densely in intestines. Subsequently, it was reported that ET-3 or ET_B receptor-deficient mice exhibited aganglionic megacolon similar to the human Hirschsprung's diseases (HOSODA et al. 1994). ET_B-deficient mice also showed coat color spotting in addition to the above anomaly. ET-1-deficient mice have craniofacial and cardiac abnormalities at birth and die of respiratory failure soon after birth. ETs and ET-receptors are suggested to play crucial roles in the development of organs at early stages. Since ETs and ET receptors are distributed widely through the body, unexpected roles for ETs are likely to appear in the future.

Thus, the discovery of ETs opened a new field of research stimulating a worldwide explosion in research activity. Numerous papers over the past decade have revealed the versatile nature of ET. The physiological significance of ET other than within the cardiovascular system will continue to be revealed in the near future. As to their role in the cardiovascular system, ETs are both beneficial and harmful, and are important to normal cardiovascular homeostasis. On the other hand, ET aggravates clinical features in several pathological conditions. Nevertheless, the underlying physiological and pathophysiological mechanisms of ETs still remain to be solved. Many ET receptor antagonists and ECE inhibitors have now been developed and investigators are now searching for the clinical targets of these blockers. Such blockers should prove useful for the treatment of a range of clinical disorders.

References

Arai H, Hori H, Aramori A, Ohkubo H, Nakanishi S (1990) Cloning and expression of a cDNA encoding an endothelin receptor. Nature 348:730

Aramori I, Nirei H, Shoubo M, Sogabe K, Nakamura K, Kojo H, Notsu Y, Ono T, Nakanishi S (1993) Subtype selectivity of a novel endothelin antagonist, FR139317, for the two endothelin receptors in transfected Chinese hamster ovary cells. Mol Pharmacol 43:127–131

Douglas SA, Meek TD, Ohlstein EH (1994) Novel receptor antagonists welcome a new era in endothelin biology. Trend Pharmacol Sci 15:313–316

Douglas SA, Beck GR Jr, Elliott JD, Ohlstein EH (1995) Pharmacological evidence for the presence of three functional endothelin receptor subtypes in rabbit saphenus vein. J Cardiovasc Pharmacol 26 (Suppl 3):S163–S168

Elshourbagy NA, Adamou JE, Gagnon AW, Wu H-L, Pullen M, Nambi P (1996) Molecular characterization of a novel human endothelin receptor splice variant. J Biol Chem 271:25300–25307

Emoto N, Yanagisawa M (1995) Endothelin-converting enzyme-2 is a membrane-bound phosphoramidon-sensitive metalloprotease with acidic pH optimum. J Biol Chem 279:16262–16268

Furchgott RF, Zawadzki JV (1980) The Obligatory role of endothelial cells in the relaxation of arterial smooth muscle by acetylcholine. Nature 288:373–376

Hasegawa H, Hiki K, Sawamura T, Aoyama T, Okamoto Y, Miwa S, Shimohama S, Kimura J, Masaki T (1998) Purification of a novel endothelin-converting enzyme specific for big endothelin-3. FEBS Lett 428:304–308

Haynes WG, Ferro CJ, O'Kane KPJ, Somerville D, Lomax CC, Webb DJ (1996) Systemic endothelin receptor blockade decreases peripheral vascular resistance and blood pressure in humans. Circulation 93:1860–1870

Hickey KA, Rubanyi G, Paul RJ, Highsmith RF (1985) Characterization of a coronary vasoconstrictor produced by cultured endothelial cells. Am J Physiol 248:C550–C556

Hosoda K, Hammer RE, Richardson JA, Baynash AG, Cheung JC, Giaid A, Yanagisawa M (1994) Interaction of endothelin-3 with endothelin-B receptor is essential for development of epidermal melanocytes and enteric neurons. Cell 79:1277–1285

Ihara M, Noguchi K, Saeki T, Fukuroda T, Tsuchida S, Kimura S, Fukami T, Ishikawa K, Nishikibe M, Yano M (1991) Biological profiles of highly potent novel endothelin antagonists selective for the ET_A receptor. Life Science 50:247–255

Inoue A, Yanagisawa M, Kimura S, Kasuya Y, Miyauchi T, Goto K, Masaki T (1989) The human endothelin family: three structurally and pharmacologically distinct isopeptides predicted by three separate genes. Proc Natl Acad Sc. USA 86:2863–2867

Inui T, Ninomiya H, Sasaki Y, Makatani M, Urade Y, Masaki T, Yamamura M (1999) Selective activation of excitation-contraction coupling pathways by ET_A and ET_B in tracheal smooth muscle. Brit J Pharmacol 126:893–902

Iwamuro Y, Miwa S, Zhang XF, Minowa T, Enoki T, Okamoto Y, Hasegawa H, Furutani H, Okazawa M, Ishikawa M, Hashimoto N, Masaki T (1999) Activation of three types of voltage-independent Ca2+ channel in A7r5 cells by endothelin-1 as revealed by a novel Ca2+ channel blocker LOE 908. Br J Pharmacol 126:1107–1114

Karne S, Jayawickreme CK, Lerner MR (1993) Cloning and characterization of an endothelin-3 specific receptor (ETC receptor) from Xenopus laevis dermal melanophores. J Biol Chem 268:19126–19133

Kido T, Sawamura T, Hosikawa H, Orleans-Juste P, Denault J-B, Leduc R, Kimura J, Masaki T (1997) Processing of proendothelin-1 at the C-terminus of big endothelin-1 is essential for proteolysis by endothelin-converting enzyme-1 in vivo. Eur J Biochem 244:520–526

Koseki C, Imai M, Hirata Y, Yanagisawa M, Masaki T (1989) Autoradiographic distribution in rat tissues of binding sites for endothelin: a neuropeptide? Am J Physiol 256:R858–R866

Kurihara Y, Kurihara H, Suzuki H, Kodama T, Maemura K, Nagai R, Oda H, Kuwaki T, Cao W, Kamada N, Jishage K, Ouchi Y, Azuma S, Toyoda Y, Ishikawa T, Kumada M, Yazaki Y (1994) Elevated blood pressure and craniofacial abnormalities in mice deficient in endothelin-1. Nature 368:703–710

Landan G, Bdolah A, Wollberg Z, Kochva E, Graur D (1991) Evolution of the sarafotoxin/endothelin superfamily of proteins. Toxicon 29:237–244

Lewin ER (1995) Endothelins. N Engl J Med 333:356–363

Ling GY, Cao W-H, Onodera M, Ju KH, Kurihara H, Kurihara Y, Yazaki Y, Kumada M, Fukuda Y, Kuwaki T (1998) Renal sympathetic nerve activity in mice: comparison between mice and rats and between normal and endothelin-1 deficient mice. Brain Res 808:238–249

Masaki, Vane J, Vanhoutt P (1994) 5th International union of pharmacology nomenclature of endothelin receptors. Pharmacol Rev 46:137–142

Masaki T (1995) Possible role of endothelin in endothelial regulation of vascular tone. Ann Rev Pharmacol Toxicol 35:235–255

Miyauchi T, Masaki T (1999) Pathophysiology of endothelin in the cardiovascular system. Ann Rev Physiol 61:391–415

Ohuchi T, Kuwaki T, Ling GY, deWit D, Ju K-H, Onodera M, Cao W-H, Yanagisawa M, Kumada M (1999) Elevation of blood pressure by genetic and pharmacologic disruption of the endothelin-B receptor in mice. Am J Physiol 276:R1071–R1077

Okamoto Y, Ninomiya H, Masaki T (1998) Posttranslational modifications of endothelin receptor type B. Trend Cardiovasc Med 8:327–329

Okazawa M, Shiraki T, Ninomiya H, Kobayashi S, Masaki T (1998) Endothelin-induced apoptosis of A375 human melanoma cells. J Biol Chem 273:12584–12592

Palmer RMJ, Ferrige AG, Moncada S (1987) Nitric oxide release accounts for the biological activity of endothelium-derived relaxing factor. Nature 327:524–526

Ruffolo R Jr (1995) Endothelin receptors from the gene to the human. CRC Press, Boca Raton, Fla

Sakurai T, Yanagisawa M, Takuwa Y, Miyazaki H, Kimura S, Goto K, Maaki T (1990) Cloning of a cDNA encoding a non-isopeptide-selective subtype of the endothelin receptor. Nature 348:782

Shimada K, Takahashi M, Tanzawa K (1994) Cloning and functional expression of endothelin-converting enzyme from rat endothelial cells. J Biol Chem 268:18274–18278

Shimada K, Takahashi M, Ikeda M, Tanzawa K (1995) Identification and characterization of two isoforms of an endothelin-converting enzyme-1. FEBS Lett 371:140–144

Sugawara F, Ninomiya H, Okamoto Y, Miwa S, Masuda O, Katsra Y, Masaki T (1996) Endothelin-1-induced mitogenic responses of Chinese hamster ovary cells expressing human endothelin A: the role of a wortmannin-sensitive signaling pathway. Mol Pharmacol 48:447–457

Takagi Y, Ninomiya H, Sakamoto A, Miwa S, Masaki T (1995) Structural basis of G-protein specificity of human endothelin receptors, A study with endothelin A/B chimeras. J Biol Chem 270:10072–10078

Tanaka H, Moroi K, Iwai J, Takahashi H, Ohnuma N, Hori S, Takimoto M, Nishiyama M, Masaki T, Yanagisawa M, Sekiya S, Kimura S (1998) Novel mutations of the endothelin B receptor gene in patients with Hirschsprung's disease and their characterization. J Biol Chem 273:11378–11383

Webb DJ, Monge JC, Rabelink TJ, Yanagisawa M (1998) Endothelin: new discoveries and rapid progress in the clinic. Trend Pharmacol Sci 19:5–8

Xu D, Emoto N, Giaid A, Slaughter C, Kaw S, deWit D, Yanagisawa M (1994) ECE-1: A membrane-bound metalloprotease that catalyzes the proteolytic activation of big endothelin-1. Cell 78:473–485

Yanagisawa M, Kurihara H, Kimura S, Tomobe Y, Kobayashi M, Mitsui Y, Yazaki Y, Goto K, Masaki T (1988) A novel potent vasoconstrictor peptide produced by vascular endothelial cells. Nature 332:411–415

Zhang ZF, Komuro T, Miwa S, Minowa T, Iwamuro Y, Okamoto Y, Ninomiya H, Sawamura T, Masaki T (1998) Role of non-selective cation channels as Ca^{2+} entry pathway in endothelin-1-induced contraction and their suppression by nitric oxide. Eur J Pharmacol 352:237–245

CHAPTER 2

Sarafotoxins and Their Relationship to the Endothelin Family of Peptides

M. Sokolovsky and Z. Shraga-Levine

A. Introduction

The sarafotoxins (SRTXs), a family of peptides isolated from the venom of the snake *Atractaspis engaddensis*, show striking structural similarities to the endothelins (ETs) produced by endothelial cells. Four isoforms of SRTXs have been identified: SRTX-a, SRTX-b, SRTX-c, and SRTX-e. SRTX-d, which was previously assumed to exist, has not been found (Bdolah et al. 1989b). SRTX-e differs from the proposed SRTX-d in having a Gly instead of an Asp in position 18 (Ducancel et al. 1993). These peptides have powerful cardiotoxic effects and cause contraction of smooth muscles. They all contain 21 amino acids and two disulfide bridges, between Cys^1-Cys^{15} and Cys^3-Cys^{11}. There is a very high degree of sequence similarity, not only within each family (ETs, 71–95%; SRTXs, 81–95%), but also between families (52–67%) (Kochva et al. 1993). The ETs are produced in minute quantities in mammals (measurable in picomoles per gram of tissue), whereas the SRTXs are produced in large amounts in the venom of the snake (0.1 mmol/g). The SRTXs and ETs may be divided into two groups on the basis of their toxicity. The most lethal are ET-1 and SRTX-b; ET-3 is less toxic, and SRTX-c and SRTX-e cause death only at very high doses, if at all. For example, in ICR mice, the LD_{50} for SRTX-b and ET-1 is 15 µg/kg body weight (Bdolah et al. 1989a). ET-1 and SRTX-b were found to cause severe disturbances in heart function at the level of the A-V conduction system and the coronary vessels, ultimately leading to cardiac arrest.

A new member of the ET/SRTX family of vasoconstrictor peptides, bibrotoxin (BTX), was isolated from the venom of the burrowing asp *Atractaspis bibroni*. The amino acid sequence of BTX differs from that of SRTX-b only in the substitution of Ala for Lys in position 4, suggesting that BTX represents the peptide isoform of *Atractaspis bibroni* corresponding to SRTX-b. BTX competes with [^{125}I]-ET-1 for binding to the human ET_B-type receptor, and induces vasoconstriction of the thoracic aorta in rats (Becker et al. 1993).

B. Structure

As mentioned above, SRTX-c, SRTX-e, and ET-3 (ET-3) are less toxic and pharmacologically weaker than the isopeptides SRTX-a, SRTX-b, and ET-1. The three peptides in the first group possess a Thr instead of a Ser at position 2, and this substitution could be responsible for the observed biological differences. LAMTHANH et al. (1994) showed that a synthetic [Thr^2]SRTX-b shows lower vasoconstriction efficacy (by approximately 35%) than SRTX-b in the rabbit aorta, but is nearly as potent as SRTX-b in toxicity tests and in influencing contraction of the rat uterus. The antigenicity of the synthetic analog is comparable to that of SRTX-b, suggesting that the overall structure of the two peptides is similar, despite the substitution at position 2. The Thr^2 substitution probably contributes to the lower activity of the 'weak' peptides in some systems.

Another functionally important substitution in the SRTX is Lys^9, as evidenced by the fact that a synthetic [Glu^9]-SRTX-b shows considerable loss of activity (KITAZUMI et al. 1990; TAKASAKI et al. 1991). However, in other synthetic peptides, one with Asn^4 (instead of Lys^4) and one with Asn^{13} (instead of Tyr^{13}), the contractile effect was diminished only slightly, if at all. In both the ETs and SRTXs, the indole group of Trp^{21} is important for both binding and vasoconstrictor activities (KOSHI et al. 1991).

The disulfide bridges are an important characteristic of the ET/SRTX peptides. The importance of the disulfide bridge locations in SRTX-b was demonstrated by comparing the cardiovascular effects of two synthetic STRX-b analogs with different disulfide bridge locations, STRX-b type A (Cys^1–Cys^{15}, Cys^3–Cys^{11}) and STRX-b type B (Cys^1–Cys^{11}, Cys^3–Cys^{15}) (LIN et al. 1990). In mice, the type A peptide produced a sustained pressor effect with a transient increase in pulse pressure at low concentrations, whereas higher concentrations were accompanied by a decrease in blood pressure, heart rate, and respiratory rate, followed by death. In contrast, type B was not lethal to mice at these doses. In the rat Langendorff heart preparation, type A produced coronary vasospasm with potency about 100 times higher than that of type B. A similar potency ratio was observed for the positive inotropic effect in rat atria. Thus, the location of the disulfide bridges in SRTX-b appears to have a profound influence on the pharmacological potency.

Experiments with the C-terminal hexapeptide (amino acids 16–21 of ETs and SRTXs) showed that this hexapeptide discriminates between different ET receptors. The binding affinity of SRTX[16-21] for arterial preparations is about half as strong as that of ET hexapeptide, probably because of the difference in amino acid residues at position 17 (Gln in SRTX, Leu in ET) (DOHERTY et al. 1991; DOUGLAS and HILEY 1991).

Most of the ET/SRTX peptides have been studied by circular dichroism (CD). A method of grouping has been suggested on the basis of their helicity from CD. ET-1, ET-2, and VIC form one group, the SRTXs form a second, and ET-3 a third (TAMAOKI et al. 1992). ATKINS et al. (1994) proposed that differences in the shapes of the CD spectra of ET-1, ET-3, and SRTX are largely attributable to differences between the amino acid sequences.

Nuclear magnetic resonance (NMR) studies of the solution conformation of the ET/SRTX peptides have not yielded a consensus structure common to the whole family. However, consistent structural features have been identified, including an extended conformation for the first three or four residues and a helix between residues 9 and 15 or 16. The biologically important and highly conserved C-terminal tail was conformationally flexible in most of the NMR studies, both in the ETs and in the SRTXs. A turn-like structure was found in ET-1 between residues 5 and 8, whereas in SRTX-b it was found between residues 3 and 6 (ATKINS et al. 1995). This conformational difference was suggested by the authors to be partly responsible for the lower binding affinity of SRTX-b than of ET-1 for the ET_AR. In the ET-1 peptide two sites, residues 3–4 and 8–9, were identified as having conformational variance (ANDERSEN et al. 1992). In SRTX-b, however, the 7–8 bond rather than the 8–9 bond was found to be the second variable region (ATKINS et al. 1995).

NMR studies of the solution structure of SRTX-c have revealed a well-defined central α-helix, similar to that of SRTX-b, but different from the generally irregular helix reported for ET peptides (MILLS et al. 1994). The authors assume that this α-helix is probably the predominant conformer in the SRTXs, while the variable results obtained for the ET peptides may reflect a different ensemble of conformers. The structural differences between the central helix of the ETs and SRTXs might be important in the binding of these ligands by ET receptors.

C. Organization of SRTX cDNA

The similarity in the physiological expression of the ETs and SRTXs probably indicates that these peptides have been highly conserved during evolution. KOCHVA et al. (1993) suggested that their phylogenetic history is an early one and that the appearance of the ET/SRTX superfamily may be traced back at least to the acraniates among the vertebrates. These peptides probably originated with an exon duplication, followed by two full gene duplications, with changes occurring after each duplication (LANDAN et al. 1991a,b). While the SRTXs could have branched off from any point at the base, the lineage probably started before the first endothelin gene duplication. The ET genes of mammals are located on different chromosomes. The human ET-1 gene is located on chromosome 6, the ET-2 gene on chromosome 1, and the human ET-3 gene on chromosome 20, all in single copies (ARINAMI et al. 1991). These closely related genes are obviously not linked and apparently were dispersed during or after gene duplication. However, the genes for the different sarafotoxins are found on the same chromosome (DUCANCEL et al. 1993). They have a different cDNA organization from the ETs and their precursors also differ from ET precursors. Based on nucleotide sequencing of full-length cDNAs of the SRTXs, it was predicted that their precursors consist of a large pre-pro-polypeptide of 543 amino acids. The precursor has an unusual rosary-type organization, which comprises 12 tandem stretches of 40 residues each (39 in the first). One stretch comprises an invariant spacer of 19 residues (18 in the first stretch) and is immediately followed by a mature 21-residue SRTX. Six different sequences of SRTXs have been found along the precursor polypeptide chain; three of them are SRTX-a, SRTX-b, and SRTX-c (Ducancel et al. 1993). The arrangement of the mRNA is as follows: SRTX-a; c; b; c; c; e; a; c; c; b; a; a. The sequence for SRTX-d was not found, but that of SRTX-e is very similar to that expected for SRTX-d (Ducancel et al. 1993).

The abundance of SRTX-c sequences, which occur five times along the precursor sequence, is reflected by the greater abundance of this toxin in snake venom than that of either of the other toxins (Bdolah et al. 1991). The SRTXs are probably generated in the venom glands by suitable processing of the precursor polypeptide chains freeing them from the invariant spacers, presumably with the aid of appropriate SRTX-converting enzymes. In principle, each precursor molecule may generate 12 SRTX molecules. This amplification procedure offers a simple explanation for the overproduction of SRTXs in the venom glands of *A. engaddensis*. Even though the ETs are remarkably similar to the SRTXs in terms of primary and tertiary structure and biological activity, their precursors contain only a single ET molecule, which derives from the pro-form big-ET, and one inactive ET-like molecule (Inoue et al. 1989).

The organization of SRTX genes can be postulated by analogy with the genomic organization of the ETs. ET-3 and the ET-like molecules co-exist on the same cDNA (Opgenorth et al. 1992); however, they are encoded by two different exons separated from each other by one intron of about 1.3 kb (Arinami et al. 1991; Bloch et al. 1989; Inoue et al. 1989). If the gene encoding the SRTXs comprises 12 similar exons and 11 introns, it should have a size of approximately 20 kb.

The SRTX precursor has an uneven distribution of the different copies of SRTX. SRTX-c, which is one of the weakest toxins in mammals, has the largest number of copies in the precursor, whereas for the most potent SRTX-b there are only two sequences. Differential sensitivity toward the various SRTXs in the natural prey of these snakes could perhaps explain this discrepancy.

D. Biosynthesis and Degradation

ETs are biosynthesized as pre-pro-molecules (200 residues) that are converted into pro-forms (38 residues) called 'big ETs', which in turn are converted into Ets (see Chap. 3). The SRTXs, however, do not appear to be cleaved from a 'big SRTX' precursor (as mentioned in the previous section), but rather synthesized directly by the processing of a large protein precursor containing multiple SRTX sequences (Ducancel et al. 1993).

There are also differences in the degradation of the peptides. The ETs are degraded by neutral endopeptidase in a two-step process. First the enzyme forms a nick between positions 5 and 6 and afterwards between positions 18 and 19, resulting in the release of the C-terminal peptide, thus rendering the peptide inactive (SOKOLOVSKY et al. 1990). The SRTXs, however, are more resistant than the ETs to neutral endopeptidase. The low susceptibility of SRTXs to this enzyme might explain their relatively high toxicity: under physiological conditions, the ETs might be inactivated by neutral endopeptidase or neutral endopeptidase-like enzymes, whereas the SRTXs are not inactivated and can trigger the series of reactions leading to death.

E. Receptors

Early ligand-binding studies employing various ETs and SRTXs led to the suggestion that these peptides bind to more than one receptor subtype (KLOOG et al. 1988) (see Chap. 4). Two receptor subtypes designated ET_A receptors and ET_B receptors (ET_AR and ET_BR), whose existence has been confirmed by cloning and stable expression, have been recognized. The two subtypes share 59% identity and 78% homology. The ligand binding of ET_AR is selective. ET-1 and ET-2 bind to it with high and similar affinities; ET-3 binds with an affinity 70- to 100-fold lower than that of ET-1, and SRTX-c binds with an affinity more than 1000-fold lower than that of ET-1. However, the ET_BR binds all of the above with high and similar affinities.

Of the sarafotoxins, SRTX-c usually binds with high affinity to the ET_BR subtype only, and is used as a selective ET_BR agonist in binding studies as well as in functional studies (NAMBI et al. 1992; WILLIAMS et al. 1991). On the other hand, SRTX-b, like ET-1, does not distinguish between ET_AR and ET_BR.

The development of subtype-specific antagonists played a critical role in demonstrating the involvement of ET_AR and ET_BR in normal cellular functions as well as in various pathological states (reviewed in SOKOLOVSKY 1995). However, the use of these compounds has led to reports of ET-mediated responses that do not appear to fit the current receptor classification, and has prompted speculation that additional subtypes of endothelin receptors exist. Furthermore, recent reports have shown that certain endothelin antagonists are less effective against ET-1 than against other ET/SRTX family members (see Sect. F).

I. Binding

The distribution pattern of high-affinity binding sites for the ET/SRTX peptides is similar in human, porcine, and rat tissues (DAVENPORT et al. 1991). However, recent reports have pointed to differences in the binding profiles of the ligands. In one study, for example, ET-1 was found to bind to a significantly higher density of receptors than SRTX-b (MAGUIRE et al. 1996). This implies the existence of an additional receptor that is not sensitive to SRTX-b. Other studies have also reported a higher density of ET-1 sites than SRTX-b sites in rat cerebellum (Johns and Hiley 1990) and in rat spinal cord (Gulati and Rebello 1992). However, ET-1 and SRTX-b must have at least one binding site in common, because each is able to compete for the specific binding of the other (Maguire et al. 1996 and references therein). Also, differences in binding affinities were noted between ET-1 and SRTX-b. For example, in bovine cardiac sarcolemmal vesicles the binding affinity for SRTX-b was at least one order of magnitude lower than for ET-1 (Shannon and Hale 1994).

Differences in the binding patterns of ET-3 and SRTX-c were also seen. For example, SRTX-c was 20 times more potent in displacing [^{125}-I]-ET-3 binding from canine lung than from spleen receptors, suggesting that SRTX-c shows some selectivity for lung over spleen ET_BR (Nambi et al. 1995).

Studies of competitive binding between ET-1 and SRTX-b to homo-genates of human saphenous vein revealed additional receptor subtypes. In this preparation, in the presence of a high concentration of [^{125}I]-ET-1, SRTX-b competed with nanomolar affinity for the majority of the sites but with micromolar affinity for the remaining sites. This suggests that ET-1 binds to two receptor populations that can be distinguished by SRTX-b (Maguire et al. 1996).

One of the characteristics of binding of ETs and SRTXs in various tissues is the extremely slow rate of dissociation of receptor-ligand complexes. In many preparations, the binding is almost irreversible. In membrane prepara-tions of rat cerebellum, atrium and guinea-pig ileum, dissociation of [^{125}I]-labeled ET-1, ET-3, and SRTX-b from their preformed complexes with the receptor was significantly slower in the cerebellum than in the ileal and atrial preparations, and slower in the atrium than in the ileum (Sokolovsky 1995). Differences in dissociative behavior between ET-3 and ET-1 or SRTX-b in both cerebellum and ileum were noticed (Galron et al. 1991). These might result from differences in the nature of their receptor-ligand complexes, i.e., from different modes of binding.

In addition, functional studies of isolated blood vessels in rat renal arter-ies have also revealed marked differences in the kinetics of [^{125}I]-ET-1 and [^{125}I]-SRTX-b binding (Devadason and Henry 1997). Following a 3-h period of association of [^{125}I]-ET-1 with its receptors, no significant dissociation of receptor-bound [^{125}I]-ET-1 was observed during a 4-h washout period. In con-trast, dissociation studies revealed that specific binding of [^{125}I]-SRTX-b to ET_A receptors was reversible (t0.5 is half-life time of dissociation, 100 min). This study further supports the notion that differences exist in the kinetics of agonist binding.

Maguire et al. (1996) reported that in homogenates of human saphenous vein the dissociation rate for ET-1 and SRTX-b appeared to be equally slow during the first 20-min period. However, [^{125}I]-ET-1 did not dissociate over the next 220 min, whereas SRTX-b continued to dissociate, so that significantly more [^{125}I]-SRTX-b than [^{125}I]-ET-1 had dissociated after 4 h.

SRTX-c also highlights the differences in binding characteristics between tissues from different species. For example, the affinity of ET_BR for SRTX-c differed in human and rat left ventricle, whereas no differences in affinity were detected for ET-3 (RUSSELL and DAVENPORT 1996). SRTX-c binding was unaffected by GTP, indicating that the different receptor affinities in human and rat heart cannot be explained by differing affinity states of the ET_BR. There are also marked differences in the affinities of SRTX-c for ET_BRs in human left ventricle and canine femoral vein (MILLER and MICHENER 1995).

Differences in binding to mutated receptors were also noted between ET and SRTX. The ET_AR subtype usually binds ET-3 and SRTX-c with equally low affinity. Mutation of Asp^{126} to Ala led to a dramatic increase in the binding affinity of ET-3 for this receptor (160-fold increase); however, the mutation did not improve the binding affinity of SRTX-c (ROSE et al. 1995).

II. ET_B Receptors

In order to distinguish between the ET receptors, specific receptor subtype ligands are usually applied. BQ-123 is commonly regarded as an ET_AR selective antagonist and SRTX-c as an ET_BR-selective agonist. It seems, however, that this classification is not a simple one and should be viewed with caution, as the following reports show.

Early studies on endothelin receptors showed that both receptor subtypes, ET_AR and ET_BR, mediate vasoconstriction in different tissues (HARRISON et al. 1992; HAY 1992). However, one of the earliest studies showed that the ET_BR mediates vasodilation through ET-induced nitric oxide formation and/or release of prostaglandin I_2 (DENUCCI et al. 1988). On the basis of these observations, the existence of two ET_BR subtypes was proposed: an ET_{B1} subtype mediating vasodilation and an ET_{B2} subtype mediating vasoconstriction (WARNER et al. 1993a).

Whether SRTX-c is specific to one or both of the ET_BR subtypes is controversial. In the rabbit trachea, treatment with SRTX-c desensitized both ET_{B1} and ET_{B2} subtypes (YONEYAMA et al. 1995), although in the same study the contractile effect of SRTX-c was shown to be mediated via the ET_{B2} subtype. In the rabbit saphenous vein, STRX-c also desensitized both the ET_{B1} and the ET_{B2} subtypes (SUDJARWO et al. 1994). SRTX-c was therefore assumed to activate both ET_B subtypes.

ZUCCARELLO et al. (1998b) recently showed, however, that RES-701-1, a selective $ET_{B1}R$ antagonist (TANAKA et al. 1995), prevents SRTX-c-induced relaxation of rabbit basilar artery. This study demonstrated that SRTX-c is selective for the $ET_{B1}R$ subtype (ZUCCARELLO et al. 1998b).

On the other hand, SRTX-c was reported to stimulate the $ET_{B2}R$ subtype as well. In rabbit pulmonary resistance arteries, SRTX-c caused vasoconstriction through the $ET_{B2}R$ and was even more potent in this respect than ET-1 (MACLEAN et al. 1998).

The ET_BR is involved in the transient systemic depressor response mediated by ET-1 or SRTX-c (Cristol et al. 1993; DeNucci et al. 1988), as well as in renal vasoconstriction in the rat (Clozel et al. 1992; Cristol et al. 1993). In conscious rats SRTX-c induced a transient decrease in blood pressure, followed by a sustained increase in pressure and a decrease in renal blood flow (Gellai et al. 1994). This indicates that the varying responses to SRTX-c may result from activation of the same receptor located on different cell types. For example, the ET_BR located on the endothelium may activate the endothelium-derived relaxing factor/nitric oxide (NO) system, whereas the same receptor on vascular smooth muscle cells could mediate the contractile process.

The above hypothesis was introduced by Gellai et al. (1996) and supported by their results, which showed that the ET_BR selective agonist SRTX-c, when given as an intravenous bolus injection in conscious Sprague-Dawley rats, elicits two simultaneous responses; vasoconstriction and vasodilation. The authors postulated that these responses are mediated by two ET_BR subtypes, one of them RES-701-1-sensitive and the other RES-701-1-insensitive. The vasodilatory effect of SRTX-c completely masked its vasoconstrictor effect during the first few seconds and limited it for a sustained period. RES-701-1 blocked the depressor response, unmasked an initial pressor effect, and potentiated the sustained response to SRTX-c. In the renal bed, vasodilation was not observed after bolus injections of SRTX-c; however, its importance became evident during the sustained infusion of SRTX-c. The potent vasoconstriction that was prevalent in both vascular beds during the first 10–20min was gradually blunted with time. Infusion of RES-701-1 prevented this gradual decline in the constrictor effect, presumably by blocking a simultaneous vasodilation induced by SRTX-c. The profile of hemodynamic changes during the sustained infusion of SRTX-c indicated that vasodilation, mediated by the RES-701-1-sensitive ET_BR subtype, is slow in onset, becomes dominant with time, and gradually limits the vasoconstriction mediated by the RES-701-1-insensitive ET_BR subtype.

In the guinea-pig ileum, SRTX-c induced a biphasic effect (relaxation and contraction). SRTX-c induced strong tachyphylaxis of both components of the response. PD145065, a potent antagonist of both ET_AR and ET_BR, acted as a noncompetitive antagonist of the contractile components of SRTX-c. RES-701-1, a specific antagonist of $ET_{B1}R$, inhibited mainly the relaxant component induced by low doses of SRTX-c. These results suggest that there are at least two distinct populations of ET_BRs mediating the biphasic response: the $ET_{B1}R$, sensitive to RES-701-1 and PD145065, and the $ET_{B2}R$, less sensitive to RES-701-1 and PD145065 (Miasiro et al. 1998).

In the guinea-pig trachea a non-peptide antagonist, Ro 47–0203 (nonselective for ET_AR over ET_BR), was more effective against SRTX-c than against ET-1. This may reflect a difference between the ET_BRs on this tissue (Gater et al. 1996).

III. ET$_A$ Receptors

When ET/SRTX ligands bind to a receptor with the order of potency ET-1~SRTX-b>ET-3>SRTX-c, this can usually be taken as evidence that the receptor in question is the ET$_a$ subtype. However, various studies carried out with the SRTX peptides have shown that there are ET$_A$Rs which do not conform to the above rule.

The effects of ET-1 and SRTX-b on classical ET$_A$Rs are usually considered as equipotent; however, a number of reports indicate that these peptides may act differently on ET$_A$ receptors. For example, some studies have shown that ET-1-induced responses are more resistant than SRTX-b-induced responses to the selective ET$_A$R antagonist BQ-123 (NISHIYAMA et al. 1995 and references therein). These receptors belong to the ET$_A$R category, since ET$_B$ agonists did not induce any responses. It is conceivable that ET-1 and SRTX-b may exert their effects through two different ET$_A$Rs, a BQ-123-insensitive ET$_A$R and a BQ-123-sensitive ET$_A$R.

Although SRTX-c is usually regarded as a selective ET$_B$R agonist, some reports have shown that it can stimulate the ET$_A$R subtype as well. For example, SRTX-c was shown to activate an ET$_A$R subtype in guinea-pig pulmonary artery. It also caused a slight but distinct concentration-dependent increase in basal tone that was significantly attenuated by BQ-123 (MATSUDA et al. 1996). High concentrations of SRTX-c stimulated cGMP and cAMP production in rat cerebellar slices through the ET$_A$R (SOKOLOVSKY et al. 1994; SHRAGA-LEVINE et al. 1994). Also, it was recently proposed that the ET$_A$R has a modest role in mediating the systemic vasoconstrictor response to a high-dose of SRTX-c (RASMUSSEN et al. 1998).

ET isopeptides and SRTX-b were found to elicit dose-dependent contractions of the isolated rabbit iris sphincter, whereas SRTX-c and IRL 1620, at concentration of up to 1 μmol/l, elicited no contractile activity. Further experiments in this preparation with the various antagonists showed that ET receptors cannot be simply classified into the ET$_A$R and ET$_B$R subtypes established so far. When compared to the known receptor subtypes, the ET receptors in the isolated rabbit iris sphincter were quite different from the ET$_B$R, and apparently showed a pharmacological profile that was most similar to the ET$_A$R, suggesting the existence of heterogeneous and atypical ET$_A$Rs (ISHIKAWA et al. 1996).

Atypical ET$_A$Rs were also found in human saphenous vein. BQ-123 (1 μmol/l) causing nonparallel shifts, with lower concentrations of ET-1 being antagonized more strongly than higher concentrations. When SRTX-b was used as an agonist, BQ-123 (0.3–3.0 μmol/l) caused concentration-dependent biphasic shifts, and low concentrations of SRTX-b were not antagonized. This unusual action of BQ-123 supports the existence of subtypes of the ET$_A$R (PATE et al. 1998).

In the rabbit iris dilator muscle, ET-3 elicited lower contractile activities and SRTX-b elicited higher contractile activities than ET-1, while SRTX-c was inactive. BQ-123 antagonized ET-3 and SRTX-b more than ET-1. These functional experiments suggest that the iris dilator, like the iris sphincter, contains atypical ET_ARs (Nosaka et al. 1998).

IV. Atypical Receptors

Since only two mammalian ET receptors have been isolated, cloned, and expressed, the existence of additional ET receptors has been postulated on the basis of pharmacological, kinetic, and functional studies. The results of experiments using specific antagonists led some investigators to postulate the existence of two subtypes of ET_AR and/or ET_BR, or alternatively of a 'novel', 'atypical', or 'non-ET_A, non-ET_B' receptor.

Although SRTX-c is usually regarded as an ET_BR-specific agonist, various studies show that in certain tissues and cells this is not always the case. A recent example is the quail EDNRB2 (ET_BR), which exhibits an atypical ET_BR-type pharmacology. The three endothelin isopeptides, ET-1, -2, and -3, showed high (nanomolar) and approximately equal affinities for this receptor, characterizing the molecule as an ET_BR. However, further pharmacological investigation revealed that the quail EDNRB2 has low affinities for SRTXs, especially SRTX-c. Therefore, quail ET_BR behaves more like ET_AR in relation to SRTX-c. In this regard, EDNRB2 has a pharmacological property that is highly atypical for a type B receptor (Lecoin et al. 1998). ET_BR-like receptors insensitive to SRTX-c have also been reported in rat (Panek et al. 1992) and in *Xenopus* liver membranes (Nambi et al. 1994). Poulat et al. (1996) suggested that SRTX-c interacts with a non-ET_A/non-ET_B binding site in rat spinal cord.

Classical ET_BRs bind all the ET/SRTX peptides with equal affinity. However, in pig coronary arteries a receptor subtype was found that binds SRTX-c and ET-3, but not ET-1 or SRTX-b (Harrison et al. 1992). Accordingly, the existence of another 'atypical' receptor was proposed. Also, in granulosa cells ET-3 and SRTX-c were significantly more potent than ET-1 in suppressing basal and follicle stimulating hormone-stimulated estrogen production, suggesting that their effect is mediated by a non-ET_BR (Calogero et al. 1998).

In sheep choroid plexus, competitive binding studies revealed the presence of two binding sites with affinities for ET-1 in the picomolar range. One of these sites binds [Ala^{1,3,11,15}]ET-1 (a selective ET_BR agonist), but not SRTX-c (Angelova et al. 1997), suggesting the existence of an 'atypical' ET_B-like receptor that does not bind SRTX-c.

The endothelium of the human umbilical artery mediates relaxation in response to SRTX-b. This relaxation is antagonized by PD 142893, resulting in an enhancement of the contractile effect of SRTX-b. The endothelial receptors involved are neither ET_AR nor ET_BR and they are not activated by ET-1 (Laursen et al. 1998).

In guinea-pig bronchus, contractions induced by ET-1, ET-3, SRTX-c, or BQ-3020 appear to be predominantly mediated via stimulation of ET_BRs.

However, these receptors are relatively insensitive to the standard ET_BR antagonists BQ-788 and RES-701-1, suggesting that responses produced by these ligands in this tissue involve activation not of the classical ET_BR, but rather of an atypical ET receptor population (HAY and LUTTMANN 1997).

In rat pulmonary resistance arteries, vasoconstriction induced by ET-1, SRTX-c, or ET-3 is mediated predominantly by activation of an ET_B-like receptor. However, the lack of effect of some antagonists on ET-1-induced vasoconstriction suggests that the vasoconstriction is mediated via an atypical ET_BR (McCULLOCH et al. 1998).

ET-1 and ET-3 enhanced, in a concentration-dependent manner, the twitch response of the rat vas deferens to electrical stimulation. ET-1 was three times more potent than ET-3 and SRTX-c was at least 200 times less active than ET-1, suggesting that the receptor involved is an ET_AR subtype. However, further experiments with BQ-123 showed that the rat vas deferens contains an endothelin receptor that does not conform to the proposed classification of ET_AR/ET_BR subtypes (EGLEZOS et al. 1993).

The use of antagonists in various experiments has shown that the observed ET responses are not always conveyed by typical ET_ARs or ET_BRs. One receptor frequently described as 'atypical' is one in which ET_AR-specific antagonists such as BQ-123 and FR139317, in preparations containing predominantly ET_ARs, are more potent against SRTX-b than against ET-1. This has been described in numerous preparations (MAGUIRE et al. 1996 and references therein). It is possible that ET-1 elicits its response through a different receptor subtype than SRTX-b or that, in addition to having a common binding site with SRTX-b, ET-1 is able to activate a receptor that is insensitive to those antagonists (see next section).

F. Differences in Sensitivities to Antagonists

ETs and SRTXs differ markedly in their sensitivities to the different antagonists (see Chap. 9). This is important, as it shows that the inhibition exerted by antagonists against certain receptor subtypes is not complete. It also emphasizes the heterogeneity of the ET receptors. This section reviews recent reports that describe the differential effects of several antagonists on ET and SRTX.

In rat pulmonary arteries the potency of SB209670 (a mixed ET_AR/ET_BR antagonist) was higher against SRTX-c than against ET-1 (HAY et al. 1996), whereas in rabbit pulmonary arteries this antagonist abolished all responses to ET-1. In rabbit pulmonary arteries, BQ788 (an ET_BR-selective antagonist) did not inhibit responses to ET-1, though it inhibited SRTX-c induced contractions (HAY et al. 1996; MACLEAN et al. 1998). These findings could not be attributed to activation of the ET_AR by ET-1, as FR139317 (an ET_AR antagonist) also had no effect upon contractions mediated by ET-1.

In the human saphenous vein, ET-1 was much more resistant than SRTX-b to BQ-123 (NISHIYAMA et al. 1995). Similarly, PD142893 (an

ET_AR/ET_BR antagonist) hardly affected the contractile effect of ET-1, whereas it markedly antagonized the contractile effects of SRTX-b. The authors proposed that ET-1 and SRTX-b exert their effects through two different ET_AR subtypes, one of them highly sensitive and the other much less sensitive to BQ-123 and PD 142893.

In preparations containing predominantly ET_AR, ET_AR-specific antagonists such as BQ-123 and FR139317 were shown to be more potent against SRTX-b than against ET-1 itself (MAGUIRE et al. 1996 and references therein). For example, in human umbilical artery BQ-123 was more potent in inhibiting contractions induced by SRTX-b than by ET-1 (BODELSSON and STJERNQUIST 1993; BOGONI et al. 1996). In homogenates of human saphenous vein, the inhibitory effect of BQ-123 against vasoconstriction mediated by SRTX-b was 50 times stronger than against ET-1-mediated vasoconstriction (MAGUIRE et al. 1996). In the rabbit sphincter muscle, BQ-123 was competitive against ET-1 but non-competitive against SRTX-b (ISHIKAWA et al. 1996). In the rabbit iris dilator muscles BQ-123 was more strongly antagonistic against ET-3 and SRTX-b than against ET-1 (NOSAKA et al. 1998).

LAURSEN et al. (1998) recently showed that in human umbilical artery, both ET-1 and SRTX-b induced contractions were unaffected by the ET_BR antagonist BQ788. However, the ET_AR/ET_BR antagonist PD 142893 decreased the contraction induced by ET-1, whereas it enhanced the contraction induced by SRTX-b (LAURSEN et al. 1998).

The following reports demonstrate differences in the sensitivities of ET-1 and SRTX-c to various antagonists. WARNER et al. (1993a,b) showed that rat endothelial responses to SRTX-c were blocked by PD 142893, whereas rabbit pulmonary artery responses to ET-1 were only marginally blocked at higher concentrations of this antagonist. CLOZEL et al. (1992) demonstrated that bosentan (a non-peptide mixed ET_AR/ET_BR antagonist) was more effective at antagonizing SRTX-c-mediated endothelial responses than smooth-muscle responses in the rat trachea. In the guinea-pig trachea a non-peptide antagonist, Ro 47-0203, which is non-selective for endothelin ET_AR over endothelin ET_BR, had a greater effect against responses elicited by SRTX-c than by ET-1 (GATER et al. 1996). Similar findings were obtained in rabbit pulmonary resistance arteries with the non-selective ET_AR/ET_BR antagonist SB 209670 (DOCHERTY and MACLEAN 1998a,b). In rat pulmonary resistance arteries SRTX-c was slightly more potent than ET-1 in mediating contractions (MCCULLOCH et al. 1998). Bosentan increased the potency of ET-1, but inhibited responses to SRTX-c. The ET_BR antagonist BQ-788 did not inhibit responses to ET-1 but did inhibit responses to both SRTX-c and ET-3.

In guinea-pig bronchus, contractions induced by ET-1, ET-3, SRTX-c, or BQ-3020 appear to be mediated predominantly via stimulation of ET_BR. However, differences in antagonistic potencies were noted using the ET_BR antagonists BQ-788 and RES-701-1. This study also provides additional evidence that the potencies of ET receptor antagonists depend upon the specific ET agonist (HAY and LUTTMANN 1997).

G. Similar and Opposing Responses Mediated by ETs/SRTXs

Several studies point to similarity in the biological actions of ET-1 and SRTX-b. Both peptides were found to produce constriction of the isolated coronary or mesenteric vascular beds (HAN et al. 1990). Both significantly enhanced the electrical field stimulation-induced contraction of guinea-pig pulmonary artery, and in both cases the response could be blocked by BQ-123 (MATSUDA et al. 1996). Intra-arterial bolus injections of ET-1 or SRTX-b induced dose-dependent and long-lasting vasoconstriction in the isolated canine liver arterial circuit, and infusion of the ET_AR antagonist FR-139317 markedly reduced both ET-1-induced and SRTX-b induced vasoconstriction (FARO et al. 1995). Both peptides were able to induce phosphoinositide hydrolysis in a dose-dependent manner; however, ET-1 was more potent than SRTX-b. Both ET-1-induced and SRTX-b-induced accumulation of inositol phosphate was almost totally inhibited by BQ-123 (MONDON et al. 1995). Both peptides induced histamine release from Weibel-Palade bodies of the toad aorta (DOI et al. 1995). Centrally administered ET-1 produced a transient rise followed by a sustained decrease in blood pressure (KUMAR et al. 1996; REBELLO et al. 1995). Intravenous infusion of SRTX-b also induced a decrease in blood flow to several organs. Moreover, centrally administered SRTX-b produced significant changes in systemic and regional blood circulation, apparently mediated though the sympathetic nervous system (KUMAR et al. 1997). In conscious Long-Evans rats, 4 pmol of ET-1, ET-3 or SRTX-b caused a significant increase in cardiac output, and 40 pmol doses caused initial hypotension and increases in cardiac output, stroke volume and total peripheral conductance (GARDINER et al. 1990).

Other studies, however, indicate differences between the biological actions of ET-1 and SRTX-b. In the isolated rabbit iris sphincter SRTX-b showed much stronger contractile activity (about 30 times higher) than ET-1 (ISHIKAWA et al. 1996). In the human umbilical artery, both peptides induced contractions that were unaffected by an ET_BR antagonist. However, SRTX-b also induced relaxation in segments in which 5-hydroxytryptamine had caused contraction (LAURSEN et al. 1998). Also, in homogenates of human saphenous vein BQ123 exhibited a 50-fold higher affinity against SRTX-b-mediated vasoconstriction as compared to ET-1. These data suggest that BQ-123 is a much more potent blocker of SRTX-b than of ET-1 contractile responses (MAGUIRE et al. 1996). In another study, ET-1 and SRTX-b elicited potent concentration-dependent contractions of human saphenous vein with similar pD2 values and similar maximal responses. BQ-123 and PD 142893, alone or in combination, hardly affected the contractile effect of ET-1, while each of them markedly antagonized the effects of SRTX-b (NISHIYAMA et al. 1995). These studies suggest that ET-1 probably activates an additional population of receptors that may have a lower affinity for BQ-123, and that these receptors are not affected by SRTX-b (see Sect. F).

Desensitization experiments have shown differences in responses to ET and SRTX. In the human saphenous vein, for example, exposure to a high concentration of either ET-1 or SRTX-b rendered the preparations unresponsive to repeated additions of the same agonist. However, whereas addition of SRTX-b to the preparation unresponsive to ET-1 did not elicit any response, a small contraction was obtained with ET-1 in preparations not responding to SRTX-b (Maguire et al. 1996).

A recent study showed that in the Egyptian mongoose ET-1 and SRTX-b induce different physiological responses (Bdolah et al. 1997). The mongoose is resistant to the venom of *Atractaspis* and its most toxic component, SRTX-b. Intravenous administration of this toxin, at a dose of about 13 times the LD_{100} for mice, resulted in electrocardiographic disturbances in the mongoose, which returned to normal after several hours. SRTX-b failed, however, to induce contraction of mongoose aortal preparations. ET-1, which was shown by immunological methods to be present in mongoose tissue extracts, did induce contraction of the isolated mongoose aorta, but this effect was greatly decreased when ET-1 was applied in addition to SRTX-b. Binding studies revealed ET/SRTX-specific binding sites in brain and cardiovascular preparations from the mongoose. It was suggested that the ET/SRTX receptors in the mongoose contain structural features that enable them to differentiate between the two peptides.

ET-1 and SRTX-c usually elicit vasoconstriction. For example, the two peptides were shown to produce contraction of the esophageal muscularis mucosae in a concentration-dependent manner (Uchida et al. 1998). However, differences in the physiological responses of these two peptides have been reported. SRTX-c caused relaxation in rabbit basilar artery, whereas in the same preparation ET-1 mediated vasoconstriction (Zuccarello et al. 1998a). SRTX-c also relaxed rabbit basilar artery constricted with serotonin in situ, while ET-1 caused constriction in this preparation as well (Zuccarello et al. 1998b). The inability of SRTX-c to induce constriction was unexpected, as the ET_BR-mediated contractile efficacies of SRTX-c and ET-1 are similar. One possible explanation is that SRTX-c may induce greater release of a relaxant factor as a result of ET receptor activation, thereby preventing ET_BR-mediated constriction (Zuccarello et al. 1998a). Also, in isolated rabbit mesenteric arteries vascular response mediating endothelin ET_BR showed that ET-1 caused a concentration-dependent contraction, whereas SRTX-c had no effect (Iwasaki et al. 1999). Endothelin-1 caused graded tonic contractions and enhanced neurogenic contractions in rat electrical field-stimulated seminal vesicles, whereas SRTX-c was ineffective (Luciano et al. 1998). On the other hand, SRTX-c induced vasoconstriction in rabbit pulmonary resistance arteries and was even more potent in this regard than ET-1 (MacLean et al. 1998).

Numerous reports have shown that SRTX-c is a more potent vasoconstrictor than ET-1 in rabbit and human saphenous veins, in rabbit conduit pulmonary arteries, and in adult and rat pulmonary resistance arteries (Moreland et al. 1994; White et al. 1994; Douglas et al. 1995; Hay et al.

1996; McCulloch et al. 1998; Docherty and MacLean 1998a). In these reports the ET receptor involved is referred to as 'ET$_C$-like', 'ET$_B$-like', or 'atypical'. In adult rat and rabbit pulmonary resistance arteries, the contractile responses to SRTX-c are strongly potentiated by inhibition of nitric oxide synthesis (Docherty and MacLean 1998a).

Differences in responses and in magnitude of contractions between ET-1 and SRTX-c were also observed in tumor blood flow in female rats. SRTX-c was more potent than ET-1, as indicated by the fact that blood flow was increased by 175% with SRTX-c and by 75% with ET-1. Vasodilation was observed following administration of ET-1, but was not seen with SRTX-c (Bell et al. 1995). Estradiol attenuated ET-1-induced vasoconstriction, but had no effect upon SRTX-mediated vasoconstriction (Sudhir et al. 1997).

A recent study (Docherty and MacLean 1998b) showed that the potency of ET-1 and SRTX-c varies with developmental age. Contractile responses to ET-1 and SRTX-c were studied in rabbit pulmonary resistance arteries from fetuses and from animals aged 0–24h, 4 days and 7 days. SRTX-c and ET-1 were equally potent in the fetus, but SRTX-c was increasingly more potent than ET-1 with increasing age, and was about 1000 times more potent in 7-day-old animals. At 7 days the responses to ET-1 were also resistant to both FR139317 and BQ-788. The latter antagonist inhibited responses to SRTX-c at all ages examined. The nonselective antagonist SB 209670 inhibited responses to ET-1 and SRTX-c in rabbits aged 0–24h and 4 days. During the first week of life the potency of SRTX-c increased while that of ET-1 decreased, suggesting differential development of responses to ET-1 and SRTX-c and heterogeneity of responses mediated by ET$_A$R or 'ET$_B$-like' receptors. This divergence in the development of responses to ET-1 and SRTX-c suggests that these peptides either activate different receptor populations or have different binding domains on the same receptor (Docherty and MacLean 1998b).

Besides vasoconstriction and relaxation, there are other responses that differ when elicited by the different peptides. For example, ET-3 and SRTX-c were significantly more potent than ET-1 in suppressing basal and follicle stimulating hormone-stimulated estrogen production in granulosa cells obtained from immature, estrogen-primed female rats (Calogero et al. 1998). According to the authors, the greater potency of SRTX-c might suggest that its effect is mediated by a non-ET$_B$R. In human myometrial cells in culture, ET-1 stimulated DNA synthesis and proliferation but ET-3 and SRTX-c had no effect (Breuiller-Fouche et al. 1998). Also, in an osteoblast-like cell line, ET-1, ET-2, and SRTX-b rapidly stimulated prostaglandin E$_2$ production through a protein tyrosine kinase-dependent and protein kinase C-dependent pathway, whereas ET-3, SRTX-a, and SRTX-c had no such effect (Leis et al. 1998).

SRTX-c at low concentrations caused a significant enhancement of neutrophil migration, whereas at higher concentrations it inhibited neutrophil migration stimulated by chemotactic activators (Elferink and de Koster 1996). These authors demonstrated a similar pattern in the effect of ETs on neutrophil migration, namely stimulation by low concentrations and inhibition by high concentrations of the ligand (ELFERINK and DE KOSTER 1996 and references therein). There were some similarities and differences between the effects of SRTX-c and of ETs on neutrophil migration. The similarities were the stimulatory effect on random migration and the inhibitory effect on IL-8 or fMLP-activated chemotaxis. However, the stimulatory effect of SRTX-c was minor compared with that of ET. Also, the stimulatory and inhibitory effects of ET were totally dependent on extracellular calcium, while the effects of SRTX-c are not.

In rat cortical astrocytes the agonists ET-1, ET-3, SRTX-c, and IRL 1620 elicited phospholipase D activation in a concentration-dependent manner. The potencies of ET-1, ET-3, and SRTX-c were similar. However, the maximal effects evoked by ET-3, SRTX-c, and IRL 1620 were significantly lower than the maximal response to ET-1. The response to ET-1 (1 nmol/l) was inhibited, in a biphasic manner, by increasing concentrations of BQ-123. Increasing concentrations of BQ-788 inhibited the response to SRTX-c. BQ-788 also inhibited the effect of ET-1, although in this case two components were defined. Rapid desensitization was achieved by preincubation with ET-1 or SRTX-c. In cells pretreated with SRTX-c neither ET-3 nor SRTX-c activated phospholipase D, but ET-1 still induced approximately 40% of the response shown by non-desensitized cells (SERVITJA et al. 1998).

Differences were also found in the modulatory effects of the ETs on the neurogenically induced release of endogenous noradrenaline and the co-transmitter adenosine 5′-triphosphate (ATP) from the sympathetic nerves of endothelium-free segments of the rat isolated tail artery. ET-1 significantly reduced the overflow of both noradrenaline and ATP evoked by electrical field stimulation. The inhibitory effect of ET-1 was resistant to BQ-123 but was prevented by ET_BR desensitization. SRTX-c exerted a dual effect on transmitter release. At low concentrations the peptide significantly reduced the evoked noradrenaline overflow, whereas at higher concentrations it caused a significant increase in the evoked overflow of both ATP and noradrenaline. ET-3 did not affect the evoked overflow of either ATP or noradrenaline, but at high concentrations it significantly potentiated the release of both transmitters (MUTAFOVA-YAMBOLIEVA and WESTFALL 1998).

H. Signaling Differences Between ET and SRTX

The ETs and SRTXs activate many signal transduction cascades (SOKOLOVSKY 1995 and references therein) (see Chap. 5). Different signals have been attributed to differences in receptor activation and these signals have resulted in distinct responses, as described in the preceding section. In this section we summarize the research done in our laboratory and describe our latest study, which shows that the differences in signaling probably occur at the level of ligand-receptor–G-protein coupling.

In rat atrial slices, ET-1 both stimulated and inhibited cAMP formation, depending on its concentration (SOKOLOVSKY et al. 1994). ET-1 at picomolar concentrations induced an increase in cAMP formation while higher concentrations inhibited it. SRTX-b and SRTX-c showed a different pattern of behavior. No stimulation was observed at picomolar concentrations, but stimulation occurred, though at lower efficiency, at the nanomolar range. Also, the β-blocker propranolol inhibited cAMP formation by ET-1, but did not affect cAMP formation by SRTX-b or SRTX-c. ET-1 at 1pmol/l induced norepinephrine release from the atrial slices, while the SRTXs blocked this release.

ETs and SRTXs were also shown to stimulate the guanylate cyclase pathway in rat cerebellar and atrial slices (SHRAGA-LEVINE et al. 1994; SHRAGA-LEVINE and SOKOLOVSKY 1996). In both preparations, ET-1 and SRTX-b at picomolar concentrations (10^{-14}–10^{-12}mol/l) induced a dose-dependent increase in the production of cGMP, whereas higher concentrations caused its dose-dependent inhibition. At the nanomolar concentration range, both ET-3 and SRTX-c induced a dose-dependent increase in cGMP production. In the cerebellum, the ETs stimulated cGMP formation via the NO pathway through the ET_AR, while the SRTXs stimulated cGMP formation via the CO pathway, also through the ET_AR. In the rat atrium, ET-1 and SRTX-b stimulated cGMP production via the NO pathway through the ET_AR, while ET-3 and SRTX-c stimulated it via the CO pathway through the ET_BR. These data show that there are two different ET_ARs in the cerebellum and the atria, and/or that the ET_AR couples to different G proteins in these two tissues.

The results of the studies of the cAMP and cGMP pathways strongly indicate that the interaction between the receptors and the different ligands, ET-1, ET-3, SRTX-b, and SRTX-c, is not a simple one. Our conclusion that the ligands' signaling pathways differ even though their binding sites are mutually exclusive suggests that the ligand-receptor interaction is more complex than was first realized. Also, our studies showed that the same receptor subtype can transduce different signals in different tissues. One explanation is that different tissues may contain non-identical receptor subtypes; for example, the ET_AR in the atrium might be distinct from the ET_AR in the cerebellum and in the ventricle. Alternatively, the receptor-ligand complex might interact differently with the various G proteins in the different tissues; for example, the ligand-receptor-G-protein complex in the cerebellum is distinct from the ligand-receptor-G-protein complex in the atrium. This would also explain the different signals transmitted by the different ligands in the various tissues.

We have previously proposed a number of models to explain our results (SHRAGA-LEVINE et al. 1994): (*a*) ligand-induced coupling of ET receptors to specific G proteins; (*b*) pre-existing functional coupling of ET receptors to specific G proteins, and specific interaction between these receptors and the various ligands; and (*c*) two separate receptor subtypes, each interacting specifically with ligand and G proteins.

In our latest study (SHRAGA-LEVINE and SOKOLOVSKY, 2000) we examined the coupling of ET_AR and ET_BR to various G proteins after stimulation with

ET-1, ET-3, SRTX-b, and SRTX-c. The experiments were performed in two fibroblast cell lines, each overexpressing one subtype of the human endothelin receptor, ET_AR or ET_BR. The two cell lines express both the high-affinity and the super-high-affinity states of the receptor. In this study we show that the two receptors couple to various G proteins depending on the receptor subtype and on the ligand employed.

Stimulation with the different ET agonists yielded differences in G protein coupling to the two receptor subtypes. ET-1, ET-3, SRTX-b, and SRTX-c induced a marked increase in coupling of $G_{i3}\alpha$ to the ET_BR, whereas ET-1 had no effect on its coupling to the ET_AR, and ET-3 and SRTX-c even induced a decrease. On the other hand, each of the four ligands induced a much greater increase in the coupling of $G_q\alpha/G_{11}\alpha$ to the ET_AR than to the ET_BR. The ligand-induced coupling between the G_{i1} or G_{i2} α-subunits and the two receptor subtypes was basically similar. This was also true for coupling of the receptors to $G_o\alpha$, except that ET-3 increased the coupling of this α-subunit to the ET_BR, but decreased its coupling to the ET_AR.

In fibroblasts overexpressing the ET_BR, the different ligands appeared to have the same effect upon coupling of the receptor to a particular G protein, i.e., either they all increased or they all decreased the coupling between the receptor and the G protein α-subunit. Differences were observed, however, in the extent of their effects. ET-1 and SRTX-b behaved similarly, causing a more intense coupling than ET-3 or SRTX-c, which themselves had a similar effect on the extent of coupling.

Differences in receptor-G-protein coupling were observed after agonist stimulation in fibroblasts overexpressing the ET_AR. In these cells, one ligand induced an increase while another caused a decrease. For example, ET-1, SRTX-b, and SRTX-c increased the coupling of ET_AR to $G_o\alpha$, whereas ET-3 decreased it. Also, SRTX-b was the only ligand that strengthened the interaction between ET_AR and $G_{i3}\alpha$; all the others weakened it. Another example is in the coupling between ET_AR and $G_{i1}\alpha$,$G_{i2}\alpha$; whereas the basal level of this coupling was maintained by ET-3, SRTX-b, or SRTX-c, it was decreased by ET-1.

In conclusion, upon binding with endothelins and/or sarafotoxins, the ET receptors couple with appropriate G-proteins and transmit signals to the intracellular signal-transduction pathways. Our latest study shows that endothelin receptors have a promiscuous nature and can activate multiple types of G-proteins. This coupling of ET receptors to G proteins shows that such coupling is ligand- as well as receptor subtype-specific. This study provides new information showing that although the various ligands share the same binding sites, their modes of action differ because of the different receptor-ligand-G-protein complexes formed.

I. Conclusions

This chapter summarizes the relationship between the SRTXs and the ETs. This relationship was at first considered a simple one; the two families share a high degree of homology, bind to the same receptor(s), and stimulate the same signals and responses. However, the SRTXs have shown that, although the similarity exists, they serve as a useful tool for indicating the existence of new receptors and/or for pointing out unusual signals mediated by the ET receptor subtypes.

References

Andersen NH, Chen C, Marschner TM, Krystek SR, Bassolino DA (1992) Conformational isomerism of endothelin in acidic aqueous media: a quantitative NOESY analysis. Biochemistry 31:1280–1295

Angelova K, Puett D, Narayan P (1997) Identification of endothelin receptor subtypes in sheep choroid plexus. Endocrine 7:287–293

Arinami T, Ishikawa M, Inoue A, Yanagisawa M, Masaki T, Yoshida MC, Hamaguchi H (1991) Chromosomal assignments of the human endothelin family genes: the endothelin-1 gene (EDN1) to 6p23-p24, the endothelin-2 gene (EDN2) to 1p34, and the endothelin-3 gene (EDN3) to 20q13.2-q13.3. Am J Hum Genet 48:990–996

Atkins AR, Ralston GB, Smith R. (1994) Conformational stability of the endothelin/sarafotoxin family of peptides. Int J Peptide Protein Res 44:372–377

Atkins AR, Martin RC, Smith R (1995) ^{1}H NMR studies of sarafotoxin SRTb, a nonselective endothelin receptor agonist, and IRL 1620, an ET_B receptor-specific agonist. Biochemistry 34:2026–2033

Bdolah A, Wollberg Z, Ambar I, Kloog Y, Sokolovsky M, Kochva E (1989a) Disturbances in the cardiovascular system caused by endothelin and sarafotoxin. Biochem Pharmacol 38:3145–3146

Bdolah A, Wollberg Z, Fleminger G, Kochva E (1989b) SRTX-d, a new native peptide of the endothelin/sarafotoxin family. FEBS Lett 256:1–3

Bdolah A, Wollberg Z, Kochva E (1991) Snake Toxins. In: Harvey L (ed) Pergamon, New York, pp 415–424

Bdolah A, Kochva E, Ovadia M, Kinamon S, Wollberg Z (1997) Resistance of the Egyptian mongoose to sarafotoxins. Toxicon 35:1251–1261

Becker A, Dowdle EB, Hechler U, Kauser K, Donner P, Schleuning WD (1993) Bibrotoxin, a novel member of the endothelin/sarafotoxin peptide family, from the venom of the burrowing asp *Atractaspis bibroni*. FEBS Lett 315:100–103

Bell KM, Prise VE, Chaplin DJ, Tozer GM (1995) Effect of endothelin-1 and sarafotoxin S6c on blood flow in a rat tumor. J Cardiovasc Pharmacol 26:S222–S225

Bloch KD, Friedrich SP, Lee M-E, Eddy RL, Shows TB, Quertermous T (1989) Structural organization and chromosomal assignment of the gene encoding endothelin. J Biol Chem 264:10851–10857

Bodelsson G, Stjernquist M (1993) Characterization of endothelin receptors and localization of ^{125}I-endothelin-1 binding sites in human umbilical artery. Eur J Pharmacol 249:299–305

Bogoni G, Rizzo A, Calo G, Campobasso C, D'Orleans-Juste P, Regoli D (1996) Characterization of endothelin receptors in the human umbilical artery and vein. Br J Pharmacol 119:1600–1604

Breuiller-Fouche M, Heluy V, Fournier T, Dallot E, Vacher-Lavenu MC, Ferre F (1998) Role of endothelin-1 in regulating proliferation of cultured human uterine smooth muscle cells. Mol Hum Reprod 4:33–39

Calogero AE, Burrello N, Ossino AM (1998) Endothelin (ET)-1 and ET-3 inhibit estrogen and cAMP production by rat granulosa cells in vitro. J Endocrinol 157:209–215

Clozel M, Gray GA, Breu V, Loffler B-M, Osterwalder K (1992) The endothelin ETB receptor mediates both vasodilation and vasoconstriction in vivo. Biochem Biophys Res Commun 186:867–873

Cristol J-P, Warner T D, Thiemermaan C, Vane JR (1993) Mediation via different receptors of the vasoconstrictor effects of endothelins and sarafotoxins in the systemic circulation and renal vasculature of the anaesthetized rat. Br J Pharmacol 108:776–779

Davenport AP, Morton AJ, Brown MJ. (1991) Localization of endothelin-1 (ET-1), ET-2, and ET-3, mouse VIC, and sarafotoxin S6b binding sites in mammalian heart and kidney. J Cardiovasc Pharmacol 17:S152–S155

DeNucci G, Thomas R, D'Orleans-Juste P, Antunes E, Walder C, Warner TD, Vane JK (1988) Pressor effects of circulating endothelin are limited by its removal in the pulmonary circulation and by the release of prostacyclin and endothelium-derived relaxing factor. Proc Natl Acad Sci USA 85:9797–9800

Devadason PS, Henry PJ (1997) Comparison of the contractile effects and binding kinetics of endothelin-1 and sarafotoxin S6b in rat isolated renal artery. Br J Pharmacol 121:253–263

Doherty AM, Cody WL, Leitz NL, Depue PL, Taylor MD, Rapundalo ST, Hingorani GP, Major TC, Panek RL, Taylor DG (1991) Structure-activity studies of the C-terminal region of the endothelins and sarafotoxins. J Cardiovasc Pharmacol 17:S59–S61

Docherty C, MacLean MR (1998a) Development of endothelin receptors in perinatal rabbit pulmonary resistance arteries. Br J Pharmacol 124:1165–1174

Docherty C, MacLean MR (1998b) Endothelin_B receptors in rabbit pulmonary resistance arteries: effect of left ventricular dysfunction. J Pharmacol Exp Ther 284:895–903

Doi Y, Ozaka T, Katsuki M, Fukushige H, Toyama E, Kanazawa Y, Arashidani K, Fujimoto S (1995) Histamine release from Weibel-Palade bodies of toad aortas induced by endothelin-1 and sarafotoxin-S6b. Anat Rec 242:374–382

Douglas SA, Hiley CR (1991) Endothelium-dependent mesenteric vasorelexant effects and systemic actions of endothelin(16–21) and other endothelin-related peptides in the rat. Br J Pharmacol 104:311–320

Douglas SA, Beck GR Jr, Elliot JD, Ohlstein EH (1995) Pharmacologic evidence for the presence of three functional endothelin receptor subtypes in the rabbit lateral saphenous vein. Br J Pharmacol 114:1529–1540

Ducancel F, Matre U, Dupont C, Lajeunesse E, Wollberg Z, Bdolah A, Kochva E, Boulain JC, Menez A (1993) Cloning and sequence analysis of cDNAs encoding precursors of sarafotoxins. Evidence for an unusual rosary-type organization. J Biol Chem 268:3052–3055

Eglezos A, Cucchi P, Patacchini R, Quartara L, Maggi CA, Mizrahi J (1993) Differential effects of BQ-123 against endothelin-1 and endothelin-3 on the rat vas deferens: evidence for an atypical endothelin receptor. Br J Pharmacol 109:736–738

Elferink JGR, de Koster BM (1996) Modulation of human neutrophil chemotaxis by the endothelin-B receptor agonist sarafotoxin S6c.Chemico-Biol Int 101:165–174

Faro R, Grassikassisse DM, Donato JL, Boin I, Opgenorth TJ, Withrington PG, Zatz R, Antunes E, Denucci G (1995) Role of endothelin ET(A) and ET(B) receptors in the arterial vasculature of the isolated canine liver. J Cardiovasc Pharmacol 26:S204–S207

Galron R, Bdolah A, Kochva E, Wollberg Z, Kloog Y, Sokolovsky M (1991) Kinetic and cross-linking studies indicate different receptors for endothelins and sarafotoxins in the ileum and cerebellum. FEBS Lett 283:11–14

Gardiner SM, Compton AM, Kemp PA, Bennett T (1990) Cardiac output effects of endothelin-1, -2 and -3 and sarafotoxin S6b in conscious rats. J Auton Nerv Syst 30:143–147

Gater PR, Wasserman MA, Renzetti LM (1996) Effects of Ro 47–0203 on endothelin-1 and sarafotoxin S6c-induced contractions of human bronchus and guinea-pig trachea. Eur J Pharmacol 304:123–128

Gellai M, DeWolf R, Pullen M, Nambi P (1994) Distribution and functional role of renal ET receptor subtypes in normotensive and hypertensive rata. Kidney Int 46:1287–1294

Gellai M, Fletcher T, Pullen M, Nambi P (1996) Evidence for the existence of endothelin-B receptor subtypes and their physiological roles in the rat. Am J Physiol 271:R254–R261

Gulati A, Rebello S (1992) Characteristics of endothelin receptors in the central nervous system of spontaneously hypertensive rats. Neuropharmacology 31: 243–250

Han SP, Knuepfer MM, Trapani AJ, Fok KF, Westfall TC (1990) Cardiac and vascular actions of sarafotoxin S6b and endothelin-1. Life Sci 46:767–775

Harrison VJ, Randriantsoa A, Schoeffter P (1992) Heterogeneity of endothelin-sarafotoxin receptors mediating contraction of pig coronary artery. Br J Pharmac 105:511–513

Hay DWP (1992) Pharmacological evidence for distinct endothelin receptors in guinea-pig bronchus and aorta. Br J Pharmacol 106:759–761

Hay DWP, Luttmann MA (1997) Nonpeptide endothelin receptor antagonists. IX. Characterization of endothelin receptors in guinea-pig bronchus with SB 209670 and other endothelin receptor antagonists. J Pharmacol Exp Ther 280:959–965

Hay DWP, Luttmann MA, Beck G, Ohlstein EH (1996) Comparison of endothelin$_B$ (ET$_B$) receptors in rabbit isolated pulmonary artery and bronchus. Br J Pharmacol 118:1209–1217

Inoue A, Yanagisawa M, Kimura S, Kasuya Y, Miyauchi T, Goto K, Masaki T (1989) The human endothelin family: three structurally and pharmacologically distinct isopeptides predicted by three separate genes. Proc Natl Acad Sci USA. 86: 2863–2867

Ishikawa H, Haruno I, Harada Y, Yoshitomi T, Ishikawa S, Katori M (1996) Pharmacological characterization of endothelin receptors in the rabbit iris sphincter muscle: suggestion for the presence of atypical receptors. Curr Eye Res 15:73–78

Iwasaki T, Notoya M, Hayasaki-Kajiwara Y, Shimamura T, Naya N, Ninomiya M, Nakajima M (1999) Endothelium-independent vascular relaxation mediating ET$_B$ receptor in rabbit mesenteric arteries. Am J Physiol 276:H383–390

Johns CR, Hiley CR (1990) Heterogeneity of [^{125}I]sarafotoxin S6b binding in rat tissues. Eur J Pharmacol 183:1746

Kitazumi K, Shiba T, Nishiki K, Furukawa Y, Takasaki C, Tasaka K. (1990) Structure-activity relationship in vasoconstrictor effects of sarafotoxins and endothelin-1. FEBS Lett 260:269–272

Kloog Y, Ambar I, Sokolovsky M, Kochva E, Bdolah A, Wollberg Z (1988) Sarafotoxin, a novel vasoconstrictor peptide: phosphoinositide hydrolysis in rat heart and brain. Science 242:268–270

Kochva E, Bdolah A, Wollberg Z. (1993) Sarafotoxins and endothelins: evolution, structure and function. Toxicon 31:541–568

Koshi T, Suzuki C, Arai K, Mizoguchi T, Torii T, Hirata M, Ohkuchi M, Okabe T (1991) Syntheses and biological activities of endothelin-1 analogs. Chem Pharm Bull 39:3061–3063

Kumar A, Shahani BT, Gulati A (1996) Modification of systemic and regional circulatory effects of intracerebroventricular administration of endothelin-1 (ET-1) by propranolol in anesthetized rats. Gen Pharmacol 27:1025–1033

Kumar A, Morrison S, Gulati A (1997) Effect of ET$_A$ receptor antagonists on cardiovascular responses induced by centrally administered sarafotoxin 6b: role of sympathetic nervous system. Peptides 18:855–864.

Lamthanh H, Bdolah A, Creminon C, Grassi J, Menez A, Wollberg Z, Kochva E (1994) Biological activities of [Thr2]sarafotoxin-b, a synthetic analogue of sarafotoxin-b. Toxicon 32:1105–1114

Landan G, Bdolah A, Wollberg Z, Kochva E, Graur D (1991a) Evolution of the sarafotoxin/endothelin superfamily of proteins. Toxicon 29:237–244

Landan G, Bdolah A, Wollberg Z, Kochva E, Graur D (1991b) The evolutionary history of the sarafotoxin/endothelin/endothelin-like superfamily J Cardiovasc Pharmacol 17:S517–S519

Laursen M, Bodelsson G, Stjernquist M (1998) Atypical receptors mediate the response to endothelin-1 and sarafotoxin S6b in the human umbilical artery. Eur J Pharmacol 362:167–172

Lecoin L, Sakurai T, Ngo M-T, Abe Y, Yanagisawa M, Le Douarin NM (1998) Cloning and characterization of a novel endothelin receptor subtype in the avian class. Proc Natl Acad Sci USA 95:3024–3029

Leis HJ, Zach D, Huber E, Windischhofer W (1998) Prostaglandin endoperoxide synthase-2 contributes to the endothelin/sarafotoxin-induced prostaglandin E2 synthesis in mouse osteoblastic cells (MC3T3-E1): evidence for a protein tyrosine kinase-signaling pathway and involvement of protein kinase C. Endocrinology 139:1268–1277

Lin WW, Chen YM, Lee SY, Nishio H, Kimura T, Sakakibara S, Lee CY (1990) Cardiovascular effects of two disulfide analogues of sarafotoxin S6b. Toxicon 28:911–923

Luciano LG, D'Orleans-Juste P, Calixto JB, Rae GA (1998) Endothelin-1 selectively potentiates the purinergic component of sympathetic neurotransmission in rat seminal vesicle. J Cardiovasc Pharmacol 31:S515–S517

MacLean MR, Mackenzie JF, Docherty CC (1998) Heterogeneity of endothelin-B receptors in rabbit pulmonary resistance arteries. J Cardiovasc Pharmacol 31:S115–S118

Maguire JJ, Kuc RE, Rous BA, Davenport AP (1996) Failure of BQ123, a more potent antagonist of sarafotoxin 6b than of endothelin-1, to distinguish between these agonists in binding experiments. Br J Pharmacol 118:335–342

Matsuda H, Kawaguchi A, Uematsu M, Ohmori F, Nagata S, Miyatake K (1996) Endothelins contract guinea-pig pulmonary artery and enhance its adrenergic response via ETA receptors. Clin Exp Pharmacol Physiol 23:379–385

McCulloch KM, Docherty C, MacLean MR (1998) Endothelin receptors mediating contraction of rat and human pulmonary resistance arteries: effect of chronic hypoxia in the rat. Br J Pharmacol 123:1621–1630

Miasiro N, Karaki H, Paiva (1998) Distinct endothelin-B receptors mediate the effects of sarafotoxin S6c and IRL1620 in the ileum. J Cardiovasc Pharmacol 31:S175–S178

Miller V, Michener SR (1995) Modulation of contractions to and receptors for endothelins in canine veins. Am J Physiol 268:H345–H350

Mills, RG, Ralston, GB, King GF (1994) The solution structure of sarafotoxin-c. J Biol Chem 269:23413–23419

Mondon F, Maka F, Sabry S, Ferre F (1995) Endothelin-induced phosphoinositide hydrolysis in the muscular layer of stem villi vessels of human term placenta. Eur J Endocrinol 133:606–612

Moreland S, McMullen D, Abboa-Offei B, Seymour A (1994) Evidence for a differential location of vasoconstrictor endothelin receptors in the vasculature. Br J Pharmacol 112:704–708

Mutafova-Yambolieva VN, Westfall DP (1998) Inhibitory and facilitatory presynaptic effects of endothelin on sympathetic cotransmission in the rat isolated tail artery. Br J Pharmacol 123:136–142

Nambi P, Wu H-L, Pullen M, Aiyar N, Bryan D, Elliott J (1992) Identification of endothelin receptor subtypes in rat kidney cortex using subtype-selective ligands. Mol Pharmacol 42:336–339

Nambi P, Pullen M, Kumar C (1994) Identification of a novel endothelin receptor in Xenopus laevis liver. Neuropeptides 26:181–185

Nambi P, Pullen M, Brooks DP, Gellai M (1995) Identification of ETB receptor subtypes using linear and truncated analogs of ET. Neuropeptides 29:331–336

Nishiyama M, Takahara Y, Masaki T, Nakajima N, Kimura S (1995) Pharmacological heterogeneity of both endothelin ET_A- and ET_B-receptors in the human saphenous vein. Jpn J Pharmacol 69:391–398

Nosaka C, Ishikawa H, Haruno I, Yoshitomi T, Kase H, Ishikawa S, Harada Y (1998) Radioligand binding characteristics of the endothelin receptor in the rabbit iris. Jpn J Pharmacol 76:289–296

Opgenorth TH, Wu-Wong JR, Shiosaki K (1992) Endothelin-converting enzymes. FASEB J 6:2653–2659

Panek RL, Major TC, Hingorani GP, Doherty AMA, Taylor DG (1992) Endothelin and structurally related analogs distinguish between endothelin receptor subtypes. Biochem Biophys Res Commun 183:566–571

Pate MA, Chester AH, Brown TJ, Roach AG, Yacoub MH (1998) Atypical antagonism observed with BQ-123 in human saphenous vein. J Cardiovasc Pharmacol 31:S172–S174

Poulat P, De Champlain J, D'Orleans-Juste P, Couture R (1996) Receptor and mechanism that mediate endothelin- and big endothelin-1-induced phosphoinositide hydrolysis in the rat spinal cord. Eur J Pharmacol 315:327–334

Rasmussen TE, Jougasaki M, Supaporn T, Hallett JW Jr, Brooks DP, Burnett JC Jr (1998) Cardiovascular actions of ET-B activation in vivo and modulation by receptor antagonism. Am J Physiol 274:R131–R138

Rebello S, Roy S, Saxena PR, Gulati A (1995) Systemic hemodynamic and regional circulatory effects of centrally administered endothelin-1 are mediated though ETA receptors. Brain Res 676:141–150

Rose PM, Krystek SR Jr, Patel PS, Liu ECK, Lynch JS, Lach DA, Fisher SM, Webb ML (1995) Aspartate mutation distinguishes ETA but not ETB receptor subtypeselective ligand binding while abolishing phospholipase C activation in both receptors. FEBS Lett 361:243–249

Russell FD, Davenport AP (1996) Characterization of the binding of endothelin ET_B selective ligands in human and rat heart. Br J Pharmacol 119:631–636

Servitja JM, Masgrau R, Sarri E, Picatoste F (1998) Involvement of ET(A) and ET(B) receptors in the activation of phospholipase D by endothelins in cultured rat cortical astrocytes. Br J Pharmacol 124:1728–1734

Shannon TR, Hale CC (1994) Identification of a 65 kDa endothelin receptor in bovine cardiac sarcolemmal vesicles. Eur J Pharmacol 267:233–238

Shraga-Levine Z, Galron R, Sokolovsky M (1994) Cyclic GMP formation in rat cerebellar slices is stimulated by endothelins via nitric oxide formation and by sarafotoxins via formation of carbon monoxide. Biochemistry 33:14656–14659

Shraga-Levine Z, Sokolovsky M (1996) cGMP formation in rat atrial slices is ligand and endothelin receptor subtype specific. Circ Res 78:424–430

Shraga-Levine Z, Sokolovsky M (2000) Functional Coupling of G Proteins to Endothelin Receptors Is Ligand and Receptor Subtype Specific. Cellular and Molecular Neurobiology 20:305–317

Sokolovsky M (1995) Endothelin receptor subtypes and their role in transmembrane signaling mechanisms. Pharmacol Ther 68:435–471

Sokolovsky M, Galron R, Kloog Y, Bdolah A, Indig FE, Blumberg S, Fleminger G (1990) Endothelins are more sensitive than sarafotoxins to neutral endopeptidase: possible physiological significance. Proc Natl Acad Sci USA 87:4702–4706

Sokolovsky M, Shraga-Levine Z, Galron R (1994) Ligand-specific stimulation/inhibition of cAMP formation by a novel endothelin receptor subtype. Biochemistry 33:11417–11419

Sudhir K, Ko E, Zellner C, Wong HE, Hutchison SJ, Chou TM, Chatterjee K (1997) Physiological concentrations of estradiol attenuate endothelin 1- induced coronary vasoconstriction in vivo. Circulation 96:3626–3632

Sudjarwo SA, Hori M, Tanaka M, Matsuda T, Okada T, Karaki H (1994) Subtypes of endothelin ETA and ETB receptors mediating venous smooth muscle contraction. Biochem Biophys Res Commun 200:627–633

Takasaki C, Aimoto S, Kitazumi K, Tasaka K, Shiba T, Nishiki K, Furukawa Y, Takayanagi R, Ohnaka K, Nawata H (1991) Structure-activity relationships of sarafotoxins: chemical syntheses of chimera peptides of sarafotoxins S6b and S6c. Eur J Pharmacol 198:165–169

Tamaoki H, Kyogoku Y, Nakajima K, Sakakibara S, Hayashi M, Kobayashi Y (1992) Conformational study of endothelins and sarafotoxins with the cystine-stabilized helical motif by means of CD spectra. Biopolymers 32:353–357

Tanaka T, Ogawa T, Matsuda Y. (1995) Species differences in the binding characteristics of RES-701-1: a potent endothelin ETB receptor-selective antagonist. Biochem Biophys Res Commun 209:712–716

Uchida K, Yuzuki R, Kamikawa Y (1998) The role of receptor-operated Ca^{2+} influx in endothelin-induced contraction of the muscularis mucosae. J Cardiovasc Pharmacol 31:S504–S506

Warner TD, Allcock GK, Corder R, Vane JR (1993a) Use of the endothelin antagonists BQ-123 and PD 142893 to reveal three endothelin receptors mediating smooth muscle contraction and the release of EDRF. Br J Pharmacol 110:777–782

Warner TD, Allcock GK, Mickley EJ, Corder R, Vane JR (1993b) Comparative studies with the endothelin receptor antagonists BQ-123 and PD 142893 indicate at least three endothelin receptors J Cardiovasc Pharmacol 22:S117–S120

White DG, Garrat H, Mundin JW, Sumner MJ, Vallance PJ, Watts IS (1994) Human saphenous vein contains both endothelin ET_A and ET_B contractile receptors. Eur J Pharmacol 257:307–310

Williams DL, Jones KL, Pettibone DJ, Lis EV, Clinschmidt BV (1991) Sarafotoxin S6c: an agonist which distinguishes between endothelin receptor subtypes. Biochem Biophys Res Commun 175:556–661

Yoneyama T, Hori M, Tanaka T, Matsuda Y, Karaki H (1995) Endothelin ETA and ETB receptors facilitating parasympathetic neurotransmission in the rabbit trachea. J Pharmacol Exp Therap 275:1084–1089

Zuccarello M, Boccaletti R, Rapoport RM (1998a) Endothelin ETB receptor-mediated constriction in the rabbit basilar artery. Eur J Pharmacol 350:R7–R9

Zuccarello M, Boccaletti R, Rapoport RM (1998b) Endothelin ETB receptor-mediated relaxation of rabbit basilar artery. Eur J Pharmacol 357:67–71

Identity of Endothelin-Converting Enzyme and Other Targets for the Therapeutic Regulation of Endothelin Biosynthesis

R. CORDER

A. Biosynthesis of the Endothelins

I. Introduction

The biosynthesis of the three endothelin isoforms (ET-1, ET-2 and ET-3) follows a similar pattern with initial synthesis as precursor proteins ($\approx$200 amino acid residues) which undergo selective proteolysis to yield the biologically inactive intermediates called big endothelins (BLOCH et al. 1989a,b; OHKUBO et al. 1990; YANAGISAWA et al. 1988). The human ET-1 gene encodes a 212 amino acid precursor, preproendothelin-1 (BLOCH et al. 1989b). Removal of the signal sequence generates proendothelin-1 (proET-1, 195 a.a.) which is processed at double basic amino acid residues to release the intermediate referred to as big ET-1 (human form, 38 a.a.) (Fig. 1). The active endothelins are generated by hydrolysis of the Trp^{21}-Val^{22} bond in big ET-1 and big ET-2, or Trp^{21}-Ile^{22} in big ET-3 (Fig. 2). This cleavage is unique to the endothelins and hence the role of a specific endothelin-converting enzyme (ECE) was proposed (YANAGISAWA et al. 1988). Most studies have concluded that proendothelin-1 is processed intracellularly to ET-1 prior to its release (CORDER et al. 1995b; HARRISON et al. 1993, 1995; RUSSELL and DAVENPORT 1999; XU et al. 1994). This is also the case for ET-2 (LAMBERT et al. 2000), but studies of an ET-3 secreting cell line have yet to be described. For ET-1 there is evidence of polar secretion from the abluminal surface of endothelial cells (WAGNER et al. 1992), which indicates that ET-1 functions primarily as a paracrine modulator of vascular function. Secretion of the endothelins is mainly via constitutive secretory pathways, hence the most important factor affecting their synthesis is the level of gene expression.

II. Tissue Expression and Localisation

Studies in rats have shown that the genes for ET-1, ET-2 and ET-3 have distinct patterns of tissue expression (FIRTH and RATCLIFFE 1992). In the basal state, ET-1 and ET-3 are the most widely expressed isoforms of this peptide family. Prepro-ET-1 mRNA levels are highest in the lung and large intestine, with detectable levels in most organs including the heart and kidney. Prepro-

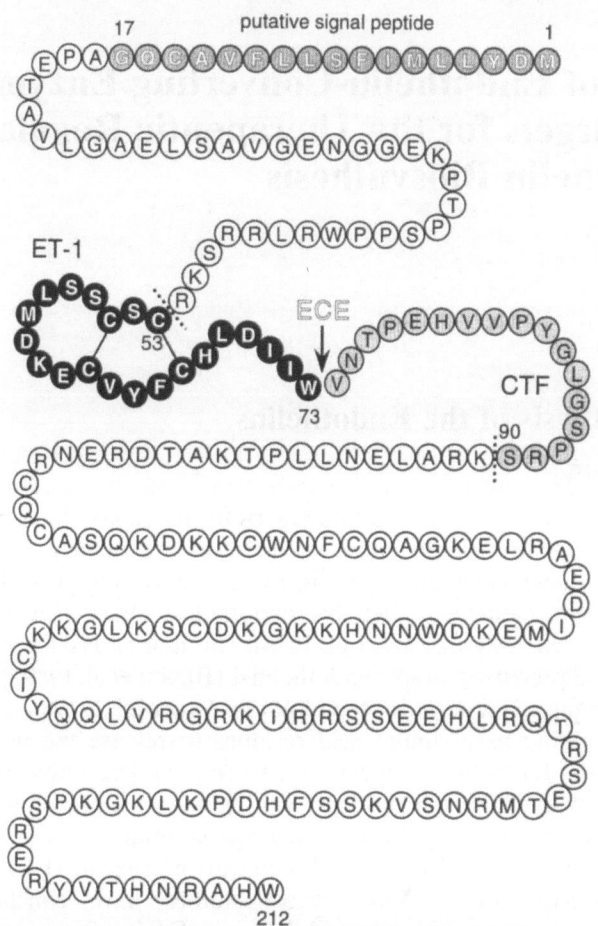

Fig. 1. Amino acid sequence of human preproendothelin-1 (Bloch et al. 1989b). Amino acid residues 53–90 represent big ET-1. ET-1 and the C-terminal fragment (CTF) of big ET-1 are indicated. The location of the Trp-Val bond hydrolysed by endothelin-converting enzyme (ECE) and the processing sites for furin or another member of the prohormone convertase family are indicated by the *dashed lines*

ET-2 mRNA is mainly localised to the small and large intestines. Prepro-ET-3 expression is moderately high in lung, kidney, stomach, small and large intestine, salivary gland and brain (Firth and Ratcliffe 1992). Although these results describe a relatively low level of ET-1 synthesis, many subsequent investigations have indicated that ET-1 is an inducible gene product whose expression is closely associated with inflammatory mechanisms and disease processes. During inflammation, or on stimulation with cytokines, ET-1 synthesis is induced in many cell types including endothelial and epithelial cells, and vascular and non-vascular smooth muscle (Carr et al. 1998; Corder et al.

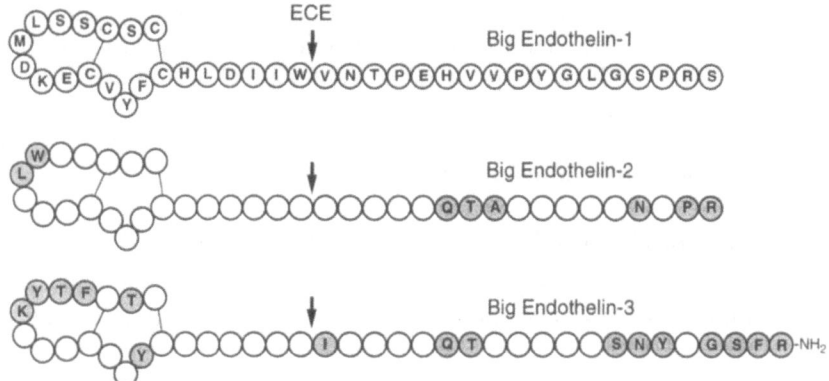

Fig. 2. Comparison of the amino acid sequences of the three big endothelin isoforms. *Open circles* indicate homology with big ET-1. Sequence differences in big ET-2 and big ET-3 are indicated by *shaded circles*

1995a; REDINGTON et al. 1997; WOODS et al. 1999). Indeed, ET-1 expression is virtually absent from normal vascular endothelium but it is markedly increased in the endothelium overlying atherosclerotic plaques (JONES et al. 1996). Similarly, vascular smooth muscle cells do not normally express ET-1 but removal of the endothelium by angioplasty leads to ET-1 synthesis (WANG et al. 1996). Such observations have played a significant part in the widely held perspective that the synthesis of ET-1 is primarily associated with disease processes. Studies of ET-2 and ET-3 synthesis under different conditions are more limited. In the kidney there seems to be an inverse relationship between ET-1 and ET-3 because levels of prepro-ET-1 mRNA were increased after ischaemia, whereas those of prepro-ET-3 were suppressed (FIRTH and RATCLIFFE 1992). This indicates the potential differences in the physiological and pathological roles of the three endothelin isoforms.

III. Regulation of Prepro-ET-1 Expression

1. Basal Synthesis

Despite intense research the factors regulating ET-1 synthesis in vivo have yet to be fully elucidated. The precise mechanisms inducing the upregulation of ET-1 synthesis may differ not only depending on the stimuli but also the cell type involved. Under basal conditions endothelial cells in culture display a relatively high level of ET-1 synthesis compared to the in vivo state where its expression is negligible in normal endothelium (JONES et al. 1996). This difference is most likely due to the inhibitory influence of shear stress on the endothelium in vivo, because application of physiological levels of shear stress to cultured endothelial cells is one of the most effective means of suppressing ET-1 gene expression (MALEK et al. 1993). Based on experiments using primary cultures of human umbilical vein endothelial cells the mechanism of this effect has been attributed to shear stress induced NO release causing the inhibition of ET-1 production (KUCHAN and FRANGOS 1993). However, it seems unlikely that this is the only explanation because the effect of NO is rapidly lost in cultured endothelial cells (BRUNNER et al. 1994; WARNER et al. 1992), yet the effect of shear stress is maintained (MALEK et al. 1993).

2. Ca^{2+} and Intracellular pH

While short periods of stimulation of cultured endothelial cells with Ca^{2+} ionophores increase the expression of prepro-ET-1 mRNA (YANAGISAWA et al. 1989; EMORI et al. 1989), prolonged stimulation suppresses ET-1 synthesis (CORDER et al. 1993; MITCHELL et al. 1992). Detailed investigations of the effect of different intracellular Ca^{2+} concentrations ($[Ca^{2+}]_i$) on ET-1 synthesis have shown a bell-shaped relationship. Thus, $[Ca^{2+}]_i$ <110nmol/l or >1000nmol/l inhibit ET-1 production by up to 50% compared to basal production with steady state $[Ca^{2+}]_i$ of 190nmol/l (BRUNNER et al. 1994). Although Ca^{2+} ionophores stimulate endothelial cells to produce NO and prostacyclin, which may in turn suppress ET-1 release, such effects do not account for the reduction in ET-1 synthesis with Ca^{2+} ionophores (CORDER et al. 1993).

Sustained increases in $[Ca^{2+}]_i$ promote mitochondrial Ca^{2+}/H$^+$ exchange with the net result that Ca^{2+} ionophores reduce intracellular pH (pH$_i$) (TRETTER et al. 1998; KITAZONO et al. 1988). Interestingly, shear stress also reduces pH$_i$ (ZIEGELSTEIN et al. 1992). Hence, changes in pH$_i$ may underlie the inhibitory effect of shear stress and Ca^{2+} ionophores on ET-1 gene expression. In agreement with this, ET-1 synthesis is inhibited in cultured endothelial cells by agents reducing pH$_i$ (Fig. 3). Amiloride inhibits Na$^+$/H$^+$ exchanger and causes small reductions in pH$_i$ in endothelial cells (ESCORBALES et al. 1990; KITAZONO et al. 1988). In comparison, the stilbene anion exchange inhibitor, 4,4'-isothiocyanostilbene-2,2'-disulfonic acid (DIDS), blocks Cl$^-$/HCO$_3^-$ exchange by inhibiting HCO$_3^-$ influx and causes marked decreases in pH$_i$ in endothelial cells (BONANNO and GIASSON 1992). Consistent with reports that Ca^{2+} ionophores reduce pH$_i$, amiloride and DIDS suppressed ET-1 synthesis and had additive effects in combination with A23187 (Fig. 3). This indicates an important role for pH$_i$ in the regulation of ET-1 synthesis and suggests that shear stress induced decreases in pH$_i$ may contribute to its mechanism of inhibition of ET-1 gene expression.

3. Protein Kinase C

Protein kinase C appears to play a key role in the basal expression of prepro-ET-1 in cultured endothelial cells. Phorbol esters cause modest increases in prepro-ET-1 mRNA levels and ET-1 synthesis (BRUNNER 1995; YANAGISAWA et al. 1989). In comparison, a role for protein kinase C is more evident from the effect of protein kinase C inhibitors as these cause a marked suppression of

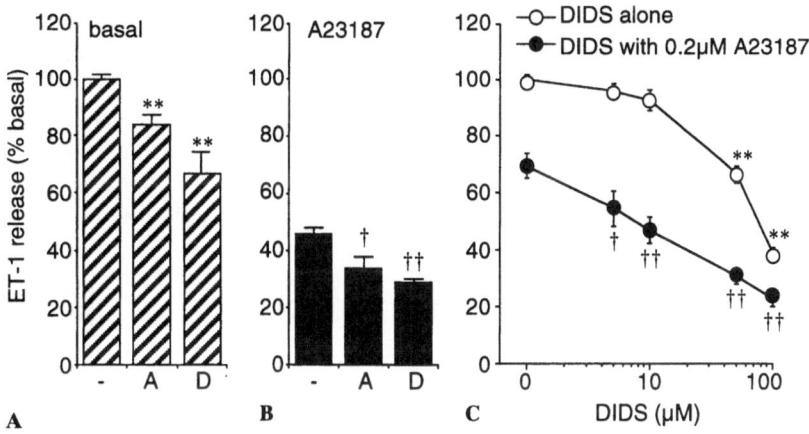

Fig. 3A–C. Effect of 100 μmol/l amiloride (*A*) or 50 μmol/l DIDS (*D*). **A** Alone. **B** In the presence of 0.2 μmol/l A23187 on the release of ET-1 over 24 h from cultured bovine aortic endothelial cells. **C** Concentration-dependent inhibition of ET-1 release by DIDS alone and in the presence of 0.2 μmol/l A23187 (**$p<0.01$ compared to control; †$p<0.05$, ††$p<0.01$ compared to release in the presence of A23187 alone)

basal prepro-ET-1 expression in endothelial cells (BRUNNER 1995; CORDER and BARKER 1999; MALEK et al. 1993). Basal regulation of ET-1 gene transcription is dependent on an AP-1 site in the promoter region regulated by the transcription factors Fos and Jun (LEE et al. 1991), but the precise relationship with constitutive activation of specific isoforms of protein kinase C has yet to be elucidated.

4. Cytokines

Cytokines, including TNFα, TGFβ and IL-1β stimulate the synthesis and secretion of ET-1 both in vitro and in vivo (CORDER et al. 1995a; KLEMM et al. 1995). Increased ET-1 synthesis correlates with raised levels of prepro-ET-1 mRNA and augmented gene transcription (KURIHARA et al. 1989; MARSDEN and BRENNER 1992). The effect of TNFα on ET-1 release is very rapid with significant increases occurring within 15 min of stimulation both in vivo (KLEMM et al. 1995) and in cultured cells (Fig. 4). The marked increase in prepro-ET-1 mRNA levels over the same time period implies that the effect on ET-1 release is mainly the result of de novo synthesis. TNFα stimulates sufficient ET-1 release to cause vasoconstriction (KLEMM et al. 1995). Hence, cytokine stimulated ET-1 synthesis may play a key role in vascular disease processes. Moreover, in common with many other effects of TNFα on endothelial cells, such as expression of adhesion molecules and tissue factor, ET-1 synthesis is induced by the p55 (TNF-R1) receptor (LEES et al. 2000).

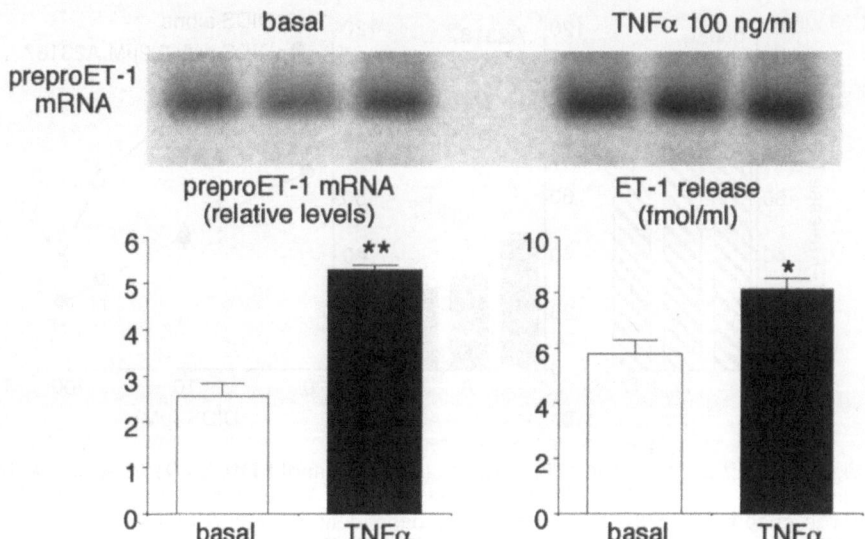

Fig. 4. TNFα induces a rapid increase in ET-1 synthesis. Bovine aortic endothelial cells were treated for 15 min with TNFα (100 ng/ml). Relative mRNA levels were assessed by semi-quantitative RT-PCR (*p<0.05, **p<0.01 compared to basal)

5. Thrombin

Thrombin was one of the first agents shown to stimulate ET-1 synthesis (YANAGISAWA et al. 1988). Subsequent studies have suggested a clear association between upregulation of ET-1 synthesis and thrombotic processes. Lysophosphatidic acid, which is released by activated platelets, has been shown to induce ET-1 expression (CHUA et al. 1998), and agents which induce ET-1 synthesis in endothelial cells also stimulate the expression of the prothrombotic protein tissue factor (MOLL et al. 1998), indicating that ET-1 synthesis is part of a co-ordinated thrombotic process.

6. Stretch

Physical forces have a variety of effects on vascular homeostasis. Increased blood pressure is generally considered as one of the factors inducing vascular remodelling in hypertension. Studies of ET-1 suggest it may play a role in this process because raised hydrostatic pressure in the jugular vein increases the synthesis of ET-1 in endothelial cells and the expression of ET$_B$ receptors in vascular smooth muscle (LAUTH et al. 2000). Similarly, application of stretch to cultured endothelial cells causes a rapid and sustained rise in ET-1 release (MACARTHUR et al. 1994).

IV. Physiological Functions of ET-1 Biosynthesis

Most research has focused on the pathophysiological roles of the endothelins, but in order to consider the consequences of inhibiting ET-1 synthesis some understanding of its physiological functions are necessary. There are two areas of particular significance: its role in development, and its contribution to normal vascular homeostasis.

1. Role of ET-1 in Development

ET-1 gene deletion leads to abnormal foetal development and the death of all ET-1$^{-/-}$ homozygous mice (KURIHARA et al. 1994) (see Chap. 6). This is the result of morphological abnormalities in the pharyngeal arch-derived craniofacial tissues which cause death at birth due to respiratory failure. Consistent with the role of ET-1 in the development of these tissues, wild type embryos show high levels of ET-1 gene expression in the pharyngeal arch at 9.5 days (KURIHARA et al. 1994). ET$_A$-receptor knockout causes an identical pattern of disrupted development (CLOUTHIER et al. 1998). More recently, the key role of ET-1 in development has been confirmed by toxicological studies of a mixed ET$_A$/ET$_B$-receptor antagonist. This revealed teratogenic effects in rats similar in pattern to the phenotype of both ET-1 and ET$_A$-receptor gene knockout mice (SPENCE et al. 1999).

2. ET-1 and Vascular Homeostasis

In adult animals the physiological functions of ET-1 are less clear. It is noteworthy that acute upregulation of ET-1 synthesis is induced by factors associated with infection including cytokines and bacterial cell wall fragments such as lipopolysaccharide (KLEMM et al. 1995). As agents inducing thrombotic processes also induce ET-1 synthesis, it seems likely that ET-1 functions as part of host-defence mechanisms to stop bleeding and to prevent the spread of infection through vasoconstrictor effects (including venoconstriction). A role for ET-1 in angiogenesis or capillary function in wound-healing has also been described (APPLETON et al. 1992). However, when ET-1 synthesis is induced inappropriately, or chronically, by these stimuli its actions contribute to the underlying changes in a variety of pathologies.

Conversely, recent toxicological studies of ET$_A$-receptor antagonists have implied a role for ET-1 in the normal homeostatic mechanisms maintaining vascular structure because high doses, or sustained treatment, caused arteriopathies (ALBASSAM et al. 1999). This type of effect is often associated with administration of high doses of vasodilator drugs, but the action of an ET$_A$-receptor antagonist indicates the balance of vasodilator and vasoconstrictor mechanisms in the vessel wall has significant dependence on the contribution from ET-1 synthesis. Hence, therapeutic regulation of ET-1 synthesis needs to aim to correct an imbalance rather than inhibiting synthesis completely.

B. Inhibition of ET-1 Synthesis vs Endothelin Receptor Antagonists

Most anti-endothelin therapeutic strategies have focused on the development of ET_A-receptor antagonists to prevent vasoconstrictor effects (see Chap. 9). However, the synthesis of ET-1 is widespread in both vascular and non-vascular cells and other actions, including promoting smooth muscle mitogenesis, may be of equal or greater importance in underlying pathological processes (Barnes 1994; Haynes and Webb 1998). Although the ET_B receptor subtype is frequently considered of most importance as a clearance receptor (Löffler et al. 1993; Fukuroda et al. 1995), under some circumstances ET_B receptors may play a key role in mediating the effects of ET-1 (Wellings et al. 1994). Hence, in some disease conditions, including those outlined below, inhibition of ET-1 synthesis may provide a more effective treatment than either selective or non-selective endothelin antagonists.

I. Lung

The acute actions of ET-1 in the airways and in the pulmonary vasculature include pulmonary arterial and venous vasoconstriction, extravasation and oedema formation, and bronchoconstriction (Barnes 1994). However, when expression of ET-1 is upregulated, its involvement in chronic changes may be just as important as any acute vasoconstrictor effect. This is particularly true for the pulmonary vasculature and airways where ET-1 may have a long-term role in remodelling by promoting vascular and tracheal smooth muscle mitogenesis, and stimulating collagen synthesis by pulmonary fibroblasts (Malarkey et al. 1995; Marini et al. 1996; Panettieri et al. 1996; Sun et al. 1997) (see Chap. 14).

1. Pulmonary Hypertension

Medical management of pulmonary hypertension cannot fully prevent the vascular remodelling and right heart failure which frequently become life-threatening (Riley 1991). Increases in vascular resistance result from both structural changes and altered vascular tone. Remodelling involves both smooth muscle proliferation and the formation of extracellular matrix proteins (Riley 1991). The growth factors precipitating these changes may originate from vascular smooth muscle cells, damaged endothelial cells, or from infiltrating leucocytes.

Current evidence suggests a key role for ET-1 in the mechanisms of pulmonary hypertension (see Chap. 15). Clinical studies have shown raised levels of ET-1 gene expression and ET-1 release in pulmonary hypertension, with an inverse relationship between the expression of endothelial nitric oxide synthase and the level of ET-1 immunoreactivity in the lungs from patients with pulmonary hypertension (Giaid and Saleh 1995; Nootens et al. 1995). These investigations also concluded that expression of ET-1 indicates localised areas of endothelial dysfunction (GIAID and SALEH 1995). Administration of endothelin receptor antagonists reduces both pulmonary vascular resistance and the medial thickening associated with pulmonary hypertension induced by chronic hypoxia in rats (CHEN et al. 1997; DI CARLO et al. 1995; UNDERWOOD et al. 1997). Hence, these observations strongly support a pivotal role for ET-1 in the pathological processes occurring in pulmonary hypertension. However, both ET_A and ET_B receptor subtypes appear to be involved in mediating the vasoconstrictor responses to ET-1 in human pulmonary vasculature (McCULLOCH et al. 1998). Therefore, inhibition of ET-1 synthesis may be more useful for attenuating all the changes induced by ET-1 in the pulmonary vasculature.

2. Airways Disease

Obliterative bronchiolitis, asthma, and chronic obstructive pulmonary disease have distinct origins, but all result in varying degrees of airways remodelling suggesting that there may be common underlying mechanisms (BARNES 1996; CHAPMAN 1996; JEFFERY et al. 1998; ROBERTS et al. 1995; WIGGS et al. 1997). Indeed, these conditions are characterised by inflammation, with smooth muscle cell proliferation and fibrosis which leads to airways obstruction. The precise location of the remodelling may be upper or lower airways depending on the disease, but generally the more advanced the disease the greater the structural changes. Increased ET-1 expression has been identified in obliterative bronchiolitis and this may be linked to the proliferative response which occurs during rejection of lung transplants (JEPPSON et al. 1998; McDERMOTT et al. 1998). Similarly, asthma is associated with elevated bronchial ET-1 synthesis (REDINGTON et al. 1997). Infection increases ET-1 synthesis in the airways and this may precipitate obliterative bronchiolitis (TAKEDA et al. 1997) or exacerbate asthma (CARR et al. 1998). ET-1 stimulates mitogenesis of airways smooth muscle and fibroblasts, induces extracellular matrix production, and sensitises the airways to bronchoconstrictor agents all of which further exacerbates these conditions (BARNES 1994). In human bronchus the endothelin receptor subtype has characteristics of an ET_B receptor subtype but it is not blocked by typical antagonists of this receptor, suggesting that it represents a novel receptor subtype (HAY et al. 1998). Because the various responses in the airways involve ET_A and ET_B receptors as well as atypical ET receptors, regulation of ET-1 production should be targeted in these conditions.

II. Atherosclerosis

Atherosclerotic lesions are characterised by accumulation of oxidised-LDL, endothelial damage, and the proliferation of arterial smooth muscle cells. The development of coronary atherosclerosis involves the action of many growth factors, cytokines, and other molecules which cause cell recruitment, migration and proliferation (see Chap. 12). Many of these growth factors and cytokines increase the synthesis of ET-1 (MATHEW et al. 1996). Consistent with its mitogenic properties, a number of studies have indicated ET-1 plays a pivotal role in remodelling after angioplasty (HAYNES and WEBB 1998; WANG et al. 1996). ET-1 synthesis and ET receptor density are also increased in smooth muscle cells migrating into the intima of arteries during remodelling and atherosclerosis (AZUMA et al. 1995; DASHWOOD et al. 1998). These findings, together with results outlined, below suggest that ET-1 is directly involved in the proliferation of vascular smooth muscle during atheroma formation. This mechanism may be dependent on both ET_A and ET_B receptors for vascular smooth muscle cells with a contractile phenotype generally expressing ET_A receptors but ET_B receptors are expressed on migrating cells and those with a synthetic phenotype (EGUCHI et al. 1994).

Clinical studies support a role for ET-1 in the evolution and progression of atherosclerosis in humans. Patients with symptomatic atherosclerosis, show higher plasma levels of ET-1 with a significant correlation with the number of sites of disease involvement (LERMAN et al. 1991). ET-1 is abundant in human coronary artery atherosclerotic lesions (DASHWOOD et al. 1998; HASDAI et al. 1997), where it has been identified in vascular smooth muscle cells as well as in endothelial cells (DASHWOOD et al. 1998; LERMAN et al. 1991). Moreover, ET-1 is localised to endothelial cells overlying atherosclerotic plaques and fatty streaks, while it is virtually absent from adjacent endothelium, indicating a clear relationship with the pathological changes (JONES et al. 1996). ET-1 has also been identified in foamy macrophages – the main inflammatory cell type in atherosclerotic lesions (IHLING et al. 1996).

The development of atherosclerotic lesions with an associated increase in ET-1 synthesis is relatively rapid as hypercholesterolaemia in pigs induced by a diet with 2% cholesterol for 4 months results in raised plasma ET-1 concentrations and increased coronary artery ET-1 immunoreactivity (LERMAN et al. 1993). Further support for ET-1 playing a role in atherosclerosis has been obtained from treatment of cholesterol-fed hamsters with a selective ET_A receptor antagonist (BMS-182874) as this decreased the area of the fatty streak by reducing the number and size of macrophage-foam cells (KOWALA et al. 1995). Despite these findings, the relative importance of ET_A or ET_B receptors in the development of human atherosclerotic lesions and the subsequent evolution of established lesions is not yet clear. The strong indications that ET-1 is driving this process suggests that the mechanisms involved in the induction of ET-1 synthesis are key targets for preventing atherosclerosis.

III. Renal Failure

The kidney is probably more sensitive to the vasoconstrictor effects of ET-1 than any other tissue (see Chap. 18). A large component of the acute vasoconstrictor response is mediated by ET_B receptors in human and rat kidney (WEITZBERG et al. 1995; WELLINGS et al. 1994). Chronic renal failure results in extensive remodelling and fibrosis with a progressive impairment of glomerular filtration and renal tubular function. Both TGFβ and ET-1 are strongly implicated in the underlying mechanisms of chronic renal disease (DOUTHWAITE et al. 1999; KOHAN et al. 1997). TGFβ and ET-1 both stimulate fibrosis, and TGFβ is a potent stimulus for ET-1 synthesis, but the interplay between these two factors has yet to be explored. It has been postulated that the inexorable process of chronic renal disease may be treated by ET antagonists. However, so far, chronic treatment of rat models of renal failure with ET_A antagonists has shown little or no benefit (NABOKOV et al. 1996; POLLOCK and POLAKOWSKI 1997). Perhaps this is because the role of ET-1 is more limited than expected, or that ET_B receptors are involved in mediating some of the chronic effects of ET-1. Treatment of renal disease with ET_B antagonists may not be appropriate because of the clearance function of this receptor, and because ET_B receptors induce beneficial effects on tubular function by increasing sodium and water excretion (KOHAN 1997). ET-3 expression is abundant in medullary collecting ducts where it may mediate natriuretic and diuretic effects (KOHAN 1997; TEREDA et al. 1993), but its relative importance compared to ET-1 for mediating these responses has not been determined. TGFβ stimulates ET-1 synthesis without altering ET-3 expression (TEREDA et al. 1993). Hence, in chronic renal disease, selective inhibition of ET-1 synthesis may be of greater value if this then preserves physiological functions of ET-3 mediated via ET_B receptors.

C. Endothelin-Converting Enzyme

I. Background to Endothelin-Converting Enzyme

The original hypothesis that a novel proteolytic activity was required during biosynthesis to generate ET-1 from the intermediate big ET-1 was proposed by Yanagisawa and colleagues, who coined the term endothelin-converting enzyme (ECE) (Yanagisawa et al. 1988). However, unequivocal identification of the physiologically relevant ECE has proved to be one of the most difficult tasks in endothelin research. In addition, it is still unclear whether hydrolysis of each of the three big ET molecules (big ET-1, big ET-2 and big ET-3) is dependent on one ECE, or whether each isoform has its own distinct ECE.

Compared with ET-1, big ET-1 is virtually devoid of vasoconstrictor activity on isolated vascular smooth muscle preparations (KASHIWABARA et al. 1989; HIRATA et al. 1990). However, systemic injections of big ET-1, at doses >0.5nmol kg^{-1}, have almost equal blood pressure effects to ET-1 (GARDINER et al. 1993). Responses to intravenous or left ventricle administration are essentially the same (GARDINER et al. 1993). Most studies of systemically administered big ET-1 have concluded that it is converted locally in the vasculature to ET-1 (GARDINER et al. 1993; HALEEN et al. 1993; CORDER and VANE 1995), probably by ECE activities present on endothelial or vascular smooth muscle cells (IKEGAWA et al. 1991; HISAKI et al. 1993).

Early investigations aimed at the identification of ECE detected a wide range of proteolytic activities which could hydrolyse big ET-1 in a relatively selective manner to generate ET-1 and the C-terminal fragment of big ET-1 (OPGENORTH et al. 1992). The starting material for these investigations included extracts from tissues such as lung as well as endothelial cells, vascular smooth muscle cells, and other cell types. A variety of serine-, cysteine-, aspartyl- and metallo-proteases were identified as candidate ECEs without any evidence of their relevance for endogenous biosynthesis of ET-1 (OPGENORTH et al. 1992).

However, it soon became widely accepted that the physiologically important ECE was inhibited by the metalloprotease inhibitor phosphoramidon. Several observations supported this opinion. The blood pressure response to big ET-1 is blocked by pre-treatment with phosphoramidon, whereas the response to ET-1 is not affected (CORDER and VANE 1995; FUKURODA et al. 1990; MATSUMURA et al. 1990; MCMAHON et al. 1991). ET-1 generated by incubation of big ET-1 with cultured endothelial or vascular smooth muscle cells is also inhibited by phosphoramidon (CORDER et al. 1995b; CORDER 1996; OPGENORTH et al. 1992). Studies of endothelial cell homogenates have shown the majority of neutral pH optimum ECE activity to be phosphoramidon-sensitive (OPGENORTH et al. 1992). Finally, high concentrations of phosphoramidon reduce the amounts of ET-1 in the conditioned medium of cultured endothelial cells and lead to an accumulation of big ET-1 without any change in the rate of synthesis (IKEGAWA et al. 1990; SAWAMURA et al. 1991; CORDER et al. 1995b).

Phosphoramidon (N-α-l-rhamnopyranosyloxyhydroxyphosphinyl-l-Leu-l-Trp) inhibits a number of metallopeptidases, so these different observations may reflect actions on several different enzymes rather than on a unique ECE. Indeed, the only conclusive evidence that endogenous biosynthesis of ET-1 is dependent on a phosphoramidon-sensitive ECE comes from the studies of endogenous synthesis by cultured endothelial cells. The inhibitory effect of phosphoramidon on ET-1 biosynthesis has also been demonstrated using epithelial and vascular smooth muscle cells (CORDER et al. 1995b; WOODS et al. 1999). Consistently, inhibition of ET-1 synthesis results in increases in big ET-1. If this effect is absent when ECE inhibitors are tested then reductions in ET-1 synthesis may be due to altered gene expression via some other mechanism, rather than by ECE inhibition.

A second aspect of the studies using phosphoramidon was that the concentrations required to inhibit endogenous synthesis by endothelial cells were 10–20 times higher than those required to prevent the hydrolysis of exogenous big ET-1 (CORDER et al. 1995b; WOODS et al. 1999). This indicated that hydrolysis of the endogenous peptide occurred intracellularly in a compartment which was relatively inaccessible to phosphoramidon. Consistent with the intracellular processing of big ET-1, constitutive secretory vesicles containing ET-1 have been identified in cultured endothelial cells (HARRISON et al. 1993, 1995). However, unequivocal identification of the relevant ECE in these vesicles has remained problematic.

II. Evidence of a Secondary Structure for Big ET-1

It has been hard to substantiate the original hypothesis that formation of ET-1 involves a novel selective endopeptidase called ECE. In large part this is due to the technical difficulties of isolating a unique ECE activity. The main confounding factor has been the numerous proteases which display apparently selective ECE activity even though they have no functional relevance for ET-1 synthesis (OPGENORTH et al. 1992). However, there may be a simple explanation for this phenomenon – namely, the conformation of big ET-1 is such that the Trp^{21}-Val^{22} bond is exposed in a manner that favours its hydrolysis. If this is the case, it may account for why so many peptidases cleave big ET-1 with apparent selectivity and give the impression of being specific ECEs. A number of separate observations indicate that big ET-1 forms a specific secondary structure (CORDER 1996).

The initial evidence for this hypothesis originated from studies of the crossreactivity of big ET-1 in radioimmunoassays which recognise the 1–15 loop-region of ET-1 (CORDER 1996). Big ET-1 showed little crossreactivity at high concentrations (<1%), but dilute solutions showed almost full crossreactivity. This may be accounted for by an equilibrium between an unfolded conformation recognised by ET-1 antibodies, and a folded conformation where the C-terminal sequence of big ET-1 folds over ET-$1_{[1-15]}$ preventing antibody access. Thus, if big ET-1 is mainly folded at physiological pH, high concentrations will have low crossreactivity, but in dilute solution the antibody binds a greater proportion of the unfolded form and drives the equilibrium in favour of further unfolding and higher crossreactivity.

The second line of evidence for a defined secondary structure was provided by studies of the actions of neutral endopeptidase 24.11 (NEP-24.11). While ET-1 and the C-terminal sequence of big ET-1 (big ET-$1_{[22-39]}$) are rapidly degraded by NEP-24.11, big ET-1 is comparatively resistant to degradation (MURPHY et al. 1994). The peptide bonds in ET-1 and the C-terminal fragment of big ET-1 which are sensitive to cleavage by NEP-24.11 are almost completely resistant to hydrolysis in big ET-1, presumably because the labile bonds become inaccessible, and therefore protected, as a result of its secondary structure. When NEP-24.11 hydrolyses big ET-1 this occurs primarily through cleavage of the Trp^{21}-Val^{22} bond (MURPHY et al. 1994), indicating that the conformation of big ET-1 not only protects against general proteolysis but also favours the specific cleavage of the Trp^{21}-Val^{22} bond.

Further support for a specific conformation came from studying the effects of chemical modifications of big ET-1 on its conversion to ET-1. This was assessed by measuring blood pressure responses in vivo and generation of ET-1 immunoreactivity in vitro. These studies showed that Lys^9-modified ET-1 molecules were essentially equipotent with ET-1 as vasopressor agents. This was in agreement with early structure-activity studies of ET-1 which showed that N-terminal acetylation abrogates its vasoconstrictor activity (NAKAJIMA et al. 1989), whereas substitution of Lys^9 by Ala or Leu has little effect on vasoconstrictor responses (Nakajima et al. 1989; Watanabe et al. 1991). Similarly, ET-1 biotinylated at Lys[9] retains high affinity for ET_A receptors (Magazine et al. 1991). In marked contrast, the pressor activities of big ET-1 molecules with the Lys[9] modifications were markedly attenuated indicating that conversion to ET-1 was reduced (Corder 1996). This effect was confirmed by studying the degree of hydrolysis of the modified molecules by bovine aortic endothelial and smooth muscle cells. Thus, both the in vivo and in vitro results show that the Lys[9] modifications to big ET-1 affect the ability of ECE activities to cleave the Trp[21]-Val[22] bond. Surprisingly, the incorporation of a single group on Lys[9] of big ET-1 caused a similar or greater decrease in the rate of hydrolysis as that observed with a linear big ET-1 molecule (Corder 1996). Consistent with these results, the truncated peptide big ET-1[19-35] was a very poor competitive substrate inhibitor of big ET-1 hydrolysis (Corder 1996). Hence, it was concluded that the secondary structure of big ET-1 is important for its hydrolysis, and that this structure is stabilised by an interaction between the C-terminal sequence of big ET-1 and either Lys[9] or an adjacent amino acid.

Whether this conformation is also present in big ET-2 and big ET-3 is unclear at present as both of these peptides are hydrolysed inefficiently by the ECE activities that have been identified so far (Shimada et al. 1994; Schmidt et al. 1994; Xu et al. 1994). Nevertheless it should be noted that, besides the high degree of sequence homology in ET-1, ET-2 and ET-3, certain structural features are conserved in the C-terminal sequence of all three big endothelin molecules (Fig. 2). The 23–26 and 31–33 sequences are fully conserved, including Pro[25] and Pro[30], which may be of particular significance for the formation of secondary structure. In addition there is an Arg residue at 37 or 38 in all big ET molecules (Bloch et al. 1989a; Ohkubo et al. 1990).

The existence of a specific conformation is consistent with one theoretical attempt to model the structure of big ET-1 (Menziani et al. 1991), but contradict results from two NMR studies (Inooka et al. 1991; Donlan et al. 1992). However, the pH at which NMR studies were performed is likely to disrupt any secondary structure.

III. Molecular Modelling of Big ET-1

The experimental data provided strong evidence that big ET-1 has a secondary structure stabilised by its disulphide bridges and an interaction of Lys[9] or an adjacent amino acid with the C-terminal sequence of big ET-1. Although this appears to be critical for optimal ECE activity, conventional modelling programs do not reveal a model which is consistent with the experimental observations. Therefore, Peto et al. (1996) used the THREADER algorithm to obtain a guide to backbone conformation for the full 212 residues of prepro-ET-1 in order to generate a molecule model for big ET-1. The sequence was threaded through all known 3-D structures of proteins in the Brookhaven Protein Data Base. This identified 1gdh (D-glycerate dehydrogenase – apo

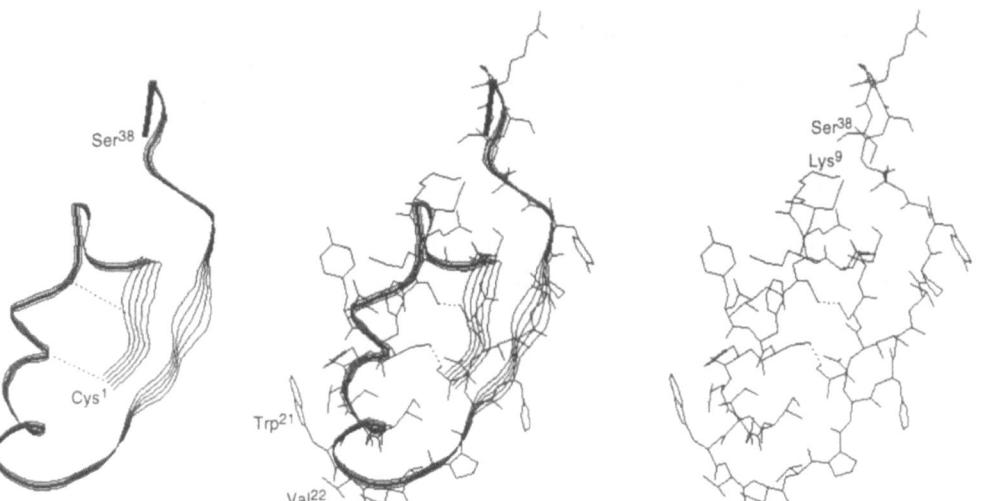

Fig. 5. Molecular model of big ET-1 demonstrating secondary structure which exposes the Trp[21]-Val[22] bond for hydrolysis by ECE. (This figure has been constructed from the model of Peto et al. 1996)

form) as the best fit. The criteria considered essential in the generation of the model were:

1. Surface exposure of basic residues in prepro-ET-1 that are cleaved during processing to release big ET-1
2. The crystal structure of ET-1 (Janes et al. 1994) could be readily inserted into the backbone structure generated by the threading algorithm
3. The Trp[21]-Val[22] bond of big ET-1 should be accessible for hydrolysis

This generated a model for big ET-1 which fitted closely with the ET-1 crystal structure (Peto et al. 1996) (Fig. 5). Consistent with the experimental data, a hydrogen bond between Lys[9] and Ser[38] stabilises a compact globular structure in this model. Other features of the model included a hydrogen-bonded loop between Asp[18] and Thr[24] causing a fold, and a parallel β-sheet between residues 1–5, 27–29. Compared to ET-1, these additional structural features are compatible with differences between ET-1 and big ET-1 observed by circular dichroism (Wallace and Corder 1997). The peptide bonds sensitive to NEP-24.11 hydrolysis are protected in this structure. More importantly, the Trp[21]-Val[22] bond is particularly well exposed in a position that would allow it access to the active sites of many different proteases (Fig. 5). Hence, this favours the efficient hydrolysis of this particular bond, but it also indicates that identification of an enzyme which generates ET-1 from big ET-1 does not have any bearing on whether it has a functional relevance for endogenous ET-1 biosynthesis.

IV. Endothelin-Converting Enzyme-1

1. Purification and Cloning of ECE-1

Once it was established that the physiologically relevant ECE was a phosphoramidon-sensitive enzyme the search for ECE focused on the purification and cloning of an ECE activity with this property. Almost simultaneously a metallopeptidase was purified from detergent-solubilised extracts of endothelial cells, lung tissue and adrenal gland. In each case cloning of these purified enzymes showed the isolated ECEs to be isoforms of the same enzyme (SHIMADA et al. 1994; SCHMIDT et al. 1994; XU et al. 1994). This phosphoramidon-sensitive type II integral membrane peptidase was called endothelin-converting enzyme-1 (ECE-1). Its identification in endothelial cells led to it being proposed as the physiologically relevant ECE for ET-1 biosynthesis (SHIMADA et al. 1994; SCHMIDT et al. 1994; XU et al. 1994). Subsequently four isoforms of this endopeptidase have been identified – namely, ECE-1a, ECE-1b, ECE-1c and ECE-1d. ECE-1a and ECE-1c are also referred to as ECE-1β and ECE-1α respectively (TURNER et al. 1998). The four isoforms are generated from the same gene by differential splicing of the 5' region containing the first four exons (YORIMITSU et al. 1995; VALDENAIRE et al. 1995, 1999; SCHWEIZER et al. 1997). The remainder of the amino acid sequence, encoded by a further 16 exons, is identical in the four isoforms (VALDENAIRE et al. 1999) (Fig. 6). A fifth ECE-1 isoform has been reported (GenBank Accession No. AF055469), but full characterisation has yet to be described. ECE-1 has a characteristic metallopeptidase HEXXH amino acid sequence in its active site which is encoded within exon 16 of the ECE-1 gene (VALDENAIRE et al. 1995). Although the various isoforms of ECE-1 are composed of 754–770 amino acid residues, on polyacrylamide gel electrophoresis native ECE-1 protein migrates with an Mr of almost 300 kDa. This is because ECE-1 protein is synthesised as a heavily glycosylated covalent homodimer (SHIMADA et al. 1996). The dimer structure is important for full enzyme activity, but whether it has other functional significance is unclear.

A second but distinct phosphoramidon-sensitive peptidase has been cloned and called ECE-2 (EMOTO and YANAGISAWA 1995). It has approximately

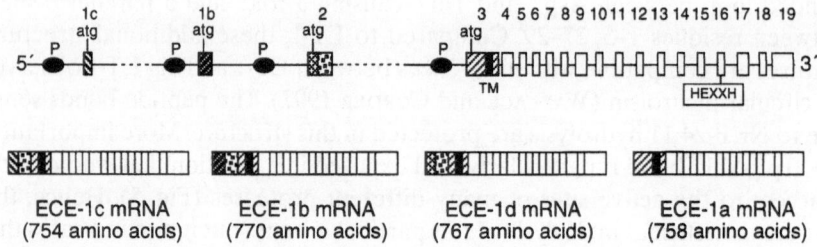

ECE-1c mRNA   ECE-1b mRNA   ECE-1d mRNA   ECE-1a mRNA
(754 amino acids)   (770 amino acids)   (767 amino acids)   (758 amino acids)

Fig. 6. Schematic representation of the human ECE-1 gene (VALDENAIRE et al. 1995, 1999; SCHWEIZER et al. 1997). The *different shading* indicates location of exons in the ECE-1 gene coding for the amino acid sequences of the four isoforms

59% homology with ECE-1, but in contrast has an acid pH optimum which may indicate a lysosomal function rather than a role in ET-1 biosynthesis. ECE-1 and ECE-2 are part of a larger family of homologous proteases which includes NEP 24.11 and the KELL and PEX proteins (TURNER and TANZAWA 1997). It is unclear whether all are inhibited by phosphoramidon or indeed how many other structurally homologous peptidases there might be in this family.

2. Expression and Localisation of ECE-1

The putative promoter sequences for the various ECE-1 isoforms encoded by the ECE-1 gene differ, suggesting they are expressed in different cell types and under different conditions. For instance, the ECE-1a promoter sequence has potential binding sites for NF-$_\kappa$B and acute phase responsive elements, whereas the start point of ECE-1b has features which are more generally associated with house-keeping genes (VALDENAIRE et al. 1995). So far, the factors regulating the expression of the different isoforms have not been systematically investigated. Studies of the cell and tissue distribution of mRNAs for ECE-1 have found them to be widespread (KORTH et al. 1999; SCHWEIZER et al. 1997; VALDENAIRE et al. 1995). However, parallel measurements of prepro-ET-1 mRNA have not been conducted to determine whether there is a relationship to ET-1 synthesis in these tissues. Where identification of mRNAs for specific ECE-1 isoforms has been made, ECE-1a appears to be abundant mainly in endothelial cells, and ECE-1c has a more extensive distribution (LAMBERT et al. 2000; SCHWEIZER et al. 1997; VALDENAIRE et al. 1995).

A number of studies have investigated the cellular expression and intracellular localisation of ECE-1. A fairly widespread distribution of ECE-1 in blood vessels has been found with it mainly localised on endothelial cells rather than vascular smooth muscle cells (BARNES et al. 1998; KORTH et al. 1999; RUSSELL et al. 1998; RUSSELL and DAVENPORT 1999). Synthesis of ECE-1 is increased in the vascular smooth muscle of diseased vessels (GRANTHAM et al. 1998) and following balloon injury (WANG et al. 1996; MINAMINO et al. 1997) indicating a role in remodelling or tissue repair. In endothelial cells human ECE-1a (ECE-1β), and rat and human ECE-1c (ECE-1α) are localised at least in part to the luminal surface with the characteristics of a classical ectoenzyme (TAKAHASHI et al. 1995; BARNES et al. 1998; BROWN et al. 1998; RUSSELL et al. 1998). ECE-1 has also been identified intracellularly with some evidence of co-localisation with ET-1, for instance in porcine lung endothelium (BARNES et al. 1998) and human coronary artery (RUSSELL et al. 1998), but there does not appear to be an extensive co-localisation of ET-1 and ECE-1. It has also been suggested that ECE-1 cycles between the cell surface and the TGN so that ECE-1 processing of big ET-1 can occur during vesicle translocation from TGN to the plasma membrane, with the ECE-1 protein undergoing recycling (BARNES et al. 1998).

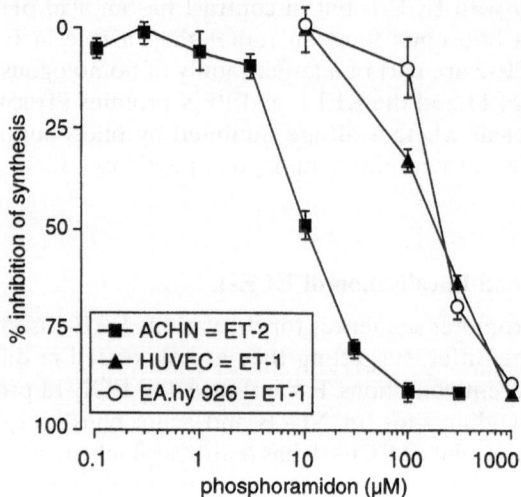

Fig. 7. Comparison of the inhibitory action of phosphoramidon on ET-1 synthesis by human umbilical vein endothelial cells or the cell line EA.hy 926 (CORDER et al. 1995) with its effect on ET-2 synthesis by ACHN cells (LAMBERT et al. 2000)

The four isoforms of ECE-1 characterised to date localise to different intracellular sites when expressed in CHO cells or Madin-Darby canine kidney (MDCK) cells (SCHWEIZER et al. 1997, 1999; AZARANI et al. 1998; BROWN et al. 1998). ECE-1a is localised to the cell surface, ECE-1b is found predominantly intracellularly with a pattern of distribution comparable to the *trans* Golgi network (TGN), ECE-1c is found mainly on the cell surface with some associated with the TGN, and ECE-1d is found both on the cell surface and intracellularly. However, the significance of ECE-1 protein identified at particular intracellular sites is frequently unclear as it may simply represent newly synthesised ECE-1 in transit from the TGN to the cell surface or to some other location where it exerts its physiological function.

Consistent with a cell surface localisation, both ECE-1a and ECE-1c isoforms expressed in CHO cells function as ectoenzymes with big ET-1 as the substrate (XU et al. 1994; TAKAHASHI et al. 1995). A number of studies have shown the various ECE-1 isoforms to hydrolyse big ET-1 more efficiently than either big ET-2 or big ET-3, casting doubts over the role ECE-1 may play in processing these intermediates (SHIMADA et al. 1994; SCHMIDT et al. 1994; XU et al. 1994). Moreover, ET-2 synthesis is considerably more sensitive to inhibition by phosphoramidon than ET-1 synthesis (Fig. 7) (LAMBERT et al. 2000). Based on these observations – low efficiency processing of big ET-2 and increased phosphoramidon sensitivity – it seems likely that ET-2 biosynthesis is dependent on a novel ECE distinct from ECE-1.

The importance of the secondary structure of big ET-1 for ECE-1 processing has been tested using site-directed mutagenesis of prepro-ET-1 followed by co-expression of mutant cDNA and ECE-1a cDNA (BROOKS and ERGUL 1998). Ala replacement of the conserved amino acid residues in the C-terminal sequence of big ET-1 showed Val^{22}, Pro^{25}, Pro^{30}, Gly^{32} and Leu^{33} mutants (numbering based on human big ET-1[1–38]; Fig. 2) to have a decreased efficiency of hydrolysis. These observations are in agreement with the studies described for modified big ET-1 molecules (CORDER 1996).

3. Implications from ECE-1 Gene Deletion Studies

Studies using gene knock-out experiments (see Chap. 6) have failed to clarify the role played by ECE-1 in ET-1 biosynthesis (YANAGISAWA et al. 1998). The phenotype generated by these gene deletion studies has the features of combined ET-1/ET-3 knock-out, suggesting that, in the absence of ECE-1, the biosynthesis of ET-1 and ET-3 are both disrupted (YANAGISAWA et al. 1998). But measurements of tissue levels of ET-1 in non-viable foetuses showed reductions of only 48% compared with the corresponding control foetuses. No increases in big ET-1 were observed, indicating that partial reductions in ET-1 synthesis were not due to inadequate processing. Therefore ET-1 biosynthesis is most likely independent of ECE-1 via a distinct ECE (YANAGISAWA et al. 1998).

One of the most perplexing questions arising from ECE-1 gene knockout is why it appears to have such a similar effect to ET-1 gene or ET$_A$ receptor gene deletion. One interpretation is that ET-1 synthesis is deficient in the pharyngeal arch during development even though ET-1 synthesis is still synthesised at other sites. But if this is not the reason, what then is the function of ECE-1? Another explanation may be that ECE-1 as an ectoenzyme functions to degrade a peptide mediator which acts as a physiological antagonist of ET-1. Of some relevance to this hypothesis are observations that phosphoramidon and other relatively non-selective inhibitors of ECE-1 have a number of pharmacological effects which are attributed to inhibition of ET-1 synthesis but may be due to other actions. For instance phosphoramidon causes vasodilatation in vivo and in vitro at concentrations (HAYNES et al. 1995; SMITH et al. 1997) below those required to inhibit ET-1 synthesis in culture (CORDER et al. 1995b). Similarly, the novel ECE inhibitor CGS 26303 protects against cerebral vasospasm in a model of subarachnoid haemorrhage (KWAN et al. 1997), and lowers blood pressure in spontaneously hypertensive rats (DE LOMBAERT et al. 1994). However, like phosphoramidon, very high concentrations of CGS 26303 are required for inhibition of ET-1 synthesis (WOODS et al. 1999). Hence, it seems more likely that at the doses CGS 26303 and other "ECE inhibitors" have been used in vivo, that these actions are due to an unrelated mechanism such as potentiation of the effects of a vasodilator peptide. Indeed, ECE-1 has been shown to degrade bradykinin (HOANG and TURNER 1997), and a number of other peptide substrates (JOHNSON et al. 1999), with similar efficiency to big ET-1. Therefore, as an ectoenzyme it may have a number of biologically important substrates. In comparison its potential intracellular functions and substrates are less clear.

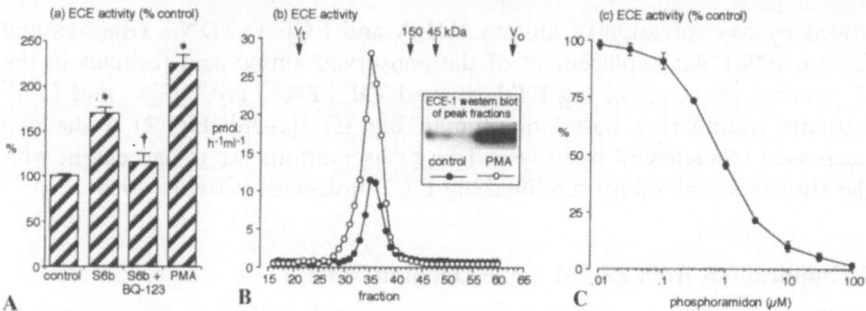

Fig. 8. A Regulation of ECE activity in bovine aortic smooth muscle cells by ET_A-receptors and phorbol ester. Treatment for 24 h with sarafotoxin S6b 1 μmol/l alone or with BQ-123 10 μmol/l, or PMA 0.1 μmol/l (*$p<0.001$ compared to control, † compared to S6b alone). **B** Sephacryl S300 gel filtration chromatography of solubilised ECE activity from control and PMA-treated cells. *Insert* compares the relative amounts of ECE-1 protein in the peak fractions determined by western blot. **C** Phosphoramidon inhibition of solubilised ECE-1

There may be further explanations for the phenotype observed after ECE-1 gene deletion. For instance, the processes of development induced via ET_A receptors may depend on induction of ECE-1 expression in order that it may exert a pivotal proteolytic function. This effect would then be downstream of ET_A receptor signalling. So, if the normal response to ET_A receptor activation was lost, this would give the impression that ET-1 synthesis was perturbed. In agreement with this idea, ET_A receptors are able to induce ECE-1 synthesis in bovine aortic smooth muscle cells (Fig. 8). Finally, ECE-1 knockout may cause the phenotype described by a mechanism completely unrelated to either ET-1 synthesis or its actions because a similar pattern of craniofacial defects is not uncommon as it also occurs with microdeletions of chromosome 22q11 (YAMAGISHI et al. 1999). Hence, the interplay between ECE-1 and many other genes may explain the phenotype of the ECE-1$^{-/-}$ mice.

V. The Search for the Physiologically Relevant ECE

The large number of enzymes identified with ECE activity appears to be explained by the structure of big ET-1 making it sensitive to specific hydrolysis by many enzymes, rather than due to any functional relevance of these enzymes. Therefore criteria need to be defined that will identify the physiologically relevant ECE. Evidence for intracellular conversion of big ET-1 can be derived from the effects of phosphoramidon, and agents such CGS 26303, on endogenous synthesis of ET-1 as the hydrolysis of exogenous big ET-1 is consistently more sensitive to inhibition by these agents (CORDER et al. 1995b; WOODS et al. 1999). However, because these inhibitors are used in high concentrations, when the selectivity for different metallopeptidases is likely to be low, these experiments provide little insight into whether the same peptidase is involved in the intracellular conversion of endogenous big ET-1 and the hydrolysis of exogenous peptide.

When ET-1 and big ET-1 were identified in constitutive secretory vesicles from bovine aortic endothelial cells, ~30% of total ET-1 immunoreactivity in the vesicle fraction was present as big ET-1 (HARRISON et al. 1995), compared to <5% in the conditioned medium (MACARTHUR et al. 1994; CORDER et al. 1995b). This indicates that processing of big ET-1 occurs during vesicle transport from the TGN to the cell surface. This bears a considerable similarity to the intracellular processing of many endocrine peptides in the regulated secretory pathway where efficient processing of peptide prohormones into bioactive peptides is dependent on the co-ordinated expression of the prohormone mRNA and the relevant processing enzymes (prohormone convertases, carboxypeptidase H) (BLOOMQUIST et al. 1991; SCHUPPIN and RHODES 1996). If the precedents of peptide precursors such as proinsulin and proopiomelanocortin are also true for ET-1 then efficient processing of proET-1 in endothelial cells should be dependent on the parallel expression, biosynthesis, and packaging into constitutive secretory vesicles of the physiologically relevant ECE.

This hypothesis was first tested by evaluating whether ET-1 synthesis and expression is co-ordinated with one of the isoforms ECE-1. Conditions were used which either stimulated or inhibited ET-1 synthesis in endothelial cells (CORDER and BARKER 1999). Levels of mRNA for ECE-1 isoforms were compared with those of prepro-ET-1 mRNA to determine whether parallel regulation occurs. In these experiments changes in prepro-ET-1 mRNA levels induced with cytokines, or inhibited by 2-chloroadenosine or staurosporine closely paralleled the changes in ET-1 synthesis. However, these studies provided no evidence for co-ordinated regulation of prepro-ET-1 with the expression of any of the ECE-1 isoforms (CORDER and BARKER 1999). This contrasts markedly with the regulation of prohormone convertases, but there may be no need for ECE-1 levels to be altered provided there is an adequate supply for the continual processing of constitutively synthesised big ET-1.

Other studies have also failed to show a relationship between the level of ET-1 synthesis and the expression of ECE-1. For instance, in congestive heart failure (KOBAYASHI et al. 1999) and pulmonary hypertension (IVY et al. 1998) prepro-ET-1 mRNA levels and ET-1 synthesis are increased, but ECE-1 levels do not change. Moreover, ECE-1 expression following balloon angioplasty of the carotid artery in rats does not display the same temporal relationship to ET-1 mRNA (WANG et al. 1996). ECE-1 levels rise rapidly in the first 24 h and then fall, whereas ET-1 levels are only increased significantly three days after angioplasty when expression of ECE-1 mRNA has returned to baseline levels (WANG et al. 1996).

Therefore, to identify an ECE activity which is synthesised in parallel with ET-1, another feature of regulated secretory pathways was examined – namely, that during release of biologically active peptides the active processing enzymes are co-secreted, albeit in very low amounts, into the extracellular medium (GUEST et al. 1992; JEAN et al. 1993). Investigations of the co-secretion of ET-1 and ECE activity using cultured bovine aortic endothelial cells and pulmonary artery smooth muscle cells have identified a secreted ECE whose synthesis is co-ordinated with that of ET-1. In contrast to the studies of changes in levels of ECE-1 mRNA (CORDER and BARKER 1999), secretion of ET-1 and ECE activity from endothelial cells increased with TNFα stimulation and decreased with 2-chloroadenosine such that peptide and ECE levels were closely correlated (CORDER et al. 1998). Furthermore, vascular smooth muscle cells, which have a low basal synthesis of ET-1, also showed very low levels of secretion of ECE activity, but TNFα stimulation increased ET-1 synthesis around threefold and ECE secretion around twofold (CORDER et al. 1998). Consistent with the physiologically relevant enzyme being phosphoramidon-sensitive (CORDER et al. 1995b; IKEGAWA et al. 1990; SAWAMURA et al. 1991; WOODS et al. 1999), the secreted ECE activities from both endothelial and vascular smooth muscle cells were inhibited by phosphoramidon.

It is still unclear whether these studies of ECE secretion have identified a specific isoform of ECE-1 or a novel ECE. The quantities of ECE measured in these experiments are relatively low, but this is in agreement with the prohormone convertases which are only produced in sufficient amounts for prohormone processing to occur efficiently. Of considerable similarity to the isolation of a physiologically relevant ECE, attempts to isolate the prohormone convertases were unsuccessful because identification of a low abundance processing enzyme against a background of other proteolytic activity was too difficult. Instead, discovery of the prohormone convertases was the result of a cloning strategy based on homology with the yeast processing enzyme Kex2 (STEINER et al. 1992).

D. Alternative Strategies for Inhibiting ET-1 Synthesis

In many cases, increased ET-1 gene expression appears to be part of a co-ordinated response to infection or inflammation (LEES et al. 2000). In the vascular endothelium, as well as ET-1, this involves the expression of adhesion molecules, chemokines, cytokines and tissue factor (MOLL et al. 1995; READ et al. 1995). Consequently, inflammation is associated with leucocyte recruitment combined with prothrombotic changes and the potential for vasoconstriction if ET-1 production exceeds counterbalancing vasodilator influences. This pattern of events may underlie many chronic pathologies including atherosclerosis, and acute exacerbation may precipitate events such as myocardial infarction, stroke, or acute renal failure. Hence treatments which attenuate the overall response of the endothelium may be more useful as therapeutic strategies than those that simply inhibit ET-1 synthesis. There are at least two potential approaches to this problem. The first of these includes inhibition of intracellular signalling mechanisms that stimulate specific kinase cascades, or induce other mechanisms that co-ordinate the upregulation of genes leading to the pro-inflammatory state. An alternative to this approach is the use of agents which induce a receptor-mediated state of refractoriness to proinflammatory stimuli and thereby attenuate responses.

I. Inhibition of Intracellular Signalling Mechanisms

Intracellular signal transduction leading to increased prepro-ET-1 gene transcription almost certainly involves a number of different pathways and kinase cascades. In the future the various steps from the cell surface to the nucleus and specific crosstalk between pathways needs to be defined so that convergent points can be targeted for inhibitor development. Initial comparisons of the effects of protein kinase C and tyrosine kinase inhibitors have confirmed the important role of protein kinase C in basal ET-1 expression in endothelial cells, and revealed that the TNFα mediated response is via a distinct pathway insensitive to protein kinase C inhibition (Fig. 9). In contrast, tyrosine kinase inhibitors cause the reduction of both basal and TNFα stimulated ET-1 synthesis (Fig. 9). Thrombin induction of ET-1 is also independent of protein kinase C but blocked by tyrosine kinase inhibitors (MARSEN et al. 1995).

Cytokine induction of protein expression has been extensively linked to NF-$_\kappa$B dependent mechanisms in endothelial cells and elsewhere (READ et al. 1995). Pertinent to this, the proteasome inhibitor, calpain inhibitor I, suppresses basal ET-1 synthesis and completely abrogates the response to TNFα (Fig. 9). Moreover, a proteasome inhibitor has been reported to lower blood pressure and inhibit ET-1 synthesis in DOCA-salt hypertension, indicating that NF-$_\kappa$B dependent responses may play a role in certain hypertensive mechanisms (OKAMOTO et al. 1998). These findings indicate future directions for research into the regulation of ET-1 synthesis.

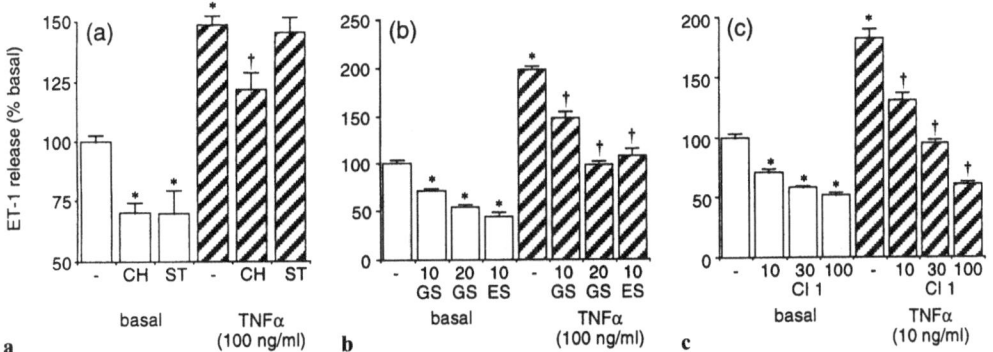

Fig. 9. Comparison of the effects on basal and TNFα-stimulated ET-1 release from cultured bovine aortic endothelial cells of: **a** protein kinase C inhibitors, chelerythrine (CH 3 μmol/l) or staurosporine (ST, 50 nmol/l); **b** tyrosine kinase inhibitors, genistein (10 or 20 μmol/l) or erbstatin (10 μmol/l); **c** the proteasome inhibitor, calpain inhibitor I (Ac-Leu-Leu-norleucinal, 10, 30 or 100 μmol/l) (*$p<0.05$ compared to basal; † $p<0.05$ compared to TNFα alone)

II. Anti-Inflammatory Mediators which Suppress ET-1 Synthesis

Endogenous anti-inflammatory mediators may play a key role in maintaining vascular homeostasis. For instance, adenosine has a number of cardioprotective effects, but the precise mechanism of these actions remains unclear. Its effects include concentration-dependent inhibition of ET-1 synthesis and decreases in prepro-ET-1 mRNA levels (CORDER and BARKER 1999). Adenosine causes endothelium-dependent vasodilatation through NO release, and inhibition of both adhesion molecule and tissue factor expression. Coronary infusion of adenosine, which is known to reduce myocardial infarct size, is reported to suppress ET-1 release induced by myocardial ischaemia and reperfusion (VELASCO et al. 1993), but the underlying mechanism of this effect has yet to be identified.

Anti-inflammatory prostaglandin molecules including 15Δ-PGJ$_2$ represent a second group of compounds which suppress ET-1 synthesis. Whether novel stable prostaglandin analogues could be used to treat endothelial dysfunction is uncertain. However, these effects may be mediated via the peroxisome proliferator-activated receptor-γ as the antidiabetic agents, thiazolidinediones, also have a similar action on ET-1 synthesis (SATOH et al 1999). Hence, the spectrum of therapeutic actions of thiazolidinediones may include some protection against vascular pathologies for patients with type 2 diabetes.

E. Summary

This chapter has provided an outline of the many different factors affecting ET-1 synthesis in health and disease, and provided an indication of specific diseases which may benefit from treatment with agents that lead to a reduction in ET-1 synthesis. The field of endothelin-converting enzyme has been critically reviewed. There is considerable doubt as to whether ECE-1 has a role in ET-1 synthesis, but until such time as a better candidate enzyme is identified as the physiologically relevant ECE this will remain an unresolved question. Finally, other approaches to modifying endothelial function and reducing ET-1 synthesis through decreased gene expression have been briefly considered.

Acknowledgements. I am indebted to Dr. M. Carrier for his contribution to Fig. 4, and Ms. N. Khan for her efforts towards Figs 3, 4, 7 and 8. Figure 5 was constructed with modelling data provided by Dr. B. Wallace and colleagues.

List of Abbreviations

DIDS	4,4'-isothiocyanostilbene-2,2'-disulfonic acid
ET	endothelin
ECE	endothelin-converting enzyme
IL-1β	interleukin-1β
$[Ca^{2+}]_i$	intracellular Ca^{2+} concentrations

pH$_i$	intracellular pH
NEP-24.11	neutral endopeptidase 24.11
NO	nitric oxide
TGFβ	transforming growth factor-β
TGN	*trans* Golgi network
TNFα	tumour necrosis factor-α

References

Albassam MA, Metz AL, Gragtmans NJ, King LM, Macallum GE, Hallak H, McGuire EJ (1999) Coronary arteriopathy in monkeys following administration of CI-1020, an endothelin A receptor antagonist. Toxicological Pathology 27:156–164

Appleton I, Tomlinson A, Chander CL, Willoughby DA (1992) Effect of endothelin-1 on croton oil-induced granulation tissue in the rat. A pharmacologic and immuno-histochemical study. Lab Invest 67:703–710

Azarani A, Boileau G, Crine P (1998) Recombinant human endothelin-converting enzyme ECE-1b is located in an intracellular compartment when expressed in polarized Madin-Darby canine kidney cells. Biochem J 333:439–448

Azuma H, Hamasaki H, Sato J, Isotani E, Obayashi S, Matsubara O (1995) Different localization of ET$_A$ and ET$_B$ receptors in the hyperplastic vascular wall. J Cardiovasc Pharmacol 25:802–809

Barnes PJ (1994) Endothelins and pulmonary disease. J Appl Physiol 77:1051–1059

Barnes PJ (1996) Pathophysiology of asthma. Brit J Clin Pharmacol 42:3–10

Barnes K, Brown C, Turner AJ (1998) Endothelin-converting enzyme: ultrastructural localization and its recycling from the cell surface. Hypertension 31:3–9

Bloch KD, Eddy RL, Shows TB, Quertermous T (1989a) cDNA cloning and chromosomal assignment of the gene encoding endothelin-3. J Biol Chem 264:18156–18161

Bloch KD, Friedrich SP, Lee M-N, Eddy RL, Shows TB, Quertermous T (1989b) Structural organization and chromosomal assignment of the gene encoding endothelin. J Biol Chem 264:10851–10857

Bloomquist BT, Eipper BA, Mains RE (1991) Prohormone-converting enzymes: regulation and evaluation of function using antisense RNA. Mol Endocrinol 5:2014–2024

Bonanno JA, Giasson C (1992) Intracellular pH regulation in fresh and cultured bovine corneal endothelium. II. Na$^+$:HCO$_3^-$ cotransport and Cl$^-$/HCO$_3^-$ exchange. Invest Ophthalmol Vis Sci 33:3068–3079

Brooks C, Ergul A (1998) Identification of amino acid residues in the C-terminal tail of big endothelin-1 involved in processing to endothelin-1. J Mol Endocrinol 21:307–315

Brown CD, Barnes K, Turner AJ (1998) Anti-peptide antibodies specific to rat endothelin-converting enzyme-1 isoforms reveal isoform localisation and expression. FEBS Letts 424:183–187

Brunner F, Stessel H, Simecek S, Graier W, Kukovetz WR (1994) Effect of intracellular Ca^{2+} concentration on endothelin-1 secretion. FEBS Letts 350:33–36

Brunner F (1995) Dependence of endothelin-1 secretion on Ca^{2+}. Biochem Pharmacol 49:1785–1791

Carr MJ, Spalding LJ, Goldie RG, Henry PJ (1998) Distribution of immunoreactive endothelin in the lungs of mice during respiratory viral infection. Eur Resp J 11:79–85

Chapman KR (1996) Therapeutic approaches to chronic obstructive pulmonary disease: an emerging consensus. Am J Med 100 (1A):5S–10S

Chen SJ, Chen YF, Opgenorth TJ, Wessale JL, Meng QC, Durand J, DiCarlo VS, Oparil S (1997) The orally active nonpeptide endothelin A-receptor antagonist A-127722 prevents and reverses hypoxia-induced pulmonary hypertension and pulmonary vascular remodeling in Sprague-Dawley rats. J Cardiovasc Pharmacol 29:713–725

Chua CC, Hamdy RC, Chua BHL (1998) Upregulation of endothelin-1 production by lysophosphatidic acid in rat aortic endothelial cells. Biochim Biophys Acta 1405: 29–34

Clouthier DS, Hosoda K, Richardson JA, Williams CA, Yangisawa H, Kuwaki T, Kumada M, Hammer RE, Yanagisawa M (1998) Cranial and cardiac neural crest defects in endothelin-A receptor-deficient mice. Development 125:813–824

Corder R, Khan N, Anggard EE, Vane JR (1993) J Cardiovasc Pharmacol 22 [Suppl 8]:S42–S45

Corder R, Carrier M, Khan N, Klemm P, Vane JR (1995a) Cytokine regulation of endothelin-1 release from bovine aortic endothelial cells. J Cardiovasc Pharmacol 26 (Suppl 3):S56–S58

Corder R, Khan N, Harrison VJ (1995b) A simple method for isolating human endothelin converting enzyme (ECE-1) free from contamination by neutral endopeptidase 24.11. Biochem Biophys Res Commun 207:355–362

Corder R, Vane JR (1995) Radioimmunoassay evidence that the pressor effect of big endothelin-1 is due to local conversion to endothelin-1. Biochem Pharmacol 49: 375–380

Corder R (1996) The conformation of human big endothelin-1 favours endopeptidase hydrolysis of the Trp^{21}-Val^{22} bond. Biochem Pharmacol 51:259–266

Corder R, Khan N, Barker S (1998) Studies of endothelin-converting enzyme in bovine endothelial cells and vascular smooth muscle cells: further characterisation of the biosynthetic process. J Cardiovasc Pharmacol 31 [Suppl 1]:S46–S48

Corder R, Barker S (1999) The expression of endothelin-1 and endothelin-converting enzyme-1 (ECE-1) are independently regulated in bovine aortic endothelial cells. J Cardiovasc Pharmacol 33:671–677

Dashwood MR, Timm M, Muddle JR, Ong ACM, Tippins JR, Parker R, McManus D, Murday AJ, Madden BP, Kaski JC (1998) Regional variations in endothelin-1 and its receptor subtypes in human coronary vasculature: pathophysiological implications in coronary disease. Endothelium 6:61–70

De Lombaert S, Ghai RD, Jeng AY, Trapani AJ, Webb RL (1994) Pharmacological profile of a non-peptidic dual inhibitor of neutral endopeptidase 24.11 and endothelin-converting enzyme. Biochem Biophys Res Comm 204:407–412

Di Carlo VS, Chen S-J, Meng QC, Durand J, Yano M, Chen Y-F, Oparil S (1995) ET_A-receptor antagonist prevents and reverses chronic hypoxia-induced pulmonary hypertension in rat. Am J Physiol 269:L690–L697

Donlan ML, Brown FK, Jeffs PW (1992) Solution conformation of human big endothelin-1. J Biomolecular NMR 2:407–420

Douthwaite JA, Johnson TS, Haylor JL, Watson P, El Nahas AM (1999) Effects of transforming growth factor-β1 on renal extracellular matrix components and their regulating proteins. J Am Soc Nephrol 10:2109–2119

Eguchi S, Hirata Y, Imai T, Kanno K, Marumo F (1994) Phenotypic change of endothelin receptor subtype in cultured rat vascular smooth muscle cells. Endocrinology 134:222–228

Emori T, Hirata Y, Ohta K, Shichiri M, Marumo F (1989) Secretory mechanism of immunoreactive endothelin in cultured bovine endothelial-cells. Biochem Biophys Res Commun 160:93–100

Emoto N, Yanagisawa M (1995) Endothelin-converting enzyme-2 is a membrane-bound, phosphoramidon-sensitive metalloprotease with acidic pH optimum. J Biol Chem 270:15262–15268

Escobales N, Longo E, Cragoe EJ, Danthuluri NR, Brock TA (1990) Osmotic activation of Na^+-H^+ exchange in human endothelial cells. Am J Physiol 259:C640–C646

Firth JD, Ratcliffe PJ (1992) Organ distribution of the three rat endothelin messenger RNAs and the effects of ischemia on renal gene expression. J Clin Invest 90: 1023–1031

Fukuroda T, Noguchi K, Tsuchida S, Nishikibe M, Ikemoto F, Okada K, Yano M (1990) Inhibition of biological actions of big endothelin-1 by phosphoramidon. Biochem Biophys Res Commun 172:390–395

Fukuroda T, Fujikawa T, Ozaki S, Ishikawa K, Yano M, Nishikibe M (1994) Clearance of circulating endothelin-1 by ET_B receptors in rats. Biochem Biophys Res Commun 199:1461–1465

Gardiner SM, Kemp PA, Bennett T (1993) Regional haemodynamic responses to intravenous and intra-arterial endothelin-1 and big endothelin-1 in conscious rats. Br J Pharmacol 110:1532–1536

Giaid A, Saleh D (1995) Reduced expression of endothelial nitric oxide synthase in the lungs of patients with pulmonary hypertension. N Engl J Med 333:214–221

Grantham JA, Schirger JA, Williamson EE, Heublein DM, Wennberg PW, Kirchengast M, Muenter K, Subkowski T, Burnett JC (1998) Enhanced endothelin-converting enzyme immunoreactivity in early atherosclerosis. J Cardiovasc Pharmacol 31 [Suppl 1]:S22–S26

Guest PC, Arden SD, Bennett DL, Clark A, Rutherford NG, Hutton JC (1992) The post-translational processing and intracellular sorting of PC2 in the Islets of Langerhans. J Biol Chem 267:22401–22406

Haleen SJ, Davis LD, Ladouceur DM, Keiser JA (1993) Why big endothelin-1 lacks a vasodilator response. J Cardiovasc Pharmacol 22 [Suppl 8]:S271–S273

Harrison VJ, Corder R, Änggård EE, Vane JR (1993) Evidence for vesicles that transport endothelin-1 in bovine aortic endothelial cells. J Cardiovasc Pharmacol 22 [Suppl 8]:S57–S60

Harrison VJ, Barnes K, Turner AJ, Wood E, Corder R, Vane JR (1995) Identification of endothelin 1 and big endothelin 1 in secretory vesicles isolated from bovine aortic endothelial cells. Proc Natl Acad Sci USA 92:6344–6348

Hasdai D, Holmes DR Jr, Garratt KN, Edwards WD, Lerman A (1997) Mechanical pressure and stretch release endothelin-1 from human atherosclerotic coronary arteries in vivo. Circulation 95:357–362

Hay DW, Luttmann MA, Pullen MA, Nambi P (1998) Functional and binding characterization of endothelin receptors in human bronchus: evidence for a novel endothelin B receptor subtype?. J Pharmacol Exp Ther 284:669–677

Haynes WG, Ferro CE, Webb DJ (1995) Physiologic role of endothelin in maintenance of vascular tone in humans. J Cardiovasc Pharmacol 26 [Suppl 3]:S183–S185

Haynes WG, Webb DJ (1998) Endothelin as a regulator of cardiovascular function in health and disease. J Hypertension 16:1081–1098

Hirata Y, Kanno K, Watanabe TX, Kumagaye S, Nakajima K, Kimura T, Sakakibara S, Marumo F (1990) Receptor binding and vasoconstrictor activity of big endothelin. Eur J Pharmacol 176:225–228

Hisaki K, Matsumura Y, Nishiguchi S, Fujita K, Takaoka M, Morimoto S (1993) Endothelium-independent pressor effect of big endothelin-1 and its inhibition by phosphoramidon in rat mesenteric artery. Eur J Pharmacol 241:75–81

Hoang MV, Turner AJ (1997) Novel activity of endothelin-converting enzyme: hydrolysis of bradykinin. Biochem J 328:23–26

Ihling C, Gobel HR, Lippoldt A, Wessels S, Paul M, Schaefer HE, Zeiher AM (1996) Endothelin-1-like immunoreactivity in human atherosclerotic coronary tissue: a detailed analysis of the cellular distribution of endothelin-1. J Pathology 179: 303–308

Ikegawa R, Matsumura Y, Tsukahara Y, Takaoka M, Morimoto S (1990) Phosphoramidon, a metalloproteinase inhibitor, suppresses the secretion of endothelin-1 from cultured endothelial cells by inhibiting a big endothelin-1 converting enzyme. Biochem Biophys Res Commun 171:669–675

Ikegawa R, Matsumura Y, Tsukahara Y, Takaoka M, Morimoto S (1991) Phosphoramidon inhibits the generation of endothelin-1 from exogenously applied big endothelin-1 in cultured vascular endothelial cells and smooth muscle cells. FEBS Letts 293:45–48

Inooka H, Endo S, Kikuchi T, Wakimasu M, Mizuta E, Fujino M (1991) Solution conformation of human big endothelin-1 (big ET-1). Peptide Chemistry 28:409–414

Ivy DD, LeCras TD, Horan MP, Abman SH (1998) Increased lung preproET-1 and decreased ET$_B$-receptor expression in fetal pulmonary hypertension. Am J Physiol 18:L535–L541

Janes RW, Peapus DH, Wallace BA (1994) The crystal structure of human endothelin. Nature Struct Biol 1:311–319

Jean F, Basak A, Rondeau N, Benjannet S, Hendy GN, Seidah NG, Chretien M, Lazure C (1993) Enzymic characterization of murine and human prohormone convertase-1 (mPC1 and hPC1) expressed in mammalian GH4C1 cells. Biochem J 292: 891–900

Jeffery PK (1998) Structural and inflammatory changes in COPD: a comparison with asthma. Thorax 53:129–136

Jeppsson A, Tazelaar HD, Miller VM, McGregor CG (1998) Distribution of endothelin-1 in transplanted human lungs. Transplantation 66:806–809

Johnson GD, Stevenson T, Ahn KH (1999) Hydrolysis of peptide hormones by endothelin-converting enzyme-1 – a comparison with neprilysin. J Biol Chem 274: 4053–4058

Jones GT, van Rij AM, Solomon C, Thomson IA, Packer SGK (1996) Endothelin-1 is increased overlying atherosclerotic plaques in human arteries. Atherosclerosis 124: 25–35

Kashiwabara T, Inagaki Y, Ohta H, Iwamatsu A, Nomizu M, Morita A, Nishikori K (1989) Putative precursors of endothelin have less vasoconstrictor activity in vitro but a potent pressor effect in vivo. FEBS Letts 247:73–76

Kitazono T, Takeshige K, Cragoe EJ, Minakami S (1988) Intracellular pH changes of cultured bovine aortic endothelial cells in response to ATP addition. Biochem Biophys Res Commun 152:1304–1309

Klemm P, Warner TD, Hohlfeld T, Corder R, Vane JR (1995) Endothelin-1 mediates ex vivo coronary vasoconstriction caused by exogenous and endogenous cytokines. Proc Natl Acad Sci USA 92:2691–2695

Kobayashi T, Miyauchi T, Sakai S, Kobayashi M, Yamaguchi I, Goto K, Sugishita Y (1999) Expression of endothelin-1, ETA and ETB receptors, and ECE and distribution of endothelin-1 in failing rat heart. Am J Physiol 45:H1197–H1206

Kohan DE (1997) Endothelins in the normal and diseased kidney. Am J Kidney Diseases 29:2–26

Korth P, Bohle RM, Corvol P, Pinet F (1999) Cellular distribution of endothelin-converting enzyme-1 in human tissues. J Histochem Cytochem 47:447–461

Kowala MC, Rose PM, Stein PD, Goller N, Recce R, Beyer S, Valentine M, Barton D, Durham SK (1995) Selective blockade of the endothelin subtype A receptor decreases early atherosclerosis in hamsters fed cholesterol. Am J Pathology 146: 819–826

Kuchan MJ, Frangos JA (1993) Shear stress regulates endothelin-1 release via protein kinase C and cGMP in cultured endothelial cells. Am J Physiol 264:H150–H156

Kurihara H, Yoshizumi M, Sugiyama T, Takaku F, Yanagisawa M, Masaki T, Hamaoki M, Kato H, Yazaki Y (1989) Transforming growth factor-β stimulates the expression of endothelin mRNA by vascular endothelial cells. Biochem Biophys Res Commun 159:1435–1440

Kurihara Y, Kurihara H, Suzuki H, Kodama T, Maemura K, Nagai R, Oda H, Kuwaki T, Cao W-H, Kamada N, Jishage K, Ouchi Y, Azuma S, Toyoda Y, Ishikawa T, Kumada M, Yazaki Y (1994) Elevated blood pressure and craniofacial abnormalities in mice deficient in endothelin-1. Nature 368:703–710

Kwan AL, Bavbek M, Jeng AY, Maniara W, Toyoda T, Lappe RW, Kassell NF, Lee KS (1997) Prevention and reversal of cerebral vasospasm by an endothelin-converting, enzyme inhibitor, CGS 26303, in an experimental model of subarachnoid hemorrhage. J Neurosurgery 87:281–286

Lambert GL, Barker S, Lees DM, Corder R (2000) Endothelin-2 synthesis is stimulated by the type-1 tumor necrosis factor receptor and cAMP: comparison with endothelin-converting enzyme-1 expression. J Mol Endocrinol 24:273–283

Lauth M, Berger M-M, Cattaruzza M, Hecker M (2000) Pressure-induced upregulation of preproendothelin-1 and endothelin-B receptor expression in the rabbit jugular vein in situ – implications for vein graft failure? Arterioscler Thromb Vasc Biol 20:96–103

Lee M-E, Dhadly MS, Temizier DH, Clifford JA, Yoshizumi M, Quertermous T (1991) Regulation of endothelin-1 by Fos and Jun. J Biol Chem 266:19034–19039

Lees DM, Pallikaros Z, Corder R (2000) The p55 tumor necrosis factor receptor (CD120a) induces endothelin-1 synthesis in endothelial and epithelial cells. Eur J Pharmacol 390:89–94

Lerman A, Edwards BS, Hallett JW, Heublein DM, Sandberg SM, Burnett JC (1991) Circulating and tissue endothelin immunoreactivity in advanced atherosclerosis. N Engl J Med 325:997–1001

Lerman A, Webster MW, Chesebro JH, Edwards WD, Wei CM Fuster V, Burnett JC (1993) Circulating and tissue endothelin immunoreactivity in hypercholesterolemic pigs. Circulation 88:2923–2928

Löffler BM, Breu V, Clozel M (1993) Effect of endothelin receptor antagonists and of the novel non-peptide antagonist Ro 46–2005 on endothelin levels in rat plasma. FEBS Letts 333:108–110

Macarthur H, Warner TD, Wood EG, Corder R, Vane JR (1994) Endothelin-1 release from endothelial cells in culture is elevated both acutely and chronically by short periods of mechanical stretch. Biochem Biophys Res Commun 200:395–400

Magazine HI, Andersen TT, Goligorsky MS, Malik AB (1991) Evaluation of endothelin receptor populations using endothelin-1 biotinylated at lysine-9 sidechain. Biochem Biophys Res Commun 181:1245–1250

Malarkey K, Chilvers ER, Lawson MF, Plevin R (1995) Stimulation by endothelin-1 of mitogen-activated protein kinases and DNA synthesis in bovine tracheal smooth muscle cells. Br J Pharmacol 116:2267–2273

Malek AM, Greene AL, Izumo S (1993) Regulation of endothelin 1 gene by fluid shear stress is transcriptionally mediated and independent of protein kinase C and cAMP. Proc Natl Acad Sci USA 90:5999–6003

Marini M, Carpi S, Bellini A, Patalano F, Mattoli S (1996) Endothelin-1 induces increased fibronectin expression in human bronchial epithelial cells. Biochem Biophys Res Commun 220:896–899

Marsden PA, Brenner BM (1992) Transcriptional regulation of the endothelin-1 gene by TNFα. Am J Physiol 262:C854–C861

Marsen TA, Simonson MS, Dunn MJ (1995) Thrombin induces the preproendothelin-1 gene in endothelial cells by a protein tyrosine kinase-linked mechanism. Circ Res 76:987–995

Matsumura Y, Hisaki K, Takaoka M, Morimoto S (1990) Phosphoramidon, a metalloproteinase inhibitor, suppresses the hypertensive effect of big endothelin-1. Eur J Pharmacol 185:103–106

Mathew V, Hasdai D, Lerman A (1996) The role of endothelin in coronary atherosclerosis. Mayo Clinic Proceedings 71:769–777

McCulloch KM, Docherty C, MacLean MR (1998) Endothelin receptors mediating contraction of rat and human pulmonary resistance arteries: effect of chronic hypoxia in the rat. Brit J Pharmacol 123:1621–1630

McDermott CD, Shennib H, Giaid A (1998) Immunohistochemical localization of endothelin-1 and endothelin-converting enzyme-1 in rat lung allografts. J Cardiovasc Pharmacol 31 [Suppl 1]:S27–S30

McMahon EG, Palomo MA, Moore WM, McDonald JF, Stern MK (1991) Phosphoramidon blocks the pressor activity of porcine big endothelin-1-(1–39) in vivo and conversion of big endothelin-1-(1–39) to endothelin-1-(1–21) in vitro. Proc Natl Acad Sci USA 88:703–707

Menziani MC, Cocchi M, De Benedetti PG, Gilbert RG, Richards WG, Zamai M, Caiolfa VR (1991) A theoretical study of the structure of big endothelin. J Chim Phys Phys-Chim Biol 88:2687–2694

Minamino T, Kurihara H, Takahashi M, Shimada K, Maemura K, Oda H, Ishikawa T, Uchiyama T, Tanzawa K, Yazaki Y (1997) Endothelin-converting enzyme expression in the rat vascular injury model and human coronary atherosclerosis. Circulation 95:221–230

Mitchell MD, Branch DW, Lamarche S, Dudley DJ (1992) The regulation endothelin production in human umbilical vein endothelial-cells – unique inhibitory-action of calcium ionophores. J Clin Endocrinol Metab 75:665–668

Moll T, Czyz M, Holzmüller H, Hoferwarbinek R, Wagner E, Winkler H, Bach FH, Hofer E (1995) Regulation of the tissue factor promoter in endothelial cells – binding of NF-$_\kappa$B-like, AP-1-like, and SP1-like transcription factors J Biol Chem 270:3849–3857

Murphy LJ, Corder R, Mallet AI, Turner AJ (1994) Generation by the phosphoramidon-sensitive peptidases, endopeptidase-24.11 and thermolysin, of endothelin-1 and C-terminal fragment from big endothelin-1. Br J Pharmacol 113:137–142

Nabokov A, Amann K, Wagner J, Gehlen F, Münter K, Ritz E (1996) Influence of specific and non-specific endothelin receptor antagonists on renal morphology in rats with surgical renal ablation. Nephrol Dial Transplant 11:514–520

Nakajima K, Kubo S, Kumagaye S, Nishio H, Tsunemi M, Inui T, Kuroda H, Chino N, Watanabe T, Kimura T, Sakakibara S (1989) Structure-activity relationships of endothelin: importance of charged groups. Biochem Biophys Res Commun 163: 424–429

Nootens M, Kaufmann E, Rector T, Toher C, Judd D, Francis GS, Rich S (1995) Neurohormonal activation in patients with right ventricular failure from pulmonary hypertension: relation to hemodynamic variables and endothelin levels. J Am Coll Cardiol 26:1581–1585

Ohkubo S, Ogi K, Hosoya M, Matsumoto H, Suzuki N, Kimura C, Ondo H, Fujino M (1990) Specific expression of human endothelin-2 (ET-2) gene in a renal adenocarcinoma cell line. Molecular cloning of cDNA encoding the precursor of ET-2 and its characterization. FEBS Letts 274:136–140

Okamoto H, Takaoka M, Ohkita M, Itoh M, Nishioka M, Matsumura Y (1998) A proteasome inhibitor lessens the increased aortic endothelin-1 content in deoxycortisone acetate-salt hypertensive rats. Eur J Pharmacol 350:R11–R12

Opgenorth TJ, Wu-Wong JR, Shiosaki K (1992) Endothelin-converting enzymes. FASEB J 6:2653–2659

Panettieri RA Jr, Goldie RG, Rigby PJ, Eszterhas AJ, Hay DW (1996) Endothelin-1-induced potentiation of human airway smooth muscle proliferation: an ETA receptor-mediated phenomenon. Brit J Pharmacol 118:191–197

Peto H, Corder R, Janes RW, Wallace BA (1996) A molecular model for human big-endothelin-1 (Big ET-1). FEBS Letts 394:191–195

Pollock DM, Polakowski JS (1997) ET$_A$ receptor blockade prevents hypertension associated with exogenous endothelin-1 but not renal mass reduction in the rat. J Am Soc Nephrol 8:1054–1060

Read MA, Neish AS, Luscinskas FW, Palombella VJ, Maniatis T, Collins T (1995) The proteasome pathway is required for cytokine-induced endothelial leukocyte adhesion molecule expression. Immunity 2:493–506

Redington AE, Springall DR, Meng QH, Tuck AB, Holgate ST, Polak JM, Howarth PH (1997) Immunoreactive endothelin in bronchial biopsy specimens: increased expression in asthma and modulation by corticosteroid therapy. J Allergy Clin Immunol 100:544–552

Riley DJ (1991) Vascular Remodeling. Chap 5.2.7 The Lung: Scientific Foundations. 1189–1198. Eds Crystal RG and West JB, Raven Press, New York

Roberts CR (1995) Is asthma a fibrotic disease? Chest 107 (3) Suppl 111S–117S

Russell FD, Skepper JN, Davenport AP (1998) Human endothelial cell storage granules. A novel intracellular site for isoforms of the endothelin-converting enzyme. Circ Res 83:314–321

Russell FD, Davenport AP (1999) Secretory pathways in endothelin synthesis. Brit J Pharmacol 126:391–398

Satoh H, Tsukamoto K, Hashimoto Y, Hashimoto N, Togo M, Hara M, Maekawa, Isoo N, Kimura S, Watanabe T (1999) Thiazolidinediones suppress endothelin-1 secretion from bovine vascular endothelial cells: a new possible role of PPARγ on vascular endothelial function. Biochem Biophys Res Commun 254:757–763

Sawamura T, Kasuya Y, Matsushita Y, Suzuki N, Shinmi O, Kishi N, Sugita Y, Yanagisawa M, Goto K, Masaki T, Kimura S (1991) Phosphoramidon inhibits the intracellular conversion of big endothelin-1 to endothelin-1 in cultured endothelial cells. Biochem Biophys Res Commun 174:779–784

Schmidt M, Kroger B, Jacob E, Seulberger H, Subkowski T, Otter R, Meyer T, Schmalzing G, Hillen H (1994) Molecular characterization of human and bovine endothelin converting enzyme (ECE-1). FEBS Letts 356:238–243

Schuppin GT, Rhodes CJ (1996) Specific co-ordinated regulation of PC3 and PC2 gene expression with that of preproinsulin in insulin-producing beta TC3 cells. Biochem J 313:259–268

Schweizer A, Valdenaire O, Nelbock P, Deuschle U, Dumas Milne Edwards JB, Stumpf JG, Löffler BM (1997) Human endothelin-converting enzyme (ECE-1): three isoforms with distinct subcellular localizations. Biochem J 328:871–877

Schweizer A, Löffler BM, Rohrer J (1999) Palmitoylation of the three isoforms of human endothelin-converting enzyme-1. Biochem J 340:649–656

Shimada K, Takahashi M, Tanzawa K (1994) Cloning and functional expression of endothelin-converting enzyme from rat endothelial cells. J Biol Chem 269: 18275–18278

Shimada K, Takahashi M, Turner AJ, Tanzawa K (1996) Rat endothelin-converting enzyme-1 forms a dimer through Cys^{412} with a similar catalytic mechanism and a distinct substrate binding mechanism compared with neutral endopeptidase-24.11. Biochem J 315:863–867

Smith RM, Brown TJ, Roach AG, Williams KI, Woodward B (1997) Evidence for endothelin involvement in the pulmonary vasoconstrictor response to systemic hypoxia in the isolated rat lung. J Pharmacol Exp Ther 283:419–425

Spence S, Anderson C, Cukierski M, Patrick D (1999) Teratogenic effects of the endothelin receptor antagonist L-753,037 in the rat. Reproductive Toxicology 13: 15–29

Steiner DF, Smeeken SP, Ohagi S, Chan SJ (1992) The new enzymology of precursor processing endoproteases. J Biol Chem 267:23435–23438

Sun G, Stacey MA, Bellini A, Marini M, Mattoli S (1997) Endothelin-1 induces bronchial myofibroblast differentiation. Peptides 18:1449–1451

Takahashi M, Fukuda K, Shimada K, Barnes K, Turner AJ, Ikeda M, Koike H, Yamamoto Y, Tanzawa K (1995) Localization of rat endothelin-converting enzyme to vascular endothelial cells and some secretory cells. Biochem J 311:657–665

Takeda S, Sawa Y, Minami M, Kaneda Y, Fujii Y, Shirakura R, Yanagisawa M, Matsuda H (1997) Experimental bronchiolitis obliterans induced by in vivo HVJ-liposome-mediated endothelin-1 gene transfer Ann Thoracic Surg 63:1562–1567

Terada Y, Tomita K, Nonoguchi H, Yang T, Marumo F (1993) Expression of endothelin-3 mRNA along rat nephron segments using polymerase chain reaction. Kidney Int 44:1273–1280

Tretter L, Chinopoulos C, Adam-Vizi V (1998) Plasma membrane depolarization and disturbed Na^+ homeostasis induced by the protonophore carbonyl cyanide-p-trifluoromethoxyphenyl-hydrazon in isolated nerve terminals. Mol Pharmacol 53:734–741

Turner AJ, Barnes K, Schweizer A, Valdenaire O (1998) Isoforms of endothelin-converting enzyme: why and where? TIPS 19:483–486

Turner AJ, Tanzawa K (1997) Mammalian membrane metallopeptidases: NEP, ECE, KELL, and PEX. FASEB J 11:355–364

Underwood DC, Bochnowicz S, Osborn RR, Luttmann MA, Hay DW (1997) Non-peptide endothelin receptor antagonists. X. Inhibition of endothelin-1- and hypoxia-induced pulmonary pressor responses in the guinea pig by the endothelin receptor antagonist, SB 217242. J Pharmacol Exp Ther 283:1130–1137

Valdenaire O, Rohrbacher E, Mattei MG (1995) Organization of the gene encoding the human endothelin-converting enzyme (ECE-1). J Biol Chem 270:29794–29798

Valdenaire O, Lepailleur-Enouf D, Egidy G, Thouard A, Barret A, Vranckx R, Tougard C, Michel J-P (1999) A fourth isoform of endothelin-converting enzyme (ECE-1) is generated from an additional promoter. Molecular cloning and characterization. Eur J Biochem 264:341–349

Velasco CE, Jackson EK, Morrow JA, Vitola JV, Inagami T, Forman MB (1993) Intravenous adenosine suppresses cardiac release of endothelin after myocardial ischaemia and reperfusion. Cardiovasc Res 27:121–128

Wagner OF, Christ G, Wojta J, Vierhapper H, Parzer S, Nowotny PJ, Schneider B, Waldhäusl W, Binder BR (1992) Polar secretion of endothlein-1 by cultured endothelial cells. J Biol Chem 267:16066–16068

Wallace BA, Corder R (1997) Circular dichroism studies of human big-endothelin-1 (big ET-1). J Peptide Res 49:331–335

Wang X, Douglas SA, Louden C, Vickery-Clark LM, Feuerstein GZ, Ohlstein EH (1996) Expression of endothelin-1, endothelin-3, endothelin-converting enzyme-1, and endothelin-A and endothelin-B receptor mRNA after angioplasty-induced neointimal formation in the rat. Circ Res 78:322–328

Warner TD, Schmidt HHHW, Murad F (1992) Interactions of endothelins and EDRF in bovine native endothelial-cells – selective effects of endothelin-3. Am J Physiol 262: H1600–H1605

Watanabe T, Itahara Y, Nakajima K, Kumagaye S, Kimura T, Sakakibara S (1991) The biological activity of endothelin-1 analogues in three different assay systems. J Cardiovasc Pharmacol 17 [Suppl 7]:S5–S9

Weitzberg E, Hemsen A, Lundberg JM, Ahlborg G (1995) ET-3 is extracted by and induces potent vasoconstriction in human splanchnic and renal vasculatures. J Applied Physiol 79:1255–1259

Wellings RP, Corder R, Warner TD, Cristol J-P, Thiemermann C, Vane JR (1994) Evidence from receptor antagonists of an important role for ET_B receptor mediated vasoconstrictor effects of endothelin-1 in the rat kidney. Br J Pharmacol 111: 515–520

Wiggs BR, Hrousis CA, Drazen JM, Kamm RD (1997) On the mechanism of mucosal folding in normal and asthmatic airways. J Appl Physiol 83:1814–1821

Woods M, Mitchell JA, Wood EG, Barker S, Walcot NR, Rees GM, Warner TD (1999) Endothelin-1 is induced by cytokines in human vascular smooth muscle cells: Evidence for intracellular endothelin-converting enzyme. Mol Pharmacol 55: 902–909

Xu D, Emoto N, Giaid A, Slaughter C, Kaw S, deWit D, Yanagisawa M (1994) ECE-1: a membrane-bound metalloprotease that catalyzes the proteolytic activation of big endothelin-1. Cell 78:473–485

Yamagishi H, Garg V, Matsuoka R, Thomas T, Srivastava D (1999) A molecular pathway revealing a genetic basis for human cardiac and craniofacial defects. Science 283:1158–1161

Yanagisawa H, Yanagisawa M, Kapur RP, Richardson JA, Williams SC, Clouthier DE, de Wit D, Emoto N, Hammer RE (1998) Dual genetic pathways of endothelin-mediated intercellular signaling revealed by targeted disruption of endothelin converting enzyme-1 gene. Development 125:825–836

Yanagisawa M, Kurihara H, Kimura S, Tomobe Y, Kobayashi M, Mitsui Y, Yazaki Y, Goto K, Masaki T (1988) A novel potent vasoconstrictor peptide produced by vascular endothelial cells. Nature 332:411–415

Yanagisawa M, Inoue A, Takuwa Y, Mitsui Y, Kobayashi M, Masaki T (1989) The human preproendothelin-1 gene – possible regulation by endothelial phosphoinositide turnover signaling. J Cardiovasc Pharmacol 13 [Suppl 5]:S13–S17

Yorimitsu K, Moroi K, Inagaki N, Saito T, Masuda Y, Masaki T, Seino S, Kimura S (1995) Cloning and sequencing of a human endothelin converting enzyme in renal adenocarcinoma (ACHN) cells producing endothelin-2. Biochem Biophys Res Commun 208:721–727

Ziegelstein RC, Cheng L, Capogrossi MC (1992) Flow-dependent cytosolic acidification of vascular endothelial cells. Science 258:656–659

CHAPTER 4

Endothelin Receptors

P.J. HENRY and R.G. GOLDIE

A. Introduction

Receptors are cellular macromolecules that combine with hormones, neuro-transmitters, drugs, or intracellular messengers to initiate a change in cell function, and are thereby concerned directly and specifically in chemical signalling between and within cells. In this article we review the progress that has been made over the past decade in our understanding of various aspects of endothelin receptor (ET receptor) biology. These G protein-coupled receptors are stimulated by the endothelin superfamily of peptides, as well as the sarafotoxins, which are constituents of a rare snake venom (see Chap. 2). There has emerged an enormous body of literature describing the locations of ET receptors within various cell, tissue, and organ systems as well as the myriad of effects produced following their activation. In large part, the great interest in ET receptor biology is generated in response to the likelihood that their activation is linked to the initiation, progression, and symptomology of various disease processes. These aspects of ET receptor biology have been excellently covered in earlier works (HUGGINS et al. 1993; RUBANYI and POLOKOFF 1994; GRAY and WEBB 1996), and detailed, up-to-date accounts of recent findings are described in Chaps. 11–19 in this volume. The focus of this chapter will be to describe some of the important developments that have been made in the fields of ET receptor classification, structure, and ligand binding. Furthermore, it will highlight the important role ET receptors play in fetal development and the extent to which ET receptor expression and function can be modulated by various cytokines, drugs, and in disease.

B. Pharmacological Classification of ET Receptors

I. General

The International Union of Pharmacology Committee for Receptor Nomenclature and Drug Classification (NC-IUPHAR) has defined four major criteria that must be fulfilled before a receptor can be accepted as functionally significant for inclusion in a pharmacological classification. These are a known protein structure, a defined link to a transduction system, evidence of endogenous expression and established operational characteristics from determination of potencies and selectivities of agonists and antagonists (Vanhoutte et al. 1998). All of these criteria have been satisfied for two endothelin receptors, the ET_A receptor and ET_B receptor, which mediate the actions of three endogenous isoforms of the endothelin family, endothelin-1, endothelin-2, and endothelin-3. The ET_A receptor has a very high affinity for endothelin-1 and endothelin-2, in the subnanomolar range, and a 100-fold lower affinity for endothelin-3 (Saeki et al. 1991). The ET_B receptor has a high and equal affinity for each of the endothelin isoforms (Saeki et al. 1991). Prior to the cloning of ET_A and ET_B receptors in 1990, evidence supporting the existence of endothelin receptor subtypes had come from several sources, including the discovery of varying agonist potencies in different tissues (Douglas and Hiley 1990; Huggins et al. 1993), reports of receptor-radioligand cross-linking (Masuda et al. 1989), and use of two-site binding models to describe adequately binding curves (Galron et al. 1989; Bousso-Mittler et al. 1989). The nomenclature of ET_A and ET_B receptors was originally introduced to distinguish sites in rat aorta (A) and guinea-pig bronchus (B) which were respectively sensitive and insensitive to the C-terminal hexapeptide of endothelin-1 (Maggi et al. 1989).

II. ET_A and ET_B Receptor-Selective Agonists and Antagonists

The classification of ET_A and ET_B receptors has been facilitated greatly by the discovery and development of receptor-selective ligands (see Chap. 9). ET_B receptors are selectively stimulated by a wide range of compounds, including sarafotoxin S6c, which is 30,000 times more potent as an inhibitor of $[^{125}I]$-endothelin-1 binding at ET_B than ET_A receptors (Williams et al. 1991). In addition, linear analogues of endothelin-1 such as $[Ala^{1,3,11,15}]$endothelin-1 (Hiley et al. 1990), BQ-3020 (N-acetyl-$[Ala^{11,15}]$endothelin-1(6–21)) (Ihara et al. 1992), and IRL1620 (N-succinyl-$[Glu^9,Ala^{11,15}]$endothelin-1(8–21)) (Takai et al. 1992) are ET_B receptor-selective agonists. Despite the wide availability of agonists that display selectivity for ET_B receptors, there are still no known selective agonists at ET_A receptors. In the absence of ET_A receptor-selective agonists, the pharmacological characterization of ET_A receptors is currently reliant upon the use of potent, but non-selective agonists such as endothelin-1 and endothelin-2 in concert with selective ET_A receptor antagonists. The first ET_A receptor antagonists were peptidic analogues of endothelin-1 ([diaminoproprionic acid^1-Asp^{15}]endothelin-1; (Spinella et al. 1991)), but the most significant development came from modifications of the cyclic pentapeptides BE18257A (cyclo[-D-Trp^1-D-Glu^2-Ala^3-D-Val^4-Leu^5-]) and BE18257B (cyclo[-D-Trp^1-D-Glu^2-Ala^3-D-alloIle^4-Leu^5-]; (Kojiri et al. 1991; Nakajima et al. 1991)), discovered in the fermentation broth of the microbe, *Streptomyces misakiensis*. From these lead compounds, several potent and highly ET_A receptor-selective compounds were produced including BQ-123 (cyclo[-D-Trp-D-Asp-L-Pro-D-Val-L-Leu-]; (IHARA et al. 1991)). Other ET_A receptor-selective antagonists include the tripeptide FR139317 (SOGABE et al. 1993) and nonpeptide, orally active compounds such as BMS-182874 (STEIN et al. 1994), PD155080 (DOHERTY et al. 1995), SB 234551 (OHLSTEIN et al. 1998), and T-0201 (HOSHINO et al. 1998). As a result of investigations into the development of tripeptide ET_A antagonists (e.g., BQ-610), it was discovered that some structural modifications promoted ET_B receptor affinity. Further studies led to the development of the potent and selective ET_B receptor antagonist BQ-788 (ISHIKAWA et al. 1994). At the same time, another ET_B receptor-selective antagonist, the cyclic, 16 amino acid peptide RES-701-1, was isolated from the culture broth of *Streptomyces misakiensis* (MORISHITA et al. 1994).

III. Characterization of ET_A and ET_B Receptor Populations

Our current understanding of the density, distribution, and function of ET_A and ET_B receptors in cells, tissues, and organs has principally been obtained from functional and radioligand binding studies using ET receptor-selective ligands. Some tissues contain a predominance of either ET_A or ET_B receptors whereas others contain a variable mixture of ET_A and ET_B receptors. These data are often confirmed by molecular biological evidence demonstrating the presence or otherwise of ET_A and ET_B receptor mRNA.

Endothelin-1 is a potent spasmogen in many vascular preparations, and the vasoconstrictor response is most commonly mediated by ET_A receptors. The principal involvement of the ET_A receptor subtype was typically determined on the basis that (a) endothelin-1-induced effects were potently inhibited by ET_A receptor-selective antagonists such as BQ-123 or FR139317, but not by ET_B receptor-selective antagonists, (b) ET_B receptor-selective agonists were inactive, and (c) endothelin-1 was more potent than endothelin-3. For example, in human isolated pulmonary artery preparations, endothelin-1 elicited concentration-dependent contractions that were potently inhibited by an ET_A receptor-selective antagonist BQ-123, whereas the ET_B receptor-selective agonist sarafotoxin S6c was inactive (FUKURODA et al. 1994). Consistent with these functional data, radioligand binding experiments have established that greater than 90% of the specific ET receptors labelled by [^{125}I]-endothelin-1 in human pulmonary artery were of the ET_A receptor subtype (FUKURODA et al. 1994; DAVENPORT et al. 1995). The ET_A receptor appears to be the predominant ET receptor in many human vascular preparations including aorta, coronary artery, internal mammary artery, and saphenous vein (DAVENPORT et al. 1995), guinea-pig aorta (SCHOEFFTER and RANDRIANTSOA 1993), rat aorta (SUMNER et al. 1992), as well as in some nonvascular tissues such as sheep tracheal smooth muscle (GOLDIE et al. 1994).

Functional evidence for a homogeneous population of ET_B receptors was typically provided on the basis that (a) endothelin-1-induced effects were inhibited by ET_B receptor-selective antagonists such as BQ-788 or RES-701 but not by ET_A receptor-selective antagonists, (b) ET_B receptor-selective agonists, such as sarafotoxin S6c and BQ3020, were active, and (c) endothelin-1 and endothelin-3 had similar potencies. ET_B receptors are widely distributed throughout vascular tissue, mediating endothelium-dependent vasodilation and vascular smooth muscle contraction (Gray and Webb 1996). ET_B receptors are also present in many nonvascular tissues, including the lung where they mediate a variety of responses including bronchial smooth muscle contraction. In human bronchus, endothelin-1 elicited BQ-123-insensitive contractions, endothelin-1 and endothelin-3 were equipotent, and sarafotoxin S6c and BQ3020 were both potent spasmogens. Consistent with this, quantitative autoradiographic studies revealed that approximately 90% of specific $[^{125}I]$-endothelin-1 binding to human bronchial smooth muscle could be inhibited by sarafotoxin S6c (Goldie et al. 1995). Together, these functional and autoradiographic data are indicative of the presence of an essentially homogeneous population of ET_B receptors on human bronchial smooth muscle. However, a recent study by Hay and coworkers demonstrated that contractions of human isolated bronchial smooth muscle induced by endothelin-1, endothelin-3 and sarafotoxin S6c were not inhibited by BQ-788 or RES-701, and thus the ET_B receptors present may be a novel subtype (Hay et al. 1998).

Rather than containing a homogeneous population of ET_A or ET_B receptors, the majority of ET receptor-expressing cells and tissues appear to possess mixed populations of ET_A and ET_B receptors. Within blood vessels, coexisting ET_A and ET_B receptors on vascular smooth muscle both appear to mediate contraction, whereas ET_B receptors present on nearby vascular endothelial cells can functionally antagonize the contraction through the production and release of nitric oxide. Functional evidence for a mixed population of ET_A and ET_B receptors was generally established on the basis that (a) ET_B receptor-selective agonists were at least partial agonists, (b) endothelin-1-induced responses were only weakly or partially inhibited by either an ET_A or ET_B receptor-selective antagonist, and (c) endothelin-1-induced responses were inhibited in the combined presence of an ET_A and an ET_B receptor antagonist or by a non-selective antagonist. For example, endothelin-1-induced contractions of mouse isolated tracheal smooth muscle were not inhibited by either BQ-123 or BQ-788 when used alone, but were significantly inhibited when the antagonists were used in combination (Carr et al. 1996). Autoradiographic analysis of $[^{125}I]$-endothelin-1 binding to murine tracheal smooth muscle confirmed the presence of both ET_A and ET_B receptors in approximately equal densities (Carr et al. 1996). Mixed populations of ET_A and ET_B receptors have been reported in many other tissues (Bax and Saxena 1994), including internal mammary artery (Tschudi and Luscher 1994; Seo et al. 1994; Liu et al. 1996) and jugular vein (Sumner et al. 1992; Lodge and Halaka 1993; Calo et al. 1996). Interestingly, regional variations in the proportion of ET_A and ET_B receptors have been reported; for example, as one moves down the porcine respiratory tract the proportion of ET_A:ET_B receptors on airway smooth muscle changes from 30:70 in trachea to 70:30 in bronchus (GOLDIE et al. 1996).

C. Molecular Biology of ET Receptors

I. Cloning of ET Receptor Genes

The first reports of the cloning and characterization of the genes encoding ET_A (ARAI et al. 1990) and ET_B (SAKURAI et al. 1990) receptors appeared simultaneously in 1990, just two years after the initial description of the endogenous agonist endothelin-1 (YANAGISAWA et al. 1988) and before the definition of subtype-selective agonists or antagonists (HUGGINS et al. 1993). Expression-cloning techniques were used to isolate ET_A receptor cDNA from a bovine lung cDNA library following expression in *Xenopus oocytes* and screening with a voltage clamp technique (ARAI et al. 1990). Similarly, for the ET_B receptor study, a rat lung cDNA library was screened in COS-7 cells (SAKURAI et al. 1990). The bovine lung ET_A receptor cDNA predicted a 427 amino acid protein (ARAI et al. 1990), whereas rat lung ET_B receptor cDNA encoded a 415 amino acid sequence (SAKURAI et al. 1990). A single copy of the human ET_A and ET_B receptor genes localize to chromosomes 4 (HOSODA et al. 1992) and 13 (ARAI et al. 1993), respectively, and cDNAs encoding the human ET_A and ET_B receptors predict 427 (ADACHI et al. 1991) and 442 (NAKAMUTA et al. 1991) amino acids, respectively. There exists considerable homology, approximately 60% at the amino acid level, between human ET_A and ET_B receptors (ARAI et al. 1993), with the most conserved regions occurring within the hydrophobic transmembrane spanning domains (90% homology). Substantial homologies (about 90%) also exist between either ET receptor subtype across mammalian species (HAENDLER et al. 1992). It has been deduced from the nucleotide sequences of the cDNAs that ET_A and ET_B receptors are G protein-coupled receptors containing seven transmembrane domains.

The genes for ET_A and ET_B receptors are large. The 40kb of DNA which spans the human ET_A receptor gene contains eight exons and seven introns (HOSODA et al. 1992), and the 24kb of DNA spanning the human ET_B receptor gene contains seven exons and six introns (ARAI et al. 1993). Other than an extra intron present in the 5'-noncoding region of the ET_A receptor gene, the exon-intron splice sites are conserved between the two receptors (RUBANYI and POLOKOFF 1994). Introns typically occur within the coding region immediately before or after one of the transmembrane helix domains, suggesting that the corresponding exons may encode functional units (RUBANYI and POLOKOFF 1994). Within the human ET_A receptor gene, exon 1 is in a part of the 5'-noncoding region, exon 2 encodes transmembrane domains I and II, and the other exons encode one transmembrane domain each. Similarly, exon 1 of the human ET_B receptor gene encodes transmembrane domains I and II, and exons 2–6 encode transmembrane domains III–VII, respectively. Altered expression of ET receptor genes in response to a variety of stimuli is discussed in detail below.

II. Mutations of ET Receptor Genes

To determine the developmental role of the ET_A receptor, several studies have used gene targeting to generate an ET_A receptor-null mouse (CLOUTHIER et al. 1998; YANAGISAWA et al. 1998) (see Chap. 6). Although these mice are born alive, they suffer severe craniofacial and cardiovascular defects and die soon after birth (CLOUTHIER et al. 1998). A similar profile of effect is also seen in endothelin-1- (KURIHARA et al. 1994) and endothelin-1 converting enzyme- (YANAGISAWA et al. 1998) deficient mice. Thus, stimulation of ET_A receptors by endogenously produced endothelin-1 plays an essential role in the process of aortic arch patterning, probably by affecting the postmigratory cardiac neural crest cell development (YANAGISAWA et al. 1998). These effects observed in ET_A receptor-deficient mice resemble the human conditions collectively termed CATCH 22 or velocardiofacial syndrome (CLOUTHIER et al. 1998).

The ET_B receptor appears to be critical for the developmental regulation of neural crest cells that become enteric ganglia and melanocytes. In mice, targeted disruption of the ET_B receptor gene results in aganglionic megacolon associated with coat colour spotting and premature death at about 3–4 weeks of age (HOSODA et al. 1994). This phenotype closely resembles hereditary syndromes observed in mice (Piebald-lethal mouse (HOSODA et al. 1994)), rats (spotting lethal rat (GARIEPY et al. 1996)), and horses (lethal white foal syndrome (SANTSCHI et al. 1998; METALLINOS et al. 1998). In humans, mutation of the ET_B receptor gene appears to be associated with Hirschsprung's disease, which affects 1 in 5000 children and is characterized by the absence of ganglion cells in the distal portion of the intestinal tract and consequent failure of innervation of the colon (PUFFENBERGER et al. 1994). In a large inbred kindred with a high incidence of Hirschsprung's disease (Mennonite pedigree), a G → A missense mutation in exon 4 of the ET_B receptor gene resulted in the substitution of the highly conserved Trp-276 residue in transmembrane domain V with a Cys residue (PUFFENBERGER et al. 1994). Many other ET_B receptor gene alterations have since been identified in sporadic and familial Hirschsprung's disease (see KUSAFUKA et al. 1997). In Piebald-lethal mice the entire ET_B receptor gene is deleted (HOSODA et al. 1994), in spotting lethal rats there is a 301 bp deletion in the 3' end of the first exon of the ET_B receptor gene (CECCHERINI et al. 1995), and in lethal white foals the gene contained a 2-bp nucleotide change leading to a missense mutation (I118K) in transmembrane domain I of the receptor (METALLINOS et al. 1998). In the spotting lethal rat, transgenic expression of the ET_B receptor gene prevented the intestinal defect but not the coat colour spotting, suggesting that the critical time for ET_B receptor gene expression in the development of the enteric nervous system begins after separation of the melanocyte lineage and the interstinal lineage and their common precursor (GARIEPY et al. 1998). Not unexpectedly, mutations in the gene encoding endothelin-3, the likely endogenous activator of the ET_B receptor, is also associated with sporadic Hirschsprung's disease in humans and lethal spotting in mice.

D. Subclassification of ET_A and ET_B Receptors

ET_A and ET_B receptor genes have similar structural organization suggesting that they originated from the same structural gene (GRAY and WEBB 1996), and analysis of human genomic DNA with probes specific for human ET_A and ET_B receptors detected just two hybridising fragments (SAKAMOTO et al. 1991). These findings suggest that if other ET receptor genes are present in the human genome then they must have low sequence similarities to ET_A and ET_B receptors (GRAY and WEBB 1996).

Nevertheless, in their recent review, Bax and Saxena brought into focus the fact that a number of effects induced by endothelin-1 and related peptides do not readily fit the present criteria for ET_A and ET_B receptors, and posed the question as to whether it was time to reconsider the classification of ET receptors (BAX and SAXENA 1994). Indeed, there are many instances within the literature of references to atypical ET receptors, and somewhat more specifically to subtypes of ET_A (e.g., ET_{A1}, ET_{A2}) and ET_B (e.g., ET_{B1}, ET_{B2}) receptors. However, much of the evidence in support of subtypes of ET_A and ET_B receptors has come from functional data, and additional molecular biological, operational, and structural information is clearly required (KENAKIN et al. 1992). Some of the evidence for and against the existence of such subtypes will be presented below.

I. ET_A Receptor Subtypes

In many tissues, ET_A receptor-selective antagonists more potently inhibit contractile responses to endothelin-3 (or sarafotoxin S6b) than to endothelin-1. For example, BQ-123 and FR139317 were both several orders of magnitude more potent in inhibiting contractions of rabbit iris sphincter muscle induced by either endothelin-3 or sarafotoxin S6b than by endothelin-1 (ISHIKAWA et al. 1996). Similar findings have been reported in rat aorta (SUMNER et al. 1992), vas deferens (EGLEZOS et al. 1993; WARNER et al. 1993), renal artery (CLARK and PIERRE 1995), human small omental vein (RIEZEBOS et al. 1994), coronary artery (GODFRAIND 1993; BAX et al. 1994), and saphenous vein (BAX et al. 1993; NISHIYAMA et al. 1995; PATE et al. 1998). One possible explanation for these data is that there exist two ET_A receptor subtypes. According to this postulate, one of these receptors, termed ET_{A1}, is activated by endothelin-1, endothelin-3, and sarafotoxin S6b and inhibited by BQ-123 and FR139317, whereas the other, termed ET_{A2}, is activated by endothelin-1 and is insensitive to blockade by the antagonists. However, radioligand binding data have not generally supported the existence of such ET_A receptor subtypes

(DEVADASON and HENRY 1997; NOSAKA et al. 1998). Competition binding experiments using [^{125}I]-endothelin-1, [^{125}I]-sarafotoxin S6b, and [^{3}H]-BQ-123 detected only typical (i.e., BQ-123-sensitive) ET$_A$ receptors in the rat renal artery (DEVADASON and HENRY 1997) and the rabbit iris (PATE et al. 1998). However, significant differences have been found in the binding kinetics of sarafotoxin S6b and endothelin-1 in both rat renal artery (DEVADASON and HENRY 1997) and ovine trachea (HENRY and KING 1999). The binding of [^{125}I]-sarafotoxin S6b was reversible whereas [^{125}I]-endothelin-1 binding was irreversible. Thus, an alternate explanation for the greater potency of BQ-123 against sarafotoxin S6b-induced responses is that BQ-123 is better able to inhibit the actions of a reversibly binding agonist such as sarafotoxin S6b than an irreversibly binding ligand such as endothelin-1. Consistent with this postulate, Gresser and coworkers have established that BQ-123 is intrinsically more potent in inhibiting the actions of endothelin-3 than endothelin-1 in cloned ET$_A$ receptors of bovine, rat, and human origins, and that there is no need to postulate the existence of multiple ET$_A$ receptor subtypes to account for the singular actions of BQ-123 (GRESSER et al. 1996).

II. ET$_B$ Receptor Subtypes

The bulk of the evidence favouring the existence of subtypes of ET$_B$ receptors has come from studies showing that the potencies of several ET receptor antagonists (e.g., PD142893, BQ-788, IRL1038, and RES-701) varied considerably depending upon the agonist used and the tissue studied. With regard to PD142893, Warner and coworkers reported that this non-peptidic antagonist blocked endothelin-1- and sarafotoxin S6c-induced vasodilatation of rat isolated perfused mesentery to a greater extent than it inhibited endothelin-1- and sarafotoxin S6c-induced constriction of rabbit pulmonary artery and rat stomach strip (WARNER et al. 1993). PD142893 inhibited the endothelin-3-induced endothelium-dependent vasorelaxation of rabbit saphenous vein, but not vasoconstriction induced by endothelin-1, endothelin-3, or sarafotoxin S6c (DOUGLAS et al. 1995). In addition, PD142893 blocked sarafotoxin S6c-induced relaxations of mouse aorta more potently than contractions of mouse gastric fundus induced by endothelin-1, sarafotoxin S6c, or IRL1620 (MIZUGUCHI et al. 1997). Furthermore, PD142893 inhibited the contractions of rabbit saphenous vein induced by IRL1620, but not those induced by endothelin-1, endothelin-3, or sarafotoxin S6c (NISHIYAMA et al. 1995). A similar profile was observed in human saphenous vein (NISHIYAMA et al. 1995).

In many instances, ET$_B$ receptor-mediated responses are differentially inhibited by the ET$_B$ receptor antagonists IRL1038, RES-701, and BQ-788, providing some prima facie, but circumstantial, evidence for the existence of multiple ET$_B$ receptor subtypes. For example, Sudjarwo and colleagues reported that IRL1038 inhibited endothelin-3-induced relaxations in swine pulmonary artery but not the contractions of swine pulmonary vein induced by endothelin-1, endothelin-3, sarafotoxin S6c, or IRL1620 (SUDJARWO et al. 1993). IRL1038 had a similar inhibitory profile in rabbit pulmonary artery and rabbit saphenous vein (SUDJARWO et al. 1994). Despite the limitations associated with the use of IRL1038 (URADE et al. 1992, 1994), similar effects were observed with another ET_B receptor antagonist RES-701. In rabbit pulmonary artery, BQ-788 inhibited responses to sarafotoxin S6c>endothelin-3=BQ-3020, but RES-701 was inactive (HAY et al. 1996). In addition, BQ-788 inhibited the ability of IRL1620, sarafotoxin S6c and endothelin-3 to augment electrical field stimulated contractions in rabbit tracheal preparations, whereas RES-701 was inactive against sarafotoxin-induced responses (YONEYAMA et al. 1995). Furthermore, in rabbit isolated bronchus, BQ-788, but not RES-701, inhibited BQ-3020-induced contractions (HAY et al. 1996) and in human isolated bronchus, contractions induced by IRL1620 and BQ3020 were inhibited by BQ-788, but not by RES-701 (HAY et al. 1998). In addition, several in vivo and radioligand binding studies have proposed the existence of multiple ET_B receptors (WELLINGS et al. 1994; GELLAI et al. 1996; HAY et al. 1998).

Together, these and other data have been used to suggest that two ET_B receptor subtypes exist, termed ET_{B1} and ET_{B2}. It is purported that BQ-788 inhibits responses mediated by both receptors (KARAKI et al. 1994), whereas RES-701 (and IRL1038) selectively blocks ET_{B1} receptors. Recent data suggests that the ET_B receptors responsible for the heterogeneous responses described above are derived from the same gene. For example, in ET_B gene knockout mice (MIZUGUCHI et al. 1997) and in Piebald-lethal mice, a naturally occurring mutant with deletion of the known ET_B receptor gene (GILLER et al. 1997), sarafotoxin S6c-induced vasodilator responses in the thoracic aorta (MIZUGUCHI et al. 1997) and in the intact anaesthetized animal (GILLER et al. 1997) were absent, as were contractile responses in the stomach (GILLER et al. 1997; MIZUGUCHI et al. 1997). Furthermore, there was no evidence of any ET_B receptor binding and function in the congenital aganglionosis rat, which has a mutant ET_B receptor gene (KARAKI et al. 1996). These findings suggest that if subtypes of ET_B receptors exist, then they are derived from the same gene (KARAKI et al. 1996). One mechanism through which this may occur is alternative splicing of the ET_B receptor gene; however the ET_B receptor splice variants described thus far either do not function as signal transducers (ELSHOURBAGY et al. 1996), or have indistinguishable binding and signal transducing activities compared to the wild-type ET_B receptor (SHYAMALA et al. 1994). Interestingly, the ET_B receptors on human endothelial and smooth muscle cells have recently been shown to exhibit equivalent binding pharmacology (FLYNN et al. 1998). Although these latter findings do not discount the possibility that more than one ET_B receptor subtype exists, further studies are clearly necessary to reconcile the functional findings, which suggest the existence of subtypes, with much of the currently available molecular biological and radioligand binding data which supports the existence of a single ET_B receptor subtype.

III. ET$_C$ and Other Putative Subtypes

In several functional and radioligand binding studies, responses induced by endothelin-1 and related peptides were not readily classified as having been mediated via either ET$_A$ and/or ET$_B$ receptors. In rat cerebellar homogenates, an ET$_B$ receptor-selective radioligand [^{125}I]-BQ3020 recognized only one third of the receptors identified using [^{125}I]-endothelin-1 and only 2% of [^{125}I]-endothelin-1 binding was sensitive to inhibition by the ET$_A$ receptor-selective ligand BQ-123 (WIDDOWSON and KIRK 1996). Furthermore, whereas typical ET$_B$ receptor ligands had only a moderate affinity for the rat cerebellar endothelin receptor, endothelin-3 had a higher affinity than endothelin-1. Similarly, whereas endothelin-3 effectively inhibited prolactin secretion from pituitary lactotrophs (SAMSON et al. 1990) and stimulated an increase in intracellular Ca^{2+} in bovine carotid artery endothelial cells (EMORI et al. 1990), endothelin-1 was inactive. It has been suggested that these responses, which are very sensitive to endothelin-3, may be mediated by a third subtype of ET receptor, termed the ET$_C$ receptor. However, the existence of the ET$_C$ receptor is still controversial, being primarily based on the rank order of affinity of endothelin-3>endothelin-1 in radioligand binding and functional studies and no ET$_C$ receptor has hitherto been cloned from mammalian sources. Nevertheless, an endothelin-3-preferring receptor, consistent with the ET$_C$ subtype has been cloned from *Xenopus* dermal melanophores (KARNE et al. 1993). The 424 amino acid sequence of the predicted mature ET$_C$ receptor is 47% and 52% homologous to ET$_A$ and ET$_B$ receptors, respectively (KARNE et al. 1993). Other studies using *Xenopus* tissues have revealed novel ET receptors. A cDNA was isolated from *Xenopus* heart that encoded a 415 amino acid protein (ET$_{AX}$ receptor) with a 74% homology to the human ET$_A$ receptor (KUMAR et al. 1994). Competition binding studies of the cloned ET$_{AX}$ receptor indicated that it was ET$_A$-like, but the ET$_A$ receptor-selective antagonist BQ-123 was inactive (KUMAR et al. 1994). In related studies, membranes prepared from *Xenopus* liver displayed high densities of high affinity sites for [^{125}I]-endothelin-1 and [^{125}I]-endothelin-3, suggestive of an ET$_B$-like receptor (ET$_{BX}$ receptor), but the binding was not inhibited by an ET$_B$ receptor-selective ligand sarafotoxin S6c (NAMBI et al. 1994). It is possible that ET$_{AX}$ and ET$_{BX}$ receptors represent the amphibian counterparts of mammalian ET$_A$ and ET$_B$ receptors. Novel variants of the ET$_A$ receptor, which had low affinity for BQ-123, have also been identified in the gills of the rainbow trout (LODHI et al. 1995). More recently, during the search for ET-like receptors, a novel gene that encoded a putative ET$_B$ receptor-like protein (hET$_B$R-LP) was cloned and characterized from human hippocampus tissue (ZENG et al. 1997). However, the 614 amino acid peptide encoded by the gene, which was 52% similar and 27% identical to the human ET$_B$ receptor, did not bind radiolabeled endothelin-1 or endothelin-3 (ZENG et al. 1997). A related peptide, termed ET$_B$R-LP-2, is also strongly expressed in the human central nervous system, but did not bind endothelin-1, endothelin-2, endothelin-3, bombesin, cholecystokinin-8, or gastrin-releasing peptide (VALDENAIRE et al. 1998). The endogenous ligand(s) for these orphan receptors remain to be determined.

IV. Splice Variants of ET Receptors

In addition to full-size ET_A receptor mRNA, the human ET_A receptor gene gives rise to at least three alternatively spliced ET_A receptor transcripts, termed ET_A-RΔ3 (BOURGEOIS et al. 1997), ET_A-RΔ4, and ET_A-RΔ3,4 (MIYAMOTO et al. 1996) corresponding to the exclusion of exon 3, exon 4, and exons 3 and 4, respectively. Translation of each of these transcripts generates a truncated form of the ET_A receptor. For example, translation of ET_A-RΔ3,4 mRNA generates an ET_A receptor protein with an extracellular N-terminal tail, five (instead of seven) membrane-spanning domains, two (instead of three) cytoplasmic loops, and a cytoplasmic C-terminal tail (MIYAMOTO et al. 1996). Similarly, the ET_A-RΔ3 (BOURGEOIS et al. 1997) and ET_A-RΔ4 transcripts generate very truncated ET_A receptor proteins with only two and three membrane-spanning domains, respectively. In view of the highly truncated nature of the ET_A receptor proteins, it is unlikely that they act as receptors. Indeed, COS-7 cells transiently transfected with ET_A-RΔ3,4 cDNA did not exhibit any specific binding affinity for $[^{125}I]$-endothelin-1 (MIYAMOTO et al. 1996). It has been suggested that the splicing mechanism of ET_A receptor mRNA may constitute a post-transcriptional means of controlling the levels of active ET_A receptor (BOURGEOIS et al. 1997).

Novel human splice variants of the ET_B receptor have also been reported (SHYAMALA et al. 1994; ELSHOURBAGY et al. 1996). One variant (termed ET_{B1}), was found in low abundance in human brain, placenta, lung, and heart by reverse transcriptase PCR, but was absent in bovine, rat, and porcine tissue. ET_{B1} encodes a protein that contains an extra ten amino acids in the second cytoplasmic domain of the ET_B receptor, but that had similar ligand binding and signal transducing (cAMP and inositol phosphate turnover) activities as the ET_B receptor (SHYAMALA et al. 1994). More recently, an ET_B receptor splice variant (termed ET_B-SVR), which shared a 91% identity to the human ET_B receptor, was identified from a human placental DNA library (ELSHOURBAGY et al. 1996). ET_B-SVR was identical to the ET_B receptor from the 5'-untranslated region through the 7 transmembrane domains, but differed completely in the 52 amino acids of the carboxy-terminal tail region. Cloned ET_B-SV and ET_B receptors expressed in COS cells had similar ligand binding profiles, but endothelin-1 increased inositol phosphate accumulation and intracellular acidification rates only in ET_B receptor-transfected cells, suggesting that elements within the carboxy tail region of ET_B receptor are critical for signal transduction but not agonist binding of these receptors (ELSHOURBAGY et al. 1996). Northern blots established that ET_B-SVR mRNA expression is generally low, representing less than 10% of total ET_B receptor mRNA expression in heart, brain, lung, and liver, and being undetectable in smooth muscle and endothelial cells. It has been postulated that ET_B-SVR, which binds

endothelin-1 but does not participate in signal transduction, might function as
a clearance binding site for the endothelins.

E. ET Receptor Structure

I. Structure

The amino acid sequences encoded by cloned ET receptor cDNA predict
a protein containing seven stretches of 20–27 hydrophobic amino acids,
consistent with the heptahelical membrane spanning structure that is common to the G-protein-coupled receptors of the rhodopsin-type receptor superfamily. The seven transmembrane domains and cytoplasmic loops of ET
receptors are highly conserved (Ogawa et al. 1991; Elshourbagy et al. 1992),
whereas the N-terminal and other extracellular domains differ between
receptors with respect to both length and amino acid sequence. Although
the extracellular NH_2-terminal domain is unusually long for a peptide G
protein-coupled receptor (75–100 amino acids in length), proteolytic truncation studies reveal that only those residues near the first transmembrane
domain are essential for endothelin-1 binding (Kozuka et al. 1991; Hashido
et al. 1992).

As expected, the transmembrane domains of the ET receptors contain
seven membrane-spanning helices (I–VII) joined by three intracellular and
three extracellular loops (Rubanyi and Polokoff 1994). Sakamoto et al.
(1993), using chimeric ET_A/ET_B receptors, have proposed that the ET receptor can be divided into two parts, one important for ligand receptor binding
(domains I, II, III, and VII) and another that determines isopeptide selectivity (domains IV, V, and VI) (Sakamoto et al. 1993). Thus, there are at least two
separable ligand interaction domains within the ET receptors; binding of
endothelin-1 to ET_A receptors involves binding of the C-terminal head region
of endothelin-1 to a region associated with transmembrane helices I, II, III,
and VII and binding of the N-terminal end to a region associated with transmembrane helices VI, V, and VI. In contrast, binding to the ET_B receptor
requires only the C-terminal end of endothelin-1. ET_A and ET_B receptor-like
binding characteristics can be interchanged by substituting transmembrane
domains IV, V, and VI and their intervening loops (Adachi et al. 1993;
Sakamoto et al. 1993). Site-directed mutagenesis studies have revealed that
Tyr[129] (Krystek et al. 1994; Lee et al. 1994) and Lys[140] (Adachi et al. 1994)
in transmembrane domain II of the ET_A receptor are important for
endothelin-1 binding. For example, mutation of Tyr[129], which is located about
half way down transmembrane helix II, had no effect on the binding of
endothelin-1 or endothelin-2, or Ro 46–2005 but decreased affinity of BQ-123
and BMS182874, and increased affinity of endothelin-3 and STX6c (Krystek
et al. 1994). Furthermore, Asp residues located one helical turn above (Asp[133])
and below (Asp[126]) Tyr[129], which have their side chains directed towards the
putative binding cavity, are important for ET_A receptor binding (Krystek

et al. 1994). Asp[126] in the ET_A receptor (and Asp[147] in the ET_B receptor) correspond to the highly conserved Asp in transmembrane helix II of many G-protein coupled receptors critical for agonist efficacy, and mutagenesis of these Asp residues attenuated PLC activity (measure of efficacy). In other mutagenesis studies, replacement of Lys[181] by aspartic acid in the third transmembrane region of the ET_B receptor reduced its affinity to endothelin peptides and sarafotoxin 6c without affecting G protein coupling (ZHU et al. 1992). The COOH-terminal region of the ET_A receptor is not involved in the binding of endothelin-1 but rather appears to be important in anchoring the receptor properly in the lipid bilayer in order to maintain the extracellular binding site (RUBANYI and POLOKOFF 1994).

Several studies suggest that ET receptors are post-translationally modified by glycosylation of the NH_2 terminus and by palmitoylation and phosphorylation of the cytoplasmic surface.

II. Post-Translational Modification of ET Receptors

1. Glycosylation

Evidence provided by Sokolovsky and colleagues suggest that some ET receptors in rat cerebellular and atrial membranes exist as glycoproteins (BOUSSO-MITTLER et al. 1991; SHRAGA-LEVINE and SOKOLOVSKY 1998). Moreover, N-glycosylation of the ET_A receptor appears to play a role in receptor function since deglycosylation of the carbohydrate chains of the ET_A receptor by endo H caused a marked decrease in [125I]-endothelin-1 binding to the nanomolar-affinity state (SHRAGA-LEVINE and SOKOLOVSKY 1998). The observed decrease in the number of binding sites (B_{max}), in the absence of an observable change in ligand affinity, suggests that the carbohydrate moiety is necessary for stabilization and full expression of the ET_A receptor rather than for ligand binding (SHRAGA-LEVINE and SOKOLOVSKY 1998). Glycosylation appears to play a lesser role in the super high affinity (picomolar)-state of the ET_A receptor and the ET_B receptor (SHRAGA-LEVINE and SOKOLOVSKY 1998).

2. Palmitoylation

Many G protein-coupled receptors, including ET receptors, undergo a post-translational modification termed palmitoylation, a covalent attachment of palmitic acid to one or more cysteine residues located in the carboxyl-terminal cytoplasmic tail (HORSTMEYER et al. 1996; OKAMOTO et al. 1997; ROOS et al. 1998). Palmitoylation at carboxyl terminal sites is known to modify the signal transducing activities of G protein-coupled receptors. ET_B receptors isolated from bovine lung are palmitoylated at two Cys residues (ROOS et al. 1998). Removal of several potential palmitoylation sites in the carboxyl terminus of the human ET_B receptor by site-directed mutagenesis did not affect ligand binding but abolished the ability of the receptor to transmit either an inhibitory effect on adenylate cyclase or a stimulatory effect on phospholipase C (OKAMOTO et al. 1997). Interestingly, a mutant ET_B receptor containing only the most proximal of the palmitoylation sites, and lacking the downstream carboxyl terminus, transmitted an effect on phospholipase C but not adenylate cyclase, suggesting a differential requirement for the carboxyl terminus downstream to the palmitoylation site in the coupling with G protein subtypes (OKAMOTO et al. 1997). Palmitoylation also appears to modulate differentially ET_A receptor signal transduction activity (HORSTMEYER et al. 1996). Mutated non-palmitoylated ET_A receptors transmitted an endothelin-1-induced increase in adenylate cyclase activity but failed to transmit an increase in phosphotidylinositol hydrolysis by phospholipase C (HORSTMEYER et al. 1996). Although there is almost no homology in the carboxyl terminal region between ET_A and ET_B receptors, amongst the very few conserved amino acids are two cysteine residues (palmitoylation sites), indicating the potential importance of palmitoylation in ET receptor signal transduction (Roos et al. 1998), and a couple of serine residues. These highly conserved serine residues plus another five have been demonstrated to be phosphorylated on ET_B receptors isolated from bovine lung (Roos et al. 1998). It has been suggested that one of these phosphorylated serine residues, Ser^{304}, may be important, since a single site mutation in the ET_B receptor gene from patients with Hirschsprung's disease resulted in the replacement of this serine residue with Asn (AURICCHIO et al. 1996; Roos et al. 1998). Phosphorylation of serine and threonine residues appears to be an important mechanism in receptor desensitization and this is discussed in Sect. G.I.

III. Binding Characteristics of ET Receptors

The binding of endothelin-1 to its membrane receptors is almost an irreversible process. In studies of membrane preparations from rat heart, rat lung, rat brain, and porcine vascular smooth muscle, Waggoner and coworkers estimated that the dissociation half-life for bound $[^{125}I]$-endothelin-1 was in excess of 30h for each tissue examined (WAGGONER et al. 1992). Similarly, $[^{125}I]$-endothelin-1 dissociated only very slowly from human cloned ET_B receptors (NAMBI et al. 1996; CHIOU et al. 1997), aortic ET_A receptors (MAGUIRE et al. 1996), and lung (NAMBI et al. 1994), as well as various animal tissues including rat lung (NAMBI et al. 1994), rat renal artery (DEVADASON and HENRY 1997), rat cardiac myocytes (HILAL-DANDAN et al. 1997), and rat cerebellum (NAMBI et al. 1994), dog lung (NAMBI et al. 1996), dog cerebellum (NAMBI et al. 1994), and pig lung (NAMBI et al. 1996). The pseudoirreversibility of the binding of endothelin-1 to its receptors has several important implications.

First, it suggests that the use of Scatchard analysis, which is based on the equilibrium process of binding of a ligand with a receptor to form a dissociable receptor-ligand complex, may not be valid for estimating dissociation equilibrium constants (K_d) and receptor densities (B_{max}) for endothelin-1 (WAGGONER et al. 1992). At the low concentrations of $[^{125}I]$-endothelin-1 typically employed in radioligand binding studies, the rate of binding is very slow and equilibrium will be approached only after incubation times that far exceed the half-life for the dissociation process (>>30h). In general, long incubation periods cannot be reliably employed, in part due to degradation of [^{125}I]-endothelin-1. If true equilibrium conditions are not used, then this may lead to the estimation of artificially high values of K_d. Indeed, the use of Scatchard analysis has generated K_d values that have varied by up to a 1000-fold, from a few picomolars to a few nanomolars. The threshold concentrations of endothelin-1 eliciting a functional response also varies markedly between preparations, leading to the suggestion that several subtypes of ET receptor exists (with superhigh, high, and low affinities for endothelin-1). As an alternative, DESMARETS et al. (1996) have performed kinetic experiments and saturation experiments performed under quasi equilibrium conditions and have identified a single high affinity state for the endothelin-1/cloned bovine ET_A receptor complex, with an estimated K_d value of 20pmol/l (DESMARETS et al. 1996). This corresponds to the lower range of previously reported K_d values and to the sites described as superhigh affinity sites (SOKOLOVSKY et al. 1992; DESMARETS et al. 1996).

Second, for irreversibly binding ligands such as endothelin-1, the rate of receptor binding is concentration-dependent (MARSAULT et al. 1991; DESMARETS et al. 1996; DEVADASON and HENRY 1997). Because of these time-limited second-order rate conditions, subnanomolar rather than picomolar concentrations of endothelin-1 are necessary to occupy an important fraction of picomolar sites within the time-frame of most functional experiments (DESMARETS et al. 1996), and offer an explanation as to why endothelin-1-induced actions often require concentrations 100-fold higher than the estimated K_d value of 20pmol/l. Under these conditions, endothelin-1-induced actions can occur at picomolar concentrations, but would require the presence of an amplification system such as a receptor reserve (DESMARETS et al. 1996). Together, these data suggest that the sensitivity of any preparation to endothelin-1 is only remotely related to the K_d values of endothelin-1/ET receptor complexes, and should be viewed as a cellular property rather than as a receptor property (DESMARETS et al. 1996).

Third, although the rate of endothelin-1 binding at high concentrations is fast (at least as fast as that of two other vasoconstrictors, angiotensin II and vasopressin), the rate of tension development in response to endothelin-1 is very much slower than the response to angiotensin II and vasopressin (MARSAULT et al. 1991). The slowly-developing contractile action of the endothelins is therefore unlikely to be due to a slow rate of binding, but may be a result of a late intracellular event involving slowly-developing pharmacomechanical coupling mechanism (MARSAULT et al. 1991). In addition to being slowly-developing, endothelin-1-induced contractions are typically long-lasting and difficult to reverse by washout. Studies by Marsault and coworkers suggest that irreversibility of binding and irreversibility of actions are not related in rat aortic strips, and it has been proposed that recycling of endocytosed ET receptors contributes to the sustained contractile action of endothelin-1 in vascular smooth muscle (Marsault et al. 1991, 1993). Indeed, in rat aortic rings, BQ-123 could still reverse an endothelin-1-induced contraction by 85% when given 40 min after endothelin-1 (Warner et al. 1994). Nevertheless, data obtained from cardiac myocytes indicate that the long-lasting effects of endothelin-1 are directly attributable to the irreversible nature of proximal aspects of endothelin action, such as binding and activation of G protein-linked signalling pathways (Hilal-Dandan et al. 1997). In these experiments, the ET_A receptor-selective antagonist BQ-123 competed for binding and biochemical effects when added simultaneously with endothelin-1, but was completely inactive if added later than 5 min after endothelin-1 (Hilal-Dandan et al. 1997). Interestingly, Chun et al. (1995) have demonstrated that $[^{125}I]$-endothelin-1 remains bound to the ET_A receptor for up to 2 h after endocytosis, possibly in plasma membrane caveolae (Chun et al. 1994), and continues to activate a signal transducing G protein, thus accounting for the sustained contraction that occurs in smooth muscle cells after a single addition of endothelin-1.

Fourth, although the binding of endothelin-1 is essentially irreversible, the binding of many other agonists and antagonists to ET receptors is reversible. In a series of studies of ET_A receptors (from rat pituitary tumour cells, human smooth muscle cells) and ET_B receptors (human astrocytoma, porcine cerebellum, human cloned expressed in CHO cells), Wu-Wong and coworkers have established that the irreversible nature of endothelin binding has profound influences on the observed potency of ET receptor antagonists (Wu-Wong et al. 1994a,b). In each case, endothelin-1 binding to ET receptors was less reversible than that of antagonists including BQ-123, FR139317, Ro46–2005, and PD142893. It has been suggested that the greater reversibility of antagonist binding accounts for the diminished potency of antagonists in inhibiting $[^{125}I]$-endothelin-1 binding (Wu-Wong et al. 1995) and endothelin-1-induced responses (Wu-Wong et al. 1994). For example, although BQ-123 (5 nmol/l), FR139317 (1 nmol/l), and Ro46–2005 (0.5 μmol/l) inhibited $[^{125}I]$-endothelin-1 binding to membranes prepared from rat pituitary cells by greater than 80% after 15 min of incubation, the extent of inhibition decreased to less than 20% after 24 h of incubation (Wu-Wong et al. 1995). A similar profile of effect was seen against $[^{125}I]$-endothelin-1 binding in porcine cerebellum membranes (Wu-Wong et al. 1994) and endothelin-1-induced mitogenesis in human smooth muscle cells (Wu-Wong et al. 1994).

Although the binding of endothelin-1 to ET receptors is essentially irreversible, a range of different kinetic profiles have been reported for other ET receptor agonists. Among the natural ligands, $[^{125}I]$-endothelin-3 binding to cultured rat aortic smooth muscle cells was more reversible than that of $[^{125}I]$-endothelin-1 and $[^{125}I]$-endothelin-2 (Roubert et al. 1991). In contrast, the binding of $[^{125}I]$-endothelin-2 to Swiss 3T3 fibroblasts was more reversible than that of $[^{125}I]$-endothelin-1 (Devesly et al. 1990) and $[^{125}I]$-endothelin-3 binding to porcine lung membranes was almost irreversible (Watakabe et al. 1992).

Furthermore, [^{125}I]-sarafotoxin S6b binding was reversible in rat renal artery (DEVADASON and HENRY 1997), but essentially irreversible in human aorta (MAGUIRE et al. 1996). This collection of contradictory findings highlight the fact that the characteristics of agonist binding may vary considerably depending upon the species investigated (NAMBI and PULLEN 1995) and the type of preparation used (membranes vs intact cells, HARA et al. 1998), and thus caution must be exercised when making comparisons of agonist binding kinetics across species and experimental conditions. Of the linear truncated analogues of endothelin-1, the binding of [^{125}I]-IRL-1620 in porcine lung (WATAKABE et al. 1992), dog cerebellum, human lung, and human ET$_B$ receptor clone (NAMBI et al. 1994) was readily reversible, whereas in rat tissues (cerebellum, lung) binding was irreversible (NAMBI et al. 1994). A similar profile of binding was seen with the longer analogue BQ3020, except the magnitude of dissociation of [^{125}I]-BQ3020 was less than that of [^{125}I]-IRL-1620 (NAMBI and PULLEN 1995).

The binding of many ET receptor antagonists studied to date is reversible. For example, the binding of [^{3}H]-BQ-123 to SK-N-MC cells was rapidly reversible, with a half time of dissociation in the order of a few minutes (IHARA et al. 1995). In other studies, extensive washing of bound BQ-123, FR139317, and Ro46–2005 from MMQ cells restored [^{125}I]-endothelin-1 binding by 62, 29, and 48%, respectively (WU-WONG et al. 1995). These findings suggest that these bound antagonists were easier to dissociate than bound endothelin-1, although different antagonists displayed different degrees of reversible binding (WU-WONG et al. 1995). One ET receptor antagonist that displays irreversible binding kinetics is CGS27830 (CHIN et al. 1996). CGS27830 appears to bind initially in a competitive manner before irreversible binding occurs, perhaps resulting from covalent modification of ET receptors by the anhydride structure of this nonpeptidic antagonist (CHIN et al. 1996).

1. Differences Between Membranes and Intact Cells

One typical means of characterization of ET receptors involves performing radioligand binding studies on either intact cells (cell cultures, tissue sections, etc.) or cell membrane preparations. However, recent studies have demonstrated that ET$_B$ receptors associated with intact cells and cell membranes exhibit different binding profiles (HARA et al. 1998), suggesting that the destruction of intact cells in order to generate a cell membrane preparation may lead to changes in ET$_B$ receptor binding properties.

F. ET Receptor Coupling

Activation of ET receptors produces a wide range of biological effects in many different tissues. The signal transduction systems that link ET receptors to cellular response are also diverse. ET receptors are able to couple to various types of G protein, including G_q, G_{11}, G_s, and G_{i2} (TAKIGAWA et al. 1995), suggesting that ET receptors may stimulate multiple effectors via several types of G protein simultaneously, depending upon the level of expression of each G-protein subtype. The downstream signal transduction pathways which have been reported to be activated following stimulation of ET receptors include enzyme systems (e.g., phospholipases A_2, C, and D, protein kinase C, protein tyrosine kinase, NO synthase, adenylate cyclase, and guanylate cyclase), ion channels (e.g., calcium channels, chloride channels), and ion transporters. These important systems will be described in detail in Chap. 5.

G. Alterations in ET Receptor Expression and Density

I. Desensitization and Internalization

Receptor desensitization, often defined as the diminished or abolished response to an agonist after repeated or continued stimulation, is an important feature of G-protein-coupled receptors to prevent overstimulation and potential damage of activated cells (CYR et al. 1993; CRAMER et al. 1998). Although endothelin-1 can elicit prolonged physiologic responses, it is well-established that complete and prolonged ET receptor desensitization occurs quickly in tissues expressing endogenous ET receptors (CYR et al. 1993) and in cells expressing cloned human ET receptors. For example, homologous, or receptor-specific, desensitization occurred within 4 min both in ET_A receptor-expressing A10 cells and in 293 cells transfected with either human ET_A or ET_B receptor (FREEDMAN et al. 1997). Less well established is the underlying mechanism of ET receptor desensitization. For many G-protein-coupled receptors, phosphorylation of serine and threonine residues in the third cytosol-facing loop and the carboxyl-terminal segment of the receptor reduces the ability of the receptor to interact with G proteins and appears to be an important mechanism for rapid receptor desensitization (CHUN et al. 1995; CRAMER et al. 1998). In 293 cells transfected with the human ET_A or ET_B receptors, endothelin-1-induced desensitization corresponded temporally with agonist-induced receptor phosphorylation (FREEDMAN et al. 1997), and appeared to involve the action primarily of G protein-coupled receptor kinase-2 (GRK2) rather than other GRKs or protein kinase C (FREEDMAN et al. 1997). A result that somewhat surprised Freedman et al. (1997) was that, despite the dissimilarity of their cytoplasmic C-terminal domains, ET_A and ET_B receptors were regulated indistinguishably by GRK-initiated desensitization. In contrast, Cramer and colleagues have reported subtype-specific desensitization of human ET_A and ET_B receptors which reflected differential receptor phosphorylation (CRAMER et al. 1997, 1998). Results obtained from Sf9, CHO, and COS cells expressing ET_A or ET_B receptors and from Rat-1 cells endogenously expressing ET_A receptors indicated that stimulation of ET_A receptors by endothelin-1 caused a sustained activation, retaining >30% of its initial activity even 20 min after endothelin-1 addition, whereas the ET_B receptor quickly deactivated losing >80% of its initial activity within 5 min (CRAMER et al. 1997, 1998). Consistent with these differences in receptor inactivation, ET_A receptors failed to undergo endothelin-1-induced phosphorylation, whereas the ET_B receptor was rapidly phosphorylated (CRAMER et al. 1997). The subtype-specific modulation of endothelin receptors has been proposed by Cramer and colleagues possibly to account for the short-term hypotensive effects of endothelins via rapidly down-regulating ET_B receptors and the long-lasting hypertensive effects due to sustained ET_A receptor activation.

In addition to receptor desensitization, internalization of the receptor bound ligand is a common method of signal termination for many peptidic ligands (BHOWMICK et al. 1998). The process of receptor-mediated endocytosis (receptor internalization) typically involves the sequential internalization of the ligand-receptor complex via clathrin-coated pits, degradation of ligands in lysosomes, and recycling or degradation of the receptor (CHUN et al. 1995). It is known that endothelin-1 binding to the ET_A receptor promotes internalization, perhaps via caveolae, with subsequent degradation of at least a portion of the bound ligand. BHOWMICK et al. (1998) report that occupancy of the ET_A receptor, rather than signalling, was sufficient to promote rapid endocytosis (t1/2 for internalization, 5 min). Chimeric ET_A receptors, which bound endothelin-1 but did not signal (as determined by IP_3 formation), internalized [^{125}I]-endothelin-1, albeit at a slower rate than wild type, signalling ET_A receptors (BHOWMICK et al. 1998). Somewhat surprisingly, the ET_A receptor-selective antagonist BQ-123 also promoted rapid internalization in cells expressing ET_A receptors. Following endocytosis, approximately 30% of internalized [^{125}I]-endothelin-1 remains intact and tightly bound to the ET_A receptor for up to 2 h (CHUN et al. 1995). Similar findings have been reported in cellular lysates of smooth muscle A-10 cells (BERMEK et al. 1996) and HG108–15 cells (ANGELOVA et al. 1997). Chun hypothesized that internalized endothelin-1/ET_A receptor complexes continued to activate a transducing G protein. Thus, receptor internalization may serve two roles; as a mechanism to terminate cell surface signalling and, in some cases, to prolong potentially the intracellular signal. No significant differences in the kinetics of endothelin-1-induced internalization were observed between ET_A and ET_B receptors (CRAMER et al. 1997), suggesting perhaps a subtype-specific inactivation of human ET receptors involving fast receptor phosphorylation of ET_B receptors and a slow receptor internalization of ET_A receptors.

Although they are activated by ET receptors, and would thus seem likely candidates for regulating ET receptors, protein kinase C isoforms have not in general been implicated in ET receptor down-regulation (COZZA et al. 1990; FREEDMAN et al. 1997). Nevertheless, phorbol esters, which are established activators of protein kinase C, promoted the sustained down-regulation of ET receptors in human vascular smooth muscle cells (RESINK et al. 1990). In addition to reducing the levels of existing ET receptors, endothelin-1 may reduce the levels of ET receptor mRNA by decreasing the stability of mRNA molecules (SAKURAI et al. 1992).

II. Drugs

1. Glucocorticoids

Dexamethasone induced a concentration- and time-dependent reduction in the density of ET receptors on cultured vascular endothelial (Stanimirovic et al. 1994) and smooth muscle (Nambi et al. 1992; Roubert et al. 1993) cells. A similar profile has also been observed in rat brain following in vivo administration of dexamethasone (Shibata et al. 1995). In rat vascular smooth muscle cells, dexamethasone was more effective than prednisolone and hydrocortisone (Nambi et al. 1992). Both ET_A (Nambi et al. 1992; Stanimirovic et al. 1994) and ET_B (Shibata et al. 1995) receptors can be down-regulated by dexamethasone and this effect was abolished by a glucocorticoid receptor antagonist cortexolone (Stanimirovic et al. 1994). Dexamethasone-induced down-regulation of ET receptors was associated with a marked reduction in the steady state levels of ET receptor mRNA (Nambi et al. 1992; Shibata et al. 1995), and was preceded by an increase in endothelin-1 content (Shibata et al. 1995). Consistent with this, dexamethasone induced a 2- to 3-fold increase in endothelin-1 levels in rat aortic smooth muscle cells (Roubert et al. 1993). Thus, it has been suggested that the regulatory effects of dexamethasone on ET receptors are mediated by endothelin-1 production (Roubert et al. 1993; Shibata et al. 1995) and is consistent with an autocrine control of ET receptors by endothelin-1 (Clozel et al. 1993).

2. Cyclosporine

The use of the immunosuppressive agent cyclosporine A is associated with potentially severe side-effects including renal vasoconstriction and systemic hypertension (see Chap. 18). Evidence has accumulated that activation and upregulation of the endothelin system may contribute to these hemodynamic alterations. Cyclosporine A stimulates endothelin-1 release both in vitro (Bunchman and Brookshire 1991; Takeda et al. 1994; Haug et al. 1995; Yokokawa et al. 1998) and in vivo (Kon et al. 1990). In addition to the increase in endothelin-1 production, cyclosporine A appears to differentially upregulate the expression of ET receptors in a tissue specific manner. For example, in aortic smooth muscle cells isolated from rats pretreated with cyclosporine A for 4 weeks, ET_A receptor mRNA expression was elevated twofold whereas ET_B receptor mRNA expression was unchanged (Iwai et al. 1995). In contrast, in rat glomerular mesangial cells, cyclosporine A caused a twofold increase in ET_B receptor mRNA expression but no change in ET_A mRNA (Takeda et al. 1994). Consistent with these latter findings, acute intravenous administration of cyclosporine to rats caused little change in mRNA expression of ET_A receptor in the kidney medulla, but was associated with an increased expression of ET_B receptors (Iwasaki et al. 1994). In addition to increased mRNA expression, enhanced binding of [125I]-endothelin-1, indicative of elevated ET receptor levels, have been reported in glomeruli, renal medulla, and cardiac membranes (NAYLER et al. 1989; AWAZU et al. 1991). Interestingly, renal transplant recipients on cyclosporine A had selective downregulation of ET_A receptor mRNA in cortical biopsy tissue (KARET and DAVENPORT 1996). The underlying mechanisms for cyclosporine A-induced changes in ET receptor expression are presently unclear.

3. Phosphoramidon

The administration of the ECE inhibitor phosphoramidon (see Chaps. 3 and 7) to cultured cells is frequently associated with increased ET receptor density (FUJITANI et al. 1992; WU-WONG et al. 1993; CLOZEL et al. 1993; ROUBERT et al. 1993; FLYNN et al. 1998). Increases in both ET_A (CLOZEL et al. 1993; ROUBERT et al. 1993) and ET_B receptors (CLOZEL et al. 1993; FLYNN et al. 1998) have been demonstrated, although the magnitude of the increase is cell dependent. For example, phosphoramidon evoked a 16-fold increase in ET_B receptor density in human umbilical vein endothelial cells but only a 1.1-fold increase in human aortic smooth muscle cells (FLYNN et al. 1998). Interestingly, phosphoramidon had a greater stimulatory effect on ET receptor density in cells with high endothelin-1 production (CLOZEL et al. 1993). One possible explanation is that autocrine production of endothelin, either by binding or by downregulation, decreases the number of ET receptors. Phosphoramidon, by inhibiting the conversion of big endothelin-1 to endothelin-1, inhibits the production of endothelin-1 and thereby attenuates ET receptor down-regulation (CLOZEL et al. 1993; FLYNN et al. 1998). This postulate is supported by the findings that other neutral metalloproteases without significant ECE activity, such as thiorphan, did not affect either endothelin-1 production or ET receptor density in these cells (WU-WONG et al. 1993; CLOZEL et al. 1993). An additional mechanism is that phosphoramidon inhibits the protease responsible for degrading the ET receptor (WU-WONG et al. 1993).

III. Cytokines and Other Peptides

Although it is well established that cytokines and endothelins can modulate the others synthesis and release, relatively little is known of the influence of cytokines on ET receptor expression. In A617 cells, a vascular smooth muscle-derived cell line, ET_A receptor density was increased in a concentration- and time-dependent manner by basic fibroblast growth factor (bFGF), reduced by transforming growth factor-beta (TGF-β), and unaffected by platelet-derived growth factor (PDGF), interleukin-6 (IL-6), tumour necrosis factor-alpha (TNF-α), and fetal bovine serum (CRISTIANI et al. 1994). In human vascular smooth muscle cells, IL-1β stimulated ET receptor mRNA expression (NEWMAN et al. 1995). More recently, Smith and colleagues demonstrated that the influence of bFGF and TNF-α on ET_B receptor mRNA expression by cultured human umbilical vein endothelial cells was dependent on whether cells were grown in the presence of an angiogenic (fibrin matrix) or non-angiogenic (plastic) stimuli (SMITH et al. 1998). ET_B receptor mRNA expression was enhanced by bFGF and TNF-α in cells grown on the fibrin matrix but was inhibited in cells grown on plastic, suggesting a role for ET_B receptors in vascular tube formation (SMITH et al. 1998).

Vasoactive peptides such as the C-type natriuretic peptides and angiotensin II also modulate ET receptor expression. Exposure of vascular smooth muscle cells derived from human pulmonary artery to the vasoconstrictor agent angiotensin II was associated with increased expression of ET receptor mRNA and increased ET_A receptor density (HATAKEYAMA et al. 1994). The potent vasodilator C-type natriuretic peptides (ANP, BNP, CNP) also stimulated an increase in the density of ET_A and ET_B receptors in rat vascular smooth muscle cells (EGUCHI et al. 1994). In rat cardiomyocytes, angiotensin II, via AT1 receptors, increased ET_B receptor density (KANNO et al. 1993). Insulin, another peptide which may influence vascular function, increased ET_B but not ET_A receptor mRNA in retinal microvascular pericytes (McDONALD et al. 1995).

IV. Cell and Tissue Culture

The process of cell culture is frequently associated with profound changes in the levels and proportions of ET_A and ET_B receptors. Eguchi and coworkers reported that serial passage of cultured rat vascular smooth muscle cells was associated with a phenotypic change in the ET receptor subtype from a predominantly ET_A receptor in early passage (10th–15th) cells to a predominantly ET_B receptor subtype in later passage (30th–35th) cells (EGUCHI et al. 1994). Northern blot analysis also demonstrated the principal expression of ET_A receptor mRNA in the early passage and ET_B receptor mRNA in the late passage (EGUCHI et al. 1994). Similarly, growth of rat and lamb tracheal smooth muscle cells in culture (6th to 14th passages) was associated with substantial increases in ET_B receptor density, with little change in ET_A receptor density (MAXWELL et al. 1998a,b). Interestingly, these changes could be prevented by a two-day period of serum deprivation (MAXWELL et al. 1998a,b). Furthermore, the MEG-01 human megakaryoblastic cell line, which in early passages expresses only functional ET_A receptors, at later passages expresses ET_B receptor mRNA that is translated, processed, and targeted to the cell membrane as a functional receptor (HAMROUN et al. 1998). In stark contrast, the culture of rat cerebellar neurons was associated with time-related increases in ET_A receptor density and concomitant reductions in ET_B receptor density (LYSKO et al. 1995), which mirrored changes in ET receptor mRNA expression.

Organ culture of vascular segments also appears to be associated with significant modulation of ET receptor expression. A significant increase in both ET_B receptor mRNA as well as contractile responsiveness to sarafotoxin S6c was observed after organ culture of human omental artery (ADNER et al. 1996) and rat mesenteric artery (MOLLER et al. 1997). Responses to sarafotoxin S6c were abolished by the transcriptional inhibitor, actinomycin D, and the translational inhibitor cycloheximide. Thus, although changes in ET_B receptor density were not directly measured in these studies, it appears that organ culture induces transcription and the subsequent translation of contractile ET_B receptors (ADNER et al. 1996; MOLLER et al. 1997). ET_B receptor plasticity was more apparent in small as compared to large arteries, more marked in veins than in arteries, and was most pronounced in the mesenteric region (ADNER et al. 1998).

V. Development

Mice deficient in the ET_A receptor gene develop defects in craniofacial structures, great vessels, and cardiac outflow tract (CLOUTHIER et al. 1998; YANAGISAWA et al. 1998), whereas targeted inactivation of the ET_B receptor gene causes colonic agangliogenesis associated with skin pigmentation abnormalities (HOSODA et al. 1994). These, and other, data suggest that endothelin-1/ET_A receptor and endothelin-3/ET_B receptor interactions are important in embryonic development, especially in the migration/proliferation/differentiation of neural crest cells. In a study of the human embryo, Brand and colleagues have recently demonstrated that ET receptor mRNA expression in neural crest cells starts at 3 weeks of gestation and continues until at least 6 weeks of gestation (BRAND et al. 1998). During this period, the distribution of ET_A receptor mRNA was more widespread in the embryo than that of ET_B receptor mRNA. Despite its first appearance in the neural crest within the neuroepithelium, ET_A receptor mRNA was often restricted to the mesodermal component of structures and organs (e.g., head and axial skeleton). ET_B receptor mRNA was most commonly present in the neural tube, sensory and sympathetic ganglia, and the endothelium. A not too dissimilar profile of ET_B mRNA expression has also been demonstrated in avian embryo (NATAF et al. 1996). In embryonic chick brain, temporal differences in the expression of ET receptors were observed during development, with the ET_A receptor subtype synthesized early and the ET_B receptor later (JENG et al. 1996).

Significant changes in ET receptor density also occur postnatally. For example, between birth and day 30, ET_A receptor density in rat kidney decreased by 65–70%, whereas ET_B receptor density was unaltered (ABADIE et al. 1996). Similarly, the number of ET receptors in kidneys from neonate rats was nearly twice the levels observed in 6- and 12-week-old animals (BECKER et al. 1996). Moreover, maturation was associated with a shift in the ET_A:ET_B ratio towards ET_B receptors in this tissue (HOCHER et al. 1995; BECKER et al. 1996). During the postnatal period in pigs, changes in the density, location, and type of ET receptor were observed in pulmonary vasculature;

however, no changes were observed in airway smooth muscle (HISLOP et al. 1995).

VI. Pregnancy/Menstrual Cycle

Profound changes in uterine ET receptor density (KUBOTA et al. 1995; COLLETT et al. 1996) and ET receptor mRNA (O'REILLY et al. 1992) occur during the menstrual cycle (O'REILLY et al. 1992; COLLETT et al. 1996), especially within the endometrium. For example, ET_A receptor expression in stroma throughout the endometrium was higher in the proliferative phase compared to the secretory and menstrual phases (COLLETT et al. 1996). ET_A receptor mRNA expression was also elevated during the proliferative phase (O'REILLY et al. 1992). In contrast, ET_B receptor expression in epithelial glands was lowest in the proliferative endometrium, higher in secretory endometrium, and highest in menstrual endometrium (COLLETT et al. 1996). A similar pattern was observed for ET_B receptor mRNA expression (O'REILLY et al. 1992). Consistent with the rise in ET_B receptors in the latter phase of the cycle and its localization within the endometrial stroma, Kohnen and coworkers suggested that ET_B receptors are involved in the onset of decidualization in the stromal cells, under the influence of progesterone (KOHNEN et al. 1998). Despite these overt changes in endometrial ET receptor expression, expression of ET receptors in the myometrium changed little during the menstrual cycle (COLLETT et al. 1996). Nevertheless, significant changes in the density of myometrial ET_A and ET_B receptors have been reported to occur during pregnancy.

Pregnancy appears to be associated with important regional changes in the proportions of both ET_A and ET_B receptors. During normal pregnancy the levels of ET_A receptor mRNA in the corpus, which develops into the upper, active uterine segment, were significantly elevated compared to non-pregnant women (WOLFF et al. 1996). Consistent with this, increased ET_A receptor density has also been observed in uterine membrane preparations from pregnant women (OSADA et al. 1997). Interestingly, the levels of ET_B receptor mRNA in the isthmus, the lower passive uterine segment that expands and thins, was also increased during pregnancy (WOLFF et al. 1996). These findings suggest that during human parturition, increased expression of ET_A receptors in the upper uterine segments mediates endothelin-1-induced contraction of the myometrium and thus participates in the movement of the fetus downward (WOLFF et al. 1996). The process of fetal movement may be facilitated by stimulation of the upregulated myometrial ET_B receptors in the lower reaches of the uterus, which mediate relaxation and dilatation of these structures (WOLFF et al. 1996).

VII. Disease

As highlighted in the latter chapters of this handbook, the endothelin receptor-effector system has been implicated in the pathophysiology of many

Table 1. Changes in ET receptor density and mRNA expression in human disease and in animal models of disease

Disease model	Species	Tissue	Effect	Reference
Adenomas	Human	Cerebral cortex	$\downarrow ET_B$	Rossi et al. (1995)
Alzheimer disease	Human	Vasculature of synovium	No ΔET	Kohzuki et al. (1995)
Arthritis	Human	Bronchial smooth muscle	No ΔET_A	Wharton et al. (1992)
Asthma	Human		No ΔET_B	Goldie et al. (1995)
Asthma	Human	Peripheral lung	No ΔET_A No ΔET_B	Knott et al. (1995)
Atherosclerosis	Human	Coronary artery	$\uparrow ET_B$	Dagassan et al. (1996)
Atherosclerosis	Human	Cultured VSMCs	$\downarrow ET$ mRNA	Winkles et al. (1993)
Atherosclerosis	Human	Coronary artery & aorta		Bacon et al. (1996)
		Media	No ΔET_A, no ΔET_B	
		Intima	$\downarrow ET$	
Brain damage, cold injury	Rat	Astrocytes	$\downarrow ET_B$ mRNA	Hama et al. (1997)
Brain tumours	Human	Meninges	$\uparrow ET_A$ mRNA	Pagotto et al. (1995)
Cardiac hypertrophy, aortovenacaval fistula	Rat	Cardiac membranes	$\uparrow ET_A$	Brown et al. (1995)
Cardiac ischemia	Rat	Cardiac membranes	$\uparrow ET$	Liu et al. (1990)
Cardiomyopathy	Hamster	Inner medullary collecting ducts	$\downarrow ET_B$	Wong et al. (1998)
Cardiomyopathy	Human	Ventricular myocytes	No ΔET_A mRNA	Hasegawa et al. (1996)
Circulatory arrest, deep hypothermic	Pig	Pulmonary parenchyma	$\downarrow ET_B$	Kirshbom et al. (1997)
Cirrhosis	Human	Hepatic tissue	$\uparrow ET_A$ mRNA $\uparrow ET_B$ mRNA	Leivas et al. (1998)
Cirrhosis, carbon tetrachloride	Rat	Liver	$\uparrow ET$	Gandhi et al. (1996)
Cirrhosis, carbon tetrachloride	Rat	Renal medulla	No ΔET_A $\uparrow ET_B$	Hocher et al. (1996)
		Renal cortex	No ΔET_A no ΔET_B	
Diabetes	Rat	Ureter	$\uparrow ET_A$ $\uparrow ET_A$ mRNA	Nakamura et al. (1997)
Diabetes, genetic	BB/W rat	Retina	$\uparrow ET_B$ mRNA	Chakrabarti et al. (1998)

Table 1. (continued)

Disease model	Species	Tissue	Effect	Reference
Diabetes, alloxan-induced	Rabbit	Corpus cavernosum	$\uparrow ET_B$ No ΔET_A	Sullivan et al. (1997)
Diabetes, streptozotocin-induced	Rat	Corpus cavernosum	$\uparrow ET_A$	Bell et al. (1995)
Diabetes, streptozotocin-induced	Rat	Vas deferens	$\uparrow ET_A$	Saito et al. (1996)
Endometrial cancer	Human	Endometrium	$\downarrow ET_A$ mRNA $\downarrow ET_B$ mRNA	Pekonen et al. (1995)
Endometrial carcinoma	Human	Endometrium	$\uparrow ET_R$	Ben-Baruch et al. (1993)
Fibrotic lung disease, scleroderma	Human	Lung	$\uparrow ET_R$, $\uparrow ET_B$, $\downarrow ET_A$	Abraham et al. (1997)
Glioblastomas	Human	Capillaries	$\uparrow ET_A$	Tsutsumi et al. (1994)
Glomerular sclerosis, puromycin, aminonucleoside	Rat	Glomeruli	$\uparrow ET_A$ mRNA $\uparrow ET_B$ mRNA	Nakamura et al. (1995)
Heart failure	Human	Atria, ventricles	No ΔET_A	Ponicke et al. (1998)
Heart failure, aortic valvular insufficiency	Rabbit	Kidney, heart	$\downarrow ET$	Loffler et al. (1993)
Heart failure, congestive	Rat	Lung	$\downarrow ET_B$ mRNA, No ΔET_A mRNA $\downarrow ET_B$ no ΔET_A	Kobayashi et al. (1998)
Heart failure, coronary artery ligation	Rat	Cardiac membranes	$\uparrow ET$	Miyauchi et al. (1995)
Heart failure, coronary artery ligation	Rat	Cardiac tissue	No ΔET_A mRNA No ΔET_B mRNA	Tonnessen et al. (1997)
Heart failure, coronary artery ligation	Rat	Mesenteric artery	$\downarrow ET$	Fu et al. (1993)
Heart failure, coronary artery ligation	Rat	Ventricle	No ΔET	Fu et al. (1993)

Condition	Species	Tissue	Result	Reference
Heart failure, coronary artery ligation	Rat	Left ventricle	↓ET_A mRNA	PICARD et al. (1998)
Hirshsprung's disease	Human		↓ET_B mRNA ↓ET_B (mutation)	PUFFENBERGER et al., (1994)
Hydatidiform moles	Human	Placenta	No ΔET_A mRNA	FAXEN et al. (1997)
Hypercholesterolaemia	Watanabe rabbit	Corpus cavernosum	↑ET_B No ΔET_A	SULLIVAN et al. (1998)
Hypertension	Human	Int. mam. artery Media Intima Heart	ET_A mRNA high ET_A mRNA low ↓ET_A mRNA	HASEGAWA et al. (1994)
Hypertension, various	Rat		↓ET_B mRNA	HAYZER et al. (1994)
Hypertension, cyclosporine-induced	Rat	Aortic SMCs	↑ET_A mRNA	IWAI et al. (1995)
Hypertension, genetic	SH rat	Heart chambers	No ΔET_A no ΔET_B	THIBAULT et al. (1995)
Hypertension, genetic	SH rat	Kidney cortex	↓ET_A ↑ET_B	GELLAI et al. (1994)
Hypertension, genetic	SH rat	Glomeruli Intrarenal art.	↑ET_A ↑ET_B ↑ET_A	HOCHER et al. (1996)
Hypertension, genetic	SPSH rat	Mesangial cells	↑ET_A mRNA	HIRAOKA et al. (1995)
Hypertrophic prostate	Human	Prostate	↑ET	KONDO et al. (1994)
Hypoxia, hypobaric (50.8kPa)	Pig	Pulmonary vasculature	↑ET_A	NOGUCHI et al. (1997)
Hypoxia, 10% O_2, 4 weeks	Rat	Lung, heart & aorta	↑ET_A mRNA ↑ET_B mRNA	LI et al. (1994a,b)
Metastatic melanoma	Human	Cultured cells	↓ET_B mRNA	KIKUCHI et al. (1996)
Myocardial infarction, coronary artery ligation	Rat	Myocardium	No ΔET	KOHZUKI et al. (1996)
Myocardial infarction, ischemia	Rat	Adrenal glands Aorta Kidney	↓ET No ΔET ↑ET_A mRNA	VESCI et al. (1994)
Nephritis	NZB/W F1 mouse		↑ET_B mRNA	NAKAMURA et al. (1993)
Nephritis, menangial proliferative	Rat	Glomeruli	No ΔET_A mRNA ↑ET_B mRNA	YOSHIMURA et al. (1995)
Nephrosis, aminonucleoside-induced	Rat	Glomeruli	No ΔET_A mRNA ↑ET_B mRNA	NAKAMURA et al. (1995)

Table 1. (continued)

Disease model	Species	Tissue	Effect	Reference
Nephrotoxicity, FK506-induced	SH rat	Kidney	No ΔET_B mRNA	Uchida et al. (1998)
Nephrotoxicity, cyclosporine-induced	Rat	Kidney	$\uparrow ET$	Nambi et al. (1990)
Neural lesions, kainic acid-induced	Rat	Brain	$\uparrow ET_B$	Sakurai-Yamashita et al. (1997)
Neuron cell death, ischemia	Rat	Various	$\uparrow ET$	Kohzuki et al. (1995)
Polycystic kidney disease	Han: SPRD rat	Kidney	$\downarrow ET_A \downarrow ET_B$; $\uparrow ET_A$ mRNA	Hocher et al. (1998)
Polycystic kidney disease	Cpk mice	Renal tissue	$\uparrow ET_B$ mRNA	Nakamura et al. (1993)
Portal hypertension, portal vein ligation	Rat	Superior mesenteric artery	$\uparrow ET_A$ mRNA	Cahill et al. (1998)
Postobstructive pulmonary vasculopathy	Rat	Pulmonary artery	$\uparrow ET_B$ mRNA; $\uparrow ET_A$	Shi et al. (1997)
Prostate cancer	Human	Prostate tissue and cell lines	$\downarrow ET_B$ mRNA	Nelson et al. (1996)
Pulmonary hypertension	Sheep	Fetal lung	$\downarrow ET_B$ mRNA	Ivy et al. (1998)
Pulmonary hypertension, hypoxia-induced	Piglet	Pulmonary artery	$\uparrow ET_A \downarrow ET_B$	Gosselin et al. (1997)
Pulmonary hypertension, monocrotaline-induced	Rat	Lung tissue	$\downarrow ET_B$ mRNA	Yorikane et al. (1993)
Raynaud's syndrome	Human	Microvessels of skin	No ΔET	Knock et al. (1993)
Remnant kidney	Rat	Renal tissue	No ΔET_A mRNA; $\uparrow ET_B$ mRNA	Benigni et al. (1996)
Renal failure, hypertonic glycerol	Rat	Cortex and medulla	$\uparrow ET_A \uparrow ET_B$	Roubert et al. (1994)

Condition	Species	Tissue	Change	Reference
Renal failure, ischemia-induced	Rat	Cortical membranes	No ΔET (affinity change)	Nambi et al. (1993)
Renal failure, glycerol-induced	Rat	Kidney	No ΔET_A mRNA; ↑ET_B mRNA	Shimizu et al. (1998)
Renal failure, acute ischemic	Rat	Glomeruli	No ΔET	Wilkes et al. (1991)
Renal hypertrophy, left nephrectomy	Rat	Renal cortex	No ΔET_A mRNA; ↑ET_B mRNA	Nakamura et al. (1995)
		Glomeruli	No ΔET_A mRNA; No ΔET_B mRNA	
Stroke, forebrain ischemia	Gerbil	Hippocampus	↓ET	Willette et al. (1993)
Subarachnoid hemorrhage	Monkey	Cerebral artery	↑ET_B mRNA; No ΔET_B	Hino et al. (1996)
Subarachnoid hemorrhage	Dog	Basilar artery	↑ET_A mRNA	Itoh et al. (1994)
Syndrome X	Human	Blood vessels	↓ET_A (*)	Newby et al. (1998)
Systemic sclerosis	Human	Microvessels of skin	↑ET	Knock et al. (1993)
Transplanted lungs	Dog	Bronchi & parenchyma	↓ET_A, ↓ET_B	Kim et al. (1997)
Varicose veins	Human	Veins	↓ET_R, ↓ET_B	Barber et al. (1997)
Vascular wall injury, angioplasty-induced	Rat	Carotid arteries	↑ET_A mRNA; ↑ET_B mRNA	Wang et al. (1996)
Vein grafts	Pig	Vein	↑ET_A	Dashwood et al. (1998)
Vein grafts	Rabbit	Vein	↑ET_B	Eguchi et al. (1997)
Ventricular hypertrophy, aortic coarctation	Rat	Ventricles	↓ET_A mRNA; ↓ET_B mRNA	Sirvio et al. (1995)
Virus infection, influenza A	Mouse	Trachea	↓ET_B	Henry and Goldie (1994); Carr et al. (1996)
Virus infection, measles, canine distemper	Rat	C6 astrocytoma cells	↓ET_A mRNA	Meissner and Koschel (1995)

human diseases. Certainly, the local production and/or circulating levels of endothelin are elevated in many diseases, which might contribute to vasoconstrictor, proliferative, or other pathologically relevant cellular responses to endothelin. An increased expression of ET receptors might also be expected to produce an increased sensitivity, and perhaps an increased maximal tissue response, to the actions of endothelin. Indeed, in various animal models of disease, and in tissue samples from patients, increased densities of ET receptors and/or expression of ET receptor mRNA have been reported (see Table 1). However, disease states are in many instances associated with reductions in ET receptor expression. One possible explanation of reduced ET receptor density is receptor down-regulation in response to over-expression of endothelin. This is consistent with the well established propensity for ET receptors to down-regulate in response to elevated levels of activating ligand. It is also likely that the myriad of cytokines and mediators that are involved in the disease process can also significantly modulate the cellular expression of ET receptors. Elucidating the underlying cellular mechanisms for the observed changes in ET receptor number during disease, which presently are almost universally unknown, is a major challenge to investigators in this field.

H. Summary

ET receptors are widely expressed in human tissues and play a pivotal role in mediating the physiological and pathophysiological effects of endothelin-1 and related peptides. ET_A and ET_B receptors are each members of the superfamily of monomeric heptahelical G protein-coupled receptors, and are linked to a diverse range of signal transduction processes. The realization that the binding of endothelin-1 to its receptors is essentially irreversible has impacted significantly on a variety of issues, including the estimation of ET receptor density, the subclassification of ET receptors, and the determination of ET receptor antagonist potency. ET receptor expression is markedly influenced by numerous substances including the endogenous ligand endothelin-1, anti-inflammatory drugs such as cyclosporine and the glucocorticoids, and a variety of cytokines and growth factors. Significant alterations in ET receptor expression were also observed during prenatal and postnatal development, during the menstrual cycle and pregnancy, and in response to cell and tissue culture. In addition, there now exists an enormous bank of literature demonstrating that variable combinations of positive and negative changes in ET_A and ET_B receptor expression occur in diseased tissues. A major challenge now facing researchers is elucidating the cellular mechanisms that control the expression of ET receptors and determining the functional effect of these changes on the disease process.

Acknowledgements. The authors wish to acknowledge the financial support of the National Health and Medical Research Council of Australia

References

Abadie L, Blazy I, Roubert P, Plas P, Charbit M, Chabrier PE, Dechaux M (1996) Decrease in endothelin-1 renal receptors during the 1st month of life in the rat. Pediatr Nephrol 10:185–189

Abraham DJ, Vancheeswaran R, Dashwood MR, Rajkumar VS, Pantelides P, Xu SW, Black CM (1997) Increased levels of endothelin-1 and differential endothelin type A and B receptor expression in scleroderma-associated fibrotic lung disease. Am J Pathol 151:831–841

Adachi M, Furuichi Y, Miyamoto C (1994) Identification of a ligand-binding site of the human endothelin-A receptor and specific regions required for ligand selectivity. Eur J Biochem 220:37–43

Adachi M, Hashido K, Trzeciak A, Watanabe T, Furuichi Y, Miyamoto C (1993) Functional domains of human endothelin receptor. J Cardiovasc Pharmacol 22 [Suppl 8]:S121–S124

Adachi M, Yang YY, Furuichi Y, Miyamoto C (1991) Cloning and characterization of cDNA encoding human A-type endothelin receptor. Biochem Biophys Res Commun 180:1265–1272

Adner M, Cantera L, Ehlert F, Nilsson L, Edvinsson L (1996) Plasticity of contractile endothelin-B receptors in human arteries after organ culture. Br J Pharmacol 119:1159–1166

Adner M, Uddman E, Cardell LO, Edvinsson L (1998) Regional variation in appearance of vascular contractile endothelin-B receptors following organ culture. Cardiovasc Res 37:254–262

Angelova K, Ergul A, Peng KC, Puett D (1997) Metabolism of endothelin-1 by neuroblastoma x glioma hybrid (NG108-15) cells. Neurosci Lett 225:1–4

Arai H, Hori S, Aramori I, Ohkubo H, Nakanishi S (1990) Cloning and expression of a cDNA encoding an endothelin receptor. Nature 348:730–732

Arai H, Nakao K, Takaya K, Hosoda K, Ogawa Y, Nakanishi S, Imura H (1993) The human endothelin-B receptor gene. Structural organization and chromosomal assignment. J Biol Chem 268:3463–3470

Auricchio A, Casari G, Staiano A, Ballabio A (1996) Endothelin-B receptor mutations in patients with isolated Hirschsprung disease from a non-inbred population. Hum Mol Genet 5:351–354

Awazu M, Parker RE, Harvie BR, Ichikawa I, Kon V (1991) Down-regulation of endothelin-1 receptors by protein kinase C in streptozotocin diabetic rats. J Cardiovasc Pharmacol 17 [Suppl 7]:S500–S502

Bacon CR, Cary NR, Davenport AP (1996) Endothelin peptide and receptors in human atherosclerotic coronary artery and aorta. Circ Res 79:794–801

Barber DA, Wang X, Gloviczki P, Miller VM (1997) Characterization of endothelin receptors in human varicose veins. J Vasc Surg 26:61–69

Bax WA, Aghai Z, van Tricht CL, Wassenaar C, Saxena PR (1994) Different endothelin receptors involved in endothelin-1- and sarafotoxin S6b-induced contractions of the human isolated coronary artery. Br J Pharmacol 113:1471–1479

Bax WA, Bos E, Saxena PR (1993) Heterogeneity of endothelin/sarafotoxin receptors mediating contraction of the human isolated saphenous vein. Eur J Pharmacol 239:267–268

Bax WA, Saxena PR (1994) The current endothelin receptor classification: time for reconsideration? Trends Pharmacol Sci 15:379–386

Becker K, Erdbrugger W, Heinroth-Hoffmann I, Michel MC, Brodde OE (1996) Endothelin-induced inositol phosphate formation in rat kidney. Studies on receptor subtypes, G-proteins and regulation during ontogenesis. Naunyn Schmiedebergs Arch Pharmacol 354:572–578

Bell CR, Sullivan ME, Dashwood MR, Muddle JR, Morgan RJ (1995) The density and distribution of endothelin 1 and endothelin receptor subtypes in normal and diabetic rat corpus cavernosum. Br J Urol 76:203–207

Ben-Baruch G, Schiff E, Galron R, Menczer J, Sokolovsky M (1993) Impaired binding properties of endothelin-1 receptors in human endometrial carcinoma tissue. Cancer 72:1955–1958

Benigni A, Zola C, Corna D, Orisio S, Facchinetti D, Benati L, Remuzzi G (1996) Blocking both type A and B endothelin receptors in the kidney attenuates renal injury and prolongs survival in rats with remnant kidney. Am J Kidney Dis 27:416–423

Bermek H, Peng KC, Angelova K, Ergul A, Puett D (1996) Endothelin degradation by vascular smooth muscle cells. Regul Pept 66:155–162

Bhowmick N, Narayan P, Puett D (1998) The endothelin subtype A receptor undergoes agonist- and antagonist-mediated internalization in the absence of signaling. Endocrinology 139:3185–3192

Bourgeois C, Robert B, Rebourcet R, Mondon F, Mignot TM, Duc-Goiran P, Ferre F (1997) Endothelin-1 and ET_A receptor expression in vascular smooth muscle cells from human placenta: a new ET_A receptor messenger ribonucleic acid is generated by alternative splicing of exon 3. J Clin Endocrinol Metab 82:3116–3123

Bousso-Mittler D, Galron R, Sokolovsky M (1991) Endothelin/sarafotoxin receptor heterogeneity: evidence for different glycosylation in receptors from different tissues. Biochem Biophys Res Commun 178:921–926

Bousso-Mittler D, Kloog Y, Wollberg Z, Bdolah A, Kochva E, Sokolovsky M (1989) Functional endothelin/sarafotoxin receptors in the rat uterus. Biochem Biophys Res Commun 162:952–957

Brand M, Le Moullec JM, Corvol P, Gasc JM (1998) Ontogeny of endothelins-1 and -3, their receptors, and endothelin converting enzyme-1 in the early human embryo. J Clin Invest 101:549–559

Brown LA, Nunez DJ, Brookes CI, Wilkins MR (1995) Selective increase in endothelin-1 and endothelin A receptor subtype in the hypertrophied myocardium of the aorto-venacaval fistula rat. Cardiovasc Res 29:768–774

Bunchman TE, Brookshire CA (1991) Cyclosporine-induced synthesis of endothelin by cultured human endothelial cells. J Clin Invest 88:310–314

Cahill PA, Hou MC, Hendrickson R, Wang YN, Zhang S, Redmond EM, Sitzman JV (1998) Increased expression of endothelin receptors in the vasculature of portal hypertensive rats: role in splanchnic hemodynamics. Hepatology 28:396–403

Calo G, Gratton JP, Telemaque S, D'Orleans-Juste P, Regoli D (1996) Pharmacology of endothelins: vascular preparations for studying ET_A and ET_B receptors. Mol Cell Biochem 154:31–37

Carr MJ, Goldie RG, Henry PJ (1996) Time course of changes in ET_B receptor density and function in tracheal airway smooth muscle during respiratory tract viral infection in mice. Br J Pharmacol 117:1222–1228

Ceccherini I, Zhang AL, Matera I, Yang G, Devoto M, Romeo G, Cass DT (1995) Interstitial deletion of the endothelin-B receptor gene in the spotting lethal (sl) rat. Hum Mol Genet 4:2089–2096

Chakrabarti S, Gan XT, Merry A, Karmazyn M, Sima AA (1998) Augmented retinal endothelin-1, endothelin-3, endothelinA and endothelinB gene expression in chronic diabetes. Curr Eye Res 17:301–307

Chin MH, Cioffi CL, Garay M, Neale RF, Shetty SS, DelGrande D, Mugrage B, Sills MA, Lipson KE (1996) The unusual binding properties of the endothelin receptor antagonist CGS 27830 distinguishes receptor/agonist interactions. J Pharmacol Exp Ther 276:74–83

Chiou WJ, Magnuson SR, Dixon D, Sundy S, Opgenorth TJ, Wu-Wong JR (1997) Dissociation characteristics of endothelin receptor agonists and antagonists in cloned human type-B endothelin receptor. Endothelium 5:179–189

Chun M, Lin HY, Henis YI, Lodish HF (1995) Endothelin-induced endocytosis of cell surface ET_A receptors. Endothelin remains intact and bound to the ET_A receptor. J Biol Chem 270:10855–10860

Chun M, Liyanage UK, Lisanti MP, Lodish HF (1994) Signal transduction of a G protein-coupled receptor in caveolae: colocalization of endothelin and its receptor with caveolin. Proc Natl Acad Sci U S A 91:11728–11732

Clark KL, Pierre L (1995) Characterization of endothelin receptors in rat renal artery in vitro. Br J Pharmacol 114:785–790

Clouthier DE, Hosoda K, Richardson JA, Williams SC, Yanagisawa H, Kuwaki T, Kumada M, Hammer RE, Yanagisawa M (1998) Cranial and cardiac neural crest defects in endothelin-A receptor-deficient mice. Development 125:813–824

Clozel M, Loffler BM, Breu V, Hilfiger L, Maire JP, Butscha B (1993) Downregulation of endothelin receptors by autocrine production of endothelin-1. Am J Physiol 265:C188–C192

Collett GP, Kohnen G, Campbell S, Davenport AP, Jeffers MD, Cameron IT (1996) Localization of endothelin receptors in human uterus throughout the menstrual cycle. Mol Hum Reprod 2:439–444

Cozza EN, Vila MC, Gomez-Sanchez CE (1990) ET-1 receptors in C-6 cells: homologous down-regulation and modulation by protein kinase C. Mol Cell Endocrinol 70:155–164

Cramer H, Muller-Esterl W, Schroeder C (1997) Subtype-specific desensitization of human endothelin ET_A and ET_B receptors reflects differential receptor phosphorylation. Biochemistry 36:13325–13332

Cramer H, Muller-Esterl W, Schroeder C (1998) Subtype-specific endothelin-A and endothelin-B receptor desensitization correlates with differential receptor phosphorylation. J Cardiovasc Pharmacol 31 [Suppl 1]:S203–S206

Cristiani C, Volpi D, Landonio A, Bertolero F (1994) Endothelin-1-selective binding sites are downregulated by transforming growth factor-beta and upregulated by basic fibroblast growth factor in a vascular smooth muscle-derived cell line. J Cardiovasc Pharmacol 23:988–994

Cyr CR, Rudy B, Kris RM (1993) Prolonged desensitization of the human endothelin A receptor in *Xenopus oocytes*. Comparative studies with the human neurokinin A receptor. J Biol Chem 268:26071–26074

Dagassan PH, Breu V, Clozel M, Kunzli A, Vogt P, Turina M, Kiowski W, Clozel JP (1996) Up-regulation of endothelin-B receptors in atherosclerotic human coronary arteries. J Cardiovasc Pharmacol 27:147–153

Dashwood MR, Mehta D, Izzat MB, Timm M, Bryan AJ, Angelini GD, Jeremy JY (1998) Distribution of endothelin-1 (ET) receptors (ET(A) and ET(B)) and immunoreactive ET-1 in porcine saphenous vein-carotid artery interposition grafts. Atherosclerosis 137:233–242

Davenport AP, O'Reilly G, Kuc RE (1995) Endothelin ET_A and ET_B mRNA and receptors expressed by smooth muscle in the human vasculature: majority of the ETA sub-type. Br J Pharmacol 114:1110–1116

Desmarets J, Gresser O, Guedin D, Frelin C (1996) Interaction of endothelin-1 with cloned bovine ET_A receptors: biochemical parameters and functional consequences. Biochemistry 35:14868–14875

Devadason PS, Henry PJ (1997) Comparison of the contractile effects and binding kinetics of endothelin-1 and sarafotoxin S6b in rat isolated renal artery. Br J Pharmacol 121:253–263

Devesly P, Phillips PE, Johns A, Rubanyi G, Parker-Botelho LH (1990) Receptor kinetics differ for endothelin-1 and endothelin-2 binding to Swiss 3T3 fibroblasts. Biochem Biophys Res Commun 172:126–134

Doherty AM, Patt WC, Edmunds JJ, Berryman KA, Reisdorph BR, Plummer MS, Shahripour A, Lee C, Cheng XM, Walker DM (1995) Discovery of a novel series of orally active non-peptide endothelin-A (ETA) receptor-selective antagonists. J Med Chem 38:1259–1263

Douglas SA, Beck GRJ, Elliott JD, Ohlstein EH (1995) Pharmacologic evidence for the presence of three functional endothelin receptor subtypes in rabbit saphenous vein. J Cardiovasc Pharmacol 26 [Suppl 3]:S163–S168

Douglas SA, Hiley CR (1990) Endothelium-dependent vascular activities of endothelin-like peptides in the isolated superior mesenteric arterial bed of the rat. Br J Pharmacol 101:81–88

Eglezos A, Cucchi P, Patacchini R, Quartara L, Maggi CA, Mizrahi J (1993) Differential effects of BQ-123 against endothelin-1 and endothelin-3 on the rat vas deferens: evidence for an atypical endothelin receptor. Br J Pharmacol 109:736–738

Eguchi D, Nishimura J, Kobayashi S, Komori K, Sugimachi K, Kanaide H (1997) Down-regulation of endothelin B receptors in autogenous saphenous veins grafted into the arterial circulation. Cardiovasc Res 35:360–367

Eguchi S, Hirata Y, Imai T, Kanno K, Marumo F (1994b) Phenotypic change of endothelin receptor subtype in cultured rat vascular smooth muscle cells. Endocrinology 134:222–228

Eguchi S, Hirata Y, Imai T, Marumo F (1994a) C-type natriuretic peptide upregulates vascular endothelin type B receptors. Hypertension 23:936–940

Elshourbagy NA, Adamou JE, Gagnon AW, Wu HL, Pullen M, Nambi P (1996) Molecular characterization of a novel human endothelin receptor splice variant. J Biol Chem 271:25300–25307

Elshourbagy NA, Lee JA, Korman DR, Nuthalaganti P, Sylvester DR, Dilella AG, Sutiphong JA, Kumar CS (1992) Molecular cloning and characterization of the major endothelin receptor subtype in porcine cerebellum. Mol Pharmacol 41:465–473

Emori T, Hirata Y, Marumo F (1990) Specific receptors for endothelin-3 in cultured bovine endothelial cells and its cellular mechanism of action. FEBS Lett 263:261–264

Faxen M, Wihman I, Lunell NO, Blanck A (1997) Messenger RNA expression of endothelin-1, endothelin-A receptor and endothelial constitutive nitric oxide synthase in hydatididorm moles. Gynecol Obstet Invest 44:221–223

Flynn MA, Haleen SJ, Welch KM, Cheng XM, Reynolds EE (1998) Endothelin B receptors on human endothelial and smooth-muscle cells show equivalent binding pharmacology. J Cardiovasc Pharmacol 32:106–116

Freedman NJ, Ament AS, Oppermann M, Stoffel RH, Exum ST, Lefkowitz RJ (1997) Phosphorylation and desensitization of human endothelin A and B receptors. Evidence for G protein-coupled receptor kinase specificity. J Biol Chem 272:17734–17743

Fu LX, Sun XY, Hedner T, Feng QP, Liang QM, Hoebeke J, Hjalmarson A (1993) Decreased density of mesenteric arteries but not of myocardial endothelin receptors and function in rats with chronic ischemic heart failure. J Cardiovasc Pharmacol 22:177–182

Fujitani Y, Oda K, Takimoto M, Inui T, Okada T, Urade Y (1992) Autocrine receptors for endothelins in the primary culture of endothelial cells of human umbilical vein. FEBS Lett 298:79–83

Fukuroda T, Kobayashi M, Ozaki S, Yano M, Miyauchi T, Onizuka M, Sugishita Y, Goto K, Nishikibe M (1994) Endothelin receptor subtypes in human versus rabbit pulmonary arteries. J Appl Physiol 76:1976–1982

Galron R, Kloog Y, Bdolah A, Sokolovsky M (1989) Functional endothelin/sarafotoxin receptors in rat heart myocytes: structure-activity relationships and receptor subtypes. Biochem Biophys Res Commun 163:936–943

Gandhi CR, Sproat LA, Subbotin VM (1996) Increased hepatic endothelin-1 levels and endothelin receptor density in cirrhotic rats. Life Sci 58:55–62

Gariepy CE, Cass DT, Yanagisawa M (1996) Null mutation of endothelin receptor type B gene in spotting lethal rats causes aganglionic megacolon and white coat color. Proc Natl Acad Sci U S A 93:867–872

Gariepy CE, Williams SC, Richardson JA, Hammer RE, Yanagisawa M (1998) Transgenic expression of the endothelin-B receptor prevents congenital intestinal aganglionosis in a rat model of Hirschsprung disease. J Clin Invest 102:1092–1101

Gellai M, DeWolf R, Pullen M, Nambi P (1994) Distribution and functional role of renal ET receptor subtypes in normotensive and hypertensive rats. Kidney Int 46:1287–1294

Gellai M, Fletcher T, Pullen M, Nambi P (1996) Evidence for the existence of endothelin-B receptor subtypes and their physiological roles in the rat. Am J Physiol 271:R254–R261

Giller T, Breu V, Valdenaire O, Clozel M (1997) Absence of ET(B)-mediated contraction in Piebald-lethal mice. Life Sci 61:255–263

Godfraind T (1993) Evidence for heterogeneity of endothelin receptor distribution in human coronary artery. Br J Pharmacol 110:1201–1205

Goldie RG, D'Aprile AC, Cvetkovski R, Rigby PJ, Henry, PJ (1996) Influence of regional differences in ET_A and ET_B receptor subtype proportions on endothelin-1-induced contractions in porcine isolated trachea and bronchus. Br J Pharmacol 117:736–742

Goldie RG, Grayson PS, Knott PG, Self GJ, Henry PJ (1994) Predominance of endothelinA (ET_A) receptors in ovine airway smooth muscle and their mediation of ET-1-induced contraction. Br J Pharmacol 112:749–756

Goldie RG, Henry PJ, Knott PG, Self GJ, Luttmann MA, Hay DW (1995) Endothelin-1 receptor density, distribution, and function in human isolated asthmatic airways. Am J Respir Crit Care Med 152:1653–1658

Gosselin R, Gutkowska J, Baribeau J, Perreault T (1997) Endothelin receptor changes in hypoxia-induced pulmonary hypertension in the newborn piglet. Am J Physiol 273:L72–L79

Gray GA, Webb DJ (1996) The endothelin system and its potential as a therapeutic target in cardiovascular disease. Pharmacol Ther 72:109–148

Gresser O, Chayard D, Herbert D, Cousin MA, Le Moullec JM, Bouattane F, Guedin D, Frelin C (1996) Et-1 and Et-3 actions mediated by cloned ET_A endothelin receptors exhibit different sensitivities to BQ-123. Biochem Biophys Res Commun 224:169–171

Haendler B, Hechler U, Schleuning WD (1992) Molecular cloning of human endothelin (ET) receptors ET_A and ET_B. J Cardiovasc Pharmacol 20 (Suppl 12):S1–S4

Hama H, Kasuya Y, Sakurai T, Yamada G, Suzuki N, Masaki T, Goto K (1997) Role of endothelin-1 in astrocyte responses after acute brain damage. J Neurosci Res 47:590–602

Hamroun D, Mathieu MN, Chevillard C (1998) Change of endothelin receptor subtype in the MEG-01 human megakaryoblastic cell line. Eur J Pharmacol 344:307–312

Hara M, Tozawa F, Itazaki K, Mihara S, Fujimoto M (1998) Endothelin ET(B) receptors show different binding profiles in intact cells and cell membrane preparations. Eur J Pharmacol 345:339–342

Hasegawa K, Fujiwara H, Doyama K, Inada T, Ohtani S, Fujiwara T, Hosoda K, Nakao K, Sasayama S (1994) Endothelin-1-selective receptor in the arterial intima of patients with hypertension. Hypertension 23:288–293

Hasegawa K, Fujiwara H, Koshiji M, Inada T, Ohtani S, Doyama K, Tanaka M, Matsumori A, Fujiwara T, Shirakami G, Hosoda K, Nakao K, Sasayama S (1996) Endothelin-1 and its receptor in hypertrophic cardiomyopathy. Hypertension 27:259–264

Hashido K, Gamou T, Adachi M, Tabuchi H, Watanabe T, Furuichi Y, Miyamoto C (1992) Truncation of N-terminal extracellular or C-terminal intracellular domains of human ET_A receptor abrogated the binding activity to ET-1. Biochem Biophys Res Commun 187:1241–1248

Hatakeyama H, Miyamori I, Yamagishi S, Takeda Y, Takeda R, Yamamoto H (1994) Angiotensin II up-regulates the expression of type A endothelin receptor in human vascular smooth muscle cells. Biochem Mol Biol Int 34:127–134

Haug C, Duell T, Voisard R, Lenich A, Kolb HJ, Mickley V, Hombach V, Grunert A (1995) Cyclosporine A stimulates endothelin release. J Cardiovasc Pharmacol 26 [Suppl 3]:S239–S241

Hay DW, Luttmann MA, Beck G, Ohlstein EH (1996) Comparison of endothelin B
(ET$_B$) receptors in rabbit isolated pulmonary artery and bronchus. Br J Pharmacol 118:1209–1217
Hay DW, Luttmann MA, Pullen MA, Nambi P (1998) Functional and binding characterization of endothelin receptors in human bronchus: evidence for a novel
endothelin B receptor subtype? J Pharmacol Exp Ther 284:669–677
Hayzer DJ, Cicila G, Cockerham C, Griendling KK, Delafontaine P, Ng SC, Runge MS
(1994) Endothelin A and B receptors are down-regulated in the hearts of hyper-
tensive rats. Am J Med Sci 307:222–227
Henry PJ, Goldie RG (1994) ET$_B$ but not ET$_A$ receptor-mediated contractions to
endothelin-1 attenuated by respiratory tract viral infection in mouse airways. Br J
Pharmacol 112:1188–1194
Henry PJ, King SH (1999) Typical endothelin ET$_A$ receptors mediate atypical
endothelin-1-induced contractions in sheep isolated tracheal smooth muscle. J
Pharmacol Exp Ther 289:1385–1390
Hilal-Dandan R, Ramirez MT, Villegas S, Gonzalez A, Endo-Mochizuki Y, Brown JH,
Brunton LL (1997a) Endothelin ET$_A$ receptor regulates signaling and ANF gene
expression via multiple G protein-linked pathways. Am J Physiol 272:H130–H137
Hilal-Dandan R, Villegas S, Gonzalez A, Brunton LL (1997b) The quasi-irreversible
nature of endothelin binding and G protein-linked signaling in cardiac myocytes.
J Pharmacol Exp Ther 281:267–273
Hiley CR, Jones CR, Pelton JT, Miller RC (1990) Binding of [^{125}I]-endothelin-1 to rat
cerebellar homogenates and its interactions with some analogues. Br J Pharmacol
101:319–324
Hino A, Tokuyama Y, Kobayashi M, Yano M, Weir B, Takeda J, Wang X, Bell GI, Macdonald RL (1996) Increased expression of endothelin B receptor mRNA following subarachnoid hemorrhage in monkeys. J Cereb Blood Flow Metab 16:688–697
Hiraoka J, Arai H, Yoshimasa T, Takaya K, Miyamoto Y, Yamashita J, Suga S, Ogawa
Y, Shirakami G, Itoh H (1995) Augmented expression of the endothelin-A receptor gene in cultured mesangial cells from stroke-prone spontaneously hypertensive rats. Clin Exp Pharmacol Physiol Suppl 95:S191–S192
Hislop AA, Zhao YD, Springall DR, Polak JM, Haworth SG (1995) Postnatal changes
in endothelin-1 binding in porcine pulmonary vessels and airways. Am J Respir
Cell Mol Biol 12:557–566
Hocher B, Rohmeiss P, Diekmann F, Zart R, Vogt V, Schiller S, Bauer C, Koppenhagen
K, Distler A, Gretz N (1995) Distribution of endothelin receptor subtypes in the
rat kidney. Renal and haemodynamic effects of the mixed (A/B) endothelin
receptor antagonist bosentan. Eur J Clin Chem Clin Biochem 33:463–472
Hocher B, Rohmeiss P, Zart R, Diekmann F, Vogt V, Metz D, Fakhury M, Gretz N,
Bauer C, Koppenhagen K, Neumayer HH, Distler A (1996a) Function and expres-
sion of endothelin receptor subtypes in the kidneys of spontaneously hyperten-
sive rats. Cardiovasc Res 31:499–510
Hocher B, Zart R, Diekmann F, Rohmeiss P, Distler A, Neumayer HH, Bauer C,
Gross P (1996b) Paracrine renal endothelin system in rats with liver cirrhosis. Br
J Pharmacol 118:220–227
Hocher B, Zart R, Schwarz A, Vogt V, Braun C, Thone-Reineke C, Braun N, Neumayer
HH, Koppenhagen K, Bauer C, Rohmeiss P (1998) Renal endothelin system in
polycystic kidney disease. J Am Soc Nephrol 9:1169–1177
Horstmeyer A, Cramer H, Sauer T, Muller-Esterl W, Schroeder C (1996) Palmitoyla-
tion of endothelin receptor A. Differential modulation of signal transduction
activity by post-translational modification. J Biol Chem 271:20811–20819
Hoshino T, Yamauchi R, Kikkawa K, Yabana H, Murata S (1998) Pharmacological
profile of T-0201, a highly potent and orally active endothelin receptor antagonist.
J Pharmacol Exp Ther 286:643–649
Hosoda K, Hammer RE, Richardson JA, Baynash AG, Cheung JC, Giaid A,
Yanagisawa M (1994) Targeted and natural (piebald-lethal) mutations of

endothelin-B receptor gene produce megacolon associated with spotted coat color in mice. Cell 79:1267–1276

Hosoda K, Nakao K, Tamura N, Arai H, Ogawa Y, Suga S, Nakanishi S, Imura H (1992) Organization, structure, chromosomal assignment, and expression of the gene encoding the human endothelin-A receptor. J Biol Chem 267:18797–18804

Huggins JP, Pelton JT, Miller RC (1993b) The structure and specificity of endothelin receptors: their importance in physiology and medicine. Pharmacology and Therapeutics 59:55–123

Huggins JP, Pelton JT, Miller RC (1993a) The structure and specificity of endothelin receptors: their importance in physiology and medicine. Pharmacol Ther 59:55–123

Ihara M, Fukuroda T, Saeki T, Nishikibe M, Kojiri K, Suda H, Yano M (1991) An endothelin receptor (ET$_A$) antagonist isolated from Streptomyces misakiensis. Biochem Biophys Res Commun 178:132–137

Ihara M, Saeki T, Fukuroda T, Kimura S, Ozaki S, Patel AC, Yano M (1992) A novel radioligand [125I]BQ-3020 selective for endothelin (ET$_B$) receptors. Life Sci 51:PL47–PL52

Ihara M, Yamanaka R, Ohwaki K, Ozaki S, Fukami T, Ishikawa K, Towers P, Yano M (1995) [^{3}H]BQ-123, a highly specific and reversible radioligand for the endothelin ET$_A$ receptor subtype. Eur J Pharmacol 274:1–6

Ishikawa H, Haruno I, Harada Y, Yoshitomi T, Ishikawa S, Katori M (1996) Pharmacological characterization of endothelin receptors in the rabbit iris sphincter muscle: suggestion for the presence of atypical receptors. Curr Eye Res 15:73–78

Ishikawa K, Ihara M, Noguchi K, Mase T, Mino N, Saeki T, Fukuroda T, Fukami T, Ozaki S, Nagase T (1994) Biochemical and pharmacological profile of a potent and selective endothelin B-receptor antagonist, BQ-788. Proc Natl Acad Sci USA 91:4892–4896

Itoh S, Sasaki T, Asai A, Kuchino Y (1994) Prevention of delayed vasospasm by an endothelin ET$_A$ receptor antagonist, BQ-123: change of ET$_A$ receptor mRNA expression in a canine subarachnoid hemorrhage model. J Neurosurg 81:759–764

Ivy DD, Le Cras TD, Horan MP, Abman SH (1998) Chronic intrauterine pulmonary hypertension increases preproendothelin-1 and decreases endothelin B receptor mRNA expression in the ovine fetal lung. Chest 114:65S

Iwai J, Kanayama Y, Negoro N, Okamura M, Takeda T (1995) Gene expression of endothelin receptors in aortic cells from cyclosporine-induced hypertensive rats. Clin Exp Pharmacol Physiol 22:404–409

Iwasaki S, Homma T, Kon V (1994) Site specific regulation in the kidney of endothelin and its receptor subtypes by cyclosporine. Kidney Int 45:592–597

Jeng AY, Wass KS, Hsu L (1996) Developmental changes in endothelin A and endothelin B receptor subtypes in embryonic chick brains. Neurosci Lett 208:208–210

Kanno K, Hirata Y, Tsujino M, Imai T, Shichiri M, Ito H, Marumo F (1993) Upregulation of ETB receptor subtype mRNA by angiotensin II in rat cardiomyocytes. Biochem Biophys Res Commun 194:1282–1287

Karaki H, Mitsui-Saito M, Takimoto M, Oda K, Okada T, Ozaki T, Kunieda T (1996) Lack of endothelin ETB receptor binding and function in the rat with a mutant ETB receptor gene. Biochem Biophys Res Commun 222:139–143

Karaki H, Sudjarwo SA, Hori M (1994) Novel antagonist of endothelin ET$_{B1}$ and ET$_{B2}$ receptors, BQ-788: effects on blood vessel and small intestine. Biochem Biophys Res Commun 205:168–173

Karet FE, Davenport AP (1996) Selective downregulation of ET$_A$ receptor mRNA in renal transplant recipients on cyclosporin A revealed by quantitative RT-PCR. Nephrol Dial Transplant 11:1976–1982

Karne S, Jayawickreme CK, Lerner MR (1993) Cloning and characterization of an endothelin-3 specific receptor (ET$_C$ receptor) from Xenopus laevis dermal melanophores. J Biol Chem 268:19126–19133

Kenakin TP, Bond RA, Bonner TI (1992) Definition of pharmacological receptors. Pharmacol Rev 44:351–362

Kikuchi K, Nakagawa H, Kadono T, Etoh T, Byers HR, Mihm MC, Tamaki K (1996) Decreased ET(B) receptor expression in human metastatic melanoma cells. Biochem Biophys Res Commun 219:734–739

Kim HK, Severson SR, Ricagna F, Barber DA, Tazelaar HD, Miller VM, McGregor CG (1997) Characteristics of endothelin receptors in acutely rejecting transplanted lungs. Transplantation 64:209–214

Kirshbom PM, Page SO, Jacobs MT, Tsui SS, Bello E, Ungerleider RM, Schwinn DA, Gaynor JW (1997) Cardiopulmonary bypass and circulatory arrest increase endothelin-1 production and receptor expression in the lung. J Thorac Cardiovasc Surg 113:777–783

Knock GA, Terenghi G, Bunker CB, Bull HA, Dowd PM, Polak JM (1993) Characterization of endothelin-binding sites in human skin and their regulation in primary Raynaud's phenomenon and systemic sclerosis. J Invest Dermatol 101:73–78

Knott PG, D'Aprile AC, Henry PJ, Hay DW, Goldie RG (1995) Receptors for endothelin-1 in asthmatic human peripheral lung. Br J Pharmacol 114:1–3

Kobayshi T, Miyauchi T, Sakai S, Maeda S, Yamaguchi I, Goto K, Sugishita Y (1998) Down-regulation of ET(B) receptor, but not ET(A) receptor, in congestive lung secondary to heart failure. Are marked increases in circulating endothelin-1 partly attributable to decreases in lung ET(B) receptor-mediated clearance of endothelin-1? Life Sci 62:185–193

Kohnen G, Campbell S, Irvine GA, Church HJ, MacLachlan F, Titterington M, Cameron IT (1998) Endothelin receptor expression in human decidua. Mol Hum Reprod 4:185–193

Kohzuki M, Kanazawa M, Yoshida K, Kamimoto M, Wu XM, Jiang ZL, Yasujima M, Abe K, Johnston CI, Sato T (1996) Cardiac angiotensin converting enzyme and endothelin receptor in rats with chronic myocardial infarction. Jpn Circ J 60:972–980

Kohzuki M, Onodera H, Yasujima M, Itoyama Y, Kanazawa M, Sato T, Abe K (1995) Endothelin receptors in ischemic rat brain and Alzheimer brain. J Cardiovasc Pharmacol 26 [Suppl 3]:S329–S331

Kojiri K, Ihara M, Nakajima S, Kawamura K, Funaishi K, Yano M, Suda H (1991) Endothelin-binding inhibitors, BE-18257A and BE-18257B. I. Taxonomy, fermentation, isolation and characterization. J Antibiot (Tokyo) 44:1342–1347

Kon V, Sugiura M, Inagami T, Harvie BR, Ichikawa I, Hoover RL (1990) Role of endothelin in cyclosporine-induced glomerular dysfunction. Kidney Int 37:1487–1491

Kondo S, Morita T, Tashima Y (1994) Endothelin receptor density in human hypertrophic and non-hypertrophic prostate tissue. Tohoku J Exp Med 172:381–384

Kozuka M, Ito T, Hirose S, Lodhi KM, Hagiwara H (1991) Purification and characterization of bovine lung endothelin receptor. J Biol Chem 266:16892–16896

Krystek SRJ, Patel PS, Rose PM, Fisher SM, Kienzle BK, Lach DA, Liu EC, Lynch JS, Novotny J, Webb ML (1994) Mutation of peptide binding site in transmembrane region of a G protein-coupled receptor accounts for endothelin receptor subtype selectivity. J Biol Chem 269:12383–12386

Kubota T, Taguchi M, Kamada S, Imai T, Hirata Y, Marumo, Aso T (1995) Endothelin synthesis and receptors in human endometrium throughout the normal menstrual cycle. Hum Reprod 10:2204–2208

Kumar C, Mwangi V, Nuthulaganti P, Wu HL, Pullen M, Brun K, Aiyar H, Morris RA, Naughton R, Nambi P (1994) Cloning and characterization of a novel endothelin receptor from Xenopus heart. J Biol Chem 269:13414–13420

Kurihara Y, Kurihara H, Suzuki H, Kodama T, Maemura K, Nagai R, Oda H, Kuwaki T, Cao WH, Kamada N (1994) Elevated blood pressure and craniofacial abnormalities in mice deficient in endothelin-1. Nature 368:703–710

Kusafuka T, Wang Y, Puri P (1997) Mutation analysis of the RET, the endothelin-B receptor, and the endothelin-3 genes in sporadic cases of Hirschsprung's disease. J Pediatr Surg 32:501–504

Lee JA, Elliott JD, Sutiphong JA, Friesen WJ, Ohlstein EH, Stadel JM, Gleason JG, Peishoff CE (1994) Tyr-129 is important to the peptide ligand affinity and selectivity of human endothelin type A receptor. Proc Natl Acad Sci U S A 91:7164–7168

Leivas A, Jimenez W, Bruix J, Boix L, Bosch J, Arroyo V, Rivera F, Rodes J (1998) Gene expression of endothelin-1 and ET(A) and ET(B) receptors in human cirrhosis: relationship with hepatic hemodynamics. J Vasc Res 35:186–193

Li H, Chen SJ, Chen YF, Meng QC, Durand J, Oparil S, Elton TS (1994a) Enhanced endothelin-1 and endothelin receptor gene expression in chronic hypoxia. J Appl Physiol 77:1451–1459

Li H, Elton TS, Chen YF, Oparil S (1994b) Increased endothelin receptor gene expression in hypoxic rat lung. Am J Physiol 266:L553–L560

Liu J, Chen R, Casley DJ, Nayler WG (1990) Ischemia and reperfusion increase ^{125}I-labeled endothelin-1 binding in rat cardiac membranes. Am J Physiol 258:H829–H835

Liu JJ, Chen JR, Buxton BF (1996) Unique response of human arteries to endothelin B receptor agonist and antagonist. Clin Sci (Colch) 90:91–96

Lodge NJ, Halaka NN (1993) Endothelin receptor subtype(s) in rabbit jugular vein smooth muscle. J Cardiovasc Pharmacol 22 [Suppl 8]:S140–S143

Lodhi KM, Sakaguchi H, Hirose S, Hagiwara H (1995) Localization and characterization of a novel receptor for endothelin in the gills of the rainbow trout. J Biochem (Tokyo) 118:376–379

Loffler BM, Roux S, Kalina B, Clozel M, Clozel JP (1993) Influence of congestive heart failure on endothelin levels and receptors in rabbits. J Mol Cell Cardiol 25:407–416

Lysko PG, Elshourbagy NA, Pullen M, Nambi P (1995) Developmental expression of endothelin receptors in cerebellar neurons differentiating in culture. Brain Res 88:96–101

Maggi CA, Giuliani S, Patacchini R, Santicioli P, Rovero P, Giachetti A, Meli A (1989) The C-terminal hexapeptide, endothelin-(16–21), discriminates between different endothelin receptors. Eur J Pharmacol 166:121–122

Maguire JJ, Kuc RE, Rous BA, Davenport AP (1996) Failure of BQ123, a more potent antagonist of sarafotoxin 6b than of endothelin-1, to distinguish between these agonists in binding experiments. Br J Pharmacol 118:335–342

Marsault R, Feolde E, Frelin C (1993) Receptor externalization determines sustained contractile responses to endothelin-1 in the rat aorta. Am J Physiol 264:C687-C693

Marsault R, Vigne P, Breittmayer JP, Frelin C (1991) Kinetics of vasoconstrictor action of endothelins. Am J Physiol 261:C987–C993

Marsault R, Vigne P, Frelin C (1991) The irreversibility of endothelin action is a property of a late intracellular signalling event. Biochem Biophys Res Commun 179:1408–1413

Masuda Y, Miyazaki H, Kondoh M, Watanabe H, Yanagisawa M, Masaki T, Murakami K (1989) Two different forms of endothelin receptors in rat lung. FEBS Lett 257:208–210

Maxwell MJ, Goldie RG, Henry PJ (1998a) Altered ET_B- but not ET_A-receptor density and function in sheep airway smooth muscle cells in culture. Am J Physiol 274:L951–L957

Maxwell MJ, Goldie RG, Henry PJ (1998b) Ca^{2+}-signalling by endothelin receptors in rat and human cultured airway smooth muscle cells. Br J Pharmacol 125:1768–1788

McDonald D, Bailie J, Archer D, Chakravarthy U (1995) Molecular characterization of endothelin receptors and the effect of insulin on their expression in retinal microvascular pericytes. J Cardiovasc Pharmacol 26 [Suppl 3]:S287–S289

Meissner NN, Koschel K (1995) Downregulation of endothelin receptor mRNA synthesis in C6 rat astrocytoma cells by persistent measles virus and canine distemper virus infections. J Virol 69:5191–5194

Metallinos DL, Bowling AT, Rine J (1998) A missense mutation in the endothelin-B receptor gene is associated with Lethal White Foal Syndrome: an equine version of Hirschsprung disease. Mamm Genome 9:426–431

Miyamoto Y, Yoshimasa T, Arai H, Takaya K, Ogawa Y, Itoh H, Nakao K (1996) Alternative RNA splicing of the human endothelin-A receptor generates multiple transcripts. Biochem J 313:795–801

Miyauchi T, Sakai S, Ihara M, Kasuya Y, Yamaguchi I, Goto K, Sugishita Y (1995) Increased endothelin-1 binding sites in the cardiac membranes in rats with chronic heart failure. J Cardiovasc Pharmacol 26 [Suppl 3]:S448–S451

Mizuguchi T, Nishiyama M, Moroi K, Tanaka H, Saito T, Masuda Y, Masaki T, de Wit D, Yanagisawa M, Kimura S (1997) Analysis of two pharmacologically predicted endothelin B receptor subtypes by using the endothelin B receptor gene knockout mouse. Br J Pharmacol 120:1427–1430

Moller S, Edvinsson L, Adner M (1997) Transcriptional regulated plasticity of vascular contractile endothelin ET(B) receptors after organ culture. Eur J Pharmacol 329:69–77

Morishita Y, Chiba S, Tsukuda E, Tanaka T, Ogawa T, Yamasaki M, Yoshida M, Kawamoto I, Matsuda Y (1994) RES-701-1, a novel and selective endothelin type B receptor antagonist produced by Streptomyces sp. RE-701. I. Characterization of producing strain, fermentation, isolation, physico-chemical and biological properties. J Antibiot (Tokyo) 47:269–275

Nakajima S, Niiyama K, Ihara M, Kojiri K, Suda H (1991) Endothelin-binding inhibitors, BE-18257A and BE-18257B II. Structure determination. J Antibiot (Tokyo) 44:1348–1356

Nakamura I, Saito M, Fukumoto Y, Yoshida M, Nishi K, Weiss RM, Latifpour J (1997) Experimental diabetes upregulates the expression of ureteral endothelin receptors. Peptides 18:1091–1093

Nakamura T, Ebihara I, Fukui M, Osada S, Tomino Y, Masaki T, Goto K, Furuichi Y, Koide H (1993) Increased endothelin and endothelin receptor mRNA expression in polycystic kidneys of cpk mice. J Am Soc Nephrol 4:1064–1072

Nakamura T, Ebihara I, Fukui M, Osada S, Tomino Y, Masaki T, Goto K, Furuichi Y, Koide H (1993) Renal expression of mRNAs for endothelin-1, endothelin-3 and endothelin receptors in NZB/W F1 mice. Ren Physiol Biochem 16:233–243

Nakamura T, Ebihara I, Fukui M, Osada S, Tomino Y, Masaki T, Goto K, Furuichi Y, Koide H (1995b) Modulation of glomerular endothelin and endothelin receptor gene expression in aminonucleoside-induced nephrosis. J Am Soc Nephrol 5:1585–1590

Nakamura T, Ebihara I, Fukui M, Tomino Y, Koide H (1995a) Modulation of endothelin family gene expression in renal hypertrophy. Nephron 73:228–234

Nakamura T, Fukui M, Ebihara I, Osada S, Takahashi T, Tomino Y, Koide H (1995) Effects of a low-protein diet on glomerular endothelin family gene expression in experimental focal glomerular sclerosis. Clin Sci (Colch) 88:29–37

Nakamuta M, Takayanagi R, Sakai Y, Sakamoto S, Hagiwara H, Mizuno T, Saito Y, Hirose S, Yamamoto M, Nawata H (1991) Cloning and sequence analysis of a cDNA encoding human non-selective type of endothelin receptor. Biochem Biophys Res Commun 177:34–39

Nambi P, Pullen M (1995) [125I]-BQ3020, a potent ET$_B$-selective agonist, displays species differences in its binding characteristics. Neuropeptides 28:191–196

Nambi P, Pullen M, Aiyar N (1996) Correlation between guanine nucleotide effect and reversible binding property of endothelin analogs. Neuropeptides 30:109–114

Nambi P, Pullen M, Contino LC, Brooks DP (1990) Upregulation of renal endothelin receptors in rats with cyclosporine A-induced nephrotoxicity. Eur J Pharmacol 187:113–116

Nambi P, Pullen M, Jugus M, Gellai M (1993) Rat kidney endothelin receptors in ischemia-induced acute renal failure. J Pharmacol Exp Ther 264:345–348

Nambi P, Pullen M, Kumar C (1994) Identification of a novel endothelin receptor in Xenopus laevis liver. Neuropeptides 26:181–185

Nambi P, Pullen M, Spielman W (1994) Species differences in the binding characteristics of [^{125}I]IRL-1620, a potent agonist specific for endothelin-B receptors. J Pharmacol Exp Ther 268:202–207

Nambi P, Pullen M, Wu HL, Nuthulaganti P, Elshourbagy N, Kumar C (1992) Dexamethasone down-regulates the expression of endothelin receptors in vascular smooth muscle cells. J Biol Chem 267:19555–19559

Nataf V, Lecoin L, Eichmann A, Le Douarin NM (1996) Endothelin-B receptor is expressed by neural crest cells in the avian embryo. Proc Natl Acad Sci U S A 93:9645–9650

Nayler WG, Gu XH, Casley DJ, Panagiotopoulos S, Liu J, Mottram PL (1989) Cyclosporine increases endothelin-1 binding site density in cardiac cell membranes. Biochem Biophys Res Commun 163:1270–1274

Nelson JB, Chan-Tack K, Hedican SP, Magnuson SR, Opgenorth TJ, Bova GS, Simons JW (1996) Endothelin-1 production and decreased endothelin B receptor expression in advanced prostate cancer. Cancer Res 56:663–668

Newby DE, Flint LL, Fox KA, Boon NA, Webb DJ (1998) Reduced responsiveness to endothelin-1 in peripheral resistance vessels of patients with syndrome X. J Am Coll Cardiol 31:1585–1590

Newman P, Kakkar VV, Kanse SM (1995) Modulation of endothelin receptor expression in human vascular smooth muscle cells by interleukin-1 beta. FEBS Lett 363:161–164

Nishiyama M, Moroi K, Shan LH, Yamamoto M, Takasaki C, Masaki T, Kimura S (1995c) Two different endothelin B receptor subtypes mediate contraction of the rabbit saphenous vein. Jpn J Pharmacol 68:235–243

Nishiyama M, Shan LH, Moroi K, Masaki T, Kimura S (1995a) Heterogeneity of endothelin ET$_A$ receptor-mediated contractions in the rabbit saphenous vein. Eur J Pharmacol 286:209–212

Nishiyama M, Takahara Y, Masaki T, Nakajima N, Kimura S (1995b) Pharmacological heterogeneity of both endothelin ET$_A$- and ET$_B$-receptors in the human saphenous vein. Jpn J Pharmacol 69:391–398

Noguchi Y, Hislop AA, Haworth SG (1997) Influence of hypoxia on endothelin-1 binding sites in neonatal porcine pulmonary vasculature. Am J Physiol 272:H669–H678

Nosaka C, Ishikawa H, Haruno I, Yoshitomi T, Kase H, Ishikawa S, Harada Y (1998) Radioligand binding characteristics of the endothelin receptor in the rabbit iris. Jpn J Pharmacol 76:289–296

O'Reilly G, Charnock-Jones DS, Davenport AP, Cameron IT, Smith SK (1992) Presence of messenger ribonucleic acid for endothelin-1, endothelin-2, and endothelin-3 in human endometrium and a change in the ratio of ET$_A$ and ET$_B$ receptor subtype across the menstrual cycle. J Clin Endocrinol Metab 75:1545–1549

Ogawa Y, Nakao K, Arai H, Nakagawa O, Hosoda K, Suga S, Nakanishi S, Imura H (1991) Molecular cloning of a non-isopeptide-selective human endothelin receptor. Biochem Biophys Res Commun 178:248–255

Ohlstein EH, Nambi P, Hay DW, Gellai M, Brooks DP, Luengo J, Xiang JN, Elliott JD (1998) Nonpeptide endothelin receptor antagonists. XI. Pharmacological characterization of SB 234551, a high-affinity and selective nonpeptide ET$_A$ receptor antagonist. J Pharmacol Exp Ther 286:650–656

Okamoto Y, Ninomiya H, Tanioka M, Sakamoto A, Miwa S, Masaki T (1997) Palmitoylation of human endothelinB. Its critical role in G protein coupling and a differential requirement for the cytoplasmic tail by G protein subtypes. J Biol Chem 272:21589–21596

Osada K, Tsunoda H, Miyauchi T, Sugishita Y, Kubo T, Goto K (1997) Pregnancy increases ET-1-induced contraction and changes receptor subtypes in uterine smooth muscle in humans. Am J Physiol 272:R541–R548

Pagotto U, Arzberger T, Hopfner U, Weindl A, Stalla GK (1995) Cellular localization of endothelin receptor mRNAs (ET$_A$ and ET$_B$) in brain tumors and normal human brain. J Cardiovasc Pharmacol 26 [Suppl 3]:S104–S106

Pate MA, Chester AH, Brown TJ, Roach AG, Yacoub MH (1998) Atypical antagonism observed with BQ-123 in human saphenous vein. J Cardiovasc Pharmacol 31 [Suppl 1]:S172–S174

Pekonen F, Nyman T, Ammala M, Rutanen EM (1995) Decreased expression of messenger RNAs encoding endothelin receptors and neutral endopeptidase 24.11 in endometrial cancer. Br J Cancer 71:59–63

Picard P, Smith PJ, Monge JC, Rouleau JL, Nguyen QT, Calderone A, Stewart DJ (1998) Coordinated upregulation of the cardiac endothelin system in a rat model of heart failure. J Cardiovasc Pharmacol 31 [Suppl 1]:S294–S297

Ponicke K, Vogelsang M, Heinroth M, Becker K, Zolk O, Bohm M, Zerkowski HR, Brodde OE (1998) Endothelin receptors in the failing and nonfailing human heart. Circulation 97:744–751

Puffenberger EG, Hosoda K, Washington SS, Nakao K, deWit D, Yanagisawa M, Chakravart A (1994) A missense mutation of the endothelin-B receptor gene in multigenic Hirschsprung's disease. Cell 79:1257–1266

Resink TJ, Scott-Burden T, Weber E, Buhler FR (1990) Phorbol ester promotes a sustained down-regulation of endothelin receptors and cellular responses to endothelin in human vascular smooth muscle cells. Biochem Biophys Res Commun 166:1213–1219

Riezebos J, Watts IS, Vallance PJ (1994) Endothelin receptors mediating functional responses in human small arteries and veins. Br J Pharmacol 111:609–615

Roos M, Soskic V, Poznanovic S, Godovac-Zimmermann J (1998) Post-translational modifications of endothelin receptor B from bovine lungs analyzed by mass spectrometry. J Biol Chem 273:924–931

Rossi G, Belloni AS, Albertin G, Zanin L, Biasolo MA, Nussdorfer GG, Palu G, Pessina AC (1995) Endothelin-1 and its receptors A and B in human aldosterone-producing adenomas. Hypertension 25:842–847

Roubert P, Gillard-Roubert V, Pourmarin L, Cornet S, Guilmard C, Plas P, Pirotzky E, Chabrier PE, Braquet P (1994) Endothelin receptor subtypes A and B are up-regulated in an experimental model of acute renal failure. Mol Pharmacol 45:182–188

Roubert P, Gillard V, Plas P, Chabrier PE, Braquet P (1991) Binding characteristics of endothelin isoforms (ET-1, ET-2, and ET-3) in vascular smooth muscle cells. J Cardiovasc Pharmacol 17 [Suppl 7]:S104–S108

Roubert P, Viossat I, Lonchampt MO, Chapelat M, Schulz J, Plas P, Gillard-Roubert V, Chabrier PE, Braquet P (1993) Endothelin receptor regulation by endothelin synthesis in vascular smooth muscle cells: effects of dexamethasone and phosphoramidon. J Vasc Res 30:139–144

Rubanyi GM, Polokoff MA (1994) Endothelins: molecular biology, biochemistry, pharmacology, physiology, and pathophysiology. Pharmacol Rev 46:325–415

Saeki T, Ihara M, Fukuroda T, Yamagiwa M, Yano M (1991) [Ala^{1,3,11,15}]endothelin-1 analogs with ET$_B$ agonistic activity. Biochem Biophys Res Commun 179:286–292

Saito M, Nishi K, Fukumoto Y, Weiss RM, Latifpour J (1996) Characterization of endothelin receptors in streptozotocin-induced diabetic rat vas deferens. Biochem Pharmacol 52:1593–1598

Sakamoto A, Yanagisawa M, Sakurai T, Takuwa Y, Yanagisawa H, Masaki T (1991) Cloning and functional expression of human cDNA for the ET$_B$ endothelin receptor. Biochem Biophys Res Commun 178:656–663

Sakamoto A, Yanagisawa M, Sawamura T, Enoki T, Ohtani T, Sakurai T, Nakao K, Toyo-oka T, Masaki T (1993) Distinct subdomains of human endothelin receptors determine their selectivity to endothelinA-selective antagonist and endothelinB-selective agonists. J Biol Chem 268:8547–8553

Sakurai-Yamashita Y, Niwa M, Yamashita K, Kataoka Y, Himeno A, Shigematsu K, Tsutsumi K, Taniyama K (1997) Endothelin receptors in kainic acid-induced neural lesions of rat brain. Neuroscience 81:565–577

Sakurai T, Morimoto H, Kasuya Y, Takuwa Y, Nakauchi H, Masaki T, Goto K (1992) Level of ET$_B$ receptor mRNA is down-regulated by endothelins through decreasing the intracellular stability of mRNA molecules. Biochem Biophys Res Commun 186:342–347

Sakurai T, Yanagisawa M, Takuwa Y, Miyazaki H, Kimura S, Goto K, Masaki T (1990) Cloning of a cDNA encoding a non-isopeptide-selective subtype of the endothelin receptor. Nature 348:732–735

Samson WK, Skala KD, Alexander BD, Huang FL (1990) Pituitary site of action of endothelin: selective inhibition of prolactin release in vitro. Biochem Biophys Res Commun 169:737–743

Santschi EM, Purdy AK, Valberg SJ, Vrotsos PD, Kaese H, Mickelson JR (1998) Endothelin receptor B polymorphism associated with lethal white foal syndrome in horses. Mamm Genome 9:306–309

Schoeffter P, Randriantsoa A (1993) Differences between endothelin receptors mediating contraction of guinea-pig aorta and pig coronary artery. Eur J Pharmacol 249:199–206

Seo B, Oemar BS, Siebenmann R, von Segesser L, Luscher TF (1994) Both ET$_A$ and ET$_B$ receptors mediate contraction to endothelin-1 in human blood vessels. Circulation 89:1203–1208

Shi W, Cernacek P, Hu F, Michel RP (1997) Endothelin reactivity and receptor profile of pulmonary vessels in postobstructive pulmonary vasculopathy. Am J Physiol 273:H2558–H2564

Shibata K, Komatsu C, Misumi Y, Furukawa T (1995) Dexamethasone down-regulates the expression of endothelin B receptor mRNA in the rat brain. Brain Res 692:71–78

Shimizu T, Kuroda T, Ikeda M, Hata S, Fujimoto M (1998) Potential contribution of endothelin to renal abnormalities in glycerol-induced acute renal failure in rats. J Pharmacol Exp Ther 286:977–983

Shraga-Levine Z, Sokolovsky M (1998) Functional role for glycosylated subtypes of rat endothelin receptors. Biochem Biophys Res Commun 246:495–500

Shyamala V, Moulthrop TH, Stratton-Thomas J, Tekamp-Olson P (1994) Two distinct human endothelin B receptors generated by alternative splicing from a single gene. Cell Mol Biol Res 40:285–296

Sirvio ML, Uhlenius N, Stewen P, Metsarinne K, Fyhrquist F (1995) The effect of aortic coarctation on expression of endothelin-1 and endothelin receptors in heart and lungs. Blood Press 4:320–323

Smith PJ, Teichert-Kuliszewska K, Monge JC, Stewart DJ (1998) Regulation of endothelin-B receptor mRNA expression in human endothelial cells by cytokines and growth factors. J Cardiovasc Pharmacol 31 [Suppl 1]:S158–S160

Sogabe K, Nirei H, Shoubo M, Nomoto A, Ao S, Notsu Y, Ono T (1993) Pharmacological profile of FR139317, a novel, potent endothelin ET$_A$ receptor antagonist. J Pharmacol Exp Ther 264:1040–1046

Sokolovsky M, Ambar I, Galron R (1992) A novel subtype of endothelin receptors. J Biol Chem 267:20551–20554

Spinella MJ, Malik AB, Everitt J, Andersen TT (1991) Design and synthesis of a specific endothelin 1 antagonist: effects on pulmonary vasoconstriction. Proc Natl Acad Sci U S A 88:7443–7446

Stanimirovic DB, McCarron RM, Spatz M (1994) Dexamethasone down-regulates endothelin receptors in human cerebromicrovascular endothelial cells. Neuropeptides 26:145–152

Stein PD, Hunt JT, Floyd DM, Moreland S, Dickinson KE, Mitchell C, Liu EC, Webb ML, Murugesan N, Dickey J (1994) The discovery of sulfonamide endothelin antagonists and the development of the orally active ET$_A$ antagonist 5-(dimethylamino)-N-(3,4-dimethyl-5-isoxazolyl)-1-naphthalenesulf onamide. J Med Chem 37:329–331

Sudjarwo SA, Hori M, Takai M, Urade Y, Okada T, Karaki H (1993) A novel subtype of endothelin B receptor mediating contraction in swine pulmonary vein. Life Sci 53:431–437

Sudjarwo SA, Hori M, Tanaka T, Matsuda Y, Okada T, Karaki H (1994) Subtypes of endothelin ET_A and ET_B receptors mediating venous smooth muscle contraction. Biochem Biophys Res Commun 200:627–633

Sullivan ME, Dashwood MR, Thompson CS, Mikhailidis DP, Morgan RJ (1998) Down-regulation of endothelin-B receptor sites in cavernosal tissue of hypercholestero-laemic rabbits. Br J Urol 81:128–134

Sullivan ME, Dashwood MR, Thompson CS, Muddle JR, Mikhailidis DP, Morgan RJ (1997) Alterations in endothelin B receptor sites in cavernosal tissue of diabetic rabbits: potential relevance to the pathogenesis of erectile dysfunction. J Urol 158:1966–1972

Sumner MJ, Cannon TR, Mundin JW, White DG, Watts IS (1992) Endothelin ET_A and ET_B receptors mediate vascular smooth muscle contraction. Br J Pharmacol 107:858–860

Takai M, Umemura I, Yamasaki K, Watakabe T, Fujitani Y, Oda K, Urade Y, Inui T, Yamamura T, Okada T (1992) A potent and specific agonist, Suc-[Glu9,Ala11,15]-endothelin-1(8–21), IRL 1620, for the ET_B receptor. Biochem Biophys Res Commun 184:953–959

Takeda M, Iwasaki S, Hellings SE, Yoshida H, Homma T, Kon V (1994) Divergent expression of EtA and EtB receptors in response to cyclosporine in mesangial cells. Am J Pathol 144:473–479

Takigawa M, Sakurai T, Kasuya Y, Abe Y, Masaki T, Goto K (1995) Molecular identification of guanine-nucleotide-binding regulatory proteins which couple to endothelin receptors. Eur J Biochem 228:102–108

Thibault G, Arguin C, Garcia R (1995) Cardiac endothelin-1 content and receptor subtype in spontaneously hypertensive rats. J Mol Cell Cardiol 27:2327–2336

Tonnessen T, Christensen G, Oie E, Holt E, Kjekshus H, Smiseth OA, Sejersted OM, Attramadal H (1997) Increased cardiac expression of endothelin-1 mRNA in ischemic heart failure in rats. Cardiovasc Res 33:601–610

Tschudi MR, Luscher TF (1994) Characterization of contractile endothelin and angiotensin receptors in human resistance arteries: evidence for two endo-thelin and one angiotensin receptor. Biochem Biophys Res Commun 204:685–690

Tsutsumi K, Niwa M, Kitagawa N, Yamaga S, Anda T, Himeno A, Sato T, Khalid H, Taniyama K, Shibata S (1994) Enhanced expression of an endothelin ETA recep-tor in capillaries from human glioblastoma: a quantitative receptor autoradi-ographic analysis using a radioluminographic imaging plate system. J Neurochem 63:2240–2247

Uchida J, Miura K, Yamanaka S, Kim S, Iwao H, Nakatani T, Kishimoto T (1998) Renal endothelin in FK506-induced nephrotoxicity in spontaneously hypertensive rats. Jpn J Pharmacol 76:39–49

Urade Y, Fujitani Y, Oda K, Watakabe T, Umemura I, Takai M, Okada T, Sakata K, Karaki H (1992) An endothelin B receptor-selective antagonist: IRL 1038, [Cys11-Cys15]-endothelin-1(11–21). FEBS Lett 311:12–16

Urade Y, Fujitani Y, Oda K, Watakabe T, Umemura I, Takai M, Okada T, Sakata K, Karaki H (1994) An endothelin B receptor-selective antagonist: IRL 1038, [Cys11-Cys15]-endothelin-1(11–21). FEBS Lett 342:103

Valdenaire O, Giller T, Breu V, Ardati A, Schweizer A, Richards JG (1998) A new family of orphan G protein-coupled receptors predominantly expressed in the brain. FEBS Lett 424:193–196

Vanhoutte PM., Humphrey PPA, Spedding M (1998) NC-IUPHAR recommendations for nomenclature of receptors. In: The IUPHAR compendium of receptor char-acterization and classification. pp31–33, IUPHAR Media, London

Vesci L, Mattera GG, Botarelli M, Tobia P, Corsico N, Martelli EA (1994) Decreased density of endothelin binding sites in adrenal glands but not in aorta, of long term infarcted rats. Life Sci 55:L421–L424

Waggoner WG, Genova SL, Rash VA (1992) Kinetic analyses demonstrate that the equilibrium assumption does not apply to [^{125}I]endothelin-1 binding data. Life Sci 51:1869–1876

Wang X, Douglas SA, Louden C, Vickery-Clark LM, Feuerstein GZ, Ohlstein EH (1996) Expression of endothelin-1, endothelin-3, endothelin-converting enzyme-1, and endothelin-A and endothelin-B receptor mRNA after angioplasty-induced neointimal formation in the rat. Circ Res 78:322–328

Warner TD, Allcock GH, Corder R, Vane JR (1993) Use of the endothelin antagonists BQ-123 and PD 142893 to reveal three endothelin receptors mediating smooth muscle contraction and the release of EDRF. Br J Pharmacol 110:777–782

Warner TD, Allcock GH, Mickley EJ, Vane JR (1993) Characterization of endothelin receptors mediating the effects of the endothelin/sarafotoxin peptides on autonomic neurotransmission in the rat vas deferens and guinea-pig ileum. Br J Pharmacol 110:783–789

Warner TD, Allcock GH, Vane JR (1994) Reversal of established responses to endothelin-1 in vivo and in vitro by the endothelin receptor antagonists, BQ-123 and PD 145065. Br J Pharmacol 112:207–213

Watakabe T, Urade Y, Takai M, Umemura I, Okada T (1992) A reversible radioligand specific for the ET$_B$ receptor: [^{125}I]Tyr13-Suc-[Glu9,Ala11,15]-endothelin-1(8–21), [^{125}I]IRL 1620. Biochem Biophys Res Commun 185:867–873

Wellings RP, Corder R, Warner TD, Cristol JP, Thiemermann C, Vane JR (1994) Evidence from receptor antagonists of an important role for ETB receptor-mediated vasoconstrictor effects of endothelin-1 in the rat kidney. Br J Pharmacol 111:515–520

Wharton J, Rutherford RA, Walsh DA, Mapp PI, Knock GA, Blake DR, Polak JM (1992) Autoradiographic localization and analysis of endothelin-1 binding sites in human synovial tissue. Arthritis Rheum 35:894–899

Widdowson PS, Kirk CN (1996) Characterization of [^{125}I]-endothelin-1 and [^{125}I]-BQ3020 binding to rat cerebellar endothelin receptors. Br J Pharmacol 118:2126–2130

Wilkes BM, Pearl AR, Mento PF, Maita ME, Macica CM, Girardi EP (1991) Glomerular endothelin receptors during initiation and maintenance of ischemic acute renal failure in rats. Am J Physiol 260:F110–F118

Willette RN, Ohlstein EH, Pullen M, Sauermelch CF, Cohen A, Nambi P (1993) Transient forebrain ischemia alters acutely endothelin receptor density and immunoreactivity in gerbil brain. Life Sci 52:35–40

Williams DLJ, Jones KL, Pettibone DJ, Lis EV, Clineschmidt BV (1991) Sarafotoxin S6c: an agonist which distinguishes between endothelin receptor subtypes. Biochem Biophys Res Commun 175:556–561

Winkles JA, Alberts GF, Brogi E, Libby P (1993) Endothelin-1 and endothelin receptor mRNA expression in normal and atherosclerotic human arteries. Biochem Biophys Res Commun 191:1081–1088

Wolff K, Faxen M, Lunell NO, Nisell H, Lindblom B (1996) Endothelin receptor type A and B gene expression in human nonpregnant, term pregnant, and preeclamptic uterus. Am J Obstet Gynecol 175:1295–1300

Wong WH, Wong BP, Wong EF, Huang MH, Wong NL (1998) Downregulation of endothelin B receptors in cardiomyopathic hamsters. Cardiology 89:195–201

Wu-Wong JR, Chiou WJ, Dixon DB, Opgenorth TJ (1995) Dissociation characteristics of endothelin ETA receptor agonists and antagonists. J Cardiovasc Pharmacol 26 [Suppl 3]:S380–S384

Wu-Wong JR, Chiou WJ, Huang ZJ, Vidal MJ, Opgenorth TJ (1994b) Endothelin receptors in human smooth muscle cells: antagonist potency differs on agonist-evoked responses. Am J Physiol 267:C1185-C1195

Wu-Wong JR, Chiou WJ, Naugles KEJ, Opgenorth TJ (1994a) Endothelin receptor antagonists exhibit diminishing potency following incubation with agonist. Life Sci 54:1727–1734

Wu-Wong JR, Chiou WJ, Opgenorth TJ (1993) Phosphoramidon modulates the number of endothelin receptors in cultured Swiss 3T3 fibroblasts. Mol Pharmacol 44:422–429

Yanagisawa H, Hammer RE, Richardson JA, Williams SC, Clouthier DE, Yanagisawa M (1998b) Role of Endothelin-1/Endothelin-A receptor-mediated signaling pathway in the aortic arch patterning in mice. J Clin Invest 102:22–33

Yanagisawa H, Yanagisawa M, Kapur RP, Richardson JA, Williams SC, Clouthier DE, de Wit D, Emoto N, Hammer RE (1998a) Dual genetic pathways of endothelin-mediated intercellular signaling revealed by targeted disruption of endothelin converting enzyme-1 gene. Development 125:825–836

Yanagisawa M, Kurihara H, Kimura S, Tomobe Y, Kobayashi M, Mitsui Y, Yazaki Y, Goto K, Masaki T (1988) A novel potent vasoconstrictor peptide produced by vascular endothelial cells. Nature 332:411–415

Yokokawa K, Kohno M, Minami M, Yasunari K, Mandal AK, Yoshikawa J (1998) Heparin suppresses cyclosporine-induced endothelin-1 synthesis in rat endothelial cells. J Cardiovasc Pharmacol 31 [Suppl 1]:S460–S463

Yoneyama T, Hori M, Tanaka T, Matsuda Y, Karaki H (1995) Endothelin ET_A and ET_B receptors facilitating parasympathetic neurotransmission in the rabbit trachea. J Pharmacol Exp Ther 275:1084–1089

Yorikane R, Miyauchi T, Sakai S, Sakurai T, Yamaguchi I, Sugishita Y, Goto K (1993) Altered expression of ET_B-receptor mRNA in the lung of rats with pulmonary hypertension. J Cardiovasc Pharmacol 22 (Suppl 8):S336–S338

Yoshimura A, Iwasaki S, Inui K, Ideura T, Koshikawa S, Yanagisawa M, Masaki T (1995) Endothelin-1 and endothelin B type receptor are induced in mesangial proliferative nephritis in the rat. Kidney Int 48:1290–1297

Zeng Z, Su K, Kyaw H, Li Y (1997) A novel endothelin receptor type-B-like gene enriched in the brain. Biochem Biophys Res Commun 233:559–567

Zhu G, Wu LH, Mauzy C, Egloff AM, Mirzadegan T, Chung FZ (1992) Replacement of lysine-181 by aspartic acid in the third transmembrane region of endothelin type B receptor reduces its affinity to endothelin peptides and sarafotoxin 6c without affecting G protein coupling. J Cell Biochem 50:159–164

CHAPTER 5

Cell Signaling by Endothelin Peptides

M.S. SIMONSON

A. Introduction and Overview

Endothelin (ET) peptides are surprisingly multifunctional. They control diverse physiological events such as vascular tone, cardiac, pulmonary, and renal function. Endothelins also direct phenotypic functions including cell growth, development, and differentiation. ET receptors control these events by evoking a complex, integrated network of signaling pathways. ET signaling has emerged as an important model for understanding how G protein-coupled receptors function in physiology and pathophysiology, particularly for those signals that regulate the genome.

The chapter reviews ET post-receptor signals linked to short-term cytosolic actions (i.e., contraction) and long-term events requiring regulation of the genome (i.e., cell growth and differentiation). Rather than discussing each cell signaling pathway in isolation, the emphasis is on discussing these signals in the context of specific cell functions (i.e., contraction or growth). This chapter is by necessity brief and selective; the reader is referred to several reviews for detailed information (MIYAUCHI and MASAKI 1999; POLLOCK and HIGHSMITH 1998; YANAGISAWA 1994; SIMONSON 1993; FORCE 1998).

B. Contractile Signaling by ET-1

ET-1 controls contractile elements in the cytosol (i.e., actin-myosin) of vascular smooth muscle, cardiac muscle, and in non-muscle cells such as glomerular mesangial cells and fibroblasts (Table 1). Although specific mechanisms governing actin-myosin interactions are cell type-specific, contraction in all cells requires elevation of cytosolic free [Ca^{2+}]. ET-1 is a potent contractile peptide largely by virtue of its ability to produce robust and sustained increases in cytosolic [Ca^{2+}]. Figure 1 summarizes the signaling network underlying elevation of [Ca^{2+}] in response to ET-1.

I. Activation of Phospholipase C by Endothelin-1

ET receptors initiate Ca^{2+} and contractile signaling by activating phosphotidylinositol-specific phospholipase C (PLC) (MASAKI 1995). ET peptides

Table 1. Cells in which ET-1 stimulates contractile signaling

Cell type	Physiological actions
Vascular smooth muscle	Vascular tone
Cardiac myocytes	Strength and rate of cardiac contractions
Glomerular mesangial cells	Ultrafiltration coefficient and glomerular filtration
Fibroblasts	Wound healing and collagen remodeling

rapidly induce a dose-dependent increase in phosphotidylinositol (PtdIns) turnover in contractile cells including vascular smooth muscle cells (MARSDEN et al. 1989; HIGHSMITH et al. 1989; RESINK et al 1988; REYNOLDS et al. 1989; VANRENTERGHEM et al. 1988; DANTHULARI and BROCK 1998), cardiac myocytes (VIGNE et al 1989; GALRON et al 1989), and glomerular mesangial cells (SIMMMONSON et al. 1989; BADR et al. 1989a,b). PtdIns turnover has also been demonstrated in intact tissues such as aorta (OHLSTEIN et al. 1989) and coronary artery strips (KASUYA et al. 1989). Experiments with stably transfected CHO cells demonstrate that both ET_A and ET_B receptors activate PLC (ARAMORI and NAKANISHI 1992). The pattern of PtdIns metabolites formed after activation by ET depends on the phosphatase/kinase activities of target cells (SIMONSON and DUNN 1990a). ET receptors activate primarily the PLCβ isoform by both Gαq- and G$\beta\gamma$-dependent mechanisms.

Activation of phospholipase C by ET produces at least two second messengers (Fig. 1): (i) the water soluble Ins(1,4,5)P_3, which diffuses to specific receptors on specialized compartments of the endoplasmic reticulum to release intracellular Ca^{2+}; and (ii) the neutral 1,2-diacylglycerol, which remains in the plasma membrane and with cofactors Ca^{2+} and phosphatidylserine activates protein kinase C (PKC). 1,2-Diacylglycerol can also be formed by the phospholipase C-mediated hydrolysis of other phospholipids such as phosphatidylcholine and phosphatidylethanolamine (RESINK et al. 1990a; MACNULTY et al. 1990), but the exact role of this putative signaling pathway remains unclear. ET-1 contractile signaling, as discussed below, is initiated primarily by elevation of cytosolic Ca^{2+}, but the intensity of the contractile response is modified by PKC.

II. Ca^{2+} Signaling and ET-1-Induced Contraction

Ca^{2+} signaling is a nearly universal response to ET receptor activation (POLLOCK et al. 1995). Experiments in many laboratories have shown that ET-induced Ca^{2+} signaling is surprisingly complex and utilizes multiple mechanisms in concert to produce an integrated Ca^{2+} signal. For example, in glomerular mesangial cells loaded with the fluorescent Ca^{2+} indicator, fura-2, ET isopeptides and sarafotoxin S6b evoke a biphasic increase in intracellular free $[Ca^{2+}]$ consisting of a rapid (2–5 s) spike increase followed by a lesser but sustained phase (SIMONSON et al. 1989, 1990; SIMONSON and DUNN 1991). Lower doses of ET-1 only stimulate a monophasic, sustained increase in $[Ca^{2+}]$, con-

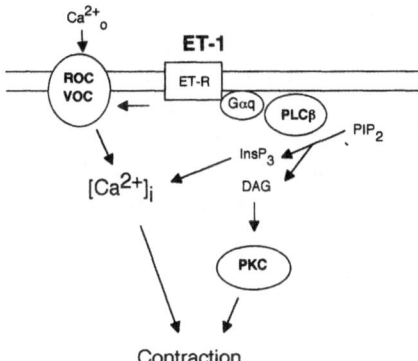

Contraction

Fig. 1. Summary of the major post-receptor signals by which ET-1 stimulates contraction in muscle and non-muscle cells. ET-1 binds to a cognate G protein-coupled receptor and activates phospholipase $C\beta$ (PLCβ) mainly through the Gαq GTP binding protein. PLC catalyzes formation of two soluble mediators: Ins(1,4,5)P$_3$ (InsP$_3$) which diffuses to specific receptors on specialized compartments of the endoplasmic reticulum to release intracellular Ca^{2+}; and 1,2-diacylglycerol (DAG), which activates protein kinase C (PKC). Cytosolic free [Ca^{2+}] is also elevated by influx of extracellular Ca^{2+} through both receptor-operated or voltage-operated Ca^{2+} channels. As described in the text, both Ca^{2+} and PKC regulate ET-1-induced contraction. Note that the specific pathway involved can vary depending on the cell type and the specific contractile response (i.e., vascular tone vs. cardiac contractility)

sistent with different Ca^{2+} signaling pathways coupled to distinct ET receptor isoforms (SIMONSON et al. 1989; SIMONSON and DUNN 1991). Ca^{2+} signaling induced by ET-1 requires conversion of big ET-1 to ET-1 (SIMONSON and DUNN 1991), suggesting that mature ET-1 is constrained in an inactive conformation within the proET-1 molecule. The duration of the sustained phase of Ca^{2+} is especially pronounced with ET-1 and ET-2 (SIMONSON and DUNN 1991; SIMONSON et al. 1990), which might contribute to the prolonged contractile activity of these ET isoforms. The peak increase in Ca^{2+} is dose-dependent with ET-2=ET-1>S6b>ET-3 (SIMONSON and DUNN 1991; SIMONSON et al. 1990). Thus, ET isopeptides and sarafotoxins evoke similar elevations of Ca^{2+} but with dissimilar potencies and kinetics. Two distinct but interconnected mechanisms contribute to ET-1 induced increases in Ca^{2+}. ET peptides release Ca^{2+} from intracellular stores, presumably by the action of Ins(1,4,5)P3 to gate a Ca^{2+} channel on specialized compartments of the endoplasmic reticulum (BERRIDGE 1997). More importantly, as discussed below, ET increases Ca^{2+} influx from the extracellular space by activating multiple types of Ca^{2+} ion channels in the plasma membrane.

III. Ca^{2+} Channels and ET-1

Ca^{2+} influx occurs through two main classes of Ca^{2+} channels (BERRIDGE 1997). Voltage-operated calcium channels (VOC) are gated principally by depolarization of membrane potential, but Ca^{2+} conductance can be modulated by receptors via diffusible second messenger systems or via G proteins. Receptor-operated calcium channels (ROG) are a diverse set of channel proteins ranging from Ca^{2+}-selective voltage-insensitive channels to non-selective cation channels capable of conducting Ca^{2+} ions. Although ROC remain poorly characterized, they appear to be regulated by ligands, second messenger/ protein kinase systems, or by Ca^{2+} itself. Current evidence demonstrates that ET peptides activate both VOC and ROC to increase Ca^{2+}. Indirect evidence of ROC activation by ET comes from the observations that vasoconstriction or ET-mediated increases in Ca^{2+} are insensitive to dihydropyridine or phenylalkylamine Ca^{2+} channel blockers (OHLSTEIN et al. 1989; D'ORLEANS-JUSTE et al. 1989; EDWARDS et al. 1990; MITSUHASHI et al. 1989; KORBMACHER et al. 1989; SIMPSON and ASHLEY 1989; AUGUET et al. 1988; BORIC et al. 1990; KIOWSKI et al. 1991; WILKES et al. 1991). ET-1-induced Ca^{2+} signaling is also blocked by a selective inhibitor of ROC (CHAN and GREENBURG 1991). Further evidence for the involvement of ROC comes from studies utilizing Mn^{2+} as a probe of Ca^{2+}-permeable channels in mesangial cells (SIMONSON and DUNN 1991), medullary interstitial cells (WILKES et al. 1991), and A10 vascular smooth muscle cells (SIMPSON et al. 1990).

Recent studies in cultured vascular smooth muscle cells demonstrate that ET-1 activates at least two types of Ca^{2+}-permeable, non-selective cation channels (IWAMURO et al. 1998). Interestingly, the channels were differentially activated by high (>10 nmol/l) or low (<1 nmol/l) concentrations of ET-1. Additional studies in intact aortic rings show that Ca^{2+} entry through these ROC is required for smooth muscle contraction (ZHANG et al. 1998). These studies also demonstrated that inhibition by nitric oxide of ET-1-induced contraction resulted from inhibition of Ca^{2+} entry through these ROC Ca^{2+} channels (ZHANG et al. 1998). The mechanism by which nitric oxide blocks ROC activation is presently unknown, but these studies amplify the importance of these Ca^{2+} channels in the physiological actions of ET-1 and control of vascular tone.

It became clear early on, however, that VOC can also contribute to ET-induced Ca^{2+} signaling (Fig. 1). For instance, in vascular smooth muscle vasoconstriction and Ca^{2+} signaling are reduced by dihydropyridine channel blockers (HIGHSMITH et al. 1989; VANRENTERGHEM et al. 1988; VIGNE et al. 1989; EDWARDS et al. 1990; SIMPSON and ASHLEY 1989; YANAGISAWA et al. 1988; GOTO et al. 1989; MADEDDU et al. 1990; MAHER et al. 1991). In humans nifedipine not only blocks ET-induced vasoconstriction in the forearm but unmasks the vasodilatory action of ET (KIOWSKI et al. 1991). But these results were puzzling as ET clearly failed to bind directly to L-type VOC (see above). Direct measurements of Ca^{2+} conductance by patch-clamp techniques reveal that ET-1 indeed increases VOC activity but by an indirect, modulatory mechanism (GOTO et al. 1989; SILBERBERG et al. 1989; INOUE et al. 1990). In addition, ET-1 does not activate a partially purified, reconstituted VOC from porcine cardiac muscle (NAITOH et al. 1990). Although the molecular mechanism by which ET activates VOC remains unclear, in patch clamp experiments the ability of bath-applied ET-1 in the cell attached mode to activate VOC strongly argues that ET acts via a diffusible second messenger system (SILBERBERG et al. 1989; INOUE et al. 1990). The ability of ET to modulate VOC activity has important consequences for electrically excitable cells in the vasculature and nervous system (MASAKI et al. 1990). For instance, the ability of ET to increase VOC potentiates excitation-contraction coupling in coronary artery strips when the membrane is depolarized by high KCl_o (GOTO et al. 1989).

IV. Unresolved Issues in ET-1 Ca^{2+} Signaling

There are still many issues regarding ET-induced Ca^{2+} signaling that remain unresolved. For example, there is preliminary evidence that ET affects Ca^{2+} signaling through mechanisms regulated by membrane depolarization but distinct from dihydropyridine-sensitive, L-type VSCC (RASMUSSEN and PRINTZ 1989; BLACKBURN and HIGHSMITH 1990). There are also observations suggesting that the major role of ET is to increase the sensitivity of pharmacomechanical coupling to Ca^{2+} (MARSAULT et al. 1990; REMBOLD 1990). It is also possible that ET opens a non-selective cation channel in vascular smooth muscle (VANRENTERGHEM et al. 1988), which in turn gates a VSCC similar to arginine vasopressin in A7r5 cells (VANRENTERGHEM et al. 1989). Taken together, these data suggest that multiple, interacting pathways of Ca^{2+} influx contribute to ET-induced vasoconstriction. Moreover, the coexistence or tissue-specific expression of dihydropyridine-sensitive and -insensitive pathways of Ca^{2+} influx probably explains the divergent effects of Ca^{2+} channel blockers on ET-induced vasoconstriction. In human mesangial cells and vascular smooth muscle cells, ET also stimulates periodic oscillations of $[Ca^{2+}]$ following the spike increase (SIMONSON et al. 1990; SIMPSON and ASHLEY 1989). These oscillations are especially interesting in that they represent synchronized Ca^{2+} signaling in a population of cells as opposed to the more commonly observed oscillations of Ca^{2+} at a single cell level (BERRIDGE and IRVINE 1990). Similar Ca^{2+} oscillations have been observed in monolayers of ciliated tracheal epithelium (SANDERSON et al. 1988) and cardiac trabeculae (MULDER et al. 1989), and it seems likely that Ca^{2+} oscillations by ET recruit a local population of cells to function as a syncitium (SIMONSON et al. 1990). It is also important to note that at least three mechanisms have been proposed to down-regulate ET-induced Ca^{2+} signaling including protein kinase C-mediated attenuation of Ca^{2+} signaling, protein-kinase C-independent desensitization by ET ligands, and an ET-dependent Ca^{2+} efflux pathway (SIMONSON and DUNN 1991).

V. PKC and ET-1-Induced Contraction

Although a role for protein kinase C in ET-induced biological events remains unclear, accumulating evidence suggests that protein kinase C mediates both short- and long-term events. In cultured cells ET rapidly stimulates a dose-dependent, sometimes biphasic increase in 1,2 diacylglycerol that remains elevated for 20 min or longer (DANTHULURI and BROCK 1990; RESINK et al. 1990a; GRIENDLING et al. 1989; LEE et al. 1989; MULDOON et al. 1989, 1990; SUNAKO et al. 1990). Phosphorylation of an 80-kDa protein kinase C target increases rapidly after addition of ET to vascular smooth muscle cells, and down-modulation of protein kinase C with phorbol esters attenuates ET-induced phosphorylation (GRIENDLING et al. 1989). In addition, fractionation of a Ca^{2+}/phospholipid-dependent histone phosphorylation activity is consistent with translocation of protein kinase C from the cyotsol to plasma membrane (LEE et al. 1989). Activation of protein kinase C serves as an important negative feedback signal decreasing phospholipase C activity (MULDOON et al. 1990; RESINK et al. 1990b) and Ca^{2+} signaling by ET peptides (SIMONSON and DUNN 1991; SIMONSON et al. 1990; MULDOON et al. 1990). Protein kinase C activation also appears to mediate, in part, the tonic contraction of vascular smooth muscle cells to ET (DANTHULURI and BROCK 1990; SUGIURA et al. 1989; KODAMA et al. 1989). These observations are consistent with the view that protein kinase C mediates tonic vasoconstriction in the intact vasculature (BERK et al. 1988; RASMUSSEN et al. 1987). The role of protein kinase C in ET-mediated renal vasoconstriction remains unclear. Indirect evidence also implicates protein kinase C in the promitogenic response to ET peptides (SIMONSON and DUNN 1990a; YANAGISAWA and MASAKI 1989; VIGNE et al. 1990), and down-modulation of protein kinase C in Rat-1 fibroblasts nearly abolishes the mitogenic response to ET (MULDOON et al. 1990).

C. Genomic Signaling by ET: Cell Growth and Development

In contrast to cytosolic signals that regulate contraction, ET-1 post-receptor signals also control differential expression of genes. The first evidence that ET regulates gene expression came from the unexpected discovery that ET-1 is a mitogen for vascular smooth muscle and mesangial cells in culture (SIMONSON et al. 1989; BADR et al. 1989a; KOMURO et al. 1988). In most cells the increase in DNA synthesis precedes an increase in cell number (i.e., hyperplasia), but in some cells a hypertrophic response ensues (see BATTISTINI et al. 1993 for review). Although first demonstrated in mesangial and vascular smooth muscle cells, it is now clear that the mitogenic effects of ET-1 are widespread and occur in fibroblasts, epithelial cells, endothelial cells, osteoclasts, and subpopulations of glial cells (BATTISTINI et al. 1993). ET-1-induced proliferation is regulated by commonly used cell cycle genes including cyclin D1 and the retinoblastoma gene product (TERADA et al. 1998).

Another important area of genomic signaling by ET-1 involves normal embryonic development of neural crest cell lineage. Studies with targeted deletions of ET system genes in mice demonstrate that ET-1/ET_A signals direct the complex process of craniofacial and aortic arch patterning altering differentiation of postmigratory cardiac neural crest cells (see Chap. 6). Gene targeting of the ET-3/ET$_B$ axis demonstrates of role for these signals in differentiation of epidermal melanocytes and enteric neurons in the distal gut. As discussed below, relatively little is known about the ET post-receptor signals regulating these events.

I. Mitogenic Signaling by ET

Accumulating evidence suggests that ET-1 is an important mitogen in vivo, and the major pathways involved in ET-1 mitogenic signaling are illustrated in Fig. 2. Elevated secretion of ET-1 is a hallmark of fibroproliferative cardiovascular, pulmonary, and renal diseases such as atherosclerosis, glomerulosclerosis, coronary artery restenosis, transplantation-associated vascular sclerosis, and heart failure (SIMONSON 1993; SAKAI et al. 1996; KOHAN 1997) (see Chaps. 11–19). A functional role in these disorders is demonstrated by the ability of ET receptor antagonists to block smooth muscle cell proliferation and neointima formation in restenosis and atherosclerosis; mesangial cell proliferation in glomerulonephritis; and myocardial hypertrophy and fibrosis in cardiac failure (SAKAI et al. 1996; BENNIGNI et al. 1993, 1996; DOUGLAS et al. 1994; LERMAN et al. 1991; GIAD et al. 1993; FERRER et al. 1995). A unique transgenic rat model has been established in which renal ET-2 expression is greatly

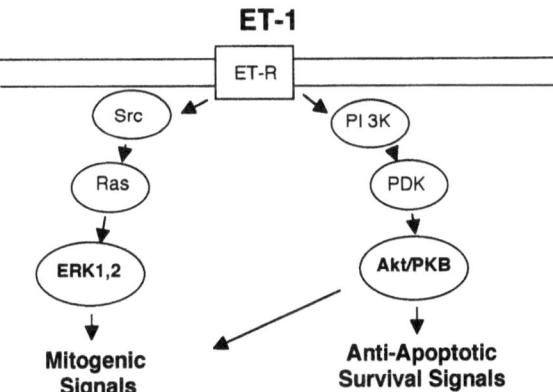

Fig. 2. Pathways of mitogenic signaling by ET-1. Binding of ET-1 to either ET$_A$ or ET$_B$ receptors stimulates at least two major signaling pathways that control cell growth. Activation of Src and Ras leads to stimulation of ERK1,2, which then activates transcription of genes (i.e., immediate early genes) required for cell cycle progression. In the other pathway, activation of phosphatidylinositol 3-kinase (PI 3K) initiates a pathway involving PDK and the Akt/PKB kinase. The PI 3K pathway in other cells is linked to both anti-apoptotic, survival events and to induction of cell cycle genes. However, events downstream of PI 3K have not been characterized in ET-1-treated cells

elevated in glomeruli (HOCHER et al. 1996a). In these rats glomeruli develop extensive glomerulosclerosis with mesangial matrix expansion and apparent increases in mesangial cell proliferation. Transgenic mice that overexpress ET-1 develop pronounced glomerulosclerosis and interstitial fibrosis (HOCHER et al. 1996b). The experiments summarized above have been performed so far only in animal models, but every indication suggests that ET-1 plays a similar role in humans.

In contrast to attention paid to growth stimulation by ET-1, relatively little research has focused on feedback mechanisms that counterbalance ET-1-induced mitogenesis in vitro or in vivo. ET-1 activates atrial natriuretic peptide (ANP) release from cardiac myocytes, and these peptides potently antagonize ET-1-induced contraction (FUKADA et al. 1988; HU et al. 1988; SANDOK et al. 1992). ANP also inhibits ET-1-stimulated growth of vascular smooth muscle cells and glomerular mesangial cells in culture (SIMONSON 1993; NEUSER et al. 1990), but it is not yet clear how or whether release of ANP by cardiac myocytes would antagonize ET-1-stimulated vascular cell growth in peripheral vessels. Nitric oxide (NO) is another candidate for a physiologically relevant counterbalance to ET-1-stimulated vascular cell growth. ET-1 potently activates NO release by endothelial cells, which in turn antagonizes ET-1-induced vasoconstriction in vivo and cell growth in culture (SIMONSON 1993) (see Chap. 11). cGMP is most likely the common mediator of the antimitogenic actions of both ANP and NO. However, given the short half-life of NO, it is uncertain whether NO would effectively regulate ET-1-induced mitogenesis in vivo.

Retinoic acid is another candidate inhibitor of ET-1-induced cell growth in vivo. In cultured glomerular mesangial cells, retinoic acid completely inhibits [^{3}H]thymidine uptake in response to ET-1 (SIMONSON 1994), and in cardiac myocytes retinoic acid inhibits hypertrophy and changes in gene expression associated with ET-1 expression and cardiac failure (ZHOU et al. 1995; WU et al. 1996). It is not yet clear how retinoic acid blocks ET-1-stimulated cell growth, but retinoic acid does block activation of the AP-1 transcription factor by ET-1 (SIMONSON 1994). AP-1 is activated by most growth factors, and AP-1 cis-elements are found in several genes associated with cell cycle progression.

Given the mitogenic actions of ET-1, it is perhaps not surprising that ET-1 has also been implicated in oncogenesis and cancer. Anchorage-independent growth of cells in culture is a hallmark of the transformed phenotype. Early studies revealed that ET-1 potentiates anchorage-independent growth of Rat-1 fibroblasts and NRK 49F cells in the presence of epidermal growth factor (MULDOON et al. 1990; KUSUHARA et al. 1992). The ability of ET-1 to increase anchorage-independent growth and mitogenesis requires active protein kinase C.

Several studies implicate ET-1 in the development of human tumors. Cells derived from human tumors – particularly tumors of the breast, pancreas, colon, and prostate – secrete abundant amounts of ET-1 (KUSUHARA et al.

1990; Kar et al. 1995; Nelson et al. 1995). Circulating levels of ET-1 are elevated in patients with hepatocellular carcinoma and metastatic prostate cancer (Kar et al. 1995; Nelson et al. 1995). ET-1 derives not only from the extensive tumor vasculature but also from transformed cells themselves (Kar et al. 1995; Nelson et al. 1995). At least two roles have been envisioned for ET-1 in the growth and survival of neoplasms. First, ET-1 is a mitogen for several cancer cell lines and might increase tumor growth in an autocrine fashion or stromal cell growth by a paracrine mechanism. Second, ET-1 might also participate in the osteoblastic response of bone to metastatic prostate and other cancers (Nelson et al. 1995).

Collectively, the experiments summarized above suggest that ET-1 plays an important role in adaptive proliferative responses in cardiovascular and renal disease, and in cancer. Investigation of the signals by which ET-1 evokes the G_o to G_1 cell cycle transition has become the major model to study how ET-1 regulates gene expression. Several signaling cascades – including mitogen-activated protein kinases (MAPK), non-receptor protein tyrosine kinases (PTK), and phosphatidylinositol-kinase (PI 3K) – have been implicated in mitogenic signaling by ET-1 and are discussed in detail later in this chapter. ET-1 serves as an important model for mitogenic signaling by other growth factors that bind to the general class of G protein-coupled receptors (Van Biesen et al. 1996).

1. ET-1 and MAPK Cascades

The mitogen-activated protein kinase cascades (MAPK) play a pivotal role in mitogenic and non-mitogenic signaling by ET-1 (Force 1998). Mammalian cells have at least five conserved MAPK signaling cascades that convert short-term, transmembrane signals into long-term phenotypic events requiring transcription (Lewis et al. 1998). MAPK pathways include the mitogenic ERK1/2 cascade, the stress-activated JNK and p38 MAPK cascades, and two poorly understood pathways, the ERK3 and ERK 5 cascades. MAPK pathways are typically organized into a "three kinase architecture" consisting of an MAPK, an MAPK activator (MAPK kinase, i.e., MEK), and an activator of the MAPK kinase (MAPK kinase kinase, i.e., Raf). Signals are transmitted by sequential phosphorylation and activation of components specific to an MAPK cascade, for example, Raf activating MEK, which in turn activates ERK1/2 (Lewis et al. 1998).

ET-1 is a potent and sustained activator of ERK1/2 (Fig. 2). ERK1/2 is activated by ET-1 in numerous cell types including glomerular mesangial cells (Wang et al. 1992), cardiac myocytes (Bogoyevitch et al. 1994; Fuller et al. 1997), and smooth muscle cells (Whelchel et al. 1997). ET-1 induces rapid activation of ERK1/2 phosphorylation; ERK1/2 remains activated for at least 15–20 min. Activation of ERK1/2 by ET-1 is not inhibited by pertussis toxin and most likely occurs through $G\alpha q$ (Wang et al. 1992). Both ET_A and ET_B receptors can activate the ERK1/2 cascade (Wang et al. 1994), and the mech-

anisms by which ET receptors induce MAPK have recently been reviewed
(FORCE 1998).

A functional role for ERK1/2 in ET-1-induced mitogenesis has recently
been demonstrated using MKP-1 (MAPK phosphatase-1), which dephospho-
rylates and inactivates ERK1/2 (LEWIS et al. 1998). When ectopically expressed
in cardiac myocytes, MKP-1 attenuated ET-1-induced hypertrophy and transcriptional activation of genes linked to hypertrophy (FULLER et al. 1997). A
catalytically inactive MKP-1 mutant failed to block ET-1-induced hypertrophy. Further evidence for ERK1/2 in ET-1 mitogenic signaling comes from
experiments with an inhibitor of the MAPK activator MEK. PD98059 blocks
ERK1/2 in airway smooth muscle cells and also inhibits ET-1-induced proliferation (WHELCHEL et al. 1997). Similar studies in cardiac myocytes show that
PD98059 blocks ET-1-stimulated hypertrophy. Taken together, these experiments provide strong evidence that ERK1/2 link ET-1 receptors to transcriptional activation of gene involved in cell cycle regulation.

Accumulating evidence also suggests a role for JNK and p 38 MAPKs in
regulation of cell growth by ET-1. ET-1 is unusual in that it activates both JNK
(CADWALLER et al. 1997; ISONO 1998; ARAKI et al. 1997; CHOUKROUN et al. 1998)
and p38 (CLERK et al. 1998) in addition to ERK1/2. Force and coworkers have
conclusively demonstrated a role for JNK in ET-1-induced cardiac myocyte
hypertrophy (CHOUKROUN et al. 1998). Dominant negative mutants of the JNK
activator protein specifically blocked ET-1 induced hypertrophy without
blocking activation of ERK1/2 or p 38 by ET-1. Thus multiple MAPK pathways might contribute to ET-1-induced cell growth in these cells. P38 MAPK
is not linked to the immediate induction of hypertrophy in myocyctes but
might be necessary over a longer period to maintain the hypertrophic response
(CLERK et al. 1998).

2. ET and Non-Receptor Protein Tyrosine Kinases in Mitogenic Signaling

When our laboratory and others began investigating transduction of signals
from ET-1 receptors to the nucleus, one question stood out: do ET-1 receptors
activate PTKs, and if so how do these kinases propagate signals to the nucleus?
This question was intriguing because at the time it was well-established that
most receptors associated with phenotypic control (i.e., growth and/or differentiation) required PTK activity (CANTLEY et al. 1991). Receptors for platelet-
derived growth factor and related growth factors have intrinsic PTK activity
that phosphorylates specific tyrosine residues on downstream effectors
required for mitogenic signaling (i.e., phosphatidyl inositol-3-kinase, ras GAP,
phospholipase Cγ, and Raf-1). Some of these effectors are themselves PTKs
that set in motion a complicated cascade of PTK signaling. Even cytokine
receptors, which lack intrinsic PTK activity, were known to recruit non-
receptor PTKs such as JAK (Janus kinases) or c-Src to propagate signals to
the nucleus. Although two early studies suggested that G protein-coupled
receptors might somehow activate PTKs (HUCKLE et al. 1992; NASMITH et al.

1989), is was generally unclear whether G protein-coupled receptors communicated with PTKs.

The first suggestion that ET-1 activates PTKs was the finding that ET-1 elevates tyrosine phosphorylation (P-Tyr) of cellular proteins. When added to quiescent cells, ET-1 increases P-Tyr of several proteins in whole cell lysates of mesangial cells, fibroblasts, and vascular smooth muscle cells (FORCE et al. 1991; SIMONSON and HERMAN 1993; ZACHARY et al. 1991a, 1992; WEBER et al. 1994; SIMONSON et al. 1996a). The increase in P-Tyr is apparently biphasic: between 5–10 min after adding ET-1 two to three proteins demonstrate increased P-Tyr, whereas by 20 min six to eight additional proteins show elevated P-Tyr (SIMONSON et al. 1996a). The P-Tyr proteins increased by ET-1 show a surprising degree of cell specificity, although proteins around pp60 kDa, pp90, pp125, and pp225 are commonly observed. In mesangial cells, the biphasic time course of ET-1-stimulated P-Tyr accumulation contrasts with the rapid and transient induction of P-Tyr by platelet-derived growth factor (SIMONSON et al. 1996a), suggesting that different mechanisms are involved. It is important to note that ET-1 stimulates accumulation of P-Tyr proteins with molecular mass similar to P-Tyr proteins that accumulate after treatment with platelet-derived growth factor and epidermal growth factor (FORCE et al. 1991; SIMONSON et al. 1996a), whose receptors possess intrinsic PTK activity. These results imply that ET-1 receptors target many of the same P-Tyr downstream effectors utilized by receptor PTKs.

Focal adhesions are intracellular, oligomeric complexes of proteins that form when integrin receptors link components of the extracellular matrix to cytoskeletal elements. Focal adhesions are a particularly rich source of P-Tyr proteins that regulate cell cycle, and P-Tyr of focal adhesion proteins is commonly observed following addition of growth factors that bind to receptor PTKs (RICHARDSON and PARSONS 1995). Immunocytochemical analysis of P-Tyr proteins in quiescent mesangial cells reveals a marked increase in focal adhesion by P-Tyr proteins in cells treated with mitogenic concentrations of ET-1 (SIMONSON et al. 1996a). Consistent with activation of focal adhesion proteins by growth factors, ET-1 also increases actin stress fiber formation (SIMONSON and DUNN 1990b). ET-1 stimulates P-Tyr of the focal adhesion-associated protein paxillin (ZACHARY et al. 1993). Paxillin is a 68-kDa protein that binds to the actin-capping protein vinculin in focal adhesions, and it is a major P-Tyr protein in cells transformed with v-src or subjected to integrin activation (RICHARDSON and PARSONS 1995). These results suggest that P-Tyr of paxillin might play an important role in cell cycle control by ET-1. Other focal adhesion-associated proteins that are P-Tyr in ET-1-treated cells have yet to be identified.

ET-1 also elevates P-Tyr in cytoplasmic proteins (SIMONSON et al. 1996a). With the exception of c-Src and focal adhesion kinase (FAK) (see below), the specific cytosolic proteins that display elevated P-Tyr following ET-1 are unknown. This is an important gap in our current knowledge of mitogenic signaling by ET-1.

P-Tyr can occur by increasing PTK activity or by reducing PTPase activity. It is also possible that growth factors increase both PTK and PTPase activity resulting in differential P-Tyr of specific substrates. To determine if ET-1-stimulated P-Tyr results from PTK activity, PTKs have been immunoprecipitated in ET-1-treated cells using anti-P-Tyr antibodies (i.e., almost all PTKs are themselves P-Tyr proteins). Using this technique, it was found that ET-1 does indeed increase PTK activity (ZACHARY et al. 1991a,b; SIMONSON et al. 1996a). The time course of activation was biphasic and paralleled accumulation of P-Tyr proteins observed in concurrent experiments (SIMONSON et al. 1996a). The important question of whether ET-1 also increases PTPase activity has not yet been addressed. However, as discussed in the next section, we have made some progress identifying specific PTKs activated by ET-1.

To identify specific PTKs activated by ET-1 receptors, immunoprecipitation/in vitro kinase assays were used with antisera for different families of PTKs. Most PTKs can be grouped into one of three families: receptor PTKs in the plasma membrane (e.g., platelet-derived growth factor receptor), non-receptor PTKs attached to the plasma membrane (e.g., c-Src, c-Yes), and diffusable PTKs (e.g., Janus kinases, FAK). Attention initially focused on non-receptor PTKs attached to the plasma membrane (i.e., c-Src) because of their close proximity to ET-1 receptors.

The first evidence that ET-1 activates a specific non-receptor PTK was the finding that ET-1 increases pp60 c-Src activity in quiescent mesangial cells (SIMONSON and HERMAN 1993; FORCE and BONVENTRE 1992). Immunoprecipitation/in vitro kinase assays with anti-v-Src antibodies revealed rapid activation of c-Src autophosphorylation and PTK activity by ET-1. Mesangial cells also express the Src family member c-Yes, but c-Yes activity is unaffected by ET-1, suggesting that activation of Src-family kinases is tightly regulated (M.S. Simonson et al., unpublished; SIMONSON and HERMAN 1993). Autophosphorylation of c-Src, which typically accompanies c-Src activation, was rapid and transient whereas Src PTK activity was sustained (SIMONSON and HERMAN 1993). The dose-response curve for ET-1 stimulated c-Src activity was identical to that for ET-1-stimulated mitogenesis. Depending on the cell type, both ET_A (SIMONSON and HERMAN 1993) and ET_B (M.S. Simonson and W.H. Herman, unpublished results) receptors stimulate c-Src activity. These experiments were among the first to demonstrate cross-talk between a G protein-coupled receptor and a non-receptor PTK.

Relatively little detailed information is known about signaling mechanisms that link ET-1 receptors to c-Src. Increased c-Src activity in ET-1-stimulated cells requires Ca^{2+} influx but is apparently independent of protein kinase C. Activation of protein kinase C by phorbol ester increases P-Tyr accumulation, suggesting that protein kinase C-dependent pathways can activate PTKs. However, inhibition or depletion of protein kinase C has no effect on P-Tyr accumulation in ET-1-treated mesangial cells (FORCE et al. 1991; SIMONSON and HERMAN 1993). In contrast, the Ca^{2+} ionophore, ionomycin, mimics ET-1-stimulated PTK activity, which is also inhibited by chelation of extracellular Ca^{2+} influx (SIMONSON et al. 1996a). Similar dependence on Ca^{2+} influx is also observed for ET-1-stimulated c-Src activation in mesangial cells (SIMONSON et al. 1996a). Ca^{2+}-dependent activation of c-Src in mesangial cells is reminiscent of Ca^{2+}-activated c-Src in a genetic program of differentiation in keratinocytes (ZHAO et al. 1992). Although the available evidence point to Ca^{2+}-dependent activation of c-Src by ET-1 receptors, the mechanisms and Ca^{2+}-dependent effectors involved have been difficult to identify.

Another PTK activated by ET receptors is FAK (ZACHARY et al. 1992; HANEDA et al. 1995), a p125 kDa cytosolic PTK that forms stable complexes with c-Src and is a major P-Tyr protein in v-src transformed fibroblasts. It is unknown whether ET-1-activated c-Src is responsible for activation of FAK; in addition, unlike activation of c-Src by ET-1, activation of FAK requires protein kinase C (HANEDA et al. 1995). The role of FAK in mitogenic signaling has not been characterized in detail, but recent experiments with a dominant negative FAK mutant suggest that FAK is required for anchorage-dependent growth of fibroblasts (RICHARDSON and PARSONS 1996). Given the potential importance of FAK in cell cycle control, FAK is an attractive target PTK activated by ET-1.

Simply demonstrating that ET-1 receptors activate PTKs is not evidence that the PTKs function in nuclear signaling by ET-1. Relatively selective PTK inhibitors (i.e., genistein, herbimycin A) block ET-1-induced c-fos mRNA induction and mitogenesis in mesangial cells (SIMONSON and HERMAN 1993). These results are consistent with a role for PTKs in ET-1 nuclear signaling, but these inhibitors block a variety of PTKs and might also have non-specific effects that are difficult to detect. Thus the results with PTK inhibitors need to be corroborated by techniques that are more specific for particular families of PTKs.

To inhibit specifically c-Src activity, we used a dominant negative c-Src mutant (Src K–) in which the conserved lysine residue (Lys 295) in the ATP-binding kinase domain is mutated to methionine (TWAMLEY-STEIN et al. 1993; ROCHE et al. 1995). In Src K–, the domains that link c-Src to upstream and downstream effectors, the SH2 and SH3 domains, are wild type whereas the domain responsible for phosphorylating effectors, the kinase domain, is inactive. Thus Src K– can interact with proteins that contribute to Src signaling but cannot transduce signals by phosphorylation. Overexpression of Src K– can in theory titrate the action of endogenous c-Src and function as a dominant negative mutant (HERSKOWITZ 1987). Indeed Courtneidge and coworkers (TWAMLEY-STEIN et al. 1993; ROCHE et al. 1995) have shown that Src K– blocks gene expression and cell growth induced by platelet-derived growth factor, which convincingly establishes a role for c-Src in some pathways of nuclear and mitogenic signaling.

Co-transfection of mesangial cells with a plasmid expressing Src K– blocked activation of the c-fos promoter by ET-1 whereas a plasmid expressing wild type Src K+ had no effect (SIMONSON et al. 1996b). By using point mutants of the c-fos promoter we determined that ET-1 requires at least two *cis*-elements to activate the promoter: the serum response element and the Ca²⁺/cAMP response element (SIMONSON et al. 1996b). The ET-1–c-Src signaling pathway apparently targets the serum response element but not the Ca²⁺/cAMP response element, suggesting a divergence of the pathway upstream of Src (SIMONSON et al. 1996b).

To obtain independent evidence that c-Src propagates an ET-1 signal to the nucleus, we transfected mesangial cells with a plasmid expressing COOH-terminal Src kinase (Csk), which phosphorylated the COOH-terminal tyrosine in Src (i.e., Tyr 527) and repressed Src activity (SABE et al. 1992). Gene targeting studies with Csk⁻/⁻ mice recently confirmed that Csk negatively regulates c-Src in vivo (IMAMOTO and SORIANO 1993). We found that overexpression of Csk also blocked activation of the *c-fos* promoter by ET-1 (SIMONSON et al. 1996b), consistent with a role for c-Src in ET-1 nuclear signaling. One caveat in interpreting the results with Src K– and Csk is that both proteins probably also inhibit closely related members of the Src family of PTKs such as Yes and Fyn. Thus it is difficult to rule out formally involvement of other Src-related kinases in ET-1 nuclear signaling. However, we have been unable to demonstrate activation by ET-1 of other Src family kinases expressed in mesangial cells (i.e., c-Yes). Taken together, the results with Src K– and Csk support a role for c-Src in induction of the *c-fos* immediate early gene by ET-1. The results also suggest a wider role for non-receptor PTKs in nuclear signaling by G protein-coupled receptors. It is important to note that a similar approach was recently used to support a role for Src in endothelin-induced hypertrophy in cardiac myocytes (KOVACIC et al. 1998). Thus ET-1-Src interactions are important in a variety of cell systems in which ET-1 differentially regulates gene expression.

We are now in a position to suggest some working models of how Src and other PTKs function in ET-1 signaling. First, ET-1-activated Src apparently sends signals to Ras, which in turn transmits signals to a variety of Ras-dependent effectors of gene expression such as the Raf-1/MAPK cascade. Second, accumulating evidence suggests that Src potentiates activation of the phosphoinositide (PI) cascade by ET-1, which by virtue of elevating intracellular free [Ca²⁺]ᵢ and activating protein kinase C induces new patterns of gene expression.

In addition to activating c-Src, ET-1 also activates p21 Ras in mesangial cells, and expression of a dominant negative Ras mutant (Asn17 c-Ha-Ras) blocks activation of the *c-fos* promoter by ET-1 (HERMAN and SIMONSON 1995). c-Src acts upstream of c-Ras in the same ET-1 signaling pathway as dominant negative Ras blocks Src activation of the *c-fos* promoter but dominant negative Src mutants fail to block activation by Ras (SIMONSON et al. 1996b; HERMAN and SIMONSON 1995) (Fig. 1). Similar to systems in which Src functions upstream of Ras, ET-1-activated Src probably stimulates Ras by promoting P-Tyr and Grb2 association of the Shc adapter protein (CAZAUBON et al. 1994). The ability of ET-1 to activate Ras undoubtedly accounts for the potent stimulation of p42,44 MAPK by ET-1 (WANG et al. 1992). Activation of a Src-Ras signaling cassette by ET-1 probably contributes to induction of genes in addition to *c-fos*.

Compelling evidence also points to a role for Src in the unusually extended activation of the PI cascade by ET-1. The PI cascade and its two major effectors, $[Ca^{2+}]_i$ and protein kinase C, are critical for ET-1 signal transduction. The ET-1-stimulated PI cascade is greatly amplified in fibroblasts transfected with v-Src (MATTINGLY et al. 1992). The ability of Src to increase P-Tyr and activity of G_q/G_{11} G protein α subunits, which mediate ET-1 stimulation of phospholipase $C\beta$ (see SIMONSON 1993 for review), accounts for much of the potentiation of the PI cascade by Src and ET-1 (LIU et al. 1996). In fact, the ability of ET-1-Src to activate the same G protein α subunit that couples to ET-1 receptors provides a general mechanism for amplifying ET-1 signal transduction. Several other proteins relevant to nuclear signaling directly associate with c-Src and are tyrosine phosphorylated: phosphatidyl inositol-3-kinase, ras GAP, phospholipase $C\gamma$, and Raf-1 (BROWN and COOPER 1996; COURTNEIDGE et al. 1993). It remains to be determined if these effectors participate in ET-1-Src signaling pathways.

3. ET and Survival, Anti-Apoptotic Signaling

Growth factors can increase cell growth by stimulating cell cycle progression or by inhibiting apoptosis, the controlled disassembly of a cell. Apoptosis contrasts with necrosis, which usually results from a severe cell insult and results in emptying of cell contents into the microenvironment resulting in inflammation. When apoptosis occurs correctly in normal physiology, the whole organism is not affected. However, dysregulation of apoptosis is observed in many pathophysiological states involving either premature cellular degeneration (neurodegeneration) or escape from normal inhibitory constraints on the cell cycle and hyperproliferation (i.e., cancer). Recent evidence from several laboratories suggests that ET-1 is a potent antiapoptotic peptide that promotes growth of vascular smooth muscle cells by inhibiting normal apoptotic constraints on the cell cycle (Fig. 2).

ET-1 blocks apoptosis in cultured vascular smooth muscle cells (WU-WONG et al. 1997), endothelial cells (SCHICHIRI et al. 1997), and fibroblasts (SCHICHIRI et al. 1998). ET-1 attenuates apoptosis induced by paclitaxel (an anticancer agent that causes apoptosis) in smooth muscle cells whereas in the same experiments angiotensin II did not inhibit apoptosis (WU-WONG et al. 1997). Binding of ET-1 to ET_B receptors in endothelial cells protects against apoptosis in serum-free medium (SCHICHIRI et al. 1997), and in fibroblasts ET-1 protected against *c-myc*-induced apoptosis (SCHICHIRI et al. 1998). In a particularly important finding, it appears that the antiapoptotic effect of ET-1 contributes adversely to vascular remodeling in vivo (SHARIFI and SCHIFFRIN 1997). ET_A receptor antagonists enhance vascular cell apoptosis in the rat model of deoxycorticosterone acetone-salt hypertension, suggesting that the antiapoptotic actions of ET-1 contribute to growth and remodeling in

vivo of vascular cells in hypertension. It seems likely that the paracrine/
autocrine antiapoptotic actions of ET-1 function in concert with the growth
factor actions to stimulate compensatory growth of vascular cells in response
to injury. However, additional experiments are clearly needed to determine
whether the antiapoptotic effect of ET-1 contributes to vascular remodeling
in other systems.

At present, little is known about the molecular mechanisms of anti-
apoptotic signaling by ET-1. The viability of many cells requires a constant or
intermittent survival signal conveyed by growth factors or cytokines. In the
absence of these apoptosis-suppressing signals, cells may undergo apoptosis.
A recently described model for antiapoptotic signaling suggests that growth
factors activate PI 3kinase (phosphatidylinositol 3-kinase), which stimulates
a signaling pathway involving Akt/PKB and the *bcl-2* family member Bad
(FRANKE et al. 1997). Activation of this pathway causes Bad phosphorylation,
sequestration of Bad in the cytosol by a 14–3-3 protein and inhibition of apoptosis. When the survival signals are absent, the phosphorylation state of Bad
shifts to the dephosphorylated form with subsequent stimulation of caspases
and apoptosis. Several groups have shown that ET-1 activates PI 3kinase
(SUGAWARA et al. 1996; FOSCHI et al. 1997; SU et al. 1999; SUZUKI et al. 1999),
and blockade of PI 3kinase blocks growth in cells stimulated by ET-1
(SUGAWARA et al. 1996; SUZUKI et al. 1999). To our knowledge, however, effects
of ET-1 downstream of PI 3kinase have not been investigated (see Fig. 2).

II. ET Post-Receptor Signals in Development and Differentiation

Gene targeting experiments have firmly established that ET peptides regulate
gene expression in development and differentiation of several tissues, particularly in cells derived from the neural crest (see Chap. 6). Targeted disruption
of the ET-1 gene in mice causes massive malformation of pharyngeal arch-
derived craniofacial tissues and organs (KURIHARA et al. 1994; YANIGASAWA et
al. 1998a). ET-1$^{-/-}$ mice suffer anoxia and die at birth from respiratory failure.
The phenotypic hallmarks are craniofacial abnormalities in the mandible,
zygomatic and temporal bones, tympanic rings, hyoid, thyroid cartilage, tongue,
and in soft tissues of the neck, palate, and ears (KURIHARA et al. 1994). All of
these cells derive from the pharyngeal arches, which in turn derive primarily
from neural crest ectomesenchymal cells. Thus the ET-1 and ET$_A$ genes direct
differentiation of neural crest-derived cells that form craniofacial developmental fields. ET-1 knockout mice also display abnormal development of
the thyroid, thymus, heart and great vessels (YANIGASAWA et al. 1998a;
KURIHARA et al. 1995). The molecular mechanisms and genes responsible for
the developmental abnormalities in ET-1 knockout mice are unclear.

Gene targeting studies of the ET-3 and ET$_B$ genes in mice reveal a requirement for normal development and migration of neural crest cells to the enteric
nervous system (myenteric ganglion neurons) and skin (i.e., epidermal
melanocytes). The lack of myenteric ganglion neurons and epidermal

melanocytes results in aganglionic megacolon and coat color spotting, respectively. These mutations do not complement in crossbreeding studies (HOSODA et al. 1994), suggesting that both ET-3 and ET_B are required (HOSODA et al. 1994). These findings in mice apparently have important implications in humans. Mutations in human ET-3 or ET_B genes predispose afflicted individuals to the genetic disorder of Hirschsprung's disease and similar neuro-cristopathies (PUFFENBERGER et al. 1994; CHAKRAVARTI 1996). The major phenotype in Hirschsprung's disease is a lack of enteric neurons in the colon leading to megacolon and a lack of melanocytes leading to abnormal pigmentation. ET-3 is a potent mitogen for early neural crest cell precursors, the majority of which give rise to melanocytes (LAHAV et al. 1996). In patients with Hirschsprung's disease, a G to T missense mutation in *EDNRB* exon 4 substitutes the highly conserved Trp-276 residue in the fifth transmembrane helix of the receptor with a Cys residue that renders the receptor less sensitive to ligand activation (PUFFENBERGER et al. 1994; CHAKRAVARTI 1996).

Relatively little is known regarding ET-1 signaling and differentiation and development (Fig. 3). One of the most interesting aspects of ET differentiation signaling is that ET-1/ETA and ET-3/ETB signaling function independently (YANAGISAWA et al. 1998b). The independence of these signaling cascade was demonstrated by targeted disruption of the endothelin converting enzyme-1 gene. The goosecoid gene is depleted in the pharyngeal arch of ET_A-null mice, suggesting that this transcription factor is downstream of ET_A post-receptor signals (CLOUTHIER et al. 1998). However, a functional role for goosecoid in ET-1 signaling remains to be shown, and the mechanisms by which ET_A receptors induce goosecoid are unknown. Preliminary evidence suggests a role for the basic helix-loop-helix transcription factor dHand in the

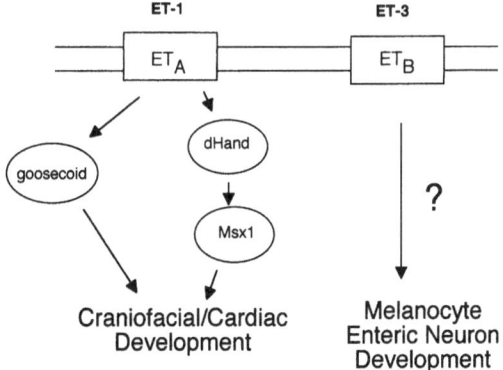

Fig. 3. Independent pathways of developmental signaling by ET. ET-1/ET_A and ET-3/ET_B activate independent signaling cascades leading to different steps in development of neural crest cells. The post-receptor signals are poorly understood, but the ET-1/ET_A axis apparently involves the goosecoid transcription factor and a cascade involving the dHand transcription factor and the Msx1 homeobox gene

ET-1/ET_A signaling axis (THOMAS et al. 1998). dHand is down-regulated in the branchial and aortic arches of ET-1$^{-/-}$ mice. A homeobox gene, Msx1, is also apparently downstream of ET-1 and dHand (THOMAS et al. 1998), but again a functional role in ET-1 signaling is at present lacking (Fig. 3).

Preliminary data also suggest that ET-1 contributes to the maturation and stabilization of blood vessels. Development of the vasculature depends critically on paracrine interactions between endothelial cells and underlying cells such as vascular smooth muscle, pericytes, and fibroblasts (FOLKMAN and D'AMORE 1996). In immature vessels, endothelial cells secrete signaling molecules that recruit neighboring cells to proliferate and form surrounding cellular and matrix structures within the vessel. In cell culture systems ET-1 is at least one of the peptides responsible for endothelial cell-dependent proliferation of vascular pericytes (CHAKRAVARTHY et al. 1992; YAMAGISHI et al. 1993). Endothelial cells enhance pericyte survival and growth in a co-culture system, which is blocked by neutralizing antibodies against ET-1. Exogenous ET-1 can in part replace the endothelial cell layer by activating ET_A receptors in underlying pericytes (YAMAGISHI et al. 1993). ET-1 has also been implicated in endothelium-directed formation of myofibroblasts in wound healing (GUIDRY and HOOK 1991; VILLASCHI and NICOSIA 1994). In fibroblasts, ET-1 induces genes for smooth muscle α-actin and collagen type I (VILLASCHI and NICOSIA 1994), two essential steps in the transformation of vascular fibroblasts to myofibroblasts that contract a newly synthesized matrix during wound repair. ET-1 secreted by endothelial cells promotes collagen gel contraction by fibroblasts (GUIDRY and HOOK 1991), an important step in wound healing.

One pathophysiological correlate of the ability of ET-1 to regulate blood vessel formation appears to be development of cerebral arteriovenous malformations (AVM) in humans (RHOTEN et al. 1997). In this disorder, which occurs in 0.5% of humans, a tangle of blood vessels form with abnormal vascular cell phenotypes (SPETZLER et al. 1978; SPETZLER and MARTIN 1986). In particular, these vessels typically lack vascular smooth muscle cells and pericytes. We recently demonstrated that the prepro-ET-1 gene is specifically repressed in 17/17 patients with cerebrovascular AVMs (RHOTEN et al. 1997). In contrast, ET-1 expression was normal in vessels feeding and draining the AVMs as well as elsewhere in the body. It seems possible that local repression of the prepro-ET-1 gene in AVMs contributes to the lack of vascular smooth muscle cells and pericytes in AVMs (RHOTEN et al. 1997). The lack of ET-1 secretion and concomitant vasoconstriction might also account for the inability of AVMs to autoregulate blood flow (SPETZLER et al. 1978).

D. Summary and Perspective

Future studies to elucidate mechanisms of signaling by ETs might focus on the following questions. What mechanisms control global and spatial aspects of ET-1 Ca^{2+} signaling? What PTKs in addition to c-Src and FAK are stimulated by ET-1 and how are they activated? What are the relevant P-Tyr target proteins in ET-1 signaling and what role, if any, do they play in induction of gene expression? What are the important targets of PI-3K in ET-1 signaling and how do they suppress apoptosis? What pathways are responsible for the independent ET-1/ET_A and ET-3/ET_B effects on differentiation of neural crest cells. Finally, a major challenge is to understand how ET post-receptor signals are organized to elicit cell- and tissue-specific responses.

Acknowledgments. The author thanks William Herman, Yuan Wang, Allison Rooney, Jennifer Jones, Patrick Rhoten, and Michael Dunn. Thanks also go to Aravinda Chakravarti, George Dubyak, and Hsing-Jien Kung for many helpful suggestions. This work was supported by a grant from the National Institutes of Health, DK-46939.

References

Araki S-I, Haneda M et al. (1997) Endothelin-1 activates c-Jun NH_2-terminal kinase in mesangial cells. Kidney Int 51:631–639

Aramori I, Nakanishi S (1992) Coupling of two endothelin receptor subtypes to differing signal transduction in transfected Chinese hamster ovary cells. J Biol Chem 267:12468–12474

Auguet M, Delaflotte S et al. (1988) Endothelin and Ca++ agonist BAY K 8644: Different vasoconstrictive properties. Biochem Biophys Res Comm 156:186–192

Badr KF, Murray JJ et al. (1989a) Mesangial cell, glomerular, and renal vascular responses to endothelin in the kidney. J Clin Invest 83:336–342

Badr KF, Munger KA et al. (1989b) High and low affinity binding sites for endothelin on cultured rat glomerular mesangial cells. Biochem Biophys Res Comm 161: 776–781

Battistini B, Chailler P et al. (1993) Growth regulatory properties of endothelins. Peptides 14:385–399

Benigni A, Zoja C et al. (1993) A specific endothelin subtype A receptor antagonist protects against functional and structural injury in a rat model of renal disease progression. Kidney Int 44:440–444

Benigni A, Zoja C et al. (1996) Blocking both type A and B endothelin receptors in the kidney attenuates renal injury and prolongs survival in rats with remnant kidney. Am J Kid Dis 27:416–423

Berk BC, Canessa M et al. (1988) Agonist-mediated changes in intracellular pH: role in vascular smooth muscle cell function. J Cardiovasc Pharmacol 12(S5):S104–S114

Berridge MJ (1997) Elementary and global aspects of calcium signaling. J Physiol. 499: 291–306

Berridge MJ, Irvine RF (1990) Inositol phosphates and cell signaling. Nature 341: 197–205

Blackburn K, Highsmith RF (1990) Nickel inhibits endothelin-induced contractions of vascular smooth muscle. Am J Physiol 258:C1025–C1030

Bogoyevitch MA, Glennon PE et al. (1994) Endothelin-1 and fibroblast growth factors stimulate the mitogen-activated protein kinase signaling cascade in cardiac myocytes. J Biol Chem 269:1110–1119

Boric MP, Donoso V et al. (1990) Endothelin reduces microvascular blood flow by acting on arterioles and venules of the hamster cheek pouch. Eur J Pharmacol 190:123–133

Brown MT, Cooper JA (1996) Regulation, substrates, and functions of src. Biochim Biophys Acta 1287:121–149

Cadwaller K, Beltman J et al. (1997) Differential regulation of extracellular signal-regulated protein kinase 1 and Jun N-terminal kinase 1 by Ca^{2+} and protein kinase C in endothelin-stimulated Rat-1 cells. Biochem J 321:795–804

Cantley LC, Auger KR et al. (1991) Oncogenes and signal transduction. Cell 64: 281–302

Cazaubon SM, Ramos-Morales F et al. (1994) Endothelin induces tyrosine phosphorylation and GRB2 association of Shc in astrocytes. J Biol Chem 269: 24805–24809

Chakravarthy U, Gardiner TA et al. (1992) The effect of endothelin 1 on the retinal microvascular pericyte. Microvasc Res 43:241–254

Chakravarti A (1996) Endothelin receptor-mediated signaling in Hirschsprung disease. Human Mol Genetics 5:303–307

Chan J, Greenburg DA (1991) SK&F 96365, a receptor-mediated calcium entry inhibitor, inhibits calcium responses to endothelin-1 in NG108–15 cells. Biochem Biophys Res Comm 177:1141–1146

Choukroun G, Hajjar R et al. (1998) Role of stress-activated protein kinases in endothelin-induced cardiomyocyte hypertrophy. J Clin Invest 102:1311–1320

Clerk A, Michael A et al. (1998) Stimulation of the p38 mitogen-activated protein kinase pathway in neonatal rat ventricular myocytes by the G protein-coupled receptor agonists, endothelin-1 and phenylephrine: A role in cardiac myocyte hypertrophy? J Cell Biol 142:523–535

Clouthier DE, Hosoda K et al. (1998) Cranial and cardiac neural crest defects in endothelin-A receptor-deficient mice. Development 125:813–824

Courtneidge SA, Fumagalli S et al. (1993) The Src family of protein tyrosine kinases: regulation and functions. Development Suppl 57–64

Danthuluri NR, Brock TA (1990) Endothelin receptor-coupling mechanisms in vascular smooth muscle: A role for protein kinase C. J Pharmacol Exp Ther 254: 393–399

D'Orleans-Juste P, deNucci G et al. (1989) Endothelin-1 contracts isolated vessels independently of dihydropyridine-sensitive Ca^{2+} channel activation. Eur J Pharmacol 165:289–295.

Douglas SA, Louden C et al. (1994) A role for endogenous endothelin-1 in neointimal formation after rat carotid artery balloon angioplasty. Circ Res 75:190–197

Edwards RM, Trizna WT et al. (1990) Renal microvascular effects of endothelin. Am J Physiol 259:F217–F221

Ferrer P, Valentine M, et al. (1995) Orally active endothelin receptor antagonist BMS-182874 suppresses neointimal development in balloon-injured rat carotid arteries. J Cardiovasc Pharmacol 26:908–915

Folkman J, D'Amore PA (1996) Blood vessel formation: What is its molecular basis? Cell 87:1153–1155

Force TL, Mechanisms of endothelin-induced mitogenesis and activation of stress response protein kinases, in Endothelin receptors and signaling mechanisms, D.M. Pollock and R.F. Highsmith, Editors. 1998, Springer-Verlag: Berlin. p. 177–215

Force T, Bonventre JV (1992) Endothelin activates Src Tyrosine kinase in glomerular mesangial cells. J Am Soc Nephrol 3:491–490

Force T, Kyriakis JM et al. (1991) Endothelin, vasopressin, and angiotensin II enhance tyrosine phosphorylation by protein kinase C-dependent and -independent pathways in glomerular mesangial cells. J Biol Chem 266:6650–6656

Foschi M, Chari S et al. (1997) Biphasic activation of p21ras by endothelin-1 sequentially activates the ERK cascade and phosphatidylinositol 3-kinase. EMBO J 16: 6439–6451

Franke TF, Kaplan DR et al. (1997) PI3K: Downstream AKTion blocks apoptosis. Cell 88:435–437

Fukada Y, Hirata Y et al. (1988) Endothelin is a potent secretagogue for atrial natriuretic peptide in cultured rat atrial myocytes. Biochem Biophys Res Comm 155: 167–171

Fuller SJ, Davies EL et al. (1997) Mitogen-activated protein kinase phosphatase I inhibits the stimulation of gene expression by hypertrophic agonists in cardiac myocytes. Biochem J 323:313–319

Galron R, Kloog Y et al. (1989) Functional endothelin/sarafotoxin receptors in rat heart myocytes: structure-activity relationships and receptor subtypes. Biochem Biophys Res Comm 163:936–943

Giad A, Yanagisawa M et al. (1993) Expression of endothelin-1 in the lungs of patients with pulmonary hypertension. New Engl J Med 328: 1732–1739

Goto K, Kasuya Y et al. (1989) Endothelin activates the dihydropyridine-sensitive, voltage-dependent Ca^{2+} channel in vascular smooth muscle. Proc Natl Acad Sci 86:3915–3918

Griendling KK, Tsuda T et al. (1989) Endothelin stimulates diacylglycerol accumulation and activates protein kinase C in cultured vascular smooth muscle cells. J Biol Chem 264:8237–8240

Guidry C, Hook M (1991) Endothelins produced by endothelial cells promote collagen gel contraction by fibroblasts. J Cell Biol 115:873–880

Haneda M, Kikkawa R et al. (1995) Endothelin-1 stimulates tyrosine phosphorylation of p125 focal adhesion kinase in mesangial cells. J Am Soc Nephrol 6:1504–1510

Herman WH, Simonson MS (1995) Nuclear signaling by endothelin-1: A Ras pathway for activation of the c-fos serum response element. J Biol Chem 270:11654–11661

Herskowitz I (1987) Functional inactivation of genes by dominant negative mutations. Nature 329:219–222

Highsmith RF, Pang DC et al. (1989) Endothelial cell-derived vasoconstrictors: mechanisms of action in vascular smooth muscle. J Cardiovasc Pharmacol 13(S5): S36–S44

Hocher B, Liefeldt L et al. (1996a) Characterization of the renal phenotype of transgenic rats expressing the human endothelin-2 gene. Hypertension 28:196–201

Hocher B, Thone-Reineke C et al. (1996b) Endothelin-1 transgenic mice develop glomerulosclerosis, interstitial fibrosis, and renal cysts in an age and gender dependent manner. J Am Soc Nephrol 12:1633–1630

Hosoda K, Hammer RE et al. (1994) Targeted and natural (Piebald-Lethal) mutations of endothelin-B receptor gene produce megacolon associated with spotted coat color in mice. Cell 79:1267–1276

Hu JR, Berninger UG et al. (1988) Endothelin stimulates atrial natriuretic peptide (ANP) release from rat atria. Eur J Pharmacol 158:177–180

Huckle WR, Dy RC et al. (1992) Calcium-dependent increase in tyrosine kinase activity stimulated by angiotensin II. Proc Natl Acad Sci 89:8837–8841

Imamoto A, Soriano P (1993) Disruption of the csk gene, encoding a negative regulator of Src family tyrosine kinases, leads to neural tube defects and embryonic lethality in mice. Cell 73:1117–1124

Inoue Y, Oike M et al. (1990) Endothelin augments unitary calcium channel currents on the smooth muscle cell membrane of guinea-pig portal vein. J Physiol 423: 171–191

Isono M (1998) Atrial natriuretic peptide inhibits endothelin-1-induced activation of JNK in glomerular mesangial cells. Kidney Int. 53:1133–1142

Iwamuro Y, Miwa S, et al. (1998) Activation of two types of Ca^{2+}-permeable nonselective cation channels by endothelin-1 in A7r5 cells. Brit J Pharmacol 124: 1541–1549

Kar S, Yousem SA et al. (1995) Endothelin-1 expression by human hepatocellular carcinoma. Biochim Biophys Res Comm 216:514–519

Kasuya Y, Takuwa Y et al. (1989) Endothelin-1 induces vasoconstriction through two functionally distinct pathways in porcine artery-contribution of phosphoinositide turnover. Biochem Biophys Res Comm 161:1049–1055

Kiowski W, Luscher TF et al. (1991) Endothelin-1-induced vasoconstriction in humans. Reversal by calcium channel blockade but not by nitrovasodilators or endothelium-derived relaxing factor. Circulation 83:469–475

Kodama M, Kanaide H et al. (1989) Endothelin-induced Ca-independent contraction of the porcine coronary artery. Biochem Biophys Res Comm 160:1302–1308

Kohan DE (1997) Endothelins in the normal and diseased kidney. J Kidney Dis. 29: 2–26

Komuro I, Kurihara H et al. (1988) Endothelin stimulates c-fos and c-myc expression and proliferation of vascular smooth muscle cells. FEBS Lett 238:249–252

Korbmacher C, Helbing H et al. (1989) Endothelin depolarized membrane voltage and increases intracellular calcium concentration in human ciliary muscle cells. Biochem Biophys Res Comm 164:1031–1039

Kovacic B, Ilic D et al. (1998) c-Src activation plays a role in endothelin-dependent hypertrophy of the cardiac myocyte. J Biol Chem 273:35185–35193

Kurihara Y, Kurihara H et al. (1994) Elevated blood pressure and craniofacial abnormalities in mice deficient in endothelin-1. Nature 368:703–710

Kurihara Y, Kurihara H et al. (1995) Aortic arch malformations and ventricular septal defect in mice deficient in endothelin-1. J Clin Invest 96:293–300

Kusuhara M, Yamaguchi K et al. (1990) Production of endothelin in human cancer cell lines. Cancer Res 50: 3257–3261

Kusuhara M, Yamaguchi K et al. (1992) Stimulation of anchorage-independent cell growth by endothelin in NRK 49F cells. Cancer Res 52:3011–3014

Lahav R, Ziller C et al. (1996) Endothelin 3 promotes neural crest cell proliferation and mediates a vast increase in melanocyte number in culture. Proc Natl Acad Sci 93:3892–3897

Lee T-E, Chao T et al. (1989) Endothelin stimulates a sustained 1,2 diacylglycerol increase and protein kinase C activation in bovine aortic smooth muscle cells. Biochem Biophys Res Comm 162:381–386

Lerman A, Edwards BS et al. (1991) Circulating and tissue endothelin immunoreactivity in advanced atherosclerosis. New Engl J Med 325:997–1001

Lewis TS, Shapiro PS et al. (1998) Signal transduction through MAP kinase cascades. Adv Cancer Res 74:49–139

Liu W, Mattingly RR et al. (1996) Transformation of Rat-1 fibroblasts with the v-src oncogene increases the tyrosine phosphorylation state and activity of the a subunit of Gq/G11. Proc Natl Acad Sci 93:8258–8263

MacNulty EE, Plevin R et al. (1990) Stimulation of the hydrolysis of phosphatidylinositol 4,5- bisphosphate and phosphatidylcholine by endothelin, a complete mitogen for Rat-1 fibroblasts. Biochem J 272:761–766

Madeddu P, Yang X et al. (1990) Efficacy of nifedipine to prevent systemic and renal vasoconstrictor effects of endothelin. Am J Physiol 259:F304–F311

Maher E, Bardequez A et al. (1991) Endothelin- and oxytocin-induced calcium signaling in cultured human myometrial cells. J Clin Invest 87:1251–1258

Marsault, R, Vigne, P, et al. (1990) The effect of extracellular calcium on the contractile action of endothelin. Biochem Biophys Res Comm 171: 301–305

Marsden PA, Danthuluri NR et al. (1989) Endothelin action on vascular smooth muscle involves inositol trisphosphate and calcium mobilization. Biochem Biophys Res Comm 158:86–93

Masaki T (1995) Possible role of endothelin in endothelial regulation of vascular tone. Ann Rev Pharmacol Toxicol 35: 235–255

Masaki T, Yanagisawa M et al. (1990) Cellular mechanism of vasoconstriction induced by endothelin. Adv Second Messenger Phosphoprotein Res 24:425–428

Mattingly RR, Wasilenko WJ et al. (1992) Selective amplification of endothelin-stimulated inositol 1,4,5-trisphosphate and calcium signaling by v-src transformation of Rat-1 fibroblasts. J Biol Chem 267:7470–7477

Mitsuhashi T, Morris RC et al. (1989) Endothelin-induced increases in vascular smooth muscle Ca^{2+} do not depend on dihydropyridine-sensitive Ca^{2+} channels. J Clin Invest 84:635–639

Miyauchi T, Masaki T (1999) Pathophysiology of endothelin in the cardiovascular system. Ann Rev Physiol 61:391–415

Mulder BJ, deTombe PP et al. (1989) Spontaneous and propagated contractions in rat cardiac trabeculae. J Gen Physiol 93:943–961

Muldoon L, Rodland KD et al. (1989) Stimulation of phosphatidylinositol hydrolysis, diacylglycerol release, and gene expression in response to endothelin, a potent new agonist for fibroblasts and smooth muscle cells. J Biol Chem 264:8529–8536

Muldoon LL, Pribnow D, et al. (1990) Endothelin-1 stimulates DNA synthesis and anchorage-independent growth of rat-1 fibroblasts through a protein kinase C-dependent mechanism. Cell Reg 1:379–390

Naitoh T, Toyo oka T, et al. (1990) An endogenous Ca^{2+} channel agonist, endothelin-1, does not directly activate partially purified dihydropyridine-sensitive Ca^{2+} channel from cardiac muscle in a reconstituted system. Biochem Biophys Res Commun 171:1205–1210

Nasmith PE, Mills GB et al. (1989) Guanine nucleotides induce tyrosine phosphorylation and activation of the respiratory burst in neutrophils. Biochem J 257:893–897

Nelson JB, Hedican SP et al. (1995) Identification of endothelin-1 in the pathophysiology of metastatic adenocarcinoma of the prostate. Nature Med 1:944–949

Neuser D, Knorr A et al. (1990) Mitogenic activity of endothelin-1 and –3 on vascular smooth muscle cells is inhibited by atrial natriuretic peptides. Artery 17:311–324

Ohlstein EH, Horohonich S et al. (1989) Cellular mechanisms of endothelin in rabbit aorta. J Pharmacol Exp Ther 250:548–555

Pollock DM, Highsmith RF (1998) Endothelin receptors and signaling mechanisms. Berlin: Springer p. 1–224

Pollock DM, Keith TL et al. (1995) Endothelin receptors and calcium signaling. FASEB J 9:1196–1204

Puffenberger EG, Hosoda K et al. (1994) A missense mutation of the endothelin-B receptor gene in multigenic Hirschsprung's disease. Cell 79:1257–1266

Rasmussen H, Takuwa Y et al. (1987) Protein kinase C in the regulation of smooth muscle contraction. FASEB J 1:177–185

Rasmussen K Printz MP (1989) Depolarization potentiates endothelin-induced effects on cytosolic calcium in bovine adrenal chromaffin cells. Biochem Biophys Res Comm 165:306–311

Rembold CM (1990) Modulation of the $[Ca^{2+}]$ sensitivity of myosin phosphorylation in intact swine arterial smooth muscle. J Physiol 429:77–94

Resink TJ, Scott-Burden T et al. (1988) Endothelin stimulates phospholipase C in cultured vascular smooth muscle cells. Biochem Biophys Res Comm 157:1360–1368

Resink TJ, Scott-Burden T et al. (1990a) Activation of multiple signal transduction pathways by endothelin in cultured human vascular smooth muscle cells. Eur J Biochem 189:415–421

Resink TJ, Scott-Burden T et al. (1990b) Phorbol ester promotes a sustained down-regulation of endothelin receptors and cellular responses to endothelin in human vascular smooth muscle cells. Biochem Biophys Re Comm 166:1213–1219

Reynolds EE, Mok LL et al. (1989) Phorbol ester dissociates endothelin-stimulated phosphoinositide hydrolysis and arachidonic acid release in vascular smooth muscle cells. Biochem Biophys Res Comm 160:868–873

Rhoten RLP, Comair YG et al. (1997) Specific repression of the preproendothelin-1 gene in intracranial arteriovenous malformations. J Neurosurg 86: 101–108

Richardson A, Parsons JT (1995) Signal transduction through integrins: a central role for focal adhesion kinase. BioEssays 17:229–236

Richardson A, Parsons JT (1996) A mechanism for regulation of the adhesion-associated protein tyrosine kinase pp125FAK. Nature 380:538–540

Roche S, Koegl M et al. (1995) DNA synthesis induced by some but not all growth factors requires Src family protein tyrosine kinases. Molec Cell Biol 15: 1102–1109

Sabe H, Knudsen B et al. (1992) Molecular cloning and expression of chicken C-terminal Src kinase: lack of stable association with c-Src protein. Proc Natl Acad Sci USA 89:2190–2194

Sakai S, Miyauchi T et al. (1996) Inhibition of myocardial endothelin pathway improves long-term survival in heart failure. Nature 384:353–355

Sanderson MJ, Chow I et al. (1988) Intercellular communication between ciliated cells in culture. Am J Physiol 254:C63–C74

Sandok EK, Lerman A et al. (1992) Endothelin in a model of acute ischemic renal dysfunction: Modulating action of atrial natriuretic factor. J Am Soc Nephrol 3: 196–202

Schichiri M, Kato H et al. (1997) Endothelin-1 as an autocrine/paracrine apoptosis survival factor for endothelial cells. Hypertension 30:1198–1203

Schichiri M, Sedivy JM et al. (1998) Endothelin-1 is a potent survival factor for c-Myc dependent apoptosis. Mol Endo 12:172–180

Sharifi AM, Schiffrin EL (1997) Apoptosis in aorta of deoxycorticosterone acetate-salt hypertensive rats: effects of endothelin receptor antagonism. J Hypertens 15: 1441–1448

Silberberg SD, Poder TC et al. (1989) Endothelin increases single-channel calcium currents in coronary arterial smooth muscle cells. FEBS Lett 247:68–72

Simonson MS (1993) Endothelins: Multifunctional renal peptides. Physiol Rev 73: 375–411

Simonson MS (1994) Anti-AP-1 activity of all-trans retinoic acid in glomerular mesangial cells. Am J Physiol 267:F805–F815

Simonson MS, Dunn MJ (1990a) Cellular signaling by peptides of the endothelin gene family. FASEB J 4:2989–3000

Simonson MS, Dunn MJ (1990b) Endothelin-1 stimulates contraction of rat glomerular mesangial cells and potentiates β-adrenergic-mediated cyclic adenosine monophosphate accumulation. J Clin Invest 85:790–797

Simonson MS, Dunn MJ (1991) Ca^{2+} signaling by distinct endothelin peptides in glomerular mesangial cells. Exp Cell Res 192:148–156

Simonson MS, Herman WH (1993) Protein kinase C and protein tyrosine kinase activity contribute to mitogenic signaling by endothelin-1: Cross-talk between G protein-coupled receptors and pp60c-src. J Biol Chem 268:9347–9357

Simonson MS, Wann S et al. (1989) Endothelin stimulates phospholipase C, Na+/H+ exchange, c-fos expression, and mitogenesis in rat mesangial cells. J Clin Invest 83:708–712

Simonson MS, Osanai T et al. (1990) Endothelin isopeptides evoke Ca^{2+} signaling and oscillations of cytosolic free $[Ca^{2+}]$ in human mesangial cells. Biochim Biophys Acta 1055:63–68

Simonson MS, Wang Y et al. (1996a) Ca^{2+} channels mediate protein tyrosine kinase activation by endothelin-1. Am J Physiol 270:F790–F797

Simonson MS, Wang Y et al. (1996b) Nuclear signaling by endothelin-1 requires Src protein tyrosine kinases. J Biol Chem 271:77–82

Simpson AW, Ashley CC (1989) Endothelin evoked Ca^{2+} transients and oscillations in A10 vascular smooth muscle cells. Biochem. Biophys Res Comm 163:1223–1229

Simpson AW, Stampfl A et al. (1990) Evidence for receptor-mediated bivalent-cation entry in A10 vascular smooth-muscle cells. Biochem J 267:277–280

Spetzler RF, Martin NA (1986) A proposed grading system for arteriovenous malformations. J Neurosurg 65:476–483

Spetzler RF, Wilson CB et al. (1978) Normal perfusion pressure breakthrough theory. Clin Neurosurg 25:651–672

Su X, Wang P et al. (1999) Differential activation of phosphoinositide 3-kinase by endothelin and ceramide in colonic smooth muscle cells. Am J Physiol 276:G853–G861

Sugawara, F, Ninomiya, H, et al. (1996) Endothelin-1-induced mitogenic responses of Chinese hamster ovary cells expressing human endothelinA: The role of a wortmannin-sensitive signaling pathway. Mol Pharmacol 49: 447–457

Sugiura M, Inagami T et al. (1989) Endothelin action: inhibition by protein kinase C inhibitor and involvement of phosphoinositols. Biochem Biophys Res Comm 158: 170–176

Sunako M, Kawahara Y et al. (1990) Mass analysis of 1,2-diacylglycerol in cultured rabbit vascular smooth muscle cells. Hypertension 15:84–88

Suzuki E, Nagata D et al. (1999) Molecular mechanisms of endothelin-1-induced cell cycle progression. Circ Res 84:611–619

Terada Y, Inoshita S et al. (1998) Cyclin D1, p16, and retinoblastoma gene regulate mitogenic signaling of endothelin in rat mesangial cells. Kidney Int 53:76–83

Thomas T, Kurihara H et al. (1998) A signaling cascade involving endothelin-1, dHand, and Msx1 regulates development of neural-crest-derived branchial arch mesenchyme. Development 125:3005–3014

Twamley-Stein GM, Pepperkok R et al. (1993) The Src family tyrosine kinases are required for platelet-derived growth factor-mediated signal transduction in NIH 3T3 cells. Proc Natl Acad Sci U.S.A. 90:7696–7700

Van Biesen T, Luttrell LM et al. (1996) Mitogenic signaling via G protein-coupled receptors. Endocrine Rev 17:698–714

VanRenterghem C, Vigne P et al. (1988) Molecular mechanism of action of the vasoconstrictor peptide endothelin. Biochem Biophys Res Comm 157:977–985

VanRenterghem C, Romey G et al. (1989) Vasopressin modulates spontaneous electrical activity in aortic cells (A7r5) by acting on three different types of ionic channels. Proc Natl Acad Sci 85:9365–9369

Vigne P, Lazdunski M et al. (1989) The inotropic effect of endothelin-1 on rat atria involves hydrolysis of phosphatidylinositol. FEBS Lett 249:143–146

Vigne P, Marsault R et al. (1990) Endothelin stimulates phosphatidylinositol hydrolysis and DNA synthesis in brain capillary endothelial cells. Biochem J 266:415–420

Villaschi S, Nicosia RF (1994) Paracrine interactions between fibroblasts and endothelial cells in a serum-free coculture model. Lab Invest 71:291–299

Wang Y, Simonson MS et al. (1992) Endothelin rapidly stimulates mitogen-activated protein kinase activity in rat mesangial cells. Biochem J 287:589–594

Wang Y, Rose PM et al. (1994) Endothelins stimulate mitogen-activated protein kinase cascade through either ETA or ETB. Am J Physiol 267:C1130–1135

Weber H, Webb ML et al. (1994) Endothelin-1 and angiotensin-II stimulate delayed mitogenesis in cultured rat aortic smooth muscle cells: Evidence for common signaling mechanisms. Mol Endocrinol 8:148–158

Whelchel A, Evans J et al. (1997) Inhibition of ERK activation attenuates endothelin-stimulated airway smooth muscle cell proliferation. Am. J. Respir. Cell Mol Biol 16:589–596

Wilkes BM, Ruston AS et al. (1991) Characterization of endothelin 1 receptor and signal transduction mechanisms in rat medullary interstitial cells. Am J Physiol 260:F579–F589

Wu J, Garami M et al. (1996) 1,25 (OH)2 Vitamin D3 and retinoic acid antagonize endothelin-stimulated hypertrophy of neonatal rat cardiac myocytes. J Clin Invest 97:1577–1588

Wu-Wong JR, Chiou WJ et al. (1997) Endothelin attenuates apoptosis in human smooth muscle cells. Biochem J 328:733–737

Yamagishi S, Hsu C-C et al. (1993) Endothelin-1 mediates endothelial cell-dependent proliferation of vascular pericytes. Biochim Biophys Res Comm 191:840–846.

Yanagisawa M (1994) The endothelin system: A new target for therapeutic intervention. Circulation 89:1320–1322

Yanagisawa M, Masaki T (1989) Endothelin, a novel endothelium-derived peptide. Biochem Pharmacol 38:1877–1883

Yanagisawa M, Kurihara H, et al. (1988) A novel potent vasoconstrictor peptide produced by vascular endothelial cells. Nature 332:411–415

Yanigisawa H, Hammer RE, et al. (1998a) Role of endothelin-1/endothelin-A receptor-mediated signaling pathway in the aortic arch patterning in mice. J Clin Invest 102:22–33

Yanagisawa H, Yanagisawa M, et al. (1998b) Dual genetic pathways of endothelin-mediated intercellular signaling revealed by targeted disruption of endothelin converting enzyme-1 gene. Development 125:825–836

Zachary I, Gil J et al. (1991a) Bombesin, vasopressin, and endothelin rapidly stimulate tyrosine phosphorylation in intact Swiss 3T3 cells. Proc Natl Acad Sci U.S.A. 88: 4577–4581

Zachary I, Sinnett-Smith J et al. (1991b) Stimulation of tyrosine kinase activity in anti-phosphotyrosine immune complexes of Swiss 3T3 cell lysates occurs rapidly after addition of bombesin, vasopressin, and endothelin to intact cells. J Biol Chem 266: 24126–24133

Zachary I, Sinnett-Smith J et al. (1992) Bombesin, vasopressin, and endothelin stimulation of tyrosine phosphorylation in Swiss 3T3 cells. J Biol Chem 267: 19031–19034

Zachary Is, Sinnett-Smith J et al. (1993) Bombesin, vasopressin, and endothelin rapidly stimulate tyrosine phosphorylation of the focal adhesion-associated protein paxillin in Swiss 3T3 cells. J Biol Chem 268:22060–22065

Zhang X-F, Komuro T et al. (1998) Role of nonselective cation channels as Ca^{2+} entry pathway in endothelin-1-induced contraction and their suppression by nitric oxide. Eur J Pharmacol 352:237–245

Zhao Y, Sudol M et al. (1992) Increased tyrosine kinase activity of c-Src during calcium-induced keratinocyte differentiation. Proc Natl Acad Sci 89:8298–8302

Zhou M, Sucov HM et al. (1995) Retinoid-dependent pathways suppress myocardial cell hypertrophy. Proc Natl Acad Sci 92:7391–7395

CHAPTER 6

Lessons from Gene Deletion of Endothelin Systems

H. Kurihara, Y. Kurihara, and Y. Yazaki

A. Introduction

Gene targeting approach in mice sometimes reveals unexpected aspects of biology. Before we found in 1992 that deletion of the endothelin (ET)-1 gene produced embryonic anomalies, few people anticipated a developmental role for ET-1, which was commonly regarded as a potent vasoconstrictor and pressor peptide mainly produced by vascular endothelial cells (Masaki 1995; Levin 1995). At that time only a few signaling molecules had been demonstrated identified to mediate morphogenesis in vertebrates and it was hard to imagine that one peptide could have dual roles in embryogenesis and cardiovascular regulation. Indeed, ET-1 was the first vasoactive substance acting through G protein-coupled receptors shown to be involved in embryonic development. Subsequently the genes for ET-3, the two ET receptors and ET converting enzyme-1 (ECE-1) were also knocked out providing further evidence of the developmental roles of the ET system. Following these studies mutations of the ET-3 and ET-B receptor genes were identified in patients with human Hirschsprung's disease. Thus, gene knockout studies paved the way to a new era of research not only on the ET system but also in the broader field of developmental biology and clinical science. Here we review the current knowledge of the developmental and pathogenetic roles of the ET system revealed by gene knockout studies.

B. Developmental Defects in Endothelin-Knockout Mice

I. Gene Knockout of ET-1/ET-A Receptor Pathway

Involvement of the ET system in embryonic development was first revealed by gene targeting of ET-1 (Kurihara et al. 1994). Mice homozygous for the ET-1 null mutation demonstrate conspicuous morphological abnormalities in the branchial arch-derived craniofacial structures, including micrognathia, microglossia, microtia, cleft palate, and thymic hypoplasia. All the ET-1 knockout homozygotes die of respiratory failure just after birth because of upper airway obstruction. In terms of skeletal development, the first and second branchial arch-derived components (i.e., Meckel's cartilage, mandibular bone, zygomatic and temporal bones, auditory ossicles, tympanic ring, hyoid and thyroid cartilages) are selectively affected. In addition, ET-1 knockout mice demonstrate cardiovascular defects including aortic arch interruption or hypoplasia, aberrant subclavian arteries, and ventricular septal defect with outflow tract misalignment, which seem to be due to impaired development of branchial arch arteries and outflow tract (KURIHARA et al. 1995, 1997). Hypoplastic fourth and sixth arch arteries and enlarged third arch artery demonstrable at early developmental stages are considered to be responsible for these great vessel anomalies in ET-1 knockout mice.

The frequency and extent of the cardiovascular abnormalities of ET-1 knockout mice are increased by treatment with neutralizing monoclonal antibodies or a selective ET-A receptor antagonist BQ123 (KURIHARA et al. 1995). This result indicates that the ET-A receptor may mediate the developmental effect of ET-1 and that maternally derived ET-1 may cause partial rescue of the cardiovascular phenotype of ET-1 null mutation. In accordance with the former hypothesis, gene knockout of the ET-A receptor in mice has been reported to produce morphological abnormalities identical to ET-1 knockout (CLOUTHIER et al. 1998; YANAGISAWA et al. 1998a). Inactivation of the ET-A receptor by an ET-A-selective antagonist, RU69986, was also shown to produce dysmorphogenesis in the hypobranchial skeleton as well as heart and aortic arch derivatives in the chick embryo, which resembles the phenotype of ET-1/ET-A receptor knockout mice (KEMPF et al. 1998). Thus, the ET-1/ET-A-mediated signaling pathway is essential for the normal craniofacial and cardiovascular development in a range of species.

II. Gene Knockout of ET-3/ET-B Receptor Pathway

In contrast to the craniofacial and cardiovascular phenotype of ET-1 and ET-A receptor knockout mice, ET-3 and ET-B receptor knockout mice exhibit a different and non-overlapping developmental phenotype (BAYNASH et al. 1994; HOSODA et al. 1994). These mice are viable at birth, but eventually die at 2–8 weeks of age with both ET-3 and ET-B receptor knockout mice showing an identical phenotype characterized by megacolon and coat color spotting. Histological examination disclosed the complete absence of Auerbach ganglion neurons in the distal part of the colon in both ET-3 and ET-B receptor knockout mice. In the skin of these mice, melanin-containing pigment cells were absent in color-affected areas. These findings indicate that the ET-1/ET-A receptor pathway and ET-3/ET-B receptor pathway are non-redundantly involved in the development of distinct cell lineages (discussed below).

The combined phenotype of aganglionic megacolon and pigmentary disorder resembles symptoms found in spontaneous mutant animals such as Piebald-lethal (sl) mice (LYON and SEARLE 1989) and Spotting lethal (sl) rats (LANE 1966). In both of these animal types mutations in the genomic locus encoding the ET-B receptor were demonstrated (HOSODA et al. 1998;

CECCHERINI et al. 1995; GARIEPY et al. 1996). Furthermore, transgenic expression of the ET-B receptor gene driven by the dopamine-β-hydroxylase promoter compensated for deficient ET-B receptor expression within enteric neurons and so prevented the development of aganglionic megacolon in a gene-dose-dependent manner (GARIEPY et al. 1998).

III. Gene Knockout of Endothelin-Converting Enzyme

Two proteolytic isoenzymes (ECE-1 and ECE-2) have been shown to catalyze the conversion of big ETs to mature active peptides (SHIMADA et al. 1994; XU et al. 1994; EMOTO and YANAGISAWA 1995) (see Chap. 3). Targeted null mutation of the mouse ECE-1 gene has been reported to reproduce the combined phenotype of the defects in the ET-1/ET-A receptor and ET-3/ET-B receptor pathways; namely, ECE-1 knockout mice show craniofacial and cardiovascular abnormalities, aganglionic megacolon and pigment disorder (YANAGISAWA et al. 1998b). This finding is a direct evidence for the essential role of ECE-1 in the proteolysis of ET-1 and ET-3 in vivo. However, large amounts of mature ET-1 are detected in ECE-1-null embryos, suggesting that other protease(s) may activate ET-1 processing whereas this is not sufficient to compensate the defect in ECE-1 (see Chap. 3).

C. Neural Crest and the Endothelin System

I. Outline of Neural Crest Development

As described above, deletion of the ET system in mice greatly affects the development of a number of tissues including craniofacial structures, great vessels, enteric neurons, and melanocytes. It has been well established that these tissues stem from a common ancestor, the neural crest. To understand the developmental roles of the ET system we will outline below the processes of neural crest development.

The neural crest is an embryonic structure which originates at the dorsalmost region of the neural tube and gives rise to multiple neural and non-neural cell lineages (KIRBY and WALDO 1995; GILBERT 1997; BAKER and BRONNER-FRASER 1997a,b; CREAZZO et al. 1998). Because of its developmental importance, the neural crest is often called 'the fourth germ layer.' Pluripotent stem cells arising from this structure migrate throughout the body and differentiate into the neurons and glial cells of the sensory and autonomic nervous systems, catecholamine-producing cells of the adrenal medulla, epidermal melanocytes, and ectomesenchymal cells constituting skeletal and connective tissues of the head. During cardiovascular development, neural crest cells from the caudal hindbrain (rhombomeres 6, 7, and 8) migrate into branchial arches 3, 4, and 6, and then a subset of these cells further migrate toward the outflow of the heart and form the aorticopulmonary septum and media of the great

Table 1. Major neural crest domains and their derivatives

Neural crest domain	Derivative
Cranial neural crest	Cranial ectomesenchyme forming branchial arches[a]
	Cartilage
	Bone
	Connective tissue
	Cranial nerve ganglia[a]
Trunk neural crest	Melanocytes[b]
	Dorsal root ganglia
	Sympathetic ganglia
	Adrenal medulla
Vagal and sacral neural crest	Parasympathetic (enteric) ganglia[b]
Cardiac neural crest	Connective tissue and smooth muscle in arch arteries[a]
	Aorticopulmonary septum[a]

[a] Affected by ET-1/ET-A null mutation
[b] Affected by ET-3/ET-B null mutation

vessels (KIRBY and WALDO 1995; GILBERT 1997; BAKER and BRONNER-FRASER 1997a,b; CREAZZO et al. 1998). The fate of neural crest cells is largely determined by the site of migration and settlement (GILBERT 1997; SIEBER-BLUM and ZHANG 1997) (see Table 1 for the four major functional domains of the neural crest and their derivatives). The contribution of the neural crest to cardiovascular development, for example, has been extensively studied by use of the avian neural crest ablation model (KURATANI and KIRBY 1991; KIRBY and WALDO 1995; CREAZZO et al. 1998). Removal of the cardiac neural crest before migration results in defective cardiovascular development, with defects including persisting truncus arteriosus, overriding aorta, ventricular septal defect, and variable great vessel anomalies such as aortic arch interruption. The thymic, thyroid, and parathyroid glands are also affected. Replacement of the removed cardiac neural crest with anterior cranial or trunk neural crests cannot rescue these phenotypes, indicating that the fate of the cardiac neural crest is already determined before migration (KURATANI and KIRBY 1991). Apparently, the cardiovascular phenotype of ET-1/ET-A receptor knockout mice is highly similar to a set of the phenotypes of the avian neural crest ablation model. Together with the distribution of tissues affected, the phenotypic manifestation of ET-1/ET-A receptor knockout mice indicates that ET-1/ET-A receptor-mediated signaling is essential for the development of cranial and cardiac neural crest-derived tissues and organs. Similarly, the phenotypes of ET-3/ET-B receptor knockout mice suggest that the ET-3/ET-B receptor-mediated signaling is indispensable for the development of melanocytes and enteric neurons derived from the trunk and vagal neural crest. Thus, two distinct pathways of the ET system have been demonstrated to participate in the normal development of different neural crest lineages.

II. Developmental Expression of the ET System

As outlined above, the developmental gene expression patterns of the ET system fit with the idea that the ET system is involved in neural crest development. ET-1 is expressed in the endothelium of the arch arteries and outflow tract of the heart as well as in the epithelium and paraxial mesoderm of the branchial arches (KURIHARA et al. 1994, 1995; CLOUTHIER et al. 1998; MAEMURA et al. 1996). These sites of ET-1 expression are not neural crest derivatives. In contrast, the ET-A receptor is intensely expressed in neural crest-derived ectomesenchyme of the branchial arches (CLOUTHIER et al. 1998; YANAGISAWA et al. 1998a; BRAND et al. 1998; KEMPF et al. 1998). These patterns of ET-1/ET-A receptor expression lead us to hypothesize that ET-1 may act as a paracrine factor on post-migratory neural crest cells at their destination. For the ET-3/ET-B receptor pathway, the ligand is expressed in mesenchyme of the gut wall and in the surrounding melanocyte precursors (NATAF et al. 1998; YANAGISAWA et al. 1998b; BRAND et al. 1998; KEMPF et al. 1998), whereas ET-B receptors are expressed in the neural tube as well as in melanocytes and enteric neuroblasts derived from the trunk and vagal neural crest (NATAF et al. 1996; BRAND et al. 1998; YANAGISAWA et al. 1998b; KEMPF et al. 1998) suggesting that the ET-3/ET-B receptor pathway could also act in a paracrine manner. Finally, the expression of ECE-1, which is essential for the activation of both ET-1/ET-A and ET-3/ET-B pathways, is strongly detected in surface epithelium and mesenchyme of the branchial arches and endocardium as well as enteric and epidermal mesenchyme (YANAGISAWA et al. 1998b).

III. Involvement of the ET System in Neural Crest Development

Clearly the most important question is whether the ET system is essential for migration, proliferation, and/or differentiation of neural crest-derived cells and what kind of signal pathways are employed by the ETs in producing their effects. Together with the expression patterns of the ET system, several findings support the idea that the ET-1/ET-A signaling pathway does not act on migrating neural crest cells but affects postmigratory neural crest-derived mesenchymal cells. For example, in the chick embryo, the ET-A-mediated signal was shown to be required only for a short period after neural crest cell migration (KEMPF et al. 1998). In addition, identification of migratory cranial neural crest cells using appropriate markers in ET-1 and ET-A receptor knockout mice indicated that the migration process was not affected (YANAGISAWA et al. 1998 and our unpublished data). As for the question of proliferation vs differentiation, no definite conclusion has been drawn as yet. In mouse neural crest cell cultures ET-1 and ET-3 stimulate the proliferation and differentiation of melanocyte progenitors (LAHAV et al. 1996; REID et al. 1996) and other neural crest lineages (STONE et al. 1997). The proliferative and differentiating effect of ETs on melanocyte progenitors are mediated by the ET-B receptor and are synergistic with Steel factor/c-Kit ligand (REID et al. 1996). Involvement of the ET-3/ET-B receptor pathway in enteric nerve development may be more complicated. Chimeras between normal and sl/sl mice (ET-B receptor-natural mutant) showed no aganglionic megacolon and sl/sl-derived enteric neurons were found in the distal colon, suggesting that the effect of ET-B receptor-null mutation is non-cell autonomous, that is through changes in the environment supportive of enteric nerve development (Kapur et al. 1993, 1995). However, ET-3 and GDNF additively stimulate the proliferation of post-migratory enteric neuron precursor cells, whereas ET-3 inhibits GDNF-induced proliferation and differentiation of enteric neurons, so maintaining the proliferating precursor cell pool (Hearn et al. 1998). These latter authors speculated that the apparent non-cell autonomous effect of ET-B receptor-null mutation may be due to a low number of precursor cells, which are necessary for extensive cell migration throughout the gut (Hearn et al. 1998).

The effect of ET-1/ET-A receptor pathway on cranial/cardiac neural crest development has not been clearly dissected in terms of proliferation and differentiation. It is also unknown whether the ET-1/ET-A-mediated signaling interacts with other factors involved in the development of neural crest-derived ectomesenchymal and/or smooth muscle cells, such as members of the TGFβ superfamily (Shah et al. 1996).

Some hints on the molecular mechanism driving the developmental effect of ET-mediated signaling may be derived from marker studies. For example, whole mount in situ hybridization in 9–11 days wild-type compared to knockout mouse embryos revealed that the expression of dHAND and eHAND, bHLH transcriptional factors involved in neural crest and heart development (Srivastava et al. 1995; Cserjesi et al. 1995) was diminished in the branchial arches and great vessels of ET-1 knockout mice, whereas the dHAND expression in the heart and limb bud was not affected (Thomas et al. 1998). Because dHAND knockout mice also exhibit branchial arch and cardiovascular defects (Srivastava et al. 1997; Thomas et al. 1998), downregulation of dHAND may be partly responsible for the branchial and great vessel anomalies of ET-1/ET-B knockout mice. In the branchial arches of dHAND knockout mice, the expression of the homeobox gene, Msx1 (Satokata and Maas 1994), is undetectable (Thomas et al. 1998). Thus, a signaling cascade involving ET-1/ET-A receptor, dHAND and Msx1 may participate in the regulation of cranial neural crest development.

Goosecoid, a homeobox gene, is also downregulated in the branchial arches of ET-1/ET-A receptor knockout mice (Clouthier et al. 1998). In goosecoid knockout mice, craniofacial structures including the tongue, nasal cavity and pits, auditory ossicles, tympanic ring, and external auditory meatus were severely affected as observed in ET-1/ET-A receptor knockout mice (Yamada G et al. 1995; Rivera-Perez et al. 1995). In addition, organ culture experiments demonstrated that the expression of Goosecoid in the branchial arches was maintained by ET-1 in conjunction with FGF-8 (Tucker et al. 1999). Taken together these findings indicate that the signaling pathway in

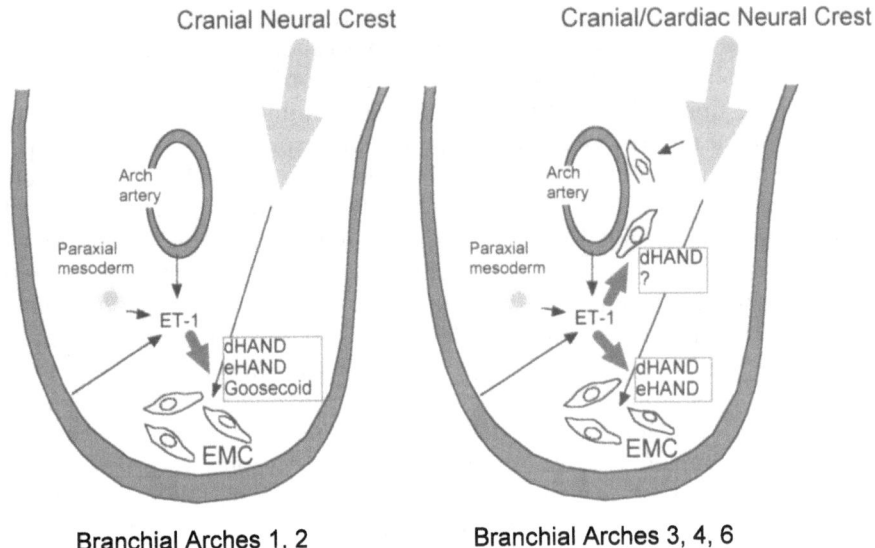

Fig. 1. Proposed role of ET-1 in the epithelial-mesenchymal interactions involved in cranial neural crest development. ET-1 produced by branchial epithelium, arch arterial endothelium, and paraxial mesoderm may act on cranial/cardiac neural crest cells to induce critical transcriptional factors and to promote their proliferation and/or differentiation into ectomesenchymal cells (*ECM*) and vascular smooth muscle cells (*SMC*)

volving ET-1 and Goosecoid may regulate the development of these structures in parallel with the possible ET-1–dHAND–Msx1 pathway. By utilizing these pathways, ET-1, which is mainly produced by epithelial and aortic endothelial cells as well as paraxial mesoderm-derived mesenchymal cells, may serve as a paracrine mediator of epithelial-mesenchymal interaction to promote craniofacial and cardiovascular morphogenesis in the branchial arches (Fig. 1).

D. Endothelin-Knockout Mice and Human Diseases

I. ET-3/ET-B Receptor Knockout Mice and Hirschsprung's Disease

The resemblance of some of the phenotypes produced in ET-knockout mice to some human diseases lead us to search for possible human gene mutations underlying these same diseases. Aganglionic megacolon in ET-3/ET-B receptor knockout mice is very similar to human Hirschsprung's disease and one autosomal dominant form of Hirschsprung's disease has been ascribed to mutations in the genomic locus encoding the receptor tyrosine kinase RET, which is mapped to 10q11.1 (EDERY et al. 1994). However, this disease is considered to be a multigenic disorder and other candidate loci have been suggested. As soon as the phenotype of ET-B receptor knockout mice appeared, Hosoda, the first author of this paper, enthusiastically searched for and eventually found a report that another type of Hirschsprung's disease is mapped to chromosome 13 (PUFFENBERGER et al. 1994a), where the ET-B receptor gene is also located (ARAI et al. 1993). Immediately these two groups started a collaboration and identified a G→T missense mutation in the exon 4 of the ET-B receptor gene that caused a W276C amino acid substitution in the fifth transmembrane domain of the receptor protein, on the haplotype associated with Hirschsprung's disease in a Mennonite kindred (PUFFENBERGER et al. 1994b). They also confirmed that the W276C substitution caused a significant reduction in ET-B-mediated signaling in transfected cells (PUFFENBERGER et al. 1994b). In the Mennonite population, the proportion of the C/C homozygotes was incomplete (74%) and some of the heterozygotes were also affected (21%). There was a higher incidence of megacolon in males than in females as is generally found in forms of Hirschsprung's disease. In addition, a small percentage of Hirschsprung patients in the Mennonites studied did not carry the W276C mutation. These facts suggest the presence of additional genes predisposing to the incidence of Hirschsprung's disease. It should also be noted that some of the affected Mennonites had pigmentary disorder and sensorineural hearing loss, which is characteristic of Waardenburg syndrome. Hirschsprung's disease and Waardenburg syndrome are regarded as neurocristopathies resulting from neural crest disorders and their association is often called Shah-Waardenburg syndrome. Subsequently, other mutations of the ET-B receptor gene were identified in both isolated Hirschsprung's disease (AMIEL et al. 1996; KUSAFUKA et al. 1996) and Shah-Waardenburg syndrome (ATTIE et al. 1995). Overall, ET-B receptor and RET mutations account for approximately 5% and 50%, respectively, of Hirschsprung's disease (CHAKRAVARTI 1996). ET-B receptor mutations were also found in lethal white foal syndrome (LWFS), a horse variant of Hirschsprung's disease (METALLINOS et al. 1998; YANG et al. 1998).

Mutations of the ET-3 gene have been also identified in patients with Hirschsprung's disease and Shah-Waardenburg syndrome (EDERY et al. 1996; HOFSTRA et al. 1996; BIDAUD et al. 1997), indicating that the ET-3-ET-B signaling pathway is essential for the development of neural crest-derived melanocytes and enteric neurons in humans, as in mice. Recently, Sox10, a member of the Sry-related HMG box-containing nucleoprotein family, has emerged as another predisposing gene for Waardenburg-Hirschsprung's disease. Mutations of the Sox10 gene was found both in patients with this disease (PINGAULT et al. 1998) and in Dom mice (SOUTHARD-SMITH et al. 1998; HERBARTH et al. 1998), whose phenotype is coat-spotting and aganglionic megacolon similar to that seen in ET-3/ET-B knockout mice. Mechanistic interactions among the Hirschsprung-predisposing genes, RET, ET-3/ET-B, Sox10, remains to be clarified.

II. ET-1/ET-A Receptor Knockout Mice and Craniofacial/Cardiovascular Disease

The phenotype of ET-1/ET-A receptor knockout mice shares apparent characteristics with human craniofacial and cardiovascular diseases. For example, the craniofacial phenotype is quite similar to first branchial arch syndrome, including Pierre-Robin syndrome and Treacher-Collins syndrome; the cardiovascular phenotype is reminiscent of congenital heart diseases such as aortic arch interruption and tetralogy of Fallot. These anomalies are often associated in humans and constitute a disease complex known as DiGeorge and velocardiofacial syndrome (PIKE and SUPER 1997; GOLDMUNTZ and EMANUEL 1997). DiGeorge and velocardiofacial syndrome is a combination of multiple anomalies including cleft palate, cardiac defects, learning difficulties, speech disorder, and characteristic facial features, with an estimated incidence of 1 in 5000. The majority of cases have a microdeletion of chromosome 22q11.2, conferring the acronym CATCH22 (Cardiac defect, Abnormal facies, Thymic hypoplasia, Cleft palate, Hypocalcemia, Chromosome 22) on this disease complex (PIKE and SUPER 1997; GOLDMUNTZ and EMANUEL 1997). The morphological resemblance between DiGeorge/velocardiofacial syndrome and ET-1/ET-A receptor knockout mice leads us to speculate that defects in ET-1/ET-A receptor signaling pathway may be involved in the pathogenesis of this congenital disease. However, no mutations of the ET-1 or ET-A receptor gene have as yet been identified in patients with this syndrome. Indeed, the ET-1 and ET-A receptor genes are not located in human chromosome 22. However, there may be interactions between ET-1 and ET-A receptor-mediated signaling and the disease-predisposing gene(s) in chromosome 22q11.2. This viewpoint may give us new insights leading to clarification of the molecular mechanisms underlying craniofacial and cardiovascular development and the pathogenesis of DiGeorge/velocardiofacial syndrome. Searches for possible ET-1/ET-A receptor mutations in patients with DiGeorge/velocardiofacial syndrome and related diseases without abnormalities in chromosome 22 are also under way.

The chromosomal location and relationship to human disease states of each component of the ET system are listed in Table 2.

E. Perspectives

Cell differentiation and morphogenesis are fundamental processes leading to the structural and functional integrity of individuals. Recent advances in genetic and cellular technology in the field of developmental biology have provided a lot of information about the molecular mechanisms underlying these developmental processes during embryogenesis. For example, the classical concept of epithelial-mesenchymal interactions inducing organogenesis has been substantiated by the discovery of signal mediators such as TGFβ superfamily members, FGFs, Shh, and so on. Now the ET system has also emerged

Table 2. Chromosomal location and relationship to human disease of genes encoding the components of the endothelin system

Gene	Chromosomal location (reference)	Related human disease
EDN1 (ET-1)	6p23–24 (Arinami et al. 1991)	CATCH22-like anomaly?
EDN2 (ET-2)	1p34 (Arinami et al. 1991)	Unknown
EDN3 (ET-3)	20q13.2–13.3 (Arinami et al. 1991; Gopal Rao et al. 1991)	Hirschsprung's disease (Shah-Waardenburg syndrome)
EDNRA (ET-A R)	4 (Hosoda et al. 1992)	CATCH22-like anomaly?
EDNRB (ET-B R)	13q22 (Puffenberger 1994b)	Hirschsprung's disease (Shah-Waardenburg syndrome)
ECE1	1p36.1 (Matsuoka et al. 1996)	Unknown

as an important participant in the molecular networks essential for neural crest development. The next important issues to be addressed include the interactions and hierarchies among individual signaling systems during embryogenesis and, more generally, the relationship between cellular events such as differentiation and morphogenesis.

The identification of mutations of the ET-3/ET-B receptor genes as underlying causes of Hirschsprung's disease is one of the most valuable products of gene targeting research into the ET system. This finding surely contributes to the understanding of the pathogenesis of Hirschsprung's disease and provides clues to the development of novel diagnostic and therapeutic approaches for this disease. Although no diseases have been attributed to mutations of the ET-1/ET-A receptor genes, further investigations on this signaling pathway are expected to pave the way to clarification of some of molecular bases of those congenital diseases associated with defects in cranial/cardiac neural crest development.

Finally, one must remember that the lethal phenotypes of the ET and ET receptor knockout mice may prevent other important biological functions of the ET system from emerging. Although some novel roles of the ET system in the cardiopulmonary system have been suggested by elaborate studies using heterozygous mice (Kuwaki et al. 1996, 1997; Ling et al. 1998; Nagase et al. 1998), the interpretation of experimental results is sometimes ambiguous due to a lack of viable homozygotes. To overcome this difficulty, tissue-specific knockout of the ET system will be awaited with interest for the future.

References

Amiel J, Attie T, Jan D, Pelet A, Edery P, Bidaud C, Lacombe D, Tam P, Simeoni J, Flori E, Nihoul-Fekete C, Munnich A, Lyonnet S (1996) Heterozygous endothelin receptor B (EDNRB) mutations in isolated Hirschsprung disease. Hum Mol Genet 5:355–357

Arai H, Nakao K, Takaya K, Hosoda K, Ogawa Y, Nakanishi S, Imura H (1993) The human endothelin-B receptor gene: structural organization and chromosomal assignment. J Biol Chem 268:3463–3470

Arinami T, Ishikawa M, Inoue A, Yanagisawa M, Masaki T, Yoshida MC, Hamaguchi H (1991) Chromosomal assignments of the human endothelin family genes: the endothelin-1 gene (EDN1) to 6p23-p24, the endothelin-2 gene (EDN2) to 1p34, and the endothelin-3 gene (EDN3) to 20q13.2-q13.3. Am J Hum Genet 48:990–996

Attie T, Till M, Pelet A, Amiel J, Edery P, Boutrand L, Munnich A, Lyonnet S (1995) Mutation of the endothelin-receptor B gene in Waardenburg-Hirschsprung disease. Hum Mol Genet 4:2407–2409

Baker CVH, Bronner-Fraser M (1997a) The origins of the neural crest. Part I: embryonic induction. Mech Dev 69:3–11

Baker CVH, Bronner-Fraser M (1997b) The origins of the neural crest. Part II: an evolutionary perspective. Mech Dev 69:13–29

Baynash AG, Hosoda K, Giaid A, Richardson JA, Emoto N, Hammer RE, Yanagisawa M (1994) Interaction of endothelin-3 with endothelin-B receptor is essential for development of epidermal melanocytes and enteric neurons. Cell 79:1277–1285

Bidaud C, Salomon R, Van Camp G, Pelet A, Attie T, Eng C, Bonduelle M, Amiel J, Nihoul-Fekete C, Willems PJ, Munnich A, Lyonnet S (1997) Endothelin-3 gene mutations in isolated and syndromic Hirschsprung disease. Eur J Hum Genet 5:247–251

Brand M Le Moullec JM, Corvol P, Gasc JM (1998) Ontogeny of endothelins-1 and –3, their receptors, and endothelin converting enzyme-1 in the early human embryo. J Clin Invest 101:549–559

Chakravarti A (1996) Endothelin receptor-mediated signaling in Hirschsprung disease. Hum Mol Genet 5:303–307

Clouthier DE, Hosoda K, Richardson JA, Williams SC, Yanagisawa H, Kuwaki T, Kumada M, Hammer RE, Yanagisawa M (1998) Cranial and cardiac neural crest defects in endothelin-A receptor-deficient mice. Development 125:813–824

Creazzo TL, Godt RE, Leatherbury L, Conway SJ, Kirby ML (1998) Role of cardiac neural crest cells in cardiovascular development. Ann Rev Physiol 60:267–286

Cserjesi P, Brown D, Lyons GE, Olson EN (1995) Expression of the novel basic helix-loop-helix gene eHAND in neural crest derivatives and extraembryonic membranes during mouse development. Dev Biol 170:664–678

Edery P, Lyonnet S, Mulligan LM, Pelet A, Dow E, Abel L, Holder S., Nihoul-Fekete C, Ponder BA, Munnich A (1994) Mutations of the RET proto-oncogene in Hirschsprung's disease. Nature 367:378–380

Edery P, Attie T, Amiel J, Pelet A, Eng C, Hofstra RMW, Martelli H, Bidaud C, Munnich A, Lyonnet S (1996) Mutation of the endothelin-3 gene in the Waardenburg-Hirschsprung disease (Shah-Waardenburg syndrome). Nature Genet12:442–444

Emoto N, Yanagisawa M (1995) Endothelin-converting enzyme-2 is a membrane-bound, phosphoramidon-sensitive metalloprotease with acidic pH optimum. J Biol Chem 270:15262–15268

Gariepy CE, Cass DT, Yanagisawa M (1996) Null mutation of endothelin receptor type B gene in spotting lethal rats causes aganglionic megacolon and white coat color. Proc Nat Acad Sci USA 93:867–872

Gariepy CE, Williams SC, Richardson JA, Hammer RE, Yanagisawa M (1998) Transgenic expression of the endothelin-B receptor prevents congenital intestinal aganglionosis in a rat model of Hirschsprung disease. J Clin Invest 102:1092–1101

Gilbert SF (1997) Developmental biology, 5th edn. pp 284–297, Sinauer Associates Inc., Sunderland

Goldmuntz E, Emanuel BS (1997) Genetic disorders of cardiac morphogenesis. The DiGeorge and velocardiofacial syndromes. Circ Res 80:437–443

Gopal Rao VVN, Loffler C, Hansmann I (1991) The gene for the novel vasoactive peptide endothelin 3 (EDN3) is localized to human chromosome 20q13.2-qter. Genomics 10:840–841

Hearn CJ, Murphy M, Newgreen D (1998) GDNF and ET-3 differentially modulate the numbers of avian enteric neural crest cells and enteric neurons in vitro. Dev Biol 197:93–105

Herbarth B, Pingault V, Bondurand N, Kuhlbrodt K, Hermans-Borgmeyer I, Puliti A, Lemort N, Goossens M, Wegner M (1998) Mutations of the Sry-related Sox10 gene in Dominant megacolon, a mouse model for Hirschsprung disease. Proc Nat Acad Sci USA 95:5161–5165

Hofstra RMW, Osinga J, Tan-Sindhunata G, Wu Y, Kamsteeg E,-J, Stulp RP, van Ravenswaaij-Arts C, Majoor-Krakauer D, Angrist M, Chakravarti A, Meijers C, Buys CHCM (1996) A homozygous mutation in the endothelin-3 gene associated with a combined Hirschsprung type 2 and Hirschsprung phenotype (Shah-Waardenburg syndrome). Nature Genet 12:445–447

Hosoda K, Nakao K, Tamura N, Arai H, Ogawa Y, Suga S, Nakanishi S, Imura H (1992) Organization, structure, chromosomal assignment, and expression of the gene encoding the human endothelin-A receptor. J Biol Chem 267:18797–18804

Hosoda K, Hammer RE, Richardson JA, Baynash AG, Cheung JC, Giaid A, Yanagi-sawa M (1994) Targeted and natural (Piebald-lethal) mutations of endothelin-B receptor gene produce megacolon associated with spotted coat color in mice. Cell 79:1267–1276

Kapur RP, Yost C, Palmiter RD (1993) Aggregation chimeras demonstrate that the primary defect responsible for aganglionic megacolon in lethal spotted mice is not neuroblast autonomous. Development 117:993–999

Kapur RP, Sweetser DA, Doggett B, Siebert JR, Palmiter RD (1995) Intercellular signals downstream of endothelin receptor-B mediate colonization of the large intestine by enteric neuroblasts. Development 121:3787–3795

Kempf H, Linares C, Corvol P, Gasc JM (1998) Pharmacological inactivation of the endothelin type A receptor in the early chick embryo: a model of mispatterning of the branchial arch derivatives. Development 125:4931–4941

Kirby ML, Waldo KL (1995) Neural crest and cardiovascular patterning. Circ Res 77:211–215

Kuratani SC, Kirby ML (1991) Initial migration and distribution of the cardiac neural crest in the avian embryo: an introduction to the concept of the circumpharyngeal crest. Am J Anat 191:215–227

Kurihara Y, Kurihara H, Suzuki H, Kodama T, Maemura K, Nagai R, Oda H, Kuwaki T, Cao W-H, Kamada N, Jishage K, Ouchi Y, Azuma S, Toyoda Y, Ishikawa T, Kumada M, Yazaki Y (1994) Elevated blood pressure and craniofacial abnormalities in mice deficient in endothelin-1. Nature 368:703–710

Kurihara Y, Kurihara H, Oda H, Maemura K, Nagai R, Ishikawa T, Yazaki Y (1995) Aortic arch malformations and ventricular septal defect in mice deficient in endothelin-1. J Clin Invest 96:293–300

Kurihara H, Kurihara Y, Maemura K, Yazaki Y (1997) The role of endothelin-1 in cardiovascular development. Ann NY Acad Sci 811:168–177

Kusafuka T, Wang Y, Puri P (1996) Novel mutations of the endothelin-B receptor gene in isolated patients with Hirschsprung's disease. Hum Mol Genet 5:347–349

Kuwaki T, Cao W-H, Kurihara Y, Kurihara H, Ling G-Y, Onodera M, Ju K-H, Yazaki Y, Kumada M (1996) Impaired ventilatory responses to hypoxia and hypercapnia in mutant mice deficient in endothelin-1. Am J Physiol270:R1279–R1286

Kuwaki T, Kurihara H, Cao WH, Kurihara Y, Unekawa M, Yazaki Y, Kumada M (1997) Physiological role of brain endothelin in the central autonomic control: from neuron to knockout mice. Prog Neurobio. 51:545–579

Lahav R, Ziller C, Dupin E, Le Douarin NM (1996) Endothelin 3 promotes neural crest cell proliferation and mediates a vast increase in melanocyte number in culture. Proc Nat Acad Sci USA 93:3892–3897

Lane PW (1966) Association of megacolon with two recessive spotting genes in the mouse. J Hered 57:29–31

Levin ER (1995) Endothelins. N Engl J Med 333:356–363

Ling GY, Cao WH, Onodera M, Ju KH, Kurihara H, Kurihara Y, Yazaki Y, Kumada M, Fukuda Y, Kuwaki T (1998) Renal sympathetic nerve activity in mice: comparison between mice and rats and between normal and endothelin-1 deficient mice. Brain Res 808:238–249

Lyon MF, Searle AG (1989) Genetic variants and strains of the laboratory mouse. 2nd edn. pp 319–320, Oxford University Press, Oxford.

Maemura K, Kurihara H, Kurihara Y, Oda H, Ishikawa T, Copeland NG, Gilbert DJ, Jenkins NA, Yazaki Y (1996) Sequence analysis, chromosomal location and developmental expression of the mouse preproendothelin-1 gene. Genomics 31:177–184

Masaki T (1995) Possible role of endothelin in endothelial regulation of vascular tone. Ann Rev Pharmacol Toxicol 35:235–255

Matsuoka R, Sawamura T, Yamada K, Yoshida M, Furutani Y, Ikura T, Shiraki T, Hoshikawa H, Shimada K, Tanzawa K, Masaki T (1996) Human endothelin converting enzyme gene (ECE1) mapped to chromosomal region 1p36.1. Cytogenet Cell Genet 72:322–324

Metallinos DL, Bowling AT, Rine J (1998) A missense mutation in the endothelin-B receptor gene is associated with lethal white foal syndrome: an equine version of Hirschsprung disease. Mammal Genome 9:426–431

Nagase T, Kurihara H, Kurihara Y, Aoki T, Fukuchi Y, Yazaki Y, Ouchi Y (1998) Airway hyperresponsiveness to methacholine in mutant mice deficient in endothelin-1. Am J Respir Crit Care Med 157:560–564

Nataf V, Lecoin L, Eichmann A, Le Douarin NM (1996) Endothelin-B receptor is expressed by neural crest cells in the avian embryo. Proc Nat Acad Sci USA 93:9645–9650

Nataf V, Amemiya A, Yanagisawa M, Le Douarin NM (1998) The expression pattern of endothelin 3 in the avian embryo. Mech Dev 73:217–220

Pike AC, Super M (1997) Velocardiofacial syndrome. Postgrad Med J 73:771–775

Pingault V, Bondurand N, Kuhlbrodt K, Goerich DE, Prehu MO, Puliti A, Herbarth B, Hermans-Borgmeyer I, Legius E, Matthijs G, Amiel J, Lyonnet S, Ceccherini I, Romeo G, Smith JC, Read AP, Wegner M, Goossens M (1998) Sox10 mutations in patients with Waardenburg-Hirschsprung disease. Nature Genet 18:171–173

Puffenberger EG, Kauffman ER, Bolk S, Matise TC, Washington SS, Angrist M, Weissenbach J, Garver KL, Mascari M, Ladda R, Chakravarti A (1994a) Identity-by-descent and association mapping of a recessive gene for Hirschsprung disease on human chromosome 13q22. Hum Mol Genet 3:1217–1225

Puffenberger EG, Hosoda K, Washington SS, Nakao K, deWit D, Yanagisawa M, Chakravarti A (1994b) A missense mutation of the endothelin-B receptor gene in multigenic Hirschsprung's disease. Cell 79:1257–1266

Reid K, Turnley AM, Maxwell GD, Kurihara Y, Kurihara H, Bartlett PF, Murphy M (1996) Multiple roles for endothelin 3 in melanocyte development: Regulation of progenitor number and stimulation of differentiation. Development 122:3911–3919

Rivera-Perez JA, Mallo M, Gendron-Maguire M, Gridley T, Behringer RR (1995) Goosecoid is not an essential component of the mouse gastrula organizer but is required for craniofacial and rib development. Development. 121:3005–3012

Satokata I and Maas R (1994) Msx1 deficient mice exhibit cleft palate and abnormalities of craniofacial and tooth development. Nature Genet 6:348–356

Shah NM, Groves AK, Anderson DJ (1996) Alternative neural crest cell fates are instructively promoter by TGFb superfamily members. Cell. 85:331–343

Shimada K, Takahashi M, Tanzawa K (1994) Cloning and functional expression of endothelin-converting enzyme from rat endothelial cells. J Biol Chem 269:18275–18278

Sieber-Blum M, Zhang J-M (1997) Growth factor action in neural crest diversification. J Anat 191:493–499

Southard-Smith EM, Kos L, Pavan WJ (1998) Sox10 mutation disrupts neural crest development in Dom Hirschsprung mouse model. Nature Genet 18:60–64

Srivastava D, Cserjesi P, Olson EN (1995) A subclass of bHLH proteins required for cardiac morphogenesis. Science 270:1995–1999

Srivastava D, Thomas T, Lin Q, Kirby ML, Brown D, Olson EN (1997) Regulation of cardiac mesodermal and neural crest development by the bHLH transcription factor, dHAND. Nature Genet 16:154–160

Stone JG, Spirling LI, Richardson MK (1997) The neural crest population responding to endothelin-3 in vitro includes multipotent cells. J Cell Sci 110:1673–1682

Thomas T, Kurihara H, Yamagishi H, Kurihara Y, Yazaki Y, Olson EN, Srivastava D (1998) A signaling cascade involving endothelin-1, dHAND and Msx1 regulates development of neural crest-derived branchial arch mesenchyme. Development 125:3005–3014

Tucker AS, Yamada G, Grigoriou M, Pachnis V, Sharpe PT (1999) Fgf-8 determines rostral-caudal polarity in the first branchial arch. Development 126:51–61

Xu D, Emoto N, Giaid A, Slaughter C, Kaw S, deWit D, Yanagisawa M (1994) ECE-1: a membrane-bound metalloprotease that catalyzes the proteolytic activation of big endothelin-1. Cell 78:473–485

Yamada G, Mansouri A, Torres M, Stuart ET, Blum M, Schultz M, De Robertis EM, Gruss P (1995) Targeted mutation of the murine goosecoid gene results in craniofacial defects and neonatal death. Development. 121:2917–2922

Yanagisawa H, Hammer RE, Richardson JA, Williams SC, Clouthier DE, Yanagisawa M (1998a) Role of endothelin-1/endothelin-A receptor-mediated signaling pathway in the aortic arch patterning in mice. J Clin Invest 102:22–33

Yanagisawa H, Yanagisawa M, Kapur RP, Richardson JA, Williams SC, Clouthier DE, de Wit D, Emoto N, Hammer RE (1998b) Dual genetic pathways of endothelin-mediated intercellular signaling revealed by targeted disruption of endothelin converting enzyme-1 gene. Development 125:825–836

Yang GC, Croaker D, Zhang AL, Manglick P, Cartmill T, Cass D (1998) A dinucleotide mutation in the endothelin-B receptor gene is associated with lethal white foal syndrome (LWFS); a horse variant of Hirschsprung disease (HSCR). Hum Mol Genet 7:1047–1052

CHAPTER 7

Endothelin-Converting Enzyme Inhibitors and their Effects

B. BATTISTINI and A.Y. JENG

A. Introduction

Previous chapters in this handbook have reviewed extensively the biochemistry and pharmacology of the family of three endothelin (ET) isopeptides and their respective precursors. Pathways responsible for the synthesis of ETs have also been introduced. In this chapter, these subjects as well as the existence and cellular/tissue localization of endothelin-converting enzyme (ECE) isoforms are nevertheless reviewed briefly as they are important in understanding the development of inhibitors of the synthesis of ET isopeptides. The major focus of the present chapter remains on the design and pharmacological properties of various types of ECE inhibitors and the effectiveness of these compounds in isolated cells, organs, tissues, and in animal models of human pathophysiology. An overall view of the comparative advantages and disadvantages of the use of ET receptor antagonists vs ECE inhibitors for the inhibition of the ET-mediated effects is also presented.

I. Mature ET Peptides and their Precursors

The ETs, originally isolated from conditioned medium of porcine aortic endothelial cells (PAECs), are the most potent peptidic vasoconstrictors (YANAGISAWA et al. 1988) and bronchoconstrictors (UCHIDA et al. 1988) discovered to date. Subsequent studies have shown that ET-1 is also produced by a variety of cell types including epithelial cells, smooth muscle cells, and macrophages (reviewed by BATTISTINI et al. 1993).

There are three isoforms in the human ET family of peptides, namely ET-1, ET-2, and ET-3. They all have 21 amino acid residues and 2 intramolecular disulfide bonds at the 1–15 and 3–11 positions, with the Trp^{21} residue at the C-terminus. However, ET-2 and ET-3 differ from ET-1 by two and six amino acids (ET-2=[Trp^6,Leu^7]ET-1; ET-3=[Thr^2,Phe^4,Thr^5,Tyr^6, Lys^7,Tyr^{14}]ET-1), respectively (INOUE et al. 1989). Such characteristic variations are responsible for the observed differences in their selective binding to ET receptors (see Chap. 4). These mature active ET isopeptides have the same amino acid sequence regardless of the mammalian species examined, e.g., human, porcine, canine, rat, mouse, or bovine.

Table 1. Comparison of the amino acid sequences of the C-terminal fragments of the big ETs

Precursor	Species	Sequence prepro-ET	# Amino acid	21	22	23	24	25	26	27	28	29	30	31
Big ET-1	Human	212	38	Trp	Val	Asn	Thr	Pro	Glu	His	Val	Val	Pro	Tyr
	Bovine	202	39											
	Porcine	203	39								Ile			
	Dog	56	26											
	Rabbit	202	39						Gly		Ile			
	Rat	202	39							Arg				
	Mouse	51	25											
Big ET-2	Human	178	37							Gln	Thr	Ala		
	Human		38							Gln	Thr	Ala		
	Dog	56	26						Gly					
	Rat	51	25					Ala						
	Mouse	160	37					Ala	Gly	Gln	Thr	Ala		
Big ET-3	Human	238	41		Ile					Gln	Thr			
	Human	224	42		Ile					Gln	Thr			
	Rabbit	114	26		Ile									
	Rat	167	41		Ile					Gln	Thr			
	Mouse	214	40		Ile					Gln	Thr			

Human ET-1 is initially synthesized as a 212-amino acid prepro-ET-1, whereas the porcine prepro-ET-1 has 203 amino acid residues (see Chap. 3). In this latter molecule, the 21 amino acid ET-1 is located between Cys^{53} and Trp^{73} (YANAGISAWA et al. 1988). The prepro-ET-1 is then sequentially hydrolyzed by dibasic amino acid endopeptidases and furin convertase to generate a precursor named big ET-1 (reviewed by TURNER and MURPHY 1996; TURNER and TANZAWA 1997). The human ET-1 precursor is a 38-amino acid peptide, whereas porcine or rat big ET-1 has 39 amino acid residues (Table 1). In comparison, human big ET-2 has either 37 or 38 amino acids while human big ET-3 has either 41 or 42 amino acids and is amidated at the C-terminus (BLOCH et al. 1989, 1991; ONDA et al. 1990) (Table 1). In addition to these differences in the length of the precursors within the same species, there are variations in the C-terminal amino acid sequence amongst both the big ETs and species (Table 1). Most importantly the cleavage site of big ET-3 is constituted of a Trp^{21}-Ile^{22} peptide bond, whereas all other precursors are cleaved between Trp^{21} and Val^{22}. These differences most probably define the varied nature and selectivity of ECEs that catalyze the conversion of big ETs to mature ETs. For instance, structure-activity relationship studies revealed that human big ET-1_{1-31} (i.e., truncated by 7 amino acid residues at the C-terminal end) and human big ET-1_{17-26} were not specifically hydrolyzed by ECE purified from bovine endothelial cells (BECs) (OKADA et al. 1991). Conversely, human big ET-1_{16-38}, a linear truncated analogue of big ET-1_{1-38}, was cleaved at a rate 4-times or 2.7-times greater than the authentic substrate using ECE purified from BAECs (OKADA et al. 1991) or PAECs (OHNAKA et al. 1993), respectively.

32	33	34	35	36	37	38	39	40	41	42		References
Gly	Leu	Gly	Ser	Pro	Arg	Ser					COOH	Itoh et al. (1988)
					Ser	Arg	Ser				COOH	Imai et al. (1992)
					Ser	Arg	Ser				COOH	Itoh et al. (1988); Takaoka et al. (1990a)
											COOH	Kimura et al. (1989)
					Ser	Arg	Ser				COOH	Marsden et al. (1992)
					Ser	Arg	Ser				COOH	Shiba et al. (1992); Nunez et al. (1991); Sakurai et al. (1991)
											COOH	Saida et al. (1989)
		Asn		Pro							COOH	Onda et al. (1991); Ohkubo et al. (1990)
		Asn		Pro	Arg						NH₂?	Suzuki et al. (1991)
											COOH	Itoh et al. (1989)
											COOH	Bloch et al. (1991)
		Asn		Pro							COOH	Saida and Mitsui (1991)
	Ser	Asn	Tyr			Gly	Ser	Phe	Arg		NH₂	Bloch et al. (1989)
	Ser	Asn	Tyr			Gly	Ser	Phe	Arg	Gly	COOH	Onda et al. (1990)
											COOH	Ohkubo et al. (1990)
	Ser	Asn	His			Gly	Ser	Leu	Arg		NH₂	Shiba et al. (1992)
	Ser	Asn	Tyr			Glu	Ser	Leu			COOH	Baynash et al. (1994)

Likewise, big ET-2$_{18-34}$, the linear truncated portion of human big ET-2$_{1-37}$, was converted at a 5-times greater rate than the authentic precursor using ECE purified from PAECs (Ohnaka et al. 1993). These results suggest that the Trp^{21} residue and the carboxy-terminal sequence between His^{27} and Gly^{34} of big ET-1 are essential for recognition by ECE while the presence of the amino-terminal disulfide loop structure may reduce the rate of hydrolysis, e.g., the access of big ET-1 to ECE (Okada et al. 1991).

II. ECE: Cloning, Classification, and Cellular Localization

In 1988, only one type of ECE was postulated for the final post-translational processing of ET-1, a step which is essential for the full expression of its biological activity (Kimura et al. 1989). At present, seven isoforms of ECEs, namely human, bovine, rat ECE-1a (also called ECE-1β in rats) (Xu et al. 1994; Ikura et al. 1994; Shimada et al. 1995), human, bovine, rat ECE-1b (also called ECE-1α in rats) (Shimada et al. 1994; Schmidt et al. 1994; Yorimitsu et al. 1995; Valdenaire et al. 1995), human ECE-1c (Schweizer et al. 1997), human and rat ECE-1d (Valdenaire et al. 1999), human and bovine ECE-2 (Emoto and Yanagisawa 1995), which is now divided as mouse/bovine ECE-2a (the original bovine ECE-2) and ECE-2b spliced variants (Nakahara et al. 1999), and bovine ECE-3 (Hasegawa et al. 1998) have been identified and cloned (reviewed by Battistini and Botting 1995; Turner and Murphy 1996; Turner and Tanzawa 1997). Mouse SEP, a novel soluble secreted metalloprotease (Ikeda et al. 1999) and XCE (Schweizer et al. 1999), are also new members of the ECE and neutral endopeptidase families (NEP, EC 3.4.24.11, also named enkephalinase, a zinc metalloprotease). SEP is not a physiologically relevant ECE but is involved in subsequent degradation of ET, while XCE is preferentially expressed in specific areas of the central nervous system. Thus, it is now recognized that the conversion of big ETs to mature ETs is catalyzed by various ECEs. Criteria for their classification, which was presented and regrouped following the Fourth International Conference on Endothelin in London, UK (1995), include: (a) a distinct N-terminal amino acid transmembrane domain sequence, (b) optimal pH for enzyme activity, (c) cellular localization, (d) substrate specificity for the processing of the three big ETs, and (e) sensitivity to inhibition by phosphoramidon (Table 2) (see BAT-TISTINI and BOTTING 1995).

ECEs are type II integral membrane-bound proteases, with a single trans-membrane-spanning domain, a conserved zinc-binding motif, N-glycosylation sites, a short N-terminal cytoplasmic tail, and a large putative extracellular domain containing the catalytic site (KORTH et al. 1997; SANSOM et al. 1995). Human ECE-1a and -1b show considerable overall amino acid sequence identity (>91%) (YORIMITSU et al. 1995). Rat ECE-1a has 37% and 31% homology with rat NEP and Kell blood group protein, respectively. The sequence similarity to NEP/Kell protein is highest within the C-terminal third of the protein that includes the zinc-binding motif. Within this region, ECE-1a exhibits 58% and 36% homology to NEP and Kell protein, respectively (XU et al. 1994). In addition, ECE-1a/-1b (deduced molecular weight: 85–86KDa), ECE-1c (deduced molecular weight: 87KDa), and ECE-1d possess comparable catalytic mechanism to NEP, but with distinct substrate binding sites. Unlike NEP, ECE-1 exist as dimers through a single Cys^{412} intermolecular disulfide bond (dimeric form of ECE-1: 250KDa; SHIMADA et al. 1996). The four ECE-1 isoforms differ only in their N-terminal regions and are derived from a single gene through alternative splicing (generated through alternative promoters) (SCHWEIZER et al. 1997). The overall sequence of bovine ECE-2 (ECE-2a) (deduced molecular weight: 89KDa) has only 53–59% homology to human ECE-1a, whereas rat and human ECE-1b are 71% identical (VALDENAIRE et al. 1999).

The four ECE-1s have an optimum activity at neutral pH of 6.8–7.2. ECE-2a is optimally active at an acidic pH of 5.5 (EMOTO and YANAGISAWA 1995), while ECE-3 operates best at around pH 6.6 (HASEGAWA et al. 1998). The ECE-1a, -1b, and -1c isoenzymes and ECE-2a convert big ET-1 more rapidly than big ET-2 and big ET-3 (TURNER and MURPHY 1996), whereas ECE-3 converts big ET-3 specifically (HASEGAWA et al. 1998).

These isoenzymes have distinct subcellular localizations. While human ECE-1a (758 amino acid residues), -1c (754 residues), and -1d (767 residues) are localized at the cell surface as ectoenzymes, ECE-1b (753–754 or 770 residues) is found to be intracellular, close to the *trans*-Golgi network (SCHWEIZER et al. 1997; CAILLER et al. 1999; VALDENAIRE et al. 1999). ECE-2a (765–787 residues) is also found to be intracellular (EMOTO and YANAGISAWA

Table 2. List of proteases that have been shown to convert ET precursors, the big ETs, into mature and active ETs

	pH	Reference
Metalloprotease		
ECE-1a	6.8–7.2	Xu et al. (1994); Ikura et al. (1994); Shimada et al. (1995)
ECE-1b	6.8–7.2	Schmidt et al. (1994); Shimada et al. (1995); Valdenaire et al. (1995)
ECE-1c	6.8–7.2	Schweizer et al. (1997)
ECE-1d	6.8	Valdenaire et al. (1999)
ECE-2 (ECE-2a)	5.5	Emoto and Yanagisawa (1995)
ECE-2b	NA	Nakahara et al. (1999)
ECE-3	6.6	Hasegawa et al. (1998)
SEP	7.0	Ikeda et al. (1999)
XCE	NA	Schweizer et al. (1999)
Aspartyl protease		
Pepsin-like	2.3	Takaoka et al. (1990c)
Cathepsin D, pepstatin A-sensitive	3.5	Sawamura et al. (1990a,b)
Cathepsin D-like, pepstatin A-sensitive	4.0	Ikegawa et al. (1990a)
Cathepsin E	4.7	Lees et al. (1990)
Pepstatin A-sensitive	3.0	Ohnaka et al. (1990)
Pepsin	NA	Pons et al. (1991)
Metal ion aspartic protease, pepstatin A-sensitive	4.0	Wu-Wong et al. (1990, 1991); Shiosaki et al. (1993)
Chymotrypsin, pepsin or chymosin-like, pepstatin A-sensitive	NA	Savage et al. (1993)
Pepstatin A-sensitive, unrelated to cathepsin D or renin	4.0	Knap et al. (1993)
Thiol protease		
p-Hydroxymercuribenzoate-sensitive	7.0–7.5	Deng et al. (1992)
Serine protease		
Chymotrypsin	NA	McMahon et al. (1989)
Chymotrypsin-like	8.0	Takaoka et al. (1990b)
α-Chymotrypsin	NA	Pons et al. (1991)
Rat lung mast cell chymase I, Chymostatin-sensitive	8.5	Wypij et al. (1992)
Human mast cell chymase	NA	Nakano et al. (1997); Kido et al. (1998)
Monkey chymase, chymostatin-sensitive	NA	Takai et al. (1998)

NA, not available

1995). The cellular localization of ECE-3 (140 Kd, number of residues not known) is not yet identified (Hasegawa et al. 1998). The cellular distributions may suggest that ECE-1b and ECE-2, the intracellular isoenzymes, are mostly responsible for big ET-1 conversion via a constitutive secretory pathway whereas ECE-1a, -1c, and -1d, as ectoenzymes, may also have an important post-secretory processing role (RUSSELL et al. 1998; RUSSELL and DAVENPORT 1999).

ECE-1a, -1b, and -1d are present in human cultured umbilical vein ECs (HUVECs) (SHIMADA et al. 1995; BROWN et al. 1998; VALDENAIRE et al. 1999). ECE-1b was also shown by western blot analysis and immunofluorescence to be present in rat cultured lung vascular ECs (TRLEC-03), and to be sensitive to phosphoramidon (SCHMIDT et al. 1994). ECE-2a is mostly expressed in the nervous system and adrenal medulla (EMOTO and YANAGISAWA 1995). ECE-3, identified in bovine iris microsomes, is also present in choroid (HASEGAWA et al. 1998).

III. Alternative Pathways for ET Production and Perspectives Provided by Gene-Targeting Studies

The results obtained from physiopharmacological approaches using ET precursors (see below) and the recent cloning of the ECEs reveal that the physiological relevant ECEs constitute a distinct phosphoramidon-sensitive metalloendoprotease family-type of enzymes, related to NEP and Kell blood group proteins (OPGENORTH et al. 1992; BATTISTINI and BOTTING 1995; TURNER and MURPHY 1996). Furthermore, the phosphoramidon-sensitive ECE that was initially reported in cytosolic fractions of cultured ECs isolated from bovine carotid artery endothelial cells (BCAECs) (TAKADA et al. 1991) could now be associated to ECE-1b and/or ECE-2. However, the cloning of ECEs have not completely excluded the participation of alternative pathways in the production of ETs (see Table 2).

Enzymes that catalyze the conversion and/or the degradation of peptides can be classified into six families: serine I, II, cysteine, aspartic, and metalloprotease I, II (BOND and BEYNON 1987). Besides the type II metalloprotease, other classes of proteases (aspartyl-, thiol-, and serine proteases) have been reported to convert or process big ETs into ETs (see TURNER and MURPHY 1996; TURNER and TANZAWA 1997). In addition, the post-translational processing of ET-1 can be catalyzed by other metalloproteases. For example, a phosphoramidon-insensitive metalloprotease capable of converting exogenous big ET-1_{1-39} was detected in cultured ECs isolated from porcine thoracic aortas (PTAECs) (MATSUMURA et al. 1991c). Recently, mouse SEP was reported to cleave specifically the exogenous big ET-1 at the Trp^{21}-Val^{22} peptidic bond to generate mature ET-1, but it also subsequently degraded ET-1 at multiple sites into inactive fragments (IKEDA et al. 1999). Thus, SEP is an endothelin-degrading enzyme (EDE) rather than a physiologically relevant ECE. The Kell blood group protein was shown to cleave big ET-3 into ET-3, and to a lesser extent big ET-1 and -2 (RUSSO et al. 1999).

1. Aspartyl Proteases

ECE-like activities have been associated with aspartyl proteases. For instance, treatment with pepsin converted porcine big ET-1_{1-39} to the 21 amino acid

ET-1 in vitro with no subsequent degradation of the formed ET-1 under those conditions (TAKAOKA et al. 1990c). An ECE-like activity was demonstrated in the soluble fraction of cultured BAECs, corresponding to the elution profile of cathepsin D and sensitive to pepstatin A (SAWAMURA et al. 1990a). Pepstatin A also abolished the converting activity in the homogenate of cultured BAECs (OHNAKA et al. 1990). Another aspartic protease, also identified as cathepsin D, was purified from bovine adrenal chromaffin granules and shown to be capable of converting big ET-1 into ET-1 (SAWAMURA et al. 1990b). Evidence for a pepstatin A-sensitive conversion of porcine big ET-1 to ET-1 was likewise reported in an extract of cultured PAECs (IKEGAWA et al. 1990a). Further experiments have suggested that aspartic proteases from an extract of PAECs presented an ECE-like activity, cleaving big ET-1 to ET-1 in vitro (KNAP et al. 1993). A metal ion aspartic protease, acting optimally at pH 4.0 and sensitive to pepstatin A, likewise converted exogenous big ET-1 into ET-1 when incubated with a membrane fraction isolated from rat lungs (WU-WONG et al. 1990, 1991; SHIOSAKI et al. 1993). Notably, human purified cathepsin E, but not cathepsin D, cleaved human big ET-1 into ET-1 in vitro and was not associated with any further degradation of the mature peptide (LEES et al. 1990). The ability of cathepsin D, chymosin, and pepsin to produce ET-1 from big ET-1 in vitro was also reported (SAVAGE et al. 1993). The product of chymosin treatment caused the contraction of isolated rabbit aortic rings, and pre-incubation of chymosin with pepstatin A abolished the contractile response. Similarly, when big ET-1 was incubated in vitro with pepsin and injected into guinea pigs at a dose of 1 nmol/kg, marked rapid bronchoconstrictor and pressor responses were observed (PONS et al. 1991). Interestingly, pepstatin A caused a significant decrease in the basal secretion of ET-1 from HUVECs (PLUMPTON et al. 1994). However, it was not associated with an increase in big ET-1.

Conversely, aspartic protease inhibitors (such as pepstatin A) did not affect the basal release of ET-1 from PAECs (NICHOLS et al. 1991). In addition, pepstatin A did not alter the ratio of ET-1(1–21) to big ET-1$_{1-39}$, nor did it affect the rate of secretion of either peptide from PAECs (SHIELDS et al. 1991). Furthermore, SQ 32056, a potent inhibitor of cathepsin E, inhibited the specific pressor response to exogenous big ET-1 in anesthetized rats in a non-specific manner since it also blocked ET-1-induced response (BIRD et al. 1992). Since cathepsin D is only active at an acidic pH and it cleaves big ET-1 in a rather non-specific manner, e.g., between the Asn[18]-Ile[19] peptide bond as well as between the proposed Trp[21]-Val[22] peptide bond (SAWAMURA et al. 1990c; TAKAOKA et al. 1990a), this enzyme is not a physiologically relevant ECE.

2. Thiol (Cysteine) Proteases

Other experiments have suggested that a thiol protease could act as an ECE by cleaving big ET-1 to ET-1 in vitro, as found in a soluble extract from PAECs (DENG et al. 1992). The product that had a retention time on HPLC identical to ET-1 also contracted rabbit aortic rings. The conversion was blocked by thiol protease inhibitors such as Z-phe-pheCHN$_2$ or p-hydroxymercuribenzoate, but not by serine protease inhibitors.

3. Serine Proteases

Incubation of big ET-1 with chymotrypsin activated porcine big ET-1$_{1-39}$ to produce peptides that induced contractions of isolated rat aortic rings and increased blood pressure in rats with potencies similar to that of ET-1 (McMahon et al. 1989). Similarly, when big ET-1 was incubated in vitro with alpha-chymotrypsin and injected into guinea pigs at a dose of 1 nmol/kg, marked and rapid bronchoconstrictive and pressor responses were observed (Pons et al. 1991). In addition, in a rat lung particulate fraction, a chymostatin-inhibitable mast cell chymase I was reported to process exogenous big ET-1 into ET-1 (Wypij et al. 1992).

The cleavage of the big ET-1 molecule by chymotrypsin is initially at the Tyr^{31}-Gly^{32} peptide bond followed by the Trp^{21}-Val^{22} link and then at the Tyr^{13}-Phe^{14} peptide bond (Takaoka et al. 1990b; Ikegawa et al. 1990a). Thus, the chymotrypsin-like endopeptidase may not physiologically contribute to the processing of big ET-1. However, recent reports showed that ET-1, -2, and -3$_{1-31}$, cleaved from their respective precursors by chymase in human mast cells, did not produce any further degradation products. These peptides are active contractile mediators in rat trachea and porcine coronary arteries, and their activities are unaffected by phosphoramidon (Nakano et al. 1997; Kido et al. 1998). Similarly, purified monkey chymase was shown to process big ET-1/-2 to ET-1/-2$_{1-31}$ that induces contractile responses in the isolated monkey trachea (Takai et al. 1998). Chymases from other species such as human, cathepsin G, and porcine alpha-chymotrypsin all degrade big ETs (Kido et al. 1998).

4. Gene Targeting

Homozygous ECE-1 gene knockout (ECE-1$^{-/-}$) mice display craniofacial and cardiac abnormalities virtually identical to those observed in ET-1- and ET$_A$ receptor-deficient animals (Yanagisawa et al. 1998) and have the developmental phenotype similar to that exhibited in the ET-3- and ET$_B$ receptor-deficient animals (see Chap. 6). Most surprisingly, no developmental abnormalities have been found in ECE-2$^{-/-}$ homozygous mice (Yanagisawa et al. 1999), suggesting that ECE-2 is not the primary enzyme for the generation of ET-1.

Intravenous administration of human big ET-1$_{1-38}$ into ECE-1$^{-/-}$ homozygous knockout mice (75% C57BL/6: 25% J129) produced similar pressor responses as those observed in wild-type anesthetized mice (N. Berthiaume et al., personal communication). These results suggest that ECE-1 may not be the only enzyme involved in the conversion of big ET-1 to ET-1 in vivo. Furthermore, the plasma levels of ET-1 in the ECE-1$^{-/-}$ mice are reduced by 30% when compared with the wild-type controls, suggesting again that ECE-1 may not be the only enzyme responsible for the endogenous production of the ETs and that other ECE(s) and non-ECE protease(s) also convert big ET-1 in vivo. These observations are not unexpected when considering that there are thus far seven distinct ECE isoforms cloned. In addition, there are non-selective, ECE-like proteases, such as SEP, XCE and Kell blood group protein.

IV. Modulation of ECE Expression and Activity

Increases in the production and release of ETs in vivo and/or in isolated cells in vitro or in various pathological states (see BATTISTINI et al. 1993) are likely to be due to the activation of ECE activities to convert big ETs into ETs. However, it is not always the case. For instance, physiologic shear stress, which dilates blood vessels, suppresses the expression of mRNA for ECE-1 in cultured BCAECs and HUVECs (MASATSUGU et al. 1998). Conversely, steady shear stress has little or no effect on the expression of ECE-1 isoforms, but enhances ET-1 gene expression in EAhy 926 cells (HARRISON et al. 1998), suggesting that these two genes are differentially regulated by the same stimulus in this cell type. ECE activity, in primary rat astrocytes, was reported to be modulated by ET_B receptor ligands (EHRENREICH et al. 1999). It was suggested that this action was induced by binding to ET_B receptors which then affected ECE-1 glycosylation. Higher basal concentrations of ET-1 were found in the media of astrocytes from ET_B-deficient than wildtype rats, suggesting an altered ECE activity related to the absence of ET_B receptors. Similarly, ET-1 itself was reported to inhibit ECE-1 mRNA expression and suppress ECE-1 protein levels through ET_B-receptor activation in rat pulmonary ECs in vitro (NAOMI et al. 1998). To that effect, angiotensin (Ang II) was reported to increase functional ECE activity in VSMCs through ET_A receptors in WKY rats in vivo (BARTON et al. 1997).

V. ECE: Substrate Specificity and pH Dependence

ECEs cleave the Trp^{21}-Val^{22} or Trp^{21}-Ile^{22} peptide bond found in big ETs, converting the precursors into active mature ET peptides. However, another physiological peptide, bradykinin (BK), a vasodilator, has been identified as a substrate for ECE (HOANG and TURNER 1997). Acting as a peptidyl dipeptidase, the rat ECE-1a expressed in CHO cells cleaved BK at Pro^7-Phe^8, producing BK (1–7) and BK (8–9). The conversion was sensitive to phosphoramidon as well. Angiotensin I, substance P, neurotensin, and the oxidized insulin B chain are also hydrolyzed as efficiently as big ET-1 by ECE-1a (JOHNSON et al. 1999).

ECE-1 activity is greatly affected by slight variations in pH. It was reported that phosphoramidon inhibited ECE-1 in a remarkably pH-dependent manner (AHN et al. 1998). With a small change of pH (1.4 pH unit), the potency of phosphoramidon varied between 49- and 1000-fold for ECE-1a inhibition. Similar results were also observed for the ECE inhibitors, PD 069185 and CGS 31447 (AHN et al. 1999). It should be noted that ECE-1s are either localized at the plasma membrane (ECE-1a, c, d) or intracellularly (ECE-1b) where the pH environment differs (neutral or acidic).

These observations may aid in the design of new and more specific ECE inhibitors. They also suggest the need to identify other physiological substrates for ECEs.

B. Development of ECE Inhibitors: Chemical Identity, Selectivity, Potency, and Oral Bioavailability

Phosphoramidon, a potent inhibitor of neprilysin, or NEP, was the first compound identified as a non-selective dual NEP/ECE inhibitor (reviewed by OPGENORTH et al. 1992). This finding, which preceeded the cloning of ECEs, is not surprising today since it has now been established that ECE-1 shares a 37% amino acid sequence identity with NEP (JENG 1997; JENG and DE LOMBAERT 1997). However, phosphoramidon was quickly found to be inadequate for the development of an orally active drug due to the presence of an acid-labile phosphorus-nitrogen bond. Thus, from the intial identification of phosphoramidon as a weak but nevertheless effective ECE inhibitor, extensive efforts have been focused on the structural modification of this compound in order to produce more potent and orally active analogues. Interestingly, thiorphan, a sulfhydryl-containing selective NEP inhibitor (see Fig. 1), was found to be inactive as an inhibitor of ECE, both in vitro (IKEGAWA et al. 1990b) and in vivo (MCMAHON et al. 1991b), suggesting that the synthesis of selective ECE inhibitors may be achievable.

I. Chemical Identities of ECE Inhibitors

ECE inhibitors reported to date can be classified in four categories: (1) metal chelators, (2) peptidic inhibitors, (3) natural compound inhibitors, and (4) low molecular weight molecules.

Structure-activity relationship studies based on NEP inhibitors such as phosphoramidon, thiorphan, CGS 24592, CGS 25015, CGS 26129 (see Figs. 1 and 2) and CGS 25452 (structure not shown) were initially pursued by pharmaceutical companies (Novartis, KUKKOLA et al. 1995; Servier/Adir, DESCOMBES et al. 1995, etc.). In general, metalloprotease inhibitors contain one of four different types of zinc-binding groups: (1) a phosphorus-containing functionality, (2) a sulfhydryl, (3) a hydroxamic acid, or (4) a carboxylic acid. Depending on the oxidation state and the substitution of the phosphorus atom, the first group of compounds (i.e., the phosphorus-containing compounds) can be further subdivided into four classes: phosphoramidates, phosphonamides (which are more potent ECE inhibitors than phosphoramidon, but also non-selective toward ACE, and are also chemically labile), phosphinic acid (with enhanced chemical stability but a greater molecular weight), and aminophosphonic acids (see Figs. 2–6). To date, the majority of published

Structure	Company
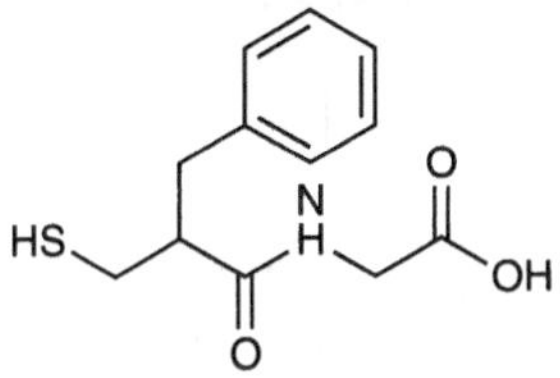 **_Thiorphan_**	generic
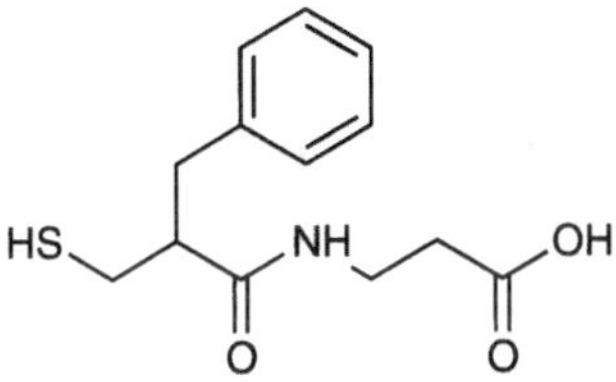 **_SQ 28603 (BMS 182309)_**	Bristol-Myers Squibb
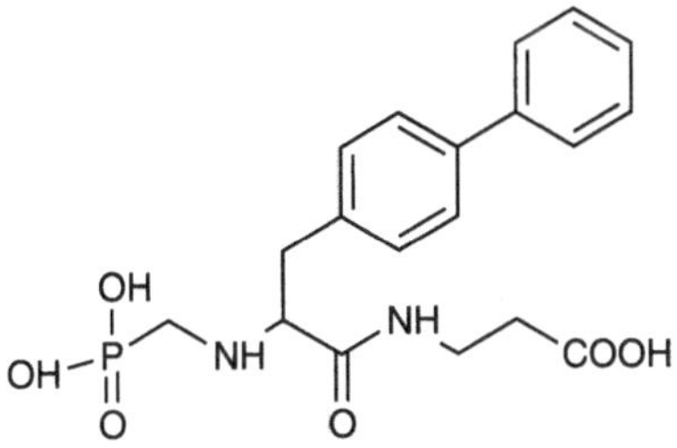 **_CGS 24592_**	Novartis
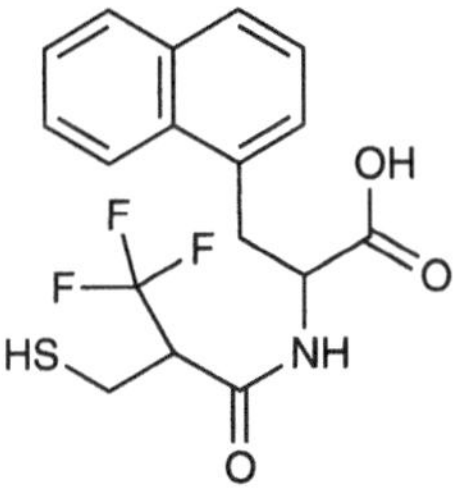 **_CGS 25015_**	Novartis
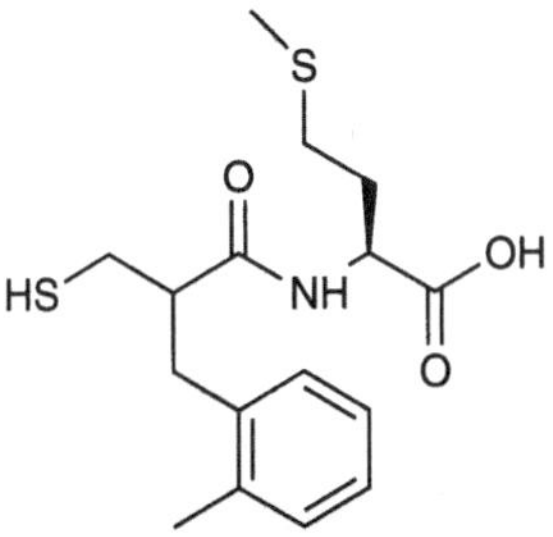 **_CGS 26129_**	Novartis

Fig. 1. Structures of selective NEP inhibitors

Class	Structure and Potency against NEP / ECE	Company
Phosphoramidate	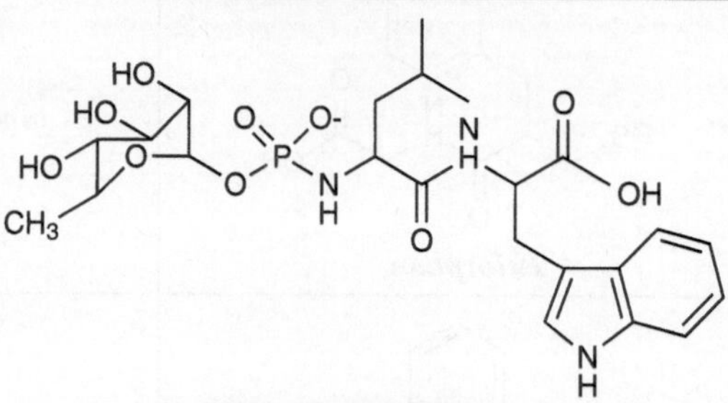 **Phosphoramidon** IC$_{50}$: 4.2 / 3 100 nmol/l	generic
Phosphoramidate	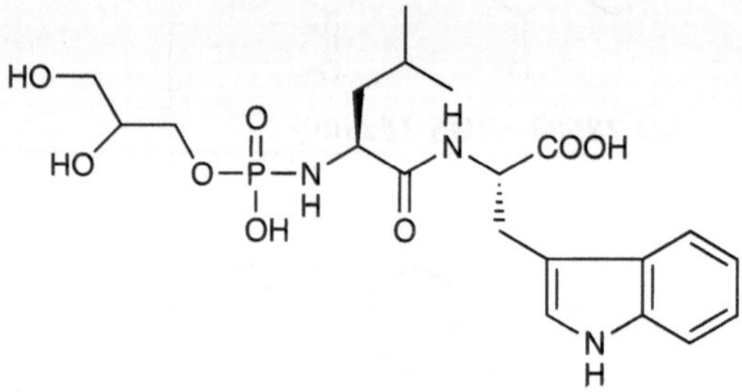 **S 17162** IC$_{50}$: NA / NA	Adir/Servier
Non-peptidic aminophosphonic acid	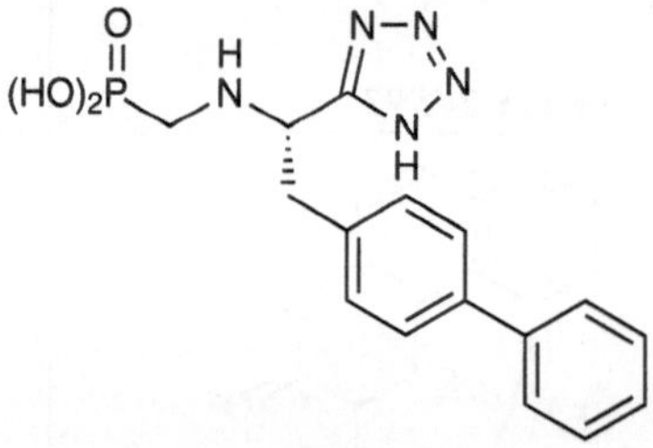 **CGS 26303** IC$_{50}$: 0.9 / 410 nmol/l	Novartis
Diphenyl prodrug of CGS 26303	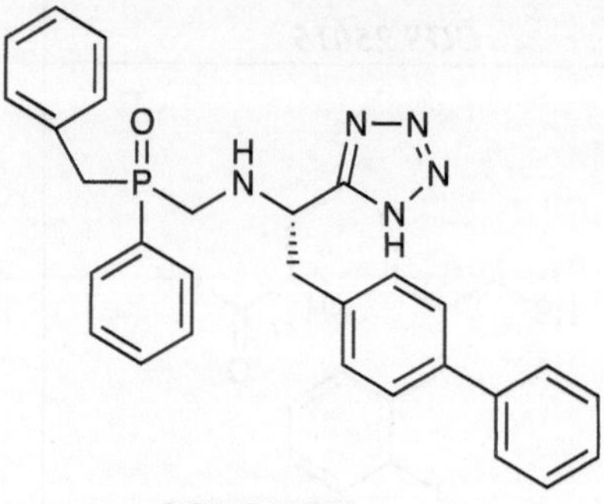 **CGS 26393**	

Fig. 2. Structures of dual NEP/ECE inhibitors

Non-peptidic aminophosphonic acid	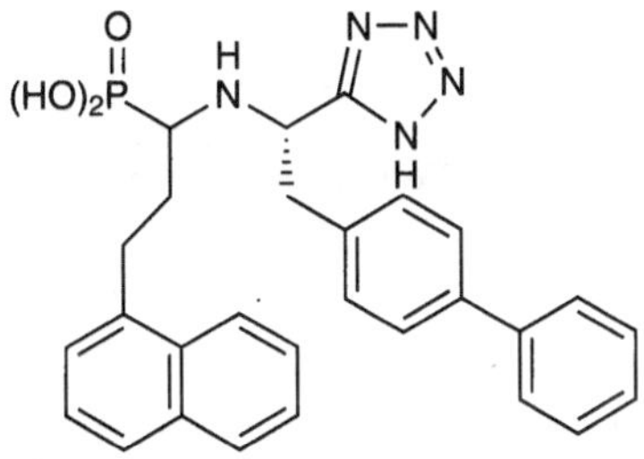 CGS 31447 IC$_{50}$: 4.8 / 17 nmol/l	Novartis
Aminophosphonic acid	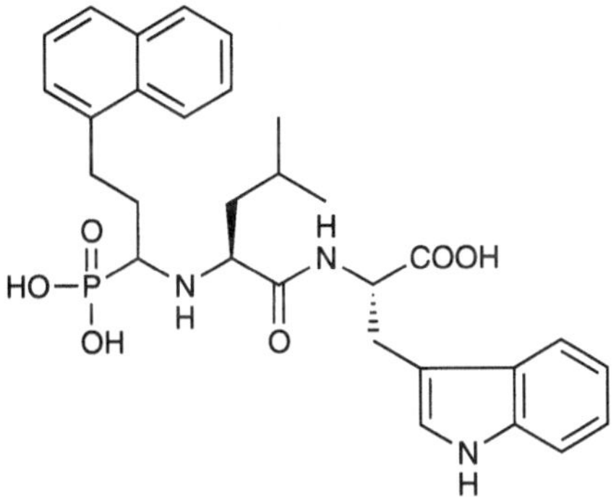 IC$_{50}$: 140 / 260 nmol/l	Banyu
Carboxylic acid	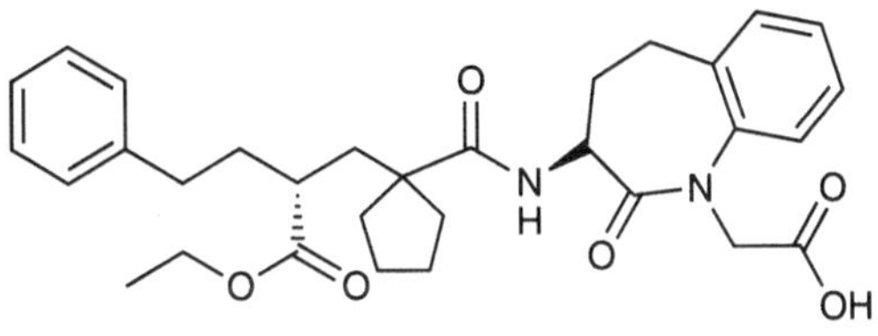 SLV-306 (KC 12615) IC$_{50}$: NA / NA	Solvay Pharma
Thiol	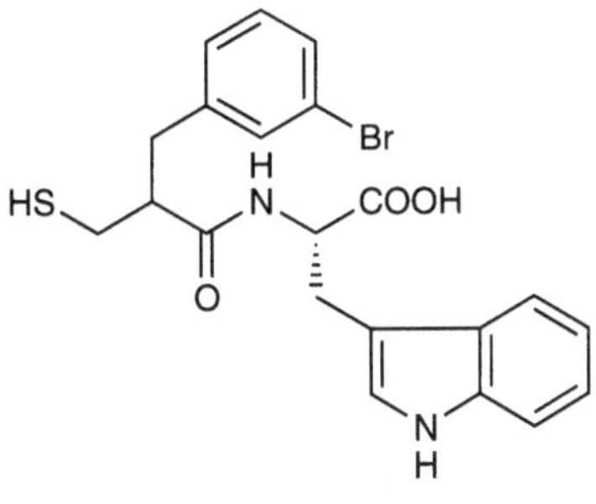 RU 69738 IC$_{50}$: NA / 20 nmol/l	Roussel-Uclaf

Fig. 2. (continued)
NA, not available.

Class	Structure Potency against ECE / NEP	Company
Phosphonamide	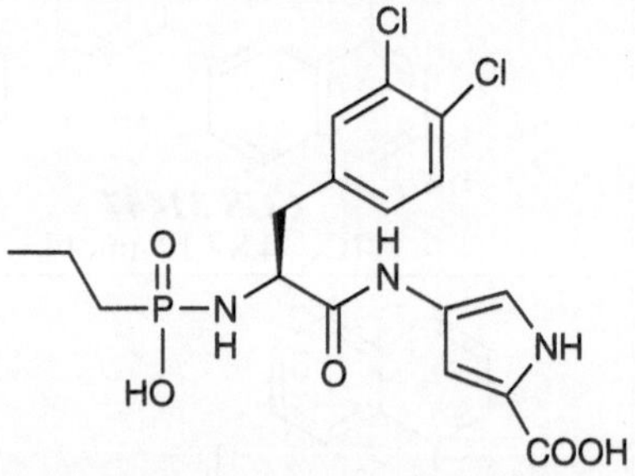 IC$_{50}$: 3 / 6 nmol/l	Zambon
Aminophosphonic acid	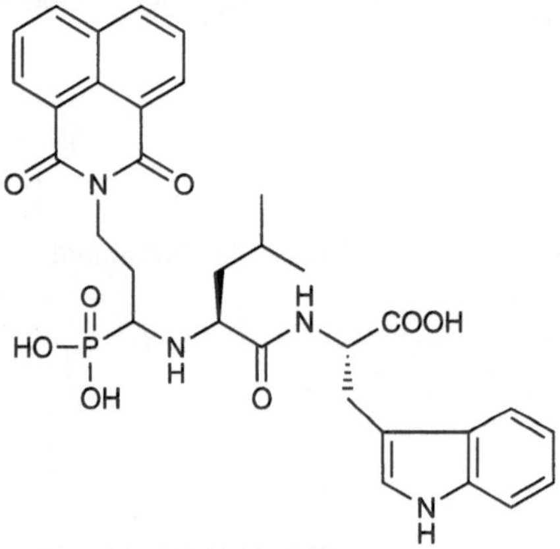 IC$_{50}$: 52 / 1 000 nmol/l	SmithKline Beecham
Natural product from *Saccharothrix sp.*	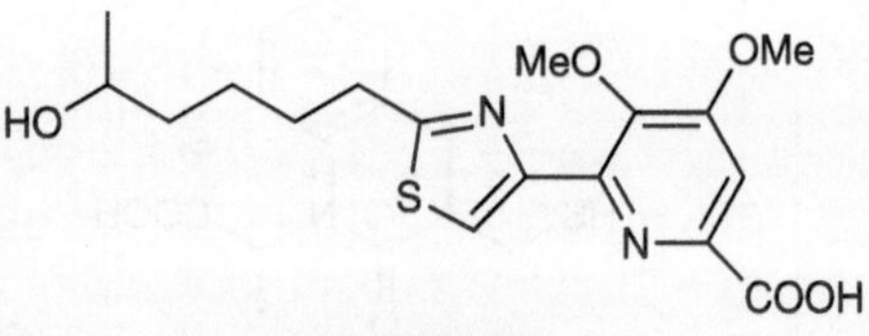 **WS 75624B** IC$_{50}$: 79 / 3 300 nmol/l	Fujisawa

Fig. 3. Structures of dual ECE/NEP inhibitors

Natural product from *Blastobacter sp.*	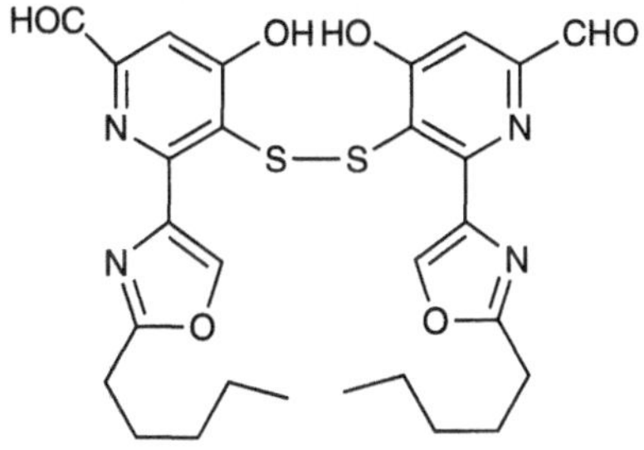 **B90063** IC$_{50}$: 900 / 70 000 nmol/l	Sankyo
Non-peptidic aminophosphonic acid	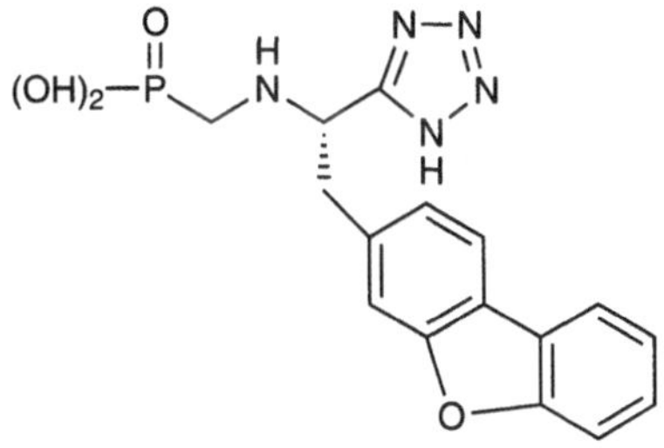 **CGS 34043** IC$_{50}$: 6 / 110 nmol/l	Novartis

Fig. 3. (continued)

ECE-1 inhibitors contain a phosphorus atom for zinc-binding at the active site of the enzyme.

Concomittantly, the search for ECE inhibitors also included natural compounds such as the WS series of compounds isolated from *Saccharothrix sp.* and *Streptosporangium roseum* by Fujisawa (TSURUMI et al. 1994, 1995a,b; YOSHIMURA et al. 1995), aspergillomarasmine B by Sapporo (MATSUURA et al. 1993), B90063 from *Blastobacter sp.* by Sankyo (SATO et al. 1996; TAKAISHI et al. 1998), daleformis from the roots of *Dalea filiciformis*, halistanol disulfate B from the sponge of *Pachastrella sp.* by SmithKline Beecham (PATIL et al. 1996, 1997), and soya saponin from soya bean by Nisshin (see Tables 3 and 4 and Figs. 2–6).

In addition, companies (e.g., Parke-Davis) also screened their libraries of synthetic compound for leads. The potency of all these protease inhibitors have not been compared to phosphoramidon in this chapter, but are reported elsewhere (JENG and DE LOMBAERT 1997).

Class	Structure Potency against ACE / NEP / ECE	Company
Phosphinic acid	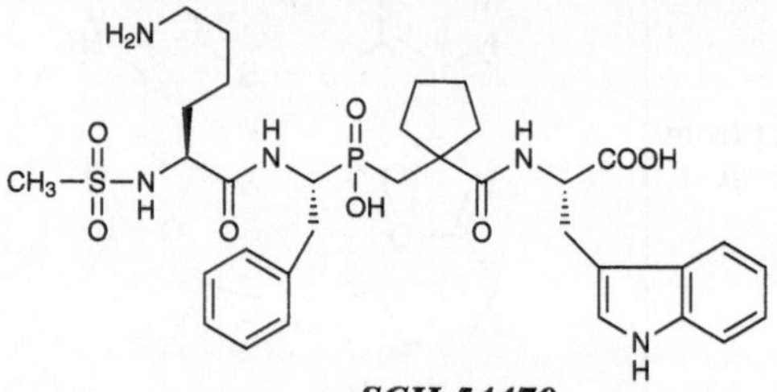 **SCH 54470** IC$_{50}$: 2.5 / 90 / 70 nmol/l	Schering-Plough
Phosphinic acid	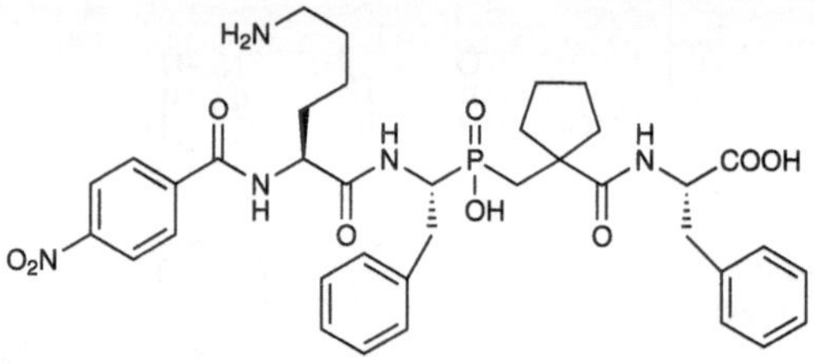 IC$_{50}$: 1.5 / 55 / 50 nmol/l	Schering-Plough
Carboxylic	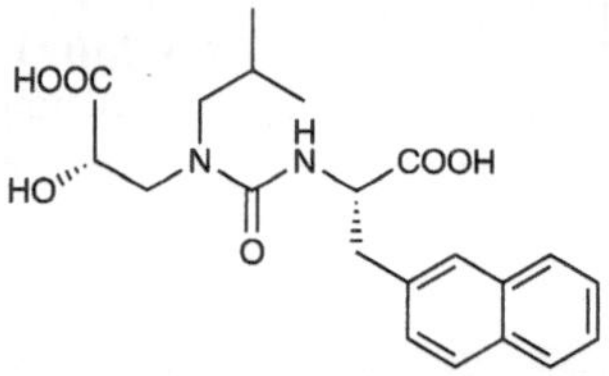 **SA 6817** IC$_{50}$: 15 / 4.4 / 910 nmol/l	Santen
Phosphonamide	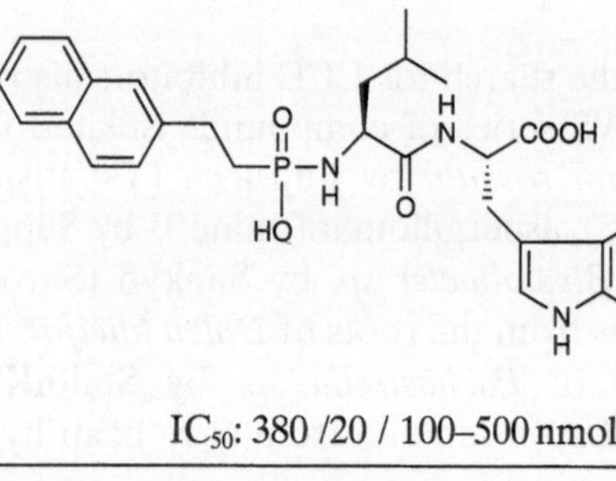 IC$_{50}$: 380 /20 / 100–500 nmol/l	Takeda

Fig. 4. Structures of triple ACE/NEP/ECE inhibitors

Phosphonamide	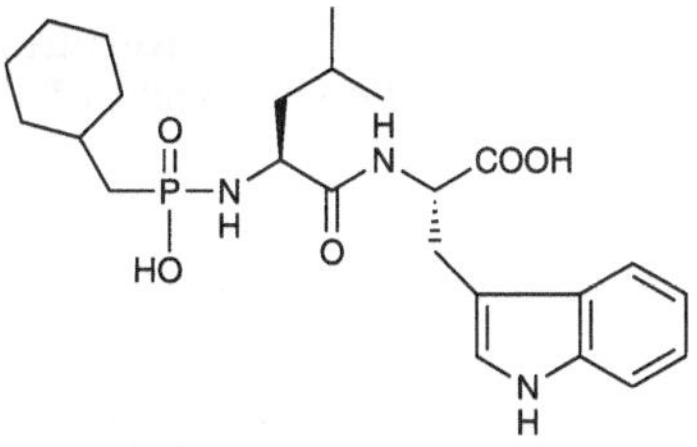 IC$_{50}$: 220 / 33 / 400–4 800 nmol/l	Monsanto
Non-peptidic Aminophosphonic acid	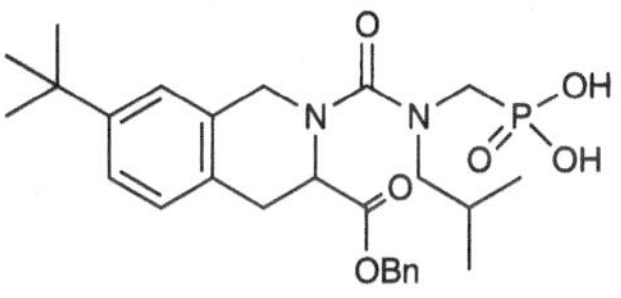 IC$_{50}$: 500 / 300 / 6 000 nmol/l	Wakamoto
Natural product from *Streptomyces galisaeus*	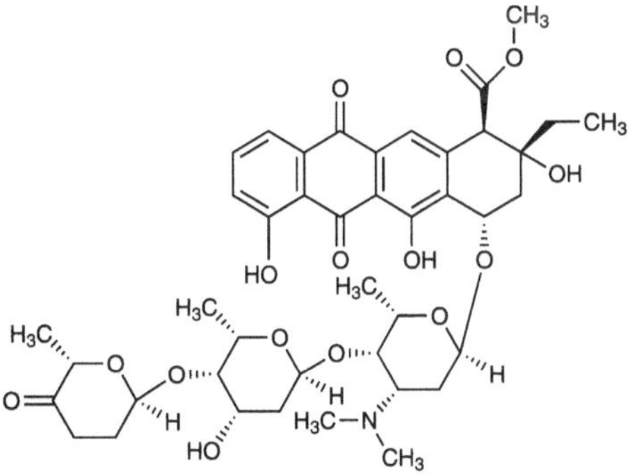 *Aclarubicin* IC$_{50}$: 30 / >100 / >100 µmol/l	Nisshin
Benzofused macrocyclic thiol lactam	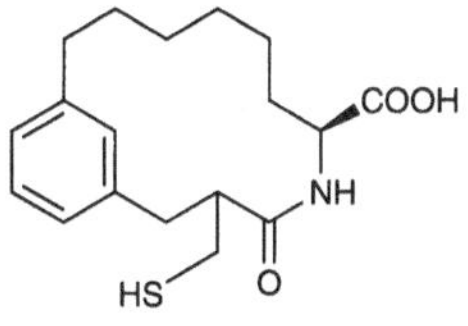 *CGS 26582* IC$_{50}$: 175 / 4 / 620 nmol/l	Novartis

Fig. 4. (continued)

II. Non-Specific ECE Inhibitors: Metal Chelators

Since ECEs are metalloproteases that require a zinc ion for their catalytic activity, depletion of this ion by metal chelators is expected to inhibit the enzyme activity. Thus, 2,3-dimercapto-1-propanol, toluene-3,4-dithiol, 8-mercaptoquinoline, and EDTA were reported to inhibit the activity of ECE

Class	Structure Potency against ECE / NEP	Company
Quinazoline	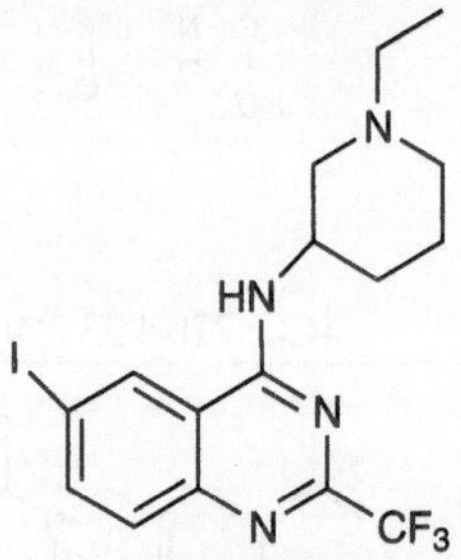 ***PD 159790*** IC$_{50}$: 2 600 / NA nmol/l	Parke-Davis
Quinazoline	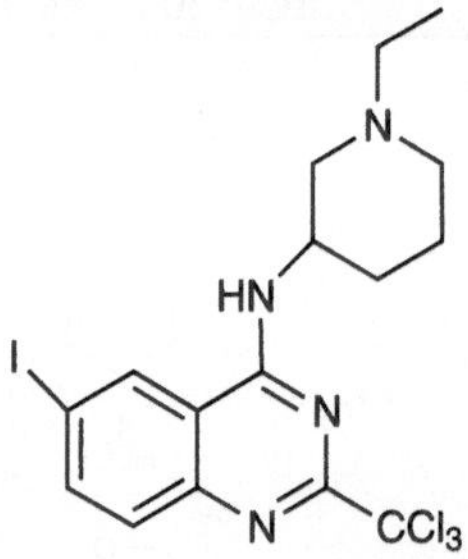 ***PD 069185*** IC$_{50}$: 900 / >100 000 nmol/l	Parke-Davis
Natural product from *Streptosporangium* *roseum*	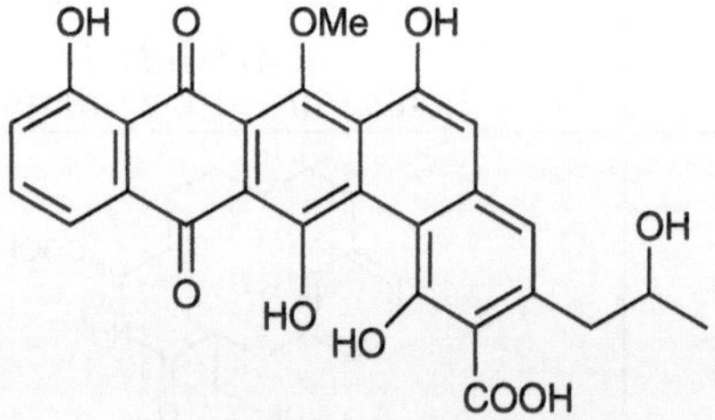 ***WS 79089B (FR901532/3)*** IC$_{50}$: 140 / >100 000 nmol/l	Fujisawa
Non-peptidic Aminophosphonic acid	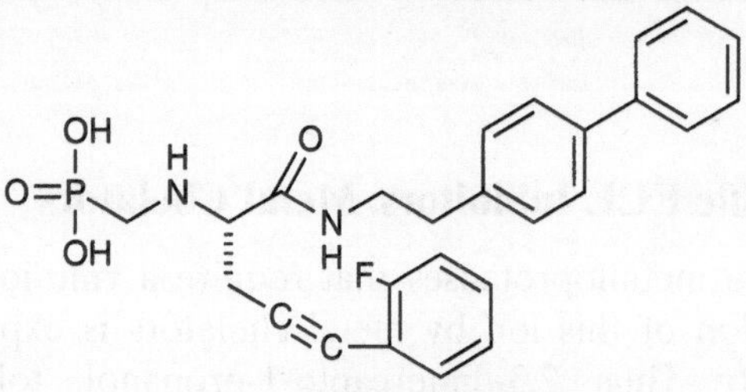 ***CGS 36112*** IC$_{50}$: 33 / 6 500 nmol/l	Novartis

Fig. 5. Structures of selective ECE inhibitors

Non-peptidic Aminophosphonic acid	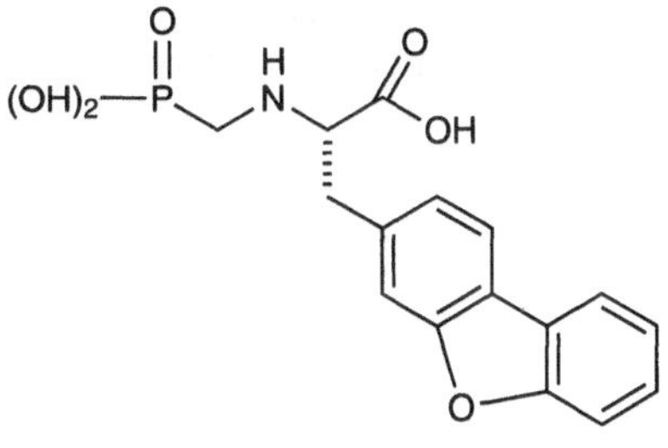 **CGS 35066** IC_{50}: 22 / 2 300 nmol/l	Novartis
Diphenyl prodrug of CGS 35066	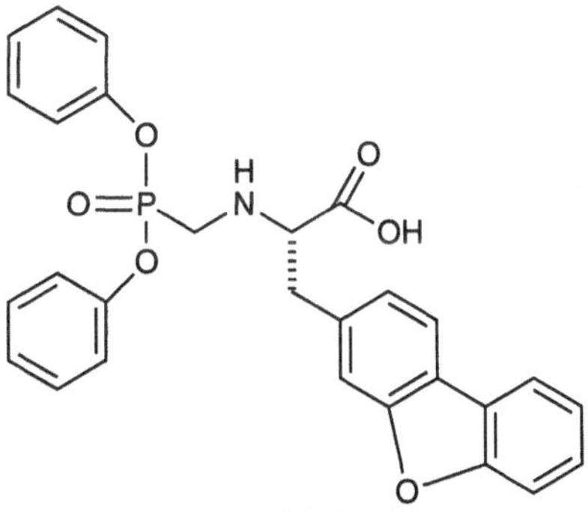 **CGS 35339**	Novartis

Fig. 5. (continued)

purified from a microsomal fraction of BAEC (ASHIZAWA et al. 1994b). The inhibitory effect is significantly less potent (IC_{50}: 0.5–82 µmol/l) than phosphoramidon (for a review, see JENG and DE LOMBAERT 1997). When injected into mice, the compounds decreased the mortality induced by big ET-1 (ASHIZAWA et al. 1994a). However, many other enzymes and proteins require divalent cations for activities. Treatment with metal chelators in vivo may cause side-effects. Thus, the use of metal chelators is non-specific (e.g., inhibiting many enzymes other than NEP or ECE) and therefore not desirable.

The effect of aspergillomarasmine B, a natural product isolated from an unidentified fungus (MATSUURA et al. 1993) (Tables 3 and 4, Fig. 6), on the inhibition of ECE activity in BAECs was reversed in the presence of 10 µmol/l Zn^{2+}, suggesting that this compound may act as a non-specific metal chelator.

III. Non-Selective Dual NEP/ECE Inhibitors

1. More Potent Against NEP

a) Phosphoramidates

S-17162 (Servier/Adir) was produced via structure-activity relationship studies. It contains a dihydroxylpropyl group acting as a truncated saccharide

Class	Structure Potency against ECE	Company
Thiol	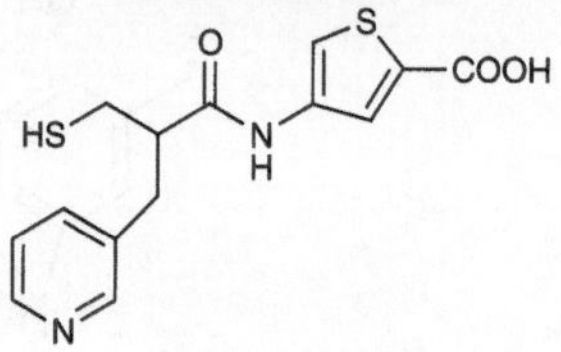 IC$_{50}$: 1 nmol/l	Zambon
Hydroxamic acid	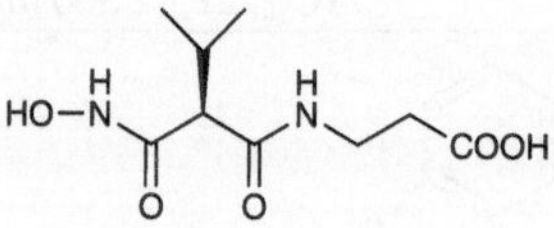 IC$_{50}$: 0.01 nmol/l	Berlex
Hydroxamic acid	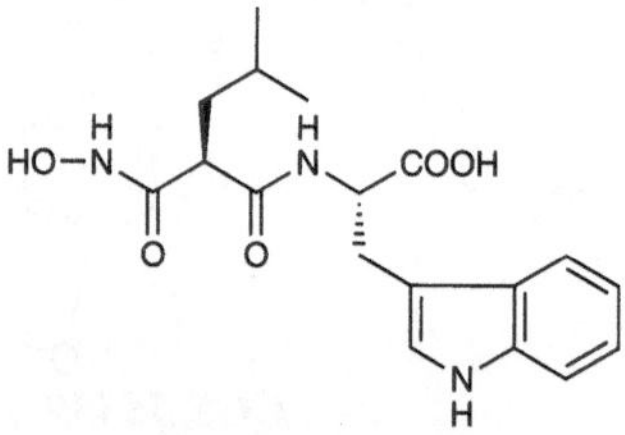 IC$_{50}$: 13 nmol/l	Yamanouchi
Carboxylic acid	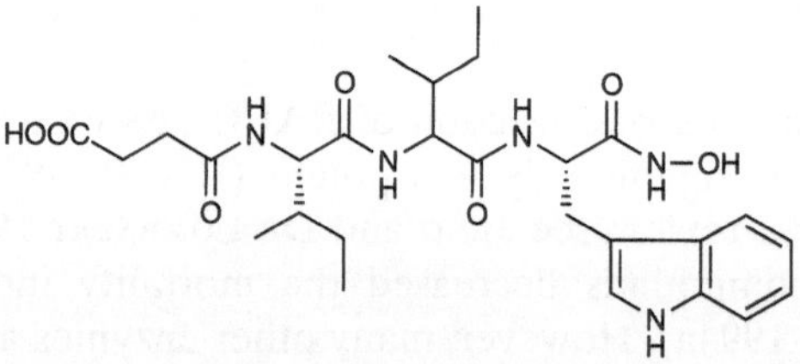 IC$_{50}$: 4700 nmol/l	Parke-Davis
Carboxylic acid	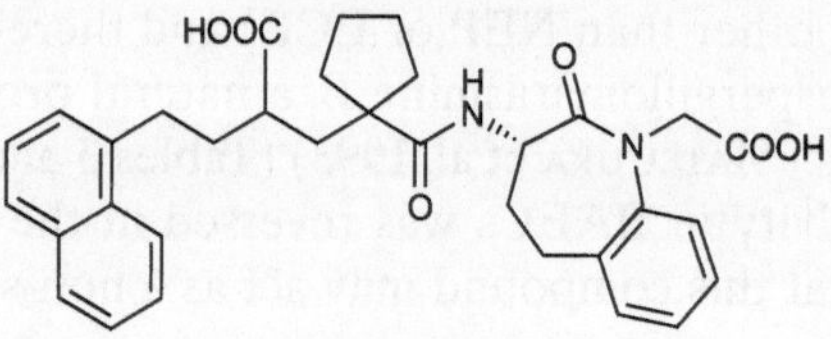 IC$_{50}$: 520 nmol/l	Kali-Chemie

Fig. 6. Structures of compounds derived from natural compounds or containing thiol, hydroxamic acid, or carboxylic acid zinc-binding groups as hydroxamic acid, ECE inhibitors. The potency of these compounds against NEP or ACE was not revealed.

Natural product from an unidentified fungus	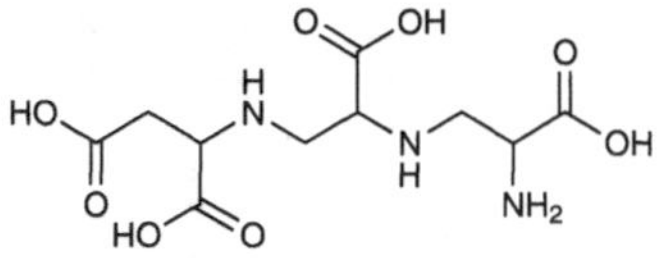 **Aspergillomarasmine A** IC$_{50}$: 3 400 nmol/l	Sapporo
Natural product from the roots of *Dalea filiciformis*	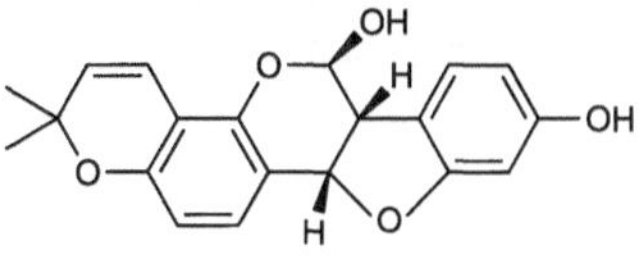 **Daleformis** IC$_{50}$: 9 000 nmol/l	SmithKline Beecham
Natural product from the sponge of *Pachastrella sp.*	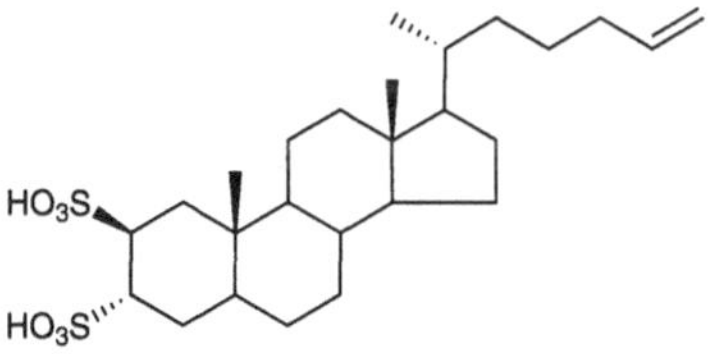 **Halistanol disulfate B** IC$_{50}$: 2 100 nmol/l	SmithKline Beecham
Natural product from Soya bean	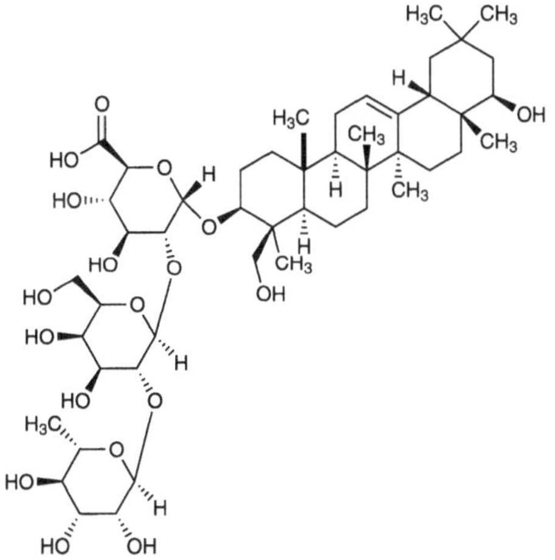 **Soya saponin** IC$_{50}$: 100 000 nmol/l	Nisshin

Fig. 6. (continued)

Table 3. List of selective and non-selective ECE inhibitors

Year	Inhibitor	IC$_{50}$ (µmol/l)[a] ACE/NEP/ECE	Selectivity compared to ECE	Company	Reference and/or Patent
1991	Phosphoramidon	NA/0.0092/0.49 78/0.034/3.5 NA/0.0042/3.1 NA/0.04/3.5 NA/0.0037/0.018	53-fold for NEP 103-fold for NEP 740-fold for NEP 88-fold for NEP 49-fold for NEP	Generic	Matsumura et al. (1991b) Kukkola et al. (1995) Fukami et al. (1994) Tsurumi et al. (1995a)
1992	A phosphonamide	0.38/0.02/0.55	28-fold for NEP	Takeda	EP-518299-A (1992)
1993	A hydroxamic acid	NA/NA/0.1 NA/NA/0.00001	ND	Berlex	Kukkola et al. (1995) WO9311154 (1993) Bihovsky et al. (1995)
	A hydroxamic acid	NA/NA/0.013	ND	Yamanouchi	JP-5039266-A (1993) Ueki et al. (1993)
	Aspergillomarasmine A	NA/NA/3.4	ND	Sapporo	Matsuura et al. (1993b) Arai et al. (1993)
	CGS 25015	NA/NA/17	ND	Novartis	Trapani et al. (1993) Balwierczak et al. (1995) Savage et al. (1998)
	CGS 26129	NA/NA/33	ND	Novartis	Trapani et al. (1993) Balwierczak et al. (1995) Savage et al. (1998)
	A sulfhydryl A phosphonamide	NA/NA/12 0.22/0.033/0.4, 4.4, 4.8	ND 12–145-fold for NEP	Monsanto Monsanto	Bertenshaw et al. (1993b) Bertenshaw et al. (1993a) Kukkola et al. (1995)
1994	WS79089A WS79089B (=FR 901533)	NA/>100/0.73 NA/>100/0.14	>137-fold for ECE >714-fold for ECE	Fujisawa	Tsurumi et al. (1994) Tsurumi et al. (1995a,b)
	WS79089C WS75624A WS75624B An aminophosphonic acid	NA/>100/3.42 NA/1/0.03 µg/ml NA/3.3/0.079 NA/0.14/0.26	>29-fold for ECE 33-fold for ECE 42-fold for ECE 2-fold for NEP	Fujisawa Banyu/Merck	Yoshimura et al. (1995) GB-2272435-A (1994) Tsumuri et al. (1995b) Fukami et al. (1994)

Year	Compound		Selectivity	Company	References
	CGS 24592	NA/0.0009/20	22,200-fold for NEP	Novartis	EP-623625-A (1994)
		NA/0.0019/>100	52,600-fold for NEP		De Lombaert et al. (1994a,b)
	A carboxylic acid	NA/NA/4.7	ND	Parke-Davis/Warner-Lambert	WO428012 (1994)
1995	S-17162	NA/NA/NA	ND	Servier	Descombes et al. (1995)
	An aminophosphonic acid	NA/1/0.052	19-fold for NEP	Adir & Co.	EP-636630-A (1995)
	A thiol	NA/NA/NA	ND	SmithKline Beecham	WO9513817 (1995); Keller et al. (1996)
	A hydroxamic acid	NA/NA/NA	ND	Zambon	EP-636621-A (1995)
	A phosphonamide	NA/0.006/0.003	2-fold for ECE	Glycomed	WO9519965 (1995)
	Aclarubicin	>100/>100/30	3-fold for ECE	Zambon	EP-636630-A (1995)
	Soya saponin	>1000/NA/100	ND	Nisshin	JP-7188034-A (1995)
1996	Halistanol disulfate B	NA/NA/2.1	ND	Nisshin / SmithKline Beecham	JP-7188033-A (1995); Patil et al. (1996)
	B-90063	NA/70/0.9	78-fold for ECE	Sankyo	JP-8208646 (1996); Takaishi et al. (1998)
	SCH 54470	0.0025/0.09/0.07	34-fold for ACE	Schering-Plough	WO9600732 (1996); McKittrick et al. (1996b)
	A carboxylic acid	NA/NA/0.52	ND	Kali-Chemie	EP-733642-A (1996)
	An aminophosphonic acid	0.5/0.3/6	20-fold for NEP	Wakamoto	JP-8092268-A (1996)
	RU 69738	NA/NA/0.02	ND	Roussel-Uclaf	Deprez et al. (1996)
	CGS 26303 (CGS 26393 as pro-drug)	NA/0.0009/0.41	455-fold for NEP	Novartis	WO9626729 (1996)
	CGS 31447	NA/0.005/0.017	3.5-fold for NEP	Novartis	De Lombaert et al. (1995); Trapani et al. (1995); Pelletier et al. (1998); Woods et al. (1999); WO9619486 (1996); De Lombaert et al. (1997); Shetty et al. (1998); Savage et al. (1998)

Table 3. (continued)

Year	Inhibitor	IC$_{50}$ (µmol/l)[a] ACE/NEP/ECE	Selectivity compared to ECE	Company	Reference and/or Patent
1997	Daleformis	NA/NA/9	ND	SmithKline Beecham	Patil et al. (1997)
	SA 6817	0.015/0.0044/0.91	207-fold for NEP	Santen	JP-9012455-A (1997) Fujita et al. (1997)
	CGS 26582	0.175/0.004/0.62	155-fold for NEP	Novartis	Jeng and De Lombaert (1997) Ksander et al. (1998)
1998	PD 069185	NA/>100/0.9 (ECE-1)	>222-fold for ECE	Parke-Davis/ Warner-Lambert	Savage et al. (1998) WO9619474 (1996) Ahn et al. (1998) Massa et al. (1998)
	PD 159790	NA/>200/2.6	>77-fold for ECE	Parke-Davis/ Warner-Lambert	Johnson et al. (1999) Ahn et al. (1998) Massa et al. (1998) Russel and Davenport (1999)
1999	SLV 306 (KC 12615)	NA/0.0042/NA	ND	Solvay Duphar	Meil et al. (1998)
	CGS 36112	NA/6.5/0.033	197-fold for ECE	Novartis	Wallace et al. (1998)
	CGS 26670	>10/0.0009/0.6	667-fold for NEP	Novartis	Ksander et al. (1998)
	Tripeptide derivative	NA/6.3/0.028	225-fold for ECE	Novartis	Wallace et al. (1998)
	Tripeptide derivative	NA/5.8/0.008	725-fold for ECE	Novartis	Wallace et al. (1998)
	CGS 34043	NA/0.11/0.006	18-fold for ECE	Novartis	Jeng et al. (1999)
	CGS 35066 (CGS 35339 as pro-drug)	NA/2.3/0.022	105-fold for ECE	Novartis	Jeng et al. (1999)

Assayed with human recombinant ECE-1 (isolated enzyme).
NA, not available; ND, not determined.
[a] Since the potencies of these inhibitors may have been determined in different assay systems using a variety of enzyme sources, the relative potency calculated in the selectivity column is more meaningful.

Table 4. Chemical names of selective NEP, non-selective NEP/ECE, non-selective triple ACE/NEP/ECE and selective ECE inhibitors

Name	Chemical Name	Company	Reference
Selective NEP inhibitors			
Thiorphan	(DL-3-mercapto-2-benzylpropanoyl)-gly-OH	Generic	SEYMOUR et al. (1991)
SQ 28603 (BMS 182309)	(N-[2-(Mercaptomethyl)-1-oxo-3-phenylpropyl]-β-alanine)	Bristol-Myers Squibb	
CGS 24592	(S)-N-[2-(Phosphonomethyl) amino)-3-(4-biphenylyl) propionyl]-3-amino-propionic acid	Novartis	DE LOMBAERT et al. (1994a, b)
CGS 25015	(α'S)-α-[[3,3,3-trifluoro-2-(mercaptomethyl)-1-oxopropyl]amino]-1-naphthalenepropanoic acid	Novartis	TRAPANI et al. (1993); BALWIERCZAK et al. (1995); SAVAGE et al. (1998)
CGS 26129	(2S)-2-[[3-Mercapto-2-[(2-methylphenyl)methyl]-1-oxopropyl]amino]-4-(methylthio)butanoic acid	Novartis	TRAPANI et al. (1993); BALWIERCZAK et al. (1995); SAVAGE et al. (1998)
Non-selective dual NEP/ECE inhibitors			
Phosphoramidon	N-[(α-L-rhamnopyranosyloxy) hydroxyphosphinyl] – L – Leu – L – Trp	generic	OPGENORTH et al. (1992)
S-17162	N-(2,3 Dihydroxy propyl phosphonyl)-(S)-Leu-(S)-Trp-OH, disodium salt	Servier	DESCOMBES et al. (1995)
No name SLV-306	3-(1-Naphtyl)-1-phosphonopropyl-L-leucyl-L-Trp (3S,2'R)-3-{ 1-[2'-(Ethoxycarbonyl) – 4'-phenylbutyl]-cyclopentane-1-carbonylamino]-2,3,4,5 -tetrahydro-2-oxo-1H-1-benzazepine-1-acetic acid	Banyu/Merck Solvay	FUKAMI et al. (1994) MEIL et al. (1998)
KC 12615 (Active metabolite of SLV 306):	3(S)-[[2(R)-Carboxy-4-phenyl-butyl]-1-cyclopentane carbonylamino]-2,3,4,5-tetrahydro-2-oxo-1H-1-benzazepine-1-acetic acid	Solvay	MEIL et al. (1998)
[Phe22] human big ET-1 (19–37)	H-Ile-Ile-Trp-Phe-Asn-Thr-Pro-Glu-His-Val-Val-Pro-Tyr-Gly-Leu-Ser-Pro-Arg-OH	Univ. Sherbrooke	CLAING et al. (1995)
CGS 26303	(S)-2-Biphenyl-4-yl-1-(1H-tetrazol-5-yl)-ethylamino-methyl phosphonic acid	Novartis	DE LOMBAERT et al. (1994b, 1995, 1997)
CGS 26393	Diphenyl ((S)-2-biphenyl-4-yl-1-(1H-tetrazol-5-yl)-ethylamino-methyl) phosphonate)	Novartis	DE LOMBAERT et al. (1994b, 1995, 1997)

Table 4. (continued)

Name	Chemical Name	Company	Reference
CGS 31447	[1-[[(1S)-2-[1,1'-Biphenyl]-4-yl-1-(1H-tetrazol-5-yl)ethyl]amino]-3-(1-naphthalenyl)propyl]phosphonic acid	Novartis	De Lombaert et al. (1997)
CGS 34043	[[[(1R)-2-Dibenzofuran-3-yl-1-(4H-1,2,4-triazol-3-yl)ethyl]amino]methyl]-phosphonic acid	Novartis	Jeng et al. (1999)
WS 75624B	6-[2-(5-Hydroxyhexyl)-4-thiazolyl]-4,5-dimethoxy-2-pyridinecarboxylic acid	Fujisawa	Tsumuri et al. (1995b)
RU 69738	(αS)-α-[3-(3-Bromophenyl)-2-(mercaptomethyl)-1-oxopropyl]amino]-3H-indole-3-propanoic acid	Roussel-Uclaf	Deprez et al. (1996)
B90063	Bis[6-formyl-4-hydroxy-2-(2'-n-pentyloxazol-4'-yl)-4-pyridon-3-yl]-disulfide (1a)	Sankyo	Takaishi et al. (1998)
Non-selective triple ACE/NEP/ECE inhibitors			
SCH 54470	(N-[[1(R)-([[6-Amino-2(S)methylsulfonyl]-amino)2-phenylethyl]hydroxyphosphinyl]methyl]cyclopentyl]carbonyl]-L-Trp lithium salt)	Schering-Plough	Vemulapalli et al. (1997); McKittrick et al. (1996b)
CGS 26582	(6S)-3-(Mercaptomethyl)-4-oxo-5-azabicyclo[11.3.1]heptadeca-1(17),13,15-triene-6-carboxylic acid	Novartis	Ksander et al. (1998); Savage et al. (1998)
SA 6817	(αS)-α-[[[[(2S)-2-Carboxy-2-hydroxyethyl](2-methylpropyl)amino]carbonyl]amino]-2-naphthalenepropanoic acid	Santen	Fujita et al. (1997)
Selective ECE inhibitors			
FR 901532 (WS 79089B)	5,6,8,13-Tetrahydro-1,6,9,14-tetrahydroxy-3-(2-hydroxypropyl)-7-methoxy-8,13-dioxo-benzo[a]naphthacene-2-carboxylic acid	Fujisawa	Tsurumi et al. (1994)
FR 901533	Sodium salt of FR 901532	Fujisawa	Tsurumi et al. (1994)
PD 069185	(1-Ethyl-piperidin-3-yl)-(6-iodo-2-trichloromethyl-quinazolin-4-yl)-amine	Warner-Lambert/Parke-Davis	Ahn et al. (1998)
PD 159790	N-(1-Ethyl-3-piperidinyl)-6-iodo-2-(trifluoromethyl)-4-quinazolinamine	Warner-Lambert/Parke-Davis	Ahn et al. (1998)
CGS 35066	(αR)-α[(Phosphonomethyl) amino]-dibenzofuran-3-propanoic acid	Novartis	Jeng et al. (1999)
CGS 35339	(αR)-α-[[(Diphenoxyphosphinyl) methyl]amino]-dibenzofuran-3-propanoic acid	Novartis	Jeng et al. (1999)
CGS 36112	(S)-[[1-[(2-Biphenyl-4-ylethyl) carbomoyl]-4-(2-fluorophenyl) but-3-ynyl] amino] methyl] phosphonic acid	Novartis	Wallace et al. (1998)

surrogate for the rhamnosyl moiety of phosphoramidon, the original phosphoramidate (Tables 3 and 4, Fig. 2). When co-infused with big ET-1 in pithed rats, this compound inhibited the big ET-1-induced pressor response with a potency similar to that of phosphoramidon (DESCOMBES et al. 1995). No results on the in vitro ACE/NEP/ECE inhibitory activity regarding S-17162 have been reported.

b) Aminophosphonic Acids

An approach to improve the chemical stability of phosphoramidon is to add a methylene group between the phosphorus-nitrogen bond, yielding an aminophosphonic acid. Novartis' non-peptidic CGS 26303 belongs to such a class of compounds (Tables 3–5, Fig. 2). This compound revealed a long duration of action in vivo and is about 3- and 30-fold more potent than phosphoramidon in the inhibition of ECE-1 and NEP activities, respectively (JENG 1997). This compound was shown to inhibit the activities of both human ECE-1 and ECE-2 expressed in Chinese hamster ovary cells with a similar IC_{50} of 100nmol/l (KAW et al. 1995). In contrast, phosphoramidon inhibited these enzymes with IC_{50} values of 1000nmol/l and 5nmol/l, respectively (KAW et al. 1995).

CGS 26393, a diphenyl pro-drug of CGS 26303, presented a significantly improved oral activity compared to its parent compound, CGS 26303. Furthermore, the addition of a naphthylethyl group to the carbon atom adjacent to the phosphorus of CGS 26303 has led to the discovery of another potent dual inhibitor of NEP and ECE-1, CGS 31447 (JENG 1997) (Table 4).

Banyu also reported several peptidic compounds (as phosphonoalkyl dipeptides) of this family. A representative compound displayed potent NEP and ECE inhibitory activity (IC_{50}: 140/260nmol/l, respectively; Tables 3 and 4, Fig. 2) (FUKAMI et al. 1994). In this compound a methylene group was included between the phosphorus-nitrogen bond in order to enhance its chemical stability in acidic conditions.

c) Carboxylic Acids

A structurally different dual inhibitor of NEP and ECE-1 is SLV 306 (Tables 3–5, Fig. 2). This orally active carboxylate is metabolized to generate the active molecule KC 12615, a compound with NEP and ECE-1 inhibitory activities comparable to those of phosphoramidon (MEIL et al. 1998).

d) Sulfhydryl-Containing Compounds

RU 69738 (DEPREZ et al. 1996) (Tables 3 and 4, Fig. 2), and two other compounds from Novartis and Monsanto (KUKKOLA et al. 1996; BERTENSHAW et al. 1993b, respectively; structures and names not shown) are derivatives that have greatly improved ECE inhibitory activity when compared to thiorphan, but with similar IC_{50} as phosphoramidon.

Table 5. Effects of ACE/NEP/ECE -related inhibitors in treatments of animal models of various pathophysiological conditions

Pathology	Inhibitor	Species	Model	Treatment	Observation and result	Reference
Hypertension	Phosphoramidon	Conscious SD rats	Diaspirin cross-linked hemoglobin (280 mg/kg i.v.) induced 27% increase in MABP after 15 min	At ~30 s, 5 mg/kg i.v.	↓ by 70% the increase in MABP	Schultz et al. (1993)
	Phosphoramidon	Conscious SHR vs WK rats	Genetically hypertensive	At ~10/-15 min, 30 mg/kg bolus i.v.	↓ MABP by 30 to 58%	McMahon et al. (1991a)
	Phosphoramidon	Anesthetized DOCA-salt rats	Hypertension	0.6–60 mg/kg/hr i.v. infusion for 2 hr	↓ MABP by 12 to 47%	Vemulapalli et al. (1993b)
	Phosphoramidon	Conscious SHR	Genetically hypertensive	10–40 mg/kg/hr i.v. infusion for 5 hr	↓ MABP by 9–40 mm Hg	McMahon et al. (1993)
	Phosphoramidon	Conscious SHR	Genetically hypertensive and renal artery-ligated animals	10–40 mg/kg/hr i.v. infusion for 5 hr	↓ MABP by 31–54 mm Hg	McMahon et al. (1993)
	Phosphoramidon	Conscious SHR	Genetically hypertensive	60 mg/kg/hr i.v. infusion for 2 hr	No effect on MABP	Vemulapalli et al. (1993b)
	SLV 306	DOCA-salt rats	Hypertension/ volume-loading hypertension	NA p.o.	Produced natriuretic and diuretic effects	Meil et al. (1998)
	CGS 26303	Conscious SHR	Genetically hypertensive	5 mg/kg/day s.c. minipumps for 13 days	↓ MABP by 15–20 mmHg after 1 day	De Lombaert et al. (1994)
	Phosphoramidon	Anesthetized pig	Cardiopulmonary bypass and circulatory arrest-induced pulmonary hypertension	At ~15 min, 30 mg/kg bolus i.v.	↓ PaP by 54% ↓ PVR by 67%	Kirshbom et al. (1995)
	FR 901533	Conscious rats	Monocrotaline-induced pulmonary hypertension	100 mg/kg/day minipumps for 1 day s.c.	Protect against development of right ventricular overload & tickening of pulmonary arteries	Takahashi (1998)
Cardiac ischemia/ reperfusion injury	Phosphoramidon	Isolated heart SD rat	Ischemic heart failure (30/60 min coronary occlusion and 24 hr reperfusion)	At ~10 min, 0.5 mg/kg/min for 2 min i.v. infusion + 0.25 mg/kg/min for 93 min	↓ myocardial infarct size by 74% in the 30-min and 60-min ligation group	Grover et al. (1992)
	Phosphoramidon	Conscious human	CHF patients treated with diuretics and ACE inhibitors	30 nmol/min i.v. infusion × 1 hr	↑ forearm blood flow by 52%	Love et al. (1996)
	SLV 306	Rats	CHF by aortic stenosis	NA	↓ cardiac hypertrophy as well as pulmonary congestion	Meil et al. (1998)

Category	Inhibitor	Animal	Model	Dose	Effects	Reference
	FR 901533	Anesthetized beagle dogs	CHF-induced by rapid right ventricular pacing (270 bpm, 14 days)	1, 3 mg/kg × 2 weeks	↓ MABP, SVR, pulmonary capillary wedge pressure, PVR, plasma ANP, renin activity, Ang II, aldosterone levels. ↓ CO, the urinary water, [a]FE_{Na}, GFR, renal plasma flow	WADA et al. (1999)
Renal failure	Phosphoramidon	Anesthetized CD rat	Ischemic acute renal failure by 40 min left renal artery ligation/ 1 hr reperfusion	At −30 min, 0.03 or 0.1 mg/kg/ min i.v. infusion	↓↓FE_{Na} by 63% ↓proteinuria by 97% ↑[b]GFR by 236%	VEMULAPALLI et al. (1993a)
	Phosphoramidon	CD rat	Ischemic acute renal failure by 30 min left renal artery ligation in uninephrectomized rats	At −20 min, 0.46 µmol/kg/min i.v. infusion and 30 min after	restored GFR ↓ tubular injury at 2 hr after reperfusion	BIRD et al. (1995)
	SCH 54470	Rats	Remnant kidney model of chronic renal failure	30 mg/kg, b.i.d. p.o. x 2 weeks	↓ MABP, proteinuria, FE_{Na} ↓ GFR, renal plasma flow	VEMULAPALLI et al. (1997)
Vasospasm	Phosphoramidon	Anesthetized mongrel dog	Cerebral vasospasm following SAH by autologous blood injection	At −30 min, 2 µmol i.c., on day 0 and day 2	Attenuated the decrease in the diameter of basilar artery by 53% on day 7	MATSUMURA et al. (1991a)
	CGS 26303	Anesthetized NZW rabbits	Cerebral vasospasm following SAH by autologous blood injection	At −1 day, 9 mg/kg/day minipumps i.p.	Prevented completely the decrease in the diameter of basilar artery on day 2 after autologous blood injection	CANER et al. (1996)
	CGS 26303	Anesthetized NZW rabbits	Cerebral vasospasm following SAH by autologous blood injection	At 1 hr after SAH, 3–30 mg/kg, b.i.d. bolus i.v	Prevented the decrease in the diameter of basilar artery in all groups on day 2	KWAN et al. (1997)
	CGS 26303			At 1 day after SAH, 3–30 mg/kg b.i.d. bolus i.v.	Reversed the decrease in the diameter of basilar artery in the 30 mg/kg dose group on day 2	KWAN et al. (1997)
Restenosis	CGS 24592 CGS 26303 CGS 26303	Conscious SD rats	Balloon angioplasty of the carotid artery	From day 0 – 14 10 mg/kg/day i.p. 10 mg/kg/day i.p. 2 × 10 mg/kg/daily s.c.	No effect No effect 17% ↓ neointimal proliferation	PHAM et al. (1998)
	CGS 26303			10 mg/kg/day minipumps i.v.	35% ↓ neointimal proliferation	
	CGS 26393			3 × 30 mg/kg/day p.o.	30% ↓ neointimal proliferation	
	Phosphoramidon			NA	↓ neointimal proliferation	MINAMINO et al. (1997)

[a] FE_{Na}: fractional sodium excretion.
[b] GFR: glomerular filtration rate.

2. More Potent Against ECE

B90063, a natural compound isolated from *Blastobacter sp.* was reported by Sankyo as an ECE inhibitor (Takaishi et al. 1998) (Tables 3 and 4, Fig. 3). This symmetrical aldehyde inhibited the activities of ECE in both BCAECs and NEP in rat kidney, being 78-times more selective for ECE inhibition. It is 16,000-fold more selective than phosphoramidon toward ECE inhibition.

a) Phosphonamides

The Zambon group published on a phosphonamide series of compound that possessed very potent ECE/NEP inhibitory activity (IC_{50}: 3/6 nmol/l; Tables 3 and 4, Fig. 3) (Norcini and Santangelo 1995).

b) Aminophosphonic Acids

Novartis conducted structure-activity relationship studies and optimized CGS 26303 by substituting a dibenzofuran for the biphenyl group, creating the analogue CGS 34043 (Jeng et al. 1999). SmithKline Beecham also revealed an aminophosphonic acid series of compound similar to those aminophosphonic acids (see above) from Banyu but with selectivity for ECE over NEP (IC_{50}: 52/1,000 nmol/l respectively; Tables 3 and 4, Fig. 3) (Keller et al. 1996).

Overall, in this class of inhibitors, the phosphonamide derivative from Zambon was found to be the most potent against ECE activity, whereas B90063 was found to be the most selective toward ECE vs NEP.

IV. Non-Selective Triple ACE/NEP/ECE Inhibitors

1. Phosphinic Acids

Schering-Plough synthesized a series of triple angiotensin-converting enzyme (ACE)/NEP/ECE inhibitors. SCH 54470 is the most potent such inhibitor reported to date (Tables 3 and 4, Fig. 4). This compound inhibits the activities of these respective enzymes with IC_{50} values of 2.5 nmol/l, 90 nmol/l, and 70 nmol/l (Vemulapalli et al. 1997). This phosphinic acid-related compound presented an enhanced oral activity attributed to the utilization of a lysine side-chain. Several other phosphinic acid derivatives that triply inhibit ACE/NEP/ECE were also presented, including an even more potent triple inhibitor (IC_{50}: 50 nmol/l, 55 nmol/l, 1.5 nmol/l; see Fig. 4) (McKittrick et al. 1996a,b; Chackalamannil et al. 1996; see Table 5 in Jeng and De Lombaert 1997). Bristol-Myers Squibb also came out with a similar analogue (Lloyd et al. 1996).

2. Phosphonamides

Takeda synthesized a compound of lower potency compared to SCH 54470, but still capable of inhibiting all three enzymes (380 nmol/l, 20 nmol/l, 100 nmol/l, respectively; Tables 3 and 4) (Wakimasu et al. 1992; Kukkola et al.

1995). Monsanto also presented a very similar compound (Table 3, Fig. 2) capable of inhibiting ACE and ECE activity from rabbit lung (BERTENSHAW et al. 1993a; KUKKOLA et al. 1995).

3. Aminophosphonic Acid

Wakamoto presented a non-peptidic aminophosphonic acid as a triple inhibitor (500 nmol/l, 300 nmol/l, 6,000 nmol/l, respectively; Tables 3 and 4, Fig. 4).

4. Other Compounds

Santen patented a carboxylic acid derivative (SA 6817) that exhibited potent triple inhibitory activities in vitro as well as ACE and ECE inhibitions in vivo (15 nmol/l, 4.4 nmol/l, 910 nmol/l; Tables 3 and 4, Fig. 4) (FUJITA et al. 1997). Novartis showed CGS 26582, a sulfhydryl-containing compound as a triple inhibitor in vitro (175 nmol/l, 4 nmol/l, 620 nmol/l) and in vivo (KSANDER et al. 1998).

Overall, in this class of inhibitors, Schering SCH 54470 was found to be the most potent against all three enzymatic activities combined followed closely by Santen SA 6817.

V. Selective ECE Inhibitors

1. Peptidic Compounds

Four analogues of human big ET-1_{1-38} with single D-amino acid substitutions have been examined, and [D-Val^{22}]big ET-1_{1-38} was the most potent (IC$_{50}$= 280 μmol/l) inhibitor of the activity of ECE purified from BAEC (INAGAKI et al. 1993; MORITA et al. 1994). Such an inhibitory activity is attributed to conformational changes of the authentic precursor. This analogue also suppressed big ET-1-induced dopamine release from the rat striatum, more potently (fivefold) than phosphoramidon (MORITA et al. 1994). A shorter analogue, [Phe^{22}]big ET-1_{19-37}, also dose-dependently attenuated big ET-1-induced increases in perfusion pressure in isolated rabbit kidneys and, likewise, was 30-times more potent than phosphoramidon (CLAING et al. 1995). However, as peptidic analogues, these inhibitors suffer from rapid proteolytic degradation and have a short half-life in circulation (HEMSEN et al. 1991).

2. Non-Peptidic Compounds from Natural Products

Selective ECE-1 inhibitors have been discovered from fermentation products (see Tables 3 and 4, Figs. 3–6). For example, WS75624B was isolated from *Saccharothrix sp.* (TSURUMI et al. 1995b; YOSHIMURA et al. 1995) and was associated with a 42-fold selectivity for ECE (activity in homogenate of bovine carotid artery) over NEP (extract of rat kidney membrane) inhibition. It is 2.3-fold more potent and 1900-times more selective toward ECE inhibition than phosphoramidon (Table 3). The same group from Fujisawa also isolated WS79089B, re-named as FR 901532 or FR 901533 (the sodium salt), a natural product from the fermentation broth of *Streptosporangium roseum* (Tsurumi et al. 1994, 1995a,b). It has about the same potency against ECE as phosphoramidon but it is selective for ECE (IC_{50} for NEP: >100 μmol/l). Noticeably, this compound displayed some species differences regarding ECE inhibition. It was as potent as phosphoramidon in inhibiting the conversion of exogenous human big ET-1 in both BCAECs and HUVECs in culture, but 100-fold less potent in cultured rat cells (abdominal aortic ECs or glomerular) (Tsurumi et al. 1995a,b).

In conclusion, several natural products have shown potencies similar to that of phosphoramidon in ECE inhibition but they are much weaker than phosphoramidon in NEP inhibition, and thus more ECE-selective. Further structural optimization of these compounds may lead to the discovery of potent and selective ECE inhibitors.

3. Other Non-Peptidic Compounds

The trisubstituted quinazoline derivative PD 069185 discovered by random screening of Parke-Davis compound library is not more potent than phosphoramidon as an ECE-1 inhibitor, but it is selective (Tables 3 and 4; Fig. 5) (Ahn et al. 1996). Replacement of the -CCl₃ group with -CF₃ gave PD 159790, a compound with increased solubility and less toxicity. No in vivo studies have been reported yet with this compound. It is interesting to note that neither FR 901533 nor PD 069185 has a known functional group for binding to zinc at the active site of ECE-1. Therefore, the mechanisms by which these two compounds inhibit the ECE-1 activity are not clear. Recently, CGS 35066 was introduced as a very potent non-peptidic ECE inhibitor, which was achieved by replacing the tetrazol in CGS 34043 with a carboxylic acid (Jeng et al. 1999). It has an IC_{50} of 22 nmol/l against recombinant human ECE-1 with an overall selectivity of 105-fold over NEP. CGS 35339, an orally available pro-drug of CGS 35066, possesses the best in vivo activity amongst Novartis's series of compounds (Jeng et al. 1999).

VI. Other Compounds Presenting Some ECE Inhibitory Activities

Besides the four groups (phosphoramidate, phosphonamide, phosphinic acid, and aminophosphonic acid) that contain a phosphorus functionality, some sulfhydryl, hydroxamic acid, and carboxylic acid derivatives have also been shown to be excellent zinc-binding groups. A number of patented thiol and hydroxamic acid derivatives were reported to possess ECE inhibitory activity and were highly potent in in vitro assays with IC_{50} values between 0.01 nmol/l and 13 nmol/l (see Fig. 6). The two natural compounds daleformis and halistanol disulfate B also inhibited the activity of solubilized recombinant ECE

(Tables 3 and 4, Fig. 6). However, very few or no additional pharmacological data have been reported to substantiate further ECE inhibition for these compounds in vivo.

C. Effects of Protease Inhibitors, Chelating Agents and Metal Ions on ECE Activities in Cultured Cells, Isolated Tissues, Perfused Organs, and In Vivo

Several studies have shown that phosphoramidon and other ECE-selective, dual NEP/ECE or triple ACE/NEP/ECE inhibitors can modulate the basal/stimulated release of ETs and/or the pharmacological effects of exogenous big ETs in various systems.

I. In Cultured Cells

1. In Endothelial Cells

Treatment of phosphoramidon attenuated the basal synthesis and release of ET-1 in culture media of BAECs and PAECs (OKADA et al. 1990; IKEGAWA et al. 1990b). It was suggested that phosphoramidon inhibited both the intracellular conversion of big ET-1 to ET-1 and the release of ET-1 from cultured BAECs (SAWAMURA et al. 1991). This compound was also shown to block the conversion of exogenous big ET-1 by BAECs, implying the existence of an extracellular membrane-bound ECE (OHNAKA et al. 1991).

The BAECs are less effective in converting big ET-3 to ET-3 than big ET-1 to ET-1 (OKADA et al. 1990; TAKADA et al. 1991). Similarly, the ECE activity found in PAECs shows the highest affinity for human big $ET-1_{1-38}$ amongst the three big ETs (IKEGAWA et al. 1990b). In PAECs, the conversion of human big $ET-2_{1-37}$ to ET-2 is much slower (compared to big ET-1 conversion) and the $Trp^{21}-Ile^{22}$ peptide bond of human big $ET-3_{1-41}$ is not cleaved (OHNAKA et al. 1993). In contrast, it has also been shown that human big $ET-3_{1-41}$ may be converted to its active peptide by a phosphoramidon-sensitive ECE partially purified from cultured PAECs (MATSUMURA et al. 1992). Such ECE activity present in endothelial cells are not inhibited by other endopeptidase inhibitors such as thiorphan or captopril, an ACE inhibitor (MATSUMURA et al. 1990b; OKADA et al. 1990).

The ECE activity present in PAECs can be inhibited by chelating agents such as EDTA or 1,10-phenantroline (MATSUMURA et al. 1990b). The ECE highly purified from PAECs can also be inhibited by several divalent cations (Cu^{2+}, Zn^{2+}, Co^{2+}, Fe^{2+}) but unaffected by Ca^{2+}, Mn^{2+}, or Mg^{2+} (OHNAKA et al. 1993). Similar sensitivity to divalent cations have been reported for the ACE activity previously (CUSHMAN and CHEUNG 1971). Exogenous addition of Zn^{2+} completely reverses the inactivation of ECE caused by EDTA, suggesting that the ECE present in PAECs is a zinc metalloprotease (OHNAKA et al. 1993).

2. In Airway Epithelial Cells

Both the dual NEP/ECE inhibitors CGS 26303 and phosphoramidon suppress the basal release of ET-1 whereas specific NEP inhibitors such as thiorphan, SQ 28603, or CGS 24592 inhibit the degradation of ET-1 from guinea-pig tracheal epithelial cells (YANG et al. 1997). The effect of CGS 26303, CGS 26393, and CGS 24592 on lipopolysaccharide (LPS)- or interleukin-1β (IL-1β)-stimulated production of ETs from the same cells were also examined in order to determine if the compounds could modulate the enhanced production and release of the vasoconstrictors observed during pathophysiological conditions. In these experiments, the cells were incubated in the presence or absence of LPS (10 μg/ml) or IL-1β (5 ng/ml) together with CGS 26303, CGS 26393, or CGS 24592 for a period of 24 h. The LPS- or IL-1β stimulated production of ET-1 from guinea-pig tracheal epithelial cells was attenuated dose-dependently by CGS 26303 or CGS 26393, but was unaffected by CGS 24592 (PELLETIER et al. 1998).

3. In Other Types of Cells

In cultured Swiss 3T3 fibroblasts, treatment with phosphoramidon increases the number of ET binding sites in a dose- and time-dependent manner (WU-WONG et al. 1993). Further studies demonstrate that the effect of phosphoramidon is due to inhibition of a protease responsible for degrading the ET receptors.

II. In Isolated Vascular and Non-Vascular Preparations

1. Vascular Preparations

These preparations were used to assess the presence of ECE activities by determining the exogenous big ET-induced contractions resulting from conversion of the precursors into mature peptides. By doing so, a phosphoramidon-sensitive ECE activity has been demonstrated in isolated rat pulmonary artery (HIRATA et al. 1990), rat thoracic aorta (KASHIWABARA et al. 1989; NISHIKORI et al. 1991), and porcine coronary artery (KIMURA et al. 1989). However, the contractions induced by big ETs are much less potent than those obtained using exogenous ETs. Interestingly, the response to big ET-1, not ET-1, is less potent in rabbit saphenous veins than in rabbit saphenous arteries. Furthermore, ECE activity in the vein is insensitive to phosphoramidon (AUGUET et al. 1992).

2. Non-Vascular Preparations

A phosphoramidon-sensitive ECE activity has also been demonstrated in isolated guinea-pig bronchi (NOGUCHI et al. 1991), guinea-pig gallbladder (in response to human and porcine big ET-1$_{1-38; 1-39}$ and human big ET-2, but not human big ET-3$_{1-41}$) (BATTISTINI et al. 1994), rat uterus (in response to human big ET-1_{1-38}) (RAE et al. 1993a; b), the prostatic portion of the rat vas deferens (in response to big ET-1 but not big ET-3) (TÉLÉMAQUE and D'ORLÉANS-JUSTE 1991), and human bronchi (ADVENIER et al. 1992; McKAY et al. 1996). The contraction induced by big ETs is normally less potent than that obtained using exogenous ETs, probably due to the degradation of the converted mature peptide by NEP located on epithelial cells. In addition, in all these preparations, both big ET-1 and big ET-2 induce contraction whereas big ET-3 is inactive except in human bronchi where it causes a contraction that is CGS 26303-inhibitable (McKAY et al. 1997). Thiorphan is ineffective in all of these preparations.

III. In Isolated and Perfused Organs

Big ET-1 is converted to ET-1 by a phosphoramidon-sensitive ECE in the rat lung (WYPIJ et al. 1992; HISAKI et al. 1994). A phosphoramidon-sensitive ECE, capable of converting exogenous big ET-1, was demonstrated in the pulmonary vascular bed of the guinea-pig (D'ORLÉANS-JUSTE et al. 1991), the rabbit (ISHIKAWA et al. 1992), and the newborn piglet (LIBEN et al. 1993). Interestingly, in guinea-pig perfused lungs, the ECE activity converts specifically big ET-1 but not big ET-3, as determined by increases in the perfusion pressure or the release of eicosanoids (D'ORLÉANS-JUSTE et al. 1991). S17162 was reported to suppress big ET-1-induced increase in perfusion pressure in isolated and perfused rat kidneys, as potently as phosphoramidon (DESCOMBES et al. 1995).

IV. In Semi-Purified and Purified Preparations

A phosphoramidon-sensitive ECE activity has been demonstrated in bovine (KUNDU and WILSON 1992) and porcine (SAWAMURA et al. 1993) lung membranes. Purification of a phosphoramidon-sensitive ECE has also been reported from rat lung (TAKAHASHI et al. 1993). An enzyme that catalyzes the conversion of human big ET-1_{1-38} to ET-1 has also been purified from membrane fractions of PAECs (OHNAKA et al. 1993). The ECE activity associated with PAECs has been postulated to be a monomeric and acidic glycoprotein with an isoelectric point of 4.1 (OHNAKA et al. 1993).

V. In Vivo Studies

In marked contrast to the previous observations in isolated preparations where big ET-1 is less potent than ET-1 in causing contractions, exogenous, intravenously administered big ET-1 has been shown to be almost equipotent to ET-1 in vivo. For instance, porcine big ET-1_{1-39} causes a phosphoramidon-sensitive increase in blood pressure similar to ET-1 in anesthetized guinea-pigs (D'ORLÉANS-JUSTE et al. 1990). The effect is accompanied by an elevation of the circulating plasma level of ET-1.

Phosphoramidon blocks the pressor response to exogenous big ET-1 (human big ET-3 is inactive as a pressor agent in guinea-pigs) as well as the increase in plasma ET levels, but not the increase in MABP induced by ET-1 in guinea-pigs (D'ORLÉANS-JUSTE et al. 1991). Human big ET-1 and big ET-3 also produce phosphoramidon-sensitive elevations in MABP in anesthetized rats (MATSUMURA et al. 1990a, 1993a; POLLOCK et al. 1993; KASHIWABARA et al. 1989; FUKURODA et al. 1990; OKADA et al. 1990; IKEGAWA et al. 1990b; McMAHON et al. 1991a; D'ORLÉANS-JUSTE et al. 1991).

The natural compound WS75624B (30 mg/kg i.v.) suppressed big ET-1-induced pressor responses, similar to the effect of phosphoramidon (same dose) in rats (TSURUMI et al. 1995b). S17162 (10 mg/kg i.v. – 20 mg/kg p.o.) also attenuated the effect of big ET-1 in pithed rats, comparable to the effects of phosphoramidon (DESCOMBES et al. 1995). CGS 26393, the pro-drug of CGS 26303, likewise suppressed the big ET-1-induced pressor response by 70%, 4 h after an oral dose of 30 mg/kg in anesthetized rats; the effects were sustained for at least 8 h (TRAPANI et al. 1995). Furthermore, SCH 54470 inhibited the big ET-1-induced pressor response by 43% when infused at a rate of 0.1 mg/kg/min for 30 min prior to the big ET-1 challenge. Since it is a triple ACE/NEP/ECE inhibitor, SCH 54470 significantly attenuated angiotensin I-induced pressor response at 3 mg/kg s.c. in anesthetized rats (VEMULAPALLI et al. 1997).

These results suggest that the conversion of big ET-1 to ET-1 takes place effectively in vivo which can be blocked by ECE inhibitors, but not completely. It may be due to the fact that both endothelial cells, underlying vascular smooth muscle cells (WOODS et al. 1998), and some circulating cells can catalyze the conversion of big ET-1 to ET-1.

D. Effect of ECE Inhibitors in Various Pre-Clinical Animal Disease Models and in Clinical Studies

Numerous lines of evidence strongly imply that ET-1 may play a role in the pathogenesis of a variety of cardiovascular, renal, and central nervous system diseases. Such conditions include hypertension, myocardial infarction, congestive heart failure, renal failure, nephrotoxicity, pulmonary hypertension, and cerebral vasospasm following subarachnoid hemorrhage (FERRO and WEBB 1996; KOHAN 1997; PATEL 1996).

The interest in ECE has emerged from the hypothesis that ECE inhibitors would suppress the biosynthesis of ET-1 and, thus, provide an effective way to block undesirable ET-mediated effects in diseases where an overproduction of this peptide plays a pathogenic role. Most of the observations reported below are not seen with a selective NEP inhibitor, suggesting that the effects are due to inhibition of ECE-1, but not of NEP. Generally, non-selective NEP/ECE inhibitors, especially phosphoramidon, have been utilized to show some efficacy in various animal models of cardiovascular, renal, and central nervous system diseases, since selective ECE inhibitors have only recently become available (Table 5).

I. Cerebral Vasospasm Following Subarachnoid Hemorrhage

Antagonism of the ET system is a logical approach to reduce the post-operative cerebral vasospasm and traumatic brain injury (see Chap. 13). Treatment with phosphoramidon has been demonstrated to have a beneficial effect in subarachnoid hemorrhage, reducing cerebral vasospasm in dogs (MATSUMURA et al. 1991a). Similar to phosphoramidon, treatment with CGS 26303 also prevented, as well as reversed, cerebral vasospasm in rabbits following experimental subarachnoid hemorrhage (CANER et al. 1996; KWAN et al. 1997).

II. Conditions Associated with Vascular Remodeling

1. Hypertension

Treatment with phosphoramidon has been shown to lower MABP in spontaneously hypertensive rats (SHR), DOCA-salt hypertensive rats (MCMAHON et al. 1993; YANO et al. 1991), and in renal-artery ligated hypertensive rats (VEMULAPALLI et al. 1993b) (see Chap. 16). Similar to phosphoramidon, infusion with CGS 26303, subcutaneously via implanted osmotic minipumps, also reduced MABP in conscious SHR (DE LOMBAERT et al. 1994b). No effects were seen with CGS 24592, a structural analogue of CGS 26303 and a selective NEP inhibitor. Oral administration of SLV 306 likewise produced natriuretic and diuretic effects in conscious rats after volume loading and lowered MABP in DOCA-salt hypertensive rats (MEIL et al. 1998). However, with different classes of antihypertensive drugs currently available, ECE inhibitors or ET receptors antagonists are not likely to be developed as a monotherapy.

2. Congestive Heart Failure

Phosphoramidon was reported to attenuate the extent and size of myocardial ischemia in rat isolated hearts (GROVER et al. 1992). Oral administration of SLV 306 also reduced cardiac hypertrophy as well as pulmonary congestion in rats with congestive heart failure caused by aortic stenosis (MEIL et al. 1998). Infusion of phosphoramidon increased the forearm blood flow in chronic heart failure patients already treated with ACE inhibitors and diuretics (LOVE et al. 1996). The fact that phosphoramidon has demonstrated vasodilatory effects even in patients receiving ACE inhibitors suggests that an add-on therapy may provide greater benefits.

The acute effects of a specific ECE inhibitor (FR 901533) were compared to those of FR 139317, a selective ET_A receptor antagonist, on cardiorenal and endocrine functions in dogs with rapid right ventricular pacing (270 bpm, 14 days)-induced congestive heart failure (WADA et al. 1999). Both drugs decreased a number of hemodynamic parameters and plasma levels of neurohumoral factors (see Table 5). Overall, FR 139317 exerted a greater vasodilatory effect on systemic and renal vasculature than FR 901533. However, FR 910533 reduced the activated secretion of factors in proportion to the severity of congestive heart failure, providing tangible data supporting investigations into the usefulness of long-term treatment with ECE inhibitor in this disease (WADA et al. 1999).

3. Post-Angioplastic Restenosis

Dual NEP/ECE inhibitors such as CGS 26303 and its prodrug CGS 26393 were tested against neointimal formation following balloon angioplasty of the rat carotid artery (PHAM et al. 1998). A 2-week treatment, initiated immediately following balloon angioplasty, provided significant inhibition of neointimal proliferation with i.p., s.c., or i.v. administration of CGS 26303 or with CGS 26393, p.o. CGS 24592, a selective NEP inhibitor, had no effect. Phosphoramidon was also tested and clearly shown to reduce the size of the neointima in this rat model (MINAMINO et al. 1997). These results further support a role for ETs in the pathogenesis of angioplasty-induced lesion formation and suggest that inhibitors of ECE(s) could be as effective as the ET receptor antagonists or any other agents thus far investigated in this animal model towards attenuating restenosis following percutaneous transluminal coronary angioplasty in patients.

III. Renal Failure

In a common rat model of ischemic acute renal failure (the renal artery of one kidney is clamped for 30–40min, and reperfused for 1–2hr) in which glomerular filtration rate (GFR) is decreased and fractional sodium excretion (FE_{Na}) and urinary protein are increased, pre-treatment with phosphoramidon increased GFR, and suppressed FE_{Na} and proteinuria (VEMULAPALLI et al. 1993a). Similar results were obtained with phosphoramidon in ischemic, uninephrectomized rats (BIRD et al. 1995). In the remnant kidney model of chronic renal failure in the rat, subcutaneous administration of SCH 54470 at 30mg/kg, b.i.d., for 2 weeks significantly decreased the MABP, proteinuria, and FE_{Na} with concomitant increases in the glomerular filtration rate and renal plasma flow when compared with the untreated controls (VEMULAPALLI et al. 1997). However, since similar renal protective effects have been observed with dual inhibitors of ACE and NEP, whether the ECE-1 inhibitory activity of SCH 54470 has additive/synergistic effects remains to be investigated (see Chap. 18).

IV. Pulmonary Vascular Diseases

Subcutaneous administration of FR 901533 via osmotic mini-pumps attenuated the development of right ventricular overload and medial thickening of pulmonary arteries in a rat model of monocrotaline-induced pulmonary hypertension (TAKAHASHI et al. 1998). In another study, injection of phosphoramidon decreased the cardiopulmonary bypass-induced increases in pulmonary arterial pressure and indexed pulmonary vascular resistance in pigs (KIRSHBOM et al. 1995) (see Chap. 15).

V. Pulmonary Respiratory Diseases

CGS 26303 was reported to attenuate the translocation and accumulation of cells in the bronchoalveolar space of C57BL/6 mice infected with a combination of the antigen *Sacharopolyspora rectivirgula* and the *Sendai* virus as a model of extrinsic allergic alveolitis (BEAUDOIN et al. 1999). The number of total cells in the lavage decreased significantly while the number of macrophages and lymphocytes also decreased upon treatment with CGS 26303 when compared with the untreated controls.

E. Conclusions and Perspectives

Numerous lines of evidence strongly support pathogenic roles for ETs in various cardiovascular, renal, and central nervous system diseases such as hypertension, myocardial infarction, congestive heart failure, renal failure, nephrotoxicity, pulmonary hypertension, and cerebral vasospasm following subarachnoid hemorrhage (FERRO and WEBB 1996; KOHAN 1997; PATEL 1996). Therefore, inhibition of the biological effects of ETs would be beneficial for treatment of these diseases.

Two classes of compounds have been developed for this purpose: ET receptor antagonists and ECE-related inhibitors. Most major pharmaceutical companies have chosen to target the ET receptors to inhibit ET-mediated effects (see Chap. 9). To date, two mammalian ET receptor subtypes have been identified (SAKURAI et al. 1992) (see Chap. 4). While the ET_A receptor has higher affinity for ET-1 and ET-2 than ET-3, the ET_B receptor exhibits similar affinity for all three ETs. Binding of ET-1 to ET_A or ET_B receptors in vascular smooth muscle cells causes constriction. In contrast, binding of ET-1 to ET_B receptors on endothelial cells triggers the release of nitric oxide and prostacyclin, which in turn mediate vasodilatory effects. In the respiratory system, however, binding of ET-1 to ET_A receptors in epithelial cells triggers the release of nitric oxide and prostacyclin in both guinea-pigs (FILEP et al. 1993; BATTISTINI et al. 1994) and rabbit trachea (GRUNSTEIN et al. 1991). Conversely, binding of ET-1 to the ET_A or ET_B receptors in underlying non-vascular smooth muscle cells causes bronchoconstriction in guinea-pigs (BATTISTINI et al. 1994).

Three types of ET receptor antagonists have been synthesized: dual ET_A/ET_B, selective ET_A, and selective ET_B receptor antagonists (see Chap. 9). However, pharmaceutical companies are unlikely to develop ET_B receptor antagonists as a class of new therapeutic agents for cardiovascular-related diseases since, in general, administration of selective ET_B receptor antagonists elevates systemic blood pressure in human subjects (Webb et al. 1995) (see Chap. 19). At the beginning, most of the ET receptor antagonists were peptides, which had poor oral bioavailability and were subjected to proteolytic degradation. Ro 46–2005 (Hoffmann-La Roche) presented the first non-peptidic, orally active dual ET_A/ET_B receptor antagonist (Webb and Meek 1997). The pharmacological properties of several potent, non-peptidic ET receptor antagonists have been recently reviewed in great details (Webb and Meek 1997; Douglas 1997; Battistini and Dussault 1998c).

Due to the dual vasodilatory and vasoconstrictive effects mediated by the ET_B receptors, it is still not clear if selective ET_A or dual ET_A/ET_B receptor antagonists would be more efficacious. In addition, other receptors, such as the ET_C receptor, yet to be cloned in mammalians, have been proposed. In many pathologic conditions, elevated ET levels have been found in the plasma, bronchoalveolar lavage fluid, cerebrospinal fluids, as well as in tissue samples. Further increases in the circulating ET-1 levels have been noted upon treatment with bosentan, a non-selective ET_A/ET_B receptor antagonist, in human (Weber et al. 1996). Infusion of the ET_B receptor antagonist BQ-788 reduced ET-1 clearance rate in rats, suggesting that ET_B receptors play an important role in such a process (Fukuroda et al. 1994). Whether the excessive ET generated could produce non-specific effects remains to be investigated. In this regard, ECE inhibitors offer an alternative approach to inhibit the biological actions of ETs. Some of these compounds have been derived from existing ACE and/or NEP inhibitors. Nevertheless, a number of potent and selective ECE inhibitors have now been synthesized as described above. To date, three classes of ECE inhibitors have been reported: dual NEP/ECE, triple ACE/NEP/ECE, and selective ECE inhibitors. Considering that dual inhibitors of NEP and ACE have already been shown to be excellent agents for the treatment of congestive heart failure and chronic renal failure, the addition of an ECE inhibitory activity may well be beneficial. However, no selective ECE or triple inhibitors are in clinical trials at present. These compounds are expected to suppress the biosynthesis of ET-1 and therefore should have effects similar to that obtained using the dual ET_A/ET_B receptor antagonists but without raising circulating ET levels.

Since the relative population of ET_A and ET_B receptors varies between tissues, vascular beds, and species, the use of ECE inhibitors could be more attractive for it would inhibit ET mediated-responses in pathological conditions regardless of the ET receptor subtypes involved in these responses (Battistini et al. 1994). In addition, other physiological substrates for ECE-1, in addition to big ET-1, have not been thoroughly investigated, at least in vivo. One recent report by Johnson et al. (1999) also revealed that other vasoactive mediators or hormones constitute excellent substrates for ECE. The effects of elevated levels of these substrates due to ECE inhibition are unknown. Furthermore, ECE may not be the only enzyme responsible for producing the ETs; the existence of alternative pathways has not been fully explored.

Table 6. Criteria to consider toward the development and use of ECE inhibitors

1 Alternative pathways leading to the production of ETs	How to achieve selectivity with the cloning of even more ECE isoenzymes (plasticity) and the existence of non-specific pathways
2 Other physiological substrates (ex. AII, BK, SP) for ECEs	ECE inhibition may affect the balance of the physiological substrates since the degradation of such peptides would be impaired
3 Optimal selectivity (single, double, triple therapy)	Inhibiting either the plasma membrane ECE (ectoenzyme) and/or intracellular ECE (also related to pH) and/or other converting/degrading enzyme such as ACE and NEP
4 Selective regulation	Inhibiting either the constitutive pathway (ex. pre-stored ETs) and/or the regulated pathway (*de novo* stimulated synthesis)
5 Advantage versus the use of ET-receptor antagonists	Blocking the up-regulated synthesis right at the site of biosynthesis, preventing ET-mediated responses compared to the selective inhibition at the sites of receptor binding; keeping note that ET_B receptors are involved in the clearance of circulatory ET-1

It should be noted that the alternative approach to inhibit the effects of ET by means of selective ECE-1 inhibitor may not be fully efficacious, since ECE-2a/-2b and ECE-3 are involved in the biosynthesis of ET-1. Dual inhibitors of ECE and NEP may also present additional benefits for the treatment of cardiovascular and renal dysfunction than the use of a selective ECE inhibitors. On the other hand, since NEP is shown to degrade ETs, elevation of ET levels due to an inhibition of the degradation of these peptides by NEP inhibitors may lead to further physiopharmacological effects (see Table 6).

Since ACE and/or dual ACE/NEP inhibitors have been shown to be excellent agents for the treatment of hypertension, congestive heart failure, or restenosis (DE LOMBAERT et al. 1996), compounds that simultaneously inhibit the activities of ACE, NEP, and ECE may be novel drugs for these diseases. As an example, both ACE inhibitors, blocking the production of Ang II, and losartan, an angiotensin AT_1 receptor antagonist, are efficacious in the rat model of restenosis, a model in which ECE inhibitors have also been demonstrated to have beneficial effects. Treatment with dual ACE/NEP or triple ACE/NEP/ECE inhibitors may give additive/synergistic results. This synergism can be envisaged since ET-1 increases the conversion of Ang I to Ang II in cultured pulmonary artery endothelial cells (KAWAGUCHI et al. 1990), and reciprocally, Ang II induces ET-1 production (EMORI et al. 1991).

Undoubtedly, clinical development of the ET receptor antagonists is more advanced than that of the ECE inhibitors. Could there be any benefits in using ECE inhibitors vs ET receptor antagonists? Only more pre-clinical studies and clinical trials could provide the much needed comparisons between ET receptor antagonists and ECE inhibitors as therapeutic agents. Thus far, the overall results obtained using ECE inhibitors compare favorably with those obtained using selective ET_A and/or dual ET_A/ET_B receptor antagonists.

Acknowledgements. We would like to thank Mrs. Nancy Laberge for typing the manuscript and Mrs. Stéphanie Molez and Mrs. Kathleen Lowde for editing. BB is a Junior I FRSQ Scholar. His research is funded by the Quebec Heart & Stroke Foundation (QHSF), the Fonds pour les Chercheurs et l'Aide à la Recherche (FCAR), les Fonds pour le Recherche en Santé du Québec (FRSQ), the Bégin Foundation and the Quebec Heart Institute Foundation. AYJ is a Principal Fellow at the Novartis Institute for Biomedical Research, Novartis Pharmaceuticals Corporation, Summit, NJ.

References

Advenier C, Lagente V, Zhang Y, Naline E (1992) Contractile activity of big endothelin-1 on the human isolated bronchus. Br J Pharmacol 106:883–887

Ahn K, Herman SB, Fahnoe DC (1998) Soluble human endothelin-converting enzyme-1: expression, purification, and demonstration of pronounced pH sensitivity. Arch Biochem Biophys 359:258–268

Ahn K, Knapp J, Johnson GD, Fahnoe DC (1999) Inhibitor potencies and substrate preference for endothelin converting enzyme-1 are dramatically affected by pH. 6th Int Conf ETs (Montréal, QC, Canada) p 4

Ahn K, Pan SM, Zientek MA, Guy PM, Sisneros AM (1996) Characterization of endothelin converting enzyme from intact cells of a permanent human endothelial cell line, EA,hy926. Biochem Mol Biol Int 39:573–580

Arai K, Ashikawa N, Nakakita Y, Matsumura A, Ashizawa N, Munekata M (1993) Aspergillomarasmine A and B, potent microbial inhibitors of endothelin-converting enzyme. Biosci Biotech Biochem 57:1944–1945

Ashizawa N, Okumura H, Kobayashi F, Aotsuka T, Asakura R, Arai K, Ashikawa N, Matsuura A (1994a) Inhibitory activities of metal chelators on endothelin-converting enzyme. II. In vivo studies. Biol Pharm Bull 17:212–216

Ashizawa N, Okumura H, Kobayashi F, Aotsuka T, Takahashi M, Asakura R, Arai K, Matsuura A (1994b) Inhibitory activities of metal chelators on endothelin-converting enzyme. I. In vitro studies. Biol Pharm Bull 17:207–211

Auguet M, Delaflotte S, Chabrier PE, Braquet P (1992) The vasoconstrictor action of big endothelin-1 is phosphoramidon-sensitive in rabbit saphenous artery, but not in saphenous vein. Eur J Pharmacol 224:101–102

Balwierczak JL, Kukkola PJ, Savage P, Jeng AY (1995) Effects of metalloprotease inhibitors on smooth muscle endothelin-converting enzyme activity. Biochem Pharmacol 49:291–296

Barton M, Shaw S, d'Uscio LV, Moreau P, Luscher TF (1997) Angiotensin II increases vascular and renal endothelin-1 and functional endothelin converting enzyme activity in vivo: role of ETA receptors for endothelin regulation. Biochem Biophys Res Commun 238:861–865

Battistini B, Botting R (1995) Endothelins: A Quantum Leap Forward. Drug News & Perspect 8:365–391

Battistini B, D'Orléans-Juste P, Sirois P (1993) Endothelins in physiopathology; circulating plasma levels and presence in other biological fluids. Lab Invest 68:600–628

Battistini B, Dussault P (1998a) The many aspects of endothelins in ischemia-reperfusion injury: emergence of a key mediator. J Invest Surg 11:297–313

Battistini B, Dussault P (1998b) Biosynthesis, distribution and metabolism of endothelins in the pulmonary system. Pulm Pharmacol Ther 11:79–88

Battistini B, Dussault P (1998c) Blocking of the endothelin system: the development of receptor antagonists. Pulm Pharmacol Ther 11:97–112

Battistini B, O'Donnell LJD, Warner TD, Fournier A, Farthing MJG, Vane JR (1994) Characterization of endothelin (ET) receptor subtypes in the isolated gallbladder of the guinea-pig. Br J Pharmacol 112:1244–1250

Baynash AG, Hosoda K, Giaid A, Richardson JA, Emoto N, Hammer RE, Yanagisawa M (1994) Interaction of endothelin-3 with endothelin-B receptor is essential for development of epidermal melanocytes and enteric neurons. Cell 79:1277–1285

Beaudoin CN, Israel-Assayag E, Dussault P, Cormier Y, Battistini B (1999) Correlation between levels of ET-1, big ET-1, NO, total and differential cell counts in the bronchoalveolar lavage fluid in patients with inflammatory lung diseases. 6th Int Conf ETs (Montréal, QC, Canada) p 49

Bertenshaw SR, Rogers RS, Stern MK, Norman BH, Moore WM, Jerome GM, Branson LM, McDonald JF, McMahon EG, Palome MA (1993a) Phosphorus-containing inhibitors of endothelin-converting enzyme: effects of the electronic nature of phosphorus on inhibitor potency. J Med Chem 36:173–176

Bertenshaw SR, Talley JJ, Rogers, Carter JS, Moore WM, Branson LM, Koboldt CM (1993b) Thiol and hydroxamic acid containing inhibitors of endothelin-converting enzyme. Bioorg Med Chem Lett 3:1953–1958

Bihovsky R, Levinson BL, Loewi RC, Erhardt PW, Polokoff MA (1995) Hydroxamic acids as potent inhibitors of endothelin-converting enzyme from human bronchiolar smooth muscle. J Med Chem 38:2119–2129

Bird JE, Waldron TL, Little DK, Asaad MM, Dorso CR, DiDonato G, Norman JA (1992) The effects of novel cathepsin E inhibitors on the big endothelin pressor response in conscious rats. Biochem Biophys Res Commun 182:224–231

Bird JE, Webb ML, Wasserman AJ, Liu EC, Giancarli MR, Durham SK (1995) Comparison of a novel ETA receptor antagonist and phosphoramidon in renal ischemia. Pharmacology 50:9–23

Bloch KD, Eddy RL, Shows TB, Quetermous T (1989) cDNA cloning and chromosomal assignment of the gene encoding endothelin 3. J Biol Chem 264:18156–18161

Bloch KD, Hong CC, Eddy RL, Shows TB, Quertermous T (1991) cDNA cloning and chromosomal assignment of the endothelin 2 gene: vasoactive intestinal contractor peptide is rat endothelin 2. Genomics 10:236–242

Bond JS, Beynon RJ (1987) Proteolysis and physiological regulation. Mol Aspects Med 9:173–287

Brown CD, Barnes K, Turner AJ (1998) Anti-peptide antibodies specific to rat endothelin-converting enzyme-1 isoforms reveal isoform localisation and expression. FEBS Lett 424:183–187

Cailler F, Zappulla JP, Boileau G, Crine P (1999) The N-terminal segment of endothelin-converting enzyme (ECE)-1b contains a di-leucine motif that can redirect neprilysin to an intracellular compartment in Madin-Darby canine kidney (MDCK) cells. Biochem J 341 (Pt 1):119–126

Caner HH, Kwan AL, Arthur A, Jeng AY, Lappe RW, Kassell NF, Lee KS (1996) Systemic administration of an inhibitor of endothelin-converting enzyme for attenuation of cerebral vasospasm following experimental subarachnoid hemorrhage. J Neurosurg 85:917–922

Chackalamannil S, Chung S, Stamford AW, McKittrick BA, Wang Y, Tsai H, Cleven R, Fawzi A, Czarniecki M (1996) Highly potent and selective inhibitors of endothelin converting enzyme. Bioorg Med Chem Lett 6:1257–1260

Claing A, Neugebauer W, Yano M, Rae GA, D'Orléans-Juste P (1995) [Phe22]-big endothelin-1[19–37]: a new and potent inhibitor of the endothelin-converting enzyme. J Cardiovasc Pharmacol 26 [Suppl 3]:S72–S74

Cushman DW, Cheung HS (1971) Spectrophotometric assay and properties of the angiotensin-converting enzyme of rabbit lung. Biochem Pharmacol 20:1637–1648

Deng Y, Jeng AY (1992) Soluble endothelin degradation enzyme activities in various rat tissues. Biochem Cell Biol 70:1385–1389

Deng Y, Savage P, Shetty SS, Martin LL, Jeng AY (1992) Identification and partial purification of a thiol endothelin-converting enzyme from porcine aortic endothelial cells. J Biochem (Tokyo) 111:346–351

De Lombaert S, Blanchard L, Tan J, Sakane Y, Berry C, Ghai RD (1995) Non-peptidic inhibitors of neutral endopeptidase 24.11. 1. Discovery and optimization of potency. Bioorg Med Chem Lett 5:145–150

De Lombaert S, Chatelain RE, Fink CA, Trapani AJ (1996) Design and pharmacology of dual angiotensin-converting-enzyme and neutral endopeptidase inhibitors. Curr Pharmaceutical Des 2:443–462

De Lombaert S, Erion MD, Tan J, Blanchard L, El-Chehabi L, Ghai RD, Sakane Y, Berry C, Trapani AJ (1994a) N-Phosphonomethyl dipeptides and their phosphonate prodrugs, a new generation of neutral endopeptidase (NEP, EC 3.4.24.11) inhibitors. J Med Chem 37:498–511

De Lombaert S, Ghai RD, Jeng AY, Trapani AJ, Webb RL (1994b) Pharmacological profile of a non-peptidic dual inhibitor of neutral endopeptidase 24.11 and endothelin-converting enzyme. Biochem Biophys Res Commun 204:407–412

De Lombaert S, Stamford LB, Blanchard L, Tan J, Hoyer D, Diefenbacher CG, Wei D, Wallace EM, Moskal MA, Savage P, Jeng AY (1997) Potent non-peptidic dual inhibitors of endothelin-converting enzyme and neutral endopeptidase 24.11. Bioorg Med Chem 7:1059–1064

Deprez P, Guillaume J, Dumas J, Vevert J.-P (1996) Thiol inhibitors of endothelin-converting enzyme. Bioorg Med Chem Lett 6:2317–2322

Descombes JJ, Mennecier P, Versluys D, Barou V, de Nanteuil G, Laubie M, Verbeuren TJ (1995) S 17162 is a novel selective inhibitor of big ET-1 responses in the rat. J Cardiovasc Pharmacol 26 [Suppl 3]:S61–S64

D'Orléans-Juste P, Lidbury PS, Warner TD, Vane JR (1990) Intravascular big-endothelin increases circulating levels of ET-1 and prostanoids in the rabbit. Biochem Pharmacol 39:R21–R22

D'Orléans-Juste P, Télémaque S, Claing A (1991) Different pharmacological profiles of big-endothelin-3 and big-endothelin-1 in vivo and in vitro. Br J Pharmacol 104:440–444

Douglas SA (1997) Clinical development of endothelin receptor antagonists. Trends Pharmacol Sci 18:408–412

Ehrenreich H, Loffler BM, Hasselblatt M, Langen H, Oldenburg J, Subkowski T, Schilling L, Siren AL (1999) Endothelin converting enzyme activity in primary rat astrocytes is modulated by endothelin B receptors. Biochem Biophys Res Commun 261:149–155

Emori T, Hirata Y, Kanno K, Ohta K, Eguchi S, Imai T, Schichiri M, Marumo F (1991) Endothelin-3 stimulates production of endothelium-derived nitric oxide via phosphoinositide breakdown. Biochem Biophys Res Commun 174:228–235

Emoto N, Yanagisawa M (1995) Endothelin-converting enzyme-2 is a membrane-bound phosphoramidon-sensitive metalloprotease with acidic pH optimum. J Biol Chem 270:15262–15268

Ferro CJ, Webb DJ (1996) The clinical potential of endothelin receptor antagonists in cardiovascular medicine. Drugs 51:12–27

Filep JG, Battistini B, Sirois P (1993) Induction by endothelin-1 of epithelium-dependent relaxation of guinea-pig trachea in vitro: role for nitric oxide. Br J Pharmacol 109:637–644

Fujita Y, Satoh C, Tsukahara Y, Miyawaki N (1997) The properties of enzyme inhibition by a novel dual inhibitor, SA6817, of neutral endopeptidase and angiotensin-converting enzyme. Jpn Pharmacol Soc p 28 (Abstract)

Fukami T, Hayama T, Amano Y, Nakamura Y, Arai Y, Matsuyama K, Yano M, Ishikawa K (1994) Aminophosphonate endothelin-converting enzyme inhibitors: potency-enhancing and selectivity-improving modifications of phosphoramidon. Bioorg Med Chem Lett 4:1257–1262

Fukuroda T, Fujikawa T, Ozaki S, Ishikawa K, Yano M, Nishikibe M (1994) Clearance of circulating endothelin-1 by ETB receptors in rats. Biochem Biophys Res Commun 199:1461–1465

Fukuroda T, Noguchi K, Tsuchida S, Nishikibe M, Ikemoto F, Okada K, Yano M (1990) Inhibition of biological actions of big endothelin-1 by phosphoramidon. Biochem Biophys Res Commun 172:390–395

Grover GJ, Sleph PG, Fox M, Trippodo NC (1992) Role of endothelin-1 and big endothelin-1 in modulating coronary vascular tone, contractile function and severity of ischemia in rat hearts. J Pharmacol Exp Ther 263:1074–1082

Grunstein MM, Rosenberg SM, Schramm CM, Pawlowski NA (1991) Mechanisms of action of endothelin 1 in maturing rabbit airway smooth muscle. Am J Physiol 260 (6 Pt 1):L434–L443

Harrison VJ, Ziegler T, Bouzourene K, Suciu A, Silacci P, Hayoz D (1998) Endothelin-1 and endothelin-converting enzyme-1 gene regulation by shear stress and flow-induced pressure. J Cardiovasc Pharmacol 31 (Suppl 1):S38–S41

Hasegawa H, Hiki K, Sawamura T, Aoyama T, Okamoto Y, Miwa S, Shimohama S, Kimura J, Masaki T (1998) Purification of a novel endothelin-converting enzyme specific for big endothelin-3. FEBS Lett 428:304–308

Hemsen A, Larsen O, Lundberg JM (1991) Characteristics of endothelin A and B binding sites and their vascular effects in pig peripheral tissues. Eur J Pharmacol 208:313–322

Hirata Y, Kanno K, Watanabe TX, Kumagaye SI, Nakajima K, Kimura T, Sakakibara S, Marumo F (1990) Receptor binding and vasoconstrictor activity of big endothelin. Eur J Pharmacol 176:225–228

Hisaki K, Matsumura Y, Maekawa H, Fujita K, Takaoka M, Morimoto S (1994) Conversion of Big Et-1 in the rat lung: role of phosphoramidon-sensitive endothelin-1-converting enzyme. Am J Physiol 266:H422–H428

Hoang MV, Turner AJ (1997) Novel activity of endothelin-converting enzyme: hydrolysis of bradykinin. Biochem J 327 (Pt 1):23–26

Ikeda T, Ohta H, Okada M, Kawai N, Nakao R, Spiegl PK, Kobayashi T, Maeda S, Miyauchi T, Nishikibe M (1999) Pathophysiological roles of endothelin-1 in Dahl salt-sensitive hypertension. Hypertension 34:514–519

Ikegawa R, Matsumura Y, Takaoka M, Morimoto S (1990a) Evidence for pepstatin-sensitive conversion of porcine big endothelin-1 to endothelin-1 by the endothelial cell extract. Biochem Biophys Res Commun 167:860–866

Ikegawa R, Matsumura Y, Tsukahara Y, Takaoka M, Morimoto S (1990b) Phosphoramidon, a metalloproteinase inhibitor, suppresses the secretion of endothelin-1 from cultured endothelial cells by inhibiting a big endothelin-1-converting enzyme. Biochem Biophys Res Commun 171:669–675

Ikura T, Sawamura T, Shiraki T, Hosokawa H, Kido T, Hoshikawa H, Shimada K, Tanzawa K, Kobayashi S, Miwa S, Masaki T (1994) cDNA cloning and expression of bovine endothelin-converting enzyme. Biochem Biophys Res Commun 203:1417–1422

Imai T, Hirata Y, Emori T, Yanagisawa M, Masaki T, Marumo F (1992) Induction of endothelin-1 gene by angiotensin and vasopressin in endothelial cells. Hypertension 19(6 Pt 2):753–757

Inagaki M, Hashimoto K, Takahashi H, Takeshita K (1993) Blood endothelin concentration from umbilical cord and preterm baby: perinatal factor and cranial ultrasonography in the cases of abnormally high concentration. No To Hattatsu 25:385–387

Inoue A, Yanagisawa M, Kimura S, Kasuya Y, Miyauchi T, Goto K, Masaki T (1989) The human endothelin family: three structurally and pharmacologically distinct isopeptides predicted by three separate genes. Proc Natl Acad Sci USA 86:2863–2867

Ishikawa S, Tsukada H, Yuasa H, Fukue M, Wei S, Onizuka M, Miyauchi T, Ishikawa T, Mitsui K, Goto K, Hori M (1992) Effects of endothelin-1 and conversion of big endothelin-1 in the isolated perfused rabbit lung. J Appl Physiol 72:2387–2392

Itoh Y, Kimura C, Onda H, Fujino M (1989) Canine endothelin-2: cDNA sequence for the mature peptide. Nucleic Acids Res 17:5389

Itoh Y, Yanagisawa M, Ohkubo S, Kimura C, Kosaka T, Inoue A, Ishida N, Mitsui Y, Onda H, Fujino M (1988) Cloning and sequence analysis of cDNA encoding the precursor of a human endothelium-derived vasoconstrictor peptide, endothelin: identity of human and porcine endothelin. FEBS Lett 231:440–444

Jeng AY (1997) Therapeutic potential of endothelin converting enzyme inhibitors. Exp Opin Ther Patents 7:1283–1295

Jeng AY, De Lombaert S (1997) Endothelin converting enzyme inhibitors. Curr Pharmaceutical Des 3:597–614

Jeng AY, De Lombaert S, Bruseo CW, Savage P. Chou M, Trapani AJ (1999) Design and synthesis of a potent and selective endothelin converting enzyme inhibitor, CGS 35066. 6th Int Conf ETs (Montréal, QC, Canada) p 16

Johnson GD, Stevenson T, Ahn K (1999) Hydrolysis of peptide hormones by endothelin-converting enzyme-1. A comparison with neprilysin. J Biol Chem 274:4053–4058

Kashiwabara T, Inagaki Y, Ohata H, Iwamatsu A, Nomizu M, Morita A, Nishikori K (1989) Putative precursors of endothelin have less vasoconstrictor activity in vitro but a potent pressor activity in vivo. FEBS Lett 247:73–76

Kaw S, Emoto N, Jeng AY, Yanagisawa M (1995) Pharmacological characterization of cloned human ECE-1 and ECE-2. 4th Int Conf ETs (London, UK) p C6

Kawaguchi H, Sawa H, Yasuda H (1990) Endothelin stimulates angiotensin I to angiotensin II conversion in cultured pulmonary artery endothelial cells. J Mol Cell Cardiol 22:839–842

Keller PM, Lee CP, Fenwick AE, Atkinson ST, Elliott JD, DeWolf WE Jr (1996) Endothelin-converting enzyme: substrate specificity and inhibition by novel analogs of phosphoramidon. Biochem Biophys Res Commun 223:372–378

Kido H, Nakano A, Okishima N, Wakabayashi H, Kishi F, Nakaya Y, Yashizumi M, Tamaki T (1998) Human chymase, an enzyme forming novel bioactive 31-amino acid length endothelins. Biol Chem 379:885–891

Kimura S, Kasuya Y, Sawamura T, Shinimi O, Sugita Y, Yanagisawa M, Goto K, Masaki T (1989) Conversion of big endothelin-1 to 21-residue endothelin-1 is essential for expression of full vasoconstrictor activity: structure activity relationships of big endothelin-1. J Cardiovasc Pharmacol 5:S5–S7

Kirshbom PM, Tsui SS, DiBernardo LR, Meliones JN, Schwinn DA, Ungerleider RM, Gaynor JW (1995) Blockade of endothelin-converting enzyme reduces pulmonary hypertension after cardiopulmonary bypass and circulatory arrest. Surgery 118:440–444; (discussion) 444–445

Knap AK, Soriano A, Savage P, Del Grande D, Shetty SS (1993) Identification of a novel aspartyl endothelin-converting enzyme in porcine aortic endothelial cells. Biochem Mol Biol Int 29:739–745

Kohan DE (1997) Endothelins in the normal and diseased kidney. Am J Kidney Dis 29:2–26

Korth P, Egidy G, Parnot C, LeMoullec JM, Corvol P, Pinet F (1997) Construction, expression and characterization of a soluble form of human endothelin-converting enzyme-1. FEBS Lett 417:365–370

Ksander GM, Savage P, Trapani AJ, Balwierczak JL, Jeng AY (1998) Benzofused macrocyclic lactams as triple inhibitors of endothelin-converting enzyme, neutral endopeptidase 24.11, and angiotensin-converting enzyme. J Cardiovasc Pharmacol 31 [Suppl 1]:S71–S73

Kukkola PJ, Bilci NA, Kozak WX, Savage P, Jeng AY (1996) Optimization of retro-thiorphan for inhibition of endothelin-converting enzyme. Bioorg Med Chem Lett 6:619–624

Kukkola PJ, Savage P, Sakane J, Berry C, Bilci NA, Ghai RD, Jeng AY (1995) Differential structure-activity relationships of phosphoramidon analogues for inhibition of three metalloproteases: endothelin-converting enzyme, neutral endopeptidase, and angiotensin-converting enzyme. J Cardiovasc Pharmacal 26 [Suppl 3]:S65–S68

Kundu GC, Wilson IB (1992) Identification of endothelin-converting enzyme in bovine lung using a new fluorogenic substrate. Life Sci 50:965–970

Kwan AL, Bavbek M, Jeng AY, Maniara W, Toyoda T, Lappe RW, Kassell NF, Lee KS (1997) Prevention and reversal of cerebral vasospasm by an endothelin-converting enzyme inhibitor, CGS 26303, in an experimental model of subarachnoid hemorrhage. J Neurosurg 87:281–286

Lees WE, Kalinka S, Meech J, Capper SJ, Cook ND, Kay J (1990) Generation of human endothelin by cathepsin E. FEBS Lett 273:99–102

Liben S, Stewart DJ, De Marte J, Perreault T (1993) Ontogeny of big endothelin-1 effects in new born piglet pulmonary vasculature. Am J Physiol 265(1 Pt 2):H139–H145

Lloyd J, Schmidt JB, Hunt JT, Barrish JC, Little DK, Tymiak AA (1996) Solid phase synthesis of phosphinic acid endothelin converting enzyme inhibitors. Bioorg Med Chem Lett 6:1323–1326

Love MP, Haynes WG, Gray GA, Webb DJ, McMurray JJ (1996) Vasodilator effects of endothelin-converting enzyme inhibition and endothelin ETA receptor blockade in chronic heart failure patients treated with ACE inhibitors. Circ 94:2131–2137

Marsden PA, Sultan P, Cybulsky M, Gimbrone MA Jr, Brenner BM, Collins T (1992) Nucleotide sequence of endothelin-1 cDNA from rabbit endothelial cells. Biochim Biophys Acta 1129:249–250

Masatsugu K, Itoh H, Chun TH, Ogawa Y, Tamura N, Yamashita J, Doi K, Inoue M, Fukunaga Y, Sawada N, Saito T, Korenaga R, Ando J, Nakao K (1998) Physiologic shear stress suppresses endothelin-converting enzyme-1 expression in vascular endothelial cells. J Cardiovasc Pharmacol 31 [Suppl 1]:S42–S45

Massa MA, Patt WC, Ahn K, Sisneros AM, Herman SB, Doherty A (1998) Synthesis of novel substituted pyridines as inhibitors of endothelin converting enzyme-1 (ECE-1). Bioorg Med Chem Lett 8:2117–2122

Matsumura Y, Fujita K, Takaoka M, Morimoto S (1993a) Big endothelin-3-induced hypertension and its inhibition by phosphoramidon in anesthetized rats. Eur J Pharmacol 230:89–93

Matsumura Y, Hisaki K, Takaoka M, Morimoto S (1990a) Phosphoramidon, a metalloproteinase inhibitor, suppresses the hypertensive effect of big endothelin-1. Eur J Pharmacol 185:103–106

Matsumura Y, Ikegawa R, Suzuki Y, Takaoka M, Uchida T, Kido H, Shinyama H, Hayashi K, Watanabe M, Morimoto S (1991a) Phosphoramidon prevents cerebral vasospasm following subarachnoid hemorrhage in dogs: the relationship to endothelin-1 levels in the cerebrospinal fluid. Life Sci 49:841–848

Matsumura Y, Ikegawa R, Tsukahara Y, Takaoka M, Morimoto S (1990b) Conversion of big endothelin-1 to endothelin-1 by two types of metalloproteinases derived from porcine aortic endothelial cells. FEBS Lett 272:166–170

Matsumura Y, Ikegawa R, Tsukahara Y, Takaoka M, Morimoto S (1991b) Conversion of big endothelin-1 to endothelin-1 by two-types of metalloproteinases of cultured porcine vascular smooth muscle cells. Biochem Biophys Res Commun 178:899–905

Matsumura Y, Ikegawa R, Tsukahara Y, Takaoka M, Morimoto S (1991c) N-ethylmaleimide differentiates endothelin-converting activity by two types of metalloproteinases derived from vascular endothelial cells. Biochem Biophys Res Commun 178:531–538

Matsumura Y, Tsukahara Y, Kininobu K, Takaoka M, Morimoto S (1992) Phosphoramidon-sensitive endothelin-converting enzyme in vascular endothelial cells converts big endothelin-1 and big endothelin-3 to their mature form. FEBS Lett 305:86–90

Matsumura Y, Okumura H, Asakura R, Ashizawa N, Takahashi M, Kobayashi F, Ashikawa N, Arai K (1993b) Pharmacological profiles of aspergillomarasmines as endothelin-converting enzyme inhibitors. Jpn J Pharmacol 63:187–193

Matsuura A, Okumura H, Ashizawa N, Kobayashi F (1992) Big endothelin-1-induced sudden death is inhibited by phosphoramidon in mice. Life Sci 50:1631–1638

McKay KO, Armour CL, Black JL (1996) Endothelin receptors and activity differ in human, dog, and rabbit lung. Am J Physiol 270(1 Pt 1):L37–L43

McKay KO, Battistini B, Jeng AY, Sirois P, Black JL (1997) Endothelin-converting enzyme activity in human airways. ALA/ATS Int Conf (San Francisco, CA, USA) p 8 (Abstract)

McKittrick BA, Czarniecki MF, Charckalamannil S, Chung S, Defrees S, Stamford AW (1996a) Phosphinic acid derivatives. Intern Patent Appl WO9600732A1

McKittrick BA, Stamford AW, Weng X, Ma K, Chackalamannil S, Czarniecki M, Cleven RM, Fawzi AB (1996b) Design and synthesis of phosphinic acids that triply inhibit endothelin converting enzyme, angiotensin converting enzyme and neutral endopeptidase 24.11. Bioorg Med Chem Lett 6:1629–1634

McMahon EG, Fok KF, Moore WM, Smith CE, Siegel NR, Trapani AJ (1989) In vitro and in vivo activity of chymotrypsin-activated big endothelin (porcine 1–40). Biochem Biophys Res Commun 161:406–413

McMahon EG, Palomo MA, Brown MA, Bertenshaw SR, Carter JS (1993) Effect of phosphoramidon (endothelin converting enzyme inhibitor) and BQ-123 (endothelin receptor subtype A antagonist) on blood pressure in hypertensive rats. Am J Hypertens 6:667–673

McMahon EG, Palomo MA, Moore WM (1991a) Phosphoramidon blocks the pressor activity of big endothelin(1–39) and lowers blood pressure in spontaneously hypertensive rats. J Cardiovasc Pharmacol 17 [Suppl 7]:S29–S33

McMahon EG, Palomo MA, Moore WM, McDonald JF, Stern MK (1991b) Phosphoramidon blocks the pressor activity of porcine big endothelin-1-(1–39) in vivo and conversion of big endothelin-1-(1–39) to endothelin-1-(1–21) in vitro. Proc Natl Acad Sci USA 88:703–707

Meil J et al. (1998) Naunyn-Schmiedeberg's Arch Pharmacol 358 (Suppl 1):R514

Minamino T, Kurihara H, Takahashi M, Shimada K, Maemura K, Oda H, Ishikawa T, Uchiyama T, Tanzawa K, Yazaki Y (1997) Endothelin-converting enzyme expression in the rat vascular injury model and human coronary atherosclerosis. Circulation 95:221–230

Morita A, Nomizu M, Okitsu M, Horie K, Yokogoshi H, Roller PP (1994) D-Val22 containing human big endothelin-1 analog, [D-Val22] big ET-1[16–38], inhibits the endothelin converting enzyme. FEBS Lett 353:84–88

Nakahara S, Emoto N, Matsuo M (1999) Isolation of an alternative spliced variant of bovine ECE-2. 6th Int Conf ETs (Montréal, QC, Canada) ET-6 abstract book p 3

Nakano A, Kishi F, Minami K, Wakabayashi H, Nakaya Y, Kido H (1997) Selective conversion of big endothelins to tracheal smooth muscle-constricting 31-amino acid-length endothelins by chymase from human mast cells. J Immunol 159:1987–1992

Naomi S, Iwaoka T, Disashi T, Inoue J, Kanesaka Y, Tokunaga H, Tomita K (1998) Endothelin-1 inhibits endothelin-converting enzyme-1 expression in cultured rat pulmonary endothelial cells. Circulation 97:234–236

Nichols JS, Berman J, Wypij DM, Wiseman JS (1991) Evidence against a role for aspartyl proteases in intracellular processing of big endothelin. J Cardiovasc Pharmacol 17 [Suppl 7]:S10–S12

Nishikori K, Akiyama H, Inagaki H, Ohta H, Kashiwabara T, Iwamutsu A, Nomizu M, Morita A (1991) Receptor binding affinity and biological activity of C-terminal elongated forms of endothelin-1. Neurochem Int 18:535–539

Noguchi K, Fukuroda T, Ikeno Y, Hirose H, Tsukada Y, Nishikibe M, Ikemoto F, Matsuyama K, Yano M (1991) Local formation and degradation of endothelin-1 in guinea-pig airway tissues. Biochem Biophys Res Commun 179:830–835

Norcini G, Santangelo F (1995) N-Heteroaryl substituted derivatives of propanamide useful in the treatment of cardiovascular diseases. EP0636630A1

Nunez DJ, Taylor EA, Oh VM, Schofield JP, Brown MJ (1991) Endothelin-1 mRNA expression in the rat kidney. Biochem J 275 (Pt 3):817–819

Ohkubo S, Ogi K, Hosoya M, Matsumoto H, Suzuki N, Kimura C, Ondo H, Fujino M (1990) Specific expression of human endothelin-2 (ET-2) gene in a renal adenocarcinoma cell line. Molecular cloning of cDNA encoding the precursor of ET-2 and its characterization. FEBS Lett 274:136–140

Ohnaka K, Takayanagi R, Nishikawa M, Haji M, Nawata H (1993) Purification and characterization of a phosphoramidon-sensitive endothelin-converting enzyme in porcine aortic endothelium. J Biol Chem 268:26759–26766

Ohnaka K, Takayanagi R, Yamauchi T, Okazaki H, Ohashi M, Umeda F, Nawata H (1990) Identification and characterization of endothelin-converting activity in cultured bovine endothelial cells. Biochem Biophys Res Commun 168:1128–1136

Ohnaka K, Takayanagi R, Yamauchi T, Umeda F, Nawata H (1991) Cultured bovine endothelial cells convert big endothelin isopeptides to mature endothelin isopeptides. Biochem Int 23:499–506

Okada K, Miyazaki Y, Takada J, Matsuyama K, Yamaki T, Yano M (1990) Conversion of big endothelin-1 by membrane-bound metalloendopeptidase in cultured bovine endothelial cells. Biochem Biophys Res Commun 171:1192–1198

Okada K, Takada J, Arai Y, Matsuyama K, Yano M (1991) Importance of the C-terminal region of big endothelin-1 for specific conversion by phophoramidon-sensitive endothelin-converting enzyme. Biochem Biophys Res Commun 180:1019–1023

Onda H, Ohkubo S, Kosaka T, Yasuhara T, Ogi K, Hosoya M, Matsumoto H, Suzuki N, Kitada C, Ishibashi Y (1991) Expression of endothelin-2 (ET-2) gene in a human renal adenocarcinoma cell line: purification and cDNA cloning of ET-2. J Cardiovasc Pharmacol 17 [Suppl 7]:S39–S43

Onda H, Ohkubo S, Ogi K, Kosaka T, Kimura C, Matsumoto H, Suzuki N, Fujino M (1990) One of the endothelin gene family, endothelin 3 gene, is expressed in the placenta. FEBS Lett 261:327–330

Opgenorth TJ, Wu-Wong JR, Shiosaki K (1992) Endothelin-converting enzymes. FASEB J 6:2653–2659

Patel TR (1996) Therapeutic potential of endothelin receptor antagonists in cerebrovascular disease CNS. Drugs 5:293–310

Patil AD, Freyer AJ, Breen A, Carte B, Johnson RK (1996) Halistanol disulfate B, a novel sulfated sterol from the sponge Pachastrella sp.: inhibitor of endothelin converting enzyme. J Nat Prod 59:606–608

Patil AD, Freyer AJ, Eggleston DS, Haltiwanger RC, Tomcowicz B, Breen A, Johnson RK (1997) Daleformis, a new phytoalexin from the roots of Dalea filiciformis: an inhibitor of endothelin converting enzyme. J Nat Prod 60:306–308

Pelletier S, Battistini B, Jeng AY, Sirois P (1998) Effects of dual endothelin-converting enzyme/neutral endopeptidase inhibitors, CGS 26303 and CGS 26393, on lipopolysaccharide or interleukin-1 beta-stimulated release of endothelin from guinea pig tracheal epithelial cells. J Cardiovasc Pharmacol 31 [Suppl 1]:S10–S12

Pham D, Battistini B, Gravel M-J, Nolet A, Jeng AY, Sirois P, Escher E (1998) Protective effects of ECE/NEP inhibitors in neointimal formation after rat carotid balloon angioplasty. FASEB J 12:A402

Plumpton C, Kalinka S, Martin RC, Horton JK, Davenport AP (1994) Effects of phosphoramidon and pepstatin A on the secretion of endothelin-1 and big endothelin-1 by human umbilical vein endothelial cells: measurement by two-site enzyme-linked immunosorbent assays. Clin Sci (Colch) 87:245–251

Pollock DM, Divish BJ, Milicic I, Novostad EI, Burres NS, Opgenorth TJ (1993) In vivo characterization of a phosphoramidon-sensitive endothelin-converting enzyme in the rat. Eur J Pharmacol 231:459–464

Pons F, Touvay C, Lagente V, Mencia-Huerta JM, Braquet P (1991) Bronchopulmonary and pressor effects of big-endothelin-1 in the guinea-pig. Neurochem Int 18:481–483

Rae GA, Calixto JB, D'Orléans-Juste P (1993a) Big-endothelin-1 contracts rat isolated uterus via a phosphoramidon-sensitive endothelin ETA receptor-mediated mechanism. Eur J Pharmacol 240:113–119

Rae GA, Calixto JB, D'Orléans-Juste P (1993b) Conversion of big endothelin-1 in rat uterus causes contraction mediated by ETA receptors. J Cardiovasc Pharmacol 22:S192–S195

Russell FD, Davenport AP (1999) Evidence for intracellular endothelin-converting enzyme-2 expression in cultured human vascular endothelial cells. Circ Res 84:891–896

Russell FD, Skepper JN, Davenport AP (1998) Evidence using immunoelectron microscopy for regulated and constitutive pathways in the transport and release of endothelin. J Cardiovasc Pharmacol 31:424–430

Russo DCW, Redman CM, Lee S (1999) The Kell blood group protein, that preferentially processes big endothelin-3, is widely expressed. 6th Int Conf ETs (Montréal, QC, Canada) ET-6 abstract book p 3

Saida K, Mitsui Y (1991) Structure of the precursor for vasoactive instestinal contractor (VIC): its comparison with those of endothelin-1 and endothelin-3. J Cardiovasc Pharmacol 17 [Suppl 7]:S55–S58

Saida K, Mitsui Y, Ishida N (1989) A novel peptide, vasoactive intestinal contractor, of a new (endothelin) peptide family. Molecular cloning, expression, and biological activity. J Biol Chem 264:14613–14616

Sakurai T, Yanagisawa M, Inoue A, Ryan US, Kimura S, Mitsui Y, Goto K, Masaki T (1991) cDNA cloning, sequence analysis and tissue distribution of rat preproendothelin-1 mRNA. Biochem Biophys Res Commun 175:44–47

Sakurai T, Yanagisawa M, Masaki T (1992) Molecular characterization of endothelin receptors. Trends Pharmacol Sci 13:103–108

Sansom CE, Hoang VM, Turner AJ (1995) Molecular modeling of the active site of endothelin-converting enzyme. J Cardiovasc Pharmacol 26 [Suppl 3]:S75–S77

Sato R, Iijima Y, Matsushita H, Negishi A, Takamatsu Y, Haruyama H, Kinoshita T, Kodama K, Ishii A (1996) Compound B90063. Jpn Patent JP8208646A

Savage P, De Lombaert S, Shimada K, Tanzawa K, Jeng AY (1998) Differential inhibition of wild-type endothelin-converting enzyme-1 and its mutants. J Cardiovasc Pharmacol 31 [Suppl 1]:S16–S18

Savage P, Shetty SS, Martin LL, Jeng AY (1993) Conversion of proendothelin-1 into endothelin-1 by aspartylproteases. Int J Peptide Protein Res 42:227–232

Sawamura T, Kasuya Y, Matsushita Y, Suzuki N, Shinmi O, Kishi N, Sugita Y, Yanagisawa M, Goto K, Masaki T, Kimura S (1991) Phosphoramidon inhibits the intracellular conversion of big endothelin-1 to endothelin-1 in cultured endothelial cells. Biochem Biophys Res Commun 174:779–784

Sawamura T, Kimura S, Shinmi O, Sugita Y, Kobayashi M, Mitsui Y, Yanagisawa M, Goto K, Masaki T (1990a) Characterization of endothelin-converting enzyme activities in soluble fraction of bovine cultured endothelial cells. Biochem Biophys Res Commun 169:1138–1144

Sawamura T, Kimura S, Shinmi O, Sugita Y, Yanagisawa M, Goto K, Masaki T (1990b) Purification and characterization of putative endothelin converting enzyme in bovine adrenal medulla: evidence for a cathepsin D-like enzyme. Biochem Biophys Res Commun 168:1230–1236

Sawamura T, Shinmi O, Kishi N, Sugita Y, Yanagisawa M, Goto K, Masaki T, Kimura S (1990c) Analysis of big endothelin-1 digestion by cathepsin D. Biochem Biophys Res Commun 172:883–889

Sawamura T, Shinmi O, Kishi N, Sugita Y, Yanagisawa M, Goto K, Masaki T, Kimura S (1993) Characterization of phosphoramidon-sensitive metalloproteinases with endothelin-converting enzyme activity in porcine lung membrane. Biochem Biophys Acta 1161:295–302

Schmidt M, Kröger B, Jacob E, Seulberger H, Subkowski T, Otter R, Meyer T, Schmalzing G, Hillen H (1994) Molecular characterization of human and bovine endothelin-converting enzyme (ECE-1). FEBS Lett 356:238–243

Schultz SC, Grady B, Cole F, Hamilton I, Burhop K, Malcom DS (1993) A role for endothelin and nitric oxide in the pressor response to diaspirin cross-linked hemoglobin. J Lab Clin Med 122:301–308

Schweizer A, Valdenaire O, Koster A, Lang Y, Schmitt G, Lenz B, Bluethmann H, Rohrer J (1999) Neonatal lethality in mice deficient in XCE, a novel member of the endothelin-converting enzyme and neutral endopeptidase family. J Biol Chem 274:20450–20456

Schweizer A, Valdenaire O, Nelbock P, Deuschle U, Dumas M, Edwards JB, Stumpf JG, Loffler BM (1997) Human endothelin-converting enzyme (ECE-1): three isoforms with distinct subcellular localizations. Biochem J 328:871–877

Seymour AA, Abboa-Offei B, Asaad MM, Rogers WL (1991) Evaluation of SQ 28,603, an inhibitor of neutral endopeptidase, in conscious monkeys. Can J Physiol Pharmacol 69:1609–1617

Shetty SS, Savage P, DelGrande D, De Lombaert S, Jeng AY (1998) Characterization of CGS 31447, a potent and nonpeptidic endothelin-converting enzyme inhibitor. J Cardiovasc Pharmacol 31 [Suppl 1]:S68–S70

Shiba R, Sakurai T, Yamada G, Morimoto H, Saito A, Masaki T, Goto K (1992) Cloning and expression of rat preproendothelin-3 cDNA. Biochem Biophys Res Commun 186:588–594

Shields PP, Gonzales TA, Charles D, Gilligan JP, Stern W (1991) Accumulation of pepstatin in cultured endothelial cells and its effects on endothelin processing. Biochem Biophys Res Commun 177:1006–1012

Shimada K, Matsushita Y, Wakabayashi K, Takahashi M, Matsubara A, Iijima Y, Tanzawa K (1995) Cloning and functional expression of human endothelin-converting enzyme cDNA. Biochem Biophys Res Commun 207:807–812

Shimada K, Takahashi M, Tanzawa K (1994) Cloning and functional expression of endothelin-converting enzyme from rat endothelial cells. J Biol Chem 269: 18275–18278

Shimada K, Takahashi M, Turner AJ, Tanzawa K (1996) Rat endothelin-converting enzyme-1 forms a dimer through Cys412 with a similar catalytic mechanism and a distinct substrate binding mechanism compared with neutral endopeptidase-24.11. Biochem J 315:863–867

Shiosaki K, Tasker AS, Sullivan GM, Sorensen BK, von Geldern TW, Wu-Wong JR, Marselle CA, Opgenorth TJ (1993) Potent and selective inhibitors of an aspartyl protease-like endothelin-converting enzyme identified in rat lung. J Med Chem 36:468–478

Suzuki N, Matsumoto H, Miyauchi T, Kitada C, Tsuda M, Goto K, Masaki T, Fujino M (1991) Sandwich-enzyme immunoassays for endothelin family peptides. J Cardiovasc Pharmacol 17 [Suppl 7]:S420–S422

Takada J, Okada K, Ikenaga T, Matsuyama K, Yano M (1991) Phosphoramidon-sensitive endothelin-converting enzyme in the cytosol of cultured bovine endothelial cells. Biochem Biophys Res Commun 176:860–865

Takahashi M, Matsushita Y, Iijima Y, Tanzawa K (1993) Purification and characterization of endothelin-converting enzyme from rat lung. J Biol Chem 268:21394–21398

Takahashi T, Kanda T, Inoue M, Sumino H, Kobayashi I, Iwamoto A, Nagai R (1998) Endothelin Converting enzyme inhibitor protects development of right ventricular overload and medial thickening of pulmonary arteries in rats with monocrotaline-induced pulmonary hypertension. Life Sci 63:PL137–PL143

Takai S, Shiota N, Jin D, Miyazaki M (1998) Chymase processes big-endothelin-2 to endothelin-2-(1–31) that induces contractile responses in the isolated monkey trachea. Eur J Pharmacol 358:229–233

Takaishi S, Tuchiya N, Sato A, Negishi T, Takamatsu Y, Matsushita Y, Watanabe T, Iijima Y, Haruyama H, Kinoshita T, Tanaka M, Kodama K (1998) B-90063, a novel endothelin converting enzyme inhibitor isolated from a new marine bacterium, Blastobacter sp. SANK 71894. J Antibiot (Tokyo) 51:805–815

Takaoka M, Hukumori Y, Shiragami K, Ikegawa R, Matsumura Y, Morimoto S (1990a) Proteolytic processing of porcine big endothelin-1 catalyzed by cathepsin D. Biochem Biophys Res Commun 173:1218–1223

Takaoka M, Miyata Y, Takenobu Y, Ikegawa R, Matsumura Y, Morimoto S (1990b) Mode of cleavage of pig big endothelin-1 by chymotrypsin. Production and degradation of mature endothelin-1. Biochem J 270:541–544

Takaoka M, Takenobu Y, Miyata Y, Ikegawa R, Matsumura Y, Morimoto S (1990c) Pepsin, an aspartic protease, converts porcine big endothelin to 21-residue endothelin. Biochem Biophys Res Commun 166:436–442

Télémaque S, D'Orléans-Juste P (1991) Presence of a phosphoramidon-sensitive endothelin-converting enzyme which converts big-endothelin-1, but not big-endothelin-3, in the rat vas deferens. Naunyn-Schmiedeberg's Arch Pharmacol 344:505–507

Trapani AJ, Balwierczak JL, Lappe RW, Stanton JL, Graybill SC, Hopkins MF, Savage P, Sperbeck DM, Jeng AY (1993) Effects of metalloprotease inhibitors on the conversion of proendothelin-1 to endothelin-1. Biochem Mol Biol Int 31:861–867

Trapani AJ, De Lombaert SD, Kuzmich S, Jeng AY (1995) Inhibition of big ET-1-induced pressor response by an orally active dual inhibitor of endothelin-converting enzyme and neutral endopeptidase 24.11. J Cardiovasc Pharmacol 26: S69–S71

Tsurumi Y, Fujie K, Nishikawa M, Kiyoto S, Okuhara M (1995a) Biological and pharmacological properties of highly selective new endothelin converting enzyme inhibitor WS79089B isolated from Streptosporangium roseum No. 79089. J Antibiot (Tokyo) 48:169–174

Tsurumi Y, Ohhata N, Iwamoto T, Shigematsu N, Sakamoto K, Nishikawa M, Kiyoto S, Okuhara M (1994) WS79089A, B and C, new endothelin-converting enzyme inhibitors isolated from Streptosporangium roseum. NO. 79089. J Antibiot (Tokyo) 47:619–630

Tsurumi Y, Ueda H, Hayashi K, Takase S, Nishikawa M, Kiyoto S, Okuhara M (1995b) WS75624 A and B, new endothelin converting enzyme inhibitors isolated from Saccharothrix sp. No. 75624. I. Taxonomy, fermentation, isolation, physico-chemical preoperties and biological activities. J Antibiot (Tokyo) 48:1066–1072

Turner AJ, Murphy LJ (1996) Molecular pharmacology of endothelin-converting enzymes. Biochem Pharmacol 51:91–102

Turner AJ, Tanzawa K (1997) Mammalian membrane metallopeptidases: NEP, ECE, KELL, and PEX. FASEB J 11:355–364

Uchida Y, Ninomiya H, Saotome M, Nomura A, Ohtsuka M, Yanagisawa M, Goto K, Masaki T, Hasegawa S (1988) Endothelin, a novel vasoconstrictor peptide, as potent bronchoconstrictor. Eur J Pharmacol 154:227–228

Ueki A, Rosen L, Andbjer B, Agnati LF, Hallstrom A, Goiny M, Tanganelli S, Ungerstedt U, Fuxe K (1993) Evidence for a preventive action of the vigilance-promoting drug modafinil against striatal ischemic injury induced by endothelin-1 in the rat. Exp Brain Res 96:89–99

Valdenaire O, Lepailleur-Enouf D, Egidy G, Thouard A, Barret A, Vranchkx R, Tougard C, Michel JB (1999) A fourth isoform of endothelin-converting enzyme (ECE-1) is generated from an additional promoter molecular cloning and characterization. Eur J Biochem 264:341–349

Valdenaire O, Rohrbacher E, Mattei MG (1995) Organization of the gene encoding the human endothelin-converting enzyme (ECE-1). J Biol Chem 270:29794–29798

Vemulapalli S et al. (1997) Cardiovasc Drug Rev 15:260–272

Vemulapalli S, Chiu PJ, Chintala M, Bernardino V (1993a) Attenuation of ischemic acute renal failure by phosphoramidon in rats. Pharmacology 47:188–193

Vemulapalli S, Watkins RW, Brown A, Cook J, Bernardino V, Chiu PJ (1993b) Disparate effects of phosphoramidon on blood pressure in SHR and DOCA-salt hypertensive rats. Life Sci 53:783–793

Wada A, Tsutamoto T, Ohnishi M, Sawaki M, Fukai D, Maeda Y, Kinoshita M (1999) Effects of a specific endothelin-converting enzyme inhibitor on cardiac, renal, and neurohumoral functions in congestive heart failure: comparison of effects with those of endothelin A receptor antagonism. Circulation 99:570–577

Wakimasu M, Mori M, Kawada A (1992) Phosphonic acid derivatives, production and use thereof. Eur Patent Appl EP 0518299A2

Wallace EM, Moliterni JA, Moskal MA, Neubert AD, Marcopulos N, Stamford LB, Trapani AJ, Savage P, Chou M, Jeng AY (1998) Design and synthesis of potent, selective inhibitors of endothelin-converting enzyme. J Med Chem 41:1513–1523

Webb ML, Bird JE, Liu EC, Rose PM, Serafino R, Stein PD, Moreland S (1995) BMS-182874 is a selective, nonpeptide endothelin ETA receptor antagonist. J Pharmacol Exp Ther 272:1124–1134

Webb ML, Meek TD (1997) Inhibitors of endothelin. Med Res Rev 17:17–67

Weber C, Schmitt R, Birnboeck H, Hopfgartner G, van Marle SP, Peeters PA, Jonkman JH, Jones CR (1996) Pharmacokinetics and pharmacodynamics of the endothelin-receptor antagonist bosentan in healthy human subjects. Clin Pharmacol Ther 60:124–137

Woods M, Bishop-Bailey D, Pepper JR, Evans TW, Mitchell JA, Warner TD (1998) Cytokine and lipopolysaccharide stimulation of endothelin-1 release from human internal mammary artery and saphenous vein smooth-muscle cells. J Cardiovasc Pharmacol 31 [Suppl 1]:S348–S350

Woods M, Mitchell JA, Wood EG, Barker S, Walcot NR, Rees GM, Warner TD (1999) Endothelin-1 is induced by cytokines in human vascular smooth muscle cells: evidence for intracellular endothelin-converting enzyme. Mol Pharmacol 55:902–909

Wu-Wong JR, Budzik GP, Devine EM, Opgenorth TJ (1990) Characterization of endothelin-converting enzyme in rat lung. Biochem Biophys Res Commun 171:1291–1296

Wu-Wong JR, Chiou WJ, Opgenorth TJ (1993) Prolonged treatment by phosphoramidon modulates the number of endothelin receptors in cultured Swiss 3T3 fibroblasts. J Cardiovasc Pharmacol 22:S77–S80

Wu-Wong JR, Devine EM, Budzik GP, Opgenorth TJ (1991) Characterization and partial purification of ECE in rat lung. J Cardiovasc Pharmacol 17:520–525

Wypij DM, Nichols JS, Novak PJ, Stacy DL, Berman J, Wiseman JS (1992) Role of mast cell chymase in the extracellular processing of big-endothelin-1 to endothelin-1 in the perfused rat lung. Biochem Pharmacol 43:845–853

Xu D, Emoto N, Giaid A, Slaughter C, Kaw S, deWit D, Yanagisawa M (1994) ECE-1: A membrane-bound metalloprotease that catalyses the proteolytic activation of big endothelin-1. Cell 78:473–485

Yanagisawa H, Richardson JA, Yanagisawa M (1999) The role of ECE-2 in cardiac development revealed by double targeted disruption of ECE-1 and ECE-2 genes in mice. 6th Int Conf ETs (Montréal, QC, Canada) ET-6 abstract book p 38

Yanagisawa H, Yanagisawa M, Kapur RP, Richardson JA, Williams SC, Clouthier DE, deWit D, Emoto N, Hammer RE (1998) Dual genetic pathways of endothelin-mediated intercellular signaling revealed by targeted disruption of endothelin-converting enzyme-1 gene. Development 125:825–836

Yanagisawa M, Kurihara H, Kimura S, Tomobe Y, Kobayashi M, Mitsui Y Yazaki Y, Goto K, Masaki T (1988) A novel potent vasoconstrictor peptide produced by vascular endothelial cells. Nature 332:411–415

Yang Q, Kawaguchi T, Battistini B, Sirois P (1997) Neutral endopeptidase degrades endothelins in guinea-pig tracheal epithelial cells. Inflamm Res 46:S171–S172

Yano M, Okada K, Takada J, Hioki Y, Matsuyama AK, Fukuroda T, Noguchi K, Nishi-
    kibe M, Ikemote F (1991) Endothelin-converting enzyme and its in vitro and in
    vivo inhibition. J Cardiovasc Pharmacol 17:S26–S28
Yorimitsu K, Moroi K, Inagaki N, Saito T, Masuda Y, Masaki T, Seino S, Kimura S (1995)
    Cloning and sequencing of a human endothelin-converting enzyme in renal
    adenocarcinoma (ACHN) cells producing endothelin-2. Biochem Biophys Res
    Comm 208:721–727
Yoshimura S, Tsurumi Y, Takase S, Okuhara M (1995) WS75624 A and B, new endothe-
    lin converting enzyme inhibitors isolated from Saccharothrix sp. No. 75624. II.
    Structure elucidation of WS75624 A and B. J Antibiot (Tokyo) 48:1073–1075

Endothelin Converting Enzymes and Endothelin Receptor Localisation in Human Tissues

A.P. DAVENPORT and F.D. RUSSELL

A. Introduction

ET-1 is the principal isoform of endothelin in the human cardiovascular system. ET-1 is continuously released from vascular endothelial cells, to produce intense vasoconstriction of the underlying smooth muscle and so maintain endogenous vascular tone. ET-1 released from other epithelial cells situated at the interface between biological compartments may have a range of actions in humans as well as functioning as a neuropeptide in the brain. Overexpression of ET-1 has been implicated in a number of pathophysiological conditions such as coronary artery disease (see Chaps. 12–19). Two strategies have emerged to block undesirable actions of ET-1 in diseased tissue: receptor antagonists and inhibitors of the enzymes responsible for ET-1 synthesis. Given the wide distribution and emerging multifunctional role of the peptide, determining the anatomical localisation of endothelin converting enzymes (ECE) is important in understanding the mechanism of production of ET and identification of the receptor sub-type interacting with the released peptide.

B. Endothelin Converting Enzymes

I. ECE-1

In mammals, including humans, analysis of cDNAs encoding ECE-1 revealed the existence of more than one isoform, ECE-1α and ECE-1β (SHIMADA et al. 1995a, b) and ECE-1a and ECE-1b (VALDENAIRE et al. 1995), leading to the development of two separate nomenclatures (TURNER et al. 1998) (see Chaps. 3, 7). Subsequently, SCHWEIZER et al. (1997) proposed that a third isoform may exist in humans. Combining the two nomenclatures, ECE-1α/ECE-1c consists of 754 amino acids in humans (and rats and cattle). Evidence is emerging that this isoform is the most abundant in a number of human tissues measured by immunoassays (MOCKRIDGE et al. 1998).

ECE-1b is proposed to encode a 770 amino acid protein that is identical to ECE-1α/ECE-1c, except for an additional 17 amino-acids deduced to be present at the N-terminus, replacing the first methionine of ECE-1α. ECE-1b cDNA has only been identified in humans (Schweizer et al. 1997).

ECE-1β/ECE-1a consists of 758 amino acids in humans; selective antisera have detected this isoform as well as ECE-1α in human umbilical endothelial cells (Russell et al. 1998b).

1. Localisation of mRNA Encoding Isoforms of ECE-1

Messenger RNA encoding ECE-1 is widely distributed in human tissue (Schmidt et al. 1994; Rossi et al. 1995; Valdenaire et al. 1995; Schweizer et al. 1997) although the use of homogenates in these studies precludes the identification of the precise cell type expressing ECE mRNA. ECE-1α (ECE-1c) mRNA predominated in human tissues as measured by ribonuclease protection assay (Schweizer et al. 1997). These studies revealed unexpected anomalies. Levels of mRNA encoding ECE-1 isoforms were apparently low in human brain compared with peripheral tissues such as the lungs.

2. Localisation of ECE-1 Using Immunocytochemistry

Conservation of the C-terminus has permitted the development of antisera which could cross-react with all ECE enzymes discovered to date in human tissue. Davenport et al. (1998b, c) used site-directed antisera raised against the C-terminus of mammalian ECE-1 (bECE-1$_{744-758}$) (Xu et al. 1994) that is conserved in both human ECE-1α (ECE-1c), ECE-1b and ECE-1β (ECE-1a) with the exception of a single residue, where the Pro for His substitution represents a minor change. The antisera also cross-reacted with the C-terminus of the deduced amino-acid sequence of bovine ECE-2 (bovine ECE-2$_{780-787}$) (Emoto and Yanagisawa 1995) that has four identical amino-acids to ECE-1 at the extreme C-terminus. Staining revealed using these antisera will be referred to as ECE-like immunoreactivity. Where antisera to the C-terminus have been reported not to cross-react with ECE-2 (Pupilli et al. 1997), staining is referred to as ECE-1-like immunoreactivity.

3. Quantification of ECE-1 Isoforms

A second strategy has been to develop antisera that can distinguish between the isoforms of human ECE-1. Site directed antisera to the deduced amino acids in the N-terminus of human ECE-1α$_{(2-16)}$ (STRYKRATLDEEDLVD) and ECE-1β (PLQGLGLQRNPFLQG) have been extensively characterised (Mockridge et al. 1998; Russell et al. 1998b, c). These antisera were used in a sensitive competition ELISA to measure the amount of immunoreactive enzyme in microsomal fractions prepared from homogenates of human tissue. In the heart, levels of ECE-1α were 0.88 ± 0.27 pmol/g and 0.43 ± 0.08 pmol/g wet weight in the atria and ventricles respectively and 0.38 ± 0.05 pmol/g wet weight in the lungs. These levels are comparatively low, reflecting the localisation of the enzyme to endothelial or epithelial cells, which represent only a small proportion of the cell type within the heart and lungs. The concentration of ECE-1β was below the level for detection in the competition ELISA, suggesting ECE-1α was the predominant isoform. Using more sensitive immunocytochemical techniques, both enzymes were localised to the cytoplasm of endothelial cells from the human umbilical vein (HUVECs) as well as human coronary arteries, suggesting that ECE-1β may be widely distributed in human vascular endothelium (RUSSELL et al. 1998c).

A third antisera to the deduced extended N-terminus of human ECE-1b$_{1-15}$ (MRGVWPPPVSALLSA) has been used to localise intense ECE-1b immunoreactivity within renal and pulmonary epithelial cells with lower levels of staining displayed by perivascular astrocytes and neuronal processes in the cerebral cortex from the brain. In diseased vessels, ECE-1b antisera stained macrophages infiltrating atherosclerotic plaques within coronary arteries. These results suggest ECE-1b may also be expressed in normal and diseased human tissue (DAVENPORT and KUC 2000). The physiological significance of multiple ECE isoforms in human tissue is unclear. The three ECE-1 isoforms have the same kinetic rate constants for cleaving Big ET-1 when expressed in cells lines and would be expected to synthesise comparable amounts of the mature peptide. It is possible that the isoforms may occupy different compartments within the same cell. A second possibility is that expression may vary according to cell type: this may account for the particularly intense staining with antisera to ECE-1b in epithelial cells whereas endothelial cell staining was difficult to detect (DAVENPORT and KUC 2000).

II. ECE-2

EMOTO and YANAGISAWA (1995) discovered a second bovine enzyme, ECE-2, that has 59% overall homology with bovine ECE-1. Following expression in CHO cells, the cloned enzyme was also found to be a membrane bound metalloprotease. The two enzymes can be differentiated biochemically. ECE-2 has an optimum pH for activity of 5.5 compared with 6.8 for ECE-1. Cloned ECE-2 is 250-fold more sensitive to phosphoramidon than ECE-1, but insensitive to PD069185, a novel ECE-1 selective inhibitor (AHN et al. 1998).

C. ECE and ET Receptor Sub-Types in the Peripheral Vasculature

I. Ultrastructural Localisation of ECE-1 in Endothelial Cells

ET-1, together with its precursor Big ET-1, is the predominant isoform synthesised and released from the human endothelium. The mature peptide has been localised in endothelial cells of all human vessels examined including large conduit and small resistance vessels (HEMSEN et al. 1991; HOWARD et al. 1992; ASHBY et al. 1995; PLUMPTON et al. 1996b). The processing of Big ET-1 to ET-1 has been attributed to activity of one or more converting enzymes that are located mainly on the plasma membrane or within intracellular compartments. Studies using endothelial cells isolated from animal tissues (HARRISON et al. 1995; CORDER et al. 1995; TAKAHASHI et al. 1995; BARNES et al. 1996, 1998) or transformed endothelial cell lines (WAXMAN et al. 1994) have concluded that ECE or ECE activity is localised to the cell surface and the enzyme acts mainly in a post-secretory processing role.

However, GUI et al. (1993), XU et al. (1994), DAVENPORT et al. (1998b) and RUSSELL et al. (1998b–d) provided evidence that ECE is either primarily expressed or has predominant activity within intracellular compartments. The co-localisation of the mature peptide and Big ET-1 within endothelial cells implies at least some ECE activity is located intracellularly. DAVENPORT et al. (1998b) compared the ability of permeabilised and non-permeabilised HUVECs to convert Big ET-1 to the mature peptide and found about 85% of ECE activity was located in intracellular compartments. In both cases, activity was inhibited by phosphoramidon but not thiorphan.

Scanning electron microscopy revealed, as expected, low levels of ECE-like immunoreactivity on the surface of the plasma membrane of cultured HUVECs (Fig. 1B) (RUSSELL et al. 1998b) as well as en face preparations of

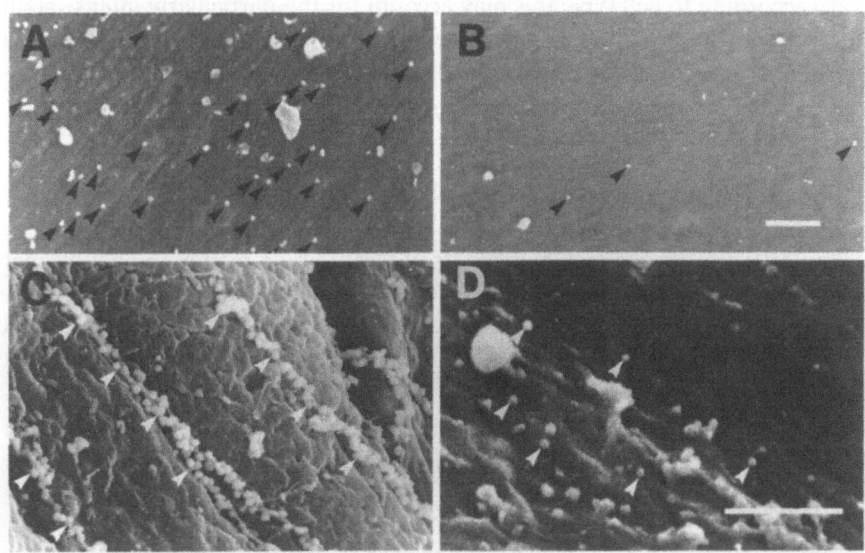

Fig. 1A–D. Scanning electron microscope images of the surface of HUVECs growing in culture and an en face preparation of endothelial cells lining human coronary artery. **A** Only moderate staining for ECE was detected on the plasma membrane of cultured HUVECS visualised using secondary antisera conjugated to colloidal gold particles (*black arrows*). **B** Pre-immune control. **C** Filamentous vWF immunoreactivity (*white arrows*) was detected on the border between coronary artery endothelial cells. **D** ECE-like immunoreactivity (*white arrows*) was moderate in comparison. *Scale bar* = $2\,\mu m$. (Modified from RUSSELL et al. 1998b)

human coronary artery (Fig. 1D), suggesting ECE-1 may function in part as an ectoenzyme to cleave circulating Big ET-1.

These authors analysed the subcellular expression of Big ET-1, ECE-1α and ECE-1β compared with von Willebrand Factor, a marker of Weibel-Palade bodies, in HUVECs that had been permeabilised to allow access of antisera to sub-cellular structures. The resulting cells were optically sectioned using confocal microscopy (Fig. 2). In agreement with the results of the scanning electron microscopy, only moderate levels of ECE-1α were detected over the plasma membrane. ECE-1α, ECE-1β and Big ET-1 were found to co-localise with von Willebrand factor in the Weibel-Palade bodies. Co-localisation of ECE isoforms to Weibel-Palade bodies was confirmed by immunoelectron microscopy in ultra-thin sections of human coronary artery. These numerous rod-shaped structures, about $0.2\,\mu m$ in diameter and $2–3\,\mu m$ in length, are located beneath the plasma membrane and are specific to endothelial cells. Stimulation by the calcium ionophore released ET-1 from cultured HUVECs. These results suggest ET-1 is synthesised by the regulated pathway and released in response to external stimuli (RUSSELL et al. 1998b).

In addition, intense staining with antisera to ECE was also discovered in smaller punctate vesicles (Fig. 2), establishing that ET is also synthesised via the constitutive secretory pathway. These results are in agreement with the ultrastructural localisation of the mature peptide in human coronary artery. Quantitative immunoelectron microscopy revealed the presence of ET-like immunoreactivity in the secretory vesicles as well as the Weibel-Palade bodies (RUSSELL et al. 1998d). The combined results demonstrate that ET is released from human endothelial cells via two distinct pathways. It is proposed that ET is continuously transported in and released from secretory vesicles by the constitutive secretory pathway, contributing to the maintenance of normal vascular tone. Continuous release from this pathway accounts for the rise in the concentration of plasma ET following systemic administration of ET receptor antagonists in volunteers (PLUMPTON 1996a). In addition, ET stored in Weibel-Palade bodies may be released following a physiological or patho-physiological stimulus by the regulated pathway, to cause additional local vasoconstriction (RUSSELL and DAVENPORT 1999b).

II. Evidence for Smooth Muscle Cell ECE and Upregulation in Cardiovascular Disease

Big ET-1 infused into the forearm of human volunteers caused pronounced vasoconstriction that was accompanied by a corresponding increase in ET and the biologically inactive C-terminal fragment (PLUMPTON et al. 1995). This constrictor action was blocked by phosphoramidon, an inhibitor that does not normally cross the plasma membrane, implying local conversion by an ectoenzyme at the site of action. Little or no conversion of Big ET-1 to ET-1 has been detected in human blood in vitro (WATANABE et al. 1991) and Big ET-1 constricts human vessels denuded of endothelium (MOMBOULI et al. 1993;

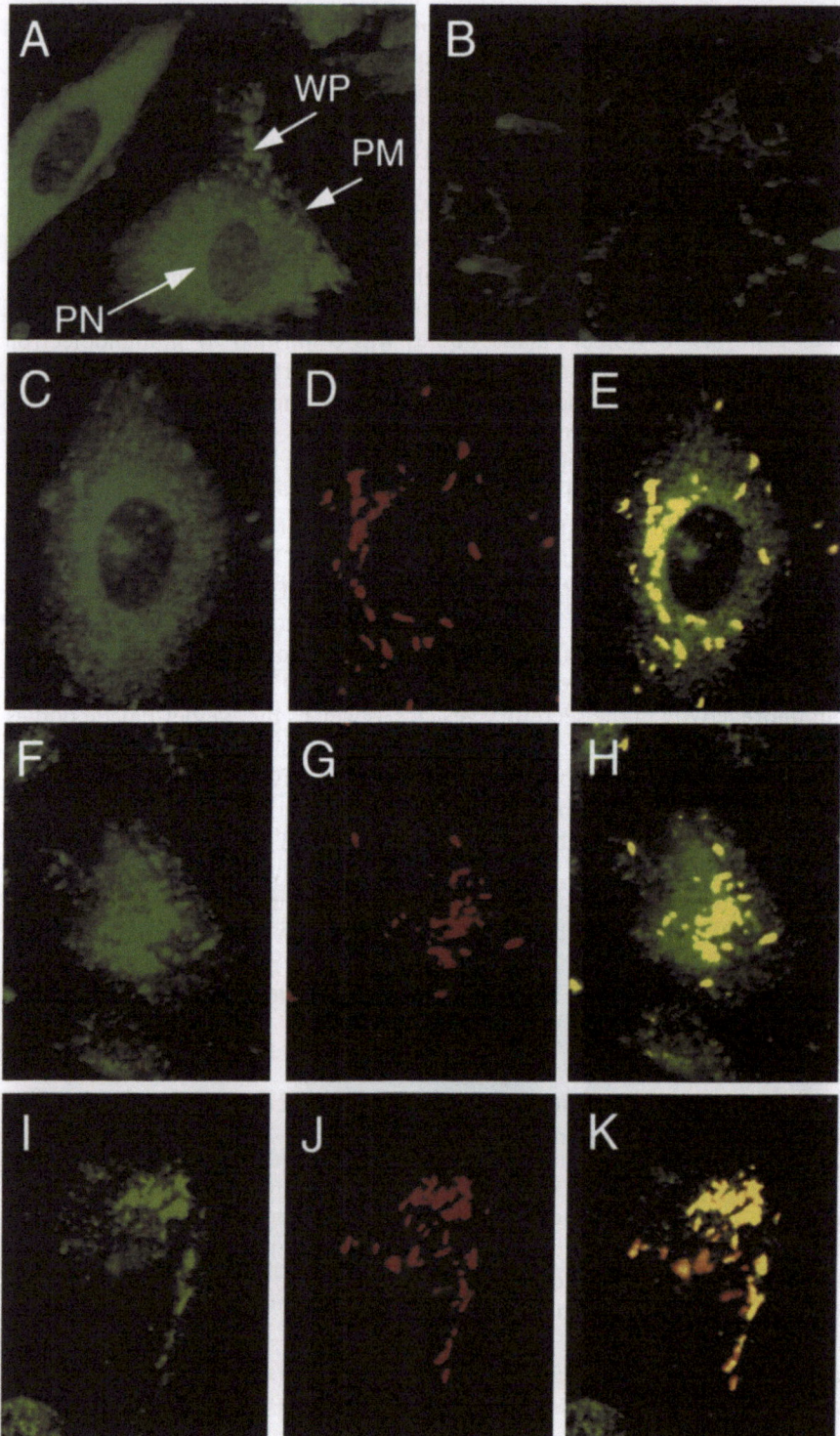

MAGUIRE et al. 1997b; MAGUIRE and DAVENPORT 1998b). Like endothelial cell ECE, this activity could be distinguished from neutral endopeptidase (that can also cleave Big ET-1) by being inhibited by phosphoramidon but not thiorphan (MAGUIRE et al. 1997b; RIZZI et al 1998) and being blocked by a selective inhibitor of ECE-1 (over ECE-2), PD159790 (MAGUIRE et al. 1999).

However, comparison of confluent cultures containing similar densities of human umbilical vein smooth muscle (HUVSMCs) and endothelial cells revealed important differences in the processing of exogenous Big ET-1 and Big ET-3. ECE activity was mainly located on the plasma membrane of smooth muscle cells which converted about 20% of Big ET-1 compared with permeabilised HUVECs (representing the combined extra and intercellular activity). Conversion of Big ET-3 was not detected in either intact or permeabilised endothelial cells, whereas the precursor was converted by a phosphoramidon-sensitive/thiorphan-insensitive enzyme in muscle cells (DAVENPORT et al. 1998b). This was not an artefact of cultured cells, since isolated vessel denuded of endothelium also efficiently converted both Big ET-1 and Big ET-3, measured by analysing the contents of the bathing medium using a radioimmunoassay (MAGUIRE et al. 1997b).

Does smooth muscle ECE have a physiological role in processing Big ET precursors at target sites? Big ET-1 does escape conversion by either the intracellular or cell-surface enzyme of the endothelium, since both ET-1 and Big ET-1 are secreted into the medium from HUVECs in a ratio of 4:1 (PLUMPTON et al. 1994) and circulates in human plasma (MATSUMOTO et al. 1994). Human endothelial cells do not synthesise ET-3 but the mature peptide and Big ET-3 can also be detected in the plasma. A further possibility is that smooth muscle ECE may be a site of conversion for Big ET-3 released for example, from secretory cells of the medulla of the human adrenal gland. This suggests that, in humans, ECE activity may be localised into two pools, one intracellular, responsible for synthesis of ET-1 and extracellular ECE, responsible for the further post-secretory processing of Big ETs. ECE staining was found in smooth muscle cells from umbilical vessels although the intensity was less than that of the endothelium but was more difficult to detect in smooth muscle cells of vessels from adults (DAVENPORT et al. 1998b; FURUKAWA et al. 1996). Combining the functional studies, the results suggest ECE activity in

◄────────────────────────────

Fig. 2A–K. Confocal microscope images showing dual-staining for von Willebrand Factor (vWF) and ECE in permeabilised human umbilical vein endothelial cells. **A** Intense ECE-1a-like immunoreactivity was detected in small vesicles within the perinuclear region (PN) and the larger Weibel-Palade bodies (WP), with only modest staining over the plasma membrane (PM), compared with the pre-immune control shown in **B**. Cells were labelled with either ECE-1α **C**, ECE-1β **F** or big ET-1 **I** as well as vWF, as a marker of Weibel-Palade bodies **D, G, J**. Digitally overlaying the two images revealed a proportion of ECE-1α **E**, ECE-1β **H** and big ET-1 **K** could be localised to Weibel-Palade bodies, with remaining ECE and Big ET-1 localised to small secretory vesicles. (From RUSSELL et al. 1998b)

smooth muscle of normal vessels is low, with conversion only seen in the presence of high, non-physiological but possibly pathophysiological concentrations of Big ET-1. Does smooth muscle ECE have a pathophysiological role? In endothelium-denuded human coronary arteries the response to Big ET-1 was significantly enhanced in vessels containing atherosclerotic lesions with a corresponding increase in mature ET formed in the bathing medium, compared with non-diseased arteries. There were no differences in responses of arteries from either group to ET-1, demonstrating upregulation of ECE activity rather than to an augmented response of the arteries to ET-1 (MAGUIRE and DAVENPORT 1998b). The identity of the ECE upregulated is unclear. MINAMINO et al. (1997) reported ECE-1 immunoreactivity was present in both smooth muscle cells and macrophages in two human coronary atherectomy samples. In occluding saphenous veins used for by-pass grafts in eight individuals retrieved 2–8 years later, intense ECE staining localised to the endothelium and infiltrating macrophages, but there was little evidence of increased ECE-like immunoreactivity within the media or proliferated smooth muscle of the occlusive lesion (Fig. 3). Both the mature peptide, Big ET-1 (BACON et al. 1996) and ET-1 mRNA (WINKLES et al. 1993) are increased in atherosclerotic vessels, but the peptide is confined to the infiltrating macrophages of the lesion and not the smooth muscle. Thus macrophages may be an additional source of the peptide and Big ET-1, and may in addition to smooth muscle ECE, locally increase conversion of Big ET-1 in the blood vessel wall.

III. ECE-2

Using the biochemical criteria outlined above, RUSSELL and DAVENPORT (1999a) detected ECE activity with an acid pH optimum in subcellular fractions of HUVECs. At a pH of 5.4 this activity was unaffected by PD159790 but was inhibited markedly by low concentrations of phosphoramidon (IC_{50} = 1.5 nmol/l), indicating ECE-2 as well as ECE-1 were co-expressed in these cells. These results were confirmed using antisera raised against the deduced amino acid sequence of bovine ECE-2 ($bECE_{697-716}$), that is distinct from the ECE-1 isoforms. Confocal microscopy revealed a punctate pattern of ECE-2-like immunoreactive staining in the cell cytosol, suggesting a possible localisation to secretory vesicles indicating a role in processing Big ET-1 whilst in transit to the cell surface via the constitutive secretory pathway, although localisation to other membranous structures cannot be excluded. In contrast to ECE-1, no staining was detected in storage granules. ECE-2-like immunoreactivity was also detected in coronary artery endothelial cells (RUSSELL et al. 1998c) suggesting the enzyme may be widely expressed in the human vasculature as well as the CNS (DAVENPORT and KUC, 2000).

Does ECE-2 have a physiological or pathophysiological role in human endothelial cells? Surprisingly, mature ET could still be measured in mouse embryos in which the ECE-1 gene had been deleted, leading YANAGISAWA et al. (1998) to suggest expression of a second ECE protease, such as ECE-2.

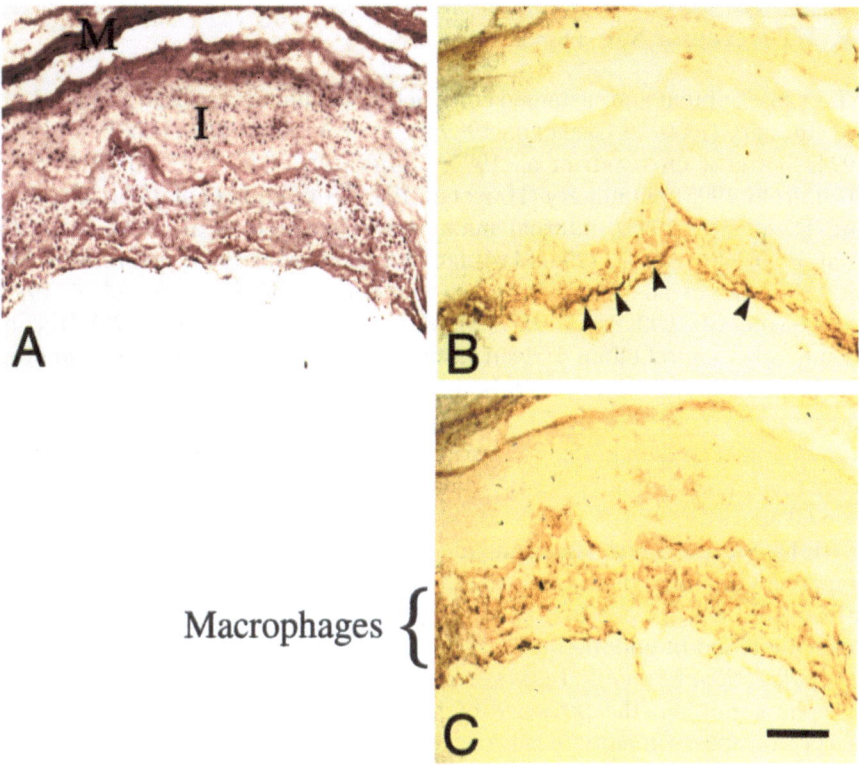

Fig. 3. A Brightfield photomicrograph showing an example of a transverse section of an occluding saphenous vein stained with haematoxylin and eosin. *M* media, *I* intima. **B** Adjacent section showing immunoreactive ECE localised to endothelial cells lining the lumen (*arrows*) and infiltrating macrophages. Little or no staining was detected within the smooth muscle of the media or thickened intima. **C** Adjacent section stained with antisera to CD68, to confirm co-localisation of immunoreactive ECE with macrophages. *Scale bar* = 250 μm. (Modified from DAVENPORT et al. 1998b)

Since ECE-2 requires an acidic pH for optimum activity, EMOTO and YANAGISAWA (1995) predicted that the enzyme would be restricted to the acid-ified environment of the trans-Golgi network or vesicles of the secretory path-way. In agreement with this hypothesis, native ECE-2-like immunoreactivity was localised to secretory vesicles in HUVECs (RUSSELL and DAVENPORT 1999a), suggesting ECE-2 could contribute to synthesis of ET under phy-siological conditions. Alternatively, synthesis of ET-1 by ECE-2 may become significant under pathophysiological conditions in which cellular pH is reduced, such as ischaemic heart disease where intracellular pH values of 5.8 have been detected in hearts subjected to global ischaemia (DOCHERTY et al. 1997), and a correlation between myocardial ischaemia and increased plasma levels of ET is now well established (TONNESSEN et al. 1993; COHN, 1996).

IV. ET Receptor Sub-Types

ET-1 causes potent and sustained constriction of human vessels in vitro including coronary (FRANCO-CERECEDA 1989; DAVENPORT et al. 1989; BAX et al. 1993, 1994; SAETRUM OPGAARD et al. 1994; GODFRAIND et al. 1993; MAGUIRE and DAVENPORT 1995) pulmonary (HAY et al. 1993; FUKURODA et al. 1994; MAGUIRE and DAVENPORT 1995), internal mammary (COSTELLO et al. 1990; LUSCHER et al. 1990; WHITE et al. 1994; MAGUIRE and DAVENPORT 1995) and renal arteries (MAGUIRE et al. 1994). Veins are also constricted (LUSCHER et al. 1990; COSTELLO et al. 1990; SEO et al. 1994; MAGUIRE and DAVENPORT 1995). In contrast to other endothelium derived vasoactive factors, a human vessel that does not respond to ET-1 has yet to be reported. The decrease in vascular resistance caused by infusion of ET antagonists in vivo has established that ET has a physiological role in humans, contributing to the maintenance of normal vascular tone (HAYNES and WEBB1994; HAYNES et al. 1996; PLUMPTON et al. 1995) (see Chap. 19).

ET_A receptors are the principle sub-type in the medial layer that is comprised mainly of smooth muscle in human arteries and veins. ET_B mRNA and a small population of receptors could be detected in the medial layer (denuded of endothelium) from certain vessels (DAVENPORT et al. 1993, 1995a, b, 1998a; BACON and DAVENPORT 1996). Electron microscope autoradiography confirmed the presence of high densities ET_A with a low density of ET_B receptors in smooth muscle cells but little or no receptors of either sub-type associated with the surrounding collagen (Fig. 4) (RUSSELL et al. 1997).

The most potent ET_B agonist, sarafotoxin S6c (see Chap. 2) does cause vasoconstriction in some, but not all, human vessels such as saphenous vein and internal mammary artery, but these responses are variable occurring in less than 50% of individuals (DAVENPORT and MAGUIRE 1994). Although the constrictor actions of sarafotoxin S6c are potent, the magnitude of the response is much less than that of ET-1. Therefore whilst some individuals may have functional constrictor ET_B receptors, the poor response to ET-3 in these vessels suggests that these receptors may have limited physiological importance in the cardiovascular system. In agreement, ET_A antagonists cause parallel and rightward shifts of the ET-1 concentration response curves in these vessels with no portion of the curve resistant to ET_A blockade (DAVENPORT and MAGUIRE 1994; MAGUIRE and DAVENPORT 1995: MAGUIRE et al. 1997a). ET_A receptors also predominate in the media of diseased arteries but are downregulated (together with ET_B receptors) in the intimal smooth muscle. The density of ET_B receptors does (BACON et al. 1996) increase in atherosclerotic coronary arteries (DAGASSAN et al. 1996) but these are localised to infiltrating macrophages and the increase in endothelial cells associated with neovascularization (BACON et al. 1996). In agreement, ET_A antagonists also fully reverse ET-1 induced vasoconstriction in diseased vessels (MAGUIRE and DAVENPORT 1998a, 1999).

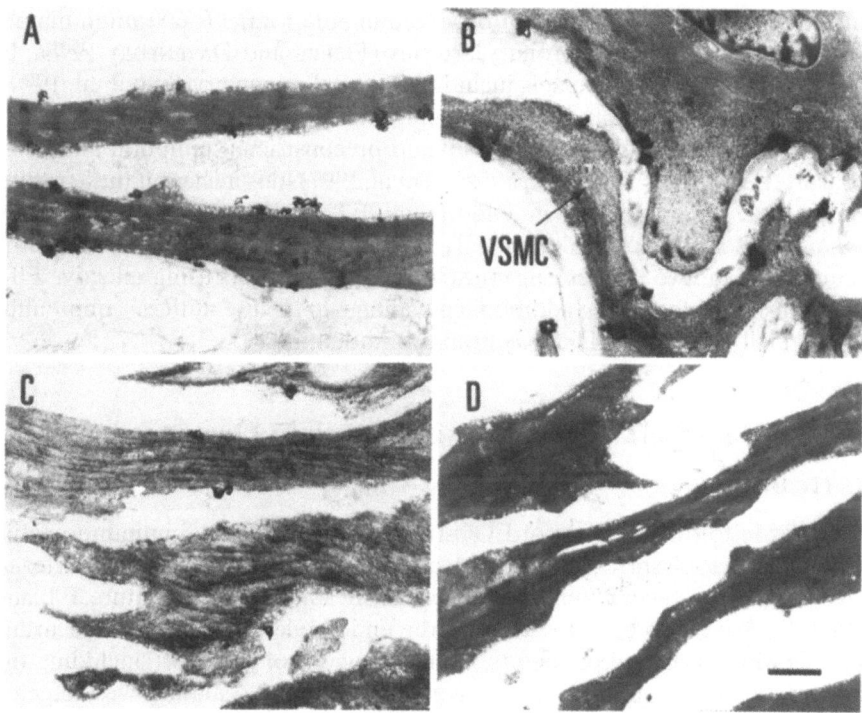

Fig. 4A–D. Electron micrographs showing the distribution of ET receptors in ultrathin sections of human coronary artery. **A** [^{125}I]-ET-1 binding to receptors was visualised as developed silver grain (*black coiled ribbons*) overlying the plasma membrane of individual vascular smooth muscle cells but not the surrounding collagen. **B** ET$_A$ receptors localised to the vascular smooth muscle cells (*VSMC*). **C** Only low densities of ET$_B$ receptors were detected. **D** Non-specific binding. *Scale bar* = 1 μm. (From RUSSELL et al. 1997)

In animals, endothelial ET$_B$ receptors are thought to be present on endothelial cells and cause vasodilatation through release of nitric oxide, prostacyclin, or an endothelium-derived hyperpolarizing factor, opposing the constrictor response (see Chap. 11). Expression of endothelial ET$_B$ receptors in human vessels may be more heterogeneous. ET$_B$ (but not ET$_A$) mRNA has been detected in HUVECs (O'REILLY et al. 1993), where the receptor (FUJITANI et al. 1992) mediates a range of functions including induction of IL6 gene expression (STANKOVA et al. 1996), proliferation and migration of the cells (ZICHE et al. 1995). The initial reduction in forearm blood flow following infusion of low doses of ET-1 into the brachial artery in vivo is thought to occur by ET$_B$ mediated vasodilatation (KIOWSKI et al. 1991) although other investigators only observed vasodilatation in response to high concentrations of ET-3 (HAYNES et al. 1995). In contrast, ET$_B$ receptors were not detected by electron microscope autoradiography in coronary artery endothelial cells (Fig. 4) (RUSSELL et al. 1997). Using sarafotoxin S6c as the agonist, ET$_B$

mediated vasodilatation was not observed in either large (~600 μm in diameter) or small (~300 μm) coronary arteries (PIERRE and DAVENPORT 1998a, b) nor in other peripheral vessels including internal mammary (SEO et al. 1994), radial (LIU et al. 1996) and small omental arteries (RIEZEBOS et al. 1994). ET$_B$ mediated relaxation did occur in isolated preconstricted temporal and cerebral arteries (LUCAS et al. 1996; NILSSON et al. 1997). It is unclear if these results reflect true heterogeneity in endothelial ET$_B$ receptor expression within human vessels. It is possible that ET$_B$ mediated vasodilator responses can be 'unmasked' in vivo by blocking constrictor ET$_A$ receptors using selective ET$_A$ antagonists, conferring an additional advantage to the use of these compounds as originally proposed (DAVENPORT and MAGUIRE 1994).

D. ECE and ET Receptor Sub-Types in Organ Systems

I. Heart

In addition to the detection of ECE-1α, ECE-1β and ECE-2 immunoreactivity in the cytoplasm of endothelial cells from epicardial coronary arteries (RUSSELL et al. 1998c), immunoreactive ECE, together with mature ET and Big ET-1, have also been localised to the endothelial cells of intramyocardial vessels with a range of diameters, from all regions of the heart, including the atria, right and left ventricle (DAVENPORT et al. 1998c). No staining was detected within the smooth muscle or myocytes of the surrounding heart muscle. Synthesis of ET-1 by one or more of these enzymes within endothelial cells throughout the coronary tree may contribute to the normal regulation of cardiac perfusion. Overexpression of ECE within atherosclerotic lesions may increase levels of ET within coronary arteries, contributing to vasospasm and the progression of the disease. ECE-1 mRNA was increased following myocardial infarction, although cells expressing increased levels of the enzyme were not identified (BOHNEMEIER et al. 1998).

Staining for ECE was also detected within endocardial endothelial cells lining the ventricle, a second major source of cardiac ET-1 (PLUMPTON 1996b). Both receptor sub-types are present in the myocardium of atria, ventricles (BAX et al. 1993; MOLENAAR et al. 1992, 1993; PETER and DAVENPORT 1995, 1996a; PÖNICKE et al. 1998) and septum of the heart although the ET$_A$ sub-type was more abundant, comprising 90% of the ET receptors in isolated myocytes. The density of this sub-type is increased by 50% in the ventricle of patients with ischaemic heart disease, whereas ET$_B$ receptors are unaltered compared to non-failing hearts (PETER and DAVENPORT 1996b). In the atrioventricular conducting system, this pattern was reversed with a higher proportion of ET$_B$ receptors (MOLENAAR et al. 1993). ET-1 is a potent positive inotropic agent, acting directly on heart muscle (MORAVEC et al. 1989; DAVENPORT et al. 1989). Synthesis of ET-1 by ECE within the endocardial endothelial cells may not only modulate the inotropic state of the heart but also exert effects on the conducting system in close proximity to endocardial cells. FUKUCHI and GIAID (1998) found immunoreactive ET and ECE-1 within macrophages that were abundant in infarcted regions of failing hearts but rare in healthy cardiac tissue. They concluded that vascular endothelial cells and macrophages were the principal sites of ET-1 synthesis in these failing hearts.

II. Lungs

In normal lung tissue, ET, Big ET-1 and ECE-like immunoreactivity were co-localised to the endothelium of pulmonary vessels having a range of lumen diameters, traversing the lung parenchyma. These included relatively thin-walled vessels in close association with airways that are characteristic of pulmonary arteries. Immunoreactive ECE was visualised within the epithelial cells lining airways of the lung including bronchioles (airways <2 mm in diameter) and small bronchi (airways 2–5 mm in diameter) (DAVENPORT et al. 1998c). Airway epithelia also stained positively for Big ET-1 and Big ET-3, while staining for all three Big ET precursors was present in submucosal glands (MARCINIAK et al. 1992). Little or no staining for either enzyme of ET peptides was observed in lung parenchyma or cartilage. ECE expression may be upregulated in scarring and thickening of the lung tissue associated with idiopathic pulmonary fibrosis: strong staining for the enzyme was observed in airway epithelium, proliferating type II pneumocytes, endothelial and inflammatory cells (SALEH et al. 1997). Similarly in the upper respiratory tract, immunostaining for ET-1 and ECE-1 was seen in the surface epithelium, endothelial cells, submucosal glands, and infrequently in vascular smooth muscle cells of the nasal mucosa with ECE-like immunoreactivity increased in inflammatory cells in chronic rhinitis (FURUKAWA et al. 1996).

Human lungs contains one of the highest densities of ET receptors (MCKAY et al. 1991; HENRY et al. 1990; MARCINIAK et al. 1992; KNOTT et al. 1995; RUSSELL and DAVENPORT 1996). Autoradiography shows that the sub-types are differentially distributed. ET_A receptors were the only sub-type detectable in resistance vessels (40–90 μm in diameter) and predominated in conduit arteries, comprising 90% of the medial layer from the main pulmonary artery. ET_A receptors were present on lung parenchyma, submucosal glands, airway smooth muscle and epithelial cells. High densities of ET_B receptors localised to airway smooth muscle, with lower levels in parenchyma, airway submucosal glands and small conduit arteries. Little or no binding was detected to connective tissues, submucosal layer and cartilage (RUSSELL and DAVENPORT 1995).

ET-1 has two main actions in the lungs, long lasting vaso- and bronchoconstriction, but the relative contribution of the sub-types is unclear. Synthesis and release of ET-1 from the pulmonary vascular endothelium is thought to cause constriction of pulmonary arteries predominantly via the ET_A sub-type (HAY et al. 1993; MAGUIRE and DAVENPORT1995), although MCCULLOCH et al. (1996) have proposed a significant contribution of ET_B receptors in resistance arteries (150–200 μm in diameter). In the bronchus, the constrictor action of the peptide released from epithelia and diffusing onto the underlying airway smooth muscle is via ET_B receptors (ADNER et al. 1996; TAKAHASHI et al. 1997; HAY et al. 1998) although FUKURODA et al. (1994) detected an ET_A component. Constriction may be modulated by the release of nitric oxide from activation of ET_A receptors present on epithelial cells (NALINE et al. 1999). ET-1 levels are elevated in patients with acute asthma; ET-1 potentiates proliferative actions of airway smooth muscle mitogens via ET_A receptors (PANETTIERI et al. 1996) (see Chap. 14). ET_A receptor antagonist may be sufficient to reverse vasoconstriction associated with pulmonary hypertension without affecting ET_B-mediated contractile effects of airway smooth muscle (see Chap. 15). Pulmonary diseases associated with a range of pathologies such as asthma may require blockade of both sub-types, which may also be accomplished by ECE-inhibition.

III. Adrenal Gland

ECE-like immunoreactivity was not detected in the secretory cells of the cortex or medulla but was confined to the vascular endothelium in the human adrenal (DAVENPORT et al. 1998c). Staining was localised to the endothelial cells (but not the smooth muscle) from arteries surrounding the gland as well as the smaller resistance vessels within the capsular plexus and the endothelium of the central vein. There was no staining in the capillary sinusoids that descend from the capsular plexus through the secretory cells of the medulla to converge on the central vein. A similar pattern of staining in the adrenal vasculature was found for ET-1 and Big ET-1 (DAVENPORT et al. 1996). Synthesis and release of ET-1 by the endothelium is consistent with the proposed role for the peptide in locally controlling the flow of blood into the adrenal (HINSON et al. 1991) mediated via ET_A receptors that are present on smooth muscle cell of the arteries and the smaller vessels within the capsular plexus. ET_B receptors were not discernible on the smooth muscle.

Similar densities of ET-1 receptors have been detected in the three regions of the cortex (zona glomerulosa, fasiculata and reticularis) that secrete steroid hormones as well as the catecholamine secreting cells of the medulla (DAVENPORT et al. 1989) but the two sub-type have a differential distribution within the gland. ET_A receptors were limited to the zona glomerulosa where ET-1 was more potent than ET-3 in stimulating aldosterone which is exclusively released from this region, from isolated human adrenals (HINSON et al. 1991). ET_B receptors were present throughout the zona fasiculata, reticularis and medulla (DAVENPORT et al. 1996). ET-1 and ET-3 are equipotent in stimulating the release of cortisol, a marker of zona fasiculata and reticularis activity, consistent with activation of ET_B receptors. Thus ET-1 may function as an intraglandular factor stimulating steroid secretion via ET_B receptors but regulating blood flow through the adrenal via ET_A sub-type.

The adrenal gland may also be a source of Big ET-3. Antisera to these precursor stained secretory cells of the medulla, although mature ET-3, were not detected within homogenates of adrenal tissue. If released, further processing of Big ET-3 could occur within the vasculature, and the adrenals may be a source of the ET-3 that can be detected in human plasma. Alternatively, a homologue of an ECE isolated from bovine iris microsomes (HASEGAWA et al. 1998) that has specificity for Big ET-3 may also exist in humans.

IV. Kidney

The kidney efficiently clears ET-1 from the circulation (GASIC et al. 1992) and may be partly responsible for excreting the peptide. However, demonstrating the presence of mRNA encoding ET-1 and the detection of the peptide and its precursor by HPLC and RIA (KARET and DAVENPORT 1993, 1996a) established that ET-1 was also synthesised within the kidney and could function as a locally-acting renal peptide.

Levels of ECE-1 mRNA measured by Northern analyses were quantitatively similar in both the cortex and medulla (PUPILLI et al. 1997). These authors used an antiserum directed to the C-terminus of bovine ECE-1 (which would be expected to cross-react with all human ECE-1 isoforms) although this antisera was reported not to cross-react with ECE-2. In the cortex, ECE-1-like immunoreactivity was detected in endothelial cells throughout the renal vasculature including the arcuate arteries which give rise to numerous interlobular arteries and arterioles (Fig. 5). Weaker but specific immunoreactivity was present over some proximal and distal tubules although the distribution was not homogeneous. Few endothelial glomerular cells were stained. In the

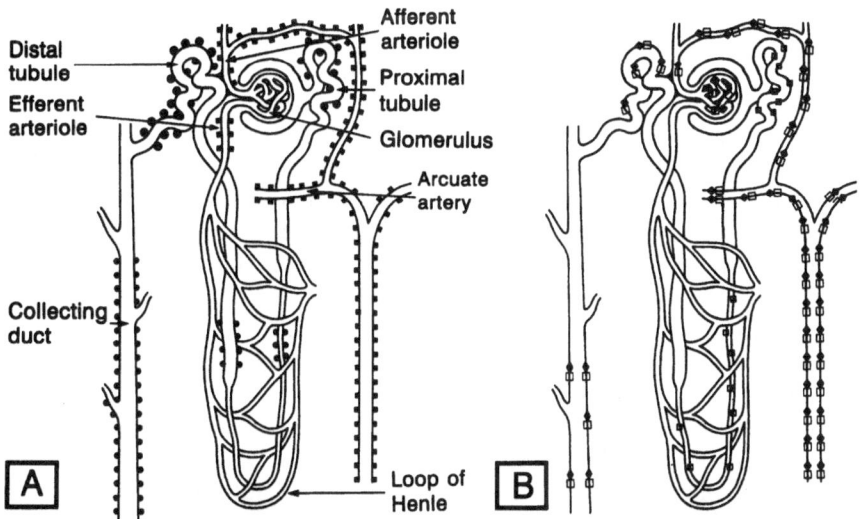

Fig. 5. A Schematic diagram summarising the distribution in human kidney of ET$_A$ (■) and ET$_B$ receptors (●). **B** ET-1 and ECE-1-like (□) immunoreactivity

medulla, ECE-1 immunoreactivity was observed in the vasa recta bundles and capillaries. ECE-1 immunostaining was also detected in the outer and inner medullary collecting ducts and thin limbs of the loop of Henle. Using a selective antisera to ECE-1α, RUSSELL et al. (1998a) found an essentially similar distribution in the endothelial cells lining the renal blood vessels in the kidney, with no detectable staining of the smooth muscle layer. However, intense staining was detected within the endothelial cells of the glomeruli, distal convoluted tubules and epithelial cells of the collecting ducts. ECE-1β-like immunoreactivity was difficult to detect in these regions.

Combining the results from a number of studies using immunocytochemistry and in situ hybridisation to localise ET, Big ET-1 or ECE, there was a good match throughout the kidney between the peptide and its synthetic enzyme in the endothelial cells of renal vasculature and epithelial cells of collecting ducts (Fig. 5). There were some discrepancies: whilst ECE-1 mRNA was localised to the thin loops of Henle little of no immunoreactive ET was detected (PUPILLI et al. 1997). At present the function, if any, of ECE-1 within the loop of Henle is unclear. The majority of studies have used normal kidney. One possible explanation (as in glial cells which express the enzyme but not the peptide) is that levels of ET precursor are upregulated in these regions during disease. In support of this hypothesis, ET immunoreactivity can be detected in these structures in pathophysiological tissue. RUSSELL et al. (1998a) found enzymatic processing of $[^{125}I]$-Big ET-1 was highest in sections of human kidney compared with other human tissue such as the heart. Binding of the resulting cleaved peptide, $[^{125}I]$-ET-1, was to ET_B receptors. If significant amounts of ECE-1 are expressed on the surface of these cells, the enzyme could contribute to clearance of Big ET-1 from the circulation (KUC et al. 1995).

The localisation of ECE-1 and ET-1 to the endothelial cells and epithelial cells is consistent with the two main actions of locally generated ET-1, vasoconstriction and natriuresis, which may be mediated by different receptor subtypes (NAMBI et al. 1992). Although the human kidney is rich in ET_B receptors (comprising 70% in both cortex and medulla), most of the ET_A receptors are localised to renal vessels of the medulla and cortex including the main renal arteries, arcuate arteries and adjacent veins at the cortico-medullary junction as well as intra-renal resistance vessels and arterioles supplying the glomerulus (DAVENPORT et al. 1994) (Fig. 5). ET_B receptors were difficult to detect in the smooth muscle layer of renal vessels but were localised to non-vascular structures such as the tubules and collecting ducts. (KARET et al. 1993; DAVENPORT et al. 1994; MAGUIRE et al. 1994).

Like other human vessels, ET-1 constriction of isolated renal arteries and veins is mediated mainly via the ET_A sub-type (MAGUIRE et al. 1994). In agreement with isolated tissue, systemic infusion of ET-1 into humans, to achieve pathophysiological levels caused pronounced vasoconstriction. ET-3, which has at least low affinity for the human ET_A receptor, infused to achieve similar levels, had no effect on blood pressure, renal haemodynamics or electrolyte excretion (KAASJAGER et al. 1997). ET_A (but not ET_B) mRNA was downregulated in renal biopsies transplant recipients receiving cyclosporin, suggesting an adaptive response to the overproduction of ET and vasoconstriction associated with this therapy (KARET and DAVENPORT 1996b). ECE may have a central role in regulating local ET production within the kidney which on release mediates responses primarily via ET_A receptor. The ET_B sub-type may also have a role as a 'clearance receptor', removing ET and locally converted Big ET (RUSSELL et al. 1998a) from the plasma.

V. Central Nervous System

1. Cerebral Vasculature

ET-1 constricts vessels from the brain including basilar (PAPADOPOLOUS et al. 1990), cerebral (ADNER et al. 1994) and pial arteries (HARDEBO et al. 1989; THORIN et al. 1998; PIERRE and DAVENPORT 1998b, 1999) and may be involved in the genesis or maintenance of cerebrovascular disorders such as the delayed vasospasm associated with subarachnoid haemorrhage or stroke (see Chap. 13). The small pial arteries and arterioles penetrating into the brain are of particular importance, playing a major role in the maintenance of cerebral blood flow (autoregulation). ET-1 does not cross the blood brain barrier in rats and ET-1 synthesised in the peripheral vasculature in humans is therefore unlikely to cross the blood brain barrier, unless this has been compromised by subarachnoid haemorrhage, stroke or head injury. Thus there are two potential targets for overexpressed ET-1 in human brain: vascular receptors mediating cerebrovasospasm that may be responsible for delayed cerebral ischaemia seen after aneurysmal subarachnoid haemorrhage and could contribute to ischaemic core volume in stroke. Second, neural receptors mediating increases in intracellular free calcium (MORTON and DAVENPORT 1992) that initiates the pathophysiological processes leading to neuronal death.

ET and ECE-like immunoreactivity was localised to the vascular endothelium of the small pial vessels (diameter $<50\,\mu m$) within sulci of the human cerebral cortex. As in the peripheral vasculature, little or no staining for either peptide or enzyme was detected in the vascular smooth muscle layer (Fig. 6). Staining was visible on the outermost surface of pial vessels which may correspond to the flattened mesothelial cells of the pia mater that is attached to the surface of the brain, continuing into the sulci and around the penetrating pial vessels. Intense ECE-like immunoreactivity was detected within endothelial cells lining the numerous smaller vessels within the grey and white matter (DAVENPORT et al. 1998b).

Raised local ET levels may elicit constriction in both large arteries and small cerebral vessels where smooth muscle cells only express the ET_A sub-type (ADNER, et al. 1994; YU et al. 1995; LUCAS et al. 1996; HARLAND et al. 1995, 1998; PIERRE and DAVENPORT 1995, 1998b, 1999). Cerebral arterioles are particularly sensitive to ET-1 compared to other agonists. Importantly, ET_A selective antagonists can relax pial arteries preconstricted with ET-1 (PIERRE and

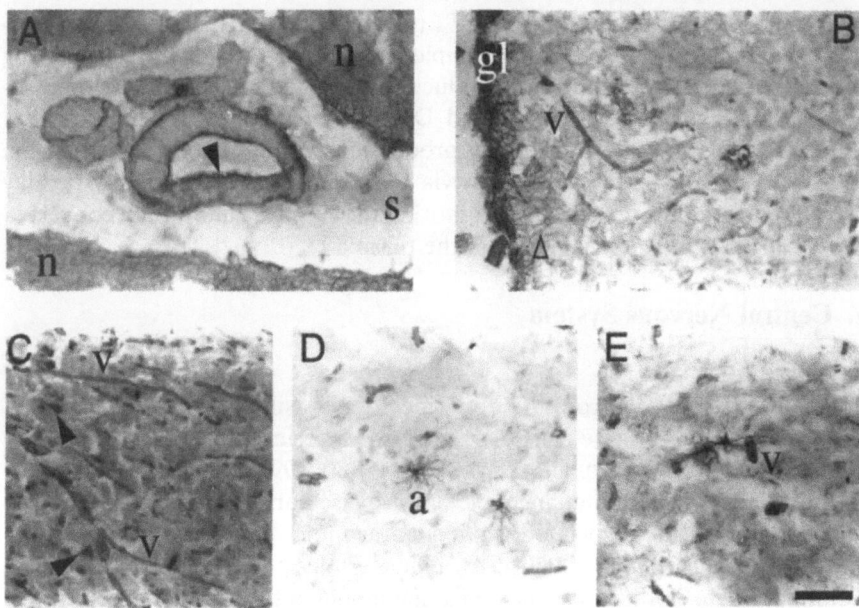

Fig. 6A–E. ECE-like immunoreactivity in the cortex of human brain. **A** Staining was visualised in the endothelium (*arrow*) of pial arteries lying within a sulcus (*S*) between neuronal tissue (*n*) of the cerebral cortex. **B** ECE-like immunoreactivity within the fibres of the glial limitans (*gl*), the outer glial membrane covering the nervous tissue and long fibres perpendicular to the pial surface in the grey matter indicated by the *open arrow*. Staining was also present in endothelial cells lining the numerous smaller vessels (*V*). **C** An example of weakly stained neuronal cell bodies (*solid arrows*) with a mixed morphology including pyramidal cells, within the deeper layers of the cortex together with penetrating blood vessels (*V*). **D** In the white matter, occasional astrocytes (*a*) were stained as well as penetrating vessels (*V*); **E** in some cases astrocytic processes terminated upon small vessels (*V*) within the white matter. (*Scale bar* = 100 μm)

DAVENPORT 1999) suggesting these compounds would reverse or prevent cerebrovasospasm.

ET$_B$ mediated relaxation has been reported to occur in isolated precon-stricted temporal arteries and small (200 μm in diameter) cerebral arteries (LUCAS et al. 1996; NILSSON et al. 1997). Intriguingly, ligand binding (YAMAGA et al. 1995) and functional evidence suggests that human brain endothelial cells isolated from capillaries (diameter <10 μm) that form the blood brain barrier and larger microvessels, express ET$_A$ receptors linked to phospholipase C and IP$_3$ accumulation (STANIMIROVIC et al. 1994; SPATZ et al. 1997). It is unclear why capillary endothelial cells have evolved to express different sub-types to peripheral endothelial cells but ET$_A$ receptors may have a role at the blood brain barrier. ET-1 acting via this sub-type has been proposed to increase capillary permeability leading to oedema (PURKISS et al. 1994).

2. Neurones

In the grey matter of human cortex, ECE-like immunoreactivity was visualised in sparse but intensely staining long fibres perpendicular to the pial surface (DAVENPORT et al. 1998c). The orientation and length of these processes is characteristic of afferent and efferent fibres of neurones. Neuronal cell bodies of mixed morphology, including pyramidal cells (Fig. 6C) were also stained although less intensely compared with the endothelial cells. The cell bodies occurred mainly in the deeper layers (layers III–VI). Less intense staining was also found in neuronal cell bodies many having the morphological features of pyramidal cells in layers III–VI. GIAID et al. (1991) localised ET-1 mRNA and ET-like immunoreactivity to fibres and neuronal cell bodies to the same layers of the human cortex. BARNES et al. (1997) detected ECE-like immunoreactivity to the pyramidal cells of the rat hippocampus. The presence of ET and ECE-like immunoreactivity within neurones provides further support for the hypothesis that ET may function as a neuropeptide in the human brain.

3. Astrocytes

Neuroglia comprise all non-neural cells and constitute nearly half of the mass of the CNS. Intense staining for ECE occurred in fibres of the glial limitans or outer glial membrane covering the nervous tissue. Within the sub-cortical white matter, occasional astrocytes (Fig. 6D) were stained, sometimes with astrocytic processes terminating upon small vessels (Fig. 6E). Staining was not detected within oligodendrocytes. Staining was also found for the ET peptides in the pial arteries, glial limitans and to a lesser extent within neurones and processes of the cortex. In agreement with GIAID et al. (1991) staining for ET was not found within astrocytes from normal human brain. Astrocytes have an important role in repair of CNS tissue following injury. Interestingly, intense ET staining was detected in reactive astrocytes surrounding metastases (ZHANG and OLSSON 1995) and following viral infections (MA et al. 1994) as well as in rat perivascular astrocytic processes in an animal model of ischaemia (GAJKOWSKA and MOSSAKOWSKI 1995).

Why has little transcription of mRNA encoding isoforms of ECE-1 been detected in human brain (SCHMIDT et al. 1994; VALDENAIRE et al. 1995) compared with peripheral tissue? This is surprising since a phosphoramidon-sensitive ECE has been detected in this tissue (WARNER et al. 1992). In bovine brain, XU et al. (1994) localised ECE-1 mRNA as expected to the endothelium of cerebral vessels, but hybridisation of the probe was not detected over neuronal cells in the cortex. High levels of ECE-2 mRNA were found in the bovine cerebral cortex by Northern blot analysis although the identity of cells expressing this enzyme was not determined (EMOTO and YANAGISAWA 1995). Analysis of individual murine Purkinje neurones and Bergmann glial cells revealed mRNA for both ECE-1 and ECE-2 (SCHMIDT-OTT et al. 1998). It is not yet known whether both enzymes are expressed in human brain but the antisera used by DAVENPORT et al. 1998c would also cross-react with the

C-terminal sequence of ECE-2 if this is conserved in the human enzyme. ECE-2 differs from ECE-1 by requiring an acidic pH optimum and is virtually inactive at neutral pH and would be expected to be associated with the acidic conditions of the trans-Golgi network (EMOTO and YANAGISAWA 1995). If ECE-2 is the predominant isoform in neural tissue, the enzyme may be activated under the acidic conditions associated with ischaemia when the pH of glial cells falls to 5.5 or less (EKHOLM et al. 1991), to increase synthesis of ET. Increased levels of ET have been reported in tissues following focal cerebral ischaemia (BARONE et al. 1995).

Although ET-1 is more abundant, ET-3 is also present in human brain (TAKAHASHI et al. 1991). The efficiency of converting Big ET-3 to the mature peptide by cloned ECE-1 is low or not detectable and the ECE activity isolated by WARNER et al. (1992) from human brain did not convert Big ET-3, suggesting the existence of an additional ECE selective for Big ET-3 as isolated by HASEGAWA et al. (1998) from bovine iris microsomes.

4. Neural Receptors

The ET_B receptor subtype accounts for about 90% of endothelin receptors in the normal cerebral cortex (HARLAND et al. 1995) being localised to neuronal regions particularly level III and IV of the cortex. The small proportion of ET_A receptors were localised in high densities to the leptomeninges, pial arteries and intracerebral vessels, with much lower but detectable levels in the grey and white matter. ET_B receptors were not detected in the vascular structures or leptomeninges. In primary cultures from rat cerebellum, single cell dynamic video imaging of intracellular free calcium ions and autoradiography showed glial cells expressed receptors characteristic of the ET_B sub-type, whereas the responses of neurones were of the ET_A (MORTON and DAVENPORT 1992; DAVENPORT and MORTON 1991). Antagonism of both receptor sub-types may be necessary to confer neuroprotection.

E. Conclusions

A consensus is emerging that ET_A receptors predominate on the smooth muscle in the peripheral vasculature of humans and ET_A receptor antagonists can fully reverse ET-1 induced constriction in both normal and diseased vessels. It is less clear whether endothelial ET_B receptors have a similar ubiquitous distribution. However, beneficial vasodilatation may possibly be 'unmasked' in the presence of ET_A selective antagonists that block the more powerful constrictor action of the peptide. Non-vascular ET_B receptors that occur in high densities in tissues such as the kidney may also have a beneficial role in removing ET from the circulation. In human brain, ET_A receptors exclusively mediate constriction in the larger cerebral vessels; activation of endothelial ET_B receptors causes significant vasodilatation. Surprisingly, ET_A and not ET_B receptors are present on the capillary endothelial cells that form the blood brain barrier. Neural tissue is rich in ET_B receptors but it remains to be determined whether neurones also express ET_A receptors that, when activated by ET-1, cause an increase in intracellular calcium as proposed in these cells from animals. If correct, blockade of ET_A receptors is predicted to reduce cerebrovasospasm whereas mixed ET_A/ET_B antagonists would be needed for neuroprotection.

ECE-like immunoreactivity is widely expressed in the human endothelium and parallels the distribution of ET-1 and Big ET-1. Some ECE activity is on the cell surface but most is intracellular, where ET synthesis appears increasingly complex. ECE-1α, ECE-1β and ECE-2 immunoreactivity have all been localised to human endothelial cells, although ECE-1α is emerging as the predominant isoform. ET-1 is unusual compared with other vasoactive peptides in being synthesised by two distinct pathways. ECE-1α and EC-1β are localised to storage granules of the regulated pathway; both isoforms are present with the possible expression of ECE-2, in vesicles of the secretory pathway.

There are discrepancies and questions. ECE activity is upregulated in coronary artery disease. How important is long range signalling by Big ET precursors and subsequent processing at target sites? Whilst human endothelial ECE is specific for Big ET-1, smooth muscle ECE can also convert Big ET-3 and may be one site for ET-3 synthesis. Do smooth muscle cells express an additional ECE, possibly more selective for Big ET-3? Glial cells within the brain appear to express the enzyme but not the peptide until these cells have been transformed to reactive astrocytes. Is this a significant source of ET in neural tissue following injury or disease? The importance of each enzyme as a target for inhibitors in pathophysiological conditions where ET levels are known to be elevated remains to be determined.

Acknowledgements. Supported by grants from the British Heart Foundation, Isaac Newton Trust, Royal Society and Medical Research Council, UK.

List of Abbreviations

Big ET-1	Big Endothelin
CHO	Chinese hamster ovary
ECE	Endothelin converting enzyme
ET	Endothelin
HUVECs	human umbilical vein endothelial cells
HUVSMCs	human umbilical vein smooth muscle cells
vWF	von Willebrand Factor

References

Adner M, Jansen I, Edvinsson L (1994) Endothelin-A receptors mediate contraction in human cerebral, meningeal and temporal arteries. J Autonomic Nervous System 49:S117–S121

Adner M, Cardell LO, Sjoberg T, Ottosson A, Edvinsson L (1996) Contractile endothelin-B (ET$_B$) receptors in human small bronchi. Eur Respir J 9:351–355

Ahn K, Sisneros AM, Herman SB, Pan SM, Hupe D, Lee C, Nikam S, Cheng X-M, Doherty AM, Schroeder RL, Haleen SJ, Kaw S, Emoto N, Yanagisawa M (1998) Novel selective quinazoline inhibitors of endothelin converting enzyme-1. Biochem Biophys Res Commun 243:184–190

Ashby, MJ Plumpton C, Teale P, Kuc RE, Houghton E Davenport AP (1995) Analysis of endogenous human endothelin peptides by high performance liquid chromatography and mass spectrophometry. J Cardiovasc Pharmacol 26 (S3):S247–S249

Bacon CR, Cary, NRB, Davenport AP (1996) Endothelin peptide and receptors in human atherosclerotic coronary artery and aorta. Circulation Research 79:794–801

Bacon CR, Davenport AP (1996) Endothelin receptors in human coronary artery and aorta. Br J Pharmacol 117:986–992

Barnes K, Shimada K, Takahashi M, Tanzawa K, Turner AJ (1996) Metallopeptidase inhibitors induce an up-regulation of endothelin-converting enzyme levels and its redistribution from the plasma membrane to an intracellular compartment. J Cell Sci 109:919–928

Barnes K, Walkden BJ, Wilkinson TC, Turner AJ (1997) Expression of endothelin-converting enzyme in both neuroblastoma and glial cells and its localization in rat hippocampus. J Neurochem 68:570–577

Barnes K, Brown C, Turner AJ (1998) Endothelin-converting enzyme. Ultrastructural localization and its recycling from the cell surface. Hyperten 31:3–9

Barone FC, Willette RN, Yue TL, Feurestein G (1995) Therapeutic effects of endothelin receptor antagonists in stroke. Neurol Res 17:259–264

Bax WA, Bruinvels AT, Van-Suylen RJ, Saxena PR, Hoyer D (1993) Endothelin receptors in the human coronary artery, ventricle and atrium. A quantitative autoradiographic analysis. Naunyn-Schmiedeberg's Arch Pharmacol 348:403–410

Bax WA, Aghai Z, van Tricht CL, Wassenaar C, Saxena PR (1994) Different endothelin receptors involved in endothelin-1- and sarafotoxin S6B-induced contractions of the human isolated coronary artery. Br J Pharmacol 113:1471–1479

Bohnemeier H, Pinto YM, Horkay F, Toth M, Juhasz Nagy A, Orzechowski HD, Bohm M, Paul M (1998) Endothelin converting-enzyme-1 mRNA expression in human cardiovascular disease. Clin Exp Hypertens 20:417–437

Cohn JN (1996) Is there a role for endothelin in the natural history of heart failure? Circulation 94:604–606

Corder R, Khan N, Harrison VJ (1995) A simple method for isolating human endothelin converting enzyme free from contamination by neutral endopeptidase 24.11. Biochem Biophys Res Commun 207:355–362

Costello KB, Stewart DJ, Baffour R (1990) Endothelin is a potent constrictor of human vessels used in coronary revascularization surgery. Eur J Pharmacol 186:311–314

Dagassan PH, Breu V, Clozel M, Kunzli A, Vogt P, Turina M, Kiowski W, Clozel JP (1996) Up-regulation of endothelin-B receptors in atherosclerotic human coronary arteries. J Cardiovasc Pharmacol 27:147–153

Davenport AP, Nunez DJ, Hall JA, Kaumann AJ, Brown MJ (1989) Autoradiographical localization of binding sites for porcine [125I] endothelin-1 in humans, pigs, and rats: functional relevance in humans. J Cardiovasc Pharmacol 13 (S5):S166–170

Davenport AP, Morton AJ (1991) Binding sites for 125I ET-1, ET-2, ET-3 and vasoactive intestinal contractor are present in adult rat brain and neurone-enriched primary cultures of embryonic brain cells. Brain Res 554:278–285

Davenport AP, O'Reilly G, Molenaar P, Maguire JJ, Kuc RE, Sharkey A, Bacon CR, Ferro A (1993) Human endothelin receptors characterised using reverse transcriptase-polymerase chain reaction, in situ hybridization and sub-type selective ligands BQ123 and BQ3020: evidence for expression of ET$_B$ receptors in human vascular smooth muscle. J Cardiovasc Pharmacol 22 (S8):22–25

Davenport AP, Maguire JJ (1994) Is endothelin-induced vasoconstriction mediated only by ET_A receptors in man? Trends Pharmacol Sci 15:9–11

Davenport AP, Kuc RE, Hoskins SL, Karet FE, Fitzgerald F (1994) [^{125}I]-PD151242: a selective ligand for endothelin ET_A receptors in human kidney which localises to renal vasculature. Br J Pharmacol 113:1303–1310

Davenport AP, O'Reilly G, Kuc, RE (1995a) Endothelin ET_A and ET_B mRNA and receptors expressed by smooth muscle in the human vasculature: majority of the ET_A sub-type. Br J Pharmacol 114:1110–1116

Davenport AP, Kuc RE, Maguire JJ, Harland SP (1995b) ET_A receptors predominate in the human vasculature and mediate constriction. J Cardiovasc Pharmacol, 26 (S3), S265–267

Davenport AP, Hoskins SL, Kuc RE, Plumpton C (1996) Differential distribution of endothelin peptides and receptors in human adrenal gland. Histochemical Journal 28:779–789

Davenport AP, Kuc RE, Ashby MJ, Patt WC, Doherty AM (1998a) Characterisation of [^{125}I]-PD164333, an ET_A-selective non-peptide radiolabelled antagonist, in normal and diseased human tissues. Br J Pharmacol 123:223–230

Davenport AP, Kuc RE, Mockridge JW (1998b) Endothelin-converting enzyme (ECE) in the human vasculature: evidence for a differential conversion of Big ET-3 by endothelial and smooth muscle cells. J Cardiovasc Pharmacol 31 (S1):S1–3

Davenport AP, Kuc RE, Plumpton C, Mockridge JW, Barker PJ, Huskisson NS (1998c) Endothelin- converting enzyme in human tissues. Histochem J 30:359–374

Davenport AP, Kuc RE (2000) Cellular expression of isoforms of endothelin-converting enzyme-1 (ECE-1c, ECE-1b and ECE-1a) and endothelin-converting enzyme-2. J Cardiovasc Pharmacol 36 (S1), 512–514

Docherty JC, Gunter HE, Kuzio B, Shoemaker L, Yang LJ, Deslauriers R (1997) Effects of cromakalim and glibenclamide on myocardial high energy phosphates and intracellular pH during ischemia-reperfusion: P-31 MRS studies. J Mol Cell Cardiol 29:1665–1673

Ekholm A, Katsura K, Siesjo BK (1991) Tissue lactate content and tissue PCO2 in complete brain ischaemia: implications for compartmentation of H+. Neurol Res 13:74–76

Emoto N, Yanagisawa M (1995) Endothelin-converting enzyme-2 is a membrane-bound, phosphoramidon-sensitive metalloprotease with acidic pH optimum. J Biol Chem 270:15262–15268

Franco-Cereceda A (1989) Endothelin- and neuropeptide Y-induced vasoconstriction of human epicardial coronary arteries in vitro. Br J Pharmacol 97:968–972

Fujitani Y, Oda K, Takimoto M, Inui T, Okada T, Urade Y (1992) Autocrine receptors for endothelins in the primary culture of endothelial cells of human umbilical vein. Febs Lett 298:79–83

Fukuchi M, Giaid A (1998) Expression of endothelin-1 and endothelin-converting enzyme-1 mRNAs and proteins in failing human hearts. J Cardiovasc Pharmacol 31 (S1):S421–423

Fukuroda T, Kobayashi M, Ozaki S, Yano M, Miyauchi T, Onizuka M, Sugishita Y, Goto K, Nishikibe M (1994) Endothelin receptor subtypes in human versus rabbit pulmonary arteries. J Appl Physiol 76:1976–1982

Furukawa K, Saleh D, Bayan F, Emoto N, Kaw S, Yanagisawa M, Giaid A (1996) Co-expression of endothelin-1 and endothelin-converting enzyme-1 in patients with chronic rhinitis. Am J Respir Cell Mol Biol 14:248–253

Gajkowska B, Mossakowski MJ (1995) Localization of endothelin in the blood-brain interphase in rat hippocampus after global cerebral ischemia. Folia Neuropathol 33:221–230

Gasic S, Wagner OF, Vierhapper H, Nowotny P, Waldhausl W (1992) Regional hemodynamic effects and clearance of endothelin-1 in humans: Renal and peripheral tissues may contribute to the overall disposal of the peptide. J Cardiovasc Pharmacol 19:176–180

Giaid A, Gibson SJ, Herrero MT, Gentleman S, Legon S, Yanagisawa M, Masaki T, Ibrahim NB, Roberts GW, Rossi ML, Polak JM (1991) Topographical localisation of endothelin mRNA and peptide immunoreactivity in neurones of the human brain. Histochem 95:303–314

Godfraind T (1993) Evidence for heterogeneity of endothelin receptor distribution in human coronary artery. Br J Pharmacol 110:1201–1205

Gui G, Xu D, Enoto N, Yanagisawa M (1993) Intracellular localization of membrane-bound endothelin-converting enzyme from rat lung. J Cardiovasc Pharmacol 22 (S8):S53–S56

Hardebo JE, Kahrstrom J, Owman C, Salford LG (1989) Endothelin is a potent constrictor of human intracranial arteries and veins. Blood Vessels 26:249–253

Harland SP, Kuc RE, Pickard JD, Davenport AP (1995) Characterisation of ET$_A$ receptors in human brain cortex, gliomas and meningiomas. J Cardiovasc Pharmacol 26 (S3):S408–S401

Harland SP, Kuc RE, Pickard JD, Davenport AP (1998) Expression of ET$_A$ receptors in human gliomas and meningiomas with high affinity for the selective antagonist, PD156707. Neurosurgery 43:890–898

Harrison VJ, Barnes K, Turner AJ, Wood E, Corder R, Vane JR (1995) Identification of endothelin 1 and big endothelin 1 in secretory vesicles isolated from bovine aortic endothelial cells. Proc Natl Acad Sci U.S.A. 92:6344–6348

Hasegawa H, Hiki K, Sawamura T, Aoyama T, Okamoto Y, Miwa S, Shimohama S, Kimura J, Masaki T (1998) Purification of a novel endothelin-converting enzyme specific for big endothelin-3. FEBS Lett 428:304–308

Hay DW, Henry PJ, Goldie RG (1993) Endothelin and the respiratory system. Trends Pharmacol Sci 14:29–32

Hay DW, Luttmann MA, Pullen MA, Nambi P (1998) Functional and binding characterization of endothelin receptors in human bronchus: evidence for a novel endothelin B receptor subtype? J Pharmacol Exp Ther 284:669–677

Haynes WG, Webb DJ (1994) Contribution of endogenous generation of endothelin-1 to basal vascular tone. Lancet 344:852–854

Haynes WG, Strachan FE, Webb DJ (1995) Endothelin ETA and ETB receptors cause vasoconstriction of human resistance and capacitance vessels *in vivo*. Circulation 92:357–363

Haynes WG, Ferro CJ, O'Kane KP, Somerville D, Lomax CC, Webb DJ (1996) Systemic endothelin receptor blockade decreases peripheral vascular resistance and blood pressure in humans. Circulation 93:1860–1870

Hemsen A, Gillis C, Larsson O, Haegerstrand A, Lundberg JM (1991) Characterization, localization and actions of endothelins in umbilical vessels and placenta of man. Acta Physiol Scand 143:395–404

Henry PJ, Rigby PJ, Self GJ, Preuss JM, Goldie RG (1990) Relationship between endothelin-1 binding site densities and constrictor activities in human and animal airway smooth muscle. Br J Pharmacol 100:786–792

Hinson JP, Vinson GP, Kapas S, Teja R (1991) The role of endothelin in the control of adrenocortical function: stimulation of endothelin release by ACTH and the effects of endothelin-1 and endothelin-3 on steroidogenesis in rat and human adrenocortical cells. J Endocrinol 128:275–280

Howard PG, Plumpton C, Davenport AP (1992) Anatomical localisation and pharmacological activity of mature endothelins and their precursors in human vascular tissue. J Hyperten 10:1379–1386

Kaasjager KA, Shaw S, Koomans HA, Rabelink TJ (1997) Role of endothelin receptor subtypes in the systemic and renal responses to endothelin 1 in humans. J Am Soc Nephrol 8:32–89

Karet FE, Davenport AP (1993) Human kidney: Endothelin isoforms revealed by HPLC with radioimmunoassay, and receptor sub-types detected using ligands BQ123 and BQ3020. J Cardiovasc Pharmacol 22 (S8):29–33

Karet FE, Kuc RE, Davenport AP (1993) Novel Ligands BQ123 and BQ3020 characterise endothelin receptor subtypes ET_A and ET_B in human kidney. Kidney Int 44:36–42

Karet FE, Davenport AP (1996a) Localization of endothelin peptides in human kidney. Kidney Int 49:382–387

Karet FE, Davenport AP (1996b) Selective down regulation of ET_A receptors mRNA in renal transplant recipients on cyclosporin A revealed by quantitative RT-PCR. Nephrol Dial Transplant 11:1976–1982

Kiowski W, Luscher TF, Linder L, Buhler FR (1991) Endothelin-1-induced vasoconstriction in humans. Reversal by calcium channel blockade but not by nitrovasodilators or endothelium-derived relaxing factor. Circulation 83:469–475

Knott PG, D'Aprile AC, Henry PJ, Hay DW, Goldie RG (1995) Receptors for endothelin 1 in asthmatic human peripheral lung. Br J Pharmacol 114:1–3

Kuc RE, Karet FE, Davenport AP (1995) Characterization of peptide and non-peptide antagonists in human kidney. J Cardiovasc Pharmacol 26 (S3):S373–S375

Liu JJ, Chen JR, Buxton BF (1996) Unique response of human arteries to endothelin B receptor agonist and antagonist. Clin Sci 90:91–96

Lucas GA, White LR, Juul R, Cappelen J, Aasly J, Edvinsson L (1996) Relaxation of human temporal artery by endothelin ET_B receptors. Peptides 17:1139–1144

Luscher TF, Yang Z, Bauer E, Tschudi M, Von-Segesser L, Stulz P, Boulanger C, Siebenmann R, Turina M, Buhler FR (1990) Interaction between endothelin-1 and endothelium-derived relaxing factor in human arteries and veins. Circ Res 66:1088–1094

Ma KC, Nie XJ, Hoog A, Olsson Y, Zhang WW (1994) Reactive astrocytes in viral infections of the human brain express endothelin-like immunoreactivity. J Neurol Sci 126:184–192

Maguire JJ, Kuc RE, O'Reilly G, Davenport AP (1994) Vasoconstrictor endothelin receptors characterised in human renal artery and vein in vitro. Br J Pharmacol 113:49–54

Maguire JJ, Davenport AP (1995) ET_A receptors mediate the constrictor responses to endothelin peptides in human blood vessels in vitro. Br J Pharmacol 115:191–197

Maguire JJ, Kuc RE, Davenport AP (1997a) Affinity and selectivity of PD156707, a novel non- peptide endothelin antagonist, for human ET_A and ET_B receptors. J Pharmacol Exp Ther 280:1102–1108.

Maguire JJ, Johnson CM, Mockridge JW, Davenport AP (1997b) Endothelin converting enzyme (ECE) activity in human vascular smooth muscle. Br J Pharmacol 122:1647–1654

Maguire JJ, Davenport AP (1998a) PD156707:a potent antagonist of ET-1 in human diseased coronary arteries and saphenous veins. J Cardiovasc Pharmacol 31 (S1):S239–240

Maguire JJ, Davenport AP (1998b) Increased response to big endothelin-1 in atherosclerotic human coronary artery: functional evidence for upregulation of endothelin-converting enzyme activity in disease. Br J Pharmacol 125:238–240

Maguire JJ, Davenport AP (1999) Endothelin receptor expression and pharmacology in human saphenous vein graft. Br J Pharmacol 126:443–450

Maguire JJ, Ahn K, Davenport AP (1999) Inhibition of big endothelin-1 (Big ET-1) responses in endothelium-denuded human coronary artery by the selective endothelin-converting enzyme (ECE) inhibitor, PD159790. Br J Pharmacol 126:193P

Marciniak SJ, Plumpton C, Barker PJ, Huskisson NS, Davenport AP (1992) Localization of immunoreactive endothelin and proendothelin in the human lung Pulm Pharmacol 5:175–182

Matsumoto H, Suzuki N, Kitada C, Fujino M (1994) Endothelin family peptides in human plasma and urine: their molecular forms and concentrations. Peptides 15:505–510

McCulloch KM, Docherty CC, Morecroft I, MacLean MR (1996) Endothelin B receptor mediated contraction in human pulmonary resistance arteries. Br J Pharmacol 119:1125–1130

McKay KO, Black JL, Diment LM, Armour CL (1991) Functional and autoradiographic studies of endothelin-1 and endothelin-2 in human bronchi, pulmonary arteries, and airway parasympathetic ganglia J Cardiovasc Pharmacol 17 (S7): S206–209

Minamino T, Kurihara H, Takahashi M, Shimada K, Maemura K, Oda H, Ishikawa T, Uchiyama T, Tanzawa K, Yazaki Y (1997) Endothelin-converting enzyme expression in the rat vascular injury model and human coronary atherosclerosis. Circulation 95:221–230

Mockridge JW, Barker PJ, Huskisson NS, Kuc RE, Davenport AP (1998) Characterisation of site directed antisera against endothelin converting enzymes. J Cardiovasc Pharmacol 31 (S1):S35–37

Molenaar P, Kuc RE, Davenport AP (1992) Characterization of two new ET_B selective radioligands, $[^{125}I]$-BQ3020 and $[^{125}I]$-$[Ala^{1,3,11,15}]ET$-1 in human heart. Br J Pharmacol 107:637–639

Molenaar P, O'Reilly G, Sharkey A, Kuc RE, Harding DP, Plumpton C, Gresham GA, Davenport AP (1993) Characterization and localization of endothelin receptor sub-types in the human atrioventricular conducting system and myocardium. Circ Res 72:526–538

Mombouli JV, Le SQ, Wasserstrum N, Vanhoutte PM (1993) Endothelins 1 and 3 and big endothelin-1 contract isolated human placental veins. J Cardiovasc Pharmacol (S8):S278–281

Moravec CS, Reynolds EE, Stewart RW, Bond M (1989) Endothelin is a positive inotropic agent in human and rat heart *in vitro*. Biochem Biophys Res Commun 28:14–8

Morton AJ, Davenport AP (1992) Neurons respond differentially to endothelins and sarafotoxin S6b in primary cultured cerebellum. Brain Res 581:299–306

Naline E, Bertrand C, Biyah K, Fujitani Y, Okada T, Bisson A, Advenier C (1999) Modulation of ET-1 induced contraction of human bronchi by airway dependent nitric oxide release via ET_A receptor activation. Br J Pharmacol 126:529–535

Nambi P, Pullen M, Wu HL, Aiyar N, Ohlstein EH, Edwards RM (1992) Identification of endothelin receptor subtypes in human renal cortex and medulla using subtype selective ligands. Endocrinology 131:1081–1086

Nilsson T, Cantera L, Adner M, Edvinsson L (1997) Presence of contractile endothelin-A and dilatory endothelin-B receptors in human cerebral arteries. Neurosurgery 40:346–51

O'Reilly G, Charnock-Jones DS, Morrison JJ, Cameron IT, Davenport AP, Smith SK (1993) Alternatively spliced mRNAs for human endothelin-2 and their tissue distribution. Biochem Biophysic Res Comm 193:834–840

Panettieri RA, Goldie RG, Rigby PJ, Eszterhas AJ, Hay DW (1996) Endothelin-1-induced potentiation of human airway smooth muscle proliferation: an ET_A receptor-mediated phenomenon. Br J Pharmacol 118:191–197

Papadopoulos SM, Gilbert LL, Webb R, D'Amato CJ (1990) Characterisation of contractile responses to endothelin in human cerebral arteries: Implications for cerebral vasospasm. Neurosurgery 26:810–815

Peter MG, Davenport AP (1995) Selectivity of $[^{125}I]$-PD151242 for the human, rat and porcine endothelin ET_A receptors in the heart. Br J Pharmacol 114:297–302

Peter MG, Davenport AP (1996a) Characterisation of endothelin receptor selective agonist BQ3020 and antagonists BQ123, FR139317, BQ788, 50235, Ro462005 and bosentan in the heart. Br J Pharmacol 117:455–462

Peter MG, Davenport AP (1996b) Upregulation of the endothelin ET_A receptors in the left ventricle from failing human hearts demonstrated using competitive binding studies with FR139317. Br J Pharmacol 117:207P

Pierre LN, Davenport AP (1995) Autoradiographic study of endothelin receptors in human cerebral arteries. J Cardiovasc Pharmacol 26 (S3):S326–S328

Pierre LN, Davenport AP (1998a) Endothelin receptor sub-types and their functional relevance in human small coronary arteries. Br J Pharmacol 124:499–506

Pierre LN, Davenport AP (1998b) Relative contribution of Endothelin-A and Endothelin-B receptors to vasoconstriction in small arteries from the human heart and brain. J Cardiovasc Pharmacol 31 (S1):S74–S76

Pierre LN, Davenport AP (1999) Blockade and reversal of ET-induced constriction in pial arteries from human brain. Stroke 30:638–643

Plumpton C, Horton J, Kalinka KS, Martin R, Davenport AP (1994) Effects of phosphoramidon and pepstatin A on the secretion of endothelin-1 and big endothelin-1 in human umbilical vein endothelial cells: Measurement by two-site ELISAs. Clin Sci 87:245–251

Plumpton C, Haynes WG, Webb DJ, Davenport AP (1995) Phosphoramidon inhibition of the *in vivo* conversion of big endothelin-1 to endothelin-1 in the human forearm. Br J Pharmacol 116:1821–1828

Plumpton C, Ferro CJ, Haynes WG, Webb DJ, Davenport AP (1996a) The increase in human plasma immunoreactive endothelin but not big endothelin-1 or its C-terminal induced by systemic administration of the endothelin antagonist TAK-044. Br J Pharmacol 119:311–314

Plumpton C, Ashby MJ, Kuc RE, O'Reilly G, Davenport AP (1996b) Expression of endothelin peptides and mRNA in the human heart. Clin Sci 90:37–46

Pönicke K, Vogelsang M, Heinroth M, Becker K, Zolk O, Bohm M, Zerkowski HR, Brodde OE (1998) Endothelin receptors in the failing and nonfailing human heart. Circulation 97:744–751

Pupilli C, Romagnani P, Lasagni L, Bellini F, Misciglia N, Emoto N, Yanagisawa M, Rizzo M, Mannelli M, Serio M (1997) Localization of endothelin-converting enzyme-1 in human kidney. Am J Physiol 273:F749–56

Purkiss JR, West D, Wilkes LC, Scott C, Yarrow P, Wilkinson GF, Boarder MR (1994) Stimulation of phospholipase C in cultured microvascular endothelial cells from human frontal lobe by histamine, endothelin and purinoceptor agonists. Br J Pharmacol 111:1041–1116

Rizzi A, Calo G, Battistini B, Regoli D (1998) Contractile activity of endothelins and their precursors in human umbilical artery and vein: identification of distinct endothelin-converting enzyme activities. J Cardiovasc Pharmacol 31 (S1):S58–61

Riezebos J, Watts IS, Vallance PJT (1994) Endothelin receptors mediating functional responses in human small arteries and veins. Br J Pharmacol 111:609–615

Rossi GP, Albertin G, Franchin E, Sacchetto A, Cesari M, Palu G, Pessina AC (1995) Expression of the endothelin-converting enzyme gene in human tissues. Biochem Biophys Res Commun 211:249–253

Russell FD, Davenport AP (1995) Characterisation of endothelin (ET) receptors in the human pulmonary vasculature using bosentan, SB209670 and 97–139. J Cardiovasc Pharmacol 26 (S3):S346–S347

Russell FD, Davenport AP (1996) Characterisation of the binding of endothelin ET_B selective ligands in human and rat heart. Br J Pharmacol 119:631–636

Russell FD, Skepper JN, Davenport AP (1997) Detection of endothelin receptors in human coronary artery vascular smooth muscle cells but not endothelial cells using electron microscope autoradiography. J Cardiovasc Pharmacol 29:820–826

Russell FD, Coppell AL, Davenport AP (1998a) *In vitro* enzymatic processing of radiolabelled big ET-1 in human kidney. Biochem Pharmacol 55:697–701

Russell FD, Skepper JN, Davenport AP (1998b) Human endothelial cell storage granules: a novel intracellular site for isoforms of the endothelin converting enzyme. Circ Res 83:314–321

Russell FD, Skepper JN, Davenport AP (1998c) Endothelin peptide and converting enzymes in human endothelium. J Cardiovasc Pharmacol 31 (S1):S19–21

Russell FD, Skepper JN, Davenport AP (1998d) Evidence using immuno-electron microscopy for the regulated and constitutive pathways in the transport and release of endothelin. J Cardiovasc Pharmacol 31:424–430

Russell FD, Davenport AP (1999a) Evidence for intracellular endothelin converting enzyme (ECE-2) expression in human vascular endothelial cells. Circ Res 84: 891–896

Russell FD, Davenport AP (1999b) Secretory pathways in endothelin synthesis. Br J Pharmacol 126:391–398

Saetrum Opgaard OS, Adner M, Gulbenkian S, Edvinsson L (1994) Localization of endothelin immunoreactivity and demonstration of constrictory endothelin-A receptors in human coronary arteries and veins. J Cardiovasc Pharmacol 23:576–583

Saleh D, Furukawa K, Tsao MS, Maghazachi A, Corrin B, Yanagisawa M, Barnes PJ, Giaid A (1997) Elevated expression of endothelin-1 and endothelin-converting enzyme-1 in idiopathic pulmonary fibrosis: possible involvement of proinflammatory cytokines. Am J Respir Cell Mol Biol 16:187–193

Schmidt M, Kroger B, Jacob E, Seulberger H, Subkowski T, Otter R, Meyer T, Schmalzing G, Hillen H (1994) Molecular characterization of human and bovine endothelin converting enzyme. (ECE-1)FEBS Lett 356:238–243

Schmidt-Ott KM, Tuschick S, Kirchhoff F, Verkhratsky A, Liefeldt L, Kettenmann H, Paul M (1998) Single cell characterization of endothelin system gene expression in the cerebellum in situ. J Cardiovasc Pharmacol 31 (S1):S364–366

Schweizer A, Valdenaire O, Nelbock P, Deuschle U, Dumas-Milne-Edwards JB, Stumpf JG, Loffler BM (1997) Human endothelin-converting enzyme (ECE-1): three isoforms with distinct subcellular localizations. Biochem J 328:871–877

Seo B, Oemar BS, Siebenmann R, Von-Segesser L, Luscher TF (1994) Both ET(A) and ET(B) receptors mediate contraction to endothelin-1 in human blood vessels. Circulation 89:1203–1208

Shimada K, Takahashi M, Ikeda M, Tanzawa K (1995a) Identification and characterization of two isoforms of an endothelin-converting enzyme-1. FEBS Lett 371: 140–144

Shimada K, Matsushita Y, Wakabayashi K, Takahashi M, Matsubara A, Iijima Y, Tanzawa K (1995b) Cloning and functional expression of human endothelin-converting enzyme cDNA. Biochem Biophys Res Commun 207:807–812

Spatz M, Kawai N, Merkel N, Bembry J, McCarron RM (1997) Functional properties of cultured endothelial cells derived from large microvessels of human brain. Am J Physiol 272:C231–239

Stanimirovic DB, Yamamoto T, Uematsu S, Spatz M (1994) Endothelin 1 receptor binding and cellular signal transduction in cultured human brain endothelial cells. J Neurochem 62:592–601

Stankova J, D'Orleans Juste P, Rola Pleszczynski M (1996) ET-1 induces IL-6 gene expression in human umbilical vein endothelial cells: synergistic effect of IL-1. Am J Physiol 271:C1073–1078

Takahashi K, Ghatei MA, Jones PM, Murphy JK, Lam HC, O'Halloran DJ, Bloom SR (1991) Endothelin in human brain and pituitary gland: presence of immunoreactive endothelin, endothelin messenger ribonucleic acid, and endothelin receptors. J Clin Endocrinol Metab 72:693–699

Takahashi M, Fukuda K, Shimada K, Barnes K, Turner AJ, Ikeda M, Koike H, Yamamoto Y, Tanzawa K (1995) Localization of rat endothelin-converting enzyme to vascular endothelial cells and some secretory cells. Biochem J 311:657–665

Takahashi T, Barnes PJ, Kawikova I, Yacoub MH, Warner TD, Belvisi MG (1997) Contraction of human airway smooth muscle by endothelin 1 and IRL 1620: effect of bosentan. Eur J Pharmacol 324:219–222

Thorin E, Nguyen TD, Bouthillier A (1998) Control of vascular tone by endogenous endothelin 1 in human pial arteries. Stroke 29:175–180

Turner AJ, Barnes K, Schweizer A, Valdenaire O (1998) Isoforms of endothelin-converting enzyme: why and where? Trend Pharmacol Sci 19:483–486

Tonnessen T, Naess PA, Kirkeboen KA, Offstad J, Ilebekk A, Christensen G (1993) Endothelin is released from the porcine coronary circulation after short-term ischemia. J Cardiovasc Pharmacol 22:S313–16

Valdenaire O, Rohrbacher E, Mattei MG (1995) Organization of the gene encoding the human endothelin-converting enzyme (ECE-1). J Biol Chem 270:29794–29798

Warner TD, Schmidt HH, Kuk J, Mitchell JA, Murad F (1992) Human brain contains a metalloprotease that converts big endothelin-1 to endothelin-1 and is inhibited by phosphoramidon and EDTA. Br J Pharmacol 106:505–506

Watanabe Y, Naruse M, Monzen C, Naruse K, Ohsumi K, Horiuchi J, Yoshihara I, Kato Y, Nakamura N, Kato M, Sugino N, Demura H (1991) Is big endothelin converted to endothelin-1 in circulating blood? Cardiovasc Pharmacol 17 (S7):S503–505

Waxman L, Doshi KP, Gaul SL, Wang S, Bednar RA, Stern AM (1994) Identification and characterization of endothelin converting activity from EAHY 926 cells: evidence for the physiologically relevant human enzyme. Arch Biochem Biophys 308:240–253

White DG, Garratt H, Mundin JW, Sumner MJ, Vallance PJ, Watts IS (1994) Human saphenous vein contains both endothelin ET(A) and ET(B) contractile receptors. Eur J Pharmacol 257:307–310

Winkles JA, Alberts GF, Brogi E, Libby P (1993) Endothelin-1 and endothelin receptor mRNA expression in normal and atherosclerotic human arteries. Biochem Biophys Res Comm 191:1081–1088

Xu D, Emoto N, Giaid A, Slaughter C, Kaw S, deWit D, Yanagisawa M (1994) ECE-1: A membrane- bound metalloprotease that catalyses the proteolytic activation of big endothelin-1. Cell 78:473–485

Yamaga S, Tsutsumi K, Niwa M, Kitagawa N, Anda T, Himeno A, Khalid H, Taniyama K, Shibata S (1995) Endothelin receptor in microvessels isolated from human meningiomas: quantification with radioluminography. Cell Mol Neurobiol 15: 327–340

Yanagisawa H, Yanagisawa M, Kapur RP, Richardson JA, Williams SC, Clouthier DE, de Wit D, Emoto N, Hammer RE (1998) Dual genetic pathways of endothelin-mediated intercellular signaling revealed by targeted disruption of endothelin converting enzyme-1 gene. Development 125:825–836

Yu JCM, Pickard JD, Davenport AP (1995) Endothelin ET_A receptor expression in human cerebrovascular smooth muscle cells. Br J Pharmacol, 116, 2441–2445

Ziche M, Morbidelli L, Donnini S, Ledda F (1995) ET_B receptors promote proliferation and migration of endothelial cells. J Cardiovasc Pharmacol 26 (S3):S284–286

Zhang M, Olsson Y (1995) Reactions of astrocytes and microglial cells around hematogenous metastases of the human brain. Expression of endothelin like immunoreactivity in reactive astrocytes and activation of microglial cells. J Neurol Sci 134:26–32

CHAPTER 9

Endothelin Receptor Antagonists

J.D. ELLIOTT and J.-N. XIANG

A. Peptide Endothelin Receptor Antagonists

I. ET_A Selective Antagonists

The earliest reported endothelin (ET) receptor antagonists were peptide in nature and arose structurally from relatively minor modifications to endothelin-1 (ET-1) (SPINELLA 1991). The most significant early departure from this paradigm came with the discovery of BE-18257B, a competitive and selective endothelin-A (ET_A) receptor antagonist isolated from the fermentation products of *Streptomyces misakiensis* (IHARA 1991). BE-18257B is a cyclic pentapeptide, cyclo(D-Trp-D-Glu-Ala-D-Allo-Ile-Leu), which binds selectively to ET_A receptors with an IC_{50} of 1.4 µmol/l (porcine aortic smooth muscle cells) and functionally antagonizes ET-1 induced vasoconstriction in the rabbit iliac artery with a pA_2 of 5.9. Shortly after the discovery of BE-18257B, a similar series of cyclic pentapeptides was identified from a strain of *Streptomyces* (MIYATA 1992). The most potent compound in this series, WS-7338B, is structurally identical to BE-18257B. *In vivo* evaluation in the spontaneously hypertensive rat showed that pretreatment with 10 mg/kg of WS-7338B reduced the pressor effect of ET-1 (3.2 µg/kg; 55% inhibition), while leaving unaffected the initial depressor response, the pressor and depressor responses being mediated by ET_A and endothelin-B (ET_B) receptors, respectively. While the selectivity profile of BE-18257B is attractive, the moderate activity and poor water solubility of the compound have restricted its use as a tool in pharmacological studies. Chemical modification of BE-18257B led to a series of antagonists which not only exhibit higher affinity for the ET_A receptor but also possess good solubility in aqueous solution. Thus BQ 123 (Fig. 1), cyclo(D-Trp-D-Asp-Pro-D-Val-Leu), emerged as a high affinity ET_A receptor antagonist (IC_{50} of 22 nmol/l) with greatly enhanced water solubility (IHARA 1992; ISHIKAWA 1992). The significantly lower affinity of this compound to the ET_B receptor (IC_{50} of 18 µmol/l) was demonstrated using porcine cerebellum, a tissue rich in this subtype, indicating that BQ-123 is almost 1000-fold selective for ET_A receptors. Functionally, BQ-123 antagonizes ET-1-induced contraction of porcine coronary arteries with a pA_2 of 7.4, a value that is in accord with its IC_{50} for binding to ET_A receptors. *In vivo*, BQ-123 produces a

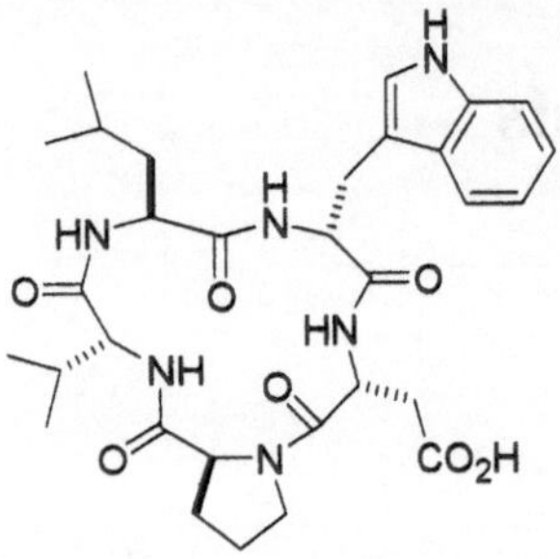

240																																																																																																																																																																																																																																																																																																																																																																																																																																																																																																																																																																																																																																																																																																																																																																																																																																																																																																																																																																																																																																																																																																																																																																																																																																																																																																																																																																																																																																																																																																																																																																																																																																																																																																																																																																																																																																																																																																																																																																																																																																																																																																																																																																																																																																																																																																																																																																																																																																																																																																																																																																																																																																																																																																																																																																																																																																																																																																																																																																																																																																																																																																																																																																																																																																																																																																																																																																																																																																																																																																																																																																																																																																																																				J.D. ELLIOTT and J.-N. XIANG

Fig. 1. ET_A-selective peptide antagonist BQ-123

dose-dependent decrease in the ET-1-induced pressor response in the conscious rat, without affecting the initial depressor response, further indicative of selectivity for the ET_A receptor (IHARA 1992). Due to its high potency, selectivity, and aqueous solubility, BQ-123 has been widely used in the investigation of the distribution of ET receptor subtypes. Furthermore, BQ-123 has been shown to be efficacious in *in vivo* models of disease, most notably hypertension (OHLSTEIN 1993) and acute renal failure (GELLAI 1994). More recently, evidence that treatment with BQ-123 markedly ameliorates the long-term survival and hemodynamic parameters in rats with chronic heart failure (SAKAI 1998) provides support for a role for ET in the etiology of this condition.

Solution conformations derived from NMR studies for BQ-123 (BEAN 1994; KRYSTE 1992) and for the related compounds cyclo[D-Trp-D-Asp-Pro-D-Val-(N-Me)Leu] [9] and cyclo[D-Trp-Cys($SO_3^-Na^+$)-Pro-D-Val-Leu], BQ-153 (BOGUSKY 1994) have been reported. In both hydrophilic and hydrophobic solvents, these compounds exhibit a highly stable backbone conformation composed of a type II β-turn containing leucine and tryptophan in the i+1 and i+2 positions, respectively, and a γ-turn centered on proline. Side-chain positions in aqueous solvent, methanol, and dimethyl sulfoxide are quite ordered, particularly about Leu-D-Trp where a strong hydrophobic interaction exists between the leucine side chain and the indole ring. In chloroform, however, the side chains of BQ-123 have a significantly higher mobility, showing no preferred orientation (BEAN 1994). The conformational stability of BQ-123 suggests a potential peptidomimetic strategy for the identification of non-peptide endothelin receptor antagonists. To date, however, this approach has only met with limited success (MURUGESAN 1995).

A number of linear tripeptide antagonists have been obtained as a result of structure-activity relationship studies based upon BQ-123. Representative examples include BQ-610, FR 139317 (ARAMORI 1993) and PD 151242 (DAVENPORT 1994) (Fig. 2). BQ-610, {N-[(hexahydro-1-azepinyl)carbonyl)]}-L-Leu-D-(1-CHO)Trp-D-Trp, inhibits ET-1 binding to ET_A receptors in porcine aortic smooth muscle ($IC_{50} = 0.7$ nmol/l), and has significantly lower affinity for ET_B receptors in porcine cerebellum ($IC_{50} = 24 \mu$mol/l). Functionally, BQ-610

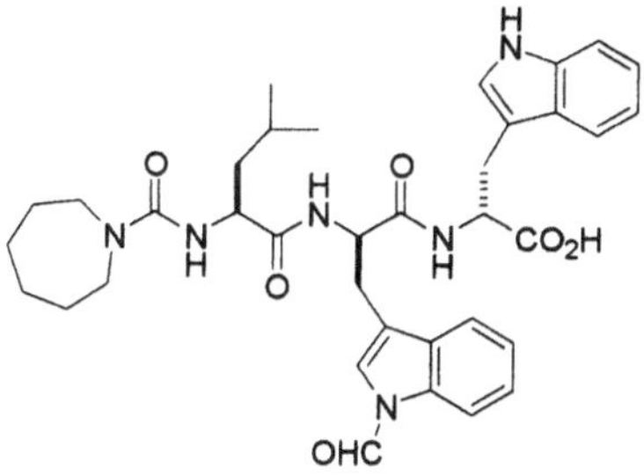

BQ 610 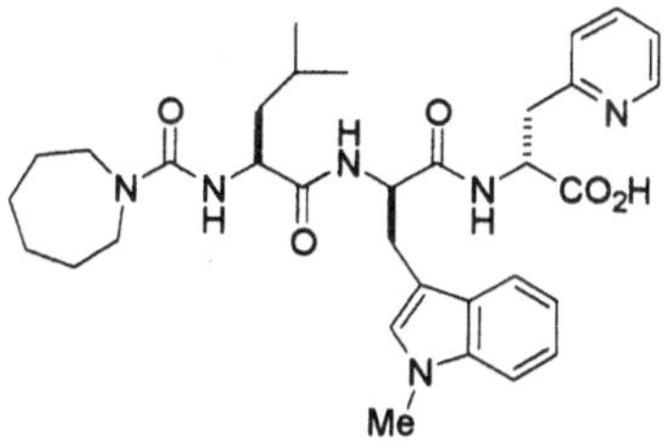   FR139317

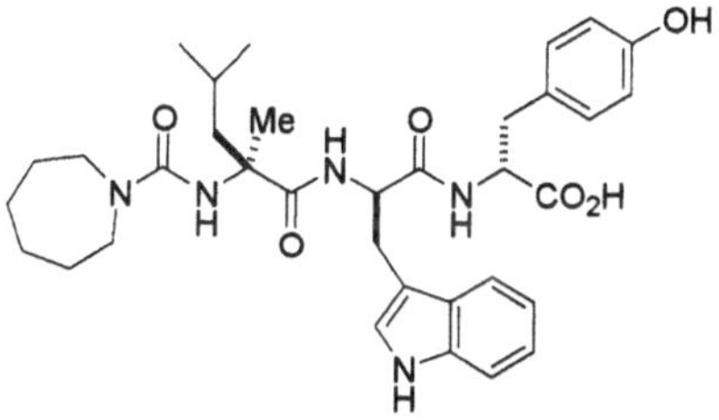

PD 151242

Fig. 2. Structures of linear tripeptide antagonists; BQ 610, FR139317, and PD151242

antagonizes ET-1 induced contraction in porcine coronary arteries with a pA_2 value of 8.2. In a study *in-vivo* in the rat, selective ET_A receptor blockade by BQ-610 improved hemodynamics in the intact circulation by causing a reduction in afterload and an increase in myocardial contractility (BEYER 1998). BQ-610 has no effect on blood pressure and heart rate, but causes vasodilatation with a corresponding increase in stroke volume, cardiac output, and ejection fraction. These positive inotropic effects may be mediated by endogenous ET-1 acting at ET_B receptors, following selective ET_A receptor blockade. Another important example of the linear tripeptide series is FR 139317, {N-[(hexahydro-1-azepinyl)carbonyl]}-Leu-D-(1-Me)Trp-D-Pya [Pya is (2-pyridyl)alanine] (ARAMORI 1993). FR 139317 is a potent, highly selective ET_A antagonist with K_i values of 1 nmol/l and 7 μmol/l, respectively, for the cloned human ET_A and ET_B receptors. FR 139317 functionally inhibits ET-1 induced contraction of the rabbit aorta (pA_2=7.2) and in conscious normotensive rats FR 139317 inhibits, in a dose-dependent manner, the ET-1 induced pressor response. In an experimental canine model of subarachnoid hemorrhage, intracisternal administration of FR 139317 significantly inhibits vasoconstriction of the basilar artery, suggesting a role for ET in this pathological condition (CARDELL 1993).

Linear hexapeptide antagonists have also been reported to have been derived from FR 139317. In common with the tripeptide series, these hexa-

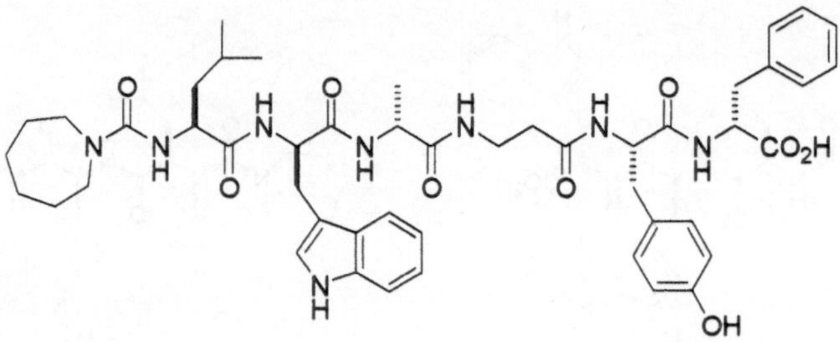

Fig. 3. Structure of TTA-386

peptides contain the alkylmethyleneiminocarbonyl group at the N-terminus, D-amino acids and an aromatic side chain at selected positions. All of these features are important for receptor binding affinity. A representative compound in this series is TTA-386 (Fig. 3), {N-[(hexahydro-1-azepinyl)carbonyl]}-Leu-D-Trp-D-Ala-β-Ala-Tyr-D-Phe) (KITADA 1993), which exhibits high affinity to ET_A receptors (IC_{50} value of 0.34 nmol/l, porcine cardiac ventricular muscle) and has no appreciable binding to ET_B receptors in bovine whole brain. This compound is a competitive antagonist of ET_A receptors in porcine cardiac ventricles as confirmed by Scatchard analysis. Functional activity for TTA-386 and selectivity for ET_A receptors was demonstrated in A-10 cells, which respond to ET-1 by an increase of cytosolic free calcium concentration. TTA-386 (10 mmol/l) completely inhibits calcium mobilization induced by ET-1 (1 nmol/l) in this cell line, while TTA-386 alone has no effect on calcium levels in these cells.

II. ET_B Selective Antagonists

In order to help define a pharmacological role for the ET_B receptor subtype, efforts have also been directed towards the discovery of ET_B selective receptor antagonists. The first peptide ET_B selective antagonist was designed based on the observations that all three ET isopeptides bind with equivalent affinity to the ET_B receptor, and that the 14 C-terminal residues are essentially conserved in ETs 1–3. Thus, IRL 1038 was prepared, which corresponds to residues 11–21 of ET-1 (URADE 1992) in which the cysteines at positions 11 and 15 are linked via a disulfide bridge not present in the intact peptides ET 1–3 {i.e., $[Cys^{11}-Cys^{15}]$-ET-1(11–21)}. IRL 1038 has much higher affinity for ET_B receptors ($K_i = 6$–11 nmol/l) than for ET_A receptors ($K_i = 0.4$–0.7 μmol/l). Functionally, IRL 1038 (3 μmol/l) antagonizes ET-3-induced contraction of guinea pig ileal and tracheal smooth muscle mediated by the ET_B receptor. The compound itself is devoid of agonist activity and has no effect on the ET_A receptor-mediated contraction of rat aortic smooth muscle.

During the extensive structure-activity relationship studies conducted on linear tripeptides (see above), it was observed that replacement of the aro-

Fig. 4. BQ-788, an ET_B selective tripeptide antagonist

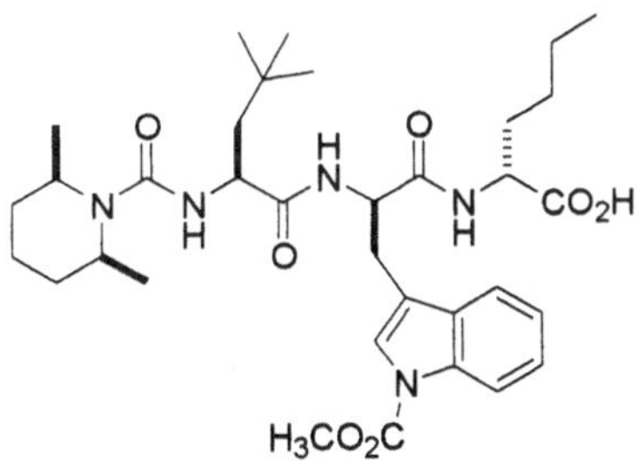

matic residue at the C-terminus (e.g., tryptophan in BQ 610) (Fig. 2) with an alkyl amino acid selectively increased affinity for the ET_B receptor. As a result of further elaboration, the potent and selective ET_B receptor antagonist, BQ-788 [N-cis-2,6-dimethylpiperidinocarbonyl-L-γ-MeLeu-D-Trp(1-CO$_2$Me)-D-Nle] (Fig. 4) was discovered (ISHIKAWA 1994). BQ-788 competitively inhibits [^{125}I]ET-1 binding to the ET_B receptor on human Girardi heart cells with an IC_{50} value of 1.2 nmol/l and shows poor affinity for ET_A receptors (IC_{50} = 1.3 μmol/l) on SK-N-MC cells, a human neuroblastoma cell line. Furthermore, BQ-788 is highly selective for the ET_B receptor as evidenced by its lack of ability to affect the binding of the physiologically relevant ligand to the angiotensin II (AII) AT_1 and calcitonin receptors. Functionally, BQ-788 antagonizes the contractions induced by a selective ET_B-agonist, BQ-3020, in isolated rabbit pulmonary arteries (a tissue rich in ET_B receptors) with a pA_2 value of 8.4. BQ-788 alone at a concentration of 10 mmol/l shows no effect on basal tension in this tissue. *In vivo* in the rat, BQ-788 (1 mg/kg, *i.v.*) completely suppresses the transient depressor response produced by ET-1, resulting in the more rapid onset of an enhanced pressor response. This result supports the contention that the transient depressor response to ET-1 in rats is mediated by the ET_B receptor and substantiates the value of receptor subtype selective antagonists for such pharmacological investigations. Contemporaneously with the discovery of BQ-788, an ET_B selective cyclic peptide antagonist, RES-701-1 (Fig. 5), was isolated from a fermentation broth of *Streptomyces* (MORISHITA 1994). RES-701-1 binds to CHO cells transfected with the human ET_B receptor with an IC_{50} value of 10 nmol/l and has no effect on [^{125}I]ET-1 binding to the human ET_A receptor. This compound selectively inhibits the ET-1-induced increase in intracellular Ca^{2+} concentration in COS-7 cells expressing the ET_B receptor, but does not inhibit the Ca^{2+} transient in ET_A-expressing cells. In anesthetized rats, RES-701-1 (250 nmol/kg) abolishes the initial depressor response to ET-1 but enhances the subsequent pressor response, corroborating the *in-vitro* data which suggest that RES-701-1 is a potent and specific antagonist for the ET_B receptor. Recent data with synthetic RES-701-1 (HE 1995) suggest a much weaker binding affinity and poorer selectivity for the ET_B receptor than those results obtained with RES-701-1 from natural sources. Further work will be needed to resolve this anomaly.

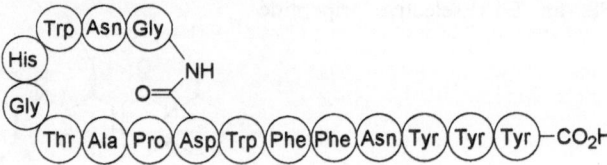

Fig. 5. RES-701-1, an ET_B selective peptide obtained from a fermentation broth

III. ET_A/ET_B Balanced Antagonists

During the course of SAR studies on ET-1 itself, certain analogs were prepared such as ET-1[Thr[18], γ-Me-Leu[19]] and ET-1[Thr[18], Leu[19]] which demonstrated non-selective ET_A/ET_B antagonism (Wakimasu 1993). The hydrophobic character of the C-terminal region of the ET isopeptides is crucial for receptor recognition, and truncation of the ET-1 sequence led to the discovery that the C-terminal hexapeptide, ET 16–21, possesses noticeable, albeit, low-affinity for both ET_A and ET_B receptors (Maggi 1989). A *mono*-D-amino acid scan of the C-terminal hexapeptide revealed that the most significant increase in binding affinities to both ET_A and ET_B receptors was achieved with a D-His residue at position 16. Thus, Ac-D-His-Leu-Asp-Ile-Ile-Trp is a non-selective antagonist with IC_{50} values for binding to ET_A (rabbit renal artery vascular smooth muscle cells) and ET_B receptors (rat cerebellum) of $8.9\,\mu mol/l$ and $9.1\,\mu mol/l$, respectively (Doherty 1993). Replacement of D-His with D-Phe at position 16 maintains the nonselective profile of the antagonist, while moderately (four-to fivefold) enhancing the affinity to both receptor subtypes. Based on these observations, extensive studies around the 16-position were carried out in which D-Phe was replaced with various D-aromatic amino acids (Doherty 1993; Cody 1992a,b, 1993). As a result it was discovered that increased steric bulk at this position provides a series of compounds with substantially increased binding affinities for both receptor subtypes. Thus, PD 142893 (Fig. 6), Ac-D-Dip-Leu-Asp-Ile-Ile-Trp (Dip = 3,3-diphenylalanine), is a functional antagonist and possesses high affinities for both ET_A (cultured rabbit renal artery vascular smooth muscle cells, $IC_{50} = 31$ nmol/l) and ET_B (rat cerebellum membranes, $IC_{50} = 54\,nmol/l$) receptors. Furthermore, PD 142893 antagonizes ET-1-induced contraction of the rabbit femoral (ET_A) and pulmonary (ET_B) arteries with pA_2 values of 5.95 and 6.31, respectively (Hingorani 1992; LaDouceur 1992; Warner 1993). Further enhancement of binding affinity at both receptors was obtained by the substitution of the unnatural tricyclic amino acid, D-Bhg, $2(R)$-10,11-dihydro-5H-dibenzo[a,d]cycloheptene glycine, at position 16, leading to PD 145065 (Fig. 6). This molecule is a linear hexapeptide which binds to ET_A and ET_B receptors with IC_{50} values of 3.5 nmol/l and 15 nmol/l, respectively. PD 145065 is also a competitive antagonist, which inhibits the ET-1- and sarafotoxin-6c-induced contractions in rabbit femoral (ET_A) and pulmonary arteries (ET_B),

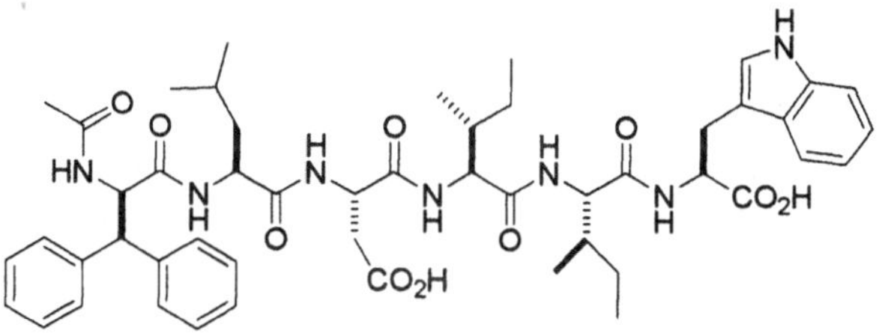

PD 142893

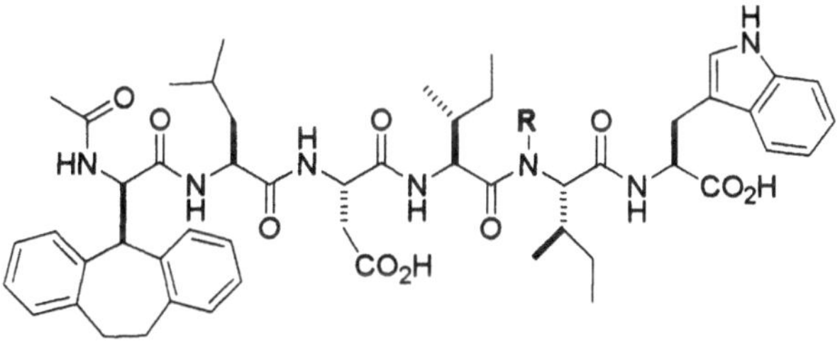

R = H, PD 145065
R = Me, PD 156252

Fig. 6. Balanced ET_A/ET_B hexapeptide endothelin receptor antagonists

with pA_2 values of 6.9 and 7.1, respectively. *In vivo*, in ganglion-blocked rats, PD 145065 inhibits, in a dose-dependent manner, both the depressor and pressor responses induced by infusion of ET-1.

While balanced (ET_A/ET_B) antagonists such as PD 142893 and PD 145065 possess quite potent receptor binding characteristics, they are however relatively unstable towards enzymatic degradation, as evidenced *in vitro* in a rat intestinal perfusate assay. This instability is probably due to carboxypeptidase activity and, indeed, co-incubation of PD 145065 with carboxypeptidase inhibitors greatly increases its half-life in the rat intestinal perfusate. An N-Me amino acid scan revealed that N-methylation of Ile20 provided a compound (Ac-D-Bhg16-Leu-Asp-Ile-[NMe]Ile-Trp21, PD 156252 [Fig. 6]) which retained full receptor affinity at both ET receptor subtypes while possessing enhanced proteolytic stability and cellular permeability (CODY 1997).

B. Non-Peptide Endothelin Receptor Antagonists

I. ET_A/ET_B Balanced Antagonists

While peptide endothelin receptor antagonists were important tools for early pharmacological studies, a very significant advance in the field came with the disclosure of the non-peptide antagonist Ro 46-2005. Ro 46-2005 (Fig. 7) was

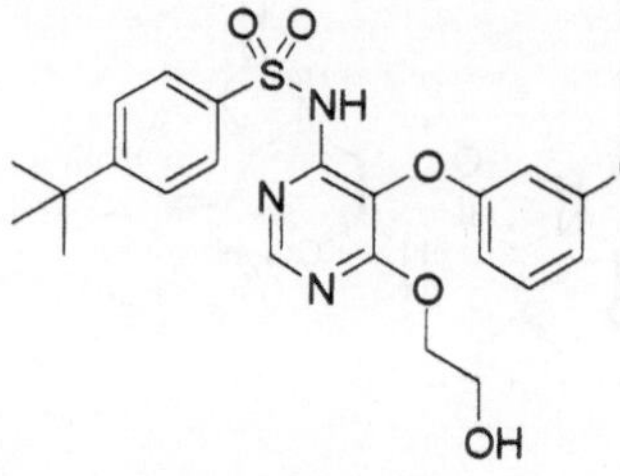

Ro 46-2005 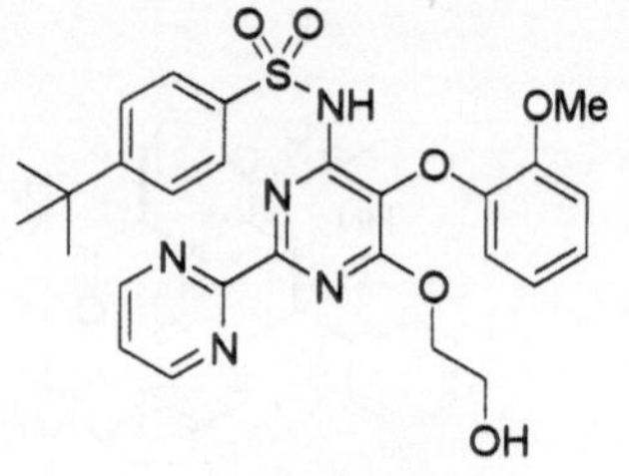   bosentan

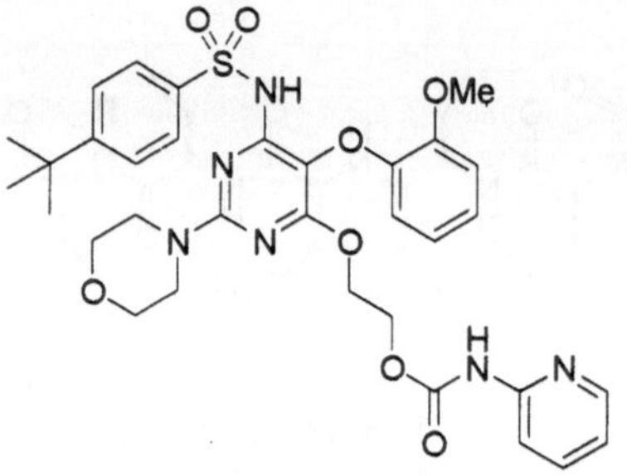

Ro 48-5695

Fig. 7. Pyrimidine-based, orally bioavailable, balanced ET_A/ET_B receptor antagonists; Bosentan, the first non-peptide agent advanced to human clinical trials

described as the first orally active endothelin receptor antagonist (CLOZEL 1993) and was the result of structural modification of a lead first synthesized as part of an antidiabetic effort. Ro 46-2005, which is devoid of hypoglycemic activity, inhibits the binding of $[^{125}I]ET$-1 to human vascular smooth muscle cells (ET_A receptors) and rat aortic endothelial cells (ET_B receptors) with IC_{50} values of 220 nmol/l and 1 μmol/l, respectively. The *in vitro* activity of Ro 46-2005 was further evaluated in functional assays for both ET_A- and ET_B-mediated effects. Thus, Ro 46-2005 produces a parallel rightward shift in the concentration-response curve to ET-1 in endothelium-denuded isolated rat aortic rings (ET_A receptors), with no effect on maximal response. The pA_2 value for Ro 46-2005 in this tissue is 6.50 and Schild analysis of the data is consistent with competitive antagonism (slope of regression = 1.05). Functional activity at the ET_B receptor was evaluated in isolated rat small mesenteric arteries which were preconstricted with methoxamine. Under these conditions, Ro 46-2005 produced a parallel rightward shift in the concentration-response curve to sarafotoxin 6c, with no depression of the maximal response, yielding a pA_2 value of 6.47. Schild analysis of the data was once again consistent with competitive antagonism. Additional experiments confirmed the selectivity of Ro 46-2005 for endothelin receptors in that, at a concentration of up to

$30\,\mu$mol/l, the compound did not affect contractions evoked by angiotensin II, serotonin, norepinephrine, potassium chloride, or prostaglandin $F_{2\alpha}$.

Subsequent structure-activity relationships in this series led to the discovery of Ro 47–0203, bosentan (Fig. 7), a more potent, balanced endothelin receptor antagonist than Ro 46-2005 (CLOZEL 1994). Bosentan binds to CHO cells separately expressing ET_A and ET_B receptors with K_i values of 6.5 nmol/l and 343 nmol/l, respectively. *In vitro*, bosentan blocks, in a competitive manner, both the ET_A mediated functional response to ET-1 in the rat aorta ($pA_2 = 7.2$) and the ET_B mediated response to sarafotoxin 6c in rat trachea ($pA_2 = 6$). Bosentan shows no inherent effects on blood pressure from *in vivo* studies in the pithed rat, but when dosed intravenously or orally does inhibit the pressor response to big ET-1, co-administered intravenously. Furthermore, it has been demonstrated clinically that bosentan reduces blood pressure in humans with essential hypertension, as effectively as enalapril, and is not associated with reflex neurohormonal activation (KRUM 1998) (see Chap. 19). Bosentan has been advanced to phase III clinical trials in congestive heart failure patients and early results suggest beneficial effects on hemodynamic parameters (KIOWSKI 1995).

Since bosentan possesses only moderate affinity for ET_A and ET_B receptors, further optimization around this scaffold has been examined, guided by a 3-D model of antagonist-receptor interaction (BREU 1995). The focus of the effort has centered around modifications of the hydroxy substituent incorporated in the 6-position, as well as the size and lipophilicity of the substituents in the 2- and 4-positions, of the central pyrimidine ring. As a result, Ro 48–5695 (Fig. 7) was identified as the most potent, balanced ET_A/ET_B antagonist in this series. Ro 48–5695 binds to ET_A and ET_B receptors separately expressed in CHO cells with IC_{50} values of 0.7 nmol/l and 5 nmol/l, respectively. Functionally, Ro 48–5695 blocks both the ET_A mediated functional response to ET-1 in rat aorta ($pA_2 = 9.3$) and the ET_B mediated response to sarafotoxin 6c in rat trachea ($pA_2 = 7.6$), showing an approximately 100-fold enhancement for both receptors relative to bosentan. This compound also inhibits intracellular calcium mobilization induced by ET-1 in both ET_A and ET_B receptor transfected HEK-293 cells, with IC_{50} values of 0.9 nmol/l and 6 nmol/l, respectively. Despite the low water solubility (sodium salt, 2 mg/l) of Ro 48–5695, the compound proved to be orally active in both dogs and rats with 40–60% bioavailability and showed good *in vivo* activity in a variety of animal models (NEIDHART 1997).

The lead structure SK&F 66861 (Fig. 8) arose through screening a selection of compounds similar in structure to known ligands of G-protein coupled receptors (ELLIOTT 1994). SK&F 66861 is a diphenylindene carboxylic acid which binds to ET_A receptors in the rat mesenteric artery but has no detectable affinity for ET_B receptors in rat cerebellum. Subsequent studies with cloned human endothelin receptors gave K_i values of $7.3\,\mu$mol/l and $>30\,\mu$mol/l, for ET_A and ET_B receptors, respectively. Significantly, SK&F 66861 was found to be a functional antagonist of endothelin-1 at the ET_A receptors

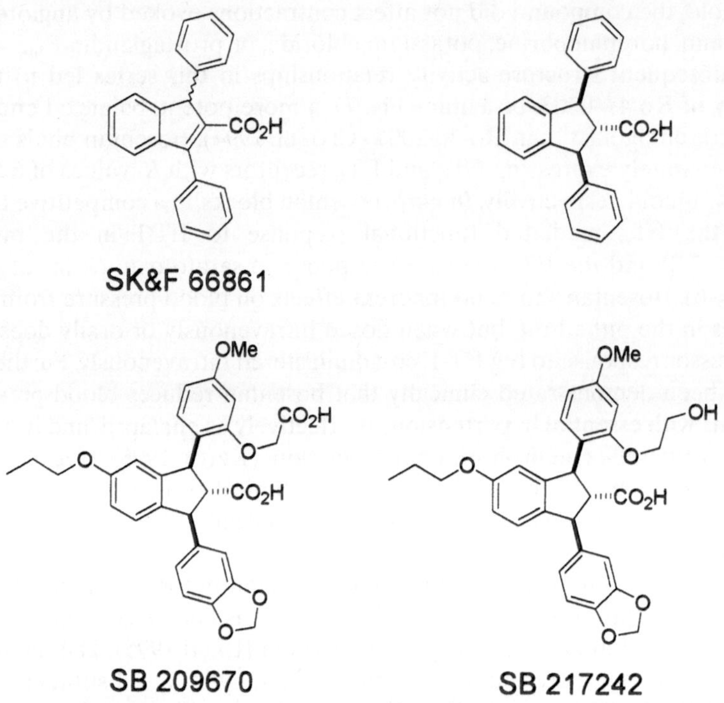

SK&F 66861

SB 209670   **SB 217242**

Fig. 8. Non-selective indane endothelin receptor antagonists

in rat aorta (K_b = 6.6 μmol/l). The oxidative lability of the indene nucleus present in SK&F 66861 led to investigation of the indane counterpart and *trans,trans*-1,3-diphenylindane-2-carboxylic acid (Fig. 8) was shown to have a receptor binding profile similar to SK&F 66861 itself.

The evolution of the indane series of ET receptor antagonists, relied heavily on peptidomimetic hypotheses and took advantage of a ^{1}H-NMR derived structure of ET-1 (ENDO 1989; MILLS 1991a,b; SAUDEK 1989). Specifically, elaboration of the indane structure was directed by the hypothesis that the 1 and 3 phenyls, and the 2 carboxylic acid are mimetics of Phe14, Tyr13, and Asp18 of ET-1, respectively. Introduction of a carboxylic acid moiety onto the 3-phenyl of the small molecule to mimic the C-terminus of ET-1 led to the very potent balanced ET_A/ET_B antagonist, SB 209670 (Fig. 8) SB 209670 was the first reported subnanomolar, non-peptide antagonist of the human ET_A receptor and the compound has substantial, albeit weaker, affinity for the ET_B receptor (K_i values: ET_A = 0.43 nmol/l; ET_B = 14.7 nmol/l) (OHLSTEIN 1994). Functionally, this ligand produces parallel rightward shifts in the concentration-response curve to ET-1 in isolated rat aorta (ET_A-mediated) with a K_b value of 0.4 nmol/l. The maximal contractile response to ET-1 is not affected by SB 209670 at concentration up to 1 μmol/l and Schild analysis of the data is consistent with competitive antagonism. Likewise, SB 209670 competitively antagonizes ET-1-induced contraction mediated by ET_B receptors in the isolated rabbit pulmonary artery with a somewhat weaker K_b value of 199 nmol/l. Furthermore, SB 209670 does not possess noticeable affinity for a range of other G-protein coupled receptors including angiotensin II AT_1, vasopressin V_1, and α-adrenergic.

When administered intravenously, SB 209670 is effective in a number of animal models of disease including hypertension (see Chap. 16), acute renal failure (GELLAI 1994; see Chap. 18), and restenosis following percutaneous transluminal coronary angioplasty (DOUGLAS 1994; see Chaps. 12, 16). Although efficacious as an intravenous agent, SB 209670 has limited oral bioavailability in the rat (4%). In an effort to enhance the oral bioavailability of compounds of this class, screening of a series of structurally related indane antagonists for enhanced permeability in intestinal tissues was conducted. As a result, SB 217242 (Fig. 8) was discovered to have substantially increased oral bioavailability in the rat (66%). SB 217242 binds with high affinity to human ET_A receptors (K_i = 1.1 nmol/l), and is somewhat more selective versus the ET_B subtype (K_i = 111 nmol/l) than SB 209670. Functionally, SB 217242 antagonizes ET-1-induced contraction of the isolated rat aorta, mediated by ET_A receptors, with a K_b value of 4.4 nmol/l. SB 217242 is currently in phase II clinical trials for the treatment of pulmonary hypertension secondary to COPD.

Another approach to identify lead ET antagonist structures relied upon the recognition of structural and functional similarities between the ET and AII receptor-ligand systems. Furthermore, these two G-protein coupled receptors exhibit similar preferences for aromatic and acidic groups in their respective agonists. Thus, screening of compounds previously prepared as AII antagonists led to the discovery of the ET_A/ET_B antagonist containing a 2,5-dibromo-3,4-dimethoxyphenoxyacetic acid, shown in Fig. 9 (WALSH 1995). This compound showed weak binding affinities to both ET_A and ET_B receptors with IC_{50} values of 11 μmol/l and 31 μmol/l, respectively. In common with other ET antagonist series, a significant potency enhancement accrues from the introduction of a methylenedioxyphenyl moiety. Replacement of the carboxylic acid present in this lead by an acylsulfonamide, in consideration of potential DMPK advantages, led to L-746072 (Fig. 9), which binds to cloned human ET_A and ET_B receptors with IC_{50} values of 26 nmol/l and 60 nmol/l, respectively. Interestingly, along with the dramatically enhanced binding affinity to ET receptors, L-746072 retains potent affinity for both AT_1 and AT_2 receptors ($IC_{50}s$ = 13 nmol/l and 32 nmol/l, respectively). Thus, L-746072 is a relatively potent antagonist of both subtypes of the ET and AII receptors and possesses an intriguing profile for a potential antihypertensive agent.

Attempts to simplify the structure of L-746072 led to discovery of L-754142 (Fig. 9), which shows high specificity for ET receptors (WILLIAMS 1995). L-754142 binds to cloned human ET_A and ET_B receptors with K_i values of 0.062 nmol/l and 2.25 nmol/l, respectively, and has no noticeable affinity for either the AT_1 or AT_2 receptors. *In vitro*, L-754142 antagonizes ET-1-induced phosphatidyl inositol hydrolysis in Chinese hamster ovary (CHO) cells sepa-

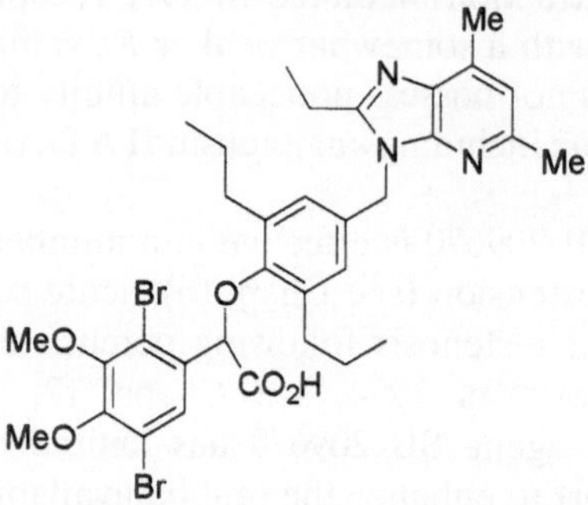

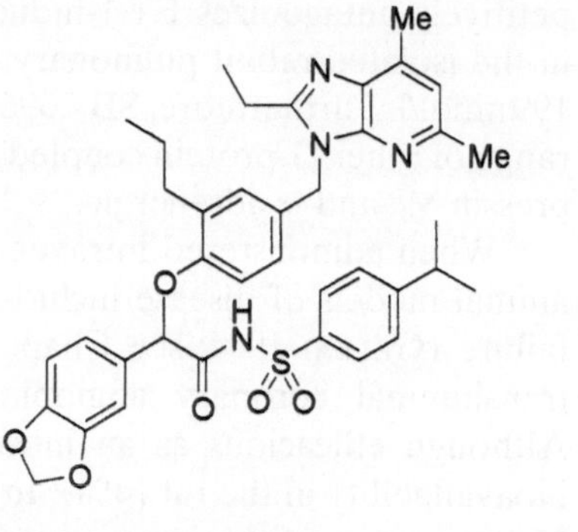

L-746072

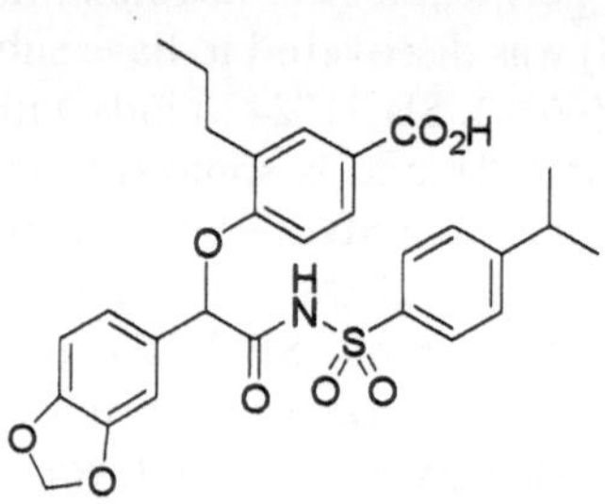

L-754142

Fig. 9. α-Phenoxyacetic acids and the corresponding acylsulfonamides, potent balanced ET_A/ET_B antagonists

rately expressing cloned human ET receptors (IC_{50}s, $ET_A = 0.35$ nmol/l, $ET_B = 26$ nmol/l). Furthermore, the compound inhibits ET-1-induced contraction of rabbit iliac artery rings ($pA_2 = 7.74$) and rat aortic rings ($pA_2 = 8.7$), in both tissues the effect being mediated via ET_A receptors. L-754142 blocks the pressor response to big ET-1 in the conscious rat with ED_{50} values of 0.3 mg/kg i.v. and 0.56 mg/kg p.o. Finally, drug metabolism and pharmacokinetic studies in the rat support a long duration of action (>12 h) for L-754142, following oral dosing at 3 mg/kg.

The prototypical pyrrolidine-3-carboxylic acid antagonists (Fig. 10) were developed based upon the indane series described earlier and they possess ET_A selectivity (see below). Additional SAR studies in this series, however, revealed that transposition of the carbonyl group of the amide in A-147627 along the side-chain led to an antagonist with a significantly diminished affinity to ET_A receptor (56-fold), thereby rendering the compound less subtype selective (IC_{50}s, $ET_A = 20$ nmol/l, $ET_B = 850$ nmol/l). Further pursuit of this finding revealed that the corresponding sulfonamides, represented by A-182086 (Fig. 10), have excellent affinities for both ET_A and ET_B receptors. A-182086 is a potent balanced ET_A/ET_B antagonist with subnanomolar affinities

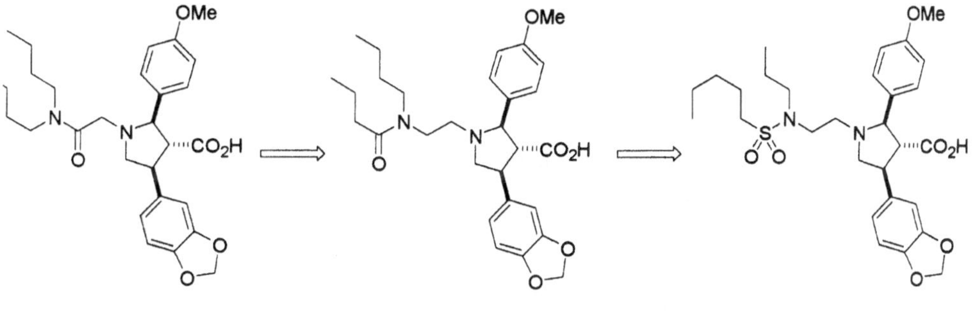

Fig. 10. Development of pyrrolidine-based ET_A/ET_B dual antagonist A-182086

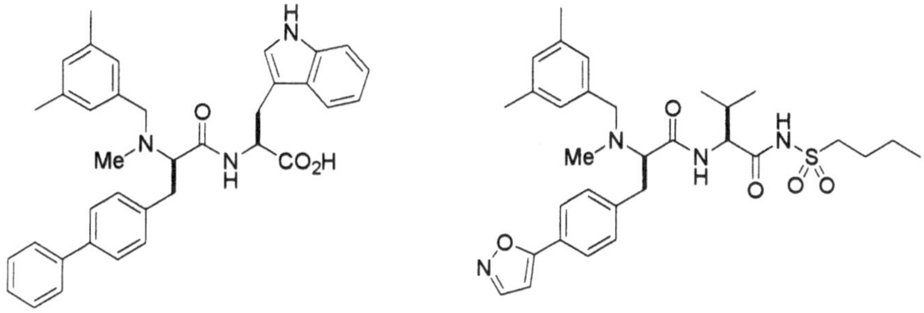

Fig. 11. IRL 3630, a balanced antagonist, resulted from modification of IRL 2500, an ET_B selective antagonist

to both receptors ($IC_{50}s$, $ET_A = 0.1\,nmol/l$, $ET_B = 0.3\,nmol/l$). The pharmacokinetic characteristics of A-182086 are also excellent with a bioavailability of 54% and $T_{1/2}$ of 8h when dosed orally in the rat (JAE 1997).

By way of contrast with A-182086 the balanced ET_A/ET_B antagonist IRL 3630 (Fig. 11) was developed based upon an earlier discovery of the ET_B selective antagonist, IRL 2500 (K_is, $ET_A = 440\,nmol/l$, $ET_B = 1\,nmol/l$) (FRUH 1996; SAKAKI 1998a,b). Structural modification of IRL 2500 indicated that the affinity to the ET_A receptor could be enhanced by replacement of the biphenyl moiety with a 4-isoxazolyl phenyl group. Subsequent replacement of the tryptophan residue by valine and the carboxylic acid present in IRL 2500 by an acylsulfonamide led to the balanced ET_A/ET_B antagonist IRL 3630 (Fig. 11, K_is, $ET_A = 1.5\,nmol/l$, $ET_B = 1.2\,nmol/l$). This molecule, despite its "peptide-like" character, is described as being resistant to degradation by plasma in rat, mouse, guinea pig and, most importantly, human.

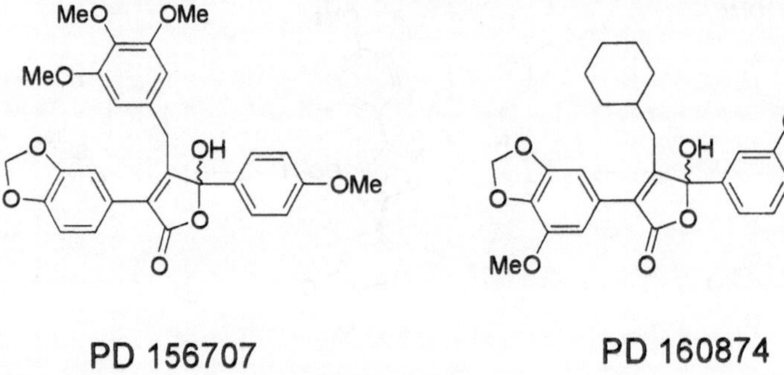

PD 156707   PD 160874

Fig. 12. PD 160874, a dual ET_A/ET_B antagonist

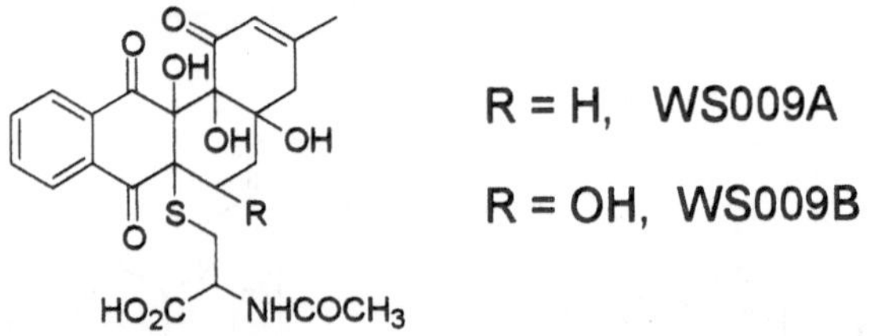

R = H,   WS009A

R = OH, WS009B

Fig. 13. WS009 A and B, early examples of non-peptide endothelin receptor antagonists

Another balanced ET_A/ET_B antagonist to be derived from an earlier ET_A selective antagonist within the chemical series is PD 160874 (Fig. 12). Thus, replacement of 3,4,5-trimethoxyphenyl group in the ET_A selective compound, PD 156707, with a cyclohexyl moiety greatly enhanced the affinity for the ET_B receptor without affecting ET_A binding. Further modification of the substituents on the methoxyphenyl ring provided PD 160874, which binds to cloned human ET_A and ET_B receptors with IC_{50}s of 3.5nmol/l (ET_A) and 8.9nmol/l (ET_B) respectively (DOHERTY 1996).

II. ET_A Selective Antagonists

The earliest identified non-peptide antagonists of ET receptors came from natural product screening and interestingly, they shared an ET_A-selective profile. WS009 A and B (Fig. 13) were isolated from a strain of *Streptomyces*. Competition binding studies using [^{125}I]ET-1 demonstrated that WS009 A and B bind with low affinity (IC_{50} values of 5.8μmol/l and 0.67μmol/l, respectively) to membranes from porcine aorta, a tissue rich in the ET_A receptor subtype (MIYATA 1992b,c). Both WS009 A and B are without effect on the ET_B receptor present in porcine brain. ET_A receptor selectivity was also demonstrated in that neither WS009 A nor B affected the appropriate ligand binding to a

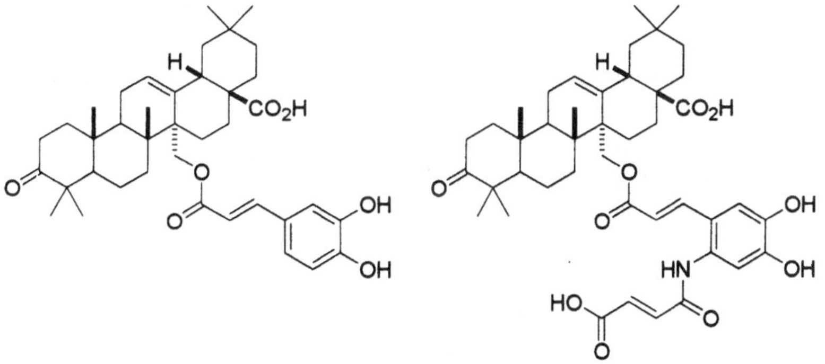

Shionogi 50-235   Shionogi 97-139

Fig. 14. Myricerone derivatives identified as endothelin receptor antagonists

range of other receptors, including the peptide receptors, AII AT_1 and substance P (MIYATA 1992a).

Lineweaver-Burk analysis of the inhibition of ET-1 binding to porcine aortic membranes is consistent with competitive antagonism. Functionally, WS009 A and B inhibit, albeit weakly, the ET-1-induced contraction of isolated rabbit aorta (MIYATA 1992a). Furthermore, in the spontaneously hypertensive rat, a 10 mg/kg, i.v. dose of WS009 A blocked the pressor, but not the depressor response to an exogenous challenge of ET-1 (3.2 μg/kg), consistent with the ET_A-selective nature of the antagonist. Finally, WS009 A competes with [^{125}I]ET-1 binding to membranes from human aorta (IC_{50} = 13 μmol/l). No more recent reports have emerged on WS009 A or B, nor has there been any description of their binding characteristics to cloned human ET receptors.

Shionogi 50–235 (Fig. 14) (FUJIMOTO 1992) is also a natural product, in this case isolated from an extract of the bayberry, *Myrica cerifera*. Shionogi 50–235 inhibits, in a concentration-dependent manner, [^{125}I]ET-1 binding to rat aortic smooth muscle A7r5 cells that express ET_A receptors (K_i = 51 nmol/l), whereas [^{125}I]ET-1 binding to Girardi heart cells (ET_B receptors) is unaffected (MIHARA 1993b). Selective ET_A antagonism was also demonstrated in a functional assay where ET-1-induced calcium mobilization was inhibited by Shionogi 50–235 in a dose-dependent fashion (IC_{50} = 11 nmol/l), while the response to potassium chloride and bombesin were unaffected. Similar functional studies in Girardi heart cells failed to show an effect of Shionogi 50–235 on the ET-1-induced response, as anticipated from the radioligand binding data (see above). The compound also inhibits the mitogenic effects of ET-1 on A7r5 cells with an IC_{50} of approximately 100 nmol/l (MIHARA 1993a). This latter result further supports the notion that Shionogi 50–235 is a functional antagonist of ET_A receptors, inasmuch as the smooth muscle proliferative effects of ET-1 have been shown to be mediated by ET_A receptors (OHLSTEIN 1992; see Chaps. 12 and 16).

While the structure of Shionogi 50–235 is relatively complex, it is amenable to some straightforward chemical modification. Thus elaboration of the caffeoyl moiety of Shionogi 50–235 gave Shionogi 97–139 (Fig. 14), a compound with some 50-fold enhanced affinity for ET_A receptors (MIHARA 1994). Shionogi 97–139 inhibits the binding of [^{125}I] ET-1 to rat A7r5 cells (ET_A receptors) with a K_i of 1 nmol/l. Unlike Shionogi 50–235, however, the compound also has some affinity, albeit very weak, for ET_B receptors on human Girardi heart cells (K_i = 1000 nmol/l). Shionogi 97–139 produces a concentration-dependent inhibition of the ET-1-induced increase in intracellular calcium levels in A7r5 cells (IC_{50} of 2.6 nmol/l). Furthermore, this compound blocks the mitogenic effects of ET-1 in A7r5 cells with an IC_{50} of 0.92 nmol/l. In isolated rat aortic ring segments, Shionogi 97–139 produces parallel rightward shifts in the ET-1 concentration-response curve (pA_2 = 8.8), without affecting the maximal contractile response. A direct comparison of BQ-123 and Shionogi 97–139 in these *in vitro* assays showed the latter compound to be nearly an order of magnitude more potent. However, this enhanced potency of Shionogi 97–139 relative to the peptide antagonist, BQ-123, was not reflected in improved efficacy in animal models. Thus, Shionogi 97–139 was approximately equipotent to BQ-123 in its ability to block the pressor response to exogenous ET-1 challenge in the pithed rat. Further evidence suggests that this discrepancy results from enhanced plasma albumin binding of Shionogi 97–139 relative to BQ-123.

In contrast to the structural complexity of the natural product lead compounds discussed thus far, the discovery that a relatively simple compound, asterric acid (Fig. 15), inhibits ET-1 binding to ET_A receptors present in A10 cells (IC_{50} = 10 μmol/l) was regarded with some interest (OHASHI 1992). The compound, a natural product from the fungus *Aspergillus*, is selective in binding to the ET_A receptors on A10 cells versus atrial natriuretic peptide or AT_1 AII receptors, which are also present in these cells.

While asterric acid contains features in common with other ET antagonists, no reports have directly connected the molecule to efforts to produce more potent antagonists.

As part of a screening program aimed at the identification of compounds capable of inhibiting [^{125}I]ET-1 binding to vascular smooth muscle A10 cells, it was discovered that the sulfonamide antibacterial sulfathiazole has low affin-

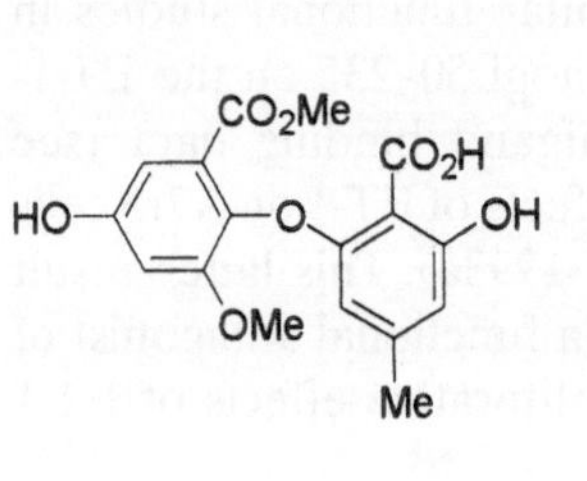

Fig. 15. Asterric acid, one of the first non-peptide endothelin receptor antagonists to possess a relatively simple chemical structure

ity for the ET_A receptor ($IC_{50} = 69 \mu mol/l$) (STEIN 1994). Subsequently, another member of this class, sulfisoxazole was found to have substantially higher affinity for the ET_A receptor ($IC_{50} = 0.78 \mu mol/l$) (STEIN 1994). Despite the encouraging binding data, sulfisoxazole is less potent in its ability to block ET-1 induced calcium mobilization in A10 cells ($IC_{50} = 40 \mu mol/l$) and, at a concentration of $100 \mu mol/l$, the compound failed to affect the concentration-response curve to ET-1 in isolated rabbit carotid artery rings. Irrespective of the disparity between binding and functional data for sulfisoxazole, it has proven to be a valuable lead structure. As was the case for bosentan and Ro 46-2005, sulfisoxazole contains a secondary sulfonamide but lacks the carboxylic acid moiety found in the earlier antagonists. The relatively acidic N-H present in sulfisoxazole is, however, critical for binding activity. Thus, substitution of the sulfonamide nitrogen with an alkyl group (e.g., methyl) or replacement of the sulfonamide with a simple amide leads to a total abrogation of binding to the ET_A receptor.

Further SAR studies in this series were focused on the benzenoid ring of the sulfonamide. It was observed that an increase in size and lipophilicity of this substituent greatly improved the binding affinity to ET_A receptors, ultimately leading to the naphthalene sulfonamide BMS 182874 (Fig. 16). BMS 182874 binds to A10 cells (ET_A receptors) with a K_i of 55nmol/l, while binding to rat cerebellar membranes (ET_B receptors) is much weaker, with a K_i of more than $200 \mu mol/l$, making the compound some 3600-fold selective in binding to the ET_A receptor. Functionally, BMS 182874 is somewhat less potent with the K_b value of 520nmol/l for the blockade of ET-1 induced contraction of the rabbit carotid artery (WU-WONG 1994). *In vivo*, BMS 182874 was demonstrated to lower blood pressure within 1h of dosing (p.o., 100 mmol/kg) to DOCA-salt rats. Similar results were obtained in a separate experiment in which an i.v. bolus of BMS 182874 was dosed to conscious, unrestrained DOCA-salt hypertensive rats and led to a reduction of blood pressure and an increase of total peripheral conductance (YU 1998).

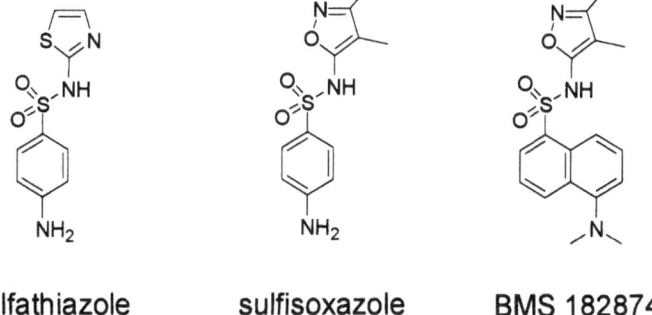

sulfathiazole   **sulfisoxazole**   **BMS 182874**

Fig. 16. Sulfonamide antagonists reported to be selective for ET_A receptors

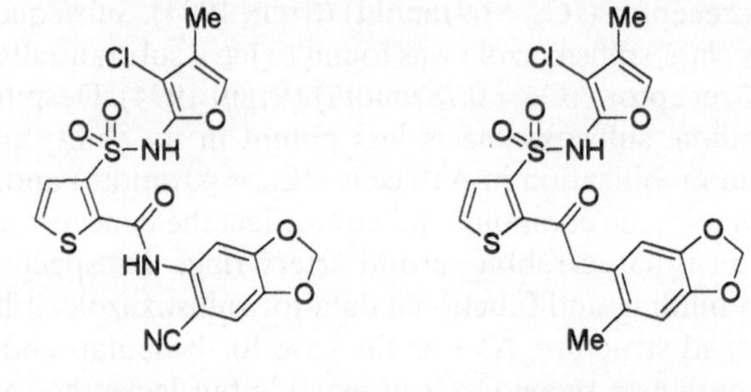

TBC11241   TBC11251

Fig. 17. TBC 11251 a potent, orally bioavailable ET_A selective antagonist

For optimal potency in this series, substitution of the isoxazole is limited to small, lipophilic groups (e.g., Me, Cl) (STEIN 1994). However, the benzenoid ring can be replaced by a variety of groups including a thiophene moiety. Modification in this manner was examined by another research group, who independently discovered that sulfisoxazole binds to the ET_A receptor. Thus, TBC11241, a member of the thiophenesulfonamide class of ET_A selective antagonists (CHAN 1996; RAJU 1997a–d) (Fig. 17), displays potent and selective binding to ET_A receptors (IC_{50}s, ET_A = 3.4 nmol/l, ET_B = 40.4 μmol/l) (WU 1997a). Although TBC11241 is a potent ET_A selective antagonist *in vitro*, it lacks oral bioavailability in the rat (8%) as well as possessing a short *in vivo* half-life (2.5 h), perhaps as a result of proteolytic cleavage of the amide bond. To circumvent this problem, the amide was replaced with a ketomethylene moiety, leading to TBC11251 (Fig. 17) (WU 1997b) which maintains the *in vitro* binding affinity of its amide congener while enhancing the oral bioavailability and duration of action. TBC11251 binds competitively to human ET_A receptors with an IC_{50} of 1.4 nmol/l (IC_{50} for ET_B = 9800 nmol/l). The compound inhibits ET-1-induced stimulation of phosphoinositol turnover with a pA_2 of 8.0, and is 60–100% orally bioavailable in both rat and dog, with a 6–7 h serum half-life in both species. TBC11251 has been advanced to phase II clinical trials for the treatment of severe congestive heart failure and initial results show a statistically significant improvement in pulmonary artery pressure versus placebo.

Further investigation of the sulfonamide BMS 182874 has been conducted using robotic synthesis, leading to the discovery that the isoxazole ring could be replaced by a pyrazine without loss of ET_A receptor affinity (Fig. 18, compound A) (BRANDBURY 1997). Replacement of the dansyl group in compound A by a biphenylcarboxamide gave a novel series of compounds exemplified by compound B (Fig. 18) (MORTLOCK 1997). In radioligand binding studies involving displacement of [^{125}I]ET-1 from membranes prepared from MEL cells transfected with cloned ET_A receptors, compound B has a pIC_{50} value of 9.3, while it is without measurable affinity for the human ET_B receptor (<50% inhibition at 10 μmol/l). When dosed at 2.5 mg/kg orally in the conscious rat,

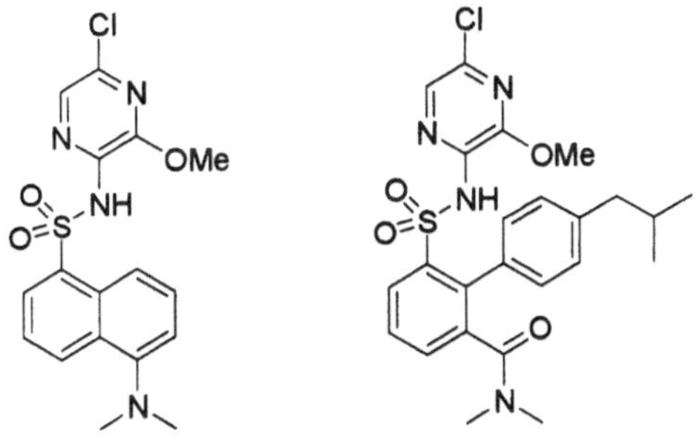

Fig. 18. Potent ET$_A$ selective sulfonamide antagonists

compound B inhibits the pressor response to exogenously administered ET-1 with a duration of action in excess of 4h, supporting the *in vitro* potency data and suggesting good metabolic stability.

The pyrrolidine series of antagonists represented by A-127722 was designed principally from the indane antagonist SB-209670. In contrast to the indane antagonist series, where additional substitution at the 2-position of the 4-methoxy bearing phenyl ring is required for optimal potency, very potent pyrrolidine antagonists have been obtained which bear a simple 4-methoxyphenyl at C-2. In the pyrrolidine series the nitrogen substituent exerts a profound effect on activity and the extremely potent compound, A-127722, has an *N,N*-dibutyl acetamide installed at this position (WINN 1996). A-127722 exhibits high affinity for cloned human ET$_A$ receptors (IC$_{50}$ = 0.082 nmol/l) and is more than 1000-fold selective in binding versus ET$_B$ receptors (IC$_{50}$ = 114nmol/l). Functionally the compound inhibits ET-1 induced contraction in the isolated rat aorta with a pA$_2$ of 9.20. Furthermore, A-127722 is a potent functional antagonist of ET-1 stimulated phosphoinositol hydrolysis in MMQ cells (IC$_{50}$ = 0.16nmol/l). The resolution of A-127722 has been achieved through formation of chiral salt with *R*-(+)-*α*-methylbenzylamine. As would be anticipated, based upon the absolute configuration of the indane SB 209670, the activity of A-127722 resides mainly in the RRS antipode (Fig. 19). Pharmacokinetic studies with A-147627 (the most active enantiomer of A-127722, shown in Fig. 19) have been performed in the rat with i.v. and oral administrations of 5mg/kg and 10mg/kg, respectively. The oral bioavailability of this compound is estimated to be 70% and plasma clearance is very low (<5ml/min per kg). SAR studies around A-127722 have provided a second generation of antagonists with much greater selectivity for ET$_A$ receptor binding. Thus, replacement of the pendant *p*-anisolyl group in A-127722 with simple aliphatic chains afforded compounds with greater ET$_A$ selectivity, as is the case with A-216546 (Fig. 19) which incorporates an *n*-pentyl substituent at this position (LIU 1998). A-216546 binds to CHO cells permanently transfected

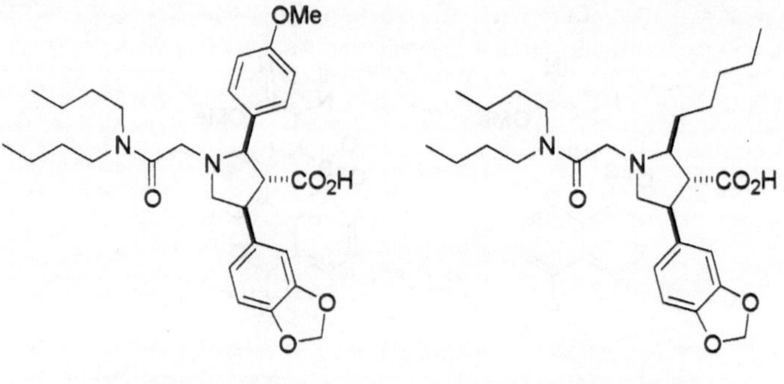

A-127722   A-216546

Fig. 19. A-127722 and A-216546, highly potent ET_A selective antagonists

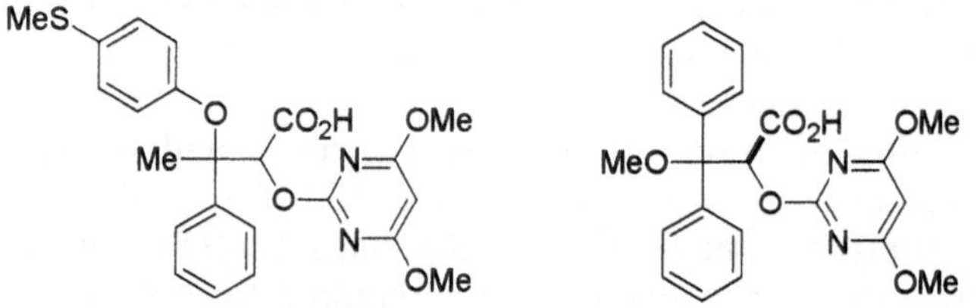

LU 110897   LU 127043

Fig. 20. LU 110897, an antagonist lead structure generated via high throughput screening and LU 127043, an potent, orally bioavailable ET_A selective antagonist

with human ET_A or ET_B receptors with IC_{50} values of 0.46 nmol/l and 13 μmol/l, respectively. Based on the overall superior affinity, high selectivity for the ET_A receptor, and good oral bioavailability (48% in rats), A-216546 has been selected as a potential clinical backup for A-147627 (the active enantiomer of A-127722).

LU 110897 (Fig. 20), a compound originally prepared as a herbicide, was discovered as a lead antagonist structure through high-throughput screening with the recombinant human ET_A receptor (RIECHERS 1996). The compound has moderate binding affinity for the ET_A receptor ($K_i = 160$ nmol/l) and is somewhat selective versus the ET_B receptor ($K_i = 4.7$ μmol/l). Since LU 110897 contains two chiral centers, the focus of chemical efforts has been to enhance binding affinity while simplifying the structure. As a result, LU 127043 (Fig. 20) was prepared which exhibits K_i values of 6 nmol/l and 1000 nmol/l for ET_A and ET_B receptor binding, respectively. The (S)-enantiomer, whose absolute configuration was confirmed by X-ray analysis and asymmetric synthesis, was found to be 50 times more active than the (R)-enantiomer in ET_A binding [(S)-enantiomer, $K_i = 3$ nmol/l]. Functionally, LU 127043 inhibits

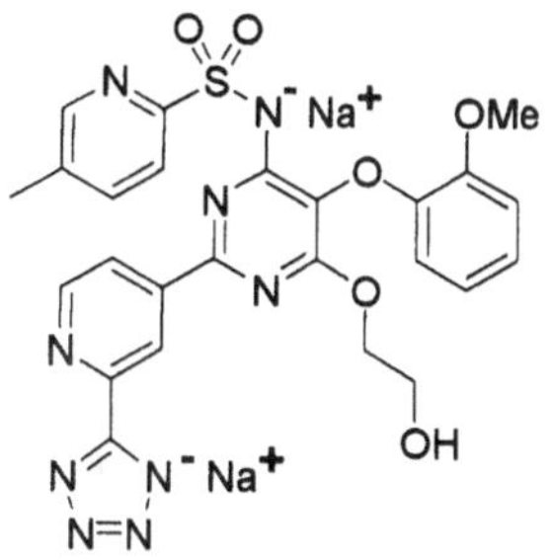

Fig. 21. Ro 61–1790, a water soluble ET_A selective antagonist derived from bosentan

ET-1 induced contraction of rabbit aortic rings, in which the endothelium is intact, with a pA_2 of 7.34. The oral bioavailability of LU 127043 was assessed in a rat model of ET-induced death. Following treatment with 30 mg/kg LU 127043 p.o., complete protection was observed after 4 h. At a higher dose (100 mg/kg p.o.), complete protection persisted to 8 h, indicating the effectiveness and long duration of action of this molecule.

Several groups have shown interest in the development of an ET_A antagonist to be administered intravenously for the treatment of subarachnoid hemorrhage. Thus, a potent member of the class of trifunctionalized heteroarylsulfonamide pyrimidines has been designed with good water solubility. Ro 61–1790 (Roux 1997) (Fig. 21) is a competitive ET antagonist with an affinity for the ET_A receptor in the subnanomolar range. It has an approximately 1000-fold selectivity for the ET_A versus the ET_B receptor as assessed by functional assays (e.g., ET-1-induced inositol-1,4,5-triphosphate release or ET-1-induced intracellular calcium mobilization). Functionally, Ro 61–1790 inhibits ET-1 induced contraction of isolated rat aorta with a pA_2 of 9.5 (ET_A receptors), while exhibiting a pA_2 of 6.4 for inhibition of sarafotoxin S6c contraction of rat trachea (ET_B receptors). *In vivo*, Ro 61–1790 inhibits the pressor response to big ET-1 in pithed rats with an ID_{50} value of 0.05 mg/kg. An i.v. bolus of Ro 61–1790 induces a long-lasting antihypertensive effect in DOCA salt rats instrumented with telemetry. In a canine double-hemorrhage model of subarachnoid hemorrhage, Ro 61–1790 both prevents and reverses cerebral vasospasm in a dose-dependent manner. In an established cerebral vasospasm, 3 mg/kg per day Ro 61–1790 dosed i.v. was only half as efficacious as intrabasilar papaverine, and when dosed at 20 mg/kg per day totally prevented the occurrence of vasospasm. These data demonstrate that Ro 61–1790 is a potent and selective ET_A receptor antagonist, suitable for parenteral use. Furthermore, the compound has considerable therapeutic potential for the prevention of delayed ischemic deficit in patients with subarachnoid hemorrhage.

Another member of the pyrimidine sulfonamide class of ET receptor antagonists, designed around the bosentan framework, is T-0201 (Fig. 22) (Ohnishi 1998). While T-0201 is more than 1000-fold selective for ET_A receptors versus ET_B receptors (K_i values of 0.015 nmol/l and 41 nmol/l, respectively), it is nonetheless quite a potent ET_B receptor antagonist. Thus in conscious

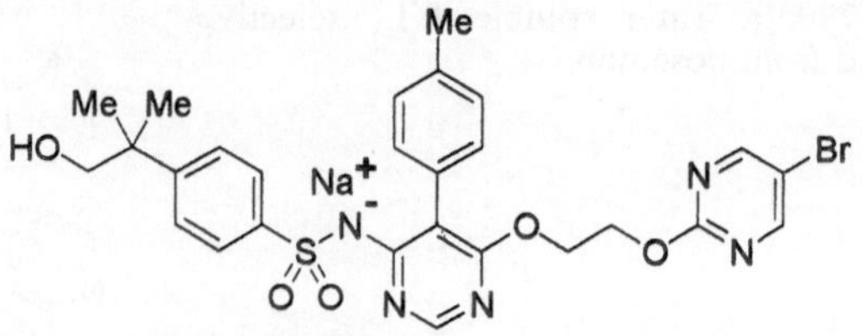

Fig. 22. T-0201, an ET$_A$ selective antagonist

normal dogs, T-0201 significantly inhibits the initial hypotension resulting from exogenous ET-1 administration and attenuates the subsequent hypertension; effects which are mediated by ET$_B$ and ET$_A$ receptors, respectively.

Considerable therapeutic interest has centered around the observation that plasma ET levels are elevated in patients suffering from congestive heart failure (CHF). To evaluate the therapeutic potential of an ET antagonist for this indication, T-0201 was dosed orally to dogs with CHF, induced by rapid right ventricular pacing (22 days, 270 beats/min). Thus, at a dose of 0.3 mg/kg per day given for 15 days, beginning 8 days after pacing, T-0201 significantly prevented the deterioration of cardiorenal function during the development of CHF, as evidenced by a decrease in cardiac pressure and an increase in cardiac and urine output. Since, at the dose given, both ET$_A$ and ET$_B$ receptors would be effectively blocked, these results suggest that chronic antagonism of ET receptors can prevent the progressive exacerbation of CHF.

Another unique series of non-peptide ET$_A$ selective antagonists to have emerged from high-throughput screening are the butenolides, which began with the micromolar antagonist PD 012527 (PATT 1997). Following a Topliss decision tree analysis this lead was parlayed into a nanomolar compound, PD 155080. Further structural modifications to the aryl substituents appended to the butenolide ring led to the subnanomolar ET$_A$ selective antagonist PD 156707 (Fig. 23), [IC$_{50}$s in binding = 0.3 nmol/l (ET$_A$) and 780 nmol/l (ET$_B$)]. PD 156707 inhibits the ET$_A$ receptor mediated release of arachidonic acid from rabbit renal artery vascular smooth muscle cells with an IC$_{50}$ of 1.1 nmol/l and also antagonizes the ET-1 induced contraction of rabbit femoral artery rings (ET$_A$ mediated) with a pA$_2$ = 7.6. This compound also displays *in vivo* functional activity in rats, i.e., inhibition of the hemodynamic response due to exogenous administration of ET-1. Pharmacokinetic parameters for PD 156707 in rats and dogs are similar, with a bioavailability of approximately 60% in fasted animals. In the dog the mean plasma terminal elimination half-life was 1.9 h following oral dosing and 0.62 h following intravenous dosing. PD 156707 has been demonstrated to be efficacious in a cat model of stroke, suggesting a possible therapeutic utility (PATEL 1995).

An intriguing feature of many of the more potent ET receptor antagonists reported to-date is the inclusion of a methylenedioxyphenyl benzodioxole moiety, within their structure, which is important for binding affinity.

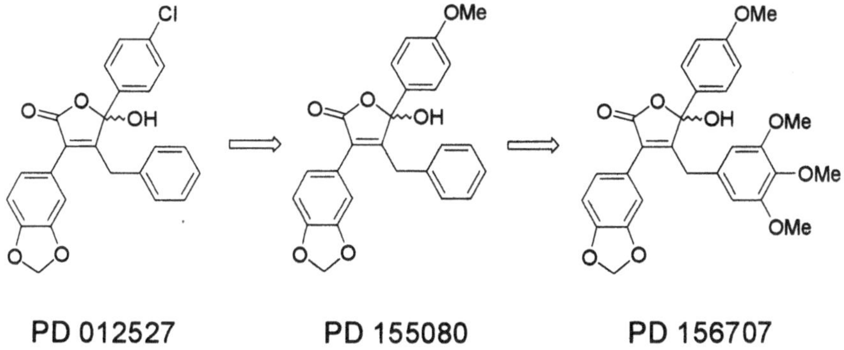

Fig. 23. Elaboration of the high throughput screening lead PD 012527 which led to the potent, orally bioavailable ET_A selective antagonist, PD 156707

While literature precedent exists to support the interaction of such a group with cytochrome P450 enzymes this has apparently not been a liability for the ET antagonists thus far advanced to human clinical trials. Nevertheless, a number of research groups have set as their objective the replacement of the methylenedioxyphenyl group, with retention of high antagonist affinity towards the endothelin receptors. Thus, in the pyrrolidine series of antagonists, exemplified by A-127722, it was demonstrated that a 2,3-dihydrobenzofuranyl moiety could replace the methylenedioxyphenyl without penalty with respect to receptor affinity (TASKER 1997). In other antagonist series such as the butenolides and the dansylsulfonamides, a benzothiadiazole has been shown to function as a methylenedioxyphenyl bio-isostere, providing compounds of equivalent or even greater potency (ANZALI 1998; MEDUSKI 1998). Thus, EMD 122801 (Fig. 24) displayed clearly superior functional activity as compared to PD 156707, with a pA_2 value of approximately 8.5 vs 7.6.

Based upon the most critical structural elements needed for receptor affinity of the ET_A/ET_B indane antagonist SB 209670, namely appropriately juxtaposed methylenedioxyphenyl and carboxyl groups, database searching led to the identification of the benzylimidazole, SK&F 107328, as a lead structure. This compound was originally prepared as part of an AII receptor antagonist program and it is a weak, non-selective antagonist of ET receptors (K_is $ET_A = 400$ nmol/l, $ET_B = 3400$ nmol/l), possessing moderate affinity for the AII AT_1 receptor ($K_i = 180$ nmol/l) (KEENAN 1992).

Molecular modeling overlays of the benzylimidazole SK&F 107328 with the indane SB 209670 suggested an analog in which a phenyl substituent corresponding to the indane 3-phenyl ring is placed at the 3-position of the acrylic acid. On the basis of known SAR in the angiotensin receptor antagonist area it was further anticipated that such a molecule would lack affinity for the AT_1 receptor. The resulting compound, SB 209834 (Fig. 25), is indeed a potent ET_A selective antagonist (K_is $ET_A = 2$ nmol/l, $ET_B = 500$ nmol/l) which is selective

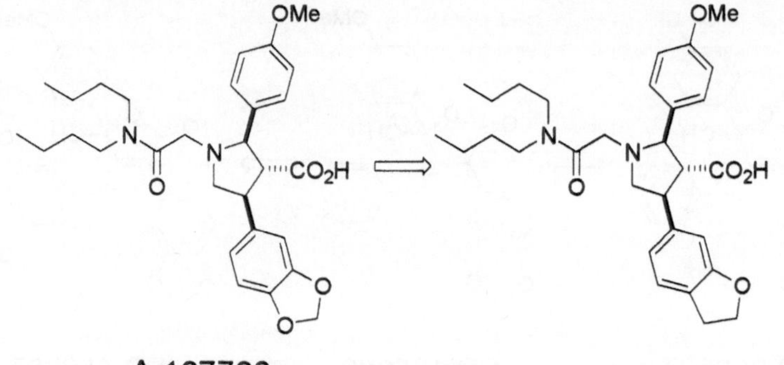

Fig. 24. Identification of surrogates of the methylenedioxyphenyl (benzodioxole) moiety

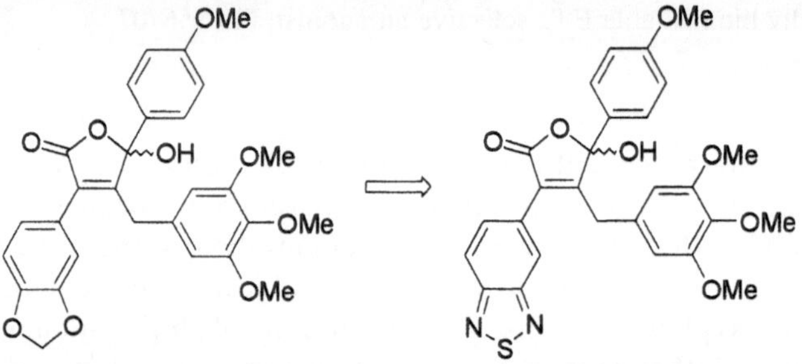

for endothelin receptors versus the AII AT_1 receptor (Elliott 1996). While SB 209834 lacks the stereochemical complexity of SB 209670, geospecific synthesis of the tetrasubstituted olefin still presents a significant challenge. Somewhat surprisingly SAR studies around SB 209834 showed that the 3 aryl substituent could be transposed to the N-1 position of the imidazole without loss of activity. Further modification of this framework led to the discovery of SB 224519 (Fig. 25), a potent selective ET_A antagonist (K_is ET_A = 0.2 nmol/l, ET_B = 700 nmol/l) (Xiang 1999) with a bioavailability of 44% in the rat and a moderate plasma clearance of 21 ml/min per kg.

Further SAR studies based upon SB 224519 revealed that the imidazole ring could be replaced with the phenyl group to provide the biphenyl compound SB 227543 (Fig. 25) (Xiang 1997) which maintains the potency and selectivity of its imidazole congener (K_is ET_A = 0.18 nmol/l, ET_B = 340 nmol/l). SB 227543 has good oral bioavailability in the rat and dog, with values of 50% and 23%, respectively and, SB 227543 displays a superior duration of action than SB 224519 in-vivo in the rat (7 h vs 2.5 h).

Both SB 224519 and SB 227543 represent attractive development candidates studies by virtue of their potency, selectivity, pharmacokinetic proper-

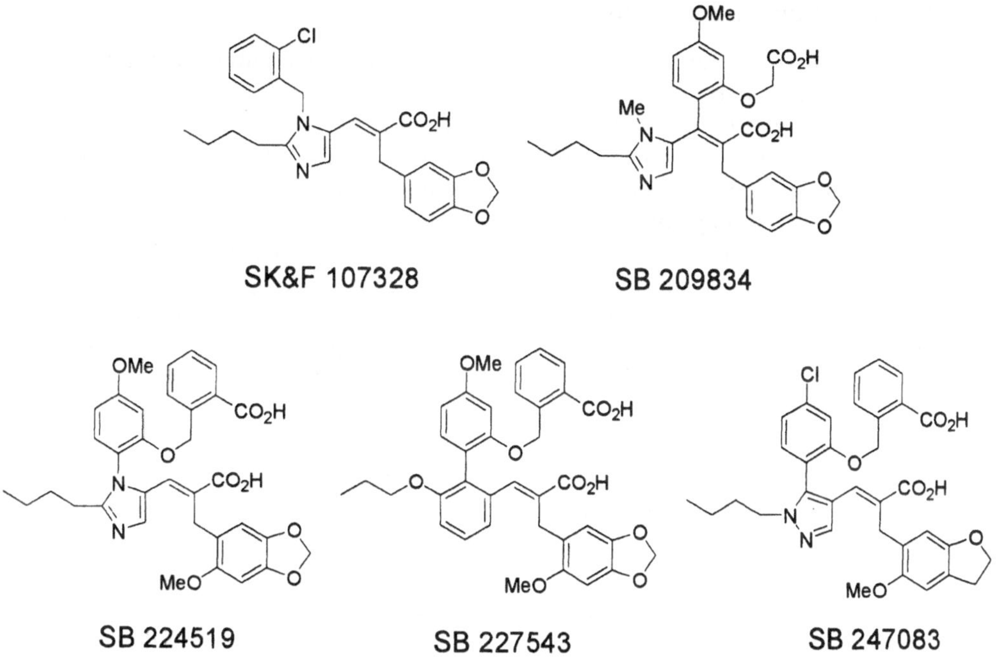

Fig. 25. SK&F 107328, a lead generated *via* screening of compounds containing key features present in the indane SB 209670. SB 209834, a potent ET_A selective antagonist designed from overlays of SK&F 107328 and indane SB 209670 (Fig. 8). SB 224519, SB 227543 and SB 247083, a series of potent orally bioavailable ET_A selective antagonists related structurally to SB 209834

ties, and simplified structures relative to the indane series of antagonists. However, as a result of steric congestion around the phenyl-imidazole C-N or biphenyl C-C bonds, atropisomeric interconversion is slow even at ambient temperature. To obviate the need for resolution of the atropisomers a search for alternative heterocycles with lower rotational barriers around the aryl-heteroaryl bond was instigated. This led to the pyrazole series of antagonists represented by SB 247083, (XIANG 1997) (Fig. 25), which also incorporates a 2,3-dihydrobenzfuranyl moiety in place of the ubiquitous methylene-dioxyphenyl group. SB 247083 is a potent ET_A selective antagonist (K_is $ET_A = 0.4$nmol/l, $ET_B = 467$nmol/l) (DOUGLAS 1998) which functionally, *in vitro*, inhibits ET-1-induced rat aortic ring contraction with a K_b of 3.5nmol/l. The compound is significantly less potent as a functional ET_B antagonist with K_b value of 0.34 μmol/l in the inhibition of sarafotoxin S6c induced rabbit pulmonary artery contraction. Pharmacodynamic and pharmacokinetic studies reveal that SB 247083 is effectively absorbed from the gastrointestinal tract. A single bolus dose inhibits the hemodynamic actions of ET-1 for up to 8h in the rat, a species in which the molecule is 46% bioavailable (DOUGLAS 1998). Thus, SB 247083 represents an attractive candidate for a therapeutic agent.

III. ET_B Selective Antagonists

In contrast to the burgeoning literature on ET_A selective antagonists, efforts to provide non-peptide ET_B selective compounds have been less well documented. The first non-peptide ET_B selective antagonist to emerge in the literature was Ro 46–8443 (Breu 1996; Clozel 1996) (Fig. 26), a member of the pyrimidine benzenesulfonamide series which includes bosentan. Ro 46–8443 competes for binding with $[^{125}I]ET-1$ to the human ET_B receptor, transfected into CHO cells, with an IC_{50} of 69 nmol/l. The selectivity of Ro 46–8443 binding to ET_B versus ET_A receptors is apparent from the micromolar IC_{50} value found for this compound in competitive binding studies with $[^{125}I]ET-1$ on cultured human vascular smooth muscle cells (ET_A receptors). Functionally, Ro 46–8443 inhibits the sarafotoxin-6c-induced contraction of rat tracheal rings with a pA_2 of 7.1.

The pyrrolidine framework has been previously discussed in connection with both balanced and ET_A selective compounds (see above). As further evidence of the utility of this structural type, A-192621 (Fig. 26) has been prepared, which is more than 1000-fold selective for the ET_B receptor (Von Geldern 1999). A-192621 binds ET_B and ET_A receptors with IC_{50} values of 4.5 nmol/l and 6.3 μmol/l, respectively. The compound is orally bioavailable in the rat (35%), with a half-life of 5 h. *In vivo*, A-192621 causes sustained and progressive hypertension in rats and should be an important tool to help define the potential therapeutic utility of an ET_B selective antagonist.

Finally, as outlined above, the potent ET_B selective antagonist IRL 2500 (Fig. 11) (Fruh 1996) was the starting point for investigations which led to a balanced ET_A/ET_B antagonist IRL 3630.

During the decade that has elapsed since the elucidation of the structure of ET-1, many receptor antagonists have been described, with selectivities encompassing the gamut of possibilities for the two characterized human receptor subtypes. Several of these agents have progressed into clinical trials

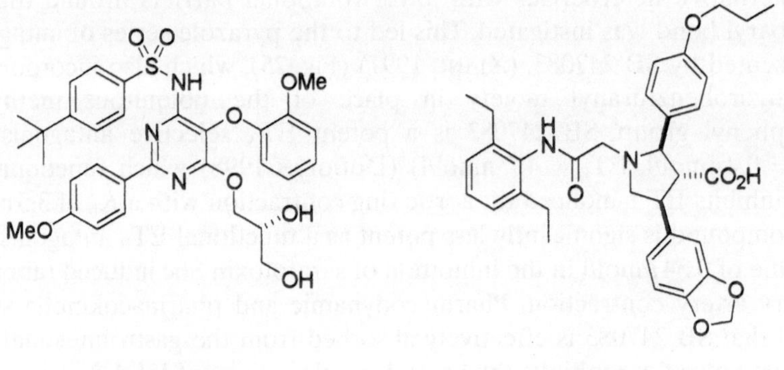

Ro 46-8443   A-192621

Fig. 26. Ro 46–8443 and A-192621, ET_B selective antagonists

and encouraging results have emerged with respect to the treatment of congestive heart failure. Further work will be needed with compounds of diverse selectivity to help define the appropriate profile for any particular therapeutic indication.

References

Anzali S, Mederski WWKR, Osswald M, Dorsch D (1998) 1. Endothelin antagonists: search for surrogates of methylenedioxyphenyl by means of a Kohonen Neural network. Bioorg Med Chem Lett 8:11–16

Aramori I, Nirei H, Shoubo M, Sogabe K, Nakamura K, Kojo H, Notsu Y, Ono T, Nakanishi S (1993) Subtype selectivity of a novel endothelin antagonist, FR139317, for the two endothelin receptors in transfected Chinese hamster ovary cells. Mol Pharmacol 43:127–131

Bean JW, Peishoff CE, Kopple KD (1994) Conformations of cyclic pentapeptide endothelin receptor antagonists. Int J Peptide Protein Res 44:223–232

Beyer ME, Slesak G, Brehm BR, Hoffmaeister HM (1998) Hemodynamic and inotropic effects of the endothelin A antagonist BQ-610 in vivo. J Cardiovasc Pharmacol 31 [Suppl 1]:S258–S261

Bogusky MJ, Brady SF, Sisko JT, Nutt RF, Smith GM (1994) Synthesis and solution conformation of c(Trp-D-Cys(SO$_3$Na$^+$)-Pro-D-Val-Leu), a potent endothelin-A receptor antagonist. Int J Peptide Protein Res 42:194–203

Brandbury RH, Bath C, Butlin RJ, Dennis M, Heys C, Hunt SJ, James R, Mortlock AA, Sumner NF, Tang EK, Telford B, Whiting E, Wilson C (1997) New non-peptide endothelin-A receptor antagonists: Synthesis, biological properties, and structure-activity relationships of 5-(dimethylamino)-N-pyridyl-, -N-pyrimidinyl, -N-pyridazinyl, and -N-pyrazinyl-1-naphthalenesulfonamides. J Med Chem 40:996–1004

Breu V, Clozel M, Burri K, Hirth G, Neidhart W, Ramuz H (1996) In vitro characterization of Ro 46–8443, the first non-peptide antagonist selective for the endothelin ET$_B$ receptor. FEBS Lett 383:37–41

Breu V, Hashido K, Broger C, Miyamoto C, Furuichi Y, Kalina B, Loffler B-M, Ramuz H, Clozel M (1995) Separable binding sites for the natural agonist endothelin-1 and the non-peptide antagonist bosentan on human endothelin-A receptors. Eur J Biochem 231:226–270

Cardell LO, Uddman R, Edvinsson L (1993) A novel ET$_A$-receptor antagonist, FR139317, inhibits endothelin-induced contractions of guinea-pig pulmonary arteries, but not trachea. Br J Pharmacol 108:448–452

Chan MF, Raju B, Kois A, Castillo RS, Verner EJ, Wu C, Hwang E, Okun I, Stavros F, Balaji VN (1996) Halogen substitution at the isoxazole ring enhances the activity of N-(Isoxazolyl)sulfonamide endothelin antagonists. Bioorg Med Chem Lett 6(20):2393–2398

Clozel M, Breu V (1996) The role of ET$_B$ receptors in normotensive and hypertensive rats as revealed by the non-peptide selective ET$_B$ receptor antagonist Ro 46–8443. FEBS Lett 383:42–45

Clozel M, Breau V, Gray GA, Kalina B, Loffler B-M, Burri K, Cassal J-M, Hirth G, Muller M, Neidhart W, Ramuz H (1994) Pharmacological characterization of bosentan, a new potent orally active non-peptide endothelin receptor antagonist. J Pharmacol Exp Ther 270:228–235

Clozel M, Breau V, Burri K, Cassal J-M, Fischli W, Gray GA, Hirth G, Loffler B-M, Muller M, Neidhart W, Ramuz H (1993) Pathophysiological role of endothelin revealed by the first orally active endothelin receptor antagonist. Nature (London) 365:759–761

Cody WL, He JX, Reily MD, Haleen SJ, Walker DM, Reyner EL, Stewart BH, Doherty AM (1997) Design of a potent combined pseudopeptide endothelin-A/

endothelin-B receptor antagonist, Ac-DBhg16-Leu-Asp-Ile-[NMe]Ile-Trp21 (PD 156252): Examination of its pharmacokinetic and spectral properties. J Med Chem 40(14):2228–2240

Cody WL, Doherty AM, He JX, DePue PL, Waite LA, Topliss JG, Haleen SJ, LaDouceur D, Flynn MA, Hill KE, Reynolds EE (1993) The rational design of an highly potent combined ET_A and ET_B receptor antagonist (PD145065) and related analogues. Med Chem Res 3:154–162

Cody WL, Doherty AM, He JX, DePue, PL, Rapundalo ST, Higorani GA, Major TC, Panek RL, Dudley D, Haleen SJ, LaDouceur D, Hill KE, Flynn MA, Reynolds EE (1992a) Design of a functional hexapeptide antagonist of endothelin. J Med Chem 35:3301–3303

Cody WL, Doherty AM, He JX, Topliss JG, Haleen SJ, LaDouceur D, Flynn MA, Hill KE, Reynolds EE (1992b) Structure-activity relationships in the C-terminus of endothelin-1 (ET-1). The discovery of potent antagonists. In: Schneider CH, Eberle AN (eds) Peptides 1992: Proceedings of the Twenty-Second European American Peptide Symposium. ESCOM Science: Leiden, The Netherlands, pp 687–688

Davenport AP, Kuc RE, Fitzgerald F, Maguire JJ, Berryman K, Doherty AM (1994) [^{125}I]-PD151241: a selective radioligand for human ET_A receptors. Br J Pharmacol 111:4–6

Doherty A, Patt W, Reisdorph B, Repine J, Walker D, Flynn M, Welch K, Reynolds E, Haleen S (1996) Design and pharmacological evaluation of a series of non-peptide endothelin ETA selective and ET_A/ET_B receptor antagonists. In: Yamazake M (ed) Med Chem: Today Tomorrow, Proc AFMC Int Med Chem Symp. Blackwell Science Ltd, London, UK, pp 255–261

Doherty AM, Cody WL, DePue PL, He JX, Waite LA, Leonard DM, Leitz NL, Dudley D, Rapundalo ST, Higorani GA, Haleen SJ, LaDouceur D, Hill KE, Flynn MA, Reynolds EE (1993a) Structure-activity relationships of C-terminal endothelin hexapeptide antagonists. J Med Chem 36:2585–2594

Doherty AM, Cody WL, He JX, DePue PL, Leonard DM, Dunbar JB, Hill KE, Flynn MA, Reynolds EE (1993b) Design of C-terminal peptide antagonists of endothelin. Structure-activity relationship of ET-1[16–21, D-His16]. Bioorg Med Chem Lett 3:497–502

Douglas SA, Nambi P, Gellai M, Luengo JI, Xiang J-N, Brooks DP, Ruffolo RR Jr, Elliott JD, Ohlstein EH (1998) Pharmacologic characterization of the novel, orally available endothelin-A-selective antagonist SB 247083, Pharmacologic characterization of the novel, orally available endothelin-A-selective antagonist SB 247083. J of Cardiovasc Pharmacol 31 [Suppl 1]:S273–S276

Douglas SA, Louden C, Vickery-Clark LM, Storer BL, Hart T, Feuerstein G, Elliott JD, Ohlstein EH (1994) A role for endogenous endothelin-1 in neointimal formation after rat carotid artery balloon angioplasty. Circ Res 75:190–197

Elliott JD, Bryan DL, Nambi P, Ohlstein EH (1996) A Novel Series of Non-Peptide Endothelin Receptor Antagonists In: Kaumana PTP, Hodges RD (eds) Peptides: chemistry, structure and biology. Mayflower Scientific Ltd, Kingswinford, pp 673–676

Elliott JD, Lago MA, Cousins RD, Gao A, Leber JD, Erhart KF, Nambi P, Elshourbagy NA, Kumar C, Lee JA, Bean JW, DeBrosse CW, Eggleston DS, Brooks DP, Feuerstein G, Ruffolo RR Jr, Weinstock J, Gleason JG, Peishoff CE, Ohlstein EH (1994) 1,3-Diarylindane-2-carboxylic acids, potent and selective non-peptide endothelin receptor antagonists. J Med Chem 37:1553–1557

Endo S, Inooka H, Ishibashi Y, Kitada C, Mizuta E, Fujino M (1989) Solution conformation of endothelin determined by nuclear magnetic resonance and distance geometry. FEBS Lett 257:149–154

Fruh T, Saika H, Svensson L, Pitterna T, Sakaki J, Okada T, Urade Y, Oda K, Fujutani Y, Takimoto M, Yamamura T, Inui T, Makatani M, Takai M, Umemura I, Teno N, Toh H, Hayakawa K, Murata T (1996) IRL 2500: A potent ET_B selective endothelin antagonist. Bioorg Med Chem Lett 6(20):2323–2328

Fujimoto M, Mihara S, Nakajima S, Ueda M, Nakamura M, Sakurai K (1992) A novel non-peptide endothelin antagonist isolated from bayberry, *Myrica cerifera*. FEBS Lett 305:41–44

Gellai M, Jugus M, Fletcher T, DeWolf R, Nambi P (1994) Reversal of post-ischemic acute renal failure with a selective endothelin A receptor antagonist in the rat. J Clin Invest 93:900–906

He JX, Cody WL, Flynn MA, Welch KM, Reynolds EE, Doherty AM (1995) RES-701-1, synthesis and a re-evaluation of its effects on the endothelin receptors. Bioorg Med Chem Lett 5(6):621–626

Hingorani G, Major T, Panek R, Flynn M, Reynolds E, He X, Cody W, Doherty A, Rapundalo S (1992) *In vitro* pharmacology of a nonselective ET_A/ET_B endothelin receptor antagonist, PD 142893 (Ac-(β-phenyl)D-Phe-Leu-Asp-Ile-Ile-Trp trifluoroacetate). FASEB J 6(Part 1, No.4):392

Ihara M, Noguchi K, Saeki T, Fukuroda T, Tsuchida S, Fukami T, Ishikawa K, Nishikibe M, Yano M (1992) Biological profiles of highly potent novel endothelin antagonists selective for the ET_A receptor. Life Sci 50:247–255

Ihara M, Fukuroda T, Saeki T, Nishikibe M, Kojiri K, Suda H, Yano M (1991) An endothelin receptor (ET_A) antagonist isolated from *Streptomyces misakiensis*. Biochem Biophys Res Commun 178:132–137

Ishikawa K, Ihara M, Noguchi K, Mase T, Mino N, Saeki T, Fukuroda T, Fukami T, Ozaki S, Nagase T, Nishikibe M, Yano M (1994) Biochemical and pharmacological profile of a potent and selective endothelin-B receptor antagonist, BQ-788. Proc Natl Acad Sci USA 91:4892–4896

Ishikawa K, Fukami T, Nagase T, Fujita K, Hayama T, Niiyama K, Mase T, Ihara M, Yano M (1992) Cyclic pentapeptide endothelin antagonists with high ET_A selectivity. Potency- and solubility-enhancing modifications. J Med Chem 35:2139–2142

Jae H-S, Winn M, Dixon, DB, Marsh KC, Nguyen B, Opgenorth TJ, von Geldern TW (1997) Pyrrolidine-3-carboxylic acids as endothelin antagonists. 2. Sulfonamide-based ET_A/ET_B mixed antagonists. J Med Chem 40:3217–3227

Keenan RM, Weinstock J, Finkelstein JA, Franz RG, Gaitanopoulos DE, Girard GR, Hill DT, Morgan TM, Samanen JM, Hempel J, Eggleston DS, Aiyar N, Griffin E, Ohlstein EH, Stack EJ, Weidley EF, Edwards RM (1992) Imidazole-5-acrylic acids: potent nonpeptide angiotensin II receptor antagonists designed using a novel peptide pharmacophore model. J Med Chem 35:3858–3872

Kiowski W, Sutsch G, Hunziker P, Muller P, Kim J, Oechslin E, Schmitt R, Jones R, Bertel O (1995) Evidence for endothelin-1-mediated vasoconstriction in severe chronic heart failure. Lancet 346:732–736

Kitada C, Ohtaki T, Masuda Y, Masuo Y, Nomura H, Asami T, Matsumoto Y, Satou M, Fujino M (1993) Design and synthesis of ET_A receptor antagonists and study of ET_A receptor distribution. J Cardiovasc Pharmacol 22 [Suppl 8]:S128–S131

Krum H, Viskoper RJ, Lacourciere Y, Budde M, Charlon V (1998) The effect of an endothelin-receptor antagonist, bosentan, on blood pressure in patients with essential hypertension. N Engl J Med 338(12):784–790

Kryste SR Jr, Bassolino DA, Bruccoleri RE, Hunt JT, Porubcan MA, Wandler CF, Anderson NH (1992) Solution conformation of a cyclic pentapeptide endothelin antagonist. Comparison of structures obtained from constrained dynamics and conformational search. FEBS Lett 299:255–261

LaDouceur DM, Davis LS, Keiser JA, Doherty AM, Cody WL, He FX, Haleen SJ (1992) Effects of the endothelin receptor antagonist PD 142893 (Ac-(β-phenyl)D-Phe-Leu-Asp-Ile-Ile-Trp trifluoroacetate) on endothelin-1 (ET-1) induced vasodilation and vasoconstriction in regional arterial beds of the anesthetized rat. FASEB J 6(Part 1, No.4):390

Liu G, Henry KJ, Szczepankiewicz BG, Winn M, Kozmina NS, Boyd SA, Wasicak J, von Geldern TW, Wu-Wong JR, Chiou WJ, Dixon DB, Nguyen B, Marsh KC, Opgenorth TJ (1998) Pyrrolidine-3-carboxylic acids as endothelin antagonists. 3.

Discovery of a potent, 2-nonaryl, highly selective ET_A antagonist (A-216546). J Med Chem 41(17):3261–3275

Maggi CA, Giuliani S, Patacchini R, Santicioli P, Giachetti A, Meli A (1989) The C-terminal hexapeptide, endothelin(16–21), discriminates between different receptors. Eur J Pharmacol 174:23–31

Mederski WWKR, Osswald M, Dorsch D, Anzali S, Christadler M, Schmitges C-J, Wilm C (1998) 2. Endothelin antagonists: Evaluation of 2,1,3-benzothiadiazole as a methylenedioxyphenyl bioisoster. Bioorg Med Chem Lett 8:17–22

Mihara S, Nakajima S, Matumura S, Kohnoike T, Fujimoto M (1994) Pharmacological characterization of a potent nonpeptide endothelin receptor antagonist 97–139. J Pharmacol Exp Ther 268:1122–1128

Mihara S, Fujumoto M (1993a) The endothelin ET_A receptor-specific effect of 50–235, a non-peptide endothelin antagonist. Eur J Pharmacol Mol Pharmacol 246:33–38

Mihara S, Sakurai K, Nakamura M, Konoike T, Fujimoto M (1993b) Structure-activity relationships of an endothelin ET_A receptor antagonist, 50–235, and its derivatives. Eur J Pharmacol Mol Pharmacol 247:219–221

Mills RG, O'Donoghue SI, Smith R, King GF (1992) Solution structure of endothelin-3 determined using NMR spectroscopy. Biochemistry 31:5640–5645

Mills RG, Atkins AR, Harvey T, Junius FK, Smith R, King GF (1991) Conformation of sarafotoxin-6b in aqueous solution determined by NMR spectroscopy and distance geometry. FEBS Lett 282:247–252

Miyata S, Fukami N, Neya M, Takase S, Kiyoto S (1992a) WS-7338, new endothelin receptor antagonists isolated from *Streptomyces* sp. No. 7338. III. Structure of WS-7338 A, B, C and D and total synthesis of WS-7338, B. J Antibiot 45:788–791

Miyata S, Ohata N, Murai H, Masui Y, Ezaki M, Takase S, Nishikawa M, Kiyoto S, Okuhara M, Kohsaka M (1992b) WS009 A and B, new endothelin receptor antagonists isolated from *Streptomyces* sp. No. 89009. I. Taxonomy, fermentation, isolation, physico-chemical properties and biological activities. J Antibiot 45: 1029–1040

Miyata S, Hashimoto M, Fujie K, Shouho M, Sogabe K, Kiyoto S, Okuhara M, Kohsaka M (1992c) WS009 A and B, new endothelin receptor antagonists isolated from *Streptomyces* sp. No. 89009. II. Biological characterization and pharmacological characterization of WS009 A and B. J Antibiot 45:1041–1046

Morishita Y, Chiba S, Tsukuda E, Tanaka T, Ogawa T, Yamasaki M, Yoshida M, Kawamoto I, Matsuda Y (1994) RES-701-1, a novel and selective endothelin type B receptor antagonist produced by Streptomyces sp. RE-701. I. Characterization of producing strain, fermentation, isolation, physico-chemical and biological properties. J Antibiotics 47:269–275

Mortlock AA, Bath C, Butlin RJ, Heys C, Hunt SJ, Reid AC, Sumner NF, Tang EK, Whiting E, Wilson C, Wright ND (1997) N-Methyl-2-[4-(2-methylpropyl) phenyl]-3-(3-methoxy-5-methylpyrazin-2-ylsulfamoyl)benzamide; One of a class of novel benzenesulphonamides which are orally-active, ET_A-selective endothelin antagonists. Bioorg Med Chem Lett 7(11):1399–1402

Murugesan N, Gu Z, Lee V, Webb ML, Liu EC-K, Hermsmeier M, Hunt JT (1995) Design and synthesis of nonpeptidal endothelin receptor antagonists based on the structure of a cyclic pentapeptide. Bioorg Med Chem Lett 5:253–258

Neidhart W, Breu V, Burri K, Clozel M, Hirth G, Klinkhammer U, Giller T, Ramuz H (1997) Discovery of Ro 48-5695: A potent endothelin receptor antagonist optimized from Bosentan. Bioorg Med Chem Lett 7:2223–2228

Ohashi H, Akiyama H, Nishikori K, Mochizuki J-I (1992) Asterric acid, a new endothelin binding inhibitor. J Antibiot 45:1684–1685

Ohlstein EH, Nambi P, Douglas SA, Edwards RM, Gellai M, Lago MA, Leber JD, Cousins RD, Gao A, Peishoff CE, Bean JW, Eggleston DS, Elshourbagy NA, Kumar C, Lee JA, Brooks DP, Ruffolo RR Jr, Feuerstein G, Weinstock J, Gleason JG, Elliott JD (1994) SB 209670, a rationally designed potent non-peptide endothelin receptor antagonist. Proc Natl Acad Sci USA 91:8052–8056

Ohlstein EH, Vickery-Clark LM, Storer BL, Douglas SA (1993) Antihypertensive effects of the endothelin receptor antagonist BQ-123 in conscious spontaneously hypertensive rats. J Cardiovasc Pharmacol 22 [Suppl 8]:S321–S324

Ohlstein EH, Arleth A, Bryan H, Elliott JD, Sung CP (1992) The selective endothelin-A receptor antagonist BQ-123 antagonizes ET-1 mediated mitogenesis in vascular smooth muscle. Eur J Pharmacol 225:347–350

Ohnishi M, Wada A, Tsutamoto T, Fukai D, Sawaki M, Maeda Y, Kinoshita M (1998) Chronic effects of a novel, orally active endothelin receptor antagonist, T-0201, in dogs with congestive heart failure. J Cardiovasc Pharmacol 31 [Suppl 1]: S236–238

Patel T, Galbraith S, McAuley M, Doherty A, Graham D, McCulloch J (1995) Therapeutic potential of endothelin receptor antagonists in experimental stroke. J Cardiovasc Pharmacol 26 [Suppl 3]:S412–S415

Patt WC, Edmunds JJ, Repine JT, Berryman KA, Reisdorph BR, Lee C, Plummer, MS, Shahripour A, Haleen SJ, Keiser JA, Flynn MA, Welch KM, Reynolds EE, Rubin R, Tobias B, Hallak H, Doherty AM (1997) Structure-activity relationships in a series of orally active γ-hydroxy butenolide endothelin antagonists. J Med Chem 40:1063–1074

Raju B, Okun I, Stavros F, Chan MF (1997a) Search for surrogates: A study of endothelin receptor antagonist structure activity relationships. Bioorg Med Chem Lett 7:933–938

Raju B, Okun I, Stavros F, Chan MF (1997b) Amide bond surrogates: A study in thiophenesulfonamide based endothelin receptor antagonists. Bioorg Med Chem Lett 7:939–944

Raju B, Wu C, Castillo RS, Okun IEJ, Stavros F, Chan MF (1997c) 2-Aryloxycarbonylthiophene-3-sulfonamides: highly potent and ET_A selective endothelin receptor antagonists. Bioorg Med Chem Lett 7(16):2093–2098

Raju B, Wu C, Kois A, Verner EJ, Okun I, Stavros F, Chan MF (1996) Thiophenesulfonamides as endothelin antagonists. Bioorg Med Chem Lett 6(22):2651–2656

Riechers H, Albrecht H-P, Amberg W, Baumann E, Bernard H, Bohm H-J, Klinge D, Kling A, Muller S, Raschack M, Unger L, Walker N, Wernet W (1996) Discovery and optimization of a novel class of orally active nonpeptidic endothelin-A receptor antagonists. J Med Chem 39:2123–2128

Roux S, Breu V, Giller T, Neidhart W, Ramuz H, Coassolo P, Clozel JP, Clozel M (1997) Ro 61–1790, a new hydrosoluble endothelin antagonist: general pharmacology and effects on experimental cerebral vasospasm. J Pharmacol Exp Ther 283(3):1110–1118

Sakai S, Miyauchi T, Kobayashi T, Yamaguchi I, Goto K, Sugishita Y (1998) Altered expression of isoforms of myosin heavy chain mRNA in the failing rat heart is ameliorated by chronic treatment with an endothelin receptor antagonist. J Cardiovasc Pharmacol 31 [Suppl 1 Endothelin V]:S302–S305

Sakaki J, Murata T,Yuumoto Y, Nakamura I, Frueh T, Pitterna T, Iwasaki G, Oda K, Yamamura T, Hayakawa K (1998a) Discovery of IRL 3461: A novel and potent endothelin antagonist with balanced ET_A/ET_B affinity. Bioorg Med Chem Lett 8:2241–2246

Sakaki J, Murata T,Yuumoto Y, Nakamura I, Hayakawa K (1998b) Stereoselective synthesis of a novel and bifunctional endothelin antagonist IRL 3630. Bioorg Med Chem Lett 8:2247–2252

Saudek V, Hoflack J, Pelton JT (1989) ^{1}H-NMR study of endothelin, sequence specific assignment of the spectrum and a solution structure. FEBS Lett 257:145–148

Spinella MJ, Malik AB, Everitt J, Anderson TT (1991) Design and synthesis of a specific endothelin-1 antagonist: effects on pulmonary vasoconstriction. Proc Natl Acad Sci USA 88:7443–7446

Stein PD, Hunt JT, Floyd DM, Moreland S, Dickinson KE Jr, Mitchell C, Liu EC-K, Webb ML, Murugesan N, Dickey J, McMullen D, Zhang R, Lee VG, Serafino R, Delany C, Schaeffer TR, Kozlowski M (1994) The discovery of sulfonamide endothelin antagonists and the development of the orally active ET_A antagonist 5-(dimethylamino)-N-(3,4-dimethyl-5-isoxazolyl)-1-naphthalenesulfonamide.   J Med Chem 37:329–331

Tasker AS, Sorensen BK, Jae H-S, Winn M, Von Geldern TW, Dixon DB, Chiou WJ, Dayton BD, Calzadila S, Hernandes L, Marsh KC, WuWong JR, Opgenorth TJ (1997) Potent and selective non-benzodioxole-containing endothelin-A receptor antagonists. J Med Chem 40:332–340

Urade Y, Fujitani Y, Oda K, Watakabe T, Umemura I, Takai M, Okada T, Ssakata K, Karaki H (1992) An endothelin B receptor-selective antagonist: IRL 1038, [Cys11-Cys15]-endothelin-1(11–12). FEBS Lett 311:12–16

Von Geldern T (1999) ABT-627 and beyond: An update on Abbott's endothelin antagonist program. In: Book of Abstracts 217th ACS National Meeting March 21–25 1999. American Chemical Society, Washington, DC, MEDI-136

Wakimasu M, Kikuchi T, Kubo K, Asami T, Ohtaki T, Fujino M (1993) Studies on endothelin antagonists. In: Testa B (ed) Perspectives in medical chemistry. Verlag Helvetica Chim Acta Basel, pp 165–177

Walsh TF, Fitch KJ, Williams DL Jr, Murphy KL, Nolan NA, Pettibone DJ, Chang RSL, O'Malley SS, Clineschmidt BV, Veber DF, Greenlee WJ (1995) Potent dual antagonists of endothelin and angiotensin II receptors derived from α-phenoxyphenylacetic acids (part III). Bioorg Med Chem Lett 5:1155–1158

Warner TD, Allcock GH, Vane JR (1993) The endothelin receptor antagonist PD 142893 inhibits endothelium-dependent vasodilatations induced by endothelin/sarafotoxin peptides. Br J Pharmacol 109:56P

Williams DL Jr, Murphy KA, Nolan NA, O'Brian DJ, Pettibone DJ, Kivlighn SD, Krause SM, Lis EV Jr, Zingaro GJ, Gabel RA, Clayton FC, Siegl PKS, Zhang K, Naue J, Vyas K, Walsh TF, Fitch KJ, Chakravaty PK, Greenlee WJ, Clineschmidt BV (1995) Pharmacology of L-754142, a highly potent, orally active, nonpeptidyl endothelin antagonist. J Pharmacol Exp Ther 275:1518–1526

Winn M, von Geldern TW, Opgenorth TJ, Jae H-S, Tasker AS, Boyd SA, Kester JA, Mantei RA, Bal R, Sorensen BK, Wu-Wong JR, Chiou WJ, Dixon DB, Novosad EI, Hernandez L, Marsh KC (1996) 2,4-Diarylpyrrolidine-3-carboxylic acids-potent ET_A selective endothelin receptor antagonists. 1. Discovery of A-127722. J Med Chem 39: 1039–1048

Wu C, Chan MF, Stavros F, Raju B, Okun I, Castillo RS (1997a) Structure-activity relationships of N^2-aryl-3-(isoxazolylsulfamoyl)-2-thiophenecarboamides as selective endothelin receptor-A antagonists. J Med Chem 40:1682–1689

Wu C, Chan MF, Stavros F, Raju B, Okun I, Mong S, Kellerm KM, Brock T, Kogan TP, Dixon RAF (1997b) Discovery of TBC11251, a potent, long acting, orally active endothelin receptor-A selective antagonist. J Med Chem 40:1690–1697

Wu-Wong JR, Chiou WJ, Naugles KE Jr, Opgenorth TJ (1994) Endothelin receptor antagonists exhibit diminishing potency following incubation with agonist. Life Sci 54:1727–1734

Xiang J-N, Luengo JI, Ohlstein EH, Elliott JD (1999) Endothelin receptor antagonists. In: Book of Abstracts, 217th ACS National Meeting March 21–25 1999 American Chemical Society, Washington, DC, MEDI-136

Xiang J-N, Atkinson ST, Gao A, Corbett DF, MacPherson DT, Nambi P, Ohlstein EH, Elliott JD (1997) A Novel series of ET_A selective antagonists. In: Book of Abstracts, 214th ACS National Meeting. Sept. 7–11 1997. American Chemical Society, Washington, DC MEDI-206

Yu M, Gopalakrishnan V, McNeill JR (1998) Hemodynamic effect of a selective endothelin-A receptor antagonist in deoxycorticosterone acetate-salt hypertensive rats. J Cardiovasc Pharmacol 31 [Suppl 1]:S262–S264

CHAPTER 10
Toxicology of Endothelin Antagonists

S.J. Morgan, P.K. Cusick, and B.A. Trela

A. Introduction

The widespread distribution of endothelin receptors and the suspected role of the endothelins are discussed thoroughly in previous and subsequent chapters. With the exception of issues of teratology, surprisingly few adverse effects secondary to administration of endothelin receptor antagonists have been reported in pre-clinical trials in experimental animals. Similarly, relatively few adverse effects have been reported in clinical trials in healthy volunteers or patients. The reader should be appropriately skeptical, however, as endothelin antagonists are relatively new therapeutic entities and, as such, their toxicity spectrum could be broader and deeper than the literature would suggest at this point in time. Only wider preclinical and clinical experience can answer such questions.

B. Cardiovascular System

A wide variety of pharmacologic agents that cause vasodilation, including minoxidil, hydralazine, and theobromine have been known to result in coronary arterial segmental medial hemorrhage, necrosis and acute inflammation (Dogterom and Zbinden 1992). These agents have been associated with a rather profound reduction in mean arterial pressure and a compensatory reflex tachycardia (Dogterom and Zbinden 1992). It is thought that the lesions are secondary to increased tension on the vascular wall as they tend to occur in the most pharmacologically responsive segment of the arterial system (Mesfin et al. 1989) rather than being the result of a direct cytotoxicity. Similar lesions have been observed in dogs that have received continuous i.v. infusions with a mixed endothelin receptor antagonist (ET_A and ET_B), SB 209670 (Louden et al. 1998). In contrast to minoxidil, hydralazine, and theobromine, physiologic changes in dogs that received SB 209670 have been limited to a slight decrease in mean arterial pressure and a minor increase in heart rate (Louden et al. 1998). For a point of reference, the threshold hemodynamic effects that are associated with cardiovascular toxicity in dogs receiving minoxidil are an increase in heart rate of at least 55 beats/min and a decrease in mean arterial pressure of at least 30 mm Hg (Mesfin et al. 1996). These are considerably greater than those changes noted with SB 209670 in which there was an increase in heart rate of 10–30 beats/min and decrease in mean arterial pressure of 10–15 mm Hg (Louden et al. 1998). Despite the lack of systemic evidence for a physiologic change which is sufficient to result in a coronary arterial lesion, the arterial changes with SB 209670 are still considered to be the result of an exaggerated pharmacologic effect (local vasodilation) rather than a direct cytotoxic effect due to the unique localization of endothelin receptors. As the right atria has the highest concentration of endothelin receptors, it is theorized that there may be significant local vasodilation with resultant increased regional blood flow and increased shear and pressure on the coronary arterial wall (Louden et al. 1998).

The cardiovascular alterations seen in dogs treated with minoxidil and other vasodilators do not appear to have clinical relevance to humans. In general, it is thought that dogs may not be the ideal model for predicting cardiovascular toxicity due to the differential sensitivity of dogs and humans to the effects of a variety of cardioactive drugs (Dogterom and Zbinden 1992). For example, if one estimates the safety margin of minoxidil based on the ratio of nontoxic dose/serum concentration in dogs to the serum concentration of the efficacious dose in humans, the safety margin would be less than 0 (Mesfin et al. 1996) in spite of the lack of cardiac lesions noted during autopsy of approximately 200 investigational patients who received minoxidil (Sobota 1989). To date, there is also no evidence of cardiovascular toxicity in humans that have received endothelin antagonists. To date, coronary arterial lesions have not been reported in other species of experimental animals that have received endothelin antagonists.

C. Teratology

In contrast to the relative paucity of adverse effects that have been reported for endothelin receptor antagonists in subacute and chronic studies in experimental animals with endothelin antagonists, the potential for extensive life-threatening effects on the developing embryo/fetus is of major concern. Both ET_A and ET_B receptors are likely to have a major regulatory role in development, and considerable evidence suggests that their disruption during critical periods is likely to have serious sequelae (see Chap. 6). The potential toxicity from blockade of the ET_A and ET_B receptors will be discussed separately.

I. ET_A Receptor

Numerous investigations indicate the importance of the ET-1, an ET_A ligand, on the epithelial cells of the pharyngeal arches, indicating that ET-1 deficiency results in disruption of normal development of mandibular and other pharyngeal arch-derived tissues. Mice deficient for the ET_A receptor mimic the human conditions collectively termed CATCH 22 or velocardiofacial syndrome, which includes severe craniofacial deformities and defects in the cardiovascular outflow tract. Craniofacial defects noted in pups include a poorly formed mandible with a variable lack of midline fusion, hypoplastic pinnae, thickening of the palate, hypoplasia of the tongue and associated musculature, and a hypoplastic and rostrally displaced thymus (CLOUTHIER et al. 1998). Similar defects have been observed in mice deficient for either ET-1 (KURIHARA et al. 1994) or endothelin converting enzyme 1 (EDF-1; YANAGISAWA et al. 1988). Other defects noted in ET_A deficient mice include aberrant middle ear development (CLOUTHIER et al. 1998).

Abnormalities detected in the heart of ET_A deficient mice included ventricular septal defect (VSD) with the aorta overriding the defective septum, double outlet-right ventricle, persistent truncus arteriosus and complete transposition of the great arteries. Abnormalities of the outflow tract of ET_A deficient mice included interruption of the aorta, absent right subclavian artery, extra arteries branching off the right and left common carotid arteries, and right dorsal aorta with a right-sided ductus arteriosus. Again, similar defects have been reported in mice deficient for ET-1, but their incidence is relatively low unless there is concurrent administration of anti-ET-1 neutralizing antibodies or a selective ET_A antagonist (BQ123) to pregnant endothelin-1 deficient mice, suggesting that it may be the loss of ET-1/ET_A interactions that is of importance in development of the defects (KURIHARA et al. 1995).

Despite the lack of current evidence that endothelin antagonists may result in cardiovascular lesions in the adult, there is concern that blocking of the ET_A receptor during critical phases of embryological development may result in significant teratologic findings in the human. Endothelin receptors have been shown to be present on the epithelial cells of both mouse and human embryonal mandibular tissues, suggesting possible developmental roles in both species (BARNI et al. 1995). Other lines of evidence that suggest that endothelin and/or its interaction with the ET_A receptor may contribute to alterations in development includes the local repression of the preproendothelin-1 gene and the low expression of ET_A receptor associated with intracranial arteriovenous malformations (RHOTEN et al. 1997).

II. ET_B Receptor

Interruption of the ET_B receptor is not without its own set of potential problems. Evidence suggests that mutations in the ET_B receptor results in congenital aganglionic megacolon with pigment abnormalities in rats, mice, humans, and horses which is due to defects in development of neural crest-derived cell lineages to enteric neurons and epidermal melanocytes (ROBERTSON et al. 1997; HOSADA et al. 1994; METALLINOS et al. 1998; KUNIEDA et al. 1996). Although there is currently no documentation to validate the existence of similar defects in teratology studies with endothelin compounds, the presence of neural lesions in a wide variety of species, including humans, with ET_B receptor defects suggests that the possibility of induction of congenital aganglionosis secondary to lack of ET_B/endothelin interactions during critical stages of embryonal development cannot be ruled out.

Although best known for its role in the neural development of the colonic region, ET_B receptors appear to be involved in neural development in other regions as well. Recent evidence indicates that ET_B receptors are expressed prenatally in the ventricular and subventricular zones as well as postnatally in the ependymal and subependymal zones. ET_B mRNA has also been detected prenatally in the dorsal root ganglia and postnatally in the cerebellar Bergmann glial cells and epithelial cells of the choroid plexus. Thus, it is possible that the ET_B receptor may have an important role in the differentiation, proliferation, or migration of a side variety of neural cells during development (TSAUR et al. 1997). However, to date, the only neural crest defects associated with ET_B receptor interruption are those described above, or colonic aganglionosis.

D. Clinical Trials

In clinical trials, endothelin antagonist receptors have been associated with relatively few adverse side effects. Administration of bosentan (Ro 47–0203), a mixed ET_A and ET_B receptor antagonist, to subjects has been associated with a slight decrease in blood pressure and heart rate (5 mm Hg and 5 beats/min). Vomiting and local irritation (thrombophlebitis or partially occluded vein) was observed at the higher intravenous doses. A mild headache was the most common adverse event, occurring with both oral intravenous routes of administration (WEBER et al. 1996). Hypertensive patients receiving bosentan orally have also exhibited headaches. Other adverse clinical signs in patients included flushing, leg edema, and more rarely, asymptomatic increases in serum alanine and aspartate aminotransferase activities (KRUM et al. 1998). It has been reported that a variety of other endothelin receptor antagonists also causes a transient headache when administered to subjects, possibly associated with enhanced vascular nitric oxide release associated with unopposed stimulation of ET_B receptors (WEBB and STRACHAN 1998).

References

Barni T, Maggi M, Fantoni G, Serio M, Tollaro I, Gloria L, Vannelli GB (1995) Identification and localization of endothelin-1 and its receptors in human fetal jaws. Dev Biol 169:373–377

Cloutheir DE, Hosoda K, Richardson JA, Williams SC, Yanaglsawa H, Kuwaki T, Kumada, M, Hammer RE, Yanagisawa M (1998) Cranial and cardiac neural crest defects in endothelin-A receptor-deficient mice. Development 125:813–824

Dogterom P, Zbinden G (1992) Cardiotoxicity of vasodilators and positive inotrophic/vasodilating drugs in dogs: an overview. Crit Rev Toxicol 22:203–241

Hosoda K, Hammer RE, Richardson JA, Baynash AG, Cheung JC, Giaid A, Yanagisawa M (1994) Targeted and natural (piebald-lethal) mutations of endothelin-B receptor gene produce megacolon associated with spotted coat color in mice. Cell (79):1267–1276

Krum H, Viskoper RJ, Lacourciere Y, Budde M, Charlon V(1998) The effect of an endothelin-receptor antagonist, bosentan, on blood pressure in patients with essential hypertension, N Engl J Med 338:784–790

Kunieda T, Kumagai T, Tsuji T, Ozaki T, Karaki H, Ikadae H (1996) A mutation in the endothelin-B receptor gene causes myenteric aganglionosis and coat color spotting in rats. DNA Res: 30(2):101–105

Kurihara Y, Kurihara H, Suzuki H, Dodama T, Maemura K, Nagai R, Oda H, Kusawki T, Cau W-H, Kamada N, Jishage K, Ouchi Y, Azuma S, Toyoda Y, Ishikawa T, Kumada M, Yazaki Y (1994) Elevated blood pressure and cranialfacial abnormalities in mice deficient in endothelin-1. Nature 368:703–710

Kurihara Y, Kurihara H, Oda H, Maemura K, Nagai, R, Ishikawa T, Yazaki Y (1995) Aortic arch malformations and ventricular septal defect in mice deficient in endothelin-1. J Clin Invest 96:293–300

Louden C, Nambi P, Branch C, Gossett K, Pullen M, Eustis S, Solleveld HA (1998) Coronary arterial lesions in dogs treated with an endothelin receptor antagonist, J Cardiovasc Pharmacol 31[Suppl 1]:S384–385

Mesfin GM, Piper RC, DuCharme DW, Carlson RG, Humphrey SJ, Zins GR (1989) Pathogenesis of cardiovascular alterations in dogs treated with minoxidil. Toxicol Pathol 17(1):164–181

Mesfin GM, Higgins MJ, Robinson FG, Wei-Zhu Z (1996) Relationship between serum concentrations, hemodynamic effects, and cardiovascular lesions in dogs treated with minoxidil. Toxicol Appl Pharamcol 140, 337–344

Metallinois DL, Bowling AT, Rine J (1998) A missense mutation in the endothelin-B receptor gene is associated with Lethal White Foal syndrome: an equine version of Hirschsprung disease. Mamm Genome 9(6):426–431

Rhoten RL, Comair YG, Shedid D, Chyatte D, Simonson MS (1997) Specific repression of the preproendothelin-1 gene in intracranial arteriovenous malformations. J Neurosurg 86(1):101–108

Robertson K, Mason I, Hall S (1997) Hirschsprung's disease: genetic mutations in mice and men. Gut 41(4):436–441

Sobota JT (1989) Review of cardiovascular findings in humans treated with minoxidil. Toxicol Pathol 17(1 Pt 2):193–202

Tsaur, ML, Wan YC, L FP, Cheng HF (1997) Expression of B-type endothelin receptor gene during neural development. FEBS Letters 417:208–212

Webb DJ and Strachan FE (1998) Clinical experience with endothelin antagonists, Am J Hypertens 11:71S–79S

Weber C, Schmitt R, Birnboeck H, Hopfgartner G, van Marle SP, Peeters PAM, Jonkman JHG, Jones, C (1996) Pharmacokinetics and pharmacodynamics of the endothelin-receptor antagonist bosentan in healthy human subjects. Clin Pharmacol Ther 60:124–137

Yanagisawa H, Yangisawa M, Kapur RP, Richardson JA, Williams SC, Clotheir DE, deWit D, Emoto N, Hammer RE (1988) Dual genetic pathways of endothelin-mediated intercellular signaling revealed by targeted disruption of endothelin converting enzyme-1 gene. Development 125:825–836

CHAPTER 11

Endothelins and the Release of Autacoids

P. D'Orléans-Juste, G. Bkaily, M. Duval, J. Labonté, M. Plante,
G. Cournoyer, and N. Berthiaume

A. Introduction

Curiously, among several well-studied pressor peptides, only endothelin-1 (ET-1) triggers at first a transient drop in blood pressure followed by a long-lasting hypertensive response when administered intravenously in most animal species (Table 1). The characteristic biphasic response to endothelin-1 was early on explained by Warner et al. (1989). That group suggested that the biphasic response to the peptide may be due to concomitant activation of two different receptor types; a receptor type which is responsible for the hypotensive response, and another type which is mainly involved in the protracted increase in vascular resistance in several animal species. This hypothesis has been consistently confirmed in the last few years (Tsuchiya et al. 1990; Adachi et al. 1991; Gratton et al. 1995a, 1997). The receptorial dichotomy involved in the vasoactive responses to endothelin-1 was also extended to man (Verhaar et al. 1998; Webb 1996).

Animals subjected to a genetic repression of the endothelin-1 but not angiotensin II or bradykinin genes (Tsuchida et al. 1998; Alfie et al. 1996) show deleterious malformations illustrating the importance of the former peptide in several developmental stages (Kurihara et al. 1994; see Chap. 6). Furthermore, the mediators and modulators released by intravenously administered ET-1 are often involved in compensatory mechanisms aimed towards the control of the peptide's induced effects in plasma extravasation, mitogenesis, angiogenesis, and even central control of blood pressure or fever (Lehoux et al. 1992; Battistini et al. 1993; Poulat et al. 1994; Fabricio et al. 1998).

The present chapter will therefore address the physiological and pharmacological consequences of endothelin-1-induced release of several mediators and modulators. In addition, the complex cross-talk between endothelins and nitric oxide, eicosanoids, cytokines and neuromediators will be addressed in various models including the mouse as well as in humans. Finally, the chapter will be concluded with some physiopathological considerations related to the role of ET-1 in the control of blood pressure.

Table 1. ET-1-induced vasodilatory effects in various animal species and in man

Species	Responses	Agonists (ED_{50} or dose range)	Receptors	References
Rat	Transitory hypotension followed by sustained hypertension	ET-1 (1 nmol/kg)	ET_B	De Nucci et al. (1988); Warner (1999)
Cat	"	Sarafotoxin 6C (0.1–1 nmol/kg)	ET_B	Clozel et al. (1996)
	"	ET-1 (0.3 nmol/kg)	ET_B	Minkes and Kadowitz (1991)
	"	Sarafotoxin 6C (3 nmol/kg)	ET_B	Minkes and Kadowitz (1991)
Pig	"	ET-1 (100 pmol/kg)	ET_B	Clement and Albertini (1996)
Dog	"	ET-1 (+50 ng/kg per min)	ET_B	Tsuchiya et al. (1990)
Cat	Cerebral vasodilatation	IRL-1620 (0.01–1 nmol/kg)	ET_B	Kobarim et al. (1994)
		ET-1 (10 nmol/l)	ET_B	Patel et al. (1996)
Goat	Increase in coronary blood flow	ET-1 (0.01–0.03 nmol/kg)	No ET_B contribution	Garcia et al. 1994
		IRL-1620 (0.01–0.03 nmol/kg)	No ET_B contribution	Garcia et al. 1994
Guinea pig	Hypotension following NO synthase inhibition (release of vasodilatory eicosanoid)	ET-1 (0.5 nmol/kg)	ET_B	Lewis et al. 1999
Man	Reduction of total peripheral vascular resistance index	–	ET_B	Strachan et al. 1999

B. Endothelin-Induced Release of Endothelial-Derived Factors

I. Nitric Oxide

The classical observation of endothelium-derived release of nitric oxide by endothelins was firstly described by DE NUCCI et al. (1988), in which a bolus administration of ET-1 triggered the previously-mentioned hypotensive effect followed by protracted pressor response. Characteristically, to trigger the release of endothelial-derived nitric oxide, endothelin-1 must activate high affinity ET_B receptors localized on the surface of the endothelial cell. These receptors not only act as membrane-bound proteins responsible for the release of NO induced by endothelin-1, but also act as an efficient clearance receptor for the same peptide. The high efficiency of this clearance mechanism is illustrated by the significant increase in plasma levels of immunoreactive endothelins following systemic administration of a selective ET_B antagonist, BQ-788, in different animal models, such as the rat, rabbit, dog as well as human (FUKURODA et al. 1994; GRATTON et al. 1997; DUPUIS et al. 1996; STRACHAN et al. 1999). Interestingly, the group of Dupuis and co-workers has shown that the selective ET_B antagonist, BQ-788, significantly interfered with the pulmonary-dependent clearance of endothelin in healthy subjects. Of course, interference with these ET_B clearance receptors will trigger sufficiently high amounts of circulating endothelin-1 to induce a significant alteration in vascular resistance following administration of the antagonist; this was illustrated in animal models by GRATTON et al. (1997) and more recently in humans by STRACHAN et al. (1999). Characteristically, the ET_B receptor-dependent release of nitric oxide, either in cell culture conditions or in vivo, is always short-lasting and highly tachyphylactic (LE MONNIER DE GOUVILLE et al. 1990, 1991). Confusing elements have arisen in the literature, as far as the nature of the transitory and tachyphylactic release of EDRF by endothelin-1 is concerned. It was first suggested that the ET_B receptor may rapidly be uncoupled from its G-protein and consequently from the subsequent activation of intracellular mechanisms (LE MONNIER DE GOUVILLE et al. 1990, 1991). The second possibility was that the activation of the ET_A receptor on the smooth muscle cell may render the tissue insensitive to subsequent challenge by nitric oxide release from the endothelium (HALEEN et al. 1993). Unfortunately, at this stage, no group has directly monitored the capacity of intact vascular endothelium to release nitric oxide following consecutive intravascular administrations of the peptide in vivo.

II. Eicosanoids

Similarly to nitric oxide, eicosanoids are efficient modulators of the vasoactive effects of endothelin and cyclooxygenase inhibitors, such as indomethacin, will markedly affect the vasoconstrictive and/or pressor effects of the peptide in animal species such as the rat (DE NUCCI et al. 1988; WALDER et al. 1989). Albeit well documented in several animal species, the contribution of endogenous eicosanoids as modulators of endothelin-1-induced increases in the vascular tone has only been remotely studied in man (STRACHAN et al. 1995) and is mostly associated to pathological events. One of these examples is related to the beneficiary effects of epoprostenol (prostacyclin analogue) in the treatment of primary pulmonary hypertension (PPH). This particular eicosanoid is not only able to partially alleviate the debilitating status of the patient, but also reduces the circulating levels of endothelin-1 in PPH (LANGLEBEN et al. 1999; see Chap. 15).

On the other hand, it is of interest that ET-1 triggers the release of eicosanoids as direct mediators of pharmacological responses to the peptide (PONS et al. 1991). A good example of this can be observed in the marked indomethacin-sensitive bronchoconstriction induced in guinea pigs by systemic application of ET-1 or a selective ET_B agonist, IRL-1620 (PONS et al. 1991; NOGUCHI et al. 1996).

Another example would be the ET-1-induced inhibition of ex vivo ADP-triggered platelet aggregation in the rabbit (MCMURDO et al. 1993). ET-1 has no direct effect on platelets per se. However, when injected systemically, ET-1 will trigger the release of prostacyclin which will influence subsequently the capacity of platelets to aggregate in ex vivo experimental conditions. Unlike ET-1-induced release of nitric oxide from the endothelium in vivo, the eicosanoid-releasing capacities of the peptide are highly reproducible in both in vitro and in vivo systems (D'ORLÉANS-JUSTE et al. 1992; LEWIS et al. 1999).

1. Receptors Involved in the Release of Eicosanoids

Although ET_B receptors have traditionally been suggested to be involved in the ET-1-induced release of prostacyclin from endothelial cells, only few functional observations have confirmed this particular concept (FILEP et al. 1991).

In contrast, our laboratory has demonstrated the importance of ET_A-dependent release of prostacyclin in the guinea pig lung, the rabbit kidney, the rat lung, and several other systems (D'ORLÉANS-JUSTE et al. 1993, 1994). The type of receptors involved in the ET-1-induced release of prostacyclin appears to be organ-specific. For example, in the rabbit, systemically-administered ET-1 will trigger inhibition of ADP-induced platelet aggregation ex vivo through mainly ET_B receptors (MCMURDO et al. 1993). Yet, results from our laboratory show that the pulmonary generation of prostacyclin from the rabbit lung involves both ET_A and ET_B receptors (GRATTON et al. 1998). Furthermore, in the rabbit the lung is a major source of PGI_2 responsible for ET-1-dependent inhibition of platelet aggregation. Interestingly, the ET_B selective antagonist, BQ-788, markedly reduces the anti-aggregatory and PGI_2 releasing

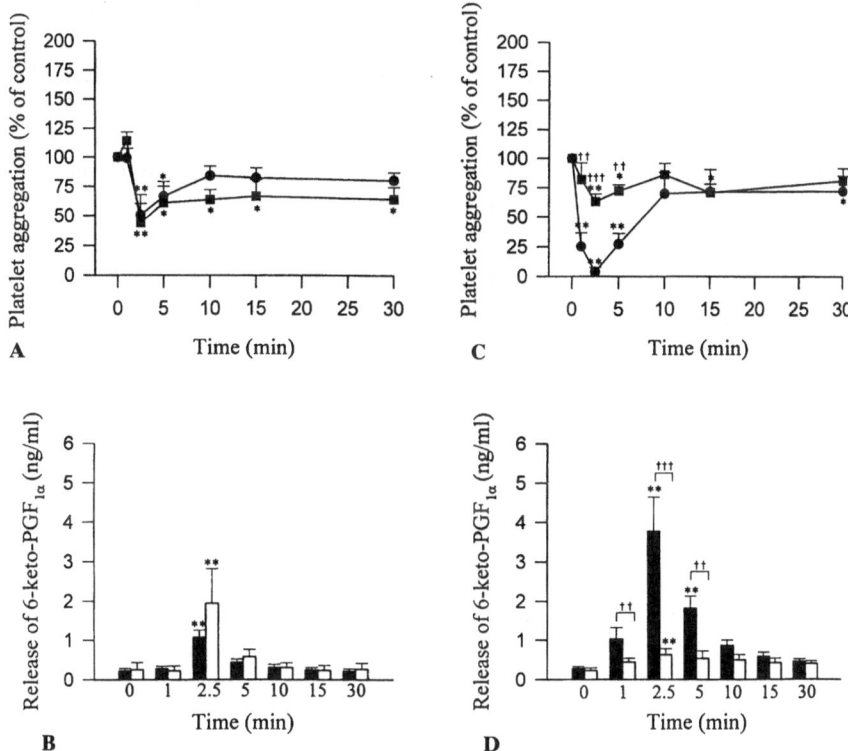

Fig. 1A–D. Time course of the effect of ET-1 (0.5 nmol/kg; intravenous; **A, B**) and IRL-1620 (3 nmol/kg; intravenous; **C, D**) on ex vivo platelet aggregation (**A, C**) and release of prostacyclin, measured as 6-keto-PGF$_{1\alpha}$ (**B, D**), either alone (and closed bars) or in the presence of BQ-788 (0.25 mg/kg; intraventricular; bolus 5 min before agonist; and opened bars). Each *point with a bar* represents the mean ± SEM of at least 5 experiments. *$P < 0.05$, **$P < 0.01$ when compared to time = 0. ††$P < 0.01$, †††$P < 0.001$ when compared to IRL-1620 alone

properties of IRL-1620, but not ET-1 in the rabbit (Fig. 1). This suggests that predominantly ET$_A$ and to a lesser extent ET$_B$ receptors are involved in the ET-1 induced release of prostacyclin that produces inhibition of platelet aggregation and systemic in the rabbit pulmonary circulations. ET$_B$ and ET$_A$-dependent release of prostacyclin from the pulmonary circulation may be explained by receptors located on the endothelium and vasculature, respectively.

This physiological dichotomy in the ET-1-induced release of eicosanoids is also demonstrated in a series of experiments performed recently in our laboratory, where systemically administered ET-1 induced the release of prostacyclin via ET$_A$ receptors and thromboxane via ET$_B$ receptors within the guinea pig pulmonary circulation. The latter receptor type is also responsible for the systemic release of vasodilatory prostanoids induced by ET-1 (LEWIS et al. 1999).

Interestingly, the prostacyclin-releasing capacities of ET-1 following ET_A receptor activation found in vivo and in perfused organs have also been correlated at the cellular level by the group of MALIK (WRIGHT and MALIK 1996), which has suggested a phospholipase D-dependent activation of the arachidonic acid cascade following receptor activation.

III. Cross-Talk Between the Endothelin, Nitric Oxide and Arachidonic Acid Pathways

The endothelium, as previously mentioned, has the capacity to release, among other factors, prostacyclin (LUSCHER and BARTON 1997), endothelin-1 (YANAG-ISAWA et al. 1998), and nitric oxide (MURAD 1998) into the circulation in an endocrine or paracrine fashion. Because of the remarkable surface area that it occupies one can truly consider the vascular endothelium to be a physio-logically important secretory gland. Secretion of one or all of the above factors will dramatically influence vascular events as well as the behavior of several blood-born cells, such as neutrophils, macrophages, and platelets (WARNER et al. 1999). This vasomodulatory ability of the endothelium is even more complex since the endothelial cell layer may upregulate or downregulate its synthesis of any one of these three factors in an autocrine fashion. For example, BOULANGER and LUSCHER (1990) has demonstrated that nitric oxide downregulates, in cultured endothelial cells in Vascular Smooth muscle cells the absolute production of mature endothelin-1.

Investigated to a lesser extent, the cross-talk between eicosanoids and the release of endothelin-1 remains unclear with the exception of an Australian study demonstrating that healthy elderly subjects treated with non-steroidal antiinflammatory drugs show a mild yet significant hypertension and a concomitant increase in immunoreactive plasma endothelin-1 (JOHNSON 1997). In a very recent study, we have also shown that endogenous nitric oxide strongly repressed the eicosanoid-releasing properties of endothelin-1 in the guinea pig. Interestingly, L-NAME sharply increased the generation of ET-1-induced thromboxane, but not prostacyclin in the pulmonary circulation of that animal species. On the other hand the release of the latter prostanoid is sharply enhanced by L-NAME in the systemic circulation of the same animal (LEWIS et al. 1999).

Interestingly, it appears that the direct bronchoconstrictive properties of eicosanoids, as illustrated with U46619 (LEWIS et al. 1999), are not potentiated in situations of NO synthase inhibition in vitro in the guinea pig perfused lung or in vivo as measured through the increase in pulmonary insufflation pres-sure in the same animal. This is one of the pieces of evidence to support the hypothesis that there is a selective potentiation of ET-induced eicosanoid release at a step beyond the direct activation of the thromboxane receptors in the NO-synthase impaired guinea pig.

The importance of the above-mentioned considerations lays in that they show a physiological interaction between ET-1, nitric oxide, and prostacyclin possibly at the level of the endothelium. In addition, this three-way cross-talk may not be limited to the endothelial layer, as other cell types such as macrophages, epithelial cells, and human bronchial cells (FILEP et al. 1993) are all able to generate these same factors.

On the other hand, in the guinea pig, it seems that only the constitutively expressed enzymes (COX-1 and eNOS), concomitantly with the ECE, are involved in the cross-modulatory properties of these three factors. This assumption is based on the fact that the inhibition of the NO synthase with L-NAME results in immediate alterations in the ET-1-induced secretion and/or synthesis of eicosanoids in the guinea pig, as illustrated in the guinea pig perfused lung (Fig. 2). These events are not observed, for example, with perfused lungs pretreated with iNOS-selective inhibitors, such as 1400 W (GARVEY et al. 1997).

Finally, conventional inducers of either iNOS or COX-2 are unnecessary to monitor this type of cross-modulation in the anesthetized rat or guinea pig (HAMILTON and WARNER 1998; LEWIS et al. 1999).

IV. Intracellular Mechanisms Involved in the Nitric Oxide or PGI$_2$-Induced Downregulation of ET-1 Production

There are complex intracellular mechanisms involved in the ET-1 autoregulatory properties of the above mentioned factors which are themselves released by the potent vasoactive peptide. At this stage, little is known of the mechanisms whereby prostacyclin inhibits endothelin release from endothelial cells. Current knowledge suggests that prostacyclin, through the IP receptor, will increase the intracellular production of cyclic AMP and so activate a cyclic AMP-dependent protein kinase (PKA). The activated PKA will then normally inhibit calcium entry and mobilization (OZAKI et al. 1996; TERTYSHNIKOVA and FEIN 1998; YAMAGISHI et al. 1994) at the vascular smooth muscle cell level. Similar events in the endothelial cell may explain the negative influence of cAMP stimulating factors on the release and/or production of ET-1. Nonetheless, little is known about the pretranscriptional influences of prostacyclin on the production of endothelin-1.

In contrast, the pre-transcriptional and post-transcriptional influences of nitric oxide on the production of endothelin are now relatively well described with the pivotal report by BOULANGER and LUSCHER (1990) who illustrated not only an inhibition of the ET-1 release, but also an inhibition of the ET-1 message, as previously mentioned.

Nitric oxide has also been shown to affect PGH synthase activity illustrating an interesting autocrine cross-talk between these two factors (MACCARRONE et al. 1997).

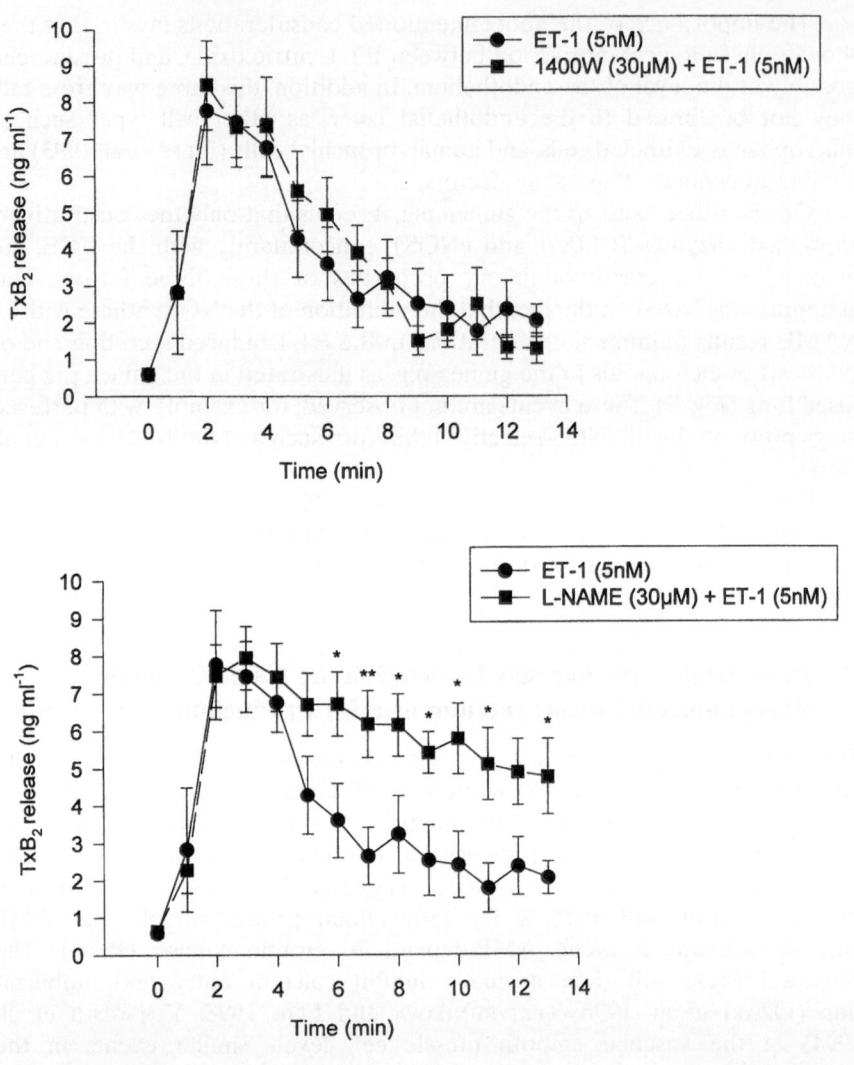

Fig. 2. Time course of the effect of ET-1 (5 nM) on thromboxane (TxB_2) release from the guinea pig lung in absence (●) or presence of 1400W or L-NAME ($30\,\mu M$, 30 min before the administration of ET-1, ■). Each *point with a bar* represents the mean ± SEM of seven experiments

V. ET-1-Induced Release of Cytokines

Different types of cytokines, growth factors, and endotoxins have been demonstrated to promote the production of endothelin either in vitro, in cell cultures of endothelial cells, or in vivo. A striking example of this cytokine-induced release of endothelin was illustrated by HOHLFELD et al. (1995), who showed that intravascular administration of TNF-α triggered an endothelin-1-dependent increase in coronary resistance in the anesthetized rat.

What is known, however, of the effect of endothelin-1 on the generation of cytokines? STANKOVA et al. (1995 and 1996) have shown, for example, that endothelin-1 through ET_B receptors acts synergistically with interleukin-1β to generate the production of interleukin-6 as well as an enhancement of its message in human umbilical cord endothelial cells. The increase of IL-6 mRNA has been suggested to be mediated through ET_B receptor activation, which both increases intracellular calcium and increases the stability of IL-6-specific mRNA (STANKOVA et al. 1995 and 1996).

C. Autacoid Mediators Generated by Endothelin-1 in Non-Vascular Cells

As previously mentioned, endothelin-1 is capable of releasing mediators, such as eicosanoids and nitric oxide in several cell types other than endothelial cells and vascular smooth muscle cells (ABDEL-LATIF et al. 1996). For example, different cell types, such as epithelial cells as well as bronchial smooth muscle cells, have been shown to be stimulated by ET-1 and to release different bronchoactive factors. Moreover, cells from the gastrointestinal tract, such as parietal cells and gastric fundus muscular cells, release HCl and PGE_2, respectively, under ET-1 stimulation (SAID and EL, MOWAFY 1998; WHITTLE and LOPEZ-BELMONTE 1993). Additional cell types, such as the hepatic Ito cells, have also been shown to release nitric oxide, when stimulated by peptides from the endothelin family (ROCKEY 1997).

Of interest is that even immortalized cells, such as LLC-PK$_1$ cells derived from porcine renal epithelial cells, release nitric oxide when stimulated by ET-1 or ET-3 through activation of ET_B receptors (WARNER et al. 1992). In renal epithelial cells lining the collecting tubule, an important physiological role has been suggested for ET_B receptors in mediating the natriuretic properties of endothelins (EDWARDS et al. 1993). Endothelin-1 has also been reported to trigger the release of atrial natriuretic factors from the cardiac atria of several species, including the mouse and the rat (FUKUDA et al. 1989). For all of the different cell types previously described, little is known of the actual intracellular mechanisms involved in their secretory responses to endothelins.

Table 2 illustrates some of the different vascular and non-vascular cells which respond to ET-1 through the release or the inhibition of modulators and/or autacoids. Some receptorial patterns may be defined through that particular table. For example, histamine, serotonin, and eicosanoids seem to be released following a predominant activation of ET_A receptors. In contrast, NO, cytokine, and acetylcholine release stimulated by ET-1 involves almost exclusively ET_B receptor activation.

Table 2. ET-1-induced modulation of autacoids released from various cell types

Autacoid	Cell type	Receptors involved	References
Histamine	Toad endothelial cells	Only ET-1 tested	Doi et al. (1995)
	Guinea pig mast cells	ET_B	Uchida et al. (1992)
	Rat and mouse peritoneal mast cells	ET_A	Yamamura et al. (1994a)
	Mouse bone marrow-derived mast cells	ET_A	Yamamura et al. (1994b)
	Mouse bone marrow derived mast cells	ET_A	Yamamura et al. (1995)
Serotonin	Mouse bone marrow derived mast cells	ET_A	Egger et al. (1995)
	Mouse peritoneal mast cells	ET_A	Yamamura et al. (1994a)
Eicosanoids	Human pericardial smooth muscle cells (A.A.)	ET_A	Wu-Wong et al. (1996)
	Mouse bone marrow derived mast cells (LTC_4)	ET_A	Yamamura et al. (1994b); Egger et al. (1995)
	Bovine aortic endothelial cells (PGI_2)	Only ET-1 tested	Filep et al. (1991)
	Dog airway cells (TxB_2,PGD_2)	Only ET-1 tested	Ninomiya et al. (1992)
	Rat aortic endothelial cells (A.A.)	ET_A	N'Diaye et al. (1997)
NO	Rat glial cells	ET_B	Murayama et al. (1998)
	Rat aorta smooth muscle cells	ET_A	Nakahashi et al. (1995)
	Rat glomerular mesangial cells	ET_B	Owada et al. (1994)
	Bovine carotid artery endothelial cells	ET_B	Hirata and Emori (1993)
	Porcine renal epithelial cells	ET_B	Warner et al. (1992)
Cytokines (IL-6)	Human umbilical vein endothelial cells	ET_B	Stankova et al. (1995, 1996)
ACh	Inhibition from cat presynaptic terminals	Only ET-3 tested	Nishimura et al. (1991)
	Guinea-pig ileum myenteric neurons	ET_B	Yoshimura et al. (1996)
	Reduction by ET-3 in dog presynaptic nerve terminals of ganglia	Only ET-3 tested	Kushiku et al. (1991)

D. Endothelin-Converting Enzyme-Dependent Release of Modulators

Kashiwabara and colleagues were the first to demonstrate the obligatory contribution of a metalloprotein which shares some catalytic homology with the neutral endopeptidase (EC 24–11) (KASHIWABARA et al. 1989) in the generation of ET-1. The endothelin-converting enzyme-1 (ECE-1), a 120 kilodalton protein, is now recognized as the main physiological entity responsible for the conversion of big-endothelins to endothelins (see Chaps. 3, 7). As described below, ECE activity can be monitored through the release of several vasoactive factors including eicosanoids and nitric oxide.

I. Eicosanoids

The stimulation of endogenous autacoids by the precursors of the endothelins, namely big-endothelin-1, 2, and 3, has only been explored scarcely. Big-endothelins have the capacity, as an intermediate metabolite, to generate eicosanoids in particular as summarized in Table 3.

As for the great majority of direct effects of big-endothelins, the eicosanoid-releasing properties of those precursors are totally dependent on a conversion through the phosphoramidon-sensitive endothelin-converting enzyme (D'ORLÉANS-JUSTE et al. 1991b). Initially it was shown that big-endothelin-1, similarly to endothelin-1, generates the release of prostacyclin from the rat perfused lung in a BQ-123-sensitive manner (D'ORLÉANS-JUSTE et al. 1992). This first series of observations prompted us to characterize fully the eicosanoid-releasing characteristics of the three precursors both in vivo and in vitro.

Consistent with the above, we found that big-ET-1 was a potent generator of prostacyclin and/or thromboxane A_2 in several organs, as depicted in Table 3.

Interestingly, we were also able to establish what we now suggest as two anatomically and functionally distinctive ECEs, one in the pulmonary circulation and the other in the systemic circulation, the former being less efficient in converting the precursor of ET-2, big-ET-2, when compared to big-ET-1. Both peptides are potent pressor agents when injected i.v. in the anesthetized rat, guinea pig and rabbit (MATSUMURA et al. 1990; FUKURODA et al. 1990; D'ORLÉANS-JUSTE et al. 1990). The precursor of ET-2, when compared to big-ET-1 is, however, incapable of either inducing a bronchoconstriction in the guinea pig or in stimulating the release of eicosanoids from perfused lungs in the guinea pig and rat (GRATTON et al. 1995a,b).

On the other hand, more recent results from our laboratory show the importance of the lung circulation to the big-ET-1-dependent inhibition of platelet aggregation ex vivo reported ten years ago in the rabbit (D'ORLÉANS-JUSTE et al. 1991a). Indeed, when administered directly into the pulmonary

Table 3. Eicosanoid-releasing properties of big-endothelins in perfused organs

Organ	Species	Big-ET subtype (100 nmol/l)	Type of eicosanoid released		Phosphoramidon-sensitive	References
			PGI_2	TxA_2		
Lung	Rabbit	1	++	0	Yes	D'Orléans-Juste et al. (1991a)
	Rat	1	++	0	Yes	D'Orléans-Juste et al. (1992); Télémaque et al. (1993)
		2	Inactive	Inactive	–	Télémaque et al. (1993); D'Orléans-Juste et al. (1991)
		3	Inactive	Inactive	–	Gratton et al. 1995a; Télémaque et al. (1993)
	Guinea pig	1	++	+++	Yes	Gratton et al. (1995b)
		2	Inactive	Inactive	–	Gratton et al. (1995b)
		3	Inactive	Inactive	–	Gratton et al. (1995b)
Kidney	Rabbit	1	++	0	Yes	Télémaque et al. (1992)
		2	++	0	Yes	
		3	Inactive	Inactive	–	

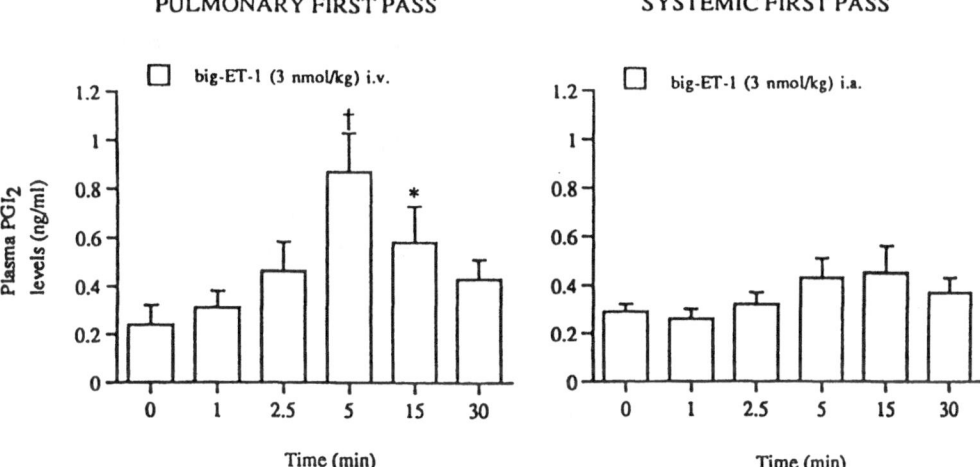

Fig. 3. Big-ET-1 (3 nmol/kg) induces a detectable raise in plasmatic prostacyclin when injected i.v. but not via the left ventricular route in the ketamine/xylazine anesthetized rabbit. *$P < 0.05$; †$P < 0.01$; $n = 5$

circulation, big-ET-1 is able to generate a detectable amount of plasmatic prostacyclin. In contrast, when administered first in the systemic circulation (through the left cardiac ventricle), big-ET-1 even at the highest dose tested remains unconverted (Fig. 3). This would suggest that the pulmonary ECE is pivotal for the first-pass conversion of big-ET-1 to ET-1 in order for the precursor to induce its prostacyclin-releasing effects. These observations prompt us to suggest that truncated analogs of big-ET-2 may be useful to interfere efficiently with the systemic ECE activity while leaving the functional characteristics of the pulmonary endothelin-converting enzyme intact. As there is an ever-growing literature suggesting a delicate balance between nitric oxide, eicosanoids, and endothelin in the pulmonary circulation (SALEH et al. 1997), such tools may be useful in understanding the role of organ-specific ECE activity in the homeostasis of the cardiovascular system.

II. Big-Endothelins and NO

Unlike endothelin-1, an intravenously-administered bolus of big-ET-1 does not generate an initial hypotensive response (D'ORLÉANS-JUSTE et al. 1991a). This suggests that big-endothelin-1 may be converted to ET-1 by an enzyme located distally from the endothelial ET_B receptors involved in the NO-releasing properties of the peptide (HALEEN et al. 1993). This, however, by no means excludes a modulatory role for endogenous nitric oxide in the produc-

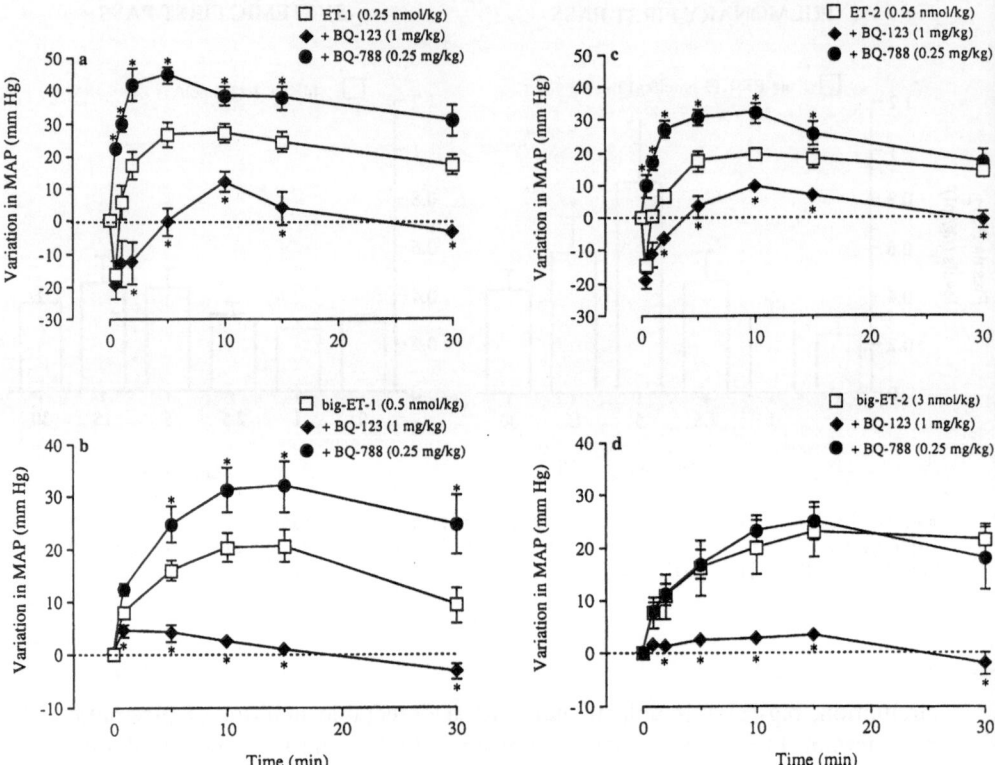

Fig. 4. a Time course of the changes in mean arterial pressure (*MAP*) induced by a ET-1 (0.25 nmol/kg). **b** Big-ET-1 (0.5 nmol/kg). **c** ET-2 (0.25 nmol/kg). **d** Big-ET-2 (3 nmol/kg). In the absence or in the presence of the selective ET_A or ET_B receptor antagonists, BQ-123 (1 mg/kg) and BQ-788, respectively, in the anesthetized rabbit. Each *point* represents the mean ± SEM of at least six experiments. *$P < 0.05$ vs control

tion (GRATTON et al. 1997) and pressor effects of big-ET-1. In series of recently published results, we have found that big-ET-1 pressor responses are sharply potentiated by an ET_B selective antagonist, BQ-788 (GRATTON et al. 2000). In contrast, the hypertensive responses to big-ET-2 are largely unaffected by the same treatment with the ET_B receptor antagonist (Fig. 4). Furthermore, intracardiac administration of big-ET-1, but not big-ET-2, induces a significant increase in plasma levels of immunoreactive endothelin (GRATTON et al. 2000). These two states of events would suggest that big-ET-1 is more efficiently converted both systemically and in the pulmonary circulation in the rabbit. This more efficient conversion would favor a more pronounced spillover of endothelin-1 than endothelin-2 in the circulation, which would consequently activate ET_B receptors to release EDRF (Fig. 4).

E. Modulator-Releasing Properties of Endothelins in Mice

This particular section is devoted to the effect of endothelin-1 in the mouse. Since this animal model is routinely used in genetic engineering and its physiological parameters are poorly explored (BROWNSTEIN, 1998; BADER, 1998), we wished to document in the present chapter the release of various modulators by endothelin in vivo and in vitro in this animal species. Indeed, we believe that it is pivotal at this stage to understand the contribution of both endothelin receptor types in the vasoactive and mediator-releasing effects of the peptide in the murine model.

I. Nitric Oxide

In a first series of experiments, BERTHIAUME et al. (1998) have demonstrated an important contribution of both ET_A and ET_B receptors to the vasoconstrictive effects of endothelins in the arterial and venous mesenteric vasculature as well as in the perfused kidney of the mouse. Indeed, both ET-1 and a selective ET_B receptor agonist were able to increase vascular resistance in the in vitro perfused models. Interestingly, both ET_A-selective and ET_B-selective antagonists, BQ-123 and BQ-788, significantly reduced the constrictive properties of ET-1; however only the latter antagonist was able to alter the responses to IRL-1620. Also of note was the apparent lack of ET_B-dependent vasodilation in the above-mentioned two perfused organs, i.e., mesentery and kidney. As a matter of fact, even when administered i.v. in the anesthetized mouse, ET-1 does not induce its initial transient depressor response, as normally seen in larger mammals and in man (GILLER et al. 1997). This would suggest that the mouse, in contrast to other animal species, does not possess a significant physiological counteracting mechanism to the potent pressor effects of endothelin-1 in the form of ET_B-dependent release of nitric oxide. Furthermore, pretreatment of the mouse perfused kidney or mesenteric vasculature with BQ-788 does not produce potentiation of the vasoconstrictor effects of either endothelin-1 or IRL-1620. This would suggest that not only is there an absence of ET_B-dependent vasodilation in high and low resistance vascular circuits in the mouse, but also that endogenous nitric oxide seems to play a relatively minor role as a modulator of the response to endothelin-1 in those vascular circuits. Nonetheless, it is worthy of mention that L-NAME per se does trigger a sustained and very significant hypertension when administered i.v. in the conscious mouse (MATTSON 1998).

II. Eicosanoids

We are currently pursuing an analysis of endothelin-1-induced eicosanoid release in mice. Our initial in vitro studies have not allowed us to detect the

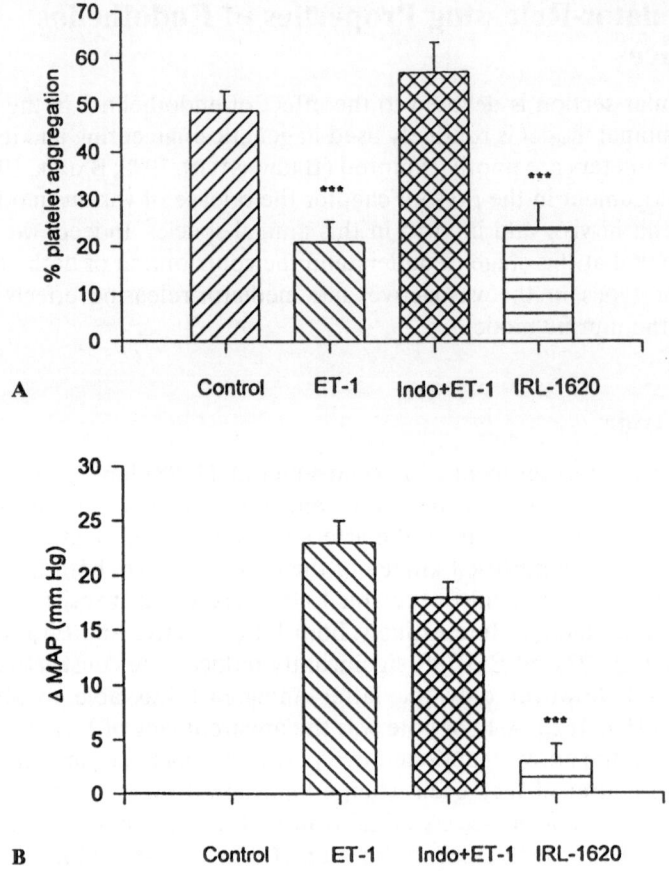

Fig. 5A,B. Profile of ET-1 (0.1 nmol/kg) and IRL-1620 (0.5 nmol/kg) induced inhibition of ADP (5 μmol/l) induced. **A** Platelet aggregation in CD-1 mice. **B** Pressor effect in CD-1 mice. Each *column* represents the mean ± SEM of at least five experiments. ***P < 0.001. *Indo* Indomethacin (10 mg/kg, administered 20 min prior to the injection of ET-1)

generation of eicosanoids from the kidney, mesenteric, pulmonary, and renal vasculatures under in vitro perfusion conditions. The situation appears, however, to be quite different in an in vivo setting, as we are currently documenting the ET-1-dependent inhibition of ADP-induced platelet aggregation in CD1 mice. Figure 5 illustrates some of the preliminary results obtained in these experiments, showing the ability of endothelin-1 to generate an indomethacin-sensitive release of circulating prostanoids via activation of ET_B receptors. Worthy of note is that even a selective ET_B agonist, IRL-1620, at a subpressor dose, is able to generate sufficient amounts of PGI_2 to alter ADP-induced platelet aggregation in the mouse.

Whether generation of circulating prostanoids will influence the pressor responses to endothelin-1, as it affects platelet aggregation, merits further experimentation. In our hands, however, indomethacin did not enhance constrictor responses to the peptide (Fig. 5). Only one study has shown that intravenously-administered endothelin-1 induces a mild but significant increase in pulmonary insufflation pressure in the murine model. However, in direct contrast to the guinea pig model (PONS et al. 1991), it appears that endogenous eicosanoids are not significantly involved in that phenomenon in the mouse (NAGASE et al. 1998).

F. Conclusion

Recent observations in healthy individuals illustrate the importance of eicosanoids as modulators of the vasoactive responses to endothelin-1; i.e., non-steroidal anti-inflammatory drugs (NSAID) significantly alter the response to endothelin-1 (JOHNSON 1997). On the other hand, ET_B receptor activation appears to be predominantly related to the release of the physiological antagonist, nitric oxide, in both healthy subjects and in patients with congestive hearth failure (HAYNES and WEBB 1998). Unexplored as yet is the cross-talk between pro-inflammatory cytokines and endothelin-1 in human subjects.

Here we have attempted to illustrate the importance of several pro-inflammatory mediators in some of the pharmacological and physiopathological effects of endothelin-1. From this chapter we must conclude that when studying the pharmacological or physiopathological effects of endogenous endothelins one should take into account their possible indirect effects potentially mediated through the activation of one or more of the autacoids described here. Such factors will indeed mask the direct effects of endothelin-1 and may lead to confusing observations when receptor antagonists or endothelin-converting enzyme inhibitors are used; e.g., as illustrated by the constrictive properties of an ET_B antagonist in both rabbits and humans (GRATTON et al. 1997; STRACHAN et al. 1999). In addition, these humoral factors may not only interfere with the direct effects of endothelin-1 but may significantly alter its production as well (GRATTON et al. 1997). Considering the complex cross-talk between the factors generated by the endothelial cell layer lining all blood vessels, further knowledge of the modulators of the physiological effects of endothelin-1 is central to an understanding of the contribution of that particular peptide, to the homeostasis of the cardiovascular system, as well as to pathophysiological changes.

Acknowledgements. The authors gratefully acknowledge Helen Morin and Pascale Martel for secretarial assistance. This project was financially supported by the Medical Research Council of Canada (MRCC) (MA-16612 and GR-13915) and the Heart and Stroke Foundation of Québec (HSFQ). PDJ is a scholar of the Fonds de la recherche en santé du Québec (FRSQ). NB is in receipt of a studentship from the Heart and Stroke Foundation of Canada.

References

Abdel-Latif AA, Yousufzai SY, El Mowafi AM, Ye Z (1996) Prostaglandins mediate the stimulatory effects of endothelin-1 on cyclic adenosine monophosphate accumulation in ciliary smooth muscle isolated from bovine, cat and other mammalian species. Investigative Ophthalmology and Visual Science 37(2):328–338

Adachi H, Shoji T, Goto K (1991) Comparison of the effects of endothelin and Bay K 8644 on cardiohemodynamics in anaesthetized pigs. Eur J Pharmacol 193:57–65

Alfie ME, Yang XP, Hess F, Carretero OA (1996) Salt-sensitive hypertension in bradykinin B$_2$ receptor knockout mice. Biochem Biophys Res Comm 224:625–630

Bader M (1998) Transgenic animal models for the functional analysis of vasoactive peptides. Braz J Med Biol Res 31(9):1171–1183

Battistini B, Chailler P, D'Orléans-Juste P, Brière N, Sirois P (1993) Growth regulatory properties of endothelins. Peptides 14(2):385–399

Berthiaume N, Yanagisawa M, Yanagisawa H, deWit D, D'Orléans-Juste P (1998) Pharmacology of endothelins in vascular circuits of normal or heterozygous ET$_A$ or ET$_B$ K.O. transgenic mice. J Cardiovasc Pharmacol 31(1):S561–S564

Boulanger C, Luscher TF (1990) Release of endothelin from the porcine aorta, inhibition by endothelion derived nitric oxide. J Clin Invest 85:587–590

Brownstein DG (1998) Genetically engineered mice: the holes in the sum of the parts. Laboratory Animal Science 48(2):121–122

Clement MG, Albertini M (1996) Differential release of prosta cyclin and nitric oxide evoked from pulmonary and systemic vascular beds of the pig by endothelin-1. Prostanglandins Leukot. Essent Fatty acids 55:279–285

Clozel M, Gray GA (1995) Are there different ET (B) receptors mediating constriction and relaxation? J Cardiovasc Pharmacol 26: S262–S264

de Nucci G, Thomas R, D'Orleans-Juste P, Antunes E, Walder C, Warner TD, Vane JR (1988) Pressor effects of circulating endothelin are limited by its removal in the pulmonary circulation and by the release of prostacyclin and endothelium derived relaxing factor. Proc Natl Acad Sci USA 85:9797–9800

Doi Y, Ozaka T, Katsuki M, Fukushige H, Toyama E, Kanazawa Y, Arashidani K, Fujimoto S (1995) Histamine release from Weibel-Palade bodies of toad aortas induced by endothelin-1 and Sarafotoxin S6b. Anatom Record 242:374–382

D'Orléans-Juste P, Claing A, Télémaque S, Maurice MC, Yano M, Gratton JP (1994) Block of endothelin-1-induced release of thromboxane A$_2$ from the guinea pig lung and nitric oxide from the rabbit kidney by a selective ET$_B$ receptor antagonist, BQ-788. Br J Pharmacol 113:1257–1262

D'Orléans-Juste P, Lidbury PS, Telemaque S, Warner TD, Vane JR (1991a) Human big endothelin releases prostacyclin in vivo and in vitro through a phosphoramidon-sensitive conversion to endothelin-1. J Cardiovasc Pharmacol 17(7):S251–S255

D'Orléans-Juste P, Lidbury PS, Warner TD, Vane JR (1990) Intravascular big endothelin increases circulating levels of endothelin-1 and prostanoids in the rabbit. Biochem Pharmacol 39(9):R21–R22

D'Orléans-Juste P, Télémaque S, Claing A (1991b) Different pharmacological profiles of big-endothelin-3 and big-endothelin-1 in vivo and in vitro. Br J Pharmacol 104(2):440–444

D'Orléans-Juste P, Télémaque S, Claing A, Ihara M, Yano M (1992) Human big-endothelin-1 and endothelin-1 release prostacyclin via the activation of ET-1 receptors in the rat perfused lung. Br J Pharmacol 105(4):773–775

D'Orléans-Juste P, Yano M, Télémaque S (1993) ET$_A$-dependent pressor effects and release of prostacyclin induced by endothelin-1 in pulmonary and renal vasculature. J Cardiovasc Pharmacol 22:S233–S238

Dupuis J, Goresky CA, Fournier A (1996) Pulmonary clearance of circulating endothelin-1 in dogs in vivo: exclusive role of ET$_B$ receptors. J Appl Physiol 81: 1510–1515

Edwards RM, Stack EJ, Pullen M, Nambi P (1993) Endothelin inhibits vasopressin action in rat inner medullary collecting duct via the ET_B receptor. JPET 267(3): 1028–1033

Egger D, Geuenich S, Denzlinger C, Schmitt E, Mailhammer R, Ehrenreich H, Dormer P, Hultner L (1995) IL-4 renders mast cells functionally responsive to endothelin-1. J Immunol 154:1830–1837

Fabricio AS, Silva CA, Rae GA, D'Orléans-Juste P, Souza GE (1998) Essential role for endothelin ET_B receptor in fever induced by LPS. Br J Pharmacol 125(3):542–548

Filep JG, Battistini B, Cote YP, Beaudoin AR, Sirois P (1991) Endothelin-1 induces prostacyclin release from bovine aortic endothelial cells. Biochem Biophys Res Comm 177:171–176

Filep JG (1993) Endothelin peptides: biological actions and pathophysiological significance in the lung. Life Sciences 52:119–133

Fukuda Y, Hirata Y, Taketani S, Kojima T, Oikawa S, Nakazato H, Kobayashi Y (1989) Endothelin stimulates accumulations of cellular atrial natriuretic peptide and its messenger RNA in rat cardiocytes. Biochem Biophys Res Comm 164(3):1431–1436

Fukuroda T, Fujikawa T, Ozaki S, Ishikawa K, Yano M, Nishikibe M (1994) Clearance of circulating endothelin-1 by ET_B receptors in rats. Biochem Biophys Res Comm 199:1461–1465

Fukuroda T, Noguchi K, Tsuchida S, Nishikibe M, Ikemoto F, Okada K, Yano M (1990) Inhibition of biological actions of big endothelin-1 by phosphoramidon. Biochem Biophys Res Comm 172(2):390–395

Garcia JL, Fernandez N, Garcia-Villalon AR, Monge L, Gomez B, Diegvez G (1996) Coronary vasoconstriction by endothelin-1 in anesthetized goats: role of endothelin receptors, nitric oxide and prostanoids. Eur J Pharmacol 315:179–186

Garvey EP, Oplinger JA, Furfine ES, Kiff RJ, Laszlo F, Whittle BJR, Knowles RG (1997) 1400 W is a slow, tight binding, and highly selective inhibitor of inducible nitric-oxide synthase in vitro and in vivo. J Biol Chem 272(8):4959–4963

Giller T, Breu V, Valdenaire O, Clozel M (1997) Absence of ET(B)-mediated contraction in Piebald-lethal mice. Life Sciences 61(3):255–263

Gratton JP, Rae GA, Bkaily G, D'Orléans-Juste P (2000) ET_B receptor blockade potentiates the pressor response to big-endothelin-1, but not big-endothelin-2, in the anesthetized rabbit. Hypertension 35:726–731

Gratton JP, Covrnoyer G, D'Orléans-Juste P (1998) Endothelin B receptor-dependent modulation of the pressor and prosta cyclin releasing properties of dynamically converted big-endothelin-1 in the anesthetized Rabbit. J Cardiovasc Pharmacol 31:S161–S163

Gratton JP, Cournoyer G, Löffler BM, Sirois P, D'Orléans-Juste P (1997) ET_B receptor and nitric oxide synthase blockade induce BQ-123-sensitive pressor effects in the rabbit. Hypertension 30 (5):1204–1209

Gratton JP, Maurice MC, Rae GA, D'Orléans-Juste P (1995a) Pharmacological properties of endothelins and big endothelins in ketamine/xylazine or urethane anaesthetized rats. Am J Hypertension 8:1121–1127

Gratton JP, Rae GA, Claing A, Télémaque S, D'Orléans-Juste P (1995b) Different pressor and bronchoconstrictor properties of human big-endothelin-1,2 (1–38) and 3 in ketamine/xylazine-anaesthetized guinea-pigs. Br J Pharmacol 114(3):720–726

Haleen SJ, Davis LS, LaDouceur DM, Keiser JA (1993) Why big-endothelin-1 lacks vasodilator response? J Cardiovasc Pharmacol 22(8):S271–S273

Hamilton LC, Warner, TD (1998) Interactions between inducible isoforms of nitric oxide synthase and cyclo-oxygenase in vivo: investigations using the selective inhibitors, 1400 W and celecoxib. Br J Pharmacol 125(2):334–340

Haynes WG, Webb DJ (1998) Endothelin as a regulator of cardiovascular function in health and disease. Hypertension 16(8):1081–1098

Hirata Y, Emori T (1993) Cellular mechanism of endothelin induced nitric oxide synthesis by cultured bovine endothelial cells. J Cardiovasc Pharmacol 22(S8): S225–S228

Hohlfeld T, Klemm P, Thiemermann C, Warner TD, Schror K, Vane JR (1995) The contribution of tumour necrosis factor-alpha and endothelin-1 to the increase of coronary resistance in hearts from rats treated with endotoxin. Br J Pharmacol 116(8):3309–3315

Johnson AG (1997) NSAIDs and increased blood pressure. What is the clinical significance? Drug Safety 17(5):277–89

Kashiwabara T, Inagaki Y, Ohta H, Iwamatsu A, Nomizu M, Morita A, Nishikori K (1989) Putative precursors of endothelin have less vasoconstrictor activity in vitro but a potent pressor effect in vivo. FEBS Letters 247(1):73–76

Kobari M, Fukuuchi Y, Tomita M, Tanahashi N, Konno S, Takeda H (1994) Cerebral vasodilatory effect of high-dose, intravascular endothelin-1: inhibition by NG-monomethyl-L-arginine. J Auton Nerv Syst 49:S111–S115

Kurihara Y, Kurihara H, Suzuki H, Kodama T, Maemura K, Nagai R, Oda H, Kuwaki T, Cao WH, Kamada N, Jishage K, Ouchi Y, Azuma S, Toyoda Y, Ishikawa T, Kumada M, Yasaki Y (1994) Elevated blood pressure and craniofacial abnormalities in mice deficient in endothelin-1. Nature 368:703–710

Kushiku K, Ohjimi H, Yamada H, Tokunaga R, Furukawa T (1991) Endothelin-3 inhibits ganglionic transmission at preganglionic sites through activation of endogenous TxA$_2$ production in dog cardiac sympathetic ganglia. J Cardiovasc Pharmacol 17(S7):S197–S199

Langleben D, Barst RJ, Badesch D, Groves BM, Tapson VF, Murali S, Bourge RC, Ettinger N, Shalit E, Clayton LM, Jobsis MM, Blackburn SD, Crow JW, Stewart DJ, Long W (1999) Continuous infusion of epoprostenol improves the net balance between pulmonary endothelin-1 clearance and release in primary pulmonary hypertension. Circulation 99(25):3266–3271

Lehoux S, Plante GE, Sirois MG, Sirois P, D'Orléans-Juste P (1992) Phosphoramidon blocks big-endothelin-1 but not endothelin-1 enhancement of vascular permeability in the rat. Br J Pharmacol 107(4):996–1000

Le Monnier de Gouville AC, Cavero I (1991) Cross tachyphylaxis to endothelin isopeptide-induced hypotension: a phenomenon not seen with preproendothelin. Br J Pharmacol 104:77–84

Le Monnier de Gouville AC, Lippton H, Cohen G, Cavero I, Hyman A (1990) Vasodilatory activity of endothelin-1 and endothelin-3: rapid development of cross-tachyphylaxis and dependence on the rate of endothelin administration. JPET 254:1024–1028

Lewis K, Cadieux A, Rae G, Gratton JP, D'Orléans-Juste P (1999) Nitric oxide modulate the thromboxane A$_2$ dependent bronchoconstriction induced by intravascularly administered endothelin-1 in the anesthetized guinea pig. Br J Pharmacol 126:93–102

Luscher TF, Barton M (1997) Biology of endothelium. Clin Cardiol 20(S11): II-3–10

Maccarrone M, Putti S, Finazzi Agro A (1997) Nitric oxide donors activate the cyclooxygenase and peroxidase activities of prostaglandin H synthase FEBS Letters 410(2–3):470–476

Matsumura Y, Hisaki K, Takaoka M, Morimoto S (1990) Phosphoramidon, a metalloproteinase inhibitor, suppresses the hypertensive effect of big endothelin-1. Eur J Pharmacol 185(1):103–106

Mattson DL (1998) Long-term measurement of arterial blood pressure in conscious mice. Am J Physiol 274(2 Pt 2):R564–R570

McMurdo L, Lidbury PS, Thiemermann C, Vane JR (1993) Mediation of endothelin-1 induced inhibition of platelet aggregation via the ET$_B$ receptor. Br J Pharmacol 109:530–534

Minkes RK, Kadowitz PJ (1991) Hemodynamic responses to sarafotoscin 6 peptides. J Cardiovasc Pharmacol 17: S293–S296

Murad F (1998) Nitric oxide signaling: Would you believe that a simple free radical could be a second messenger, autacoid, paracrine substance, neurotransmitter and hormone? Rec Prog In Hormone Res 53:43–59

Murayama T, Oda H, Sasaki Y, Okada T, Nomura Y (1998) Regulation of inducible NO synthase expression by endothelin in primary cultured glial cells. Life Sciences 62:1491–1495

Nagase T, Kurihara H, Kurihara Y, Aoki T, Fukuchi Y, Yazaki Y, Ouchi Y (1998) Airway hyperresponsiveness to methacholine in mutant mice deficient in endothelin-1. Am J Resp Critical Care Med 157(2):560–564

Nakahashi T, Fukuo K, Inoue T, Morimoto S, Hata S, Yano M, Ogihara T (1995) Endothelin-1 enhances nitric oxide induced cytotoxicity in vascular smooth muscle. Hypertension 25:744–747

Ninomiya H, Yue XY, Hasegawa S, Spannhake EW (1992) Endothelin-1 induces stimulation of prostaglandin synthesis in cells obtained from canine airways by bronchoalveolar lavage. Prostaglandins 43:401–411

Nishimura T, Krier J, Akasu T (1991) Endothelin causes prolonged inhibition of nicotinic transmission in feline colonic parasympathetic ganglia. Am J Physiol (Gastrointest Liver Physiol) 261:G628–G633

Noguchi S, Kashihara Y, Bertrand C (1996) The induction of a biphasic bronchospasm by the ET_B agonist, IRL-1620, due to thromboxane A_2 generation and endothelin-1 release in guinea-pigs. Br J Pharmacol 118:1397–1402

N'Diaye N, Pueyo ME, Battle T, Ossart C, Guédin D, Michel JB (1997) Conversion of big-endothelin-1 elicits an endothelin ET_A receptor-mediated response in endothelial cells. Eur J Pharmacol 321:387–396

Owada A, Tomita K, Terada Y, Sakamoto H, Noguchi H, Marumo F (1994) ET-3 stimulates cyclic guanosine 3′,5′-monophosphate production via ET_B receptor by producing nitric oxide in isolated rat glomerulus and in cultured rat mesangial cells. J Clinic Invest 93:556–563

Ozaki H, Abe A, Uehigashi Y, Kinoshita M, Hori M, Mitsui-Saito M, Karaki H (1996) Effects of a prostaglandin I_2 analog iloprost on cytoplasmic Ca^{2+} levels and muscle contraction in isolated guinea pig aorta. Jpn J Pharmacol 71(3):231–237

Pons F, Touvay C, Lagente V, Mencia-Huerta JM, Braquet P (1991) Bronchopulmonary and pressor activities of endothelin-1 (ET-1), ET-2, ET-3 and big ET-1 in the guinea pig. J Cardiovasc Pharmacol 17(7):S326–328

Poulat P, D'Orléans-Juste P, de Champlain J, Yano M, Couture R (1994) Cardiovascular effects of intrathecally administered endothelins and big endothelin-1 in concious rats receptor characterization and mechanism of action. Brain Res 648:239–248

Rockey D (1997) The cellular Pathogenesis of portal Hypertension: stellate cell contractility, endothelin and nitric oxide. Hepatology 25:2–5

Said SA, El-Mowafy, AM (1998) Role of endogenous endothelin-1 in stress-induced gastric mucosal damage and acid secretion. Regulatory Peptides 73(1):43–50

Saleh D, Furukawa K, Tsao MS, Maghazachi A, Corrin B, Yanagisawa M, Barnes PJ, Giaid A (1997) Elevated expression of endothelin-1 and endothelin converting enzyme-1 in idiopathic pulmonary fibrosis: possible involvement of proinflammatory cytokines. Am J Respir Cell Mol Biol 16:187–193

Stankova J, D'Orléans-Juste P, Rola-Pleszczynski M (1996) ET-1 induces IL-6 gene expression in human umbilical vein endothelial cells: synergistic effect of IL-1. Am J Physiol 271:C1073–C1078

Stankova J, Rola-Pleszczynski M, D'Orléans-Juste P (1995) Endothelin-1 and thrombin synergistically stimulate IL-6 mRNA expression and protein production in human umbilical vein endothelial cells. J Cardiovasc Pharmacol 26(S3):S505–S507

Strachan FE, Haynes WG, Webb DJ (1995) Endothelium-dependent modulation of venoconstriction to sarafotoxin S6C in human veins in vivo. J Cardiovasc Pharmacol 26(3):S180–S182

Strachan FE, Spratt JC, Wilkinson IB, Johnston NR, Gray GA & Webb DJ (1999) Systemic blockade of the endothelin B receptor increases peripheral vascular resistance in healthy men. Hypertension 33:581–585

Télémaque S, Gratton JP, Claing A, D'Orléans-Juste P (1993) Pharmacological evidence for the specificity of the phosphoramidon-sensitive endothelin-converting enzyme for big endothelin-1. J Cardiovasc Pharmacol 22(8):S85–S89

Télémaque S, Lemaire D, Claing A, D'Orléans-Juste P (1992) Phosphoramidon-sensitive effects of big endothelins in the perfused rabbit kidney. Hypertension 20(4):518–523

Tertyshnikova S, Fein A (1998) Inhibition of inositol 1,4,5-triphosphate-induced Ca^{2+} release by cAMP-dependent protein kinase in a living cell. Proc Natl Acad Sci USA 95(4):1613–1617

Tsuchida S, Matsusaka T, Chen X, Okubo S, Niimura H, Fogo A, Utsunomiya H, Inagami T, Ichikawa I (1998) Murine double nullizygotes of the angiotensin types 1_A and 1_B receptor genes duplicate severe abnormal phenotypes of angiotensinogen nullizygotes. J Clin Invest 101(4):755–760

Tsuchiya K, Naruse M, Sanaka T, Naruse K, Kato Y, Zeng ZP, Nitta K, Deura H, Shizume K, Sugino N (1990) Effects of endothelin on renal hemodynamics and excretion functions in anesthetized dog. Life Sciences 46:59–65

Uchida Y, Ninomiya H, Sakamoto T, Lee JY, Endo T, Nomura A, Hasegawa S, Hirata F (1992) ET-1 released histamine from guinea pig pulmonary but not peritoneal mast cells. Biochem Biophys Res Comm 189:1196–1201

Verhaar MC, Strachan FE, Newby DE, Cruden NL, Koomans HA, Rabelink TJ, Webb DJ (1998) Endothelin-A receptor antagonist-mediated vasodilatation is attenuated by inhibition of nitric oxide synthesis and by endothelin-B receptor blockade. Circulation 97:752–756

Walder CE, Thomas GR, Thiemermann C, Vane JR (1989) The hemodynamic effects of endothelin-1 in the pithed rat. J Cardiovasc Pharmacol 13(5):S93–S97

Warner TD, Mitchell JA, de Nucci G, Vane JR (1989) Endothelin-1 and endothelin-3 release EDRF from isolated perfused arterial vessels of the rat and rabbit. J Cardiovasc Pharmacol 13:S85–S88

Warner TD, Mitchell JA, D'Orléans-Juste P, Ishii K, Forstermann U, Murad F (1992) Characterization of endothelin-converting enzyme from endothelial cells and rat brain: detection of the formation of biologically active endothelin-1 by rapid bioassay. Mol Pharmacol 41(2):399–403

Warner TD (1999) Relationships between the endothelin and nitric oxide pathways. Clin Exp Pharmacol Physiol 26(3):247–252

Webb DJ (1996) Endothelin and blood pressure regulation. J Human Hypertension 10: 383–386

Whittle BJ, Lopez-Belmonte J (1993) Actions and interactions of endothelins, prostacyclin and nitric oxide in the gastric mucosa. J Physiol Pharmacol 44(2):91–107

Wright HM, Malik KU (1996) Prostacyclin formation elicited by endothelin-1 in rat aorta is mediated via phospholipase D activation and not phospholipase C or A_2. Circ Res 79(2):271–276

Wu-Wong JR, Dayton BD, Opgenorth TJ (1996) Endothelin-1 evoked arachidonic acid release: a Ca(2+) dependent pathway. Am J Physiol 271:C869–C877

Yamamura H, Nabe T, Kohno S, Ohata K (1994b) Endothelin-1 induces release of histamine and leukotriene C_4 from mouse bone marrow-derived mast cells. Eur J Pharmacol 257:235–242

Yamamura H, Nabe T, Kohno S, Ohata K (1994a) Endothelin-1, one of the most potent histamine releasers in mouse peritoneal mast cells. Eur J Pharmacol 265:9–15

Yamamura H, Nabe T, Kohno S, Ohata K (1995) Mechanism of histamine release by endothelin-1 distinct from that by antigen in mouse bone marrow-derived mast cells. Eur J Pharmacol 288:269–275

Yanagisawa M, Kurihara H, Kimura S, Tomobe Y, Kobayashi M, Mitsui Y, Yazaki Y, Goto K, Masaki T (1988) A novel potent vasoconstrictor peptide produced by vascular endothelial cells. Nature 332:411–415

Yoshimura M, Yamashita YS, Kan S, Niwa M, Taniyama K (1996) Localization of ET_B receptors on the myenteric plexus of guinea pig ileum and the receptor mediated release of acetylcholine. Br J Pharmacol 118:1171–1176

CHAPTER 12

The Roles of Endothelins in Proliferation, Apoptosis, and Angiogenesis

J.R. Wu-Wong and T.J. Opgenorth

A. Introduction

Endothelin (ET), originally isolated from cultured porcine aortic endothelial cells, is a highly potent vasoconstricting peptide with 21-amino acid residues (Yanagisawa et al. 1988). Three distinct members of the ET family, namely ET-1, ET-2, and ET-3, have been identified in humans through cloning (Inoue et al. 1989). The effects of ETs on mammalian organs and cells are initiated by their binding to high affinity G-protein linked receptors (see Chap. 4). ET receptors are found in various tissues, such as brain, lung, and kidney (Sokolovsky 1992). Two major types of ET receptors in the mammalian system, ET_A and ET_B, have been characterized, isolated, and their cDNAs cloned (Wada et al. 1990; Arai et al. 1990; Sakurai et al. 1990; Kozuka et al. 1991). ET_A receptors are selective for ET-1 and ET-2, while ET_B receptors bind to ET-1, ET-2, and ET-3 with equal affinity. Pharmacologically defined subtypes of ET_A and ET_B receptors have also been reported (Sudjarwo et al. 1994; Douglas et al. 1995). Various antagonists and agonists for ET receptors have been developed (Spellmeyer 1994; Warner 1994; Opgenorth 1995; Ohlstein et al. 1996; Wu-Wong et al. 1999; see Chaps 4, 9).

Endothelial cells and smooth muscle cells are integral cellular components of the blood vessels. Endothelial cells, which form a tight monolayer at the inner surface of blood vessels, play a particularly critical role in the development of new blood vessels. Formation of new blood vessels involves both vasculogenesis and angiogenesis. During embryogenesis, blood vessels develop via vasculogenesis, in which endothelial cells are born from progenitor cells, and angiogenesis, in which new capillaries arise from existing vessels. During adulthood, new vessels are formed mainly via angiogenesis, which occurs during reproduction, wound healing, chronic inflammation, tissue ischemia, and cancer. The mechanism of angiogenesis involves a complex process including degradation of the basement membrane underneath the endothelial cells, the migration and proliferation of endothelial cells, and vascular remodeling. During the final stage of new vessel formation, possibly apoptosis plays a role in the elimination of excessive endothelial cells that fail to properly fuse and connect to form the capillary tube. For more information on this subject, please refer to excellent review articles by BATTEGAY (1995), HANAHAN and FOLKMAN (1996), and BERGERS et al. (1998).

ET-1 is thought to play important roles in various pathophysiological conditions including abnormal cell accumulation such as tumor development (ASHAM et al. 1998). Disorders involving abnormal cell accumulation are usually the result of an imbalance between proliferation and apoptosis. It is now well documented that ETs regulate cell proliferation. Mitogenic effects of ETs have been demonstrated in several different cell types such as vascular smooth muscle cells, fibroblasts, mesangial, endothelial cells, and various carcinoma cells. In some cell types, ETs have also been shown to inhibit cell proliferation. Although it has only been addressed recently, and hence is less well understood, ETs may be involved in modulating apoptosis. All these studies suggest a potential role of ETs in abnormal growth disorders such as atherosclerosis, benign prostatic hyperplasia, restenosis, and tumorigenesis. It is less clear whether ETs are involved in the process of angiogenesis. In this chapter, we intend to review and discuss studies reported in the past few years regarding the modulating effects of ETs on cell proliferation and apoptosis, especially focusing on the link between these effects of ETs and angiogenesis.

B. The Role of Endothelins in Cell Proliferation

I. The Mitogenic Effect of ETs

The observation that ETs stimulate cell proliferation can be traced back to 1988, the year when ETs were first discovered. KOMURO et al. (1988) showed in a short, yet elegant study, that ET stimulates DNA synthesis in vascular smooth muscle cells in a dose-dependent manner. They promptly suggested that this mitogenic effect of ET on smooth muscle cells might be related to the development of atherosclerosis. Since then, numerous reports have shown that ETs stimulate DNA synthesis and cell proliferation in a variety of cells including fibroblasts (BROWN and LITTLEWOOD 1989; MULDOON et al. 1990), vascular smooth muscle cells, mesangial, endothelial, and various cancer cells (see Table 1 for references). Table 1 is an attempt to summarize results from reports published between 1995 and the present on the mitogenic effects of ETs as established in vitro studies. For studies on the mitogenic effects of ETs before 1995, please refer to review articles by GOTO et al. (1996) and GOLDIE et al. (1996).

From Table 1, it is obvious that ETs exhibit mitogenic effects on a wide variety of cells. One interesting observation from Table 1 is that ET_A seems to be the receptor mediating the mitogenic effect of ETs in most cells. ET_B is mainly involved in the proliferation of endothelial cells and neural crest cell precursors, in which ET_B is the predominant receptor.

In addition to the in vitro studies, the mitogenic effect of ETs has also been demonstrated in ex vivo or in vivo studies. For example, ET-1 is shown to increase the mitotic index and [^{3}H]thymidine incorporation into DNA in

Table 1. The mitogenic effect of ETs in cultured cells

Cell/tissue type	Receptor subtype	Assay used	Reference
Astrocytes (from ET_B-deficient rat)	ET_A	DNA synthesis	SASAKI et al. (1998)
CCD-18Lu (human lung)	ET_A	DNA synthesis	TAKIMOTO et al. (1996)
Cancer cells (human astrocytoma U138MG)	ET_A	DNA synthesis	WU-WONG et al. (1996a)
Cancer cells (human Meningiomas)	ET_A	DNA synthesis	PAGOTTO et al. (1995)
Cancer cells (human ovarian carcinoma)	ET_A	DNA synthesis	BAGNATO et al. (1995a, 1997)
Cancer cells (human prostatic carcinoma)	ET_A	DNA synthesis; cell number	NELSON et al. (1995, 1996, 1997)
Cardiomyocytes (rat neonatal)	NS[a]	DNA synthesis	VAN KESTEREN et al. (1997)
Chondrocytes (rat articular)	ET_A	DNA synthesis	KHATIB et al. (1998)
Endothelial cells (bovine adrenal capillaries and human umbilical veins)	ET_B	DNA synthesis	MORBIDELLI et al. (1995)
Epithelial cells (porcine airway)	ET_A	DNA synthesis; cell number	MURLAS et al. (1995)
Fibroblasts (human, IMR90)	ET_A	DNA synthesis	CHIOU et al. (1999)
Granulosa cells (human)	ET_A	DNA synthesis	KAMADA et al. (1995)
Keratinocytes (human)	ET_A	DNA synthesis	BAGNATO et al. (1995b)
Melanocytes (human)	ET_A	DNA synthesis	IMOKAWA et al. (1996)
Mesangial cells (rat kidney)	ET_A	DNA synthesis	TERADA et al. (1998)
Neural crest cell precursors (quail neural tubes).	ET_B	DNA synthesis; cell number	LAHAV et al. (1996)
Smooth muscle cells (human coronary artery)	ET_A	DNA synthesis; cell number	YOSHIZUMI et al. (1998)
Smooth muscle cells (human pericardium)	ET_A	DNA synthesis	WU-WONG et al. (1996b)
Smooth muscle cells (ovine airway)	ET_A and ET_B	Cell number	CARRATU et al. (1997)
Smooth muscle cells (human uterine)	ET_A	Cell number	BREUILLER-FOUCHE et al. (1998)
Smooth muscle cells (human corporal)	ET_A	Cell number	GIRALDI et al. (1998)
Smooth muscle cells (rat thoracic aorta)	ET_A	DNA synthesis; transactivation of EGF receptor	IWASAKI et al. (1998)

Table 1. (Continued)

Cell/tissue type	Receptor subtype	Assay used	Reference
Smooth muscle cells (human renal artery)	NS	DNA synthesis	Assender et al. (1996)
Stromal cells (ovine luteal-phase endometrium)	NS	DNA synthesis	Salamonsen et al. (1997)
Trophoblastic cells (from placentae of normal human pregnancies)	NS	DNA synthesis	Cervar et al. (1996)

ª NS: not specified.

the zona glomerulosa of the rat (Mazzocchi et al. 1997). Similarly, in a rat model of mesangial proliferative glomerulonephritis, administration of FR139317, an ET$_A$-selective receptor antagonist, results in a significant reduction in mesangial cell proliferation, suggesting that ET-1 is involved in the proliferation of mesangial cells in the kidney, possibly mediated by the ET$_A$ receptor (Fukuda et al. 1996). In the adult rat, ET-1 increases the mitotic index in the thymus cortex, which can be blocked by ET$_A$-selective receptor antagonists BQ123 and BQ610 (Malendowicz et al. 1998). In ex vivo studies, Ricagna et al. (1996) showed that incubation of dog organ cultures with ET-1 increases positive staining for proliferation cell nuclear antigen in lung parenchyma. Furthermore, in an organ culture of human saphenous vein, a model of vein graft intimal hyperplasia, an ET$_B$ selective antagonists, BQ788, significantly reduces neointima formation, suggesting that ET-1 may be mediating human vein graft intimal hyperplasia, possibly via the ET$_B$ receptor (Porter et al. 1998).

How does ET-1 stimulate cell proliferation? ET-1 is known to be coupled to various G proteins, including G$_s$, G$_i$, G$_q$, etc., and is able to activate protein kinase C (PKC), epidermal growth factor (EGF) receptor kinase, the extracellular signal-regulated kinase (ERK) pathway, and many other intracellular signaling pathways. Regarding the mechanism for the mitogenic effect of ET-1, first, it has been reported that a pertussis toxin-sensitive G-protein, possibly G$_i$, is involved in cultured human uterine smooth muscle cells (Breuiller-Fouche et al. 1998). However, in OVCA 433 cells, the mitogenic action of ET-1 is not mediated by a pertussis toxin-sensitive G protein (Bagnato et al. 1997). Therefore, the mitogenic effect of ETs is likely mediated by ET receptors coupled to more than one form of G protein, possibly depending on the cell type.

Considering the down-stream effectors of G proteins, protein kinase C seems to play a role in ET-1-mediated mitogenic effects. For example, in vascular smooth muscle cells, ET-1-induced mitogenicity is linked to the sustained, Ca^{+2}-independent activation of PKC-delta (ASSENDER et al. 1996). As a comparison, in OVCA 433 cells, PKC is necessary but not sufficient for the maximal mitogenic effect of ET-1, and a protein tyrosine kinase may be involved (BAGNATO et al. 1997). Interestingly, in some cells, ET-1 may induce DNA synthesis through transactivation of EGF receptor (IWASAKI et al. 1998), or by a wortmannin-sensitive signaling pathway (SUGAWARA et al. 1996). Last, but not least, the ERK pathway seems to play a critical role in ET-1-induced cell proliferation, since inhibition of ERK activation attenuates ET-stimulated proliferation in airway smooth muscle cells (WHELCHEL et al. 1997). Consistent with the above notion, electroporation of anti-pp60c-src and anti-p21ras antibodies into vascular smooth muscle cells blocks DNA synthesis and cell proliferation in response to ET-1 (SCHIEFFER et al. 1997). The p21ras protein is known to be an upstream signaling molecule of ERK (LIM et al. 1996).

These results collectively indicate that ETs are mitogenic and may contribute to abnormal cell accumulation in vivo.

II. The Effects of ET-1 on the Migration and Proliferation of Endothelial Cells

Because the migration and proliferation of endothelial cells are two fundamental steps in the process of angiogenesis, it is warranted to devote a separate section to the examination of whether ETs affect these two activities of the endothelial cells.

Initial reports on the possible role of ETs in the proliferation of endothelial cells appeared in 1990. First, ET-1 was shown to stimulate DNA synthesis in brain capillary endothelial cells (VIGNE et al. 1990). Second, addition of anti-ET-1 antibody was shown to inhibit DNA synthesis in human umbilical vascular endothelial cells cultured in the presence of 10% fetal calf serum (TAKAGI et al. 1990), suggesting that ETs stimulate the proliferation of endothelial cells, possibly in an autocrine manner. Furthermore, in an in vitro model of endothelial cell injury, WREN et al. (1993) showed that ET-3 enhances wound repair, suggesting that ET-3 might have a role in stimulating endothelial cell proliferation as a response to injury.

Regarding the effect of ETs on the migration of endothelial cells, CHOLLET et al. (1993) showed that ET-1 appears to be involved in bovine corneal endothelial cell proliferation and migration; the effect of ET-1 seems additive to that of basic fibroblast growth factor (bFGF), an angiogenic factor which exhibits a profound effect in stimulating the migration and proliferation of endothelial cells both in vivo and in vitro. In endothelial cells isolated from bovine adrenal capillaries and human umbilical veins, ET-1 and ET-3 stimulate cell proliferation and migration, which is blocked by IRL-1038, an ET$_B$-selective receptor antagonist, but not by BQ123, an ET$_A$-selective receptor antagonist (MORBIDELLI et al. 1995). NOIRI et al. (1997) reported that, in a model of microvascular endothelial cell transmigration using the

Boyden chemotactic apparatus, ET-1 and ET-3 are equally potent in stimulating endothelial cell migration, which is not affected by an ET_A-selective receptor antagonist, but is inhibited by an ET_B-selective receptor antagonist. These results suggest that ETs not only stimulate the proliferation of endothelial cells, but also stimulate the migration of these cells, possibly mediated by the ET_B receptor. In examining the mechanism of ET-stimulated endothelial cell migration, it is suggested that nitric oxide (NO) may be involved. In an in vitro wound healing model with microvascular endothelial cells or Chinese hamster ovary cells stably expressing ET_B receptor with or without endothelial NO synthase, an absolute requirement for functional NO synthase in ET-1-induced cell migration has been demonstrated (Noiri et al. 1997).

Studies further demonstrate that ET-1 may evoke effects on endothelial cells in an autocrine manner. It is known that endothelial cells synthesize and secrete ETs in a constitutive manner. In addition, expression of ET-1 can be up-regulated in endothelial cells under various conditions. For example, evidence clearly shows that hypoxia stimulates the expression of ET-1, and thrombin also markedly up-regulates the expression of ET-1 (Rosendorff 1997). Also, in an in vitro model with wounded endothelial cell monolayers, expression of ET-1 is upregulated (Flowers et al. 1995). Thus, endothelial cells, upon stimulation by various factors, may over-express ETs, which in turn bind to the ET_B receptor on the endothelial cells, stimulating both cell proliferation and migration. ETs, by favoring endothelial cell growth and mobilization, may contribute to neovascularization through an autocrine mechanism mediated by the ET_B-receptor (Morbidelli et al. 1995).

In our own laboratory, we have found that ET-1 is a potent mitogenic factor for smooth muscle cells and various carcinoma cells (Table 1). Figure 1A shows that, in human pericardial smooth muscle cells, ET-1 at 10 nmol/l stimulates DNA synthesis to 185% of control. As a comparision, 10% fetal calf serum (FCS) stimulates DNA synthesis by 295%, and platelet-derived growth factor-BB (PDGF-BB) at 10 ng/ml has no effect. Figure 1B shows that ET-1 at 10 nmol/l also causes an increase in the cell number to 131% of control, while 10% FCS stimulates an increase in the cell number to 173% of control. Although these smooth muscle cells express both ET_A and ET_B receptors, the mitogenic effect of ET-1 is mediated mainly by the ET_A receptor (Wu-Wong et al. 1994).

We have observed that the effect of ETs on stimulating endothelial cell proliferation is dependent on the cell type. Figure 2A shows that, in endothelial cells derived from mouse tumor, ET-1 stimulates DNA synthesis in a dose-dependent manner, with a 2.4-fold stimulation observed at 1 nmol/l ET-1, while vascular endothelial growth factor (VEGF) at 10 ng/ml stimulates DNA synthesis fivefold. As a comparison, ET-1 does not stimulate the proliferation of human dermal microvessel endothelial cells (Fig. 2B), though both bFGF and VEGF exhibit a profound mitogenic effect (10- and 8.4-fold stimulation, respectively).

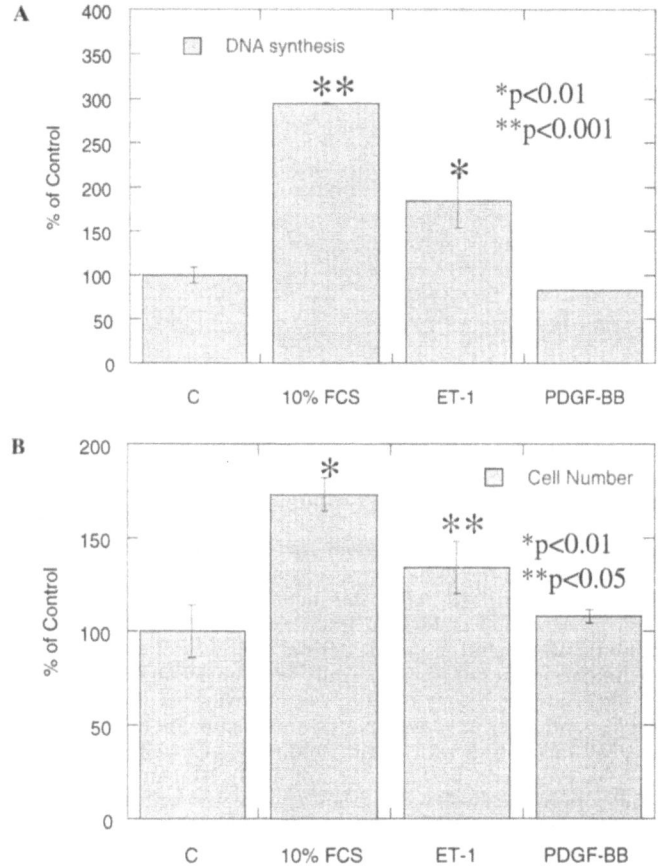

Fig. 1A,B. Effect of ET-1 on the proliferation of human pericardial smooth muscle cells. Smooth muscle cells prepared from human pericardium were provided by Dr. Maria J. Vidal (Department of Cardiology, Instituto Scientifico San Raffaele, Milano, Italy). Cells were grown in Dulbecco's modified minimal essential medium (*DMEM*) containing 10% FCS (GIBCO). **A** DNA synthesis: cells at approximately 60% confluency in 48-well plates were incubated in serum-free medium for 48h, followed by incubation with [³H]thymidine (1 μCi/well) in the presence or absence of 10% FCS or 10 nmol/l ET-1 or 10 ng/ml PDGF-BB for 24h. After the incubation, each well was washed with 1 ml of PBS, and then washed with 0.5 ml of ice-cold 10% TCA for 30 min at 4 °C. Each well was then washed again with 0.5 ml of 10% TCA. Materials not soluble in TCA were dissolved in 0.1 N NaOH for scintillation counting. Each value represents the mean ± SD of three determinations. **B** Cell number determination. Cells were treated as in **A** except that no [³H]thymidine was added, and the cell number in each well was determined by a Coulter counter at the end of the 24h incubation with FCS or ET-1 or PDGF-BB. Data are expressed as % of control (*C*, no addition of ET-1 or other growth factors). Statistical analysis was performed by an ANOVA vs control (*C*)

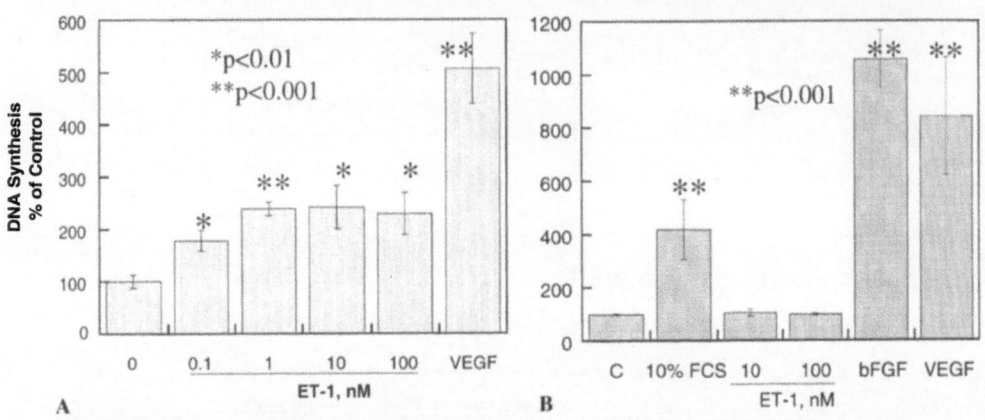

Fig. 2A,B. Effect of ET-1 on the proliferation of endothelial cells. **A** Endothelial cells derived from mouse tumor were provided by Dr. Candace Johnson (Department of Otolaryngology, University of Pittsburgh, Pittsburgh, PA). Cells were grown in DMEM containing 10% FCS (GIBCO). Cells at approximately 60% confluency in 48-well plates were incubated in serum-free medium for 48 h, followed by incubation with [³H]thymidine (1 μCi/well) in the presence or absence of ET-1 (concentration as indicated) or 10 ng/ml VEGF for 24 h. After the incubation, each well was washed and treated with TCA as described in Fig. 1A. Each value represents the mean ± SD of three determinations. **B** Human (neonatal) dermal microvessel endothelial cells were obtained from Clonetics (San Diego, CA). Cells were grown in microvessel endothelial growth medium (Clonetics) containing 5% fetal bovine serum (GIBCO). Cells at approximately 60% confluency in 48-well plates were incubated in serum-free medium for 48 h, followed by incubating with [³H]thymidine (1 μCi/well) in the presence or absence of 10% FCS, ET-1 (concentration as indicated), 10 ng/ml bFGF or VEGF for 24 h. After the incubation, each well was washed and treated with TCA as described in Fig. 1A. Each value represents the mean ± SD of three determinations. Data are expressed as % of control (*C*, no addition of ET-1 or other growth factors). Statistical analysis was performed by an ANOVA vs control (*C*)

III. The Anti-Proliferative Effect of ETs

Interestingly, ETs also exhibit anti-proliferative effects in some cell/tissue systems. For example, in myofibroblastic Ito cells, ET-1 inhibits DNA synthesis and proliferation of cells stimulated with either human serum or PDGF-BB, possibly mediated by the ET_B receptor (Mallat et al. 1995). Follow-up studies by the same group have shown that, in human hepatic stellate cells, ETs inhibit cell proliferation via the activation of ET_B (Mallat et al. 1998; Gallois et al. 1998). The anti-proliferative effect of ET-1 on these cells is likely mediated by a prostaglandin/cAMP pathway that leads to inhibition of both ERK and c-Jun kinase activities (Mallat et al. 1996). These results suggest that ET-1, through the ET_B receptor, may be involved in limiting proliferation of activated hepatic stellate cells during chronic liver diseases, and that ETs may play a key role in the negative control of liver fibrogenesis.

In conclusion, ETs are potent mitogenic factors for a variety of cells, although they also exhibit anti-proliferation effects in the hepatic Stellate cells. The involvement of ETs in regulating cell proliferation suggests that ET receptor antagonists may be useful in treating diseases with abnormal growth disorders. Indeed, it has been shown that the ET_A receptor selective antagonist, ABT-627 (A-127722, OPGENORTH et al. 1996), reduces neointimal hyperplasia in porcine iliac and coronary artery angioplasty and stent restenosis models (BURKE et al. 1997; McKENNA et al. 1998).

C. The Role of Endothelin in Apoptosis

Evidence for the involvement of ETs in modulating apoptosis started to emerge in 1997. SHICHIRI et al. (1997) showed that, in cultured endothelial cells from rat aorta, ET-1 suppresses apoptosis induced by serum starvation in a concentration-dependent manner. The ET_B-selective receptor antagonist, BQ788, but not the ET_A-selective receptor antagonist, BQ123, blocks the anti-apoptotic effect of ET-1, suggesting that ET_B may be the receptor mediating the anti-apoptosis effect of ET-1 in endothelial cells via an autocrine/paracrine manner. The same group later reported that low doses of ET-1 protect fibroblasts against serum starvation-induced apoptosis via a c-Myc-dependent process. The ET-1-induced cell survival is mediated by the ET_A receptor, and is not linked to the ability of ET-1 to induce cell proliferation. The survival function of ET-1 is abrogated by inhibiting the mitogen-activated protein kinase (MAPK) pathway (SHICHIRI et al. 1998).

We (WU-WONG et al. 1997) have found that, in both human pericardial and prostatic smooth muscle cells, addition of ET-1 reduces paclitaxel-induced DNA fragmentation and phosphatidylserine on the cell surface, two characteristics of apoptosis (Figs. 3 and 4). As a comparison, angiotensin II does not have a significant effect on apoptosis (Fig. 4). Furthermore, we have learned that ET-1 may also function as a survival factor in prostate cancer cells, especially in cells that have become androgen-independent (Joel Nelson [Johns Hopkins University] and Sookja Chung [University of Hong Kong], personal communication).

These studies suggest that ET-1 not only acts as a mitogen, but also functions as an anti-apoptotic factor for endothelial and smooth muscle cells, and possibly other cell types.

Interestingly, ET-1 may actually induce apoptosis in some cells. OKAZAWA et al. (1998) showed that ET-1 inhibits serum-dependent growth of asynchronized A375 human melanoma cells, and the growth inhibitory effect is markedly enhanced when ET-1 is applied to the cells synchronized at G1/S boundary by double thymidine blocks. Addition of ET-1 does not inhibit the cell cycle progression of cells after the release of the block, but causes a significant increase of the hypodiploid cell population that is characteristic of apoptotic cell death. ET-1-induced apoptosis is further confirmed by the

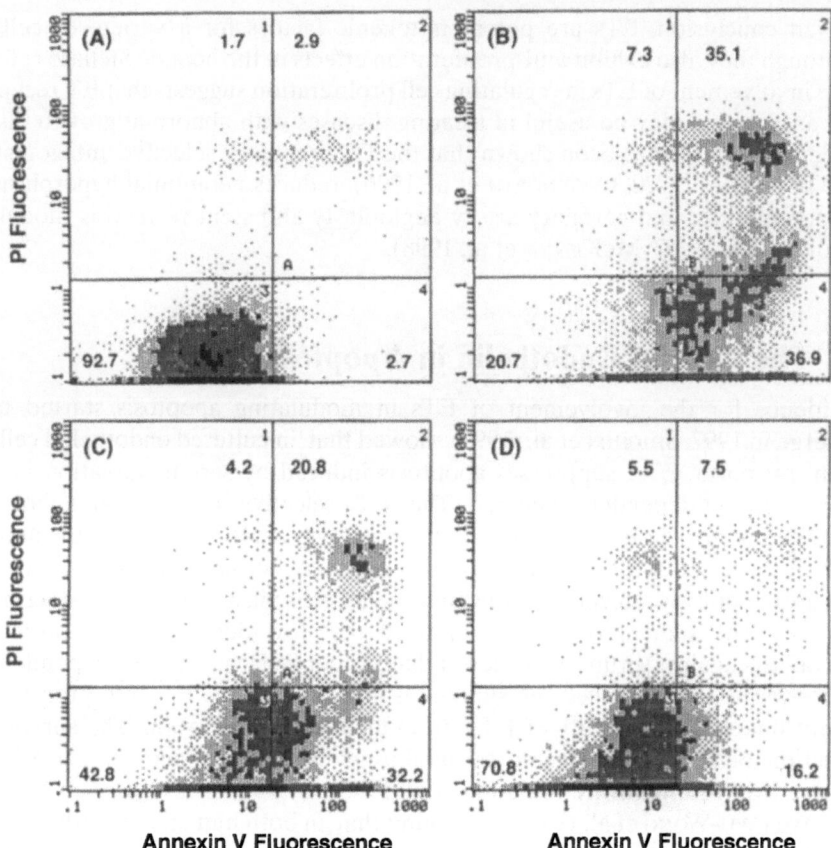

Fig. 3A–D. ET-1 attenuated paclitaxel-induced phosphatidylserine processing in human pericardial smooth muscle cells (HPSMC). HPSMC in $10\,cm^2$ petri dishes were treated with or without paclitaxel in the presence or absence of ET-1 for 48h at 37°C. Cells were then collected and double-stained with fluorescein isothiocyanate (FITC)-conjugated annexin V and propidium iodide using an apoptosis detection kit according to the manufacturer's instruction (R&D Systems, Cat. no KNX50). The kit determines the percentage of cells undergoing apoptosis in a quantitative manner based on the principle that apoptotic cells bind to annexin V, while necrotic cells only bind to propidium iodide. Samples (10,000 cells per sample) were then analyzed by flow cytometry in an Epics Elite (Coulter Corp.). The samples were excited with a 488nm Argon laser set at 25mW. The emission filters were a 515- to 535-nm band pass filter for FITC and a 565- to 585-nm band pass filter for propidium iodide. Data was analyzed using the Epics Elite software. Normal cells in *quadrant 3* of the histogram exhibit minimal (background) staining for annexin-V and propidium iodide. Cells in *quadrant 1* were annexin-V negative- and PI positive-staining cells, and were considered as necrotic cells. Cells in *quadrant 3* were stained positive for both annexin-V and PI, and possibly represented a mixture of necrotic and apoptotic cells. Annexin-V positive- and PI negative-staining cells in *quadrant 4* were apoptotic cells. Paclitaxel did not seem to have a significant effect in inducing necrosis in these cells (the % in *quadrant 1* increased from 1.7% to 7.3% after paclitaxel treatment). Therefore, cells in *quadrants 2 and 4* were scored as apoptotic. **A** Control: untreated. **B** Paclitaxel ($10\,\mu mol/l$) alone. **C** Paclitaxel ($10\,\mu mol/l$) plus ET-1 ($10\,nmol/l$). **D** Paclitaxel ($10\,\mu mol/l$) plus ET-1 ($100\,nmol/l$). Results shown are representative of five different experiments. The number of cells (in %) in each quadrant is indicated. Reprinted with permission from Wu-Wong et al. (1997)

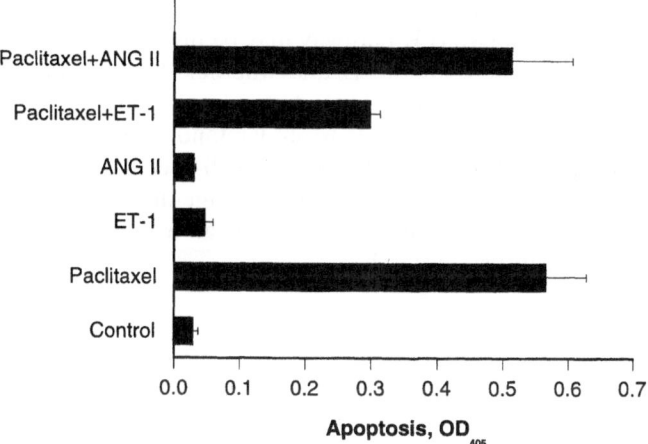

Fig. 4. ET-1 attenuated paclitaxel-induced apoptosis in human prostate smooth muscle cells. Human prostate smooth muscle cells were purchased from Clonetics (San Diego, CA) and were grown in SmGM media containing 5% fetal bovine serum. Cells in 96-well plates in SmGM media containing 5% FBS were treated with paclitaxel ($1\,\mu$mol/l) in the presence or absence of ET-1 ($10\,$nmol/l) or angiotensin II ($10\,$nmol/l) for $20\,$h at $37\,°C$. At the end of the incubation, cells were lysed in $200\,\mu$l of lysis buffer. The cell lysates were collected and centrifuged at $200 \times g$ for $10\,$min. The samples were assayed for apoptosis using an ELISA kit according to the manufacturer's instruction (Boehringer Mannheim, Cat. no 1544675). The ELISA uses monoclonal antibodies directed against DNA and histones in a quantitative sandwich-enzyme-based format. The amount of histone-associated DNA fragments (mono- and oligonucleosomes) in the cell lysates was determined at OD_{405} in a spectrophotometer. Each value represents mean $\pm$ SD of three determinations

appearance of chromatin condensation on nuclear staining and DNA fragmentation on gel electrophoresis. The pro-apoptosis effects of ET-1 appear to be mediated by the ET_B receptor via a pertussis toxin-sensitive G protein (OKAZAWA et al. 1998).

The effect of ET-1 on modulating apoptosis has also been shown in an in vivo animal study. HOCHER et al. (1998) found that ET-1 transgenic mice, in which ET-1 is over-expressed, are characterized by age-dependent development of renal cysts and interstitial fibrosis and glomerulosclerosis, leading to a progressive decrease in glomerular filtration rate. The numbers of apoptotic cells are significantly increased in the kidneys of 14-month-old ET-1 transgenic mice, whereas cell proliferation is not enhanced. Apoptotic cells are detected in the glomeruli, tubular cells, and renal interstitial cells. The study seems to suggest that ET-1 plays a pro-apoptosis role in the kidney.

Studies on the involvement of ET-1 in the regulation of apoptosis pathways are still in the early phase. So far, results seem conflicting. In some cells such as fibroblasts, endothelial, and smooth muscle cells, ET-1 functions as a survival factor. Yet in other cells, ET-1 is an apoptosis inducer. Although it has been suggested that the MAPK pathway may be involved, very little is known about how ETs modulate apoptosis, and whether the anti- vs pro-apoptosis effects of ETs in different cells are mediated by the same or different pathways. Also, it awaits further studies to understand more about whether different ET receptor subtypes with possibly cell-specific signal transduction coupling are responsible for the seemingly conflicting outcomes. The role of the ET system in regulating apoptosis remains an interesting and potentially important area for future research.

D. The Role of Endothelin in Angiogenesis

I. Is ET-1 an Angiogenic Factor?

To answer this question, we think it is necessary to compare known facts of ETs against the documented characteristics of angiogenic factors.

According to the definition described by Battegay (1995) in a review article, an angiogenic factor shall conform to at least two of the following four criteria:

1. It elicits an effect on angiogenesis in vivo upon exogenous administration or overexpression.
2. It elicits an effect on relevant endothelial cells in vitro.
3. The relevant vessels in vivo can respond to the molecule.
4. The factor's role in angiogenesis is established in the process of disease.

Some known angiogenic factors such as FGF, erythropoietin, insulin-like growth factor-I (IGF-1), PDGF, angiopoietin-1, and VEGF have all been shown to stimulate angiogenesis either in vitro, or in vivo, or both.

Regarding the role of ETs in angiogenesis, first, let us examine the data from transgenic animal studies with targeted gene inactivation of ETs or ET receptors, or overexpression of ETs. In the mouse with targeted gene inactivation of ET-1, ET-3, ET_A, or ET_B receptor, the animals exhibit severe defects in the embryonic development. The defects associated with ET-1 and ET_A inactivation are cardiovascular and craniofacial malformations (Kurihara et al. 1994; Clouthier et al. 1998), and the defects with ET-3 and ET_B receptor inactivation are colonic agangliogenesis associated with skin pigmentation abnormalities (Baynash et al. 1994; Hosoda et al. 1994; see Chap. 6). These defects are likely due to the pivotal role of ET-1 and ET-3 in the migration, proliferation, and differentiation of neural crest cells. However, these animals do not seem to suffer from abnormal vascular development, suggesting that ET-1 does not have a direct role in vasculogenesis and/or angiogenesis during embryogenesis. As a comparison, targeted gene inactivation studies in mice have shown that VEGF is necessary for the early stages of vascular development (Ferrara et al. 1996) and that angiopoietin-1 is required for the later stages of vascular remodeling (Suri et al. 1996).

Transgenic mice over-expressing the human ET-1 gene (HOCHER et al. 1997; THEURING et al. 1998) have a pathological phenotype manifested by signs such as age-dependent development of cysts, interstitial fibrosis, and glomerulosclerosis in the kidneys leading to a progressive decrease in glomerular filtration rate. Plasma and tissue ET-1 concentrations are only slightly elevated, and blood pressure is not affected even after the development of an impaired glomerular filtration rate. Despite the phenotype described above, there is no mentioning of any obvious abnormality in the vascular development and/or remodeling (HOCHER et al. 1997). Similarly, in a transgenic rat overexpressing the human ET-2 gene with high renal transgene expression, renal tissue ET-2 concentrations are significantly increased, resulting in significantly increased glomerular injury and protein excretion. Again, no obvious abnormality in vascular development and/or remodeling has been reported (HOCHER et al. 1996). As a comparison, recently it has been shown that a transgenic overexpression of angiopoietin-1 in the skin of mice produces larger, more numerous, and more highly branched vessels (SURI et al. 1998).

Angiogenic factors are generally able to induce angiogenesis in a sponge model (HU and FAN 1995) and/or in a corneal micropocket model (KENYON et al. 1996) in animals. In a rat sponge angiogenesis model, a daily dose of vasoactive intestinal peptide at 1000 pmol causes intense neovascularization, but daily doses of 10 pmol, 100 pmol, or 1000 pmol ET-3, or of 100 pmol or 1000 pmol ET-1 produce no apparent effect (HU et al. 1996). So far, no information is available about whether ET would induce neovascularization in a corneal micropocket model. However, in a study with experimentally induced corneal neovascularization, ET-1 administered topically, subconjunctivally, and intraluminally in serial concentrations ranging from $0.0005\,\mu g/ml$ to $5.0\,\mu g/ml$ in New Zealand white rabbits did not induce vasoconstriction or neovascularization (EIFERMAN et al. 1992). Besides the two in vivo models mentioned above, many angiogenic factors such as VEGF and bFGF have been shown to stimulate endothelial cells grown in 3-D collagen gels or Matrigels to develop networks of capillary-like tubes (PEPPER et al. 1992). Similar results have not been reported for ET-1. Thus, the information about the effect of ET on angiogenesis in these in vitro and in vivo models is rather limited, and available data do not support that ET-1 is directly involved in the process of angiogenesis.

Taken together, ET-1 seems to have stimulatory effects on the proliferation and migration of certain types of endothelial cells, though the in vivo studies do not seem to support a role for ET-1 as a direct angiogenic factor. However, the available information also does not exclude the possibility that ET-1 may interact with other factors to regulate angiogenesis in an indirect manner.

II. Is ET-1 Found in Diseases with Angiogenesis?

To investigate whether ET-1 plays a role in the angiogenesis process, it is of interest to examine whether ETs and ET receptors are associated with diseases in which angiogenesis is involved. Several studies have touched upon this subject.

Stiles et al. (1997) showed that, in high grade primary brain tumors including 10 glioblastoma multiforme, 10 anaplastic astrocytomas, and 10 low-grade astrocytomas, ET-1 was locally expressed in high concentrations in all cases of glioblastoma examined. Among 10 anaplastic astrocytomas, 6 tumors were positive for ET-1. In low-grade astrocytoma, 4 of 10 tumors showed weak ET-1 immunoreactivity. Endothelial cells within all tumors were positive for ET-1, and ET-1 was shown to be present in human astrocytomas and its expression correlates with tumor vascularity and malignancy.

In a different study using routine endomyocardial biopsy specimens of transplanted human hearts (72 biopsy samples), ET-1 immunoreactivity was found to be localized to vascular and endocardial endothelial cells, as well as to cardiomyocytes. The pattern of endothelial cell immunostaining with the ET-1 antiserum is similar to that of von Willebrand factor. Previous biopsy sites and areas of granulation tissue appear to have greater ET-1 immunoreactivity. There is a significant correlation between the presence of ET-1 immunoreactivity and fibrosis or granulation tissue in the biopsy specimens ($P < 0.03$). In situ hybridization with radiolabeled RNA probes reveals expression of ET-1 mRNA in endothelial cells and myocytes, also in association with granulation tissue and fibrosis (Giaid et al. 1995).

In addition, ET-1 binding sites are found in the blood vessels of human pulmonary tumors by in vitro autoradiography. Specific [^{125}I]ET-1 binding is identified in the blood vessels of all sizes in both tumor and stromal tissues. The ET-1 binding in blood vessels is specific, saturable, and time-dependent. The binding is competitively inhibited by ET-1 >> ET-2 > ET-3. The results provide evidence that ET$_A$ receptors are localized in the blood vessels of pulmonary tumors and in stromal tissues surrounding the tumor nest (Zhao et al. 1995).

The above studies suggest that ETs, and perhaps ET receptors, may be up-regulated in diseases in which progressive angiogenesis and vascular regeneration play a role. However, these studies do not provide evidence to suggest that ET-1 is directly responsible for the increased angiogenesis seen in these disease states. Because endothelial cells are a good source for ET-1, the presence of ET-1 in these pathological states is expected. Whether ET plays a role in stimulating angiogenesis in the aforementioned disease state has not been proven to date.

III. Interaction Between ETs and Angiogenic Factors

Because angiogenesis is a complex process that involves a variety of factors regulating endothelial cell proliferation and migration, degradation of basement membrane, and neovessel organization, agents that are not directly affecting the process of angiogenesis may be involved in modulating angiogenesis via modulation of the activity of "direct" angiogenic factors. Therefore, it is necessary to examine whether and how ETs may interact with known angiogenic factors.

Evidence seems to suggest that ETs and VEGF regulate each other's expression. ET-1 and ET-3 stimulate the synthesis of VEGF protein 3- to 4-fold in cultured human vascular smooth muscle cells through ET_A and ET_B receptors, and activation of protein kinase C. VEGF, secreted into the medium in cultured vascular smooth muscle cells in response to ET-1 treatment, is capable of stimulating endothelial cell proliferation and invasion of matrix. (PEDRAM et al. 1997). Furthermore, VEGF enhances prepro-ET-1 mRNA expression and ET-1 secretion in bovine aortic endothelial cells (BAECs). Similarly, in rat vascular smooth muscle cells (VSMCs), ET-1 enhances VEGF mRNA expression and stimulates VEGF secretion. ET-1-induced VEGF mRNA expression is abolished by a selective ET_A receptor antagonist, BQ485, but not by an ET_B-selective blocker, BQ788. Coculture of BAECs and VSMCs enhances both ET-1 and VEGF gene expression in these cells, and the conditioned media from BAECs and VSMCs reproduce the augmentation of gene expression of the two factors, which is partially inhibited by BQ485 or an anti-VEGF antibody. These results suggest that VEGF and ET-1 mutually enhance each other's expression, which may play an important role in regulating the proliferation of endothelial and smooth muscle cells in the vascular wall (MATSUURA et al. 1998).

It has also been shown that ET-1 enhances PDGF-induced DNA synthesis in vascular smooth muscle cells (ASSENDER et al. 1996). In Swiss 3T3 cells, ET-1 synergizes very strongly with several other growth factors including PDGF and bFGF to promote cell growth (BROWN et al. 1989). In rat mesangial cells, PDGF-AB and -BB stimulate ET-1 secretion through PDGF beta-receptors, and endogenously produced ET-1 modulates the mitogenic effect of PDGF-AB and -BB, probably via the ET_A receptor in these cells (KOHNO et al. 1994).

Human erythropoietin (HuEPO), another factor exhibiting an angiogenic effect, increases ET-1 levels in the culture supernatant of bovine pulmonary arterial endothelial cells (CARLINI et al. 1995a). In addition, in an in vitro angiogenesis assay using rat aortic rings embedded in a basement membrane matrix, HuEPO increases vessel outgrowth. The supernatant ET-1 levels of the rings incubated with HuEPO are significantly higher than controls. When rings are exposed to ET-1 alone, an increase in microvessel formation compared to control is also observed. In aortic rings co-cultured with HuEPO and an anti-ET-1 antibody, stimulation of angiogenesis by HuEPO is blunted, suggesting that ET-1 may mediate the angiogenic effect of HuEPO (CARLINI et al. 1995b).

In the light of the recent finding that NO may play a role in angiogenesis (WINK et al. 1998), perhaps it is necessary to address the relationship between ET and NO. Numerous reports have shown that ETs, by binding to the ET_B receptor on endothelial cells, are capable of stimulating NO release (reviewed by HAYNES and WEBB 1998). Conversely, endogenous NO inhibits

Table 2. The mitogenic effect of ET-1 and other growth factors on DNA synthesis in pig carotid artery smooth muscle cells

Cell Types	No ET-1	10 nmol/l ET-1
Control	100 ± 25	197 ± 44
PDGF-BB	1717 ± 284	3423 ± 262
bFGF	503 ± 50	639 ± 67
IGF-1	205 ± 23	313 ± 57

Smooth muscle cells were collected from culturing segments of carotid arteries from pigs. The arteries were cut into pieces of approximately 1 mm × 1 mm in size by a scalpel, and cultured in DMEM containing 10% fetal calf serum (FCS). Cells were collected by trypsin treatment. Cells at approximately 60% confluency in 96-well plates were incubated in serum-free medium for 48h, followed by incubating with [^{3}H]thymidine (0.5 μCi/well) in the presence or absence of ET-1 (10nmol/l) with or without the indicated growth factor for 24h. After the incubation, each well was washed and treated with TCA as described in Fig. 1A. Each value represents the mean ± SD of three determinations. Data are expressed as % of control (no addition of ET-1 or other growth factors).

endothelin-1 production through guanylyl cyclase/cGMP-dependent mechanisms (Mitsutomi et al. 1999). Furthermore, ET-1 inhibits the expression of the inducible isoform of NO synthase (iNOS) in the rat lung epithelial cell line and primary cultured rat glial cells, and in the kidney (Markewitz et al. 1997; Murayama et al. 1998; Aiello et al. 1998). Therefore, it is plausible that ETs, by modulating the expression of iNOS and the release of NO, may impact the process of angiogenesis.

From our own experience, we have observed that, in pig carotid artery smooth muscle cells, PDGF-BB and bFGF are much more effective than ET-1 in stimulating DNA synthesis (Table 2). However, ET-1 and PDGF-BB stimulate DNA synthesis in a synergistic manner, and ET-1 and bFGF or IGF-1 together exhibit an additive effect on DNA synthesis (Table 2). Figure 5 shows that, in human coronary artery smooth muscle cells, both PDGF-BB and ET-1 stimulate DNA synthesis. Again, the mitogenic effect of ET-1 is modest (twofold stimulation at 10nmol/l) in comparison to that of PDGF-BB (fourfold stimulation at 10ng/ml), and ET-1 and PDGF-BB together exhibit an additive effect on DNA synthesis (Fig. 5).

Taken together, these results suggest that ET-1 is able to interact with angiogenic factors in modulating the synthesis, secretion, and function of ET itself, or the angiogenic factor involved. This characteristic of ETs suggest that the ET system, if not directly involved in regulating angiogenesis, at least plays an indirect role in the process of angiogenesis.

E. Conclusion

In this chapter we have reviewed studies reported in the past few years regarding the effects of ETs on modulating cell proliferation, apoptosis, and possi-

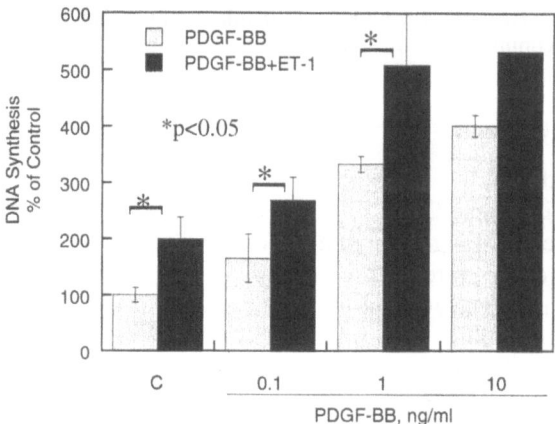

Fig. 5. Additive effect of ET-1 and PDGF-BB on the proliferation of human coronary artery smooth muscle cells. Cells purchased from Clonetics (San Diego, CA) were grown in SmGM media containing 5% fetal bovine serum (FBS). Cells at approximately 60% confluency in 48-well plates were incubated in serum-free medium for 24h, treated with ET-1 (10nmol/l) and/or PDGF-BB (concentration as indicated) for 18h, followed by incubating with [³H]thymidine (1 μCi/well) for another 8h. After the incubation, each well was washed and treated with TCA as described in Fig. 1A. Each value represents the mean ± SD of three determinations. Data are expressed as % of control (C, no addition of ET-1 or other growth factors). Statistical analysis was performed by an ANOVA

bly angiogenesis. It is well documented and widely accepted that ETs exhibit mitogenic effects in many cells and tissues, largely via stimulation of ET_A receptors. In addition, although the effect of ETs on modulating apoptosis is not as well studied as their effects on cell proliferation, ETs may function as a survival factor to further promote cell accumulation. The physiological and pathological application of these characteristics of ETs is not fully known. However, as a mitogen and a survival factor, ET-1 may play a role in diseases involving abnormal cell accumulation, such as tumorigenesis, atherosclerosis, restenosis, fibrosis, etc., and ET receptor antagonists may be useful in treating these diseases. Regarding the role of ETs in angiogenesis, data gathered so far do not provide a clear answer to the question of whether ETs are directly involved in angiogenesis. However, it is evident that ETs are capable of stimulating the proliferation and migration of endothelial cells, and can also interact with known angiogenic factors such as VEGF, bFGF, and PDGF in an additive, and perhaps synergistic, manner to regulate the activities of endothelial cells. In addition, endogenous ETs secreted from the vascular endothelium may regulate blood flow. Therefore, it is conceivable that ETs, if not directly involved in regulating angiogenesis, at least play an indirect role in the process of angiogenesis.

Acknowlegement. The author would like to thank Drs. Don Davidson and Jack Henkin for their critical comments.

List of Abbreviations

ET   endothelin
PKC   protein kinase C
ERK   extracellular signal-regulated kinase
MAPK   mitogen-activated protein kinase
EGF   epidermal growth factor
bFGF   basic fibroblast growth factor
PDGF   platelet-derived growth factor
VEGF   vascular endothelial growth factor
FCS   fetal calf serum

References

Aiello S, Remuzzi G, Noris M (1998) Nitric oxide/endothelin balance after nephron reduction. Kidney Int Suppl 65:S63–S67

Arai H, Hori S, Arimori I, Ohkubo H, Nakanishi S (1990) Cloning and expression of a cDNA encoding an endothelin receptor. Nature 348:730–732

Asham EH, Loizidou M, Taylor I (1998) Endothelin-1 and tumour development. Eur J Surg Oncol 24:57–60

Assender JW, Irenius E, Fredholm BB (1996) Endothelin-1 causes a prolonged protein kinase C activation and acts as a co-mitogen in vascular smooth muscle cells. Acta Physiol Scand 157:451–460

Bagnato A, Tecce R, Moretti C, Di Castro V, Spergel D, Catt KJ (1995a) Autocrine actions of endothelin-1 as a growth factor in human ovarian carcinoma cells. Clin Cancer Res 1:1059–1066

Bagnato A, Venuti A, Di Castro V, Marcante ML (1995b) Identification of the ETA receptor subtype that mediates endothelin induced autocrine proliferation of normal human keratinocytes. Biochem Biophys Res Commun 2091:80–86

Bagnato A, Tecce R, Di Castro V, Catt KJ (1997) Activation of mitogenic signaling by endothelin 1 in ovarian carcinoma cells. Cancer Res 57:1306–1311

Baynash AG, Hosoda K, Giaid A, Richardson JA, Emoto N, Hammer RE, Yanagisawa M (1994) Interaction of endothelin-3 with endothelin-B receptor is essential for development of epidermal melanocytes and enteric neurons. Cell 79:1277–1285

Battegay EJ (1995) Angiogenesis: mechanistic insights, neovascular diseases, and therapeutic prospects. J Mol Med 73:333–346

Bergers G, Hanahan D, Coussens LM (1998) Angiogenesis and apoptosis are cellular parameters of neoplastic progression in transgenic mouse models of tumorigenesis. Int J Dev Biol 42:995–1002

Breuiller-Fouche M, Heluy V, Fournier T, Dallot E, Vacher-Lavenu MC, Ferre F (1998) Role of endothelin-1 in regulating proliferation of cultured human uterine smooth muscle cells. Mol Hum Reprod 4:33–39

Brown KD, Littlewood CJ (1989) Endothelin stimulates DNA synthesis in Swiss 3T3 cells. Synergy with polypeptide growth factors. Biochem J 263:977–980

Burke SE, Lubbers NL, Gagne GD, Wessale JL, Dayton BD, Wegner CD, Opgenorth TJ (1997) Selective antagonism of the ET(A) receptor reduces neointimal hyperplasia after balloon-induced vascular injury in pigs. J Cardiovasc Pharmacol 30: 33–41

Carlini RG, Gupta A, Liapis H, Rothstein M (1995a) Endothelin-1 release by erythropoietin involves calcium signaling in endothelial cells. J Cardiovasc Pharmacol 26:889–892

Carlini RG, Reyes AA, Rothstein M (1995b) Recombinant human erythropoietin stimulates angiogenesis in vitro. Kidney Int 47:740–745

Carratu P, Scuri M, Styblo JL, Wanner A, Glassberg MK (1997) ET-1 induces mitogenesis in ovine airway smooth muscle cells via ETA and ETB receptors. Am J Physiol 272:L1021–L1024

Cervar M, Puerstner P, Kainer F, Desoye G (1996) Endothelin-1 stimulates the proliferation and invasion of first trimester trophoblastic cells in vitro–a possible role in the etiology of pre-eclampsia? J Investig Med 44:447–453

Chiou WJ, Wang J, Berg CE, Wu-Wong JR (1999) SV40 virus transformation down-regulates endothelin receptor, Biochem Biophys Acta 1450:35–44

Chollet P, Malecaze F, Gouzi L, Arne JL, Plouet J (1993) Endothelin 1 is a growth factor for corneal endothelium. Exp Eye Res 57:595–600

Clouthier DE, Hosoda K, Richardson JA, Williams SC, Yanagisawa H, Kuwaki T, Kumada M, Hammer RE, Yanagisawa M (1998) Cranial and cardiac neural crest defects in endothelin-A receptor-deficient mice. Development 125:813–824

Douglas SA, Beck GR Jr, Elliott JD, Ohlstein EH (1995) Pharmacological evidence for the presence of three distinct functional endothelin receptor subtypes in the rabbit lateral saphenous vein. Brit J Pharcol 114:1529–1540

Eiferman RA, Stagner J, Jumblatt M, O'Neill KP, Smith DA (1992) Endothelin does not affect experimentally induced corneal neovascularization. Ann Ophthalmol 24:236–238

Ferrara N, Carver-Moore K, Chen H, Dowd M, Lu L, O'Shea KS, Powell-Braxton L, Hillan KJ, Moore MW (1996) Heterozygous embryonic lethality induced by targeted inactivation of the VEGF gene. Nature 380:439–442

Flowers MA, Wang Y, Stewart RJ, Patel B, Marsden PA (1995) Reciprocal regulation of endothelin-1 and endothelial constitutive NOS in proliferating endothelial cells. Am J Physiol 269:H1988–997

Fukuda K, Yanagida T, Okuda S, Tamaki K, Ando T, Fujishima M (1996) Role of endothelin as a mitogen in experimental glomerulonephritis in rats. Kidney Int 49:1320–139

Gallois C, Habib A, Tao J, Moulin S, Maclouf J, Mallat A, Lotersztajn S (1998) Role of NF-κB in the antiproliferative effect of endothelin-1 and tumor necrosis factor- in human hepatic stellate cells. involvement of cyclooxygenase-2 J. Biol. Chem. 273:23183–23190

Giaid A, Saleh D, Yanagisawa M, Forbes RD (1995) Endothelin-1 immunoreactivity and mRNA in the transplanted human heart. Transplantation 59:1308–313

Giraldi A, Serels S, Autieri M, Melman A, Christ GJ (1998) Endothelin-1 as a putative modulator of gene expression and cellular physiology in cultured human corporal smooth muscle cells. J Urol. 160:1856–1862

Goldie RG, Knott PG, Carr MJ, Hay DW, Henry PJ (1996) The endothelins in the pulmonary system. Pulm Pharmacol 9: 69–93

Goto K, Hama H, Kasuya Y (1996) Molecular pharmacology and pathophysiological significance of endothelin. Jpn J Pharmacol 72:261–290

Hanahan D, Folkman J (1996) Patterns and emerging mechanims of the angiogenic switch during tumorigenesis. Cell 86:353–364

Haynes WG, Webb DJ (1998) Endothelin as a regulator of cardiovascular function in health and disease. J Hypertens 16:1081–1098

Hocher B, Liefeldt L, Thone-Reineke C, Orzechowski HD, Distler A, Bauer C, Paul M (1996) Characterization of the renal phenotype of transgenic rats expressing the human endothelin-2 gene. Hypertension 28:196–201

Hocher B, Thone-Reineke C, Rohmeiss P, Schmager F, Slowinski T, Burst V, Siegmund F, Quertermous T, Bauer C, Neumayer HH, Schleuning WD, Theuring F (1997)

Endothelin-1 transgenic mice develop glomerulosclerosis, interstitial fibrosis, and renal cysts but not hypertension. J Clin Invest 99:1380–1389

Hocher B, Rohmeiss P, Thone-Reineke C, Schwarz A, Burst V, van der Woude F, Bauer C, Theuring F (1998) Apoptosis in kidneys of endothelin-1 transgenic mice. J Cardiovasc Pharmacol 31:S554–S556

Hosoda K, Hammer RE, Richardson JA, Baynash AG, Cheung JC, Giaid A, Yanagisawa M (1994) Targeted and natural (piebald-lethal) mutations of endothelin-B receptor gene produce megacolon associated with spotted coat color in mice. Cell 79:1267–1276

Hu DE, Fan TP (1995) Suppression of VEGF-induced angiogenesis by the protein tyrosine kinase inhibitor, lavendustin A. Br J Pharmacol 114:262–268

Hu DE, Hiley CR, Fan TP (1996) Comparative studies of the angiogenic activity of vasoactive intestinal peptide, endothelins-1 and –3 and angiotensin II in a rat sponge model. Br J Pharmacol 117:545–551

Imokawa G, Yada Y, Kimura M (1996) signalling mechanisms of endothelin-induced mitogenesis and melanogenesis in human melanocytes. Biochem J 314:305–312

Inoue A, Yanagisawa M, Kimura S, Kasuya Y, Miyauchi T, Goto, Masaki T (1989) The human endothelin family: Three structurally and pharmacologically distinct isopeptides predicted by three separate genes. Proc Natl Acad Sci 86:2863–2867.

Iwasaki H, Eguchi S, Marumo F, Hirata Y (1998) Endothelin-1 stimulates DNA synthesis of vascular smooth-muscle cells through transactivation of epidermal growth factor receptor. J Cardiovasc Pharmacol 31:S182–S184

Kamada S, Blackmore PF, Kubota T, Oehninger S, Asada Y, Gordon K, Hodgen GD, and Aso T. (1995) The role of endothelin-1 in regulating human granulosa cell proliferation and steroidogenesis in vitro. J Clin Endocrinol Metab 80:3708–3714

Kenyon BM, Voest EE, Chen CC, Flynn E, Folkman J, D'Amato RJ (1996) A model of angiogenesis in the mouse cornea. Invest Ophthalmol Vis Sci 37:1625–1632

Khatib AM, Lomri A, Moldovan F, Soliman H, Fiet J, Mitrovic DR (1998) Endothelin 1 receptors, signal transduction and effects on dna and proteoglycan synthesis in rat articular chondrocytes. Cytokine 10:669–679

Kohno M, Horio T, Yokokawa K, Yasunari K, Kurihara N, Takeda T (1994) Endothelin modulates the mitogenic effect of PDGF on glomerular mesangial cells. Am J Physiol 266:F894–900

Komuro I, Kurihara H, Sugiyama T, Yoshizumi M, Takaku F, Yazaki Y (1988) Endothelin stimulates c-fos and c-myc expression and proliferation of vascular smooth muscle cells. FEBS Lett 238:249–252

Kozuka M, Ito T, Hirose S, Lodhi KM, Hagiwara H (1991) Purification and characterization of bovine lung endothelin receptor. J Biol Chem 266:16892–16896

Kurihara Y, Kurihara H, Suzuki H, Kodama T, Maemura K, Nagai R, Oda H, Kuwaki T, Cao WH, Kamada N, Jishage K, Ouchi Y, Azuma S, Toyoda Y, Ishikawa T, Kumada M, and Yazaki Y (1994) Elevated blood pressure and craniofacial abnormalities in mice deficient in endothelin-1. Nature 368:703–710

Lahav R, Ziller C, Dupin E, Le Douarin NM (1996) Endothelin 3 promotes neural crest cell proliferation and mediates a vast increase in melanocyte number in culture. Proc Natl Acad Sci 93:3892–3897

Lim L, Manser E, Leung T, Hall C (1996) Regulation of phosphorylation pathways by p21 GTPases. The p21 Ras-related Rho subfamily and its role in phosphorylation signalling pathways. Eur J Biochem 242:171–185

Malendowicz LK, Brelinska R, Decaro R, Trejer M, Nussdorfer GG (1998) Endothelin-1, acting via the A-receptor subtype, stimulates thymocyte proliferation in the rat. Life Sciences 62:1959–1963

Mallat A, Fouassier L, Preaux AM, Gal CS, Raufaste D, Rosenbaum J, Dhumeaux D, Jouneaux C, Mavier P, Lotersztajn S (1995) Growth inhibitory properties of endothelin-1 in human hepatic myofibroblastic Ito cells. An endothelin B receptor-mediated pathway. J Clin Invest 96:42–49

Mallat A, Preaux AM, Serradeil-Le Gal C, Raufaste D, Gallois C, Brenner DA, Bradham C, Maclouf J, Iourgenko V, Fouassier L, Dhumeaux D, Mavier P, Lotersztajn S (1996) Growth inhibitory properties of endothelin-1 in activated human hepatic stellate cells: a cyclic adenosine monophosphate-mediated pathway. Inhibition of both extracellular signal-regulated kinase and c-Jun kinase and upregulation of endothelin B receptors. J Clin Invest 98:2771–2778

Mallat A, Gallois C, Tao J, Habib A, Maclouf, J, Mavier P, Preaux AM, Lotersztajn S (1998) human hepatic stellate cell proliferation: antiproliferative effect of ET mediated by ET_B. J. Biol. Chem. 273:27300–27305

Markewitz BA, Michael JR, Kohan DE (1997) Endothelin-1 inhibits the expression of inducible nitric oxide synthase. Am J Physiol 272:L1078–L1083

Matsuura A, Yamochi W, Hirata K, Kawashima S, Yokoyama M (1998) Stimulatory interaction between vascular endothelial growth factor and endothelin-1 on each gene expression. Hypertension 32:89–95

Mazzocchi G, Rossi GP, Rebuffat P, Malendowicz LK, Markowska A, Nussdorfer GG (1997) Endothelins stimulate deoxyribonucleic acid synthesis and cell proliferation in rat adrenal zona glomerulosa, acting through an endothelin A receptor coupled with protein kinase C- and tyrosine kinase-dependent signaling pathways. Endocrinology 138:2333–2337

McKenna CJ, Burke SE, Opgenorth TJ, Padley RJ, Camrud LJ, Camrud AR, Johnson J, Carlson PJ, Lerman A, Holmes DR Jr, Schwartz RS (1998) Selective ET(A) receptor antagonism reduces neointimal hyperplasia in a porcine coronary stent model: ABT-627 neointimal hyperplasia in a porcine coronary artery stented injury model. Circulation 97:2551–2556

Mitsutomi N, Akashi C, Odagiri J, Matsumura Y (1999) Effects of endogenous and exogenous nitric oxide on endothelin-1 production in cultured vascular endothelial cells. Eur J Pharmacol 364:65–73

Morbidelli L, Orlando C, Maggi CA, Ledda F, Ziche M (1995) Proliferation and migration of endothelial cells is promoted by endothelins via activation of ET_B receptors. Am J Physiol 269:H686–H695

Muldoon LL, Pribnow D, Rodland KD, Magun BE (1990) Endothelin-1 stimulates DNA synthesis and anchorage-independent growth of Rat-1 fibroblasts through a protein kinase C-dependent mechanism. Cell Regul 1:379–390

Murayama T, Oda H, Sasaki Y, Okada T, Nomura Y (1998) Regulation of inducible NO synthase expression by endothelin in primary cultured glial cells. Life Sci 62:1491–1495

Murlas CG, Gulati A, Singh G, Najmabadi F (1995) Endothelin-1 stimulates proliferation of normal airway epithelial cells. Biochem Biophys Res Commun 212:953–959

Nelson JB, Hedican SP, George DJ, Reddi AH, Piantadosi S, Eisenberger MA, Simons JW (1995) Identification of endothelin-1 in the pathophysiology of metastatic adenocarcinoma of the prostate. Nat Med 1:944–949

Nelson JB, Chan-Tack K, Hedican SP, Magnuson SR, Opgenorth TJ, Bova GS, Simons JW (1996) Endothelin-1 production and decreased endothelin B receptor expression in advanced prostate cancer. Cancer Res 56:663–668

Nelson JB, Lee W-H, Nguyen SH, Jarrard DF, Brooks JD, Magnuson SR, Opgenorth TJ, Nelson WG, Bova GS (1997) Methylation of the 5' CpG island of the endothelin B receptor gene is common in human prostate cancer. Cancer Res 57:35–37

Noiri E, Hu Y, Bahou WF, Keese CR, Giaever I, Goligorsky MS (1997) Permissive role of nitric oxide in endothelin-induced migration of endothelial cells. J Biol Chem 272:1747–1752

Ohlstein EH, Elliott JD, Feuerstein G, Ruffolo RR Jr (1996) Endothelin receptors: receptor classification novel receptor antagonists, and potential therapeutic targets. Medicinal Research Reviews 16:365–390

Okazawa M, Shiraki T, Ninomiya H, Kobayashi S, Masaki T (1998) Endothelin-induced Apoptosis of A375 Human Melanoma Cells. J Biol Chem 273:12584–12592

Opgenorth TJ (1995) Endothelin receptor antagonism. Adv Pharmacol 33:1–65

Opgenorth TJ, Adler AL, Calzadilla S, Chiou WJ, Dayton BD, Dixon DB, Gehrke LJ, Hernandez L, Magnuson SR, Marsh K, Novosad EI, von Geldern TW, Wessale JL, Winn M, Wu-Wong JR (1996) Pharmacological Characterization of A-127722: An orally active and highly potent ET_A-selective receptor antagonist. J Phar Exp Ther 276:473–481

Pagotto U, Azeberger T, Hopfner U, Sauer J, Renner U, Newton CJ, Lange M, Uhl E, Weindl A, Stalla GK (1995) Expression and localization of endothelin-1 and endothelin receptors in human meningiomas. Evidence for a role in tumoral growth. J Clin Invest 96:2017–2025

Pedram A, Razandi M, Hu RM, Levin ER (1997) Vasoactive peptides modulate vascular endothelial cell growth factor production and endothelial cell proliferation and invasion. J Biol Chem 272:17097–17103

Pepper MS, Ferrara N, Orci L, Montesano R (1992) Potent synergism between vascular endothelial growth factor and basic fibroblast growth factor in the induction of angiogenesis in vitro. Biochem Biophys Res Commun 189:824–831

Porter KE, Olojugba DH, Masood I, Pemberton M, Bell PR, London NJ (1998) Endothelin-B receptors mediate intimal hyperplasia in an organ culture of human saphenous vein. J Vasc Surg 28:695–701

Ricagna F, Miller VM, Tazelaar HD, McGregor CG (1996) Endothelin-1 and cell proliferation in lung organ cultures. Implications for lung allografts. Transplantation 62:1492–1498

Rosendorff C (1997) Cardiovasc Drugs Ther 10:795–802

Salamonsen LA, Young RJ, Garcia S, Findlay JK (1997) Mitogenic actions of endothelin and other growth factors in ovine endometrium. J Endocrinol 152:283–290

Sakurai T, Yanagisawa M, Takuwa Y, Miyazaki H, Kimura S, Goto K, Masaki T (1990) Cloning of a cDNA encoding a non-isopeptide selective subtype of the endothelin receptor. Nature 348:732–735

Sasaki Y, Hori S, Oda K, Okada T, Takimoto M (1998) Both ET(A) and ET(B) receptors are involved in mitogen-activated protein kinase activation and DNA synthesis of astrocytes: study using ET(B) receptor-deficient rats. J Neurosci 10:2984–2993

Schieffer B, Drexler H, Ling BN, Marrero MB (1997) G protein-coupled receptors control vascular smooth muscle cell proliferation via pp60c-src and p21ras. Am J Physiol 272:C2019–C2030

Shichiri M, Kato H, Marumo F, Hirata Y (1997) Endothelin-1 as an autocrine/paracrine apoptosis survival factor for endothelial cells. Hypertension 30:1198–1203

Shichiri M, Sedivy JM, Marumo F, Hirata Y (1998) Endothelin-1 is a potent survival factor for c-Myc-dependent apoptosis. Mol Endocrinol 12:172–180

Sokolovsky M (1992) Structure-function relationships of endothelins, sarafotoxins, and their receptor subtypes. J Neurochem 59:809–821

Spellmeyer DC (1994) Small molecule endothelin receptor antagonists, Annual Reports in Medicinal Chemistry 29:65–71

Stiles JD, Ostrow PT, Balos LL, Greenberg SJ, Plunkett R, Grand W, Heffner RR Jr (1997) Correlation of endothelin-1 and transforming growth factor beta 1 with malignancy and vascularity in human gliomas. J Neuropathol Exp Neurol 56:435–439

Sudjarwo SA, Hori M, Tanaka T, Matsuda Y, Okada T, Karaki H (1994) Subtypes of endothelin ETA and ETB receptors mediating venous smooth muscle contraction. Biochem Biophys Res Commun 200:627–633

Sugawara F, Ninomiya H, Okamoto Y, Miwa S, Mazda O, Katsura Y, Masaki T (1996) Endothelin-1-induced mitogenic responses of Chinese hamster ovary cells expressing human endothelinA: the role of a wortmannin-sensitive signaling pathway. Mol Pharmacol 49:447–457

Suri C, Jones PF, Patan S, Bartunkova S, Maisonpierre PC, Davis S, Sato TN, Yancopoulos GD (1996) Requisite role of angiopoietin-1, a ligand for the TIE2 receptor, during embryonic angiogenesis. Cell 87:1171–1180

Suri C, McClain J, Thurston G, McDonald DM, Zhou H, Oldmixon EH, Sato TN, Yancopoulos GD (1998) Increased vascularization in mice overexpressing angiopoietin-1. Science 282:468–471

Takagi Y, Fukase M, Takata S, Yoshimi H, Tokunaga O, Fujita T (1990) Autocrine effect of endothelin on DNA synthesis in human vascular endothelial cells. Biochem Biophys Res Commun 168:537–543

Takimoto M, Oda K, Früh T, Takai M, Okada T, Sasaki Y (1996) ET_A and ET_B receptors cooperate in DAN synthesis via opposing regulations of cAMP in human lung cell line. Am J Physiol 271:L366–L373

Terada Y, Inoshita S, Nakashima O, Yamada T, Tamamori M, Ito H, Sasaki S, Marumo F (1998) Cyclin D1, P16, and retinoblastoma gene regulate mitogenic signaling of endothelin in rat mesangial cells. Kidney International 53:76–83

Theuring F, Thone-Reinecke C, Vogler H, Schmager F, Rohmeiss P, Slowinski T, Neumayer HH, Hocher B (1998) Pathophysiology in endothelin-1 transgenic mice. J Cardiovasc Pharmacol 31:S489–S491

Van Kesteren CAM, Van Heugten HAA, Lamers JMJ, Saxena PR, Schalekamp MADH, and Danser AHJ (1997) Angiotensin II-mediated growth and antigrowth effects in cultured neonatal rat cardiac myocytes and fibroblasts. J Mol Cell Cardiol 29:2147–2157

Vigne P, Marsault R, Breittmayer JP, Frelin C (1990) Endothelin stimulates phosphatidylinositol hydrolysis and DNA synthesis in brain capillary endothelial cells. Biochem J 266:415–420

Wada K, Tabuchi H, Ohba R, Satoh M, Tachibana Y, Akiyama N, Hiraoka O, Asakura A, Miyamoto C, Furuichi Y (1990) Purification of an endothelin receptor from human placenta. Biochem Biophys Res Commun 167:251–257

Warner TD (1994) Endothelin receptor antagonists. Cardiovascular Drug Reviews. 12: 105–122

Whelchel A, Evans J, Posada J (1997) Inhibition of ERK activation attenuates endothelin-stimulated airway smooth muscle cell proliferation. Am J Respir Cell Mol Biol 16:589–596

Wink DA, Vodovotz Y, Laval J, Laval F, Dewhirst MW, Mitchell JB (1998) The multi-faceted roles of nitric oxide in cancer. Carcinogenesis 19:711–721

Wren AD, Hiley CR, Fan TP (1993) Endothelin-3 mediated proliferation in wounded human umbilical vein endothelial cells. Biochem Biophys Res Commun 196: 369–375

Wu-Wong JR, Chiou W, Huang ZJ, Vidal MJ Opgenorth TJ (1994) Endothelin receptor in human pericardium smooth muscle cells: antagonist potency differs on agonist-evoked responses. Am J Physiol 267:C1185-C1195

Wu-Wong JR, Chiu W, Magnuson JR, Bianchi BR, Lin CW (1996a) Human astrocytoma U138MG cells express predominantly type-A endothelin receptors. Biochem Biophys Acta 1311:155–163

Wu-Wong JR, Dayton BD, Opgenorth TJ (1996b) Endothelin-1-evoked arachidonic acid release: a Ca^{+2}-dependent pathway. Am J Physiol 271:C869–C877

Wu-Wong JR, Chiou WJ, Dickinson R, Opgenorth TJ (1997) Endothelin attenuates apoptosis in human smooth muscle cells. Biochem J 328:733–737

Wu-Wong JR, Dixon DB, Chiou WJ, Dayton BD, Novosad EI, Adler AL, Wessale JL, Calzadilla SV, Hernandez L, Marsh KC, Liu G, Szczepankiewicz B, von Geldern TW, Opgenorth TJ (1999) Pharmacology of A-216546: a highly selective antagonist for endothelin ET(A) receptor. Eur J Pharmacol 366:189–201

Yanagisawa M, Kurihara H, Kimura S, Tomobe Y, Kobayashi Y, Mitsui M, Yazaki Y, Goto K, Masaki T (1988) A novel potent vasoconstrictor peptide produced by vascular endothelial cells. Nature 332:411–415

Yoshizumi M, Kim S, Kagami S, Hamaguchi A, Tsuchiya K, Houchi H, Iwao H, Kido H, Tamaki T (1998) Effect of endothelin-1 (1–31) on extracellular signal-regulated kinase and proliferation of human coronary artery smooth muscle cells. Br J Pharmacol 125:1019–1027

Zhao YD, Springall DR, Hamid Q, Levene M, Polak JM (1995) Localization and characterization of endothelin-1 receptor binding in the blood vessels of human pulmonary tumors. J Cardiovasc Pharmacol 26:S341–S345

CHAPTER 13

The Involvement of Endothelins in Cerebral Vasospasm and Stroke

O. Touzani and J. McCulloch

A. Introduction

Endothelin-1 (ET-1) is a 21 amino acid peptide which has a high degree of structural homology with the other members of the ET family, ET-2 and ET-3 (Inoue et al. 1989; Yanagisawa et al. 1988). ET-1 is the most potent mammalian vasoconstrictor identified to date, being approximately one order of magnitude greater in potency than either angiotensin II, vasopressin or norepinephrine. The discovery of the existence of vasoconstrictor peptides of the endothelin family in the mid- to late 1980s, let to a great deal of excitement over their potential role in cardiovascular homeostasis and their possible therapeutic potential. The isolation, purification and sequencing of the endothelins has since allowed the development of both peptide and non peptide antagonists selective for endothelin receptors (see Chap. 9).

ET was originally thought to be primarily involved in the modulation of cardiovascular tone. However, early studies have revealed that ET-like immunoreactivity, endothelin-converting enzymes (ECE), ET-mRNA and endothelin receptors are present in the central nervous system (CNS) both in vascular and non vascular tissue. Endothelin has been proposed to act as neurotransmitter or neuromodulator in the CNS.

The extremely potent and long-lasting vasoconstrictor properties of ET-1 have implicated the peptide in several pathological vascular conditions. In the CNS, a large number of recent studies strongly point to an involvement of endothelins in different pathologies such as migraine, atherosclerosis, brain trauma, subarachnoid haemorrhage (SAH) and stroke (Ohlstein et al. 1996; Patel 1996; Rogers et al. 1997; Zimmermann and Seifert 1998). The present review focuses on the recent knowledge concerning the role of ETs in normal cerebral circulation and their potential pathophysiological role in cerebral vasospasm following SAH and stroke.

B. Endothelins and Endothelin Receptors in the Brain and Cerebral Circulation

I. Endothelins in the Brain and Cerebral Circulation

Although endothelins were identified initially in endothelial cells from the aorta, early studies showed the presence of endothelins in many systems including the CNS (RUBANYI and POLOKOFF 1994). In the CNS, endothelin has not been as extensively and vigorously studied as in other systems such as the cardiovascular, renal and gastrointestinal system. In brain and spinal cord, ET mRNA and ET-like immunoreactivity are associated with endothelial cells, neurones and astrocytes (EHRENREICH et al. 1991; GIAID et al. 1989; LEE et al. 1990; MACCUMBER et al. 1990). Although endothelial cells, neurones, and glia all appear capable of producing endothelins, different types of cells may produce different isopeptides. ET-1 and ET-3 have been identified in human cerebrospinal fluid (CSF), with the concentration of ET-3 being 1.5-fold greater than ET-1 (YAMAJI et al. 1990). The distribution of ET is widespread throughout the brain, with the highest densities in the hypothalamus and posterior pituitary. ET-1 immunoreactivity has been detected in the cerebral cortex, hippocampus, midbrain, medulla, cerebellum, spinal cord and dorsal root ganglia. ET-3 immunoreactivity has been identified in the pituitary gland at much greater density than ET-1 (GIAID et al. 1991; LEE et al. 1990; MACCUMBER et al. 1989; MATSUMOTO et al. 1989; YOSHIZAWA et al. 1990). In contrast to ET-1 and ET-3, ET-2 has not been reported to be present in the brain.

II. Endothelin Receptors in the Brain and Cerebral Circulation

The combination of functional and molecular studies of endothelin receptors has demonstrated the existence of at least two distinct subtypes, namely ET_A and ET_B. The ET_A receptor is characterised by a higher affinity towards ET-1 than ET-2 and ET-3, while the ET_B receptor displays a similar affinity towards the three endothelins (HAYNES et al. 1993; SAKURAI et al. 1990; see Chap. 4). Specific ET-binding sites were observed in neurones, astrocytes, microglial cells and in cerebrovascular smooth muscle and endothelial cells (GIAID et al. 1989; MACCUMBER et al. 1990). The mammalian brain predominantly contains the ET_B receptor subtype (FERNANDEZ-DURANGO et al. 1994; HARLAND et al. 1998), which exists mainly on endothelial, neuronal, and glial cells (HARLAND et al. 1998; LYSKO et al. 1995). The ET_A receptor is mainly expressed by vascular smooth muscle cells, but it is also present on endothelial, neuronal and glial cells (GREENBERG et al. 1992; HARLAND et al. 1998; VIGNE et al. 1990).

Endothelin binding sites have been demonstrated in many brain areas (KOHZUKI et al. 1991; NAMBI et al. 1990). The widespread non-vascular distribution of endothelin binding sites suggests a role for ET receptors in nonvascular functions, such as neurotransmission and/or neuromodulation. ET binding sites have been identified in the median eminence and anterior and posterior pituitary. Binding studies with radiolabelled ET-1 have indicated the presence of binding sites in the choroidal plexus, cerebellum, subfornical organ, cerebral cortex, dentate gyrus in the hippocampus and in nuclei in the brain stem (HARLAND et al. 1998; JONES et al. 1989; NAMBI et al. 1990; NIWA et al. 1991). ET-3 specific binding sites are found in their highest density in the cerebellum compared to other brain regions (NAMBI et al. 1990).

III. Function of Endothelins in the Brain

As discussed above, ET-1, ET-3, and endothelin receptors are found throughout the brain. Both peptides and their receptors are produced by neurones and astrocytes at a relatively early stage of the foetal life in rats and therefore may have a role in the development of the brain (LEVIN 1995). The role of endothelins in the CNS is unclear, but there is increasing evidence that points towards an important role for endothelins in modulating the function of cerebral tissue. It has been suggested that ET-1 acts as a growth factor for astrocytes, regulating biological processes such as proliferation during brain development or injury (CAZAUBON et al. 1997; ISHIKAWA et al. 1997; LADEN-HEIM et al. 1993; MacCUMBER et al. 1990). ET-1 specifically stimulates the efflux of glutamate from cultures of rat astrocytes (SASAKI et al. 1997), increases glucose uptake in rat astrocytes in primary culture (TABERNERO et al. 1996), and stimulates phosphoinositol turnover in cerebellar neurones and glial cells (DAVENPORT and MORTON 1991; LIN et al. 1989). Moreover, ET-1 has been reported to increase the level of mRNA for the early gene c-fos, a marker for the activation of neuronal pathways, in neurones and glia (SULLIVAN and MORTON 1996).

In vivo and in vitro experiments have shown that endothelin stimulates the release of vasopressin, substance P, catecholamines, atrial natriuretic factor and oxytocin (STOJILKOVIC 1992). Endothelin can also induce the release of neurotransmitters from neurones. Intrastriatal injection of ET-1 or application of ET-1 to striatal slices evokes the release of dopamine via the activation of ET_B receptors located on dopaminergic neurones (WEBBER et al. 1998). Other investigations have indicated that endothelins may play a role as neuromodulators in the locomotor system or in the release of pituitary hormones (GROSS et al. 1993). However, it should be emphasised that the use of exogenous ET-1 to examine the function of endogenous ET-1 in non-vascular cerebral tissue in vivo may be misleading. Indeed, the administration of the peptide into the brain induces ischaemia that can trigger a variety of events (e.g. neurotransmitters release) which can complicate the interpretation of the in vivo data. The use of endothelin receptor antagonists allows clarification of the exact role of endothelins on neuronal function in vivo, uncomplicated by concerns that the agonist effects are the indirect consequences of induced ischaemia.

IV. Endothelins and the Cerebral Circulation

1. ET$_A$ Receptor-Mediated Effects

Cerebral arteries are among the vessels most sensitive to ET-induced constriction (Salom et al. 1995). In isolated segments of various cerebral arteries from variety of species, exogenous ET-1 elicits marked and long-lasting, concentration dependent constriction. ET-1 applied in vitro or in situ, contracts large, intermediate and small arteries and precapillary arterioles (Fig. 1) (Jansen et al. 1989; Robinson and McCulloch 1990; Saito et al. 1989; Salom et al. 1995). ET-2 shows a vasoconstrictor potency similar to that of ET-1, whereas the potency of ET-3 is much lower in different cerebral arteries and arterioles (Adner et al. 1993; Edwards et al. 1990; Feger et al. 1994). ET-1 induced vasoconstriction is primarily mediated by ET$_A$ receptors, located on the vascular smooth muscle, since it is effectively antagonised by several selective ET$_A$ receptor antagonists such as BQ123, BQ610 and PD155080 (Adner et al. 1993; Patel et al. 1996c).

The cerebral circulation is distinguished by the presence of the blood-brain barrier (BBB) that can prevent the access of peptide molecules to the adventitial surface (Egleton and Davis 1997; Pardridge 1997). In the cerebral circulation, there are conflicting reports on whether ET-1 acts from the abluminal or the luminal side. Yoshimoto et al. (1990) have reported that endothelin mRNA was expressed in cultured endothelial cells of canine cere-

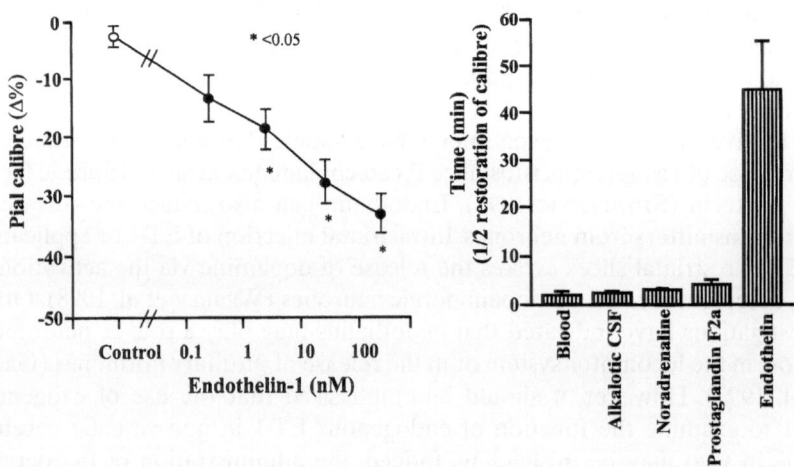

Fig. 1. *Left.* Vasomotor responses of pial arterioles to adventitial application of endothelin-1 (5 μl around individual pial arterioles in the cat. Pial arteriolar diameter was measured using the cranial window technique. Data are presented as percent alteration from baseline of arteriolar calibre (mean ± SEM). *$p < 0.05$. *Right.* The duration of action of a single microapplication of endothelin-1 (10 nmol/l). Data are expressed as time for half restoration of preinjection calibre (mean ± SEM). After Robinson and McCulloch (1990)

bral microvessels and that ET-1 produced by cultured endothelial cells of cerebral microvessels was released mainly to the basal side (corresponding to the basement membrane side). These findings indicate that ET-1 constricts arterioles locally at the same location at which it is produced by endothelial cells. Several investigations have shown that ET-1 mediates its cerebral vasoconstrictor effects from the adventitial side. In an autoradiographic study with ^{125}I-labeled ET, it has been shown that ET-1 administered intravenously to rats did not access the CNS except in those areas lacking blood-brain barrier (KOSEKI et al. 1989). Intracisternal but not intra-arterial ET-1 produces angiographically demonstrable vasoconstriction of feline and canine basilar arteries, and intravenous ET-1 has no effect on CBF or cerebral microvascular pressure in rabbits (KADEL et al. 1990; MIMA et al. 1989). These findings suggest that ET-1 is not able to cross the BBB and gain access to the cerebrovascular smooth muscle of the major resistance vessels of the cerebral circulation. In contrast, in isolated rabbit basilar artery, KAZUKI et al. (1997) have reported recently that both intra- and extraluminal application of ET-1 effected similar potent and dose-dependent vasoconstrictions. This result is consistent with the finding of GARCIA et al. (1991) who observed an increase of cerebrovascular resistance after intra-arterial injection of ET-1 in goats although this species has a history of exaggerated cerebrovascular responses to intravascular administration of vasoactive agents (EDVINSSON et al. 1993). KOBARI et al, (1994a) observed reduction of cerebral blood volume following the intracarotid injection of ET-1 in cats although some cerebral capacitance vessels are known to be unprotected by the blood-brain barrier (EDVINSSON et al. 1993). OGURA et al. (1991) also found that intraluminal endothelin induced contractions of cerebral arterioles in rats, but that the potency was not as great as that of extraluminal endothelin. The discrepancies between the above discussed studies and others may be related to the doses of exogenous ET-1 employed, species, vascular segment, and techniques used to measure cerebrovascular reactivity to ET-1. Nonetheless, the evidence overall indicates that the most potent route to evoke cerebrovascular effects of ET-1 is intracerebral (or direct adventitial) administration and not the intravascular route.

2. ET$_B$ Receptor-Mediated Effects

In many vascular beds (including cerebrovascular), endothelin induced vasoconstriction is normally attributed to activation of ET$_A$ receptors (based on the relative potency of endothelins and the ability of selective ET$_A$ receptor antagonists to attenuate the responses). The function of ET$_B$ receptors is more controversial. It is widely accepted that stimulation of ET$_B$ receptors induces vasodilatation mediated by generation of the endothelium-derived relaxing factor(s), nitric oxide and/or prostacyclin (DE NUCCI et al. 1988; KARAKI et al. 1993; SCHILLING et al. 1995). In the cat, intracarotid infusion of the ET$_B$ receptor agonist IRL 1620 elicits a vasodilatation of capacitance vessels (with increases in cerebral blood volume), suggesting the presence of ET$_B$ receptors

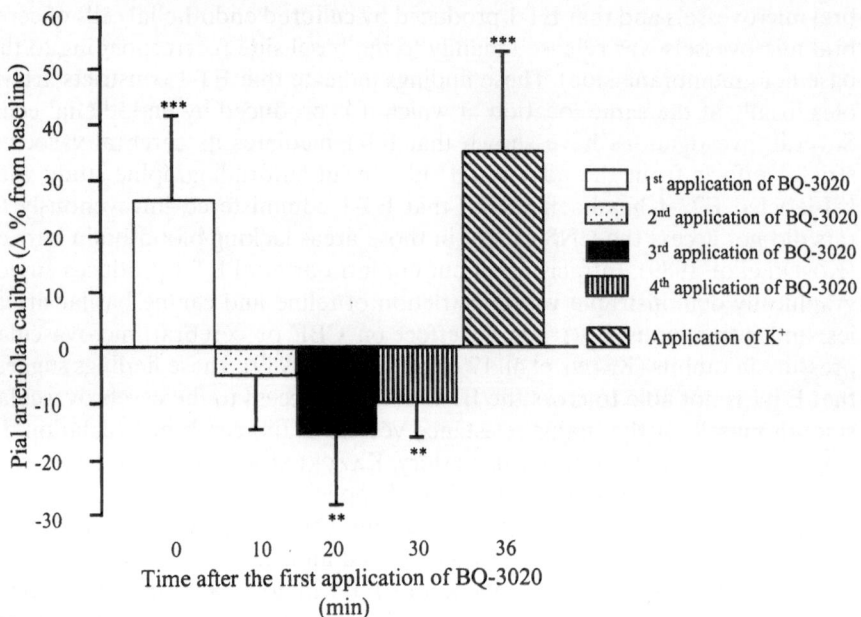

Fig. 2. Vasomotor responses of pial arterioles to repeated perivascular microapplication of BQ-3020 (ET_B receptor agonist, $1 \mu mol/l$) followed by adventitial application of potassium ($10 mmol/l$). Data are the percentage change from pre-injection baseline in pial arteriolar calibre. The first microapplication of BQ-3020 onto pial cortical arteriole elicited marked dilatation. A second application of BQ-3020 on the same arteriolar site failed to evoke any significant vasomotor response. Subsequent (third and fourth) adventitial microapplication of the ET_B receptor agonist effected a significant constriction of cerebral arterioles. Potassium-induced marked dilatation, Subsequent to repeated injection of BQ-3020, indicated that the pial arterioles were still responsive to vasodilatator agents. ** $p < 0.001$, *** $p < 0.0001$ for the comparison with CSF. Data are presented as mean $\pm$ SD (n, number of arterioles examined = 16). After TOUZANI et al. (1997)

in the vascular endothelium (KOBARI et al. 1994b). Additionally, topical application of ET_B receptor agonists (IRL 1620 and BQ 3020) onto rat basilar artery and cat cerebral resistance arterioles in vivo has been demonstrated to cause dose dependent dilatation which is attenuated, in a dose dependent manner, by the application of a selective ET_B receptor antagonist (KITAZONO et al. 1995a; PATEL et al. 1996c; TOUZANI et al. 1997). In the rat basilar artery and cat pial arterioles, the ET_B dilator response displays tachyphylaxis (Fig. 2) indicating desensitisation of the receptors (KITAZONO et al. 1995a; TOUZANI et al. 1997). This phenomenon is probably due to phosphorylation of the ET_B receptor leading to its inactivation (CRAMER et al. 1997).

    In some peripheral vascular beds, an ET_B receptor has also been linked to the contractile properties of endothelins. Indeed, it was observed that ET-1-mediated vasoconstriction was resistant to the actions of ET_A receptor antagonists, and therefore could not be solely attributed to ET_A receptor activation (GELLAI et al. 1996; McMURDO et al. 1993; SEO et al. 1994; SUMNER et al. 1992). ET_B-mediated cerebrovasoconstriction was first demonstrated by repeated selective activation of ET_B receptors on cerebral resistance arterioles in the cat (Fig. 2) (TOUZANI et al. 1997). In accordance with our findings, ZUCCARELLO et al. (1998a, b) have shown, through the use of different ET_B receptor agonists and antagonists, that ET_B receptors can mediate both vasoconstriction and vasodilatation in isolated basilar arteries. The dilator or constrictor properties of ET_B receptors may vary according to the physiological or pathological state of the vessel (DAGASSAN et al. 1996; KITAZONO et al. 1995b). For example, it has been shown that some arterial segments in culture exhibit a substantial ET_B receptor mediated contractile response, whereas such contractile activity has not been observed in fresh isolated arteries (ADNER et al. 1996; WHITE et al. 1998).

The qualitatively different effects of the endothelins at the ET_B receptor have led to subdivision of ET_B receptors into ET_{B1} receptors (on endothelial cells) that mediate vasodilatation and ET_{B2} receptors (on smooth muscle cells) that mediate vasoconstriction (MASAKI et al. 1994; WARNER et al. 1993). However, no molecular correlate of ET_B receptor subtypes has been reported (MIZUGUCHI et al. 1997). The relative function of the different endothelin receptor subtypes in cerebral vessels is unclear. Nonetheless, the demonstration of constrictor ET_A and the dilator and/or constrictor ET_B receptors has important implications for the development of endothelin receptor antagonists to be used in cerebrovascular disease.

3. Role of Endothelins in the Regulation of Cerebrovascular Tone

The role of endothelins in the regulation of cerebrovascular tone and CBF in physiological conditions remains controversial. Investigations in endothelial and vascular smooth muscle cell culture have indicated that there may be a basal release of ET-1 that could be involved in the regulation of vascular tone (YOSHIMOTO et al. 1990). The available data suggest that the cerebrovascular tone may be, at least in part, affected by physiological antagonism between the vasoconstriction induced by ET-1 and the vasodilatation caused by NO (HIROSE et al. 1995; THORIN et al. 1998). Mechanical removal of the endothelium enhances the potency and efficacy of exogenous ET-1-induced vasoconstriction in cerebral arteries (JANSEN et al. 1989; KAUSER et al. 1990; SAITO et al. 1989). These observations may imply the existence of a compensatory mechanism between ET-mediated constriction and relaxation mediated by endothelial-derived relaxing factors.

We have demonstrated that perivascular application of ET_A selective (BQ123, PD156707), ET_B selective (BQ788) or mixed ET_A/ ET_B (bosentan) receptor antagonists did not induce any change in the calibre of feline cerebral resistance arterioles (PATEL et al. 1994, 1996; TOUZANI et al. 1997). These findings suggest that, at least at the level of the adventitial surface, there is minimal tone due to activation of ET_A and ET_B receptors by endogenous endothelins. These data are in agreement with other reports in the literature that have demonstrated, by blocking the ET receptors, that endothelins are not involved in the regulation of cerebral arterial calibre or CBF during resting physiological conditions (ARMSTEAD 1996; KOEDEL et al. 1998; NIREI et al. 1993; ROUX et al. 1995). Conversely, intracisternal injection of BQ123 in dogs produced a 30% increase in the basal diameter of the basilar artery 24h after injection (HIROSE et al. 1995). The findings of this study need to be confirmed for, if true, they suggest that endothelins are major determinants of the tone of cerebral arteries under physiological conditions – a view at variance with the major body of data.

C. Pathophysiological Role of Endothelins in the Cerebral Circulation

I. Subarachnoid Haemorrhage

Cerebral vasospasm remains one of the major causes of morbidity and mortality in patients with aneurysmal SAH (PASQUALIN 1998). Angiographic vasospasm, defined as a focal or diffuse narrowing of the major cerebral arteries, is usually observed 4 to 12 days after the SAH; it almost never occurs earlier than 48h and only rarely after 2 weeks. The vasospasm may lead to the development of delayed cerebral ischaemia. This is referred to as symptomatic vasospasm. Approximately 60% of patients with SAH exhibit delayed symptomatic vasospasm. Arteriographic vasospasm refers only to the radiographically visualised constriction of cerebral arteries and may have no clinically demonstrable correlate. The term 'cerebral vasospasm' has been used loosely to denote either or both of the above concepts.

Cerebral vasospasm associated with SAH is a complex disorder involving endothelial dysfunction and damage, vasoconstriction, vascular proliferation, thrombosis and vasonecrosis (MAYBERG 1998; PEERLESS et al. 1980). Despite much active research, the pathogenesis of cerebral vasospasm after SAH remains poorly understood; as a result, effective clinical management of this entity continues to be suboptimal (FEIGIN et al. 1998; PASQUALIN 1998).

A number of experimental models have been developed to investigate the pathophysiology of cerebral vasospasm. The most widely used are dog and rabbit single or double haemorrhage, models in which arterial blood is injected into the cisterna magna. Following SAH, a very large number of putative spasmogens released from the intracisternal clot have been proposed. They may be directly spasmogenic or induce the release of vasoconstrictor and/or impair endothelium-dependent relaxation (JERIUS et al. 1998; MACDONALD and WEIR 1991; TODA et al. 1980). A considerable amount of experimental and clinical work has been published which suggest a role for ET-1 in the pathogenesis of cerebral vasospasm following SAH. Three lines of evidence point to a major role of ET-1 in the pathophysiology of vasospasm after SAH: (1) cerebrovascular reactivity to intracerebral ET-1, (2) levels of ET-1 after SAH and (3) the efficacy of ECE inhibitors or endothelin receptor antagonists in preventing development of vasospasm.

1. Cerebrovascular Reactivity to Endothelins

As discussed earlier, a characteristic of ET-1 is its ability to induce marked and long-lasting vasoconstriction. ET-1 injected intracisternally into dogs and cats initiated vasoconstriction of the basilar artery which was dose dependent and lasted over 12 h (ASANO et al. 1989; KOBAYASHI et al. 1990; MIMA et al. 1989). When administered chronically, ET-1 also caused a spectrum of histological changes that resemble the degenerative changes seen in vasospastic arteries (KOBAYASHI et al. 1991). The intracisternal administration of exogenous ET-1 in rats results in brain stem ischaemia (MACRAE et al. 1991). The topical administration of ET-1 onto a major cerebral artery, for example the middle cerebral artery (MCA), induces marked reductions in CBF in the cortex and striatum, and ischaemic damage in these areas (MACRAE et al. 1993; FUXE et al. 1992). Considering these uniquely marked and long-lasting actions of ET-1 on the cerebral vasculature, the peptide has been proposed to be an important mediator of cerebrovascular and neuronal sequelae associated with SAH.

Under physiological conditions, ET-1 may also play a role in sensitising blood vessels to the actions of other circulating vasoconstrictor substances. ET-1 can augment the contractile actions of norepinephrine, serotonin and angiotensin II (YANG et al. 1990). After SAH, it has been proposed that cerebral arteries might be sensitised by ET-1 to contract to other agonists (ALAFACI et al. 1990; KAMATA et al. 1991). However, there is no direct experimental data to demonstrate the existence of such an effect in the development of cerebral vasospasm following SAH.

The existing evidence indicates a physiological antagonism between ET-1 and NO in the normal vasculature. It has been observed in variety of vascular beds that NO is released from endothelium after exposure to ET-1 and limits the contractile response to the peptide (ALABADI et al. 1997; LERMAN et al. 1992; LUSCHER et al. 1990; THOMAS et al. 1997; see Chap. 11). Moreover, it is well known that SAH is associated with endothelial damage and impairment of endothelium-dependent relaxation (HATAKE et al. 1992; NAKAGOMI et al. 1987; ONOUE et al. 1995; ZUCCARELLO et al. 1995). This impairment of endothelial function probably accounts for the increase in cerebrovascular sensitivity to ET-1 observed after SAH (ALABADI et al. 1993, 1997; ALAFACI et al. 1990; ZUCCARELLO et al. 1995).

Oxyhaemoglobin released from the subarachnoid blood clot is considered as a major cause of vasospasm after SAH (MACDONALD and WEIR 1994; MAYBERG 1998; PLUTA et al. 1998). Different mechanisms can account for its actions in vasospasm. Oxyhaemoglobin can stimulate the production and release of ET-1 from endothelial cells (Kasuya et al. 1993; Ohlstein and Storer 1992). It can also reduce NO production due to the concomitant disappearance of NO synthase from the adventitia of the vessel or a decrease in NO availability due to the "sink effect" of oxyhaemoglobin (Goretski et al. 1988; Pluta et al. 1996). These effects of oxyhaemoglobin that result in the alteration of the normal balance between ET-1 and NO provide further evidence of a role for endothelin in the development of post-haemorrhagic arterial narrowing.

2. Levels of Endothelins

Several investigations have demonstrated increased levels of ET-1 in the CSF and plasma of patients after SAH. Nonetheless, the data are controversial regarding the possible relationship between the elevation in ET-1 levels and the arterial vasospasm that occurs. Some authors have reported a direct correlation between a significant peak in plasma and CSF ET-1 levels and the occurrence of symptomatic vasospasm (Masaoka et al. 1989; Shirakami et al. 1994; Suzuki et al. 1990). Suzuki et al. (1992) have reported that plasma ET-1-like immunoreactivity levels were increased from the onset of SAH and that the rise of ET-1-like immunoreactivity correlated with the severity of the subarachnoid clots. The patients with vasospasm had higher plasma ET-1 immunoreactivity levels than those without vasospasm during the first week. However, CSF ET-1-like immunoreactivity levels in patients with vasospasm were not elevated during the early period of SAH, but rose significantly during the second week. The CSF ET-1-like immunoreactivity levels in patients without vasospasm remained within the normal range. Similarly, a correlation between the CSF levels of, ET-1, Big-ET-1 and ET-3 and the volume of haematoma and clinical occurrence of vasospasm has been demonstrated, emphasising the important role of these peptides in the development of vasospasm (Seifert et al. 1995). In contrast, Fujimori et al. (1990) found no correlation between CSF ET levels and vasospasm, although a higher plasma ET-1 level at 8–14 days after SAH was found and related a possible "stress response" in patients showing clinical sign of vasospasm. Similar observations have been made in other studies, providing no support for the theory that CSF ET-1 levels may represent a predictive factor for clinical outcome, particularly with regard to the occurrence of vasospasm (Gaetani et al. 1994; Hamann et al. 1993b).

Increases of ET-1 in the CSF and plasma have also been reported in some, but not all, experimental models of SAH (Hino et al. 1996; Pluta et al. 1997; Roux et al. 1995; Zimmermann et al. 1996). The significance of elevated plasma and CSF ET-1 levels in SAH is unclear since the available evidence indicates that the actions of ET-1 are exerted locally at the site of synthesis or release. ET is also elevated in response to other central nervous insults such as head injury, suggesting that elevations during vasospasm could reflect a non-specific response of the brain.

In addition to the wide increases of ET-1 levels in plasma and CSF, overexpression of immunoreactive ET-1 was observed in endothelium of the spastic basilar artery days after SAH in experimental model (HIROSE et al. 1995; ROUX et al. 1995; SHIGENO et al. 1991; YAMAURA et al. 1992). However, the time course of increases in cerebral vessel endothelin immunoreactivities varies between studies. YAMAURA et al. (1992) reported that the immunoreactive ET-1 level in the wall of the basilar artery was significantly increased on day 2, but not on day 7 in the canine SAH model. The authors suggest that ET-1 could act as a trigger in the early stages of cerebral vasospasm, because vasospasm was partially reversed by topical application of monoclonal antibody against ET-1 on day 2 but was rather resistant to the same antibody on day 7. In contrast, in the same model other studies have shown that immunoreactive ET-1 level in the basilar artery was still increased 7 days after the induction of SAH (HIROSE et al. 1995; ROUX et al. 1995).

The high level of ET-1 in the CSF and vessels is probably caused by an increase of ET-1 production rather than a decrease of ET-1 turnover. This is supported by the study of ROUX et al. (1995) in which they observed an increase of the ECE activity in the basilar artery after experimental SAH. Interestingly, in the same study changes in the expression of endothelin receptors were described in vasospastic basilar artery. Using binding experiments, they found that ET_A receptors are predominantly expressed in sham-operated dogs, whereas in dogs subjected to SAH, ET_B receptors were the predominant population. These data indicate a potential role of ET_B receptors in ET-1-mediated vasospasm.

The sources of endothelin released following SAH have not been definitively established. Some evidence suggests that ET-1 associated with SAH can originate not only from endothelium but also from neurones and glia (EHRENREICH et al. 1992; PLUTA et al. 1997; YAMADA et al. 1995). A number of processes may account for the elevation of the synthesis and release of endothelins after SAH. It has been shown that the production and release of ET-1 is increased in endothelial cultures treated with haemoglobin (KASUYA et al. 1993; OHLSTEIN and STORER 1992). Following SAH, haemoglobin is released from lysed blood cells in the subarachnoid clot and can gain access to both vascular and non-vascular cells. This could lead to elevations in the levels of ET-1 (MACDONALD and WEIR 1994). However, increased ET-1 production could not be demonstrated in vitro after exposure of astrocytes and endothelial cells to oxyhaemoglobin (PLUTA et al. 1997); rather, this was associated with an inhibition of ET-1 production. In the same study, hypoxia induced an increase in ET-1 release from astrocytes but not from endothelial cells in culture. Hypoxia is often associated with post-haemorrhagic vasospasm and has been reported to enhance ET-1 and endothelin receptor gene expression in other studies (KOUREMBANAS et al. 1991; LI et al. 1994). Other putative triggers for ET-1 release are thrombin and transforming growth factor β, which can be generated from blood clots after SAH. Both of these compounds have been reported to induce the synthesis and release of ET-1 (Ehrenreich et al. 1990, 1993; Kurihara et al. 1989).

3. Treatment with Endothelin Receptor Antagonists

Considerable effort has been applied to look for corroborating evidence linking endothelins to vasospasm and to test the hypothesis that elevated endothelin levels are causally related to the development of vasospasm. The strategy in such studies has been to intervene with a variety of agents, in an attempt to block the action of endogenous endothelins. The synthesis of endothelins is controlled at the level of transcription (Inoue et al. 1989; Yanagisawa et al. 1988). Inhibition of the synthesis of endothelin at the transcriptional level by actinomycin D resulted in complete prevention of vasospasm after SAH in the dog (Shigeno et al. 1991). Although actinomycin D does not specifically inhibit ET-1 synthesis, the findings of this study are consistent with a role for ET-1 in the pathogenesis of vasospasm. The development of endothelin receptor antagonist and ECE inhibitors has been important for elucidating the role of endothelins in vascular pathology (Barnes et al. 1997; Douglas 1997; Douglas et al. 1994; Webb et al. 1998). The finding that drugs that antagonise the actions of ET-1 prevent or reverse experimental vasospasm directly supports the evidence (cited above) that endothelins are involved in cerebral vasospasm.

The efficacy of endothelin receptor antagonists (see Chap. 9) and ECE inhibitors (see Chap. 7) in SAH have been reported by different groups in variety of species (Table 1). The intracisternal administration of BQ-123, a peptide antagonist selective for the ET_A receptor, resulted in remarkable restoration of CBF in rats subjected to SAH (Clozel and Watanabe 1993). In contrast, when the same antagonist was delivered intravenously, no effect on CBF reduction was observed. These data demonstrate that BQ-123 does not cross the BBB, and emphasise the importance of the subsequent development of non-peptide endothelin receptor antagonists that do not require intracisternal administration to gain access to smooth muscle cells.

In all published studies (Foley et al. 1994; Hino et al. 1995; Hirose et al. 1995; Itoh et al. 1994; Kim et al. 1996) but one (Cosentino et al. 1993), BQ-123, administered intracisternally, before or after the induction of SAH, attenuates significantly spasm in the basilar artery. Similar effects in reducing vasospasm have been reported using other peptide antagonists (BQ-485 and FR-139317) selective for ET_A receptors (Itoh et al. 1993; Nirei et al. 1993).

An agent capable of preventing the progression of the delayed ischaemic deficit caused by vasospasm without altering blood pressure would represent a true improvement over current treatment with calcium channel blockers that lack cerebrovascular selectivity. Because of the paracrine/autocrine mode of action of ET, a systemic blockade created by the intravenous administration of an endothelin receptor antagonist might offer a unique opportunity to reverse localised vasospasm without inducing hypotension. An important goal

Table 1. Effects of endothelin receptor antagonists and ECE inhibitors in animal models of SAH

Compound	Species (model)	Administration schedule (dosage, route)	Outcome	Reference
BQ 123[a]	Dog (DH)	4 and 5 days after SAH (0.6 mg/kg, IC)	22% decrease in BA diameter at day 6 (37% decrease in control animals)	Hirose et al. (1995)
	Rat (SH)	10 min before SAH (10 μmol/l, IC)	CBF restored to baseline at 60 and 120 min	Clozel and Watanabe (1993)
	Rat (SH)	10 min before SAH (10 μmol/l, IV)	No effects on SAH-induced CBF reduction	Clozel and Watanabe (1993)
	Dog (DH)	30 min before SAH (100 μmol/l per day, IC)	No significant effect on BA calibre at day 7	Cosentino et al. (1993)
	Rabbit (DH)	4 days after 1st SAH (40 nmol/l, superfusion BA)	71% increase in diameter of vasospastic BA	Foley et al. (1994)
	Dog (DH)	2 days before SAH (5 μmol/day, IC)	Restoration of BA diameter to baseline levels at day 7	Itoh et al. (1994)
	Monkey (SH)	Continuous infusion (5 mg/kg per day, IC)	16% decrease in MCA diameter at day 7 (36% decrease in vehicle-treated animals)	Kim et al. (1996)
	Monkey (SH)	The day of SAH induction (6 mg/kg per day, IC)	7% decrease in MCA diameter at day 7 (34% decrease in vehicle-treated animals)	Hino et al. (1995)
BQ 485[a]	Dog (DH)	2 days before 1st SAH (120 mg/day, SC)	60% decrease in BA diameter at day 7 (75% decrease in control animals)	Itoh et al. (1993)
BQ 610[a]	Rabbit (DH)	4 days after 2nd SAH (1 μmol/l, superfusion BA)	45% increase in diameter of vasospastic BA at day 6	Zuccarello et al. (1994)
FR 139317[a]	Dog (DH)	0, 2 and 4 days after SAH (0.1 mg, IC)	24% decrease in BA diameter at day 7 (58% decrease in control animals)	Nirei et al. (1993)

Table 1. (Continued)

Compound	Species (model)	Administration schedule (dosage, route)	Outcome	Reference
PD 155080[a]	Rabbit (DH)	1h after the 1st SAH (20mg/kg bid, PO)	16% decrease in BA diameter at day 7 (34% decrease in control animals)	Zuccarello et al. (1996)
S 0139[a]	Dog (DH)	Immediately after the 1st SAH (0.03 mg/kg + 28ng/kg per min, IC)	Restoration of BA diameter to baseline levels at 7 days	Kita et al. (1998)
Bosentan[b]	Dog (DH)	Immediately after 1st SAH (10mg/kg bid, IV)	21% decrease in BA diameter at day 7 (49% decrease in vehicle-treated animals)	Shigeno et al. (1995)
	Rabbit (DH)	1h after the 1st SAH (20mg/kg bid, PO)	9% decrease in BA diameter at day 7 (34% decrease in control animals)	Zuccarello et al. (1996)
	Rabbit (SH)	2 days after SAH (30mg/kg, IV)	36% increase in BA calibre at 30 and 60min after drug injection	Roux et al. (1995)
	Dog (DH)	7 days after SAH (30mg/kg, IV)	10% increase in BA calibre at 30 to 120min after drug injection	Roux et al. (1995)
	Dog (DH)	The day of SAH induction (30mg/kg per day bid, PO)	13% decrease in BA diameter at day 7 (31% decrease in control animals)	Zimmermann et al. (1996)
	Monkey (SH)	The day of SAH induction (5mg/kg per day IC)	No effect on MCA diameter at 7 days compared to vehicle-treated animals	Hino et al. (1995)
Ro 46-2005[b]	Rat (SH)	10min before SAH (3mg/kg, IV)	20% increase in CBF at 30 and 60min versus vehicle-treated animals	Clozel et al. (1993)

Agent	Animal (model)	Dose/timing	Effect	Reference
SB 209670[b]	Dog (DH)	1 day before 1st SAH (360 µg/day, IC)	27% decrease in BA diameter at day 7 (68% decrease in vehicle-treated animals)	WILLETTE et al. (1994)
PD 145065[b]	Rabbit (DH)	4 days after 2nd SAH (1 µmol/l, superfusion BA)	87% increase in diameter of vasospastic BA at day 6	ZUCCARELLO et al. (1994)
BQ 788[c]	Rabbit (DH)	1 day before SAH (0.1 nmol/h, IC)	10% decrease in BA diameter at day 7 (30% decrease in control animals)	ZUCCARELLO et al. (1998c)
RES 701-1[c]	Rabbit (DH)	1 day before 1st SAH (0.1 nmol/h, IC)	10% decrease in BA diameter at day 7 (30% decrease in control animals)	ZUCCARELLO et al. (1998c)
CGS 26303[d]	Rabbit (SH)	1 day before SAH (1.2 mg/h, IP)	13% decrease in BA diameter at day 2 (58% decrease in vehicle-treated animals)	CANER et al. (1996)
	Rabbit (SH)	24h after SAH (30 mg/kg, IV)	21% decrease in BA diameter at day 2 (59% decrease in vehicle-treated animals)	KWAN et al. (1997)
Phosphoramidon[d]	Dog (DH)	30min before 1st and 2nd SAH (2 µmol, IC)	20% decrease in BA diameter at day 7 (45% decrease in vehicle-treated animals)	MATSUMURA et al. (1991)
Phosphoramidon[d]	Dog (DH)	30min before SAH (200 µmol/l/day, IC)	No significant effect on BA calibre at day 7	COSENTINO et al. (1993)

BA, basilar artery; bid, twice daily; CBF, cerebral blood flow; DH, double haemorrhage; ECE, endothelin-converting enzyme; IC, intracisternal; IP, intraperitoneal; IV, intravenous; MCA, middle cerebral artery; PO, oral; SAH, subarachnoid haemorrhage; SH, single haemorrhage.
[a] Selective ETA antagonist.
[b] Mixed ETA/ETB antagonist.
[c] Selective ETB antagonist.
[d] ECE inhibitor.

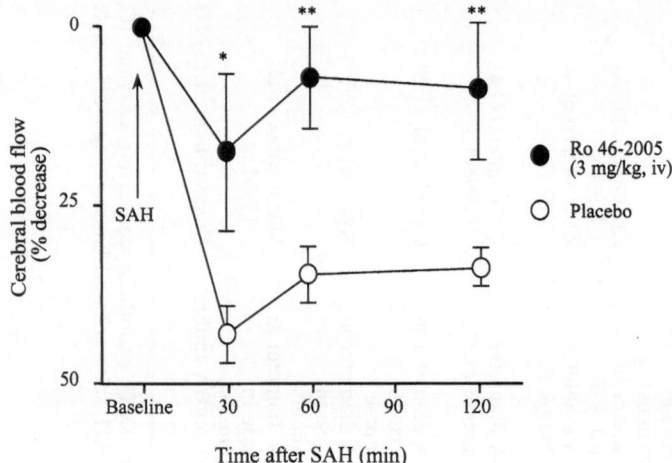

Fig. 3. Prevention by Ro 46–2005 (mixed ET_A/ET_B receptor antagonist) of the decrease in cerebral blood flow (CBF) after subarachnoid haemorrhage (SAH) in rats. Ro 46–2005 (●, 3 mg/kg; n = 9 rats) or glucose as a placebo (○, n = 10 rats) was injected intravenously 10 min before SAH. CBF was measured before and 30 min, 60 min and 120 min after SAH by the radioactive microspheres technique. Values depicted are the percent changes in CBF measured in cerebellum. *$p < 0.05$, **$p < 0.01$. After Clozel et al. (1993)

of ongoing research remains the discovery of ET-blocking agents that are effective when administered systemically.

Clozel et al. (1993) were the first to develop a non-peptide endothelin receptor antagonist, namely, Ro 46-2005. This mixed ET_A/ET_B receptor antagonist increased CBF when administered intravenously to rats subjected to SAH (Fig. 3) (Clozel et al. 1993). Subsequently, other non-peptide endothelin receptor antagonists (selective ET_A or mixed ET_A/ET_B) have been shown to elicit significant reduction of post-haemorrhagic vasospasm when administered orally or intravenously, without effects on systemic blood pressure (Kita et al. 1998; Roux et al. 1995; Shigeno et al. 1995; Zimmermann et al. 1996; Zuccarello et al. 1996). In disagreement with these findings, the study of Hino et al. (1995) showed that bosentan (a peptide mixed ET_A/ET_B antagonist) had no significant effects on arterial narrowing after SAH in the monkeys. In this study, bosentan (5 mg/kg) was administered intracisternally twice a day into an Ommaya reservoir with a catheter along the middle cerebral artery. The lack of efficacy of bosentan in these experiments, however, may be related to the inadequate CSF levels obtained by this dosing regimen. Bosentan was not detected in CSF at day seven after SAH (Hino et al. 1995).

As discussed earlier, the function of ET_B receptors in normal and pathological conditions are not clear. ET_B receptors have been reported to mediate vasodilatation and vasoconstriction (Kitazono et al. 1995a; Seo et al. 1994;

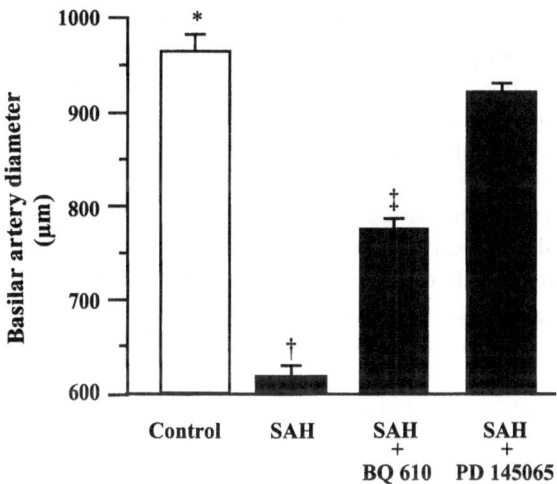

Fig. 4. Effects of the endothelin ET_A receptor antagonist, BQ 610, and endothelin ET_A/ET_B receptor antagonist, PD 145065, on subarachnoid haemorrhage (SAH)-induced vasospasm of the rabbit basilar artery. Six days after SAH induction, the basilar artery was superfused in situ with BQ 610 (1 μmol/l) and PD 145065 (1 μmol/l). Basilar artery diameter was measured using the cranial window technique. Data are presented as mean ± SEM. N, the number of rabbits was 13 for control, 5 for SAH, 5 for SAH + BQ 610 and 5 for SAH + PD 145065. *Significantly different from SAH and SAH + BQ 610. †Significantly different from all other values. ‡Significantly different from control and SAH + PD 145065. After ZUCCARELLO et al. (1994)

TOUZANI et al. 1997; ZUCCARELLO et al. 1998a). One of the key issues is whether selective ET_A or mixed ET_A/ET_B antagonists are likely to be more effective in treating vasospasm. In an elegant experiment, ZUCCARELLO et al. (1994) have shown that spasm of the basilar artery of the rabbit was only partially reversed by a selective ET_A antagonist, whereas a mixed ET_A/ET_B antagonist induced complete relaxation of the artery (Fig. 4). These results indicate that ET-1-dependent vasospasm is mediated not only by ET_A receptor activation but also by ET_B receptor activation. The most recent investigation by ZUCCARELLO et al. (1998c) suggest that ET_B activation is the more important event in vasospasm as selective blockade of ET_B receptor prevents and reverses SAH-induced cerebral vasospasm in rabbits.

The observations discussed above support the idea that use of a broad spectrum endothelin receptor antagonist would be of greater benefit in reversing the increased cerebrovascular tone driven by the release of endogenous endothelins after SAH in human. Two of the endothelin receptor antagonists (bosentan and TAK 044) have already made the transition to clinical trials in SAH patients but the limited information in the public domain suggest that whatever efficacy (if any) may be demonstrable in man, it is minimal in relation to the compelling data obtained with induced SAH in animals.

Inhibition of synthesis of ET-1 provides another approach for limiting ET-1-mediated vasospasm. The metalloprotease inhibitor, phosphoramidon, is capable of blocking ECE and thus inhibits the conversion of Big-ET-1 to ET-1 (see Chaps. 3, 7). The effects of phosphoramidon on vasospasm however, have been mixed (Table 1). When given intracisternally before SAH, the observed decrease in arterial diameter after phosphoramidon was only 20% compared with 50% in non-treated dogs (MATSUMURA et al. 1991). Phosphoramidon was also effective in inhibiting vasoconstriction and pressor effects of intracisternally applied big-ET-1 in dogs, suggesting that the physiological actions of big-ET-1 occur after conversion to ET-1. However, COSENTINO et al. (1993) did not observe any reduction of vasospasm even after continuous intracisternal administration of a higher dose of phosphoramidon in the canine model of SAH. Development of the non-peptide ECE inhibitor, CGS 26303, has allowed the authors to examine the effects of its systemic administration on experimental vasospasm (Table 1). When injected intravenously, CGS 26303 blocks the constrictor response to Big ET-1, but not ET-1, topically applied to the basilar artery of the rabbit (CANER et al. 1996). In addition, CGS 26303 prevents and reverses vasospasm in the basilar artery following SAH in the rabbit (CANER et al. 1996; KWAN et al. 1997). The observation that this ECE inhibitor was able to reverse the established vasospasm suggests that there is a continuous production of ET-1 after SAH. The development of new ECE inhibitors specific for different isoforms of the enzyme should further clarify the exact role of endothelins in SAH.

II. Cerebral Ischaemia

Cerebral ischaemia remains a major clinical problem in the world (SACCO 1997). Ischaemic stroke is often caused by thromboembolic occlusion of an artery supplying a portion of the brain. Ischaemic stroke is a complex vascular and metabolic process that evolves with time. Occlusion of a cerebral artery results in a markedly ischaemic core surrounded by a moderately hypoperfused region known as the penumbra. The penumbra is a region at risk; it can evolves towards infarction or towards viability (OBRENOVITCH 1995). A number of mechanisms have been proposed to explain the progression of the penumbra to infarction. These include, generation of cytotoxic substances, occurrence of spreading depression and formation of vasoconstrictor elements that can worsen the hypoperfusion.

The ability of exogenous ET-1 to overwhelm intrinsic autoregulatory mechanisms that maintain CBF has been demonstrated. The intracisternal administration of exogenous ET-1 in rats results in brain stem ischaemia (MACRAE et al. 1991). The topical administration of ET-1 onto a major cerebral artery, such as the MCA, induces marked and dose-dependent reductions in CBF in the cortex and striatum, and development of infarction similar to that produced by occlusion of the artery (FUXE et al. 1992; MACRAE et al. 1993; SHARKEY et al. 1993). These ET-1-mediated haemodynamic alterations and the fact that insults such as hypoxia, stress and blood products (factors associated with cerebral ischaemia) induce ET-1 expression have led to conjecture that this peptide may be involved in the pathogenesis of cerebral ischaemia.

Endothelin-1 does not appear to have a direct neurotoxic potential when applied to cortical neuronal cell cultures (LUSTIG et al. 1992; NIKOLOV et al. 1993). However, the administration of exogenous ET-1 into the cerebroventricular system exacerbated the cerebral damage associated with focal cerebral ischaemia in the mouse (NIKOLOV et al. 1993). These observations suggest that the pathologic role of ET-1 is probably due to its vasoconstrictive properties and not direct neurotoxic effects. Nonetheless, neurochemical effects of ET-1 could contribute to ischaemic cerebral injury indirectly by exacerbating cytotoxic mechanisms that operate in cerebral ischaemia. SASAKI et al. (1997) demonstrated that ET-1 can stimulate the efflux of glutamate from rat astrocytes in culture via the activation of ET_B receptors. In addition, ET-1 increased free intracellular levels in various cell cultures (CHAN and GREENBERG 1991; ZHANG et al. 1990). ET-1 can also play a pathologic role in cerebral ischaemia by its ability to evoke increase in free radical formation (KASEMSRI and ARMSTEAD 1997; PHILLIS 1994), and adhesion molecule expression (HAYASAKI et al. 1996; MCCARRON et al. 1993). All these factors have been proposed to play pivotal roles in the progression of ischaemic brain damage (HALLENBECK 1996; PHILLIS 1994).

1. Clinical Observations in Man

Currently, only limited clinical studies are available and, even in these, controversy exists in respect of the interpretation of the data. Elevation of plasma ET-1 levels has been reported in patients in the acute, subacute and the chronic stages after stroke (ESTRADA et al. 1994; ZIV et al. 1992; WEI et al. 1993). WEI et al. (1993) have showed a positive correlation between the infarct size and plasma ET-1 level in 21 stroke patients. ESTRADA et al. (1994) have reported that ET-1 concentration in the plasma is correlated with the clinical status on admission and with the final outcome, but not with the size of infarction. Conversely, HAMANN et al. (1993a) found no overall elevation in plasma Big-ET-1 in patients following stroke. Recently, plasma and CSF ET-1 levels were measured in 26 patients within the first 18 h of non-haemorrhagic stroke (LAMPL et al. 1997). In this population of patients, the mean concentration of ET-1 in the plasma was not different to that in the control group, and was not correlated with the volume of infarction. However, the CSF ET-1 level was significantly higher in stroke patients (16.1 ± 3.5 pg/ml) compared to the control group (5.1 ± 1.5 pg/ml). Moreover, the CSF levels correlated to the volume of cortical infarction as well as to the degree of neurological deficit. It is difficult to interpret results across studies because of the considerable variation in ET-1 levels found in control plasma and CSF. The question of whether the increased levels of ET-1 reflect a role of the peptide in mediating cerebral ischaemic damage, or whether they are a marker for the degree of cerebral damage, cannot be definitively answered by these observations. Indeed, as discussed earlier, the pathological significance of elevated plasma ET-1 levels in stroke as in SAH is unclear since the available evidence indicates that ET-1 does not cross the BBB. Furthermore, because of the autocrine/paracrine mode of action of endothelin, its circulating levels do not always reflect local events.

2. Induced Cerebral Ischaemia in Animals

a) Global Cerebral Ischaemia

Experimental models of global cerebral ischaemia have been developed to examine the mechanisms of slowly evolving selective neuronal death in the brain. Global cerebral ischaemia is characterised by brief, global temporary reductions of CBF, followed by transient hyperaemia and delayed hypoperfusion. Neuronal death following transient ischaemia occurs in vulnerable brain regions (hippocampal CA1 region, striatum and cortex), with regionally characteristic delays in appearance of up to 7 days.

A variety of studies have been designed to examine the role of ET-1 in haemodynamic perturbations and development of cell death after global cerebral ischaemia. Increases of forebrain ET-1 immunoreactivity have been observed in rats and gerbils subjected to global ischaemia (Barone et al. 1994; Giuffrida et al. 1992; Willette et al. 1993; Yamashita et al. 1993, 1994). Barone et al. (1994) found that the increase of ET-1 immunoreactivity in the extracellular space is observed very early (<3h) after the induction of global ischaemia in rats. Yamashita et al. (1993, 1994) reported that elevation of ET-1 immunostaining in the hippocampal CA1 cell layer correlated with the appearance of neuronal death on days 4 and 7 post-ischaemia. Interestingly, these studies also showed an increase in ET_B receptor binding in microglia in the hippocampal pyramidal cell field 7 days after global ischaemia. These authors suggested that ET-1 released from astrocytes, in response to ischaemia, could activate microglia via ET_B receptors to promote neuronal death in the hippocampus.

Pharmacological intervention with endothelin receptor antagonists has shown a possible role for ET-1 in neuronal degeneration in models of global ischaemia. Intracerebroventricular administration of BQ-123 and SB 209967, peptide and non peptide ET_A selective antagonists, respectively, resulted in significant neuroprotection in the CA1 region of the hippocampus (Fig. 5) (Feuerstein et al. 1994; Ohlstein et al. 1993). Whether this neuroprotection is a result of amelioration of cerebral perfusion is not clear. In comparison, Bosentan has failed to reverse delayed hypoperfusion after global ischaemia in rats and gerbils (Patel and McCulloch 1996; Yasuma et al. 1997), whereas, BQ 123 and Ro 61–1790 (ET_A antagonists) were effective in ameliorating post-ischaemic hypoperfusion in gerbils (Yasuma et al. 1997). The failure of bosentan to increase CBF may be related to its blockade of dilator action of ET_B receptors. Nonetheless, our results obtained following focal cerebral ischaemia

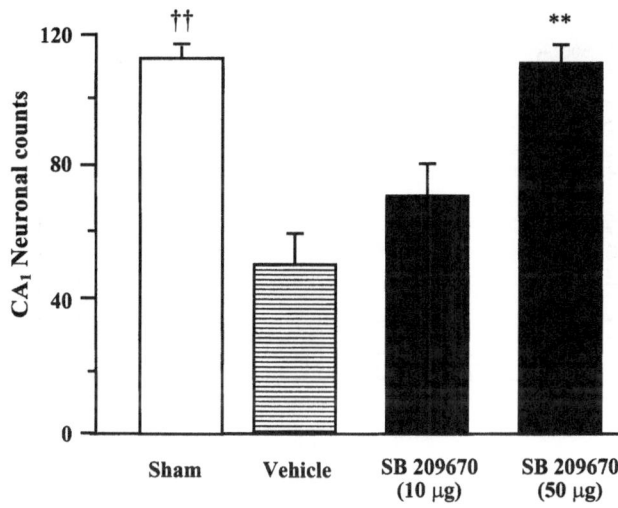

Fig. 5. Protective effect of SB 209670 (non peptide ET_A/ET_B receptor antagonist) against neuronal damage in the CA1 region of gerbil hippocampus after 6.5 min of ischaemia. SB 209670 was administered intracerebroventricularly 5 min pre- and 60 min post-ischaemia. Data represent the average of CA1 hippocampus neuronal counts of a 750 μm length of both left and right hemispheres at 7 days after occlusion. $^{\dagger\dagger}p < 0.01$ for values obtained in vehicle-treated animals compared to sham-operated animals; $^{**}p < 0.01$ for values obtained in animals treated with SB 209670 compared to vehicle-treated animals. After OHLSTEIN et al. (1994)

in the cat showed that the dilator effects of ET_B receptors are lost after 30 min of ischaemia (TOUZANI et al. 1997). Further studies employing other endothelin receptor antagonists and ECE inhibitors with optimal pharmacological properties are needed to clarify this question.

b) Focal Cerebral Ischaemia

Several lines of evidence implicate endothelins in the pathogenesis of focal cerebral ischaemia, although the precise definition of their pathophysiological role has not yet been fully elucidated. In normotensive rats subjected to permanent middle cerebral artery occlusion (MCAO), ET-1 immunoreactivity increased progressively and markedly in the ischaemic hemisphere (Fig. 6). At 3 days after the occlusion, there was a 450% increase in ET-1 immunoreactivity in the ipsilateral cortex compared to the contralateral one (DUVERGER et al. 1992; VIOSSAT et al. 1993). In these studies, the elevation of brain ET-1 immunoreactivity was related to the severity of ischaemia since transient focal ischaemia was accompanied only by modest increase in ET-1 immunoreactivity. The sources of progressive production of ET-1 do not appear to be pial vessels or arteries from the circle of Willis (DUVERGER et al. 1992; VIOSSAT et al. 1993). Similar results have been obtained by BARONE et al.

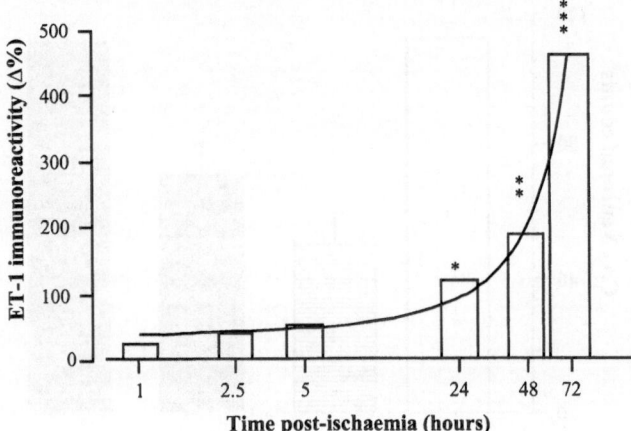

Fig. 6. Time course of increase of ET-1 immunoreactivity following permanent focal ischaemia in the rat. The results express the percentage variation between the ipsilateral and the contralateral hemispheres. Each group contains $n = 6$ animals. *$p < 0.05$; **$p < 0.01$; ***$p < 0.001$. After Duverger et al. (1992)

(1994) in models of temporary and permanent ischaemia in spontaneously hypertensive rats.

The marked production of ET-1 following a cerebral arterial occlusion may result in increase of vascular tone that jeopardises hypoperfused cerebral tissue. We have examined this issue in cats subjected to permanent occlusion of the MCA. The responses of the cortical resistance arterioles after MCAO can be divided into two groups. One group undergoes marked constriction, and the direct adventitial application of the non-peptide endothelin receptor antagonists bosentan and PD 155080 (mixed ET_A/ET_B and selective ET_A antagonists, respectively) increase the calibre of the arterioles (Patel et al. 1996b). The other group of arterioles undergoes dilatation. Direct application of bosentan, PD 155080, and BQ788 (peptide ET_B selective antagonist) elicits further increase in the calibre of these arterioles (Patel et al. 1996b; Touzani et al. 1997). The ability of these endothelin receptor antagonists to increase the arteriolar calibre of post ischaemic constricted and dilated arterioles demonstrates an increase in endothelin-mediated tone in cerebral resistance arterioles following focal cerebral ischaemia. The most compelling evidence that endothelins are involved in the evolving cerebral circulatory disturbances after cerebral ischaemia has come from analysis of the effects of systemically administered endothelin receptor antagonists on cerebral blood flow (CBF) and ischaemic pathology after middle cerebral artery occlusion. There are no reliable reports that any endothelin antagonist significantly alters CBF in normal animals under normocapnic and normotensive conditions. The intra-

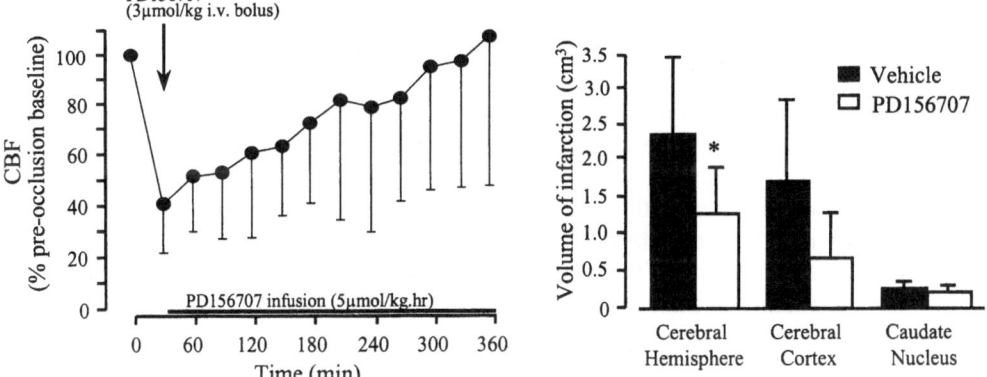

Fig. 7. Effects of PD 156707 (non peptide ET_A receptor antagonist) on cerebral blood flow (CBF) and volume of ischaemic damage after permanent cerebral ischaemia in the cat ($n = 5$ in each group). CBF was measured continuously by laser Doppler flowmetry technique. PD 156707 (3μmol/kg) was administered as an intravenous bolus injection 30 min after induction of ischaemia, followed by a continuous intravenous infusion (5μmol/kg per h). Vehicle had no effect on cerebral blood flow in the post ischaemia period. The volume of infarction was determined 6h after arterial occlusion. Data are mean ± SD. *$p < 0.05$. After PATEL et al. (1996)

venous administration of the ET_A receptor antagonist PD 156707, initiated 30 min following permanent middle cerebral artery occlusion in the cat, progressively restores CBF to pre-occlusion baseline levels in the ischaemic penumbra by 6h following occlusion (PATEL et al. 1996c). PD 156707 also reduces the volume of hemispheric infarction by 45% in the same animals (Fig. 7). The success of the study is due in part to its rigorous pharmacokinetic control, particularly the demonstration of access of the antagonist to CSF and that these concentrations were sufficient for ET_A receptor blockade. The neuroprotective effects of PD 156707 and other endothelin receptor antagonists following focal cerebral ischaemia in the rat have also been reported but the neuroprotective effect in rats was generally less impressive than in cats (BARONE et al. 1995; TATLISUMAK et al. 1998; McAULEY et al. 1996; TAKASAGO et al. 1997). Inhibition of ECE by phosphoramidon produced a significant reduction in the volume of infarction in rats subjected to MCA occlusion (DUVERGER et al. 1993). In experimental focal ischaemia there is, thus, a persuasive case that dilatation in and blood flow to the penumbra is submaximal because of the generation of endothelins and that endothelin antagonists can enhance blood flow to this zone and so prevent its infarction. Endothelin mechanisms are an increasingly interesting therapeutic target for human stroke, particularly so in view of the disappointing efficacy observed to date with therapies targeted at excitotoxic mechanisms.

D. Concluding Remarks

In the cerebral circulation, exogenous ET-1 evokes a potent and, most importantly, extremely long-lasting vasoconstriction. The role of ET-1 in maintaining cerebrovascular tone is controversial. Although some data indicate that the cerebrovascular tone may be, at least in part, affected by physiological antagonism between the vasoconstriction induced by ET-1 and the vasodilatation induced by endothelium-derived relaxing factors (e.g. nitric oxide and prostacyclin), the more compelling evidence via the use of endothelin receptor antagonists is that ET-1 exerts a minimal effect in maintaining vascular calibre in resting physiological conditions. The presence of endothelin system in non-vascular tissue in brain (i.e. neurones and glia) suggest that ET-1 has actions which might influence the function of neuronal tissue. However its role in physiological and pathophysiological conditions remains to be determined.

The available evidence indicates that ET-1 is involved in cerebrovascular pathologies where vascular homeostasis is overridden. In SAH-related vasospasm, elevated ET-1 levels have been reported in humans and animals. The majority of experimental investigations have shown the ability of endothelin receptor antagonists to attenuate, to different degree, cerebral vasospasm. These findings support the idea that ET-1 may be causal, or at least contributory, in this pathology, although, given the multifactorial features of post-haemorrhagic vasospasm, ET-1 should not be considered as the unique agent driving the development of vasospasm.

In stroke, relatively few studies have examined the contribution of ET-1 to haemodynamic and histologic alterations. The available data indicate that ET-1 is overexpressed following cerebral ischaemia and it participates significantly in the increased cerebrovascular tone that follows arterial occlusion.

The identification of endothelin receptor subtypes has important implications for the use of combined ET_A/ET_B receptor antagonists vs antagonists selective for the ET_A receptor in cerebrovascular disease. The exact contribution of ET_A and ET_B receptors in abnormalities observed in SAH and stroke is not clear. Thus the selectivity profile for the ideal endothelin receptor antagonist to be used as therapeutic agent is presently unknown. Further studies, with highly selective and brain penetrating antagonists, are required to resolve this issue. Inhibition of ECE remains an alternative approach to elucidate the exact role of ET-1 in cerebrovascular pathologies. Characterisation of different ECE subtypes in brain and development of selective, non-peptide ECE inhibitors will have great utility in the development of endothelin-targeted therapeutic strategies.

References

Adner M, Cantera L, Ehlert F, Nilsson L, Edvinsson L (1996) Plasticity of contractile endothelin-B receptors in human arteries after organ culture. Br J Pharmacol 119:1159–1166

Adner M, You J, Edvinsson L (1993) Characterization of endothelin-A receptors in the cerebral circulation. Neuroreport 4:441–443

Alabadi JA, Salom JB, Torregrosa G, Miranda FJ, Jover T, Alborch E (1993) Changes in the cerebrovascular effects of endothelin-1 and nicardipine after experimental subarachnoid hemorrhage. Neurosurgery 33:707–714

Alabadi JA, Torregrosa G, Miranda FJ, Salom JB, Centeno JM, Alborch E (1997) Impairment of the modulatory role of nitric oxide on the endothelin-1-elicited contraction of cerebral arteries: a pathogenetic factor in cerebral vasospasm after subarachnoid hemorrhage? Neurosurgery 41:245–252

Alafaci C, Jansen I, Arbab MA, Shiokawa Y, Svendgaard NA, Edvinsson L (1990) Enhanced vasoconstrictor effect of endothelin in cerebral arteries from rats with subarachnoid haemorrhage. Acta Physiol Scand 138:317–319

Armstead WM (1996) Role of endothelin in pial artery vasoconstriction and altered responses to vasopressin after brain injury. J Neurosurg 85:901–907

Asano T, Ikegaki I, Suzuki Y, Satoh S, Shibuya M (1989) Endothelin and the production of cerebral vasospasm in dogs. Biochem Biophys Res Commun 159:1345–1351

Barnes K, Turner AJ (1997a) The endothelin system and endothelin-converting enzyme in the brain: molecular and cellular studies. Neurochem Res 22:1033–1040

Barone FC, Globus MY, Price WJ, White RF, Storer BL, Feuerstein GZ, Busto R, Ohlstein EH (1994) Endothelin levels increase in rat focal and global ischemia. J Cereb Blood Flow Metab 14:337–342

Barone FC, White RF, Elliott JD, Feuerstein GZ, Ohlstein EH (1995) The endothelin receptor antagonist SB 217242 reduces cerebral focal ischemic brain injury. J Cardiovasc Pharmacol 26 Suppl 3:S404–S407

Caner HH, Kwan AL, Arthur A, Jeng AY, Lappe RW, Kassell NF, Lee KS (1996) Systemic administration of an inhibitor of endothelin-converting enzyme for attenuation of cerebral vasospasm following experimental subarachnoid hemorrhage. J Neurosurg 85:917–922

Cazaubon S, Chaverot N, Romero IA, Girault JA, Adamson P, Strosberg AD, Couraud PO (1997) Growth factor activity of endothelin-1 in primary astrocytes mediated by adhesion-dependent and -independent pathways. J Neurosci 17:6203–6212

Chan J, Greenberg DA (1991) Endothelin and calcium signaling in NG108–15 neuroblastoma x glioma cells. J Pharmacol Exp Ther 258:524–530

Clozel M, Breu V, Burri K, Cassal JM, Fischli W, Gray GA, Hirth G, Loffler BM, Muller M, Neidhart W (1993) Pathophysiological role of endothelin revealed by the first orally active endothelin receptor antagonist. Nature 365:759–761

Clozel M, Watanabe H (1993) BQ-123, a peptidic endothelin ETA receptor antagonist, prevents the early cerebral vasospasm following subarachnoid hemorrhage after intracisternal but not intravenous injection. Life Sci 52:825–834

Cosentino F, McMahon EG, Carter JS, Katusic ZS (1993) Effect of endothelinA-receptor antagonist BQ-123 and phosphoramidon on cerebral vasospasm. J Cardiovasc Pharmacol 22 Suppl 8:S332–S335

Cramer H, Muller-Esterl W, Schroeder C (1997) Subtype-specific desensitization of human endothelin ETA and ETB receptors reflects differential receptor phosphorylation. Biochemistry 36:13325–13332

Dagassan PH, Breu V, Clozel M, Kunsli A, Vogt P, Turina M, Kiowski W, Clozel JP (1996) Up-regulation of endothelin-B receptors in atherosclerotic human coronary arteries. J Cardiovasc Pharmacol 27:147–153

Davenport AP, Morton AJ (1991) Binding sites for 125I ET-1, ET-2, ET-3 and vasoactive intestinal contractor are present in adult rat brain and neurone-enriched primary cultures of embryonic brain cells. Brain Res 554:278–285

de Nucci G, Thomas R, D'Orleans-Juste P, Antunes E, Walder C, Warner TD, Vane JR (1988) Pressor effects of circulating endothelin are limited by its removal in the pulmonary circulation and by the release of prostacyclin and endothelium-derived relaxing factor. Proc Natl Acad Sci U.S.A. 85:9797–9800

Douglas SA (1997) Clinical development of endothelin receptor antagonists. Trends Pharmacol Sci 18:408–412

Douglas SA, Meek TD, Ohlstein EH (1994) Novel receptor antagonists welcome a new era in endothelin biology. Trends Pharmacol Sci 15:313–316

Duverger D, Viossat I, Chapelat M, Chabrier PE, Pirotzky E, Braquet P (1993) Effects of phosphoramidon on immunoreactive ET-1 contents in brain volume infarctus in MCA-occluded rats. J Cereb Blood Flow Metab 13 (Suppl 1):S194

Duverger D, Viossat I, Chapelat M, Chabrier PE, Pirotzky E, Braquet P (1992) Ischémie cérébrale focale chez le rat : quantification de l'endothéline tissulaire. Circ Metab Cerveau 9:85–93

Edvinsson L, Mackenzie ET, McCulloch J (1993) Cerebral Blood Flow and Metabolism, Raven Press

Edwards R, Trizna W (1990) Response of isolated intracerebral arterioles to endothelins, Pharmacology 41:149–152

Egleton RD, Davis TP (1997) Bioavailability and transport of peptides and peptide drugs into the brain, Peptides 18:1431–1439

Ehrenreich H, Anderson RW, Fox CH, Rieckmann P, Hoffman GS, Travis WD, Coligan JE, Kehrl JH, Fauci AS (1990) Endothelins, peptides with potent vasoactive properties, are produced by human macrophages. J Exp Med 172:1741–1748

Ehrenreich H, Costa T, Clouse KA, Pluta RM, Ogino Y, Coligan JE, Burd PR (1993) Thrombin is a regulator of astrocytic endothelin-1. Brain Res 600:201–207

Ehrenreich H, Kehrl JH, Anderson RW, Rieckmann P, Vitkovic L, Coligan JE, Fauci AS (1991) A vasoactive peptide, endothelin-3, is produced by and specifically binds to primary astrocytes. Brain Res 538:54–58

Ehrenreich H, Lange M, Near KA, Anneser F, Schoeller LA, Schmid R, Winkler PA, Kehrl JH, Schmiedek P, Goebel FD (1992) Long term monitoring of immunoreactive endothelin-1 and endothelin-3 in ventricular cerebrospinal fluid, plasma, and 24-h urine of patients with subarachnoid hemorrhage. Res Exp Med (Berl) 192:257–268

Estrada V, Tellez MJ, Moya J, Fernandez-Durango R, Egido J, Fernandez CA (1994) High plasma levels of endothelin-1 and atrial natriuretic peptide in patients with acute ischemic stroke. Am J Hypertens 7:1085–1089

Feger GI, Schilling L, Ehrenreich H, Wahl M (1994) Endothelin-induced contraction and relaxation of rat isolated basilar artery: effect of BQ-123. J Cereb Blood Flow Metab 14:845–852

Feigin VL, Rinkel GJ, Algra A, Vermeulen M, van Gijn J (1998) Calcium antagonists in patients with aneurysmal subarachnoid hemorrhage: a systematic review. Neurology 50:876–883

Fernandez-Durango R, de Juan JA, Zimman H, Moya FJ, Garcia de la Coba M, Fernandez-Cruz A (1994) Identification of endothelin receptor subtype (ETB) in human cerebral cortex using subtype-selective ligands. J Neurochem 62:1482–1488

Feuerstein G, Gu JL, Ohlstein EH, Barone FC, Yue TL (1994) Peptidic endothelin-1 receptor antagonist, BQ-123, and neuroprotection, Peptides 15:467–469

Foley PL, Caner HH, Kassell NF, Lee KS (1994) Reversal of subarachnoid hemorrhage-induced vasoconstriction with an endothelin receptor antagonist, Neurosurgery 34:108–112

Fujimori A, Yanagisawa M, Saito A, Goto K, Masaki T, Mima T, Takakura K, Shigeno T (1990) Endothelin in plasma and cerebrospinal fluid of patients with subarachnoid haemorrhage [letter], Lancet 336:633

Fuxe K, Kurosawa N, Cintra A, Hallstrom A, Goiny M, Rosen L, Agnati LF, Ungerstedt U (1992) Involvement of local ischemia in endothelin-1 induced lesions of the neostriatum of the anaesthetized rat. Exp Brain Res 88:131–139

Gaetani P, Rodriguez, Grignani G, Spanu G, Pacchiarini L, Paoletti P (1994) Endothelin and aneurysmal subarachnoid haemorrhage: a study of subarachnoid cisternal cerebrospinal fluid. J Neurol Neurosurg Psychiatry 57:66–72

Garcia JL, Gomez B, Monge L, Garcia-Villalon AL, Dieguez G (1991) Endothelin action on cerebral circulation in unanesthetized goats. Am J Physiol 261:R581–R587

Gellai M, Fletcher T, Pullen M, Nambi P (1996) Evidence for the existence of endothelin-B receptor subtypes and their physiological roles in the rat. Am J Physiol 271:R254–R261

Giaid A, Gibson SJ, Herrero MT, Gentleman S, Legon S, Yanagisawa M, Masaki T, Ibrahim NB, Roberts GW, Rossi ML (1991) Topographical localisation of endothelin mRNA and peptide immunoreactivity in neurones of the human brain. Histochemistry 95:303–314

Giaid A, Gibson SJ, Ibrahim BN, Legon S, Bloom SR, Yanagisawa M, Masaki T, Varndell IM, Polak JM (1989) Endothelin 1, an endothelium-derived peptide, is expressed in neurons of the human spinal cord and dorsal root ganglia. Proc Natl Acad Sci U.S.A. 86:7634–7638

Giuffrida R, Bellomo M, Polizzi G, Malatino LS (1992) Ischemia-induced changes in the immunoreactivity for endothelin and other vasoactive peptides in the brain of the Mongolian gerbil. J Cardiovasc Pharmacol 20 Suppl 12:S41–S44

Goretski J, Hollocher TC (1988) Trapping of nitric oxide produced during denitrification by extracellular hemoglobin. J Biol Chem 263:2316–2323

Greenberg DA, Chan J, Sampson HA (1992) Endothelins and the nervous system. Neurology 42:25–31

Gross PM, Beninger RJ, Shaver SW, Wainman DS, Espinosa FJ, Weaver DF (1993) Metabolic and neuroanatomical correlates of barrel-rolling and oculoclonic convulsions induced by intraventricular endothelin-1: a novel peptidergic signaling mechanism in visuovestibular and oculomotor regulation? Exp Brain Res 95:397–408

Hallenbeck JM (1996) Inflammatory reactions at the blood-endothelial interface in acute stroke. Adv Neurol 71:281–97; discussion 297–300:281–297

Hamann G, Isenberg E, Strittmatter M, Moili R, Schimrigk K (1993a) Big-endothelin in acute stroke. J Stroke Cerebrovasc Dis 3:256–260

Hamann G, Isenberg E, Strittmatter M, Schimrigk K (1993b) Absence of elevation of big endothelin in subarachnoid hemorrhage. Stroke 24:383–386

Harland SP, Kuc RE, Pickard JD, Davenport AP (1998) Expression of endothelin(A) receptors in human gliomas and meningiomas, with high affinity for the selective antagonist PD156707. Neurosurgery 43:890–898

Hatake K, Wakabayashi I, Kakishita E, Hishida S (1992) Impairment of endothelium-dependent relaxation in human basilar artery after subarachnoid hemorrhage. Stroke 23:1111–1116

Hayasaki Y, Nakajima M, Kitano Y, Iwasaki T, Shimamura T, Iwaki K (1996) ICAM-1 expression on cardiac myocytes and aortic endothelial cells via their specific endothelin receptor subtype. Biochem Biophys Res Commun 229:817–824

Haynes WG, Davenport AP, Webb DJ (1993) Endothelin: progress in pharmacology and physiology. Trends Pharmacol Sci 14:225–228

Hino A, Tokuyama Y, Kobayashi M, Yano M, Weir B, Takeda J, Wang X, Bell GI, Macdonald RL (1996) Increased expression of endothelin B receptor mRNA following subarachnoid hemorrhage in monkeys. J Cereb Blood Flow Metab 16:688–697

Hino A, Weir BK, Macdonald RL, Thisted RA, Kim CJ, Johns LM (1995) Prospective, randomized, double-blind trial of BQ-123 and bosentan for prevention of vasospasm following subarachnoid hemorrhage in monkeys. J Neurosurg 83:503–509

Hirose H, Ide K, Sasaki T, Takahashi R, Kobayashi M, Ikemoto F, Yano M, Nishikibe M (1995) The role of endothelin and nitric oxide in modulation of normal and spastic cerebral vascular tone in the dog. Eur J Pharmacol 277:77–87

Inoue A, Yanagisawa M, Kimura S, Kasuya Y, Miyauchi T, Goto K, Masaki T (1989) The human endothelin family: three structurally and pharmacologically distinct isopeptides predicted by three separate genes. Proc Natl Acad Sci USA 86: 2863–2867

Ishikawa N, Takemura M, Koyama Y, Shigenaga Y, Okada T, Baba A (1997) Endothelins promote the activation of astrocytes in rat neostriatum through ET(B) receptors. Eur J Neurosci 9:895–901

Jansen I. Fallgren B, Edvinsson L (1989) Mechanisms of action of endothelin on isolated feline cerebral arteries: in vitro pharmacology and electrophysiology. J Cereb Blood Flow Metab 9:743–747

Jerius H, Beall A, Woodrum D, Epstein A, Brophy C (1998) Thrombin-induced vasospasm: cellular signaling mechanisms. Surgery 123:46–50

Jones CR, Hiley CR, Pelton JT, Mohr M (1989) Autoradiographic visualization of the binding sites for [125I]endothelin in rat and human brain, Neurosci Lett. 97:276–279

Kadel KA, Heistad DD, Faraci FM (1990) Effects of endothelin on blood vessels of the brain and choroid plexus. Brain Res 518:78–82

Kamata K, Nishiyama H, Miyata N, Kasuya Y (1991) Changes in responsiveness of the canine basilar artery to endothelin-1 after subarachnoid hemorrhage. Life Sci 49:217–224

Karaki H, Sudjarwo SA, Hori M, Takai M, Urade Y, Okada T (1993) Induction of endothelium-dependent relaxation in the rat aorta by IRL 1620, a novel and selective agonist at the endothelin ETB receptor. Br J Pharmacol 109:486–490

Kasemsri T, Armstead WM (1997) Endothelin production links superoxide generation to altered opioid- induced pial artery vasodilation after brain injury in pigs. Stroke 28:190–196

Kasuya H, Weir BK, White DM, Stefansson K (1993) Mechanism of oxyhemoglobin-induced release of endothelin-1 from cultured vascular endothelial cells and smooth-muscle cells. J Neurosurg 79:892–898

Kauser K, Rubanyi GM, Harder DR (1990) Endothelium-dependent modulation of endothelin-induced vasoconstriction and membrane depolarization in cat cerebral arteries. J Pharmacol Exp Ther 252:93–97

Kazuki S, Ohta T, Ogawa R, Tsuji M, Tamura Y, Yoshizaki Y, Takase T (1997) Effects of intraluminal or extraluminal endothelin on perfused rabbit basilar arteries. J Neurosurg 86:859–865

Kim CJ, Bassiouny M, Macdonald RL, Weir B, Johns LM (1996) Effect of BQ-123 and tissue plasminogen activator on vasospasm after subarachnoid hemorrhage in monkeys. Stroke 27:1629–1633

Kita T, Kubo K, Hiramatsu K, Sakaki T, Yonetani Y, Sato S, Fujimoto M, Nakashima T (1998) Profiles of an intravenously available endothelin A-receptor antagonist, S-0139, for preventing cerebral vasospasm in a canine two- hemorrhage model. Life Sci 63:305–315

Kitazono T, Heistad DD, Faraci FM (1995a) Dilatation of the basilar artery in response to selective activation of endothelin B receptors in vivo. J Pharmacol Exp Ther 273:1–6

Kitazono T, Heistad DD, Faraci FM (1995b) Enhanced response of the basilar artery to activation of endothelin B receptors in stroke-prone spontaneously hypertensive rats. Hypertension 25:490–494

Kobari M, Fukuuchi Y, Tomita M, Tanahashi N, Konno S, Takeda H (1994a) Constriction/dilatation of the cerebral microvessels by intravascular endothelin-1 in cats. J Cereb Blood Flow Metab 14:64–69

Kobari M, Fukuuchi Y, Tomita M, Tanahashi N, Konno S, Takeda H (1994b) Dilatation of cerebral microvessels mediated by endothelin ETB receptor and nitric oxide in cats. Neurosci Lett 176:157–160

Kobayashi H, Hayashi M, Kobayashi S, Kabuto M, Handa Y, Kawano H (1990) Effect of endothelin on the canine basilar artery. Neurosurgery 27:357–361

Kobayashi H, Hayashi M, Kobayashi S, Kabuto M, Handa Y, Kawano H, Ide H (1991) Cerebral vasospasm and vasoconstriction caused by endothelin. Neurosurgery 28:673–678

Koedel U, Lorenzl S, Gorriz C, Arendt RM, Pfister HW (1998) Endothelin B receptor-mediated increase of cerebral blood flow in experimental pneumococcal meningitis. J Cereb Blood Flow Metab 18:67–74

Kohzuki M, Chai SY, Paxinos G, Karavas A, Casley DJ, Johnston CI, Mendelsohn FA (1991) Localization and characterization of endothelin receptor binding sites in the rat brain visualized by in vitro autoradiography. Neuroscience 42:245–260

Koseki C, Imai M, Hirata Y, Yanagisawa M, Masaki T (1989) Autoradiographic localization of [125I]-endothelin-1 binding sites in rat brain. Neurosci Res 6:581–585

Kourembanas S, Marsden PA, McQuillan LP, Faller DV (1991) Hypoxia induces endothelin gene expression and secretion in cultured human endothelium. J Clin Invest 88:1054–1057

Kurihara H, Yoshizumi M, Sugiyama T, Takaku F, Yanagisawa M, Masaki T, Hamaoki M, Kato H, Yazaki Y (1989) Transforming growth factor-beta stimulates the expression of endothelin mRNA by vascular endothelial cells. Biochem Biophys Res Commun 159:1435–1440

Kwan AL, Bavbek M, Jeng AY, Maniara W, Toyoda T, Lappe RW, Kassell NF, Lee KS (1997) Prevention and reversal of cerebral vasospasm by an endothelin-converting enzyme inhibitor, CGS 26303, in an experimental model of subarachnoid hemorrhage. J Neurosurg 87:281–286

Ladenheim RG, Lacroix I, Foignant-Chaverot N, Strosberg AD, Couraud PO (1993) Endothelins stimulate c-fos and nerve growth factor expression in astrocytes and astrocytoma. J Neurochem 60:260–266

Lampl Y, Fleminger G, Gilad R, Galron R, Sarova-Pinhas I, Sokolovsky M (1997) Endothelin in cerebrospinal fluid and plasma of patients in the early stage of ischemic stroke. Stroke 28:1951–1955

Lee ME, de la Monte SM, Ng SC, Bloch KD, Quertermous T (1990) Expression of the potent vasoconstrictor endothelin in the human central nervous system. J Clin Invest 86:141–147

Lerman A, Sandok EK, Hildebrand FLJ, Burnett JCJ (1992) Inhibition of endothelium-derived relaxing factor enhances endothelin- mediated vasoconstriction. Circulation 85:1894–1898

Levin ER (1995) Endothelins, N Engl J Med 333:356–363

Li H, Chen SJ, Chen YF, Meng QC, Durand J, Oparil S, Elton TS (1994) Enhanced endothelin-1 and endothelin receptor gene expression in chronic hypoxia. J Appl Physiol 77:1451–1459

Lin WW, Lee CY, Chuang DM (1989) Cross-desensitization of endothelin- and sarafotoxin-induced phosphoinositide turnover in neurons. Eur J Pharmacol 166:581–582

Luscher TF, Yang Z, Tschudi M, von Segesser L, Stulz P, Boulanger C, Siebenmann R, Turina M, Buhler FR (1990) Interaction between endothelin-1 and endothelium-derived relaxing factor in human arteries and veins. Circ Res 66:1088–1094

Lustig HS, Chan J, Greenberg DA (1992) Comparative neurotoxic potential of glutamate, endothelins, and platelet-activating factor in cerebral cortical cultures. Neurosci Lett 139:15–18

Lysko PG, Elshourbagy NA, Pullen M, Nambi P (1995) Developmental expression of endothelin receptors in cerebellar neurons differentiating in culture. Brain Res Dev Brain Res 88:96–101

MacCumber MW, Ross CA, Glaser BM, Snyder SH (1989) Endothelin: visualization of mRNAs by in situ hybridization provides evidence for local action. Proc Natl Acad Sci U.S.A. 86:7285–7289

MacCumber MW, Ross CA, Snyder SH (1990) Endothelin in brain: receptors, mitogenesis, and biosynthesis in glial cells. Proc Natl Acad Sci U.S.A. 87:2359–2363

Macdonald RL, Weir BKA (1991) A review of hemoglobin and the pathogenesis of cerebral vasospasm. Stroke 22:971–982

Macrae I, Robinson M, McAuley M, Reid J, McCulloch J (1991) Effects of intracisternal endothelin-1 injection on blood flow to the lower brain stem. Eur J Pharmacol 203:85–91

Macrae IM, Robinson MJ, Graham DI, Reid JL, McCulloch J (1993) Endothelin-1-induced reductions in cerebral blood flow: dose dependency, time course, and neuropathological consequences. J Cereb Blood Flow Metab 13:276–284

Masaki T, Vane JR, Vanhoutte PM (1994) International union of pharmacology nomenclature of endothelin receptors. Pharmacol Rev 46:137–142

Masaoka H, Suzuki R, Hirata Y, Emori T, Marumo F, Hirakawa K (1989) Raised plasma endothelin in aneurysmal subarachnoid haemorrhage [letter]. Lancet 2:1402

Matsumoto H, Suzuki N, Onda H, Fujino M (1989) Abundance of endothelin-3 in rat intestine, pituitary gland and brain. Biochem Biophys Res Commun 164:74–80

Matsumura Y, Ikegawa R, Suzuki Y, Takaoka M, Uchida T, Kido H, Shinyama H, Hayashi K, Watanabe M, Morimoto S (1991) Phosphoramidon prevents cerebral vasospasm following subarachnoid hemorrhage in dogs: the relationship to endothelin-1 levels in the cerebrospinal fluid. Life Sci 49:841–848

Mayberg MR (1998) Cerebral vasospasm. Neurosurg Clin N Am 9:615–627

McAuley MA, Breu V, Graham DI, McCulloch J (1996) The effects of bosentan on cerebral blood flow and histopathology following middle cerebral artery occlusion in the rat. Eur J Pharmacol 307:171–181

McCarron RM, Wang L, Stanimirovic DB, Spatz M (1993) Endothelin induction of adhesion molecule expression on human brain microvascular endothelial cells. Neurosci Lett 156:31–34

McMurdo L, Corder R, Thiemermann C, Vane JR (1993) Incomplete inhibition of the pressor effects of endothelin-1 and related peptides in the anaesthetized rat with BQ-123 provides evidence for more than one vasoconstrictor receptor. Br J Pharmacol 108:557–561

Mima T, Yanagisawa M, Shigeno T, Saito A, Goto K, Takakura K, Masaki T (1989) Endothelin acts in feline and canine cerebral arteries from the adventitial side. Stroke 20:1553–1556

Mizuguchi T, Nishiyama M, Moroi K, Tanaka H, Saito T, Masuda Y, Masaki T, de Wit D, Yanagisawa M, Kimura S (1997) Analysis of two pharmacologically predicted endothelin B receptor subtypes by using the endothelin B receptor gene knockout mouse. Br J Pharmacol 120:1427–1430

Nakagomi T, Kassell NF, Sasaki T, Fujiwara S, Lehman RM, Torner JC (1987) Impairment of endothelium-dependent vasodilation induced by acetylcholine and adenosine triphosphate following experimental subarachnoid hemorrhage. Stroke 18:482–489

Nambi P, Pullen M, Feuerstein G (1990) Identification of endothelin receptors in various regions of rat brain. Neuropeptides 16:195–199

Nikolov R, Rami A, Krieglstein J (1993) Endothelin-1 exacerbates focal cerebral ischemia without exerting neurotoxic action in vitro. Eur J Pharmacol 248:205–208

Nirei H, Hamada K, Shoubo M, Sogabe K, Notsu Y, Ono T (1993) An endothelin ETA receptor antagonist, FR139317, ameliorates cerebral vasospasm in dogs. Life Sci 52:1869–1874

Niwa M, Kawaguchi T, Yamashita K, Maeda T, Kurihara M, Kataoka Y, Ozaki M (1991) Specific 125I-endothelin-1 binding sites in the central nervous system. Clin Exp Hypertens 13:799–806

Obrenovitch TP (1995) The ischemic penumbra: twenty years on. Cerebrovasc Brain Metab Rev 7:297–323

Ogura K, Takayasu M, Dacey RGJ (1991) Differential effects of intra- and extraluminal endothelin on cerebral arterioles. Am J Physiol 261:H531–H537

Ohlstein EH, Elliott JD, Feuerstein GZ, Ruffolo RRJ (1996) Endothelin receptors: receptor classification, novel receptor antagonists, and potential therapeutic targets. Med Res Rev 16:365–390

Ohlstein EH, Nambi P, Douglas SA, Edwards RM, Gellai M, Lago A, Leber JD, Cousins RD, Gao A, Frazee JS, Peishoff CE, Bean JW, Eggleston DS, Elshoubragy NA, Kumar C, Lee JA, Yue TL, Louden C, Brooks DP, Weinstock J, Feuerstein G, Poste G, Ruffolo RR, Gleason JG, Elliott JD (1994) SB 209670, a rationally designed potent nonpeptide endothelin receptor antagonist. Proc Natl Acad Sci USA 91:8052–8056

Ohlstein EH, Storer BL (1992) Oxyhemoglobin stimulation of endothelin production in cultured endothelial cells. J Neurosurg 77:274–278

Onoue H, Kaito N, Akiyama M, Tomii M, Tokudome S, Abe T (1995) Altered reactivity of human cerebral arteries after subarachnoid hemorrhage. J Neurosurg 83:510–515

Pardridge WM (1997) Drug delivery to the brain. J Cereb Blood Flow Metab 17:713–731

Pasqualin A (1998) Epidemiology and pathophysiology of cerebral vasospasm following subarachnoid hemorrhage. J Neurosurg Sci 42:15–21

Patel TR (1996) Therapeutic potential of endothelin receptor antagonists in cerebrovascular disease. CNS drugs 5:293–310

Patel TR, Galbraith S, Graham DI, Hallak H, Doherty AM, McCulloch J (1996a) Endothelin receptor antagonist increases cerebral perfusion and reduces ischaemic damage in feline focal cerebral ischaemia. J Cereb Blood Flow Metab 16:950–958

Patel TR, Galbraith S, McAuley MA, McCulloch J (1996b) Endothelin-mediated vascular tone following focal cerebral ischaemia in the cat. J Cereb Blood Flow Metab 16:679–687

Patel TR, McAuley MA, McCulloch J (1994) Effects on feline pial arterioles in situ of bosentan, a non-peptide endothelin receptor antagonist. Eur J Pharmacol 260:65–71

Patel TR, McAuley MA, McCulloch J (1996c) Endothelin receptor mediated constriction and dilatation in feline cerebral resistance arterioles in vivo. Eur J Pharmacol 307:41–48

Patel TR, McCulloch J (1996) Failure of an endothelin antagonist to modify hypoperfusion after transient global ischaemia in the rat. J Cereb Blood Flow Metab 16:490–499

Peerless SJ, Kassel NF, Komatsu K, Hunter IG (1980) Cerebral vasospasm: acute proliferative vasculopathy? II. Morphology. In Wilkins RH (ed) Cerebral arterial vasospasm. Baltimore, Williams & Wilkins, pp 88–96

Phillis JW (1994) A "radical" view of cerebral ischemic injury. Prog Neurobiol 42:441–448

Pluta RM, Afshar JK, Boock RJ, Oldfield EH (1998) Temporal changes in perivascular concentrations of oxyhemoglobin, deoxyhemoglobin, and methemoglobin after subarachnoid hemorrhage. J Neurosurg 88:557–561

Pluta RM, Boock RJ, Afshar JK, Clouse K, Bacic M, Ehrenreich H, Oldfield EH (1997) Source and cause of endothelin-1 release into cerebrospinal fluid after subarachnoid hemorrhage. J Neurosurg 87:287–293

Pluta RM, Thompson BG, Dawson TM, Snyder SH, Boock RJ, Oldfield EH (1996) Loss of nitric oxide synthase immunoreactivity in cerebral vasospasm. J Neurosurg 84:648–654

Robinson MJ, McCulloch J (1990) Contractile responses to endothelin in feline cortical vessels in situ. J Cereb Blood Flow Metab 10:285–289

Rogers SD, Demaster E, Catton M, Ghilardi JR, Levin LA, Maggio JE, Mantyh PW (1997) Expression of endothelin-B receptors by glia in vivo is increased after CNS injury in rats, rabbits, and humans. Exp Neurol 145:180–195

Roux S, Loffler BM, Gray GA, Sprecher U, Clozel M, Clozel JP (1995) The role of endothelin in experimental cerebral vasospasm. Neurosurgery 37:78–85

Rubanyi GM, Polokoff MA (1994) Endothelins: molecular biology, biochemistry, pharmacology, physiology, and pathophysiology. Pharmacol Rev 46:325–415

Sacco RL (1997) Risk factors, outcomes, and stroke subtypes for ischemic stroke. Neurology 49:S39–S44

Saito A, Shiba R, Kimura S, Yanagisawa M, Goto K, Masaki T (1989) Vasoconstrictor response of large cerebral arteries of cats to endothelin, an endothelium-derived vasoactive peptide. Eur J Pharmacol 162:353–358

Sakurai T, Yanagisawa M, Takuwa Y, Miyazaki H, Kimura S, Goto K, Masaki T (1990) Cloning of a cDNA encoding a non-isopeptide-selective subtype of the endothelin receptor. Nature 348:732–735

Salom JB, Torregrosa G, Alborch E (1995) Endothelins and the cerebral circulation. Cerebrovasc Brain Metab Rev 7:131–152

Sasaki Y, Takimoto M, Oda K, Fruh T, Takai M, Okada T, Hori S (1997) Endothelin evokes efflux of glutamate in cultures of rat astrocytes. J Neurochem 68:2194–2200

Schilling L, Feger GI, Ehrenreich H, Wahl M (1995) Endothelin-3-induced relaxation of isolated rat basilar artery is mediated by an endothelial ETB-type endothelin receptor. J Cereb Blood Flow Metab 15:699–705

Seifert V, Loffler BM, Zimmermann M, Roux S, Stolke D (1995) Endothelin concentrations in patients with aneurysmal subarachnoid hemorrhage, Correlation with cerebral vasospasm, delayed ischemic neurological deficits, and volume of hematoma. J Neurosurg 82:55–62

Seo B, Oemar BS, Siebenmann R, von Segesser L, Luscher TF (1994) Both ETA and ETB receptors mediate contraction to endothelin-1 in human blood vessels. Circulation 89:1203–1208

Sharkey J, Ritchie IM, Kelly PA (1993) Perivascular microapplication of endothelin-1: a new model of focal cerebral ischaemia in the rat. J Cereb Blood Flow Metab 13:865–871

Shigeno T, Clozel M, Sakai S, Saito A, Goto K (1995) The effect of bosentan, a new potent endothelin receptor antagonist, on the pathogenesis of cerebral vasospasm, Neurosurgery 37:87–90

Shigeno T, Mima T, Yanagisawa M, Saito A, Goto K, Yamashita K, Takenouchi T, Matsuura N, Yamasaki Y, Yamada K (1991) Prevention of cerebral vasospasm by actinomycin D. J Neurosurg 74:940–943

Shirakami G, Magaribuchi T, Shingu K, Kim S, Saito Y, Nakao K, Mori K (1994) Changes of endothelin concentration in cerebrospinal fluid and plasma of patients with aneurysmal subarachnoid hemorrhage. Acta Anaesthesiol Scand 38:457–461

Stojilkovic SS, Catt KJ (1992) Neuroendocrine actions of endothelins. Trends Pharmacol Sci 13:385–391

Sullivan AM, Morton AJ (1996) Endothelins induce Fos expression in neurons and glia in organotypic cultures of rat cerebellum. J Neurochem 67:1409–1418

Sumner MJ, Cannon TR, Mundin JW, White DG, Watts IS (1992) Endothelin ETA and ETB receptors mediate vascular smooth muscle contraction. Br J Pharmacol 107:858–860

Suzuki H, Sato S, Suzuki Y, Takekoshi K, Ishihara N, Shimoda S (1990) Increased endothelin concentration in CSF from patients with subarachnoid hemorrhage. Acta Neurol Scand 81:553–554

Suzuki R, Masaoka H, Hirata Y, Marumo F, Isotani E, Hirakawa K (1992) The role of endothelin-1 in the origin of cerebral vasospasm in patients with aneurysmal subarachnoid hemorrhage. J Neurosurg 77:96–100

Tabernero A, Giaume C, Medina JM (1996) Endothelin-1 regulates glucose utilization in cultured astrocytes by controlling intercellular communication through gap junctions. Glia 16:187–195

Takasago T, McCulloch J (1997) Endothelin-A receptor blockade reduces infarction after focal cerebral ischaemia in the rat. J Cereb Blood Flow Metab: S661.

Tatlisumak T, Carano RA, Takano K, Opgenorth TJ, Sotak CH, Fisher M (1998a) A novel endothelin antagonist, A-127722, attenuates ischemic lesion size in rats with temporary middle cerebral artery occlusion: a diffusion and perfusion MRI study. Stroke 29:850–857

Thomas JE, Nemirovsky A, Zelman V, Giannotta SL (1997) Rapid reversal of endothelin-1-induced cerebral vasoconstriction by intrathecal administration of nitric oxide donors. Neurosurgery 40:1245–1249

Thorin E, Nguyen TD, Bouthillier A (1998) Control of vascular tone by endogenous endothelin-1 in human pial arteries. Stroke 29:175–180

Toda N, Shimizu K, Ohta T (1980) Mechanism of cerebral arterial contraction induced by blood constituents. J Neurosurg 53:312–322

Touzani O, Galbraith S, Siegl P, McCulloch J (1997) Endothelin-B receptors in cerebral resistance arterioles and their functional significance after focal cerebral ischemia in cats. J Cereb Blood Flow Metab 17:1157–1165

Vigne P, Marsault R, Breittmayer JP, Frelin C (1990) Endothelin stimulates phosphatidylinositol hydrolysis and DNA synthesis in brain capillary endothelial cells. Biochem J 266:415–420

Viossat I, Duverger D, Chapelat M, Pirotzky E, Chabrier PE, Braquet P (1993) Elevated tissue endothelin content during focal cerebral ischemia in the rat. J Cardiovasc Pharmacol 22 Suppl 8: S306–S309

Wanebo JE, Arthur AS, Louis HG, West K, Kassell NF, Lee KS, Helm GA (1998) Systemic administration of the endothelin-A receptor antagonist TBC 11251 attenuates cerebral vasospasm after experimental subarachnoid hemorrhage: dose study and review of endothelin-based therapies in the literature on cerebral vasospasm. Neurosurgery 43:1409–1417

Warner TD, Allcock GH, Corder R, Vane JR (1993) Use of the endothelin antagonists BQ-123 and PD 142893 to reveal three endothelin receptors mediating smooth muscle contraction and the release of EDRF. Br J Pharmacol 110:777–782

Webb DJ, Monge JC, Rabelink TJ, Yanagisawa M (1998) Endothelin: new discoveries and rapid progress in the clinic. Trends Pharmacol Sci 19:5–8

Webber KM, Pennefather JN, Head GA, van den Buuse M (1998) Endothelin induces dopamine release from rat striatum via endothelin-B receptors. Neuroscience 86:1173–1180

Wei GZ, Zhang J, Sheng SL, Ai HX, Ma JC, Lui HB (1993) Increased plasma endothelin-1 concentration in patients with acute cerebral infarction and actions of endothelin-1 on pial arterioles of rat. Chin Med J (Engl) 106:917–921

White LR, Leseth KH, Juul R, Adner M, Cappelen J, Aasly J, Edvinsson L (1998) Increased endothelin ETB contractile activity in cultured segments of human temporal artery. Acta Physiol Scand 164:21–27

Willette RN, Ohlstein EH, Pullen M, Sauermelch CF, Cohen A, Nambi P (1993) Transient forebrain ischemia alters acutely endothelin receptor density and immunoreactivity in gerbil brain. Life Sci 52:35–40

Willette RN, Zhang H, Mitchell MP, Sauermelch CF, Ohlstein EH, Sulpizio AC (1994) Nonpeptide endothelin antagonist. Cerebrovascular characterization and effects on delayed cerebral vasospasm. Stroke 25:2450–2455

Yamada G, Hama H, Kasuya Y, Masaki T, Goto K (1995) Possible sources of endothelin-1 in damaged rat brain. J Cardiovasc Pharmacol 26 Suppl 3:S486–S490

Yamaji T, Johshita H, Ishibashi M, Takaku F, Ohno H, Suzuki N, Matsumoto H, Fujino M (1990) Endothelin family in human plasma and cerebrospinal fluid. J Clin Endocrinol Metab 71:1611–1615

Yamashita K, Kataoka Y, Niwa M, Shigematsu K, Himeno A, Koizumi S, Taniyama K (1993) Increased production of endothelins in the hippocampus of stroke-prone spontaneously hypertensive rats following transient forebrain ischemia: histochemical evidence. Cell Mol Neurobiol 13:15–23

Yamashita K, Niwa M, Kataoka Y, Shigematsu K, Himeno A, Tsutsumi K, Nakano-Nakashima M, Sakurai-Yamashita Y, Shibata S, Taniyama K (1994) Microglia with an endothelin ETB receptor aggregate in rat hippocampus CA1 subfields following transient forebrain ischemia. J Neurochem 63:1042–1051

Yamaura I, Tani E, Maeda Y, Minami N, Shindo H (1992) Endothelin-1 of canine basilar artery in vasospasm. J Neurosurg 76:99–105

Yanagisawa M, Kurihara H, Kimura S, Tomobe Y, Kobayashi M, Mitsui Y, Yazaki Y, Goto K, Masaki T (1988) A novel potent vasoconstrictor peptide produced by vascular endothelial cells. Nature 332:411–415

Yang ZH, Richard V, von Segesser L, Bauer E, Stulz P, Turina M, Luscher TF (1990) Threshold concentrations of endothelin-1 potentiate contractions to norepinephrine and serotonin in human arteries. A new mechanism of vasospasm? Circulation 82:188–195

Yasuma Y, McCarron RM, Strasser A, Spatz M (1997) The role of endothelin receptors in postischemic cerebral perfusion, Stroke 28:138 [Abstract]

Yoshimoto S, Ishizaki Y, Kurihara H, Sasaki T, Yoshizumi M, Yanagisawa M, Yazaki Y, Masaki T, Takakura K, Murota S (1990) Cerebral microvessel endothelium is producing endothelin. Brain Res 508:283–285

Yoshimoto S, Ishizaki Y, Mori A, Sasaki T, Takakura K, Murota S (1991) The role of cerebral microvessel endothelium in regulation of cerebral blood flow through production of endothelin-1. J Cardiovasc Pharmacol 17 Suppl 7: S260–S263

Yoshizawa T, Shinmi O, Giaid A, Yanagisawa M, Gibson SJ, Kimura S, Uchiyama Y, Polak JM, Masaki T, Kanazawa I (1990) Endothelin: a novel peptide in the posterior pituitary system. Science 247:462–464

Zhang W, Sakai N, Yamada H, Fu T, Nozawa Y (1990) Endothelin-1 induces intracellular calcium rise and inositol 1,4,5- trisphosphate formation in cultured rat and human glioma cells. Neurosci Lett 112:199–204

Zimmermann M, Seifert V (1998) Endothelin and subarachnoid hemorrhage: an overview. Neurosurgery 43:863–875

Zimmermann M, Seifert V, Loffler BM, Stolke D, Stenzel W (1996) Prevention of cerebral vasospasm after experimental subarachnoid hemorrhage by RO 47-0203, a newly developed orally active endothelin receptor antagonist. Neurosurgery 38:115–120

Ziv I, Fleminger G, Djaldetti R, Achiron A, Melamed E, Sokolovsky M (1992) Increased plasma endothelin-1 in acute ischemic stroke. Stroke 23:1014–1016

Zuccarello M, Boccaletti R, Rapoport RM (1998a) Endothelin ET(B) receptor-mediated constriction in the rabbit basilar artery. Eur J Pharmacol 350:R7–R9

Zuccarello M, Boccaletti R, Rapoport RM (1998b) Endothelin ET(B1) receptor-mediated relaxation of rabbit basilar artery. Eur J Pharmacol 357:67–71

Zuccarello M, Boccaletti R, Romano A, Rapoport RM (1998c) Endothelin B receptor antagonists attenuate subarachnoid hemorrhage-induced cerebral vasospasm. Stroke 29:1924–1929

Zuccarello M, Lewis AI, Rapoport RM (1994) Endothelin ETA and ETB receptors in subarachnoid hemorrhage-induced cerebral vasospasm. Eur J Pharmacol 259:R1–R2

Zuccarello M, Romano A, Passalacqua M, Rapoport RM (1995) Decreased endothelium-dependent relaxation in subarachnoid hemorrhage- induced vasospasm: role of ET-1. Am J Physiol 269:H1009–H1015

Zuccarello M, Soattin GB, Lewis AI, Breu V, Hallak H, Rapoport RM (1996) Prevention of subarachnoid hemorrhage-induced cerebral vasospasm by oral administration of endothelin receptor antagonists. J Neurosurg 84:503–507

CHAPTER 14
Involvement of the Endothelins in Airway Reactivity and Disease

R.G. GOLDIE and P.J. HENRY

A. Introduction

In 1988, a previously uncharacterized endothelium-derived contractile factor was isolated, purified and identified as a novel 21 amino acid sequence and named endothelin-1 (ET-1) (YANAGISAWA et al. 1988; INOUE et al. 1989). Some of the pharmacological activity of ET-1 was also reported in these studies, although there soon followed an avalanche of published research data from other laboratories describing the biology of this peptide in great detail in several mammalian systems, with particular emphasis on its spasmogenic actions in vascular tissues. In addition however, the potent contractile effects of ET-1 in airway smooth muscle were also reported in 1988 and again in 1989 (TURNER et al. 1989; UCHIDA et al. 1988), predictably, followed rapidly by evidence for high densities of ET receptors in airway smooth muscle (TURNER et al. 1989; POWER et al. 1989). Autoradiographic analyses in human and animal airway tissues established the presence of significant numbers of such receptors in several airway wall cell types in addition to airway and vascular smooth muscle (HENRY et al. 1990; GOLDIE et al. 1995). Taken together, this information constituted a reasonable basis for speculating that the actions of ET-1 might be associated with obstructive airway diseases such as asthma (HAY et al. 1993a, 1996; GOLDIE et al. 1996c), as well as with pulmonary hypertension (ALLEN et al. 1993; STELZNER et al. 1992; STEWART et al. 1991; FOLKERTS et al. 1998). (The involvement of ETs in pulmonary hypertension will be dealt with in detail in Chap. 15.) Since then, there has been a constant stream of published reports demonstrating that ET-1 can mimic many of the features of asthma in addition to its powerful spasmogenic activity in airway smooth muscle, all of which add weight to the concept of a mediator role for ET-1 in this disease (GOLDIE et al. 1996c). Evidence continues to emerge implicating the ETs in this and other lung pathologies, including degenerative fibrotic diseases such as fibrosing alveolitis (SALEH et al. 1995; SEINO et al. 1995; MUTSAERS et al. 1998).

In this review, we will evaluate much of the evidence implicating the ETs in respiratory diseases, with particular emphasis on asthma for which a strong, but still circumstantial case can be made.

B. The Endothelin System

I. Is There a Link to Asthma?

Over the years, many substances have been proposed as mediators in asthma. However, only some of these, such as the cysteinyl leukotrienes, have been confirmed as significant players in this disease after years of rigorous evaluation. Similarly, the theory that ET-1 (and/or related endogenous peptides) is a significant mediator in asthma will only receive universal acceptance after various standard criteria are fulfilled. First, ET-1 must induce actions in the respiratory tract that mimic most if not all of the features and symptoms of this disease. Second, relevant receptors must be present and actively involved in mediating relevant cellular responses to these peptides in the airways. A true asthma mediator must be an endogenous substance, synthesized, released and degraded at appropriate sites in the lung. Furthermore, the levels of the mediator must be elevated in asthma, with a positive correlation existing between these levels and disease symptom severity. Finally, ET receptor antagonists or inhibitors of ET synthesis should relieve asthma symptoms and thus be of at least potential therapeutic benefit. Before exploring these aspects further, it is important to outline briefly some of the fundamental features of the endothelin system in the airways as far as they are presently understood.

II. ET Structure, Synthesis and Degradation

1. Structure

ET-1 is one of a family of 21 amino acid endogenous mammalian peptides (ET-1, ET-2 and ET-3), each of which have similar sequences. In each sequence, two disulfide bridges spanning positions 1, 15 and 3, 11 constrain their structures as seen in Fig. 1. The sarafotoxins, which are spasmogenic components of the venom of the Middle Eastern burrowing asp, *Atractaspis engaddensis* (MASAKI et al. 1992) are also 21 amino acid sequences with similar structural characteristics and sequences to the ETs (see Chap. 2). It is perhaps not surprising then that the sarafotoxins also evoke contraction of vascular smooth muscle. Recently, a 31-amino acid ET-1 sequence was also identified which was derived from prepro-ET-1 via the action of mast cell chymase (NAKANO et al. 1997) (Fig. 1). This peptide is also a directly acting spasmogen in both vascular and airway smooth muscle, i.e. biological activity is not dependent upon cleavage to the 21 amino acid sequence, although this conversion can occur (YOSHIZUMI et al. 1998a, b; KISHI et al. 1998).

2. Synthesis

The formation of the ETs is preceded by the synthesis of 212 amino acid precursors known as prepro-ETs, e.g. prepro-ET-1 (YANAGISAWA et al. 1988; see Chaps. 3, 7). These precursors for ET-1, ET-2 and ET-3 are encoded by genes found on chromosomes 6, 1 and 20, respectively (INOUE et al. 1989; BLOCH et

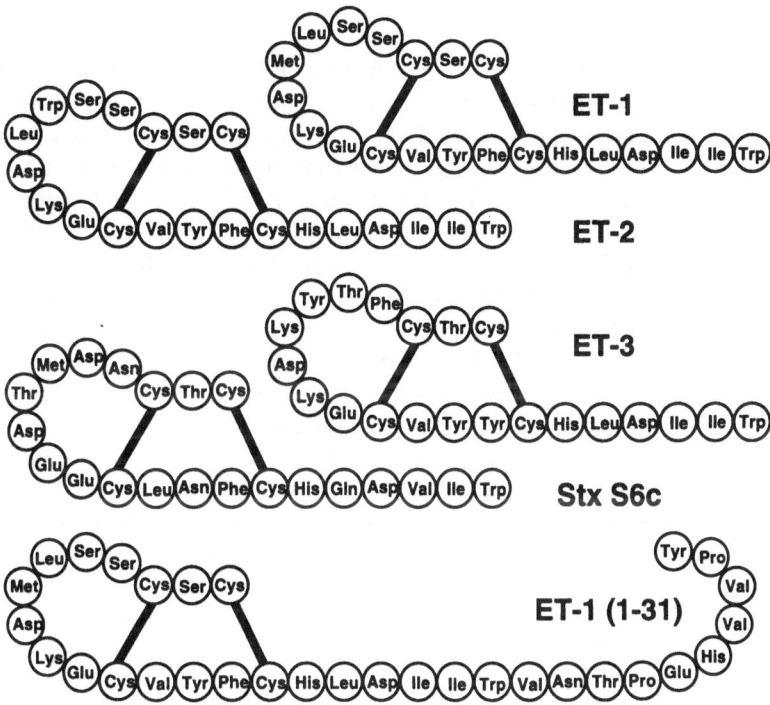

Fig. 1. Diagrammatic representation of the chemical structures of ET-1 and related peptides

al. 1989a, b, 1991) and are subsequently cleaved via dibasic amino acid residue-specific endopeptidases or by the mammalian convertase known as furin (DENAULT et al. 1995), to provide 38 amino acid residues called big ETs (INOUE et al. 1989; ITOH et al. 1988). The big ETs are not ET receptor agonists and must be cleaved to the receptor-activating 21 amino acid sequences (YANAGISAWA and MASAKI 1989; MASAKI et al. 1992; OPGENORTH et al. 1992) via ET converting enzyme (ECE) (XU et al. 1994). This enzyme is a phosphoramidon-sensitive, membrane-bound, neutral metalloprotease (EMOTO and YANAGISAWA 1995; OPGENORTH et al. 1992; VEMULAPALLI et al. 1992).

3. Degradation

Neutral endopeptidase (NEP) is found in abundance in the airways (JOHNSON et al. 1985) and the ETs have high affinity for and are actively metabolized by this enzyme (VIJAYARAGHAVAN et al. 1990; FAGNY et al. 1991). Interestingly, in activated human polymorphonuclear neutrophils, cathepsin G, rather than NEP, may be the enzyme responsible for eliminating ET-1 (FAGNY et al. 1992).

C. ET Receptors

The detection by molecular cloning of just two ET receptor subtypes in mammalian cells, designated ET_A and ET_B (ARAI et al. 1990; SAKURAI et al. 1990; SAKAMOTO et al. 1991; ADACHI et al. 1991; see Chap. 4) is entirely consistent with data derived using other approaches (SAKURAI et al. 1992; MASUDA et al. 1989; TAKAYANAGI et al. 1991) in mammalian systems including the lung (HICK et al. 1995; KONDOH et al. 1991; KOZUKA et al. 1991; HAGIWARA et al. 1992). However, in amphibian cells, an ET_C receptor has been cloned and may be functional (KARNE et al. 1993). In recent years, several studies involving vascular tissues have provided evidence for the existence of distinct subtypes within both the ET_A and ET_B receptor families (SOKOLOVSKY et al. 1992; WARNER et al. 1993). This possibility has also been raised for ET_B receptors in human bronchus, with the reporting of apparently anomalous contractile responsiveness to ET_B receptor agonists (HAY et al. 1998). Despite such functional evidence, the existence of genetic codes for such receptors has not been confirmed. Accordingly, the existence of these novel receptor subtypes must for the moment remain uncertain (BAX and SAXENA 1994). Indeed, some of the functional data for novel ET receptor subtypes may be explained in terms of differences in the kinetics of ligand interactions with ET receptors (DEVADASON and HENRY 1997).

D. ET and the Major Pathologies in Respiratory Diseases

I. Asthma

Asthma is recognized as a chronic inflammatory lung disease (BARNES et al. 1988), involving several pathologies (KAY 1991) including airway hyperreactivity (BOUSHEY et al. 1980). It is also clear that asthma is an obstructive airway disease and that a significant component of the obstruction is caused by increased airway smooth muscle tone (JAMES et al. 1989). However, occlusion of the bronchi is also a result of the hypersecretion of mucus together with reduced clearance of mucus from the airways (BEASLEY et al. 1989). Epithelial cell damage and desquamation and the addition of epithelial debris and inflammatory cells to luminal mucus further reduces the patency of the airways (NAYLOR 1962; LAITINEN et al. 1985; BEASLEY et al. 1989). Finally, airway wall restructuring as evidenced by submucosal oedema, airway smooth muscle and mucous gland hyperplasia (HEARD and HOSSAIN 1973; CARROLL et al. 1993; KNOX 1994; ROCHE 1998) and sub-epithelial fibrosis (BREWSTER et al. 1990; ROCHE et al. 1989) accompanying chronic airway inflammation further reduce lumen diameter and elevate bronchial resistance to airflow. Importantly, ET-1 has actions within the bronchial wall which mimic and potentially reproduce many of these pathologies and symptoms and these issues will be discussed in detail in later sections. It is also important to realize that treatment with anti-inflammatory glucocorticoids (TRIGG et al. 1994) and the removal of provoking stimuli, e.g. diisocyanates (SAETTA et al. 1995) has been linked to reversal of such structural changes to the bronchial wall.

II. Allergic Rhinitis

The human nasal mucosa contains the mRNA for prepro-ET-1 and expresses immunoreactive ET-1 (ir-ET-1) (MULLOL et al. 1993; CASASCO et al. 1993). Specific binding sites for ET-1 that are presumably ET receptors were found in nasal submucosal glands, venous sinusoids and small muscular arterioles and ET-1-induced stimulation of these sites in vitro caused serous and mucous cell secretions (MULLOL et al. 1993) and induced prostanoid release WU et al. 1992). Riccio and co-workers conducted the first study describing the effects of intranasal ET-1 in human rhinitics and healthy volunteers (RICCIO et al. 1995). This study demonstrated mucosal hyperresponsiveness to ET-1 in rhinitics, since sneezing frequency and the amounts of nasal secretions were increased. It has subsequently been shown that the levels of mRNAs for prepro-ET-1 and ECE were significantly increased in chronic rhinitis (FURUKAWA et al. 1996). The possibility of a mediator role for ET-1 in rhinitis requires further investigation.

III. Adult Respiratory Distress Syndrome (ARDS)

Lung injury resulting in compromised pulmonary gas exchange in ARDS is most often the result of sepsis, but may have other causes. Elevated circulating ir-ET levels have been linked to deterioration in ARDS and clinical improvement was associated with significant falls in these levels (DRUML et al. 1993; LANGLEBEN et al. 1993). Animal studies suggest that the release of ir-ET in this condition may reflect the influence of endotoxin released in sepsis (WEITZBERG et al. 1991). In rat models of respiratory distress, abnormal blood gas levels and pulmonary oedema are seen, together with increases in ir-ET levels in bronchoalveolar lavage (BAL) fluid (HERBST et al. 1995; SIMMET et al. 1992). Importantly, the abnormal blood gases and oedema were partly corrected in the presence of ET_A receptor antagonist BQ-123, suggesting the involvement of ET-1 in this model (HERBST et al. 1995).

IV. Cryptogenic Fibrosing Alveolitis (CFA)

As the name suggests, CFA is characterized by peripheral lung fibrosis involving fibroblast proliferation and collagen deposition. However, CFA is also linked to lung inflammation and type II pneumocyte proliferation. The production of ir-ET was up-regulated in these cells, and in airway epithelium (GIAID et al. 1993). Importantly, the extent of type II cell proliferation was closely correlated with the levels of ir-ET. Morphological changes were also detected in pulmonary vessels in which the endothelium over-expressed ir-ET (GIAID et al. 1993).

V. Pulmonary Fibrosis

Pulmonary fibrosis can be induced in the rat (MUTSAERS et al. 1998) and hamster by pretreatment with bleomycin (SEINO et al. 1995). In the rat, intratracheal bleomycin caused a significant elevation in ir-ET which preceded the increase in lung collagen (MUTSAERS et al. 1998). In the hamster, the ET_A receptor-selective antagonist BQ-123, attenuated bleomycin-induced alveolar fibrosis and the accompanying right ventricular hypertrophy. These data suggest that endogenous ET-1 released in this condition was mitogenic in pulmonary fibroblasts, indicating a significant link to the disease process in these models and raising the possibility of a role in the pathogenesis of human pulmonary fibrosis. Consistent with this, increased levels of ir-ET and ir-ECE-1 have been co-localized in several cell types, including proliferating type II pneumocytes, in patients with idiopathic pulmonary fibrosis (SALEH et al. 1995). Pulmonary fibrosis also occurs in systemic sclerosis where increased ir-ET levels in BAL fluid were also detected (CAMBREY et al. 1994). The levels of ir-ET in BAL fluid were high enough to induce an ET_A receptor-dependent proliferation of fibroblasts in vitro. In this disease, a significant amount of ir-ET probably comes from alveolar macrophages which have been shown to produce excessive amounts of this peptide in response to endotoxin (ODOUX et al. 1997). In other fibrotic lung conditions in the human, including that associated with scleroderma (ABRAHAM et al. 1997) and in a rat model of bronchiolitis obliterans (TAKEDA et al. 1998), significant increases in ir-ET were detected in both alveolar and peripheral bronchial epithelia. Interestingly, in scleroderma-associated lung fibrosis, total ET receptor number was increased two- to threefold. The ratio of $ET_A:ET_B$ sites also changed, with ET_B receptor numbers increased and ET_A receptor numbers reduced (ABRAHAM et al. 1997).

VI. Pulmonary Tumours

ET receptors and prepro-ET-1 mRNA have been detected in HeLa and HEp-2 human tumour cell lines and ir-ET was also released from these cells (SHICHIRI et al. 1991). ET-1 evoked increased cytosolic free Ca^{2+} and proliferation of these cells (SHICHIRI et al. 1991). In addition, ir-ET and mRNA for prepro-ET has been detected predominantly in pulmonary squamous cell carcinomas and adenocarcinomas (GIAID et al. 1990). The precise role of ET-1 in these tissues is not established, although activity as an autocrine/paracrine growth factor cannot be dismissed.

E. ET-1 and the Standard Criteria for Mediator Status in Asthma

The following is an outline of the extent to which the standard criteria for confirming the identity of an endogenous chemical mediator of disease, as described in Sect. B.I., have been fulfilled for ET-1 in asthma.

I. Synthesis, Release and Degradation of the ETs in the Lung

A true endogenous asthma mediator must be synthesized, released and degraded at appropriate sites in the lung.

1. Synthesis

The first reports describing the synthesis of the ETs established the endothelium as the primary vascular source of the peptides (YANAGISAWA et al. 1988; INOUE et al. 1989). The endothelium of small blood vessels within the airway wall is also a significant site of ET synthesis in the lung (GIAID et al. 1991; SPRINGALL et al. 1991; MACCUMBER et al. 1989). However, in relation to the status of ET-1 as a possible asthma mediator, arguably more important support for the case came in the form of several studies establishing that this peptide was also synthesized and released by the airway epithelium (ROZENGURT et al. 1990; GIAID et al. 1991; SPRINGALL et al. 1991; RENNICK et al. 1992; MACCUMBER et al. 1989). This is significant because this tissue represents a relatively large proportion of the airway mucosal volume and can potentially produce relatively large amounts of peptide which might diffuse to critical submucosal target tissues including airway smooth muscle and nerves.

As previously mentioned, ECE is a phosphoramidon-sensitive metalloprotease responsible for the conversion of Big ETs to the "mature" 21 amino acid ET peptides. This critical enzyme and has been found in guinea-pig, rabbit and human airways (BIHOVSKY et al. 1995; ISHIKAWA et al. 1992; PONS et al. 1992b). The lung contains high levels of the ETs relative to most other organs (PERNOW et al. 1989; YOSHIMI et al. 1989) and it is the airway epithelium in animal and human lung that is the richest source of these peptides (ROZENGURT et al. 1990; RENNICK et al. 1992; MACCUMBER et al. 1989; SPRINGALL et al. 1991).

2. Release

Under basal conditions, ET-1 is released abluminally from human (MATTOLI et al. 1990), porcine and canine (BLACK et al. 1989) cultured bronchial epithelial cells and from cultured tracheal epithelium from the guinea-pig (ENDO et al. 1992) and rabbit (RENNICK et al. 1993). Despite this, ETs released from vascular endothelium and some inflammatory cells (MACCUMBER et al. 1989; GIAID et al. 1991) including mast cells (EHRENREICH et al. 1992) and macrophages (EHRENREICH et al. 1990) may also play important functional roles within the airway wall in asthma.

3. Degradation

The ETs are catabolized primarily by neutral endopeptidase (NEP), a phosphoramidon-sensitive enzyme found in abundance in the airway epithelium (JOHNSON et al. 1985). Not surprisingly, both removal of the epithelium or pretreatment with phosphoramidon causes marked potentiation of the contractile actions of ET-1 in guinea-pig tracheal (HAY 1989) and human bronchial airway preparations (CANDENAS et al. 1992; YAMAGUCHI et al. 1992). These data are consistent with the notion that epithelial NEP is the major degradative enzyme for the ETs in the lung and, as such, acts as a significant modulator of the sensitivity of airway smooth muscle to this peptide (DI MARIA et al. 1992; BOICHOT et al. 1991b).

II. ET Receptor Distribution

ET receptors must be present in the lung at sites relevant to the expression of asthma symptoms.

1. Airway Smooth Muscle

Autoradiographic studies have been valuable in mapping the distribution and localization of both ET receptor subtypes in the lung. Such studies have verified the presence of high densities of ET receptors in many tissue types in the respiratory tract in the human (HEMSEN et al. 1990; HENRY et al. 1990; BRINK et al. 1991; MCKAY et al. 1991) and in animals (TURNER et al. 1989; POWER et al. 1989; TSCHIRHART et al. 1991). Consistent with the potent spasmogenic activity of the ETs in the airways, ET receptors are found in greatest density in airway smooth muscle, with many studies showing that both ET_A and ET_B receptor subtypes are expressed. There is a wide spectrum of subtype ratios detected in airway smooth muscle from different species. For example, approximately equal numbers of these subtypes are detected in mouse and rat tracheal smooth muscle (HENRY and GOLDIE 1994; HENRY 1993). However, there are some notable exceptions to this general pattern, with ET_B receptors constituting approximately 90% of the total population in human bronchial airway smooth muscle (GOLDIE et al. 1995). This proportion falls to about 70% when non-airway components of the human bronchial wall are included in the assessment (FUKURODA et al. 1996). ET_B receptors also greatly outweigh the numbers of ET_A receptors in rabbit bronchus (MCKAY et al. 1996) and also predominate in guinea-pig bronchus (GOLDIE et al. 1996a) and pig trachea (KOSEKI et al. 1989). At the other end of the spectrum, ET_A receptors exist as a homogeneous population in sheep tracheal airway smooth muscle (GOLDIE et al. 1994), strongly predominate in canine airway smooth muscle (MCKAY et al. 1996) and constitute approximately 70% of the total in pig bronchus (GOLDIE et al. 1996a).

2. Other Sites

Many cell types within the airway wall, other than airway smooth muscle, also express either or both ET_A and ET_B receptors. For example, specific [^{125}I]-ET-1 binding was also detected in epithelium and submucosal tissues, as well as in blood vessels in human bronchus (POWER et al. 1989; HENRY et al. 1990; GOLDIE et al. 1995) and in mouse, rat, guinea-pig and pig tracheal tissue

(TSCHIRHART et al. 1991; HENRY et al. 1990; GOLDIE et al. 1996a; KOSEKI et al. 1989). In sheep trachea, submucosal glands and sub-epithelial tissues expressed relatively high levels of ET_B receptors (GOLDIE et al. 1994). Peripheral lung alveoli also express very high levels of both ET_A and ET_B receptors. This has been verified in human lung (KNOTT et al. 1995) and in lung from the rat (TURNER et al. 1989; POWER et al. 1989), guinea-pig and pig (GOLDIE et al. 1996b). Neuronal tissue also contains ET receptors of both subtypes. Specific [^{125}I]-ET-1 binding has been detected in airway parasympathetic ganglia and with paravascular nerves and other neuronal pathways (McKAY et al. 1991; KOBAYASHI et al. 1993). ET receptors mostly of the ET_B subtype, have been localized to guinea-pig tracheal adrenergic and cholinergic nerve cell bodies, processes and varicosities in primary culture (TAKIMOTO et al. 1993). Recently, we used an immunofluorescence approach with confocal microscopy to show that both receptor subtypes existed in rat tracheal nerves grown in cultures (FERNANDES et al. 1998; GOLDIE et al. 1998). The actions of the ETs at these various sites may also be relevant to the airway obstruction in asthma. This will be discussed below.

In the context of assessing the role of the ETs in asthma, it is clearly important to determine whether this disease is associated with significant changes in ET receptor distribution or subtype densities in the lung. We have assessed these parameters in central (GOLDIE et al. 1995, 1996c; HAY et al. 1993a) and peripheral airways (KNOTT et al. 1995) from both asthmatic and non-asthmatic subjects. Interestingly, in asthma, no significant differences were detected in ET_A/ET_B receptor ratio in either central bronchial airway smooth muscle (non-asthmatic = 12%:88%; asthmatic = 18%:82%) (GOLDIE et al. 1995), or in alveolar wall tissue (non-asthmatic = 32%:68%; asthmatic = 29%:71%) (KNOTT et al. 1995). However, issues such as altered levels of ET synthesis and release in asthmatics compared with healthy individuals may be of importance in asthma.

III. ET Receptor-Mediated Responses Relevant to Asthma

1. Altered Bronchial Tone

ET receptors in the lung must mediate responses relevant to asthma symptoms and pathologies.

The autoradiographic detection of specific ET binding sites does not establish these sites as functional ET receptors. However, ET-1 has been shown to induce a wide range of acute and chronic effects within the bronchial wall and in peripheral lung in tissue known to contain specific binding sites for [^{125}I]-ET-1, that may be highly relevant to a mediator role in asthma. Of all the effects that an asthma mediator might be expected to induce, airway smooth muscle contraction is arguably one of the most important, since episodic and chronically elevated airway tone are cardinal features of this disease.

a) Direct Airway Smooth Muscle Contraction

The airway spasmogenic effects of ET-1 and related peptides was reported very soon after the identification of these substances in 1988 (UCHIDA et al. 1988). Subsequently, airway smooth muscle contraction has become the most widely studied action of ET-1 in the airways. ET-1 is one of the most powerful and potent spasmogens in human isolated bronchial smooth muscle preparations (UCHIDA et al. 1988; HAY et al. 1993a; TSCHIRHART et al. 1991; HEMSEN et al. 1990; MAGGI et al. 1989; ADVENIER et al. 1992; HAY 1990). Contraction to ET-1 is relatively slow to develop, but is persistent and resistant to reversal by washout. This may be explained in part by the pseudo-irreversible nature of ET-1 binding to ET receptors (NAMBI et al. 1994; WAGGONER et al. 1992; WU-WONG et al. 1994; WATAKABE et al. 1992; IHARA et al. 1995), which causes sustained receptor activation and signal transduction, consistent with the persistence of elevated bronchial tone often seen in asthma.

Although ET_B receptors greatly predominate in number in human bronchial smooth muscle, functional studies indicate that the smaller ET_A receptor population can also mediate ET-1-induced contraction (GOLDIE et al. 1995; FUKURODA et al. 1996). This is also the case in animal airway smooth muscle in which both receptors are expressed (HENRY and GOLDIE 1994; HENRY 1993; HAY et al. 1993d; GOLDIE et al. 1996a; KOSEKI et al. 1989) and in sheep tracheal smooth muscle where only ET_A receptors were detected (GOLDIE et al. 1994; ABRAHAM et al. 1993). Interestingly, in asthma, a decrease in sensitivity to the contractile effects of the ET_B receptor-selective agonist sarafotoxin S6c was demonstrated in bronchial tissue, suggesting that ET_B receptor desensitization may have occurred. It is possible that such desensitization was the result of increased synthesis and release of ET-1, with subsequent over-exposure of ET receptors to this ligand. Thus, a mediator role for ET-1 in asthma cannot be attributed in any degree to up-regulation of ET receptor function (GOLDIE et al. 1995) or density (GOLDIE et al. 1995; KNOTT et al. 1995).

b) Modulation of Neurotransmission

Neuronal pathways play an important role in bronchial wall homeostasis, including the regulation of airway tone. In asthma, the function of such systems may be perturbed, resulting in altered airway resistance or the activity of gland secretory processes. For example, airway obstruction resulting from increased airway smooth muscle tone could be induced by hyper-activity of bronchial cholinergic nerves (WARD et al. 1994; SHEPPARD et al. 1982; BEAKES 1997), or of the excitatory non-adrenergic, non-cholinergic neuronal system (BARNES et al. 1991) innervating this target tissue. Airway calibre might be similarly influenced by hypofunction of inhibitory sympathetic pathways (GOLDIE et al. 1986; GOLDIE 1990) or of inhibitory non-adrenergic, non-cholinergic nerves (ELLIS and UNDEM 1992). Thus, the balance of activities between these systems may be important in the etiology and progression of obstructive airway diseases including asthma (GOLDIE 1990; BARNES 1992). Mediator-induced modulation of such pathways is one mechanism through which this balance might be disturbed.

Recent evidence that the ETs induced potentiation of cholinergic neuro-transmission in the guinea-pig ileum (WIKLUND et al. 1989) provided a rationale for assessing the influence of these peptides on airway neuronal pathways. The first evidence for such an effect in the respiratory tract came with a report that ET-3 potentiated cholinergic nerve-evoked bronchial contraction in the rabbit (McKAY et al. 1993). This was soon followed by similar findings with ET-1 and the ET_B receptor-selective agonist sarafotoxin S6c in mouse (HENRY and GOLDIE 1995) and rat (KNOTT et al. 1996) trachea. With regard to a mediator role for ET peptides in asthma, it is particularly exciting to find that sarafotoxin S6c also caused powerful and potent potentiation of cholinergic nerve-mediated contraction in human isolated bronchial preparations (FERNANDES et al. 1996). Importantly, very recent, but as unpublished data from our laboratory indicate that ET-1 also potentiated cholinergic contraction in human bronchial ring preparations, an effect involving activation of both ET_A and ET_B receptors located on post-ganglionic nerves. These findings are consistent with results in rat tracheal tissue, where both receptor subtypes were linked to increased release of acetylcholine (KNOTT et al. 1996) and provide another mechanism through which ET peptides might influence airway tone in asthma. Ovine tracheal muscle tissue represents an interesting anomaly, in that prejunctional ET_B receptor stimulation caused inhibition of contraction due to suppression of acetylcholine release (HENRY et al. 1996). Thus, sheep airways do not provide a model which mimics the neuronal activities of ET-1 in human bronchus. We are presently evaluating the effects of exogenously applied ET-1 and similar peptides on non-cholinergic neuronal pathways in the airways.

c) Bronchoconstriction In Vivo

ET-1 caused increased airway resistance, presumably primarily as a result of bronchospasm, in several animal species, including the rat (MATSUSE et al. 1990), dog (UCHIDA et al. 1992a), guinea-pig (NOGUCHI et al. 1993; NAGASE et al. 1995; TOUVAY et al. 1990) and sheep (NOGUCHI et al. 1995). Both ET_A and ET_B receptors were involved in this effect in the guinea-pig (NOGUCHI et al. 1993; NAGASE et al. 1995) where sustained bronchoconstriction to either intravenous or aerosolized ET-1 involved the production of the secondary mediators thromboxane and platelet-activating factor (PAF) (MACQUIN-MAVIER et al. 1989; PAYNE and WHITTLE 1988; LAGENTE et al. 1989). Importantly, allergic sensitization in this model was accompanied by hyperresponsiveness to ET-1, perhaps related to reduced epithelial NEP activity (BOICHOT et al. 1990, 1991b). In addition, the early phase response to ovalbumin in this species, due largely to histamine and leukotriene release (BARNES et al. 1988), may also involve ET release early in the cascade of events, since ET_B receptor blockade inhibited allergic bronchoconstriction (UCHIDA et al. 1995). The late phase reaction to allergen may also involve ET production and the activation of ET$_A$ receptors, since inhibition at these sites attenuated this response, as well as the usual hyperresponsiveness to inhaled carbachol (NOGUCHI et al. 1995).

ET-1 may also induce bronchial hyperresponsiveness in some models. For example, ET-1 challenge in rabbits has recently been reported to enhance bronchoconstriction to inhaled histamine, an action that involved the activation of capsaicin-sensitive airway sensory nerves (DAGOSTINO et al. 1998). Hyperresponsiveness to spasmogens such as histamine and methacholine is also commonly observed in asthmatics (BOUSHEY et al. 1980). Inhaled ET-1 has been shown to induce potentiated responsiveness to such agonists in the sheep (NOGUCHI et al. 1995), although this could only be demonstrated in one (KANAZAWA et al. 1992) of five (MACQUIN-MAVIER et al. 1989; BOICHOT et al. 1991a; PONS et al. 1992a; LAGENTE et al. 1990) studies in the guinea-pig. In heterozygous ET-1 knockout mice in which ET-1 levels were abnormally low, airway responsiveness to methacholine was increased rather than reduced as might be predicted (NAGASE et al. 1998). This suggests that underproduction of this peptide might also be accompanied by anomalous production of another factor(s) such as a bronchodilator like nitric oxide (NO) (NAGASE et al. 1998).

Studies involving the actions of the ETs in human subjects, particularly asthmatic individuals, have long been awaited, since asthma is a peculiarly human condition for which there are no completely adequate animal models. The first report of the actions of inhaled ET-1 in human asthmatic and non-asthmatic subjects showed that inhaled aerosolized ET-1 had little influence on lung function in non-asthmatics, but induced severe bronchoconstriction in asthmatic patients who were also hyperresponsive to inhaled methacholine (CHALMERS et al. 1997). The lack of potency in normals may have been the result of protection of the submucosa via the degradative barrier actions of epithelial NEP. In contrast, in asthmatics, this protection might have been compromised by damaged epithelium reducing the amounts of available NEP, or by the greater permeability of the epithelium in asthmatics, allowing the penetration of ET-1 to sub-epithelial targets including airway smooth muscle and cholinergic nerves.

2. Mitogenesis

a) Fibroblasts

The pathologies which accompany chronic asthma include restructuring of various tissue elements within the bronchial wall, a phenomenon often described as bronchial remodelling. Amongst the most prominent of these changes is an increase in airway smooth muscle volume (CARROLL et al. 1993; KNOX 1994) and increased thickness of the sub-epithelial matrix caused by increased deposition of extracellular matrix protein (BREWSTER et al. 1990; ROCHE et al. 1989). The question is whether the action of ET peptides can contribute to these events. If so, then further circumstantial evidence is provided in support of a mediator role for the ETs in asthma. Indeed, the evidence from animal models clearly demonstrates the mitogenic potential of ET-1 in fibroblasts and airway smooth muscle cells. For example, ET-1 induced the proliferation of Swiss 3T3 fibroblasts in culture (TAKUWA 1993). In addition, ET-1 and ET-3 have been shown to be chemoattractants for fibroblasts and also to promote the replication of rat pulmonary artery-derived fibroblasts (PEACOCK et al. 1992).

Importantly, evidence from human cells is consistent with data from these animal studies. Fibronectin released from human bronchial epithelial cells is both an important extracellular matrix component and itself a chemotactic factor for fibroblasts. ET-1 enhanced fibronectin gene expression and fibronectin release from these cells, actions that were mediated via ET_A receptors (MARINI et al. 1996). Furthermore, in human asthmatic bronchial epithelial cells pretreated with allergen, the cytokine granulocyte/macrophage colony-stimulating factor (GM-CSF) stimulated ET-1 production, which in turn was associated with transformation of epithelial cells into myofibroblasts, cells which play a critical role in extracellular matrix deposition (SUN et al. 1997). Thus, ET-1 and epithelial cells from which it is derived, are pivotal to the promotion of sub-epithelial fibrosis. Interestingly, the mitogenic action of interleukin-1 beta (IL-1β) in porcine epithelial cells appear to be mediated in part by ET-1-induced ET_A receptor activation (MURLAS et al. 1997).

b) Airway Smooth Muscle

ET-1 has also been shown to be a potent but relatively weak mitogen in cultured tracheal airway smooth muscle cells from the guinea-pig, rabbit (NOVERAL et al. 1992) and sheep (GLASSBERG et al. 1994). Similar results were obtained in human cultured bronchial smooth muscle cells, adding considerable weight to the proposition that ET-1 might be an asthma mediator. Interestingly, although the mitogenic activity of ET-1 was modest, this action was not blunted by pretreatment with β-adrenoceptor bronchodilators such as salbutamol which had marked inhibitory effects against the proliferative effects of other mitogens including thrombin (TOMLINSON et al. 1994). Thus, ET-1 is unusual amongst mitogens in this regard and this characteristic enhances its profile as a putative asthma mediator. Perhaps more importantly, ET-1 is a potent co-mitogen in human bronchial airway smooth muscle cells, i.e. although alone, ET-1 was only weakly active, this peptide dramatically potentiated (three- to fourfold) the already powerful mitogenic effects of epidermal growth factor (EGF), an action mediated exclusively via ET_A receptors (PANETTIERI et al. 1996).

c) Mucous Glands

Bronchial mucous gland hyperplasia is also observed in chronic asthma, a phenomenon which presumably is linked to excessive production of mucus in this disease (Beasley et al. 1989; Hegele and Hogg 1996). Although not tested, given the mitogenic effects of ET-1 in bronchial smooth muscle and fibroblasts, it seems likely that mucous gland cells might also proliferate in response to ET-1.

3. Secretion of Mucus

Receptors for ET-1 have been detected in submucosal glands associated with their venous sinusoids and small muscular arterioles (Mullol et al. 1993). Studies in animal models suggest that ET-1 has effects in mucous glands consistent with the production of mucus-obstructed airways, as often reported in asthma. Namely, ET-1 increased mucous glycoprotein secretion in feline isolated tracheal submucosal glands (Shimura et al. 1992) and in ovine tracheal tissue, and reduced tracheal mucus velocity as a result of ET_A receptor activation (Sabater et al. 1996). ET-1-induced increased mucus production coupled with reduced mucus clearance are effects consistent with the promotion of airway mucus plugging as seen in asthma. Importantly, stimulation of ET receptors in human cultured nasal mucosal tissue also results in increased serous and mucous secretions (Mullol et al. 1993) and increased production of prostanoids (Wu et al. 1992). Furthermore, the vascular endothelium and venous sinusoidal tissue in human nasal mucosal tissue produced ET-1 (Casasco et al. 1993).

4. Altered Microvascular Permeability

Asthma is an inflammatory lung disease. Accordingly, the permeability of the airway microvasculature may be increased, resulting in bronchial submucosal oedema, since increased microvascular permeability is an obligatory accompaniment to airway inflammation (Persson 1991; Goldie and Pedersen 1995). In addition, oedema-associated bronchial wall swelling is a potentially important component of airway obstruction in asthma (Persson 1991). Another potentially important action of ET-1 in relation to asthma is the fact that it has been shown to increase airway microvascular permeability. This has been reported in perfused rat lung where the response was leukocyte- and plasma-dependent (Rodman et al. 1992). The formation of secondary mediator prostanoids may also be involved, although this remains uncertain (Raffestin et al. 1991; Horgan et al. 1991; Pons et al. 1991; Ercan et al. 1993). However, there is some consensus that ET-1-induced oedema formation in this model involves increased microvascular pressure (Raffestin et al. 1991; Rodman et al. 1992), mediated in part by ET_A receptor activation (Filep et al. 1993a). ET_A receptors also mediated permeability increases in the guinea-pig (Filep et al. 1995).

5. ET-1 as a Pro-Inflammatory Mediator

Given the inflammatory nature of bronchial asthma, a link between inflammatory processes in the lung and the actions of ET-1 would certainly support the case that this peptide is an asthma mediator. A large body of evidence suggests that ET-1 has significant pro-inflammatory activities in vitro, although some data suggest otherwise.

a) ET-1-Induced Pro-Inflammatory Mediator Release

The concept of ET-1 as a pro-inflammatory mediator in the airways, is largely dependent upon data from animal studies, including the description of an indomethacin-sensitive increase in the release of the prostanoid precursor arachidonic acid from feline cultured tracheal epithelial cells (Wu et al. 1993) and from epithelial membrane phospholipids (PLEWS et al. 1991). Other studies have also demonstrated ET-1-induced release of various pro-inflammatory mediators from various airway cells types. For example, ET-1 activated the release of the pro-inflammatory cytokines tumour necrosis factor alpha (TNF-α), IL-1β and IL-6 from human monocytes (HELSET et al. 1993) and of thromboxane A$_2$ from an unidentified source(s) in perfused guinea-pig lung (DE NUCCI et al. 1988). In addition, thromboxane and PGD$_2$ were detected following stimulation with ET-1 of cells from canine bronchial lavage lavage fluid (NINOMIYA et al. 1992). Furthermore, the levels of 15-HETE and of oxygen radicals in BAL fluid in the rat were raised in response to intravenous ET-1 (NAGASE et al. 1990), as was the case for oxygen radical levels in BAL fluid from the guinea-pig (FILEP et al. 1995). Guinea-pig alveolar macrophages also released arachidonic acid and thromboxane in response to ET-1 (MILLUL et al. 1991) and superoxide production was raised by ET-1 in human alveolar macrophages (HALLER et al. 1991). ET-1 also potentiated superoxide production from alveolar macrophage in response to FMLP and PAF, via an ET$_A$ receptor-mediated mechanism (FILEP et al. 1995). ET-1 caused the release of histamine from guinea-pig pulmonary mast cells (UCHIDA et al. 1992b) and of histamine, 5-HT and LTC$_4$ from IL-4-treated murine bone marrow-derived mast cells (EGGER et al. 1995).

b) ET-1-Induced Inflammatory Cell Chemotaxis

ET-1 may also be a chemotactic factor under some conditions, since it caused adhesion of leukocytes to pulmonary vascular endothelium and induced the sequestration of leukocytes from pulmonary capillaries (HELSET et al. 1994). Antigen challenge in the sensitized mouse also caused the influx of eosinphils into the respiratory tract, an effect attenuated by selective ET$_A$ receptor blockade, or pretreatment with the dual ET$_A$/ET$_B$ receptor antagonist SB 209670, but not by selective ET$_B$ receptor blockade (FUJITANI et al. 1997).

c) Some Evidence Against a Major Pro-Inflammatory Role for ET-1 in the Airways

Despite these apparently positive indications of pro-inflammatory activity and reports that ET-1 stimulated the release of prostanoids from guinea-pig trachea (HAY et al. 1993c), human bronchus (HAY et al. 1993b) and human cultured nasal mucosal tissue (WU et al. 1992), ET-1 failed to activate histamine or leukotriene release from intact airways from these sources (HAY et al. 1993b, c). Some evidence from studies in vivo also do not indicate ET-1-associated inflammatory activity. For example, in the guinea-pig, ET-1 was not a chemoattractant for inflammatory cells as assessed by histological examination of lung alveolar or vascular walls (MACQUIN-MAVIER et al. 1989; BOICHOT et al. 1991a). In addition, in this model, exposure to ET-1 was not associated with airway epithelial damage or elevated microvascular permeability (MACQUIN-MAVIER et al. 1989; PONS et al. 1992a). The chemotactic influence of ET-1 on human blood monocytes is also disputed, with one study reporting chemotaxis (ACHMAD and RAO 1992) and another failing to demonstrate this phenomenon (BATH et al. 1990).

IV. Airway Inflammation, Increased ET Levels and Asthma

Levels of ETs should be elevated in asthma and should positively correlate with disease severity.

1. Evidence from Models In Vitro

If ET-1 is involved in the chronic generation of asthma symptoms and pathologies, levels of this peptide might be expected to be significantly increased as a result of the actions of other inflammatory mediators. Indeed, it has been demonstrated that endotoxin, thrombin and other inflammatory stimuli including various cytokines, stimulate the release of ET-1 from tracheal epithelial cells in culture (NINOMIYA et al. 1991; ENDO et al. 1992; RENNICK et al. 1993; FRANCO-CERECEDA et al. 1991). In addition, in human bronchial epithelial cells in culture, the cytokines IL-1α, IL-1β and TNFα enhanced the expression of prepro-ET-1 mRNA and increased ET-1 release (NAKANO et al. 1994), providing further strong support for the proposition that inflammation and elevated ET levels in the airways were causally linked.

2. Evidence from Animal Models Ex Vivo or In Vitro

If disease-associated airway inflammation (e.g. in asthma) involves elevated ET-1 levels in the airways, it should be possible to mimic this in animal models of airway inflammation. This is indeed the case in several systems. Some investigators have used sephadex administered intra-tracheally or intravenously to model the airway inflammation of asthma, since this stimulus causes an acute airway eosinophilic inflammatory response (BJERMER et al. 1994; KUBIN et al. 1992). Results clearly demonstrate that immunoreactive ET (ir-ET) was significantly increased in BAL fluid and this response was markedly attenuated by the glucocorticoid budesonide (ANDERSSON et al. 1992, 1996). Recent work has established that the increase in lung ir-ET after intra-tracheal sephadex in the rat occurred in bronchial epithelium and macrophages and importantly, that this response preceded the airway eosinophilia, suggesting a role for ET-1 in the initiation of airway inflammation (Finsnes et al. 1997). This study also demonstrated that the ET_A/ET_B receptor antagonist bosentan blocked this inflammatory reaction (Finsnes et al. 1997). Tissue levels of ir-ET were also significantly increased in mice with airway inflammation associated with an Influenza-A respiratory tract viral infection (Carr et al. 1998).

Similarly, increased levels of ir-ET have been detected in plasma, BAL fluid and tissue in actively and passively ovalbumin-sensitized guinea-pigs which have an accompanying airway inflammatory cell infiltrate (Filep et al. 1993b; Xu and Zhong 1997). BAL fluid contained enough ir-ET to induce significant proliferation of bronchial airway smooth muscle cells in culture. It was concluded that increases in $TNF\alpha$, in response to allergic sensitization, induced the increased production of ET-1 which promoted airway smooth muscle cell proliferation (Xu and Zhong 1997). This is consistent with evidence that $TNF\alpha$ induced elevated ET-1 levels in human airway epithelial cells lines (Aubert et al. 1997). In ovalbumin-sensitized mice, the accompanying airway eosinophilia and neutrophilia were attenuated by about 50% following ET_A receptor blockade, but not by selective antagonism of ET_B receptors (Fujitani et al. 1997). Taken together, these data clearly indicate that airway inflammation results in the enhanced production and release of ETs which then has the potential to activate responses within the airway wall relevant to obstructive diseases such as asthma. However, it is important that such a link be established in human asthma.

3. Asthma-Associated Airway Inflammation and ET Levels

As previously mentioned, epithelial cells are a major potential source of ETs in the respiratory tract, although in healthy individuals, the expression of these peptides under basal conditions is very low (Vittori et al. 1992; Springall et al. 1991). It might be expected that these epithelial levels would be significantly elevated in inflamed airways in asthma. Several studies have now established that this is so. Thus, the expression of both mRNA for prepro-ET-1 and of ET-1 protein in bronchial epithelial cells from asthmatics was significantly greater than in similar tissue from healthy volunteers or from chronic bronchitics (Vittori et al. 1992). Cells derived from asthmatics have also been shown to produce increased amounts of ET-1 in response to pro-inflammatory stimuli compared with amounts produced in tissue from nonasthmatic subjects. This has been demonstrated for epithelial cells in response to IL-1, histamine (Ackerman et al. 1995) and GM-CSF (Sun et al. 1997). Peripheral blood mononuclear cells from asthmatics also released greater amounts of ET-1 than similar cells from non-asthmatics and allergen immunotherapy suppressed this response (Chen et al. 1995).

Importantly, various studies have now shown that active asthma is associated with increases in the production of ETs in the lung. In the first of these, Nomura and co-workers reported tantalizing preliminary data of a sixfold elevation in the levels of ir-ET in BAL fluid from an individual in status asthmaticus (NOMURA et al. 1989). It was subsequently shown that asthmatics have increased circulating blood (CHEN et al. 1995; AOKI et al. 1994) and BAL fluid levels of ir-ET (MATTOLI et al. 1991; SOFIA et al. 1993; BATTISTINI et al. 1991), suggesting that epithelial ET-1 levels should also be raised in asthma. Studies assessing ir-ET levels in bronchial biopsies confirmed this (SPRINGALL et al. 1991; REDINGTON et al. 1997). It is also significant that ir-ET-1 levels in lung tissue were not significantly raised in asthmatics receiving anti-inflammatory glucocorticoid therapy (REDINGTON et al. 1997). In such asthmatics demonstrating reduced ir-ET levels, the symptoms of asthma should be reduced in severity if ET-1 is a significant mediator.

4. Respiratory Tract Viral Infection, Asthma and ETs

Respiratory tract infections with viruses such as respiratory syncytial virus (RSV) have long been associated with exacerbations of asthma and bronchial hyperresponsiveness to inhaled spasmogens (BEASLEY et al. 1989; NICHOLSON et al. 1993; TEICHTAHL et al. 1997; SCHWARZE et al. 1997). Although many mechanisms for this have been proposed (BUSSE 1990; FOLKERTS et al. 1998), the phenomenon remains poorly understood. However, it is known that such infections involve airway inflammation and in the case of RSV this response involves an eosinophilia (FOLKERTS et al. 1998; SCHWARZE et al. 1997). This raises the possibility that up-regulated epithelial production of ET-1 makes a significant contribution to the induction of asthma symptoms. Consistent with this, we have recently shown that influenza-A virus infection in the mouse was linked to markedly elevated levels of ir-ET in both central and peripheral airways (CARR et al. 1998). Behera and colleagues have now established that RSV infection in human airway epithelial cell lines caused markedly increased expression of ET-1 mRNA, ET-1 protein, 5-lipoxygenase activity and cysteinyl leukotrienes (BEHERA et al. 1998). Taken together, these data suggest that ET-1 could be a significant mediator of virally-triggered asthma.

5. ET Levels and Asthma Symptoms

It has now been established that glucocorticosteroids, which are an important anti-inflammatory therapy in asthma, reduced the amount of epithelial ET-1 released in asthmatics (VITTORI et al. 1992; MATTOLI et al. 1991; REDINGTON et al. 1995). Most importantly, treatment with inhaled β-agonists and oral glucocorticoids for 15 days reduced the previously elevated BAL ET levels in asthmatics to levels approaching those detected in healthy control subjects and this was accompanied by improvements in lung function (MATTOLI et al. 1991). In a separate study, the suppressant influence of steroids on BAL ir-ET levels in asthma was confirmed (REDINGTON et al. 1995). Significantly, it was also noted that the percentage of predicted FEV_1 was lower in patients not receiving this treatment and that this correlated with the higher levels of BAL ir-ET in these patients. In contrast, in patients with nocturnal asthma, the fall in FEV_1

overnight was inversely related to the levels of ir-ET-1 in BAL fluid (KRAFT et al. 1994).

F. Conclusions

There are now many hundreds of published research papers describing the actions of the ETs in the respiratory tract, many of which support a case for ET-1 as a significant contributor to the pathologies of asthma. However, to our knowledge, there have been no studies assessing the clinical effects of ET receptor antagonists in asthmatic subjects. An alternative therapeutic approach would be to use agents that inhibit the synthesis or release of the ETs within the airway wall. Once again, studies evaluating such inhibitors have not been conducted. Until studies of this type are done and their therapeutic efficacy established, the last of the standard criteria for confirmation of ET-1 as an asthma mediator must remain unfulfilled.

Despite the fact that the case in support of this contention is circumstantial, the weight of evidence is impressive. In particular, the facts that the actions of ET-1 in human respiratory tissues mimic so many of the signal features of asthma, and that asthma/airway inflammation has been linked to increases in epithelial ET levels, provide powerful evidence suggesting that ET-1 is indeed a mediator in this disease. However, most of the data also suggest that the ETs are not initiators of this disease. Rather, their production seems to be upregulated following the establishment of airway inflammation, promoting the expression of disease symptoms. This view might have to be modified in the light of data from the study by Finsnes and co-workers, who reported a rise in ir-ET in the airways which preceded the eosinophilia in response to a pro-inflammatory stimulus (FINSNES et al. 1997).

Acknowledgements. The authors acknowledge the research funding support of the National Health and Medical Research Council (Australia).

References

Abraham DJ, Vancheeswaran R, Dashwood MR, Rajkumar VS, Pantelides P, Xu SW, du Bois RM, Black CM (1997) Increased levels of endothelin-1 and differential endothelin type A and B receptor expression in scleroderma-associated fibrotic lung disease. Am J Pathol 151:831–841

Abraham WM, Ahmed A, Cortes A, Spinella MJ, Malik AB, Andersen TT (1993) A specific endothelin-1 antagonist blocks inhaled endothelin-1-induced bronchoconstriction in sheep. J Appl Physiol 74:2537–2542

Achmad TH, Rao GS (1992) Chemotaxis of human blood monocytes toward endothelin-1 and the influence of calcium channel blockers. Biochem Biophys Res Commun 189:994–1000

Ackerman V, Carpi S, Bellini A, Vassalli G, Marini M, Mattoli S (1995) Constitutive expression of endothelin in bronchial epithelial cells of patients with symptomatic and asymptomatic asthma and modulation by histamine and interleukin-1. J Allergy Clin Immunol 96:618–27

Adachi M, Yang YY, Furuichi Y, Miyamoto C (1991) Cloning and characterization of cDNA encoding human A-type endothelin receptor. Biochem Biophys Res Commun 180:1265–1272

Advenier C, Lagente V, Zhang Y, Naline E (1992) Contractile activity of big endothelin-1 on the human isolated bronchus. Br J Pharmacol 106:883–887

Allen SW, Chatfield BA, Koppenhafer SA, Schaffer MS, Wolfe RR, Abman SH (1993) Circulating immunoreactive endothelin-1 in children with pulmonary hypertension. Association with acute hypoxic pulmonary vasoreactivity. Am J Respir Dis 148:519–522

Andersson SE, Zackrisson C, Hemsen A, Lundberg JM (1992) Regulation of lung endothelin content by the glucocorticosteroid budesonide. Biochem Biophys Res Commun 188:1116–1121

Andersson SE, Hemsen A, Zackrisson C, Lundberg JM (1996) Release of endothelin-1 into rat airways following sephadex-induced inflammation – modulation by enzyme inhibitors and budesonide. Respiration 63:111–116

Aoki T, Kojima T, Ono A, Unishi G, Yoshijima S, Kameda-Hayashi N, Yamamoto C, Hirata Y, Kobayashi Y (1994) Circulating endothelin-1 levels in patients with bronchial asthma. Ann Allergy 73:365–369

Arai H, Hori S, Aramori I, Ohkubo H, Nakanishi S (1990) Cloning and expression of a cDNA encoding an endothelin receptor Nature 348:730–732

Aubert JD, Juillerat-Jeanneret L, Leuenberger P (1997) Expression of endothelin-1 in human broncho-epithelial and monocytic cell lines: influence of tumor necrosis factor-alpha and dexamethasone. Biochem Pharmacol 53:547–552

Barnes PJ, Chung KF, Page CP (1988) Inflammatory mediators and asthma. Pharmacol Rev 40:49–84

Barnes PJ, Baraniuk JN, Belvisi MG (1991) Neuropeptides in the respiratory tract. Am J Respir Dis 144:1187–1198

Barnes PJ (1992) Neural mechanisms in asthma. Br Med Bull 48:149–168

Bath PM, Mayston SA, Martin JF (1990) Endothelin and PDGF do not stimulate peripheral blood monocyte chemotaxis, adhesion to endothelium, and superoxide production. Exp Cell Res 187:339–342

Battistini B, Filep JG, Cragoe EJ Jr, Fournier A, Sirois P (1991) A role for Na+/H+ exchange in contraction of guinea pig airways by endothelin-1 in vitro. Biochem Biophys Res Commun 175:583–588

Bax WA, Saxena PR (1994) The current endothelin receptor classification: time for reconsideration. Trends Pharmacol Sci 15:378–386

Beakes DE (1997) The use of anticholinergics in asthma. J Asthma 34:357–368

Beasley R, Roche WR, Roberts JA, Holgate ST (1989) Cellular events in the bronchi in mild asthma and after bronchial provocation. Am J Respir Dis 139:806–817

Behera AK, Kumar M, Matsuse H, Lockey RF, Mohapatra SS (1998) Respiratory syncytial virus induces the expression of 5-lipoxygenase and endothelin-1 in bronchial epithelial cells. Biochem Biophys Res Commun 251:704–709

Bihovsky R, Levinson BL, Loewi RC, Erhardt PW, Polokoff MA (1995) Hydroxamic acids as potent inhibitors of endothelin-converting enzyme from human bronchiolar smooth muscle. J Med Chem 38:2119–2129

Bjermer L, Sandstrom T, Sarnstrand B, Brattsand R (1994) Sephadex-induced granulomatous alveolitis in rat: effects of antigen manipulation. Am J Indust Med 25:73–78

Black PN, Ghatei MA, Takahashi K, Bretherton-Watt D, Krausz T, Dollery CT, Bloom SR (1989) Formation of endothelin by cultured airway epithelial cells. FEBS Lett 255:129–132

Bloch KD, Eddy RL, Shows TB, Quertermous T (1989a) cDNA cloning and chromosomal assignment of the gene encoding endothelin 3. J Biol Chem 264:18156–18161

Bloch KD, Friedrich SP, Lee ME, Eddy RL, Shows TB, Quertermous T (1989b) Structural organization and chromosomal assignment of the gene encoding endothelin. J Biol Chem 264:10851–10857

Bloch KD, Hong CC, Eddy RL, Shows TB, Quertermous T (1991) cDNA cloning and chromosomal assignment of the endothelin 2 gene: vasoactive intestinal contractor peptide is rat endothelin 2. Genomics 10:236–242

Boichot E, Lagente V, Mencia-Huerta JM, Braquet P (1990) Effect of phosphoramidon and indomethacin on the endothelin-1 (ET-1) induced bronchopulmonary response in aerosol sensitized guinea pigs. J Vasc Med Biol 2:206

Boichot E, Carre C, Lagente V, Pons F, Mencia-Huerta JM, Braquet P (1991a) Endothelin-1 and bronchial hyperresponsiveness in the guinea pig. J Cardiovasc Pharmacol 17 (Suppl 7):S329–S331

Boichot E, Pons F, Lagente V, Touvay C, Mencia-Huerta JM, Braquet P (1991b) Phosphoramidon potentiates the endothelin-1-induced bronchopulmonary response in guinea-pigs. Neurochem Int 18:477–479

Boushey HA, Holtzman MJ, Sheller JR, Nadel JA (1980) Bronchial hyperreactivity. Am J Respir Dis 121:389–413

Brewster CE, Howarth PH, Djukanovic R, Wilson J, Holgate ST, Roche WR (1990) Myofibroblasts and subepithelial fibrosis in bronchial asthma. Am J Respir Cell Mol Biol 3:507–511

Brink C, Gillard V, Roubert P, Mencia-Huerta JM, Chabrier PE, Braquet P, Verley J (1991) Effects and specific binding sites of endothelin in human lung preparations. Pulm Pharmacol 4:54–59

Busse WW(1990) Respiratory infections: their role in airway responsiveness and the pathogenesis of asthma. J Allergy Clin Immunol 85:671–683

Cambrey AD, Harrison NK, Dawes KE, Southcott AM, Black CM, du Boi RM, Laurent GJ, McAnulty RJ (1994) Increased levels of endothelin-1 in bronchoalveolar lavage fluid from patients with systemic sclerosis contribute to fibroblast mitogenic activity in vitro. Am J Respir Cell Mol Biol 11:439–445

Candenas ML, Naline E, Sarria B, Advenier C (1992) Effect of epithelium removal and of enkephalin inhibition on the bronchoconstrictor response to three endothelins of the human isolated bronchus. Eur J Pharmacol 210:291–297

Carr MJ, Spalding LJ, Goldie RG, Henry PJ (1998) Distribution of immunoreactive endothelin in the lungs of mice during respiratory viral infection. Eur Resp J 11:79–85

Carroll N, Elliot J, Morton A, James A (1993) The structure of large and small airways in nonfatal and fatal asthma. Am J Respir Dis 147:405–410

Casasco A, Benazzo M, Casasco M, Cornaglia AI, Springall DR, Calligaro A, Mira E, Polak JM (1993) Occurrence, distribution and possible role of the regulatory peptide endothelin in the nasal mucosa. Cell Tissue Res 274:241–247

Chalmers GW, Little SA, Patel KR, Thomson NC (1997) Endothelin-1-induced bronchoconstriction in asthma. Am J Respir Crit Care Med 156:382–8

Chen WY, Yu J, Wang JY (1995) Decreased production of endothelin-1 in asthmatic children after immunotherapy. J Asthma 32:29–35

Dagostino B, Filippelli A, Falciani M, Rossi F (1998) Endothelin-1 and bronchial hyperresponsiveness in the rabbit. Naunyn Schmiedebergs Arch Pharmacol 358:561–566

de Nucci G, Thomas R, D'Orleans-Juste P, Antunes E, Walder C, Warner TD, Vane JR (1988) Pressor effects of circulating endothelin are limited by its removal in the pulmonary circulation and by the release of prostacyclin and endothelium-derived relaxing factor. Proc Natl Acad Sci USA 85:9797–9800

Denault JB, Claing A, D'Orleans-Juste P, Sawamura T, Kido T, Masaki T, Leduc R (1995) Processing of proendothelin-1 by human furin convertase. FEBS Lett 362:276–280

Devadason PS, Henry PJ (1997) Comparison of the contractile effects and binding kinetics of endothelin-1 and sarafotoxin S6b in rat isolated renal artery. Br J Pharmacol 121:253–263

Di Maria GU, Katayama M, Borson DB, Nadel JA (1992) Neutral endopeptidase modulates endothelin-1-induced airway smooth muscle contraction in guinea-pig trachea. Regul Pept 39:137–145

Druml W, Steltzer H, Waldhausl W, Lenz K, Hammerle A, Vierhapper H, Gasic S, Wagner OF (1993) Endothelin-1 in adult respiratory distress syndrome. Am J Respir Dis 148:1169–1173

Egger D, Geuenich S, Denzlinger C, Schmitt E, Mailhammer R, Ehrenreich H, Dormer P, Hultner L (1995) IL-4 renders mast cells functionally responsive to endothelin-1. J Immunol 154:1830–1837

Ehrenreich H, Anderson RW, Fox CH, Rieckmann P, Hoffman GS, Travis WD, Coligan JE, Kehrl JH, Fauci AS (1990) Endothelins, peptides with potent vasoactive properties, are produced by human macrophages. J Exp Med 172:1741–1748

Ehrenreich H, Burd PR, Rottem M, Hultner L, Hylton JB, Garfield M, Coligan JE, Metcalfe DD, Fauci AS (1992) Endothelins belong to the assortment of mast cell-derived and mast cell-bound cytokines. New Biologist 4:147–156

Ellis JL, Undem BJ (1992) Inhibition by L-NG-nitro-L-arginine of nonadrenergic-non-cholinergic-mediated relaxations of human isolated central and peripheral airway. Am J Respir Dis 146:1543–1547

Emoto N, Yanagisawa M (1995) Endothelin-converting enzyme-2 is a membrane-bound, phosphoramidon-sensitive metalloprotease with acidic pH optimum. J Biol Chem 270:15262–15268

Endo T, Uchida Y, Matsumoto H, Suzuki N, Nomura A, Hirata F, Hasegawa S (1992) Regulation of endothelin-1 synthesis in cultured guinea pig airway epithelial cells by various cytokines. Biochem Biophys Res Commun 186:1594–1599

Ercan ZS, Kilinc M, Yazar O, Korkusuz P, Turker RK (1993) Endothelin-1-induced oedema in rat and guinea-pig isolated perfused lungs. Arch Int Pharmacodyn Ther 323:74–84

Fagny C, Michel A, Leonard I, Berkenboom G, Fontaine J, Deschodt-Lanckman M (1991) In vitro degradation of endothelin-1 by endopeptidase 24.11 (enkephalinase). Peptides 12:773–778

Fagny C, Michel A, Nortier J, Deschodt-Lanckman M (1992) Enzymatic degradation of endothelin-1 by activated human polymorphonuclear neutrophils. Regul Pept 42:27–37

Fernandes LB, Henry PJ, Rigby PJ, Goldie RG (1996) Endothelin$_B$ (ET$_B$) receptor-activated potentiation of cholinergic nerve-mediated contraction in human bronchus. Br J Pharmacol 118:1873–1874

Fernandes LB, Henry PJ, Spalding LJ, Cody SH, Pudney CJ, Goldie RG (1998) Immunocytochemical detection of endothelin receptors in rat cultured airway nerves. J Cardiovasc Pharmacol 31 (Suppl 1):S222–S224

Filep JG, Sirois MG, Foldes-Filep E, Rousseau A, Plante GE, Fournier A, Yano M, Sirois P (1993a) Enhancement by endothelin-1 of microvascular permeability via the activation of ET$_A$ receptors. Br J Pharmacol 109:880–886

Filep JG, Telemaque S, Battistini B, Sirois P, D'Orleans-Juste P (1993b) Increased plasma levels of endothelin during anaphylactic shock in the guinea-pig. Eur J Pharmacol 239:231–236

Filep JG, Fournier A, Foldes-Filep E (1995) Acute pro-inflammatory actions of endothelin-1 in the guinea-pig lung: involvement of ET$_A$ and ET$_B$ receptors. Br J Pharmacol 115:227–236

Finsnes F, Skjonsberg OH, Tonnessen T, Naess O, Lyberg T, Christensen G (1997) Endothelin production and effects of endothelin antagonism during experimental airway inflammation. Am J Respir Crit Care Med 155:1404–1412

Folkerts G, Busse WW, Nijkamp FP, Sorkness R, Gern JE (1998) Virus-induced airway hyperresponsiveness and asthma. Am J Respir Crit Care Med 157:1708–20

Franco-Cereceda A, Rydh M, Lou YP, Dalsgaard CJ, Lundberg JM (1991) Endothelin as a putative sensory neuropeptide in the guinea-pig: different properties in comparison with calcitonin gene-related peptide. Regul Pept 32:253–265

Fujitani Y, Trifilieff A, Tsuyuki S, Coyle AJ, Bertrand C (1997) Endothelin receptor antagonists inhibit antigen-induced lung inflammation in mice. Am J Respir Crit Care Med 155:1890–1894

Fukuroda T, Ozaki S, Ihara M, Ishikawa K, Yano M, Miyauchi T, Ishikawa S, Onizuka M, Goto K, Nishikibe M (1996) Necessity of dual blockade of endothelin ET(A) and ET(B) receptor subtypes for antagonism of endothelin-1-induced contraction in human bronchi. Br J Pharmacol 117:995–999

Furukawa K, Saleh D, Bayan F, Emoto N, Kaw S, Yanagisawa M, Giaid A (1996) Co-expression of endothelin-1 and endothelin-converting enzyme-1 in patients with chronic rhinitis. Am J Respir Cell Mol Biol 14:248–253

Giaid A, Hamid QA, Springall DR, Yanagisawa M, Shinmi O, Sawamura T, Masaki T, Kimura S, Corrin B, Polak JM (1990) Detection of endothelin immunoreactivity and mRNA in pulmonary tumours. J Pathol 162:15–22

Giaid A, Polak JM, Gaitonde V, Hamid QA, Moscoso G, Legon S, Uwanogho D, Roncalli M, Shinmi O, Sawamura T et al. (1991) Distribution of endothelin-like immunoreactivity and mRNA in the developing and adult human lung. Am J Respir Cell Mol Biol 4:50–58

Giaid A, Michel RP, Stewart DJ, Sheppard M, Corrin B, Hamid Q (1993) Expression of endothelin-1 in lungs of patients with cryptogenic fibrosing alveolitis. Lancet 341:1550–1554

Glassberg MK, Ergul A, Wanner A, Puett D (1994) Endothelin-1 promotes mito-genesis in airway smooth muscle cells. Am J Respir Cell Mol Biol 10:316–321

Goldie RG, Spina D, Henry PJ, Lulich KM, Paterson JW (1986) In vitro responsive-ness of human asthmatic bronchus to carbachol, histamine, beta-adrenoceptor agonists and theophylline. Br J Clin Pharmacol 22:669–676

Goldie RG (1990) Receptors in asthmatic airways. Am J Respir Dis 141:S151–6

Goldie RG, Grayson PS, Knott PG, Self GJ, Henry PJ (1994) Predominance of endothe-lin$_A$ (ET$_A$) receptors in ovine airway smooth muscle and their mediation of ET-1-induced contraction. Br J Pharmacol 112:749–756

Goldie RG, Henry PJ, Knott PG, Self GJ, Luttmann MA, Hay DWP (1995) Endothelin-1 receptor density, distribution and function in human isolated asthmatic airways. Am J Respir Crit Care Med 152:1653–1658

Goldie RG, D'Aprile AC, Cvetkovski R, Rigby PJ, Henry PJ (1996a) Influence of regional differences in ET$_A$ and ET$_B$ receptor subtype proportions on endothelin-1-induced contractions in porcine isolated trachea and bronchus. Br J Pharmacol 117:736–742

Goldie RG, D'Aprile AC, Self GJ, Rigby PJ, Henry PJ (1996b) The distribution and density of receptor subtypes for endothelin-1 in the peripheral lung of the rat, guinea-pig and pig. Br J Pharmacol 117:729–735

Goldie RG, Knott PG, Carr MJ, Hay DW, Henry PJ (1996c) The endothelins in the pulmonary system. Pulm Pharmacol 9:69–93

Goldie RG, Fernandes LB, Henry PJ, Spalding LJ, Pudney CJ (1998) Immunocyto-chemical detection of endothelin receptors in rat airway parasympathetic nerves. Am J Respir Crit Care Med 157:A720

Goldie RG, Pedersen KE (1995) Mechanisms of increased airway microvascular permeability: role in airway inflammation and obstruction. Clin Exp Pharmacol Physiol 22:387–396

Hagiwara H, Nagasawa T, Lodhi KM, Kozuka M, Ito T, Hirose S (1992) Affinity chro-matographic purification of bovine lung endothelin receptor using biotinylated endothelin and avidin-agarose. J Chromatog 597:331–334

Haller H, Schaberg T, Lindschau C, Lode H, Distler A (1991) Endothelin increases [Ca^{2+}]i, protein phosphorylation, and O_2^- production in human alveolar macrophages. Am J Physiol 261:L478-L484

Hay DW (1989) Guinea-pig tracheal epithelium and endothelin. Eur J Pharmacol 171:241–245

Hay DW (1990) Mechanism of endothelin-induced contraction in guinea-pig trachea: comparison with rat aorta. Br J Pharmacol 100:383–392

Hay DW, Henry PJ, Goldie RG (1993a) Endothelin and the respiratory system. Trends Pharmacol Sci 14:29–32

Hay DW, Hubbard WC, Undem BJ (1993b) Relative contributions of direct and indirect mechanisms mediating endothelin-induced contraction of guinea-pig trachea. Br J Pharmacol 110:955–962

Hay DW, Hubbard WC, Undem BJ (1993c) Endothelin-induced contraction and mediator release in human bronchus. Br J Pharmacol 110:392–398

Hay DW, Luttmann MA, Hubbard WC, Undem BJ (1993d) Endothelin receptor subtypes in human and guinea-pig pulmonary tissues. Br J Pharmacol 110:1175–1183

Hay DW, Henry PJ, Goldie RG (1996) Is endothelin-1 a mediator in asthma? Am J Respir Crit Care Med 154:1594–1597

Hay DW, Luttmann MA, Pullen MA, Nambi P (1998) Functional and binding characterization of endothelin receptors in human bronchus: evidence for a novel endothelin B receptor subtype? J Pharmacol Exp Ther 284:669–677

Heard BE, Hossain S (1973) Hyperplasia of bronchial muscle in asthma. J Pathol 110:319–331.

Hegele RG, Hogg JC (1996) The pathology of Asthma. An inflammatory disorder. In: Szefler SJ, Leung DYM (eds). Severe asthma: pathogenesis and clinical management. Marcel Dekker, New York, pp 61–76

Helset E, Sildnes T, Seljelid R, Konoski ZS (1993) Endothelin-1 stimulates human monocytes in vitro to release TNF-α, IL-1β and IL-6. Mediators Inflamm 2:417–422

Helset E, Ytrehus K, Tveita T, Kjaeve J, Jorgensen L (1994) Endothelin-1 causes accumulation of leukocytes in the pulmonary circulation. Circ Shock 44:201–209

Hemsen A, Franco-Cereceda A, Matran R, Rudehill A, Lundberg JM (1990) Occurrence, specific binding sites and functional effects of endothelin in human cardiopulmonary tissue. Eur J Pharmacol 191:319–328

Henry PJ, Rigby PJ, Self GJ, Preuss JM, Goldie RG (1990) Relationship between endothelin-1 binding site densities and constrictor activities in human and animal airway smooth muscle. Br J Pharmacol 100:786–792

Henry PJ (1993) Endothelin-1 (ET-1)-induced contraction in rat isolated trachea: involvement of ET_A and ET_B receptors and multiple signal transduction systems. Br J Pharmacol 110:435–441

Henry PJ, Goldie RG (1994) ET_B but not ET_A receptor-mediated contractions to endothelin-1 attenuated by respiratory tract viral infection in mouse airways. Br J Pharmacol 112:1188–1194

Henry PJ, Goldie RG (1995) Potentiation by endothelin-1 of cholinergic nerve-mediated contractions in mouse trachea via activation of ET_B receptors. Br J Pharmacol 114:563–569

Herbst C, Tippler B, Shams H, Simmet T (1995) A role for endothelin in bicuculline-induced neurogenic pulmonary oedema in rats. Br J Pharmacol 115:753–760

Hick S, Heidemann I, Soskic V, Muller-Esterl W, Godovac-Zimmermann J (1995) Isolation of the endothelin B receptor from bovine lung. Structure, signal sequence, and binding site. Eur J Biochem 234:251–257

Horgan MJ, Pinheiro JM, Malik AB (1991) Mechanism of endothelin-1-induced pulmonary vasoconstriction. Circ Res 69:157–164

Ihara M, Yamanaka R, Ohwaki K, Ozaki S, Fukami T, Ishikawa K, Towers P, Yano M (1995) [^{3}H]BQ-123, a highly specific and reversible radioligand for the endothelin ET_A receptor subtype. Eur J Pharmacol 274:1–6

Inoue A, Yanagisawa M, Kimura S, Kasuya Y, Miyauchi T, Goto K, Masaki T (1989) The human endothelin family: three structurally and pharmacologically distinct isopeptides predicted by three separate genes. Proc Natl Acad Sci USA 86:2863–2867

Ishikawa S, Tsukada H, Yuasa H, Fukue M, Wei S, Onizuka M, Miyauchi T, Ishikawa T, Mitsui K, Goto K et al. (1992) Effects of endothelin-1 and conversion of big endothelin-1 in the isolated perfused rabbit lung. J Appl Physiol 72:2387–2392

Itoh Y, Yanagisawa M, Ohkubo S, Kimura C, Kosaka T, Inoue A, Ishida N, Mitsui Y, Onda H, Fujino M et al. (1988) Cloning and sequence analysis of cDNA encoding the precursor of a human endothelium-derived vasoconstrictor peptide, endothelin: identity of human and porcine endothelin. FEBS Lett 231:440–444

James AL, Pare PD, Hogg JC (1989) The mechanics of airway narrowing in asthma. Am J Respir Dis 139:242–246

Johnson AR, Ashton J, Schulz WW, Erdos EG (1985) Neutral metalloendopeptidase in human lung tissue and cultured cells. Am J Respir Dis 132:564–568

Kanazawa H, Kurihara N, Hirata K, Fujiwara H, Matsushita H, Takeda T (1992) Low concentration endothelin-1 enhanced histamine-mediated bronchial contractions of guinea pigs in vivo. Biochem Biophys Res Commun 187:717–721

Karne S, Jayawickreme CK, Lerner MR (1993) Cloning and characterization of an endothelin-3 specific receptor (ET$_C$ receptor) from *Xenopus laevis* dermal melanophores. J Biol Chem 268:19126–19133

Kay AB (1991) Asthma and inflammation. J Allergy Clin Immunol 87:893–910

Kishi F, Minami K, Okishima N, Murakami M, Mori S, Yano M, Niwa Y, Nakaya Y, Kido H (1998) Novel 31-amino-acid-length endothelins cause constriction of vascular smooth muscle. Biochem Biophys Res Commun 248:387–390

Knott PG, Daprile AC, Henry PJ, Hay DWP, Goldie RG (1995) Receptors for endothelin-1 in asthmatic human peripheral lung. Br J Pharmacol 114:1–3

Knott PG, Fernandes LB, Henry PJ, Goldie RG (1996) Influence of endothelin-1 on cholinergic nerve-mediated contractions and acetylcholine release in rat isolated tracheal smooth muscle. J Pharmacol Exp Ther 279:1142–1147

Knox AJ (1994) Airway re-modelling in asthma: role of airway smooth muscle. Clin Sci 86:647–652

Kobayashi M, Ihara M, Sato N, Saeki T, Ozaki S, Ikemoto F, Yano M (1993) A novel ligand, [^{125}I]BQ-3020, reveals the localization of endothelin ET$_B$ receptors. Eur J Pharmacol 235:95–100

Kondoh M, Miyazaki H, Uchiyama Y, Yanagisawa M, Masaki T, and Murakami K (1991) Solubilization of two types of endothelin receptors, ET$_A$ and ET$_B$, from rat lung with retention of binding activity. Biomed Res 12:417–423.

Koseki C, Imai M, Hirata Y, Yanagisawa M, Masaki T (1989) Autoradiographic distribution in rat tissues of binding sites for endothelin: a neuropeptide? Am J Physiol 256:R858–R866

Kozuka M, Ito T, Hirose S, Lodhi KM, Hagiwara H (1991) Purification and characterization of bovine lung endothelin receptor. J Biol Chem 266:16892–16896

Kraft M, Beam WR, Wenzel SE, Zamora MR, O'Brien RF, Martin RJ (1994) Blood and bronchoalveolar lavage endothelin-1 levels in nocturnal asthma. Am J Respir Crit Care Med 149:946–952

Kubin R, Deschl U, Linssen M, Wilhelms OH (1992) Intratracheal application of Sephadex in rats leads to massive pulmonary eosinophilia without bronchial hyperreactivity to acetylcholine. Int Arch Allergy Immunol 98:266–272

Lagente V, Chabrier PE, Mencia-Huerta JM, Braquet P (1989) Pharmacological modulation of the bronchopulmonary action of the vasoactive peptide, endothelin, administered by aerosol in the guinea-pig. Biochem Biophys Res Commun 158:625–632

Lagente V, Boichot E, Mencia-Huerta J, Braquet P (1990) Failure of aerosolized endothelin (ET-1) to induce bronchial hyperreactivity in the guinea pig. Fundam Clin Pharmacol 4:275–280

Laitinen LA, Heino M, Laitinen A, Kava T, Haahtela T (1985) Damage of the airway epithelium and bronchial reactivity in patients with asthma. Am J Respir Dis 131:599–606

Langleben D, DeMarchie M, Laporta D, Spanier AH, Schlesinger RD, Stewart DJ (1993) Endothelin-1 in acute lung injury and the adult respiratory distress syndrome. Am J Respir Dis 148:1646–1650

MacCumber MW, Ross CA, Glaser BM, Snyder SH (1989) Endothelin: visualization of mRNAs by in situ hybridization provides evidence for local action. Proc Natl Acad Sci USA 86:7285–7289

Macquin-Mavier I, Levame M, Istin N, Harf A (1989) Mechanisms of endothelin-mediated bronchoconstriction in the guinea pig. J Pharmacol Exp Ther 250:740–745

Maggi CA, Patacchini R, Giuliani S, Meli A (1989) Potent contractile effect of endothelin in isolated guinea-pig airways. Eur J Pharmacol 160:179–182

Marini M, Carpi S, Bellini A, Patalano F, Mattoli S (1996) Endothelin-1 induces increased fibronectin expression in human bronchial epithelial cells. Biochem Biophys Res Commun 220:896–899

Masaki T, Yanagisawa M, Goto K (1992) Physiology and pharmacology of endothelins. Med Res Rev 12:391–421

Masuda Y, Miyazaki H, Kondoh M, Watanabe H, Yanagisawa M, Masaki T, Murakami K (1989) Two different forms of endothelin receptors in rat lung. FEBS Lett 257:208–210

Matsuse T, Fukuchi Y, Suruda T, Nagase T, Ouchi Y, Orimo H (1990) Effect of endothelin-1 on pulmonary resistance in rats. J Appl Physiol 68:2391–2393

Mattoli S, Mezzetti M, Riva G, Allegra L, Fasoli A (1990) Specific binding of endothelin on human bronchial smooth muscle cells in culture and secretion of endothelin-like material from bronchial epithelial cells. Am J Respir Cell Mol Biol 3:145–151

Mattoli S, Soloperto M, Marini M, Fasoli A (1991) Levels of endothelin in the bronchoalveolar lavage fluid of patients with symptomatic asthma and reversible airflow obstruction. J Allergy Clin Immunol 88:376–384

McKay KO, Black JL, Diment LM, Armour CL (1991) Functional and autoradiographic studies of endothelin-1 and endothelin-2 in human bronchi, pulmonary arteries, and airway parasympathetic ganglia. J Cardiovasc Pharmacol 17 (Suppl 7):S206–S209

McKay KO, Armour CL, Black JL (1993) Endothelin-3 increases transmission in the rabbit pulmonary parasympathetic nervous system. J Cardiovasc Pharmacol 22 (Suppl 8):S181–S184

McKay KO, Armour CL, Black JL (1996) Endothelin receptors and activity differ in human, dog, and rabbit lung. Am J Physiol – Lung Cell Mol Physiol 14:L37–L43

Millul V, Lagente V, Gillardeaux O, Boichot E, Dugas B, Mencia-Huerta JM, Bereziat G, Braquet P, Masliah J (1991) Activation of guinea pig alveolar macrophages by endothelin-1. J Cardiovasc Pharmacol 17 (Suppl 7):S233–S235

Mullol J, Chowdhury BA, White MV, Ohkubo K, Rieves RD, Baraniuk J, Hausfeld JN, Shelhamer JH, Kaliner MA (1993) Endothelin in human nasal mucosa. Am J Respir Cell Mol Biol 8:393–402

Murlas CG, Sharma AC, Gulati A, Najmabadi F (1997) Interleukin-1 beta increases airway epithelial cell mitogenesis partly by stimulating endothelin-1 production. Lung 175:117–126

Mutsaers SE, Foster ML, Chambers RC, Laurent GJ, McAnulty RJ (1998) Increased endothelin-1 and its localization during the development of bleomycin-induced pulmonary fibrosis in rats. Am J Respir Cell Mol Biol 18:611–619

Nagase T, Fukuchi Y, Jo C, Teramoto S, Uejima Y, Ishida K, Shimizu T, Orimo H (1990) Endothelin-1 stimulates arachidonate 15-lipoxygenase activity and oxygen radical formation in the rat distal lung. Biochem Biophys Res Commun 168:485–489

Nagase T, Fukuchi Y, Matsui H, Aoki T, Matsuse T, Orimo H (1995) In vivo effects of endothelin A- and B-receptor antagonists in guinea pigs. Am J Physiol 268:L846–L850

Nagase T, Kurihara H, Kurihara Y, Aoki T, Fukuchi Y, Yazaki Y, Ouchi Y. (1998) Airway hyperresponsiveness to methacholine in mutant mice deficient in endothelin-1. Am J Respir Crit Care Med 157:560–564

Nakano A, Kishi F, Minami K, Wakabayashi H, Nakaya Y, Kido H (1997) Selective conversion of big endothelins to tracheal smooth muscle constricting 31 amino acid length endothelins by chymase from human mast cells. J Immunol 159:1987–1992

Nakano J, Takizawa H, Ohtoshi T, Shoji S, Yamaguchi M, Ishii A, Yanagisawa M, Ito K (1994) Endotoxin and pro-inflammatory cytokines stimulate endothelin-1 expression and release by airway epithelial cells. Clin Exp Allergy 24:330–336

Nambi P, Pullen M, Spielman W (1994) Species differences in the binding characteristics of [^{125}I]IRL-1620, a potent agonist specific for endothelin-B receptors. J Pharmacol Exp Ther 268:202–207

Naylor B (1962) The shedding of the mucosa of the bronchial tree in asthma. Thorax 17:69–72

Nicholson KG, Kent J, Ireland DC (1993) Respiratory viruses and exacerbations of asthma in adults. Br Med J 307:982–986

Ninomiya H, Uchida Y, Ishii Y, Nomura A, Kameyama M, Saotome M, Endo T, Hasegawa S (1991) Endotoxin stimulates endothelin release from cultured epithelial cells of guinea-pig trachea. Eur J Pharmacol 203:299–302

Ninomiya H, Yu XY, Hasegawa S, Spannhake EW (1992) Endothelin-1 induces stimulation of prostaglandin synthesis in cells obtained from canine airways by bronchoalveolar lavage. Prostaglandins 43:401–411

Noguchi K, Noguchi Y, Hirose H, Nishikibe M, Ihara M, Ishikawa K, Yano M (1993) Role of endothelin ETB receptors in bronchoconstrictor and vasoconstrictor responses in guinea-pigs. Eur J Pharmacol 233:47–51

Noguchi K, Ishikawa K, Yano M, Ahmed A, Cortes A, Abraham WM (1995) Endothelin-1 contributes to antigen-induced airway hyperresponsiveness. J Appl Physiol 79:700–705

Nomura A, Uchida Y, Kameyana M, Saotome M, Oki K, Hasegawa S (1989) Endothelin and bronchial asthma. Lancet 2:747–748

Noveral JP, Rosenberg SM, Anbar RA, Pawlowski NA, Grunstein MM (1992) Role of endothelin-1 in regulating proliferation of cultured rabbit airway smooth muscle cells. Am J Physiol 263:L317–L324

Odoux C, Crestani B, Lebrun G, Rolland C, Aubin P, Seta N, Kahn MF, Fiet J, Aubier M (1997) Endothelin-1 secretion by alveolar macrophages in systemic sclerosis. Am J Respir Crit Care Med 156:1429–1435

Opgenorth TJ, Wu-Wong JR, Shiosaki K (1992) Endothelin-converting enzymes. FASEB J 6:2653–2659

Panettieri RA, Goldie RG, Rigby PJ, Eszterhas AJ, Hay DWP (1996) Endothelin-1-induced potentiation of human airway smooth muscle proliferation: an ET$_A$ receptor-mediated phenomenon. Br J Pharmacol 118:191–197

Payne AN, Whittle BJ (1988) Potent cyclo-oxygenase-mediated bronchoconstrictor effects of endothelin in the guinea-pig in vivo. Eur J Pharmacol 158:303–304

Peacock AJ, Dawes KE, Shock A, Gray AJ, Reeves JT, Laurent GJ (1992) Endothelin-1 and endothelin-3 induce chemotaxis and replication of pulmonary artery fibroblasts. Am J Respir Cell Mol Biol 7:492–499

Pernow J, Hemsen A, Lundberg JM (1989) Tissue specific distribution, clearance and vascular effects of endothelin in the pig. Biochem Biophys Res Commun 161:647–653

Persson CGA (1991) Tracheobronchial microcirculation in asthma. In: Kaliner MA, Barnes PJ, Persson CGA (eds). Asthma its pathology and treatment. Marcel Dekker, New York, pp. 209–229

Plews PI, Abdel-Malek ZA, Doupnik CA, Leikauf GD (1991) Endothelin stimulates chloride secretion across canine tracheal epithelium. Am J Physiol 261:L188–L194

Pons F, Touvay C, Lagente V, Mencia-Huerta J M, Braquet P (1991) Comparison of the effects of intra-arterial and aerosol administration of endothelin-1 (ET-1) in the guinea-pig isolated lung. Br J Pharmacol 102:791–796

Pons F, Boichot E, Lagente V, Touvay C, Mencia-Huerta JM, Braquet P (1992a) Role of endothelin in pulmonary function. Pulm Pharmacol 5:213–219

Pons F, Touvay C, Lagente V, Mencia-Huerta JM, Braquet P (1992b) Involvement of a phosphoramidon-sensitive endopeptidase in the processing of big endothelin-1 in the guinea-pig. Eur J Pharmacol 217:65–70

Power RF, Wharton J, Zhao Y, Bloom SR, Polak JM (1989) Autoradiographic localization of endothelin-1 binding sites in the cardiovascular and respiratory systems. J Cardiovasc Pharmacol 13 (Suppl 5):S50–S56

Raffestin B, Adnot S, Eddahibi S, Macquin-Mavier I, Braquet P, Chabrier PE (1991) Pulmonary vascular response to endothelin in rats. J Appl Physiol 70:567–574

Redington AE, Springall DR, Ghatei MA, Lau LC, Bloom SR, Holgate ST, Polak JM, Howarth PH (1995) Endothelin in bronchoalveolar lavage fluid and its relation to airflow obstruction in asthma. Am J Respir Crit Care Med 151:1034–1039

Redington AE, Springall DR, Meng QH, Tuck AB, Holgate ST, Polak JM, Howarth PH (1997) Immunoreactive endothelin in bronchial biopsy specimens: increased expression in asthma and modulation by corticosteroid therapy. J Allergy Clin Immunol 100:544–552

Rennick RE, Loesch A, Burnstock G (1992) Endothelin, vasopressin, and substance P like immunoreactivity in cultured and intact epithelium from rabbit trachea. Thorax 47:1044–1049

Rennick RE, Milner P, Burnstock G (1993) Thrombin stimulates release of endothelin and vasopressin, but not substance P, from isolated rabbit tracheal epithelial cells. Eur J Pharmacol 230:367–370

Riccio MM, Reynolds CJ, Hay DW P, Proud D (1995) Effects of intranasal administration of endothelin-1 to allergic and nonallergic individuals. Am J Respir Crit Care Med 152:1757–1764

Roche WR, Beasley R, Williams JH, Holgate ST (1989) Subepithelial fibrosis in the bronchi of asthmatics. Lancet 1:520–524

Roche WR (1998) Inflammatory and structural changes in the small airways in bronchial asthma. Am J Respir Crit Care Med 157:S191–S194

Rodman DM, Stelzner TJ, Zamora MR, Bonvallet ST, Oka M, Sato K, O'Brien RF, McMurtry IF (1992) Endothelin-1 increases the pulmonary microvascular pressure and causes pulmonary edema in salt solution but not blood-perfused rat lungs. J Cardiovasc Pharmacol 20:658–663

Rozengurt N, Springall DR, Polak JM (1990) Localization of endothelin-like immunoreactivity in airway epithelium of rats and mice. J Pathol 160:5–8

Sabater JR, Otero R, Abraham WM, Wanner A, O'Riordan TG (1996) Endothelin-1 depresses tracheal mucus velocity in ovine airways via ET-A receptors. Am J Respir Crit Care Med 154:341–345

Saetta M, Maestrelli P, Turato G, Mapp CE, Milani G, Pivirotto F, Fabbri LM, Di Stefano A (1995) Airway wall remodeling after cessation of exposure to isocyanates in sensitized asthmatic subjects. Am J Respir Crit Care Med 151:489–94

Sakamoto A, Yanagisawa M, Sakurai T, Takuwa Y, Yanagisawa H, Masaki T (1991) Cloning and functional expression of human cDNA for the ET_B endothelin receptor. Biochem Biophys Res Commun 178:656–663

Sakurai T, Yanagisawa M, Takuwa Y, Miyazaki H, Kimura S, Goto K, Masaki T (1990) Cloning of a cDNA encoding a non-isopeptide-selective subtype of the endothelin receptor. Nature 348:732–735

Sakurai T, Yanagisawa M, Masaki T (1992) Molecular characterization of endothelin receptors. Trends Pharmacol Sci 13:103–108

Saleh D, Yanagisawa M, Corrin B, Giaid A (1995) Co-expression of endothelin and endothelin converting enzyme-1 in the lungs of normal humans and patients with idiopathic pulmonary fibrosis. Am J Respir Crit Care Med 151:A51

Schwarze J, Hamelmann E, Bradley KL, Takeda K, Gelfand EW (1997) Respiratory syncytial virus infection results in airway hyperresponsiveness and enhanced airway sensitization to allergen. J Clin Invest 100:226–233

Seino S, Kato S, Takahashi H, Iwabuchi K, Yuki H, Nakamura H, Tomoike H (1995) BQ123, endothelin type A receptor antagonist, attenuates bleomycin-induced pulmonary fibrosis. Am J Respir Crit Care Med 151:A63

Sheppard D, Epstein J, Holtzman MJ, Nadel JA, Boushey HA (1982) Dose-dependent inhibition of cold air-induced bronchoconstriction by atropine. J Appl Physiol 53:169–174

Shichiri M, Hirata Y, Nakajima T, Ando K, Imai T, Yanagisawa M, Masaki T, Marumo F (1991) Endothelin-1 is an autocrine/paracrine growth factor for human cancer cell lines. J Clin Invest 87:1867–1871

Shimura S, Ishihara H, Satoh M, Masuda T, Nagaki N, Sasaki H, Takishima T (1992) Endothelin regulation of mucus glycoprotein secretion from feline tracheal submucosal glands. Am J Physiol 262:L208–L213

Simmet T, Pritze S, Thelen KI, Peskar BA (1992) Release of endothelin in the oleic acid-induced respiratory distress syndrome in rats. Eur J Pharmacol 211:319–322

Smart SJ, Casale TB (1998) Abnormal autonomic control in respiratory disease. In: Barnes PJ (ed) Autonomic control of the respiratory system. Harwood Academic Publishers, Amsterdam pp. 313–338

Sofia M, Mormile M, Faraone S, Alifano M, Zofra S, Romano L, Carratu L (1993) Increased endothelin-like immunoreactive material on bronchoalveolar lavage fluid from patients with bronchial asthma and patients with interstitial lung disease. Respiration 60:89–95

Sokolovsky M, Ambar I, Galron R (1992) A novel subtype of endothelin receptors. J Biol Chem 267:20551–20554

Springall DR, Howarth PH, Counihan H, Djukanovic R, Holgate ST, Polak JM (1991) Endothelin immunoreactivity of airway epithelium in asthmatic patients. Lancet 337:697–701

Stelzner TJ, O'Brien RF, Yanagisawa M, Sakurai T, Sato K, Webb S, Zamora M, McMurtry IF, Fisher JH (1992) Increased lung endothelin-1 production in rats with idiopathic pulmonary hypertension. Am J Physiol 262:L614–L620

Stewart DJ, Levy RD, Cernacek P, Langleben D (1991) Increased plasma endothelin-1 in pulmonary hypertension: marker or mediator of disease? Ann Intern Med 114:464–469

Sun G, Stacey MA, Bellini A, Marini M, Mattoli S (1997) Endothelin-1 induces bronchial myofibroblast differentiation. Peptides 18:1449–1451

Takayanagi R, Ohnaka K, Takasaki C, Ohashi M, Nawata H (1991) Multiple subtypes of endothelin receptors in porcine tissues: characterization by ligand binding, affinity labeling and regional distribution. Regul Pept 32:23–37

Takeda S, Miyoshi S, Omori K, Utsumi T, Kogaki S, Sawa Y, Yanagisawa M, Matsuda H (1998) Pulmonary disease models induced by in vivo hemagglutinating virus of Japan liposome-mediated endothelin-1 gene transfer. J Cardiovasc Pharmacol 31 (Suppl 1):S336–S338

Takimoto M, Inui T, Okada T, Urade Y (1993) Contraction of smooth muscle by activation of endothelin receptors on autonomic neurons. FEBS Lett 324:277–282

Takuwa Y(1993) Endothelin in vascular and endocrine systems: biological activities and its mechanisms of action. Endocrine J. 40:489–506

Teichtahl H, Buckmaster N, Pertnikovs E (1997) The incidence of respiratory tract infection in adults requiring hospitalization for asthma. Chest 112:591–596

Tomlinson PR, Wilson JW, Stewart AG (1994) Inhibition by salbutamol of the proliferation of human airway smooth muscle cells grown in culture. Br J Pharmacol 111:641–647

Touvay C, Vilain B, Pons F, Chabrier PE, Mencia-Huerta JM, Braquet P (1990) Bronchopulmonary and vascular effect of endothelin in the guinea-pig. Eur J Pharmacol 176:23–33

Trigg CJ, Manolitsas ND, Wang J, Calderon MA, McAulay A, Jordan SE, Herdman MJ, Jhalli N, Duddle JM, Hamilton SA et al. (1994) Placebo-controlled immunopathologic study of four months of inhaled corticosteroids in asthma. Am J Respir Crit Care Med 150:17–22

Tschirhart EJ, Drijfhout JW, Pelton JT, Miller RC, Jones CR (1991) Endothelins: functional and autoradiographic studies in guinea pig trachea. J Pharmacol Exp Ther 258:381–387

Turner NC, Power RF, Polak JM, Bloom SR, Dollery CT (1989) Endothelin-induced contractions of tracheal smooth muscle and identification of specific endothelin binding sites in the trachea of the rat. Br J Pharmacol 98:361–366

Uchida Y, Ninomiya H, Saotome M, Nomura A, Ohtsuka M, Yanagisawa M, Goto K, Masaki T, Hasegawa S (1988) Endothelin, a novel vasoconstrictor peptide, as potent bronchoconstrictor. Eur J Pharmacol 154:227–228

Uchida Y, Hamada M, Kameyama M, Ohse H, Nomura A, Hasegawa S, Hirata F (1992a) ET-1 induced bronchoconstriction in the early phase but not late phase of anesthetized dogs is inhibited by indomethacin and ICI 198615. Biochem Biophys Res Commun 183:1197–1202

Uchida Y, Ninomiya H, Sakamoto T, Lee JY, Endo T, Nomura A, Hasegawa S, Hirata F (1992b) ET-1 released histamine from guinea pig pulmonary but not peritoneal mast cells. Biochem Biophys Res Commun 189:1196–1201

Uchida Y, Nomura A, Ohse H, Sakamoto T, Ninomiya H, Iijima H, To J, Hasegawa S, Hirata F (1995) Evidence for the involvement of endothelins in antigen-induced immediate and late asthmatic reactions of guinea-pig experimental asthma. Am J Respir Crit Care Med 151:A228

Vemulapalli S, Rivelli M, Chiu PJ, del Prado M, Hey JA (1992) Phosphoramidon abolishes the increases in endothelin-1 release induced by ischemia-hypoxia in isolated perfused guinea pig lungs. J Pharmacol Exp Ther 262:1062–1069

Vijayaraghavan J, Scicli AG, Carretero OA, Slaughter C, Moomaw C, Hersh LB (1990) The hydrolysis of endothelins by neutral endopeptidase 24.11 (enkephalinase). J Biol Chem 265:14150–14155

Vittori E, Marini M, Fasoli A, De Franchis R, Mattoli S (1992) Increased expression of endothelin in bronchial epithelial cells of asthmatic patients and effect of corticosteroids. Am J Respir Dis 146:1320–5

Waggoner WG, Genova SL, Rash VA (1992) Kinetic analyses demonstrate that the equilibrium assumption does not apply to [^{125}I] endothelin-1 binding data. Life Sci 51:1869–1876

Ward MJ, Fentem PH, Smith WH, Davies D (1994) Ipratropium bromide in acute asthma. Br Med J 1981:598–600

Warner TD, Allcock GH, Mickley EJ, Corder R, Vane JR (1993) Comparative studies with the endothelin receptor antagonists BQ-123 and PD-142893 indicate at least 3 endothelin receptors. J Cardiovasc Pharmacol 22:S117–S120

Watakabe T, Urade Y, Takai M, Umemura I, Okada T (1992) A reversible radioligand specific for the ETB receptor: [^{125}I]Tyr13-Suc-[Glu9,Ala11,15]-endothelin-1(8–21), [^{125}I]IRL 1620. Biochem Biophys Res Commun 185:867–873

Weitzberg E, Lundberg JM, Rudehill A (1991) Elevated plasma levels of endothelin in patients with sepsis syndrome. Circ Shock 33:222–227

Wiklund NP, Wiklund CU, Ohlen A, Gustafsson LE (1989) Cholinergic neuromodulation by endothelin in guinea pig ileum. Neurosci Lett 101:342–346

Wu T, Mullol J, Rieves RD, Logun C, Hausfield J, Kaliner MA, Shelhamer JH (1992) Endothelin-1 stimulates eicosanoid production in cultured human nasal mucosa. Am J Respir Cell Mol Biol 6:168–174

Wu T, Rieves RD, Larivee P, Logun C, Lawrence MG, Shelhamer JH (1993) Production of eicosanoids in response to endothelin-1 and identification of specific endothelin-1 binding sites in airway epithelial cells. Am J Respir Cell Mol Biol 8:282–290

Wu-Wong JR, Chiou WJ, Magnuson SR, Opgenorth TJ (1994) Endothelin receptor agonists and antagonists exhibit different dissociation characteristics. Biochim Biophys Acta 1224:288–294

Xu D, Emoto N, Giaid A, Slaughter C, Kaw S, deWit D, Yanagisawa M (1994) ECE-1: a membrane-bound metalloprotease that catalyzes the proteolytic activation of big endothelin-1. Cell 78:473–485

Xu J, Zhong NS (1997) The interaction of tumour necrosis factor alpha and endothelin-1 in pathogenetic models of asthma. Clin Exp Allergy 27:568–573

Yamaguchi T, Kohrogi H, Kawano O, Ando M, Araki S (1992) Neutral endopeptidase inhibitor potentiates endothelin-1-induced airway smooth muscle contraction. J Appl Physiol 73:1108–1113

Yanagisawa M, Kurihara H, Kimura S, Tomobe Y, Kobayashi M, Mitsui Y, Yazaki Y, Goto K, Masaki T (1988) A novel potent vasoconstrictor peptide produced by vascular endothelial cells. Nature 332:411–415

Yanagisawa M, Masaki T (1989) Endothelin, a novel endothelium-derived peptide. Pharmacological activities, regulation and possible roles in cardiovascular control. Biochem Pharmacol 38:1877–1883

Yoshimi H, Hirata Y, Fukuda Y, Kawano Y, Emori T, Kuramochi M, Omae T, Marumo F (1989) Regional distribution of immunoreactive endothelin in rats. Peptides 10:805–808

Yoshizumi M, Inui D, Okishima N, Houchi H, Tsuchiya K, Wakabayashi H, Kido H, Tamaki T (1998a) Endothelin-1-(1–31), a novel vasoactive peptide, increases [Ca^{2+}](i) in human coronary artery smooth muscle cells. Eur J Pharmacol 348:305–309

Yoshizumi M, Kim S, Kagami S, Hamaguchi A, Tsuchiya K, Houchi H, Iwao H, Kido H, Tamaki T (1998b) Effect of endothelin-1 (1–31) on extracellular signal regulated kinase and proliferation of human coronary artery smooth muscle cells. Br J Pharmacol 125:1019–1027

CHAPTER 15

Endothelin-1 and Pulmonary Hypertension

D.J. STEWART and Y.D. ZHAO

A. Mechanisms of Pulmonary Hypertension

The pulmonary bed is distinct from the systemic beds in that it must accommodate the entire cardiac output and provide the conditions necessary for efficient gas exchange. One critical aspect of pulmonary vascular physiology is the ability of the lung circulation to adapt to large variations in cardiac output with minimal impedance to blood flow at all levels (VOELKEL and WEIR 2000). In patients with PH, this ability is compromised, usually because of a narrowing or loss of pulmonary microvessels, due to an increase in vascular tone (i.e., vasoconstriction), remodelling, with hypertrophy of medial and intimal layers and muscularization of small arterioles (VOEWKEL and WEIR 2000), or destruction of vascular channels.

The mechanism of PH is sometimes clearly evident, as in the case of PH associated with pulmonary thromboembolism. In this condition, blockage of large pulmonary arteries increases the impedance of the lung. However, in other cases of PH, the aetiology is often uncertain even when there is a clear association with a well defined pulmonary or systemic disorder. Perhaps a good example of this is PH secondary to congestive heart failure. This is usually a complication of advanced heart failure, and is found in the setting of persistent and marked elevations in left ventricular filling pressures. Thus, it has been suggested that increases in transpulmonary gradient, that is the difference between the mean pulmonary artery pressure and the left atrial pressure, result from a "reactive" increase in pulmonary arteriolar tone caused by the chronic increase in capillary pressure. However, at present it is unclear how such a response might be transduced, and what specific stimuli might interact in the development of this complication. What is certain is that the appearance of PH foreshadows a poor outcome, and may contribute negatively to the overall prognosis of patients with advanced heart failure. Moreover, the presence of this complication can preclude the use of more definitive therapeutic options in the disease, namely cardiac transplantation.

B. Endothelial Dysfunction and Pulmonary Hypertension

A role of endothelial dysfunction in the pathogenesis of primary pulmonary hypertension has been previously suggested (MENDELSOHN et al. 1992). In pulmonary hypertension associated with shunts and high blood flow and repeated exposure to endotoxin, endothelial injury or dysfunction is apparent (VOELKEL and WEIR 2000). In an animal model of pulmonary hypertension induced by monocrotaline administration (MEYRICK and REID 2000), endothelial injury was an early event, followed by endothelial proliferation and finally the development of vascular remodelling. In humans, evidence of abnormal endothelial function has often been inferred from the circulating levels of endothelium-derived substances. Increased von Willebrand factor (vWF) antigen levels have been reported in patients with primary PH (RABINOVITCH et al. 1987; GEGGEL et al. 1987) and abnormalities in circulating forms of vWF correlated well with increased immunostaining in lung biopsy tissue. More recently, an imbalance in vasoconstrictor and vasodilator prostanoids in the pulmonary circulation has been demonstrated in idiopathic PH, as assessed by the measurement of the stable metabolites of prostacyclin and thromboxane (CHRISTMAN et al. 1992). Therefore, with the discovery of the endothelin-1 by Yanagisawa et al., it was natural to consider a potential role of this potent endothelium-derived vasoconstrictor and growth factor in PH, particularly in patients with primary disease. In the following sections, the evidence in favour of a contribution of endothelin in pulmonary vascular diseases will be briefly reviewed.

C. The Search for a "Vasoconstrictor Factor" in Pulmonary Hypertension

Pulmonary hypertension (PH) remains a poorly understood disease. As a result of our lack of knowledge regarding its pathogenesis, there is a paucity of specific therapy, which is most acutely felt for primary PH (PPH) (WORLD HEALTH ORGANIZATION 1975). This rare disorder affects predominantly females and shows a bimodal age distribution, striking both the young (children to young adults) and those in late middle age (WALCOTT et al. 1970; FISHMAN 1998). PPH is a devastating disorder and, with survival periods often less than two years from initial diagnosis(FISHMAN 1998); it has a prognosis as poor as many malignant cancers. There is no single pathopneumonic feature for PPH and the clinical diagnosis relies on the careful exclusion of various secondary causes of PH, including pulmonary, cardiac and systemic disorders. Even the hallmark plexiform lesions found on histological examination of lung biopsies (WAGENVOORT and WAGENVOORT 1970) neither rule in or out the diagnosis since they are not found in all instances of PPH and they are also characteristic of severe secondary forms of PH, such as Eisenmenger's syndrome involving large congenital left to right shunts (VONGPATANASIN et al 1998).

PPH is a progressive disorder; and in its early stages is characterized by pulmonary vasoconstriction that can be reversed by various vasodilator agents, but in its later stages there is progressive pulmonary vascular remodelling or obstruction, which produces fixed increases in pulmonary vascular resistance (PVR). In the 1950s, Prof. Paul Wood developed a paradigm for the pathogenesis of PH that remains of value even today (WOOD 1950). He proposed the existence of a "vasoconstrictor factor" which might be responsible for the inappropriate increase in PVR in the early stages of PPH, and possible even contribute to pulmonary vascular remodelling.

Until the discovery of endothelin-1 (ET-1), vasoconstrictor prostaglandins such as thromboxane A_2 (TxA_2) were prime candidates for the elusive vasoconstrictor factor. The work of CHRISTMAN et al. (1992) demonstrated an imbalance in the release of TxA_2 compared to vasodilator prostaglandins [i.e., prostacyclin (PGI_2)] in patients with PPH compared to patients with secondary PH due to chronic obstructive lung disease. Indeed, consistent with this paradigm the continuous infusion of PGI_2 analogues, such as Iloprost (Flolan), was developed as a therapeutic strategy for PPH, and is currently in wide spread clinical use (BARST et al. 1996; STRICKER et al. 1999). However, despite hemodynamic improvement and modest prognostic benefit observed with its use, long-term intravenous infusion of Flolan is associated with significant morbidity, mainly related to the need for a chronic indwelling catheter. There also exists significant tachyphylaxis resulting in progressive dose escalation. Moreover, this therapy did not prove to be beneficial in secondary PH due to chronic congestive heart failure, resulting in excess of cardiac events likely due to significant systemic vasodilation. Thus, PH (both primary and secondary) largely remains an urgent target for new therapies.

The discovery of ET-1 by YANAGISAWA et al. (1988) marked the beginning of a new era in the understanding of the pathogenesis of PH. This review will focus predominantly on ET-1, since it is the most relevant isoform for vascular diseases. ET-1 is the most potent pulmonary vasoconstrictor factor yet identified (FURCHGOTT and VANHOUTTE 1989; VANHOUTTE et al. 1991). Moreover, ET-1-induced vasoconstriction is characterized by an unusually protracted nature of response, reminiscent of pathological vasoconstriction and even vasospasm. ET-1 acts by binding and activating two distinct endothelin receptors, ET_A and ET_B (ARAI et al. 1990; SAITO et al. 1991), discussed in more detail elsewhere (see Chap. 4). ET_A is selective for ET-1 (and the closely related ET-2) and is found predominantly on "target tissues", such as vascular smooth muscle cells (SMC) (SAETRUM et al. 1994; BAX and SAXENA 1994), mediating the well known vasoconstrictor effects of ET-1. In contrast, ET_B is nonselective and interacts with equal affinity with all three ET isoforms and is found predominantly on the vascular endothelium (DASHWOOD et al. 1998; KOBAYASHI et al. 1998; BAX and SAXENA 1994). As will be discussed in more detail below, binding of ET-1 to this endothelial receptor results in the release of vasodilator autacoids, such as nitric oxide (NO) and PGI_2, which counteract the direct vasoconstrictor effects of ET-1(VANE et al. 1990; STEWART and BAFFOUR 1990) (see Chap. 11). More recently it has been recognized that ET_B may also be expressed on SMCs and in this position it may mediate vasoconstriction in a manner identical to ET_A (WARNER et al. 1994). A detailed review of signal transduction and ET receptor biology is beyond the scope of the present review and the reader is referred to other chapters (Chaps. 4, 5) dealing specifically with these issues. Thus, there appears to be a remarkable parallelism between the endothelin and prostaglandin systems with respect to regulation of pulmonary vascular tone, involving the balance between vasoconstrictor and vasodilator effects. In the case of ET-1, the fulcrum of this balance resides at both the level of ET-1 production and the relative expression of endothelial (i.e., ET_B) and smooth muscle (ET_A or ET_B) receptors.

In addition to its vasoconstrictor effects, ET-1 also has other important biological actions which may be highly relevant to the pathogenesis of PH. ET-1 is a growth factor (BATTISTINI et al. 1993), and thus may contribute to SMC proliferation and hypertrophy which are important components of vascular remodelling (MOREAU et al. 1997) (see Chap. 12). Although it induces only relatively weak mitogenic actions in isolation, it is a potent co-mitogen in the presence of other growth factors (HAFIZI et al. 1999). Perhaps more importantly from the standpoint of pulmonary vascular remodelling is the ability of ET-1 to induce the production and secretion of extracellular matrix (CAMERON 1998; FALANGA et al. 1992). ET-1 is a potent secretogogue for collagen, and the importance of the endothelin system in fibrosis and scaring has been demonstrated in a number of organs (NGUYEN et al. 1998; HOCHER et al. 1997), including the lung (GIAID et al. 1993a). Recently the importance of extracellular matrix, and the various enzymes which regulate its turnover, have been convincingly demonstrated in vascular remodelling in systemic arteries (STRAUSS et al. 1994) and pulmonary arteries and arterioles in PH (RABINOVITCH 1998).

Thus, ET-1 fulfils the description of the long sought "vasoconstrictor factor" postulated to play an important role in the pathogenesis of PPH. For this reason our group was intrigued by the possibility that ET-1 production may be increased in patients with PH, particularly in those with PPH. Less than a year after its discovery, we developed a radioimmunoassay and measured plasma levels of ET-1 in various cardiovascular diseases (STEWART et al. 1991, 1992; CERNACEK and STEWART 1989). We found marked elevations in circulating ET-1 in many cardiovascular disorders, with greatest increase (up to 40-fold) in patients with cardiogenic shock following myocardial infarction (CERNACEK and STEWART 1989). However, patients with PH also exhibited a significant increase in plasma ET-1 compared to normal subjects, albeit some less marked than in patients in shock. We attempted to understand more fully the alterations in ET-1 metabolism in PH by measuring its plasma levels in the systemic venous (or mixed venous) blood flowing into the lung compared with levels in the systemic arterial (or aortic) blood representing the pulmonary effluent. This report included a group of patients with PH due to a variety of secondary causes, compared to patients with PPH by strict NHLBI criteria

(STEWART et al. 1991). Both groups showed elevated ET-1 levels in the systemic venous and arterial samples compared to normal subjects or patients with stable coronary disease and no PH. Although on average there was no significant difference in mean plasma ET-1 levels between the primary and secondary groups, there was an intriguing difference in the metabolism of ET-1 across the lung in these patients. While the control subjects showed significantly lower ET-1 levels in plasma samples of the pulmonary outflow vs inflow, patients with secondary PH demonstrated no significant gradient across the pulmonary bed. These data represented the first evidence in humans that the lung may play a crucial role in the clearance of circulating ET-1 under physiological conditions, an important physiological function which is discussed in more detail below. The lack of difference between venous and arterial ET-1 levels in secondary PH is open to a variety of interpretations, but it is consistent with either a loss of ET-1 clearance by the lung or an increase in pulmonary production or both. More sophisticated studies are needed, for example using the multiple indicator dilution method, to distinguish between these potential mechanisms, as will be discussed below. Of particular interest, however, was the observation that in patients with PPH the gradient was reversed, with ET-1 levels being significantly higher in the aortic (pulmonary effluent) than the mixed venous (inflow) blood. This result could only be explained by an increase in the local production of ET-1 within the lungs, providing the first evidence of activation of the ET system in the pulmonary vasculature of PPH patients.

D. ET-1 Homeostasis in the Pulmonary Vasculature in Health and Disease

Based on the earlier observations suggesting a gradient of plasma ET-1 across the normal lung, DUPUIS et al. (1994) examined the metabolism of ET-1 in the pulmonary vasculature using a multiple indicator dilution technique which had been widely applied to the study of catecholamine spillover (DUPUIS et al. 1994). In normal anaesthetized dogs, there was a 50% first pass clearance of ET-1 across the lung. Thus, under physiological conditions the lung appears to be the major site of clearance of circulating ET-1. However, under conditions of supraphysiological levels of ET-1, for example as a result of administration of high doses of exogenous peptide, the pulmonary clearance mechanisms may become saturated, and then other organs, particularly the kidney, take on a more important role in ET-1 clearance (GASIC et al. 1992; PERNOW et al. 1989). In subsequent studies the role of the lung in the clearance of ET-1 from the circulation was demonstrated to be entirely dependent on the ET_B receptor (DUPUIS et al. 1996), presumably expressed on the luminal surface of the pulmonary vascular endothelium. As suggested on the basis of simple transpulmonary plasma determinations, it was also found that ET-1 clearance was profoundly reduced in patients with heart failure and PH due to mitral stenosis (DUPUIS et al. 1998), confirming that this is likely an important mechanism of elevation in circulating levels in this condition.

Thus, both increased pulmonary production and decreased clearance may contribute to the changes in circulating ET-1 seen in patients with PH. In order to distinguish between these two complementary mechanisms it is necessary to study tissue expression of ET-1 peptide and mRNA in PH of various etiologies. In the first human study, tissue distribution of ET-1 was determined using imunocytochemistry and prepro-ET-1 mRNA expression was defined by in situ hybridization in specimens from patients undergoing lung transplantation for PH compared to lung tissue from patients without PH undergoing other pulmonary surgery or to normal donor lungs explanted but not used in transplant surgery (GIAID et al. 1993b). In the control lung specimens, there was little evidence of ET-1 expression at any level of the pulmonary vasculature. However, in patients with PH there was a remarkable increase in ET-1 peptide and mRNA expression that was largely localized to the vascular endothelium. Increased ET-1 expression was found at all levels of the vasculature, including large pulmonary arteries and veins, arterioles and capillaries, and even the systemic bronchial arteries. The greatest increases in expression were seen in the muscular arteries and arterioles, which also demonstrated the greatest degree of morphological abnormalities, specifically medial hyperplasia. Interestingly, increased ET-1 expression was seen to a similar degree in both secondary and primary "plexigenic" PH, with the exception of muscular pulmonary arteries which exhibited significantly greater staining in the PPH lungs. In addition, very high levels of ET-1 expression were also associated with plexiform lesions that are characteristic of PPH, and may represent an abortive attempt at recanalization of occluded vascular channels. Moreover, there was a direct and significant correlation between increased pulmonary endothelial ET-1 immunoreactivity and the degree of elevation in total pulmonary vascular resistance in a subset of patients with "plexigenic" PH for whom hemodynamic data were available. This provides evidence of a fundamental link between ET-1 overproduction and the abnormalities in vascular structure and function in PPH. In contrast, there was no such relationship for patients with PH due to various secondary causes. However, given the highly heterogeneous nature of this group and the small numbers of patients in any given category, one cannot exclude that a relationship did exist for certain subgroups (i.e. heart failure), and not for others (i.e. pulmonary thromboembolism). Indeed, Cody has shown a remarkable correlation between plasma levels of ET-1 and the degree of inappropriate elevation in pulmonary vascular resistance in patients with congestive heart failure, consistent with a role for increased ET-1 production in the pathogenesis of secondary PH in this disorder (CODY et al. 1992). Therefore, these observations provided additional support for local activation of the endothelin system in the pulmonary vasculature of patients with PH, in particular suggesting a role for ET-1 both in the medial hypertrophy of small muscular arteries and arterioles and possibly in the pathogenesis of plexiform arteriopathy. Other studies have also demonstrated elevations in plasma and tissue ET-1 in a variety of other conditions associated with pulmonary hypertension, including high altitude pulmonary edema (GOERRE et al. 1995), persistent PH of the newborn (KUMAR et al. 1996), and PH associated with congenital heart disease (LUTZ et al. 1999), among others.

Activation of the endothelin system has also been found in diverse animal models of PH, including chronic hypoxia model in the rat (LI et al. 1994; STELZNER et al. 1992) and monocrotaline rat models (FRASCH et al. 1999; MIYAUCHI et al. 1993), as well PH induced by endotoxin (SNAPPER et al. 1998), diaphgragmatic hernia (OKAZAKI et al. 1998), or air-induced chronic pulmonary hypertension (TCHEKNEVA et al. 1998), further strengthening the association between activation of the ET system and PH. In addition, alterations in the expression of ET receptors have been reported that are consistent with a role for this system in PH. In the canine rapid pacing model of heart failure, we have also shown increased expression of both ET-1 and ET_A, the predominant receptor on vascular smooth muscle (MOE et al. 1998). At the same time, the expression of ET_B, which is present on vascular endothelium and mediates clearance of ET peptides from the circulation and release of vasodilator autacoids, was profoundly down-regulated. A very similar pattern of ET receptor expression was seen in the rat myocardial infarction model of heart failure (KOBAYASHI et al. 1998, 1999) and chronic hypoxia-induced PH was found to be associated with increased ET_A relative to ET_B signalling in isolated SMCs in vitro (LI et al. 1999; SOMA et al. 1999). However, another group has reported increased ET_B-induced pulmonary vasodilation and receptor expression in the chronic hypoxic PH model (MURAMATSU et al. 1999), although the opposite has also been demonstrated (SATO et al. 1999).

These results suggest a paradigm for ET activation in PH (Fig. 1). Not only is the expression and local production of ET markedly increased, but there is also a parallel increase in expression of the predominant vasoconstrictor ET receptor, thereby amplifying the biological effect of the increased ET-1. Moreover, the endothelial ET_B receptor is profoundly down-regulated (with the possible exception of hypoxia-induced PH), and thus there is less buffering of the direct actions of ET-1 by NO and PGI_2, further potentiating its vasoconstrictor, pro-proliferative, and fibrotic effects. Finally, the clearance of ET-1 may be reduced by the same mechanism, prolonging its actions and contributing to even higher steady-state tissue and circulating ET-1 levels. Therefore, the activation of the ET system in PH appears to involve a positive feedback loop resulting in both higher levels of ET-1 and greater vascular responses, both in terms of vasoconstriction and remodelling.

Further evidence supporting a central role for the endothelin system in the pathogenesis of pulmonary hypertension has come from gene transfer experiments, producing overexpression of ET-1 in the pulmonary vasculature and/or parenchyma. Liposomes incorporating UV-inactivated hemagglutinating virus of Japan liposomes were used to encapsulate the ET-1 gene. The plasmid DNA and liposome complexes were instilled into the trachea or

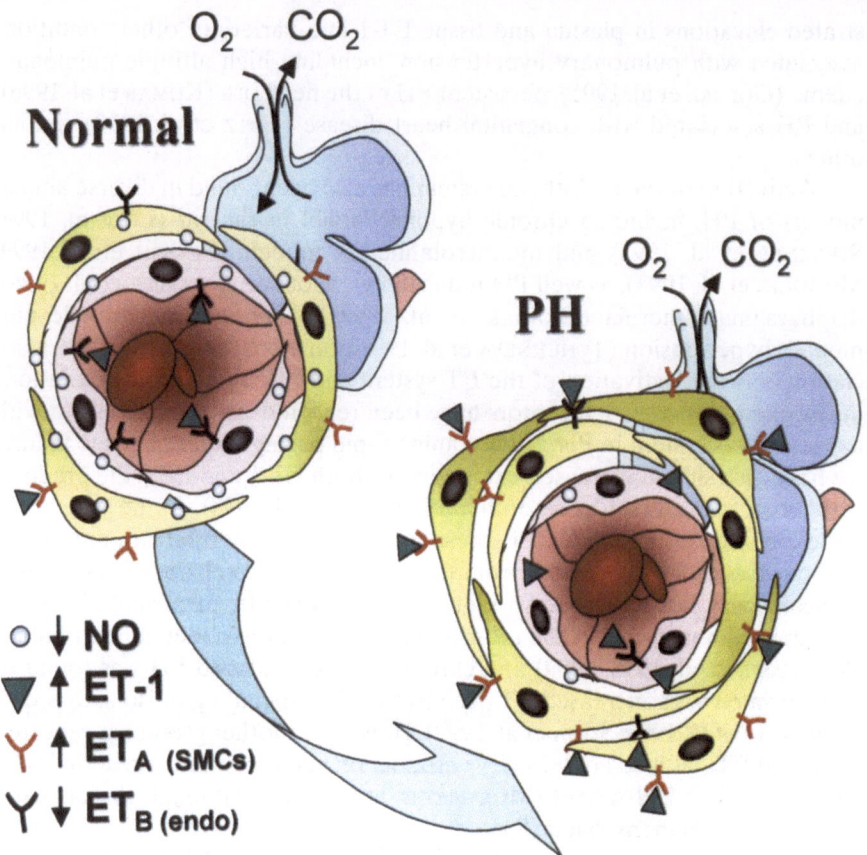

Fig. 1. Alteration in the balance of endothelial factors in pulmonary hypertension (*PH*). Under normal physiological conditions, there is a relatively low basal ET-1 production (*green triangles*) by pulmonary arteriolar endothelial cells (*EC*). These cells also express abundant ETB receptors (*black 'Y'*), which interact with both the ET-1 produced locally and that circulating in the blood. Binding of ET-1 to endothelial ETB results in the release of endothelial-derived vasodilator factors, such as NO (*blue spheres*) and represents a major pathway of clearance of this peptide, thus maintaining vascular homeostasis. However, in PH, EC activation results in increased local ET-1 production, while at the same time endothelial ETB receptor expression is reduced. This results in a decrease in ET-1 clearance and reduced NO release, and therefore an imbalance between these opposing endothelial factors in the pulmonary microcirculation. ETA (and ETB) receptors (*red 'Y'*) on pulmonary smooth muscle cells are also upregulated, potentiating the vasoconstrictor and growth promoting effects of increased ET-1 expression in PH

injected into the pulmonary artery of Wistar rats to achieve local ET-1 over-expression in the lung (Takeda et al. 1997, 1998). Tracheo-bronchial ET-1 gene transfer resulted in hyperplastic connective tissue plaques, reminiscent of bronchiolitis obliterans, whereas pulmonary arterial overexpression produced smooth muscle proliferation and medial hypertrophy, similar to that seen in pulmonary hypertension. This suggests that ET-1 may contribute to a variety of pulmonary disorders characterized by increased cell growth and matrix production, depending on the specific pattern of gene expression, and localization of the ET-1 production is critical in determining the pathological phenotype (LEE et al. 1998). This is consistent with the observation of striking ET-1 immunostaining, localized predominantly to type 2 pneumocytes, in patients with idiopathic pulmonary fibrosis, whether or not there was concomitant PH (SALEH et al. 1997). Transgenic mouse models of ET-1 overexpression under control of its own promoter have also demonstrated an intriguing phenotype dominated not by vasoconstriction and systemic hypertension as might be expected, but by chronic inflammation and fibrosis of the kidney (THEURING et al. 1998) and the lung (B. Hocher, personal communications). Again, these studies underscore the versatility of this peptide in pulmonary pathology, involving a spectrum of biological actions.

Recently, we have developed a cell-based pulmonary gene transfer approach that results in highly selective transfection of the pulmonary microcirculation at the level of the distal arteriolar bed (CAMPBELL et al. 1999). This method was used successfully to reduce monacrotaline-induced PH in the rat model, using eNOS as therapeutic transgene (CAMPBELL et al. 1999). We tested the hypothesis that overexpression of ET-1 localized to the distal pulmonary arteriolar level could reproduce the hemodynamic and histological features of PH. Pulmonary artery smooth muscle cells from Fisher 344 rats were established in culture and transfected with the full-length coding region of human prepro-ET-1, and then reintroduced into the pulmonary circulation of normal syngeneic rats. After 28 days, right ventricular systolic pressure was found to be significantly elevated (Figs. 2 and 3) compared with control animals that received only mock transfected cells. In addition, significant morphological changes of the pulmonary vasculature could be discerned on histological examination, including abnormal arteriolar muscularization and medial hypertrophy of small muscular arteries. These changes were similar but not as marked as those induced by the monacrotaline, which itself was associated with marked increase in endogenous ET-1 expression. Moreover, the combination of monocrotaline with ET-1 gene transfer did not potentiate the development of PH in response to this potent pulmonary endothelial cell toxin, again consistent with a high degree of endogenous ET-1 production in this model.

E. Role of ET Antagonists in Pulmonary Hypertension

Based on these observations, it would be predicted that inhibition of the ET system would break the vicious cycle of increased ET-1 production and activity, and be effective in the treatment of PH. BQ-123, an ET_A selective, peptide ET antagonist, was the first to be used in an animal model of PH. In the rat monocrotaline model of PH (MIYAUCHI et al. 1993) BQ123 nearly completely

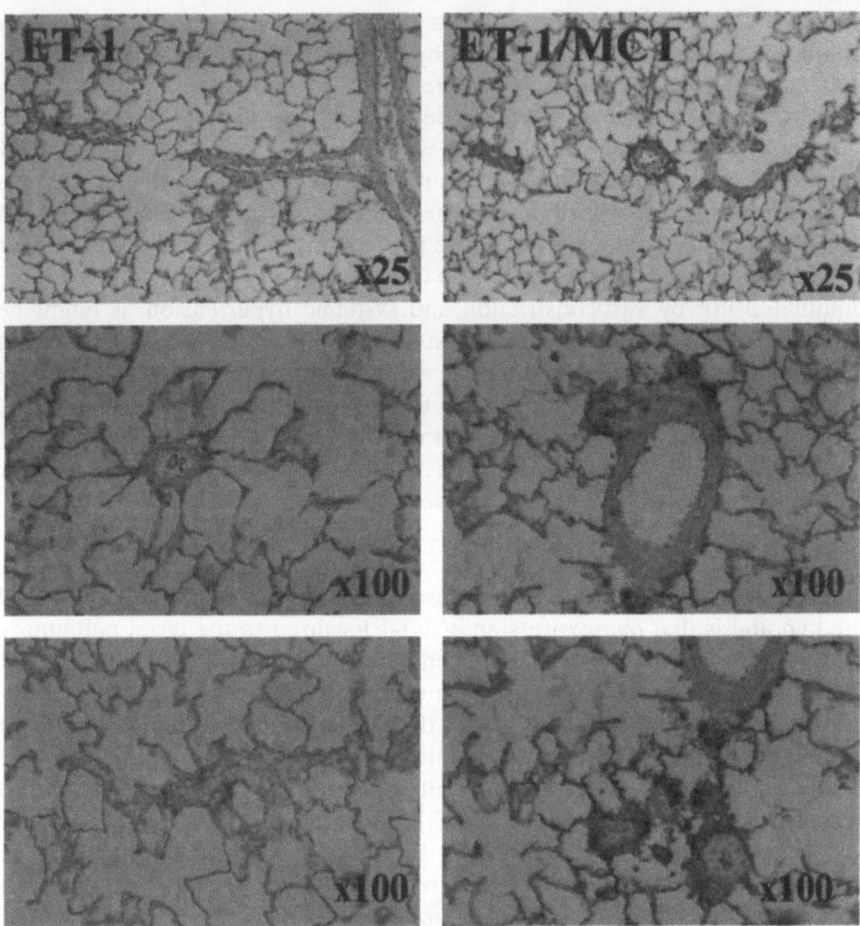

Fig. 2. Representative photomicrographs depicting the morphological abnormalities in rats receiving cell-based gene transfer of prepro-ET-1 alone (*left panels*) or together with the potent endothelial cell toxin, monocrotaline (MCT 70 mg/kg i.p., *right panels*). ET-1 gene transfer resulted in patchy increases in ET-1 expression (*brown staining*) localized mainly to the small muscular arterioles, which also showed evidence of modest increase in muscularization and medial hypertrophy. Treatment with MCT resulted in a much greater ET-1 expression and more severe pulmonary vascular abnormalities

prevented the increase in right ventricular systolic pressure, and normalized the ratio of right to left ventricular thickness, indicative of improvement in PVR. The same antagonist was also effective in both preventing and reversing PH induced by long-term exposure of rats to hypoxic air (OPARIL et al. 1995). In this study, the authors also demonstrated that the marked increase in medial hypertrophy of small muscular arteries and arterioles could be prevented by BQ-123. The same investigators have also reported that Bosen-

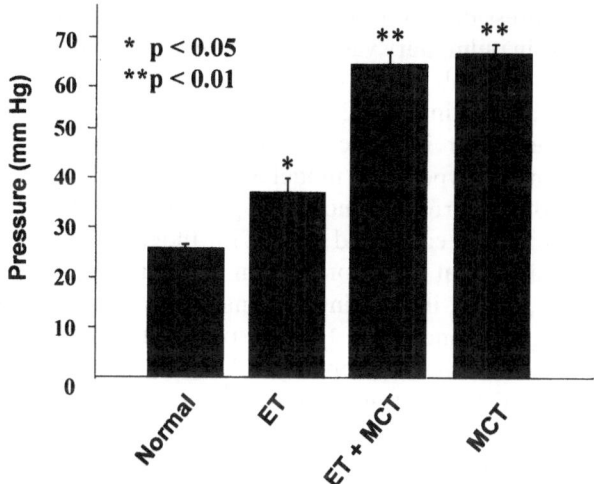

Fig. 3. Effect of ET-1 gene transfer of right ventricular (RV) systolic pressure in normal rats, rats treated with cell-based ET-1 gene transfer alone or together with MCT, compared with MCT alone. ET-1 gene transfer resulted in a significant increase in pulmonary arterial pressure compared with untreated normal rats. Addition of MCT resulted in a further substantial increase in RV pressure to levels not different from MCT alone

tan, a "balanced" ET_A and ET_B antagonist, improved PH and its sequellae in the hypoxia model (CHEN et al. 1995). A number of other reports have confirmed the efficacy ET antagonist in these (LI et al. 1999; TJEN-A-LOOI et al. 1996) and other models of PH (COPPOLA et al. 1998; SCHMECK et al. 1998). In particular, the use of a selective ET_A antagonists reduced markedly pulmonary arterial and right atrial pressures and PVR in the rat myocardial infarction model of heart failure (SAKAI et al. 1996), suggesting a role for these agents in the therapy of secondary, as well as primary forms of PH. In addition, inhibition of the endothelin converting enzyme has been shown to reduce the deleterious effects of monocrotaline-induced PH, again consistent with the view that increased ET-1 biosynthesis is an important mechanism of PH in this model.

An important question that remains to be addressed relates to the relative merits of selective ET_A vs mixed $ET_{A/B}$ receptor antagonists. Theoretical arguments can be developed to support either approach. For instance selective antagonists might offer certain advantages since they block ET_A, and therefore interrupt the vicious cycle of ET-induced vasoconstriction and vascular remodelling, without inhibiting the endothelial ET_B receptor. Thus, preserving the actions of ET-1 on the endothelium and increasing the production and release of vasodilatory autacoids might serve to amplify the beneficial effects of ET_A blockade, resulting in greater vasodilation and beneficial remodelling of the pulmonary vasculature. On the other hand, it has been proposed that the expression of ET_B receptors on smooth muscle cells might play an important role in pulmonary vasoconstriction in some species (McCULLOCH et al. 1998; MacLEAN et al. 1998; LaDOUCEUR et al. 1993), and therefore mixed ET receptor blockade using and $ET_{A/B}$ antagonist might be more effective.

The evidence so far is not conclusive. BQ-123 and Bosentan appear equally effective in the hypoxic rat model of PH (OPARIL et al. 1995). However, Bosentan may not be a truly balanced antagonist since it has about a 30-fold greater affinity for ET_A compared with ET_B. Indeed, a majority of reports support the predominant role for ET_A in the pulmonary vasoconstrictor response to ET-1, which, if anything, becomes more important in models of PH (FRANCO-CERECEDA and HOLM 1998; UNDERWOOD et al. 1999; HOLM 1997). However, more head to head comparisons between the selective and nonselective classes of ET receptor antagonists are needed to address this critical issue better.

To date there is only very limited data on the effect of ET antagonists in patients with PH, and the studies that have been reported have addressed predominantly PH secondary to congestive heart failure (CHF). SUTSCH et al. (1998) demonstrated highly favourable hemodynamic effects of short-term infusions of Bosentan in patients with severe CHF (NYHA class III or IV), with a somewhat greater decrease in pulmonary pressures and PVR than that seen in the systemic circulation. Moreover, these same investigators found that the acute effects of the ET antagonist were sustained over a two used week period of oral administration (SUTSCH et al. 1998). Indeed, the hemodynamic effects of Bosentan were greater after two weeks of treatment than on initial dosing. This was particularly true for the pulmonary circulation, with even greater relative reductions in pressures and vascular resistance compared with the systemic bed (SUTSCH et al. 1998). Even greater selectivity of action on the pulmonary vascular bed was observed in patients with CHF using a new highly selective ET_A antagonist developed by the Texas Biotechnology Corporation. Dramatic decreases in PAP and PVR were noted with the virtual absence of a systemic vasodilator response (D.L. Mann, personal communication). These early findings confirm an important role of ET in PH secondary to CHF, and strongly support the view that ET receptor antagonists will be useful in the treatment of patients with PH. Whether these agents will also be effective in patients with other secondary etiologies of PH, or perhaps more importantly in those with PPH, remains to be determined.

References

Arai H, Hori S, Aramori I, Ohkubo H, Nakanishi S (1990) Cloning and expression of a cDNA encoding an endothelin receptor. Nature 348:730–732
Barst RJ, Rubin LJ, Long WA et al. (1996) A comparison of continuous intravenous epoprostenol (prostacyclin) with conventional therapy for primary pulmonary hypertension. The Primary Pulmonary Hypertension Study Group. N Engl J Med 334:296–302

Battistini B, Chailler P, D'Orleans-Juste P, Briere N, Sirois P (1993) Growth regulatory properties of endothelins. Peptides 14:385–399

Bax WA, Saxena PR (1994) The current endothelin receptor classification: time for reconsideration? Trends Pharmacol Sci 15:379–386

Cameron IT (1998) Matrix metalloproteinases, prostaglandins and endothelins: paracrine regulators of implantation. Gynecol Endocrinol 12:415–419

Campbell AI, Kuliszewski MA, Stewart DJ (1999) Cell-based gene transfer to the pulmonary vasculature: Endothelial nitric oxide synthase overexpression inhibits monocrotaline-induced pulmonary hypertension. Am J Respir Cell Mol Biol 21:567–575

Cernacek P, Stewart DJ (1989) Immunoreactive endothelin in human plasma: marked elevations in patients in cardiogenic shock. Biochem Biophys Res Commun 161:562–567

Chen SJ, Chen YF, Meng QC, Durand J, DiCarlo VS, Oparil S (1995) Endothelin-receptor antagonist bosentan prevents and reverses hypoxic pulmonary hypertension in rats. J Appl Physiol 79:2122–2131

Christman BW, McPherson CD, Newman JH et al. (1992) An imbalance between the excretion of thromboxane and prostacyclin metabolites in pulmonary hypertension. N Engl J Med 327:70–75

Cody RJ, Haas GJ, Binkley PF, Capers Q, Kelley R (1992) Plasma endothelin correlates with the extent of pulmonary hypertension in patients with chronic congestive heart failure [published erratum appears in Circulation 1993 Mar;87(3):1064]. Circulation 85:504–509

Coppola CP, Au-Fliegner M, Gosche JR (1998) Endothelin-1 pulmonary vasoconstriction in rats with diaphragmatic hernia. J Surg Res 76:74–78

Dashwood MR, Mehta D, Izzat MB et al. (1998) Distribution of endothelin-1 (ET) receptors (ET(A) and ET(B)) and immunoreactive ET-1 in porcine saphenous vein-carotid artery interposition grafts. Atherosclerosis 137:233–242

Dupuis J, Cernacek P, Tardif JC et al. (1998)Reduced pulmonary clearance of endothelin-1 in pulmonary hypertension. Am Heart J 135:614–620

Dupuis J, Goresky CA, Fournier A (1996) Pulmonary clearance of circulating endothelin-1 in dogs in vivo: exclusive role of ETB receptors. J Appl Physiol 81:1510–1515

Dupuis J, Goresky CA, Stewart DJ (1994) Pulmonary removal and production of endothelin in the anesthetized dog. J Appl Physiol 76:694–700

Falanga V, Katz MH, Kirsner R, Alvarez AF (1998) The effects of endothelin-1 on human dermal fibroblast growth and synthetic activity. J Surg Res 53:515–519

Fishman AP (1998) Etiology and pathogenesis of primary pulmonary hypertension: a perspective. Chest 114:242S–247S

Franco-Cereceda A, Holm P (1998) Selective or nonselective endothelin antagonists in chronic hypoxic pulmonary hypertension? J Cardiovasc Pharmacol. 31 Suppl 1:S447–S452

Frasch HF, Marshall C, Marshall BE (1999) Endothelin-1 is elevated in monocrotaline pulmonary hypertension. Am J Physiol 276:L304–L310

Furchgott RF, Vanhoutte PM (1989) Endothelium-derived relaxing and contracting factors. FASEB J 3:2007–2018

Gasic S, Wagner OF, Vierhapper H, Nowotny P, Waldhausl W (1992) Regional hemodynamic effects and clearance of endothelin-1 in humans: renal and peripheral tissues may contribute to the overall disposal of the peptide. J Cardiovasc Pharmacol 19:176–180

Geggel RL, Carvalho AC, Hoyer LW, Reid LM (1987) von Willebrand factor abnormalities in primary pulmonary hypertension. Am Rev Respir Dis 135:294–299

Giaid A, Michel RP, Stewart DJ, Sheppard M, Corrin B, Hamid Q (1993a) Expression of endothelin-1 in lungs of patients with cryptogenic fibrosing alveolitis. Lancet 341:1550–1554

Giaid A, Yanagisawa M, Langleben D et al. (1993b) Expression of endothelin-1 in the lungs of patients with pulmonary hypertension. N Engl J Med 328:1732–1739

Goerre S, Wenk M, Bartsch P et al. (1995) Endothelin-1 in pulmonary hypertension associated with high-altitude exposure. Circulation 91:359–364

Hafizi S, Allen SP, Goodwin AT, Chester AH, Yacoub MH (1999) Endothelin-1 stimulates proliferation of human coronary smooth muscle cells via the ET(A) receptor and is co-mitogenic with growth factors. Atherosclerosis 146:351–359

Hocher B, Thone-Reineke C, Rohmeiss P et al. (1997) Endothelin-1 transgenic mice develop glomerulosclerosis, interstitial fibrosis, and renal cysts but not hypertension. J Clin Invest 99:1380–1389

Holm P (1997) Endothelin in the pulmonary circulation with special reference to hypoxic pulmonary vasoconstriction. Scand Cardiovasc J Suppl 46:1–40

Kobayashi T, Miyauchi T, Sakai S et al. (1999) Expression of endothelin-1, ETA and ETB receptors, and ECE and distribution of endothelin-1 in failing rat heart. Am J Physiol 276:H1197–H1206

Kobayashi T, Miyauchi T, Sakai S et al. (1998) Down-regulation of ET(B) receptor, but not ET(A) receptor, in congestive lung secondary to heart failure. Are marked increases in circulating endothelin-1 partly attributable to decreases in lung ET(B) receptor-mediated clearance of endothelin-1? Life Sci 62:185–193

Kumar P, Kazzi NJ, Shankaran S (1996) Plasma immunoreactive endothelin-1 concentrations in infants with persistent pulmonary hypertension of the newborn. Am J Perinatol 13:335–341

LaDouceur DM, Flynn MA, Keiser JA, Reynolds E, Haleen SJ (1993) ETA and ETB receptors coexist on rabbit pulmonary artery vascular smooth muscle mediating contraction. Biochem Biophys Res Commun 196:209–215

Lee KM, Tsai KY, Wang N, Ingber DE (1998) Extracellular matrix and pulmonary hypertension: control of vascular smooth muscle cell contractility. Am J Physiol 274:H76–H82

Li H, Chen SJ, Chen YF et al. (1994) Enhanced endothelin-1 and endothelin receptor gene expression in chronic hypoxia. J Appl Physiol 77:1451–1459

Li KX, Fouty B, McMurtry IF, Rodman DM (1999) Enhanced ET(A)-receptor-mediated inhibition of K(v) channels in hypoxic hypertensive rat pulmonary artery myocytes. Am J Physiol 277:H363–H370

Lutz J, Gorenflo M, Habighorst M, Vogel M, Lange PE, Hocher B (1999) Endothelin-1- and endothelin-receptors in lung biopsies of patients with pulmonary hypertension due to congenital heart disease. Clin Chem Lab Med 37:423–428

MacLean MR, Mackenzie JF, Docherty CC (1998) Heterogeneity of endothelin-B receptors in rabbit pulmonary resistance arteries. J Cardiovasc Pharmacol 31 Suppl 1:S115–S118

McCulloch KM, Docherty C, MacLean MR (1998) Endothelin receptors mediating contraction of rat and human pulmonary resistance arteries: effect of chronic hypoxia in the rat [published erratum appears in Br J Pharmacol 1998 Aug;124(7):1586]. Br J Pharmacol 123:1621–1630

Mendelsohn ME, Loscalzo J (1992) Endotheliopathies. In: Vascular Medicine: a textbook of vascular biology and disease, edited by Loscalzo J, Creager MA, Dzau YJ. Little Brown, and Co., Boston, 279–305

Meyrick B, Reid L (2000) Development of pulmonary arterial changes in rats fed crotalaria spectabilis. Am J Path 94:37

Miyauchi T, Yorikane R, Sakai S et al. (1993) Contribution of endogenous endothelin-1 to the progression of cardiopulmonary alterations in rats with monocrotaline-induced pulmonary hypertension. Circ Res 73:887–897

Moe GW, Albernaz A, Naik GO, Kirchengast M, Stewart DJ (1998) Beneficial effects of long-term selective endothelin type A receptor blockade in canine experimental heart failure. Cardiovasc Res 39:571–579

Moreau P, d'Uscio LV, Shaw S, Takase H, Barton M, Luscher TF (1997) Angiotensin II increases tissue endothelin and induces vascular hypertrophy: reversal by ET(A)-receptor antagonist. Circulation 96:1593–1597

Muramatsu M, Oka M, Morio Y, Soma S, Takahashi H, Fukuchi Y (1999) Chronic hypoxia augments endothelin-B receptor-mediated vasodilation in isolated perfused rat lungs. Am J Physiol 276:L358–L364

Nguyen QT, Cernacek P, Calderoni A et al. (1998) Endothelin A receptor blockade causes adverse left ventricular remodeling but improves pulmonary artery pressure after infarction in the rat. Circulation 98:2323–2330

Okazaki T, Sharma HS, McCune SK, Tibboel D (1998) Pulmonary vascular balance in congenital diaphragmatic hernia: enhanced endothelin-1 gene expression as a possible cause of pulmonary vasoconstriction. J Pediatr Surg 33:81–84

Oparil S, Chen SJ, Meng QC, Elton TS, Yano M, Chen YF (1995) Endothelin-A receptor antagonist prevents acute hypoxia-induced pulmonary hypertension in the rat. Am J Physiol 268:L95–L100

Pernow J, Hemsen A, Lundberg JM (1989) Tissue specific distribution, clearance and vascular effects of endothelin in the pig. Biochem Biophys Res Commun 161:647–653

Rabinovitch M, Andrew M, Thom H et al. (1987) Abnormal endothelial factor VIII associated with pulmonary hypertension and congenital heart defects. Circulation 76:1043–1052

Rabinovitch M (1998) Elastase and the pathobiology of unexplained pulmonary hypertension. Chest 114:213S–224S

Saetrum OO, Adner M, Gulbenkian S, Edvinsson L (1994) Localization of endothelin immunoreactivity and demonstration of constrictory endothelin-A receptors in human coronary arteries and veins. J Cardiovasc Pharmacol 23:576–583

Saito Y, Mizuno T, Itakura M et al. (1991) Primary structure of bovine endothelin ETB receptor and identification of signal peptidase and metal proteinase cleavage sites. J Biol Chem 266:23433–23437

Sakai S, Miyauchi T, Sakurai T et al. (1996) Pulmonary hypertension caused by congestive heart failure is ameliorated by long-term application of an endothelin receptor antagonist. Increased expression of endothelin-1 messenger ribonucleic acid and endothelin-1-like immunoreactivity in the lung in congestive heart failure in rats. J Am Coll Cardiol 28:1580–1588

Saleh D, Furukawa K, Tsao MS et al. (1997) Elevated expression of endothelin-1 and endothelin-converting enzyme-1 in idiopathic pulmonary fibrosis: possible involvement of proinflammatory cytokines. Am J Respir Cell Mol Biol 16:187–193

Sato K, Rodman DM, McMurtry IF (1999) Hypoxia inhibits increased ETB receptor-mediated NO synthesis in hypertensive rat lungs. Am J Physiol 276:L571–L581

Schmeck J, Koch T, Patt B, Heller A, Neuhof H, van Ackern K (1998) The role of endothelin-1 as a mediator of the pressure response after air embolism in blood perfused lungs. Intensive Care Med 24:605–611

Snapper JR, Thabes JS, Lefferts PL, Lu W (1998) Role of endothelin in endotoxin-induced sustained pulmonary hypertension in sheep. Am J Respir Crit Care Med 157:81–88

Soma S, Takahashi H, Muramatsu M, Oka M, Fukuchi Y (1999) Localization and distribution of endothelin receptor subtypes in pulmonary vasculature of normal and hypoxia-exposed rats. Am J Respir Cell Mol Biol 20:620–630

Stelzner TJ, O'Brien RF, Yanagisawa M, et al. (1992) Increased lung endothelin-1 production in rats with idiopathic pulmonary hypertension. Am J Physiol 262:L614–L620

Stewart DJ, Baffour R (1990) Functional state of the endothelium determines the response to endothelin in the coronary circulation. Cardiovasc Res 24:7–12

Stewart DJ, Cernacek P, Costello KB, Rouleau JL (1992) Elevated endothelin-1 in heart failure and loss of normal response to postural change. Circulation 85:510–517

Stewart DJ, Kubac G, Costello KB, Cernacek P (1991) Increased plasma endothelin-1 in the early hours of acute myocardial infarction. J Am Coll Cardiol 18:38–43

Strauss BH, Chisholm RJ, Keeley FW, Gotlieb AI, Logan RA, Armstrong PW (1994) Extracellular matrix remodeling after balloon angioplasty injury in a rabbit model of restenosis. Circ Res 75:650–658

Stricker H, Domenighetti G, Fiori G, Mombelli G (1999) Sustained improvement of performance and haemodynamics with long-term aerosolised prostacyclin therapy in severe pulmonary hypertension. Schweiz Med Wochenschr 129:923–927

Sutsch G, Kiowski W, Yan XW et al. (1998) Short-term oral endothelin-receptor antagonist therapy in conventionally treated patients with symptomatic severe chronic heart failure. Circulation 98:2262–2268

Takeda S, Miyoshi S, Omori K et al. (1998) Pulmonary disease models induced by in vivo hemagglutinating virus of Japan liposome-mediated endothelin-1 gene transfer. J Cardiovasc Pharmacol 31 Suppl 1:S336–S338

Takeda S, Sawa Y, Minami M et al. (1997) Experimental bronchiolitis obliterans induced by in vivo HVJ-liposome- mediated endothelin-1 gene transfer. Ann Thorac Surg 63:1562–1567

Tchekneva E, Quertermous T, Christman BW, Lawrence ML, Meyrick B (1998) Regional variability in preproEndothelin-1 gene expression in sheep pulmonary artery and lung during the onset of air-induced chronic pulmonary hypertension. Participation of arterial smooth muscle cells. J Clin Invest 101:1389–1397

Theuring F, Thone-Reinecke C, Vogler H et al. (1998) Pathophysiology in endothelin-1 transgenic mice. J Cardiovasc Pharmacol 31 Suppl 1:S489–S491

Tjen-A-Looi S, Ekman R, Osborn J, Keith I (1996) Pulmonary vascular pressure effects by endothelin-1 in normoxia and chronic hypoxia: a longitudinal study. Am J Physiol 271:H2246–H2253

Underwood DC, Bochnowicz S, Osborn RR et al. (1999) Effect of SB 217242 on hypoxia-induced cardiopulmonary changes in the high altitude-sensitive rat. Pulm Pharmacol Ther 12:13–26

Vane JR, Anggard EE, Botting RM (1990) Regulatory functions of the vascular endothelium. N Engl J Med 323:27–36

Vanhoutte PM, Luscher TF, Graser T (1991) Endothelium-dependent contractions. Blood Vessels 28:74–83

Voelkel NF, Weir EK (2000) Etiologic mechanisms in primary pulmonary hypertension. In: Weir EK, Reaves JT (eds) Pulmonary vascular physiology and pathophysiology. Marcel Dekker, New York, 513–539

Vongpatanasin W, Brickner ME, Hillis LD, Lange RA (1998) The Eisenmenger syndrome in adults. Ann Intern Med 128:745–755

Wagenvoort CA, Wagenvoort N. Primary pulmonary hypertension. A pathologic study of the lung vessels in 156 clinically diagnosed cases. Circulation 42, 1163. 1970

Walcott G, Burchell HB, Brown AL (1970) Primary pulmonary hypertension. Am J Med 49, 70

Warner TD, Allcock GH, Vane JR (1994) Reversal of established responses to endothelin-1 in vivo and in vitro by the endothelin receptor antagonists, BQ-123 and PD 145065. Br J Pharmacol 112:207–213

Wood P (1950) Congenital Heart Dis. Br Med J ii, 693. 2000

World Health Organization. Primary Pulmonary hypertension: report on a WHO meeting. In: Primary pulmonary hypertension, edited by Hatano S.S.T. 1975

Yanagisawa M, Kurihara H, Kimura S et al. (1988) A novel potent vasoconstrictor peptide produced by vascular endothelial cells. Nature 332:411–415

CHAPTER 16
Vascular and Cardiac Effects of Endothelin

S.A. Douglas and E.H. Ohlstein

A. Endothelin-Induced Vascular Smooth Muscle Contraction

Endothelin (ET)-1, the most potent mammalian vasoconstrictor identified to date, was originally characterized as a peptide capable of producing contractile responses that were both sustained and extremely resistant to wash-out (Yanagisawa et al. 1988). Subsequently, ET-1 has been shown to contract both resistance arterioles and venous tissue in addition to conduit arteries (venous tissue generally being more sensitive than arterial vessels; Brain et al. 1988, 1989; Borges et al. 1989b; Cocks et al. 1989; D'Orleans-Juste et al. 1989). Furthermore, since subthreshold concentrations of ET-1 augment the contractile actions of norepinephrine, 5-HT, and angiotensin II, it has been proposed that ET-1 also sensitizes vessels to the actions of other "classical" circulating neurohumoral factors (MacLean and McGrath 1990; Lerman et al. 1991; Nakayama et al. 1991).

The diverse intracellular signaling pathways utilized by ET-1 are described in detail in Chap. 5 and are reviewed elsewhere (Douglas and Ohlstein 1997). Briefly, however, ET-induced constriction results from an influx of extracellular Ca^{2+} secondary to PLC-mediated IP_3/DAG generation. Although responses are attenuated after extracellular Ca^{2+} chelation (EGTA) or inhibition of extracellular Ca^{2+} influx (dihydropyridine Ca^{2+} antagonists), ET-1 is still capable of inducing vasoconstriction (Ohlstein et al. 1989). In contrast, inhibition of the G-protein-mediated inositol phosphate generation (with pertussis toxin) or the initial release of intracellular Ca^{2+} from the sarcoplasmic reticulum (TMB-8 or caffeine) effectively abolishes the vasoconstrictor actions of ET-1 (Miasiro et al. 1988; Van Renterghem et al. 1988; Bialecki et al. 1989; Borges et al. 1989a; Kim et al. 1989). In addition to PLC, PKC may also contribute to the contractile actions of ET-1 since contractile responses are attenuated by staurosporine, phloretin and H-7 (Auguet et al. 1989; Kodama et al. 1989; Ohlstein et al. 1989; Lee et al. 1989; Sugiura et al. 1989a).

B. Modulatory Role of the Endothelium on the Contractile Actions of the Endothelin Isopeptides

The vascular endothelium plays a critical role in modulating the vasocon-strictor actions of the ET isopeptides. ET isopeptides stimulate the release of nitric oxide and both vasodilator (PGI_2, PGE_2) and vasoconstrictor (TXA_2; De Nucci et al. 1988) eicosanoids (see Chap. 11). Consequently, the concomitant release of these labile vasoactive factors influences the direct contractile and indirect hemostatic actions of this isopeptide family.

In the presence of active tone, ET-1 produces endothelium-dependent vasorelaxation in vitro via a mechanism that is abrogated by compounds which interfere with the source (CHAPS and deoxycholate), synthesis (L-NMMA and L-NAME) or actions (methylene blue and oxyhemoglobin) of nitric oxide (Hiley et al. 1989; Randall et al. 1989; Warner et al. 1989a, b; Douglas and Hiley 1990). In accord, ET-1 stimulates endothelium-dependent cGMP accumulation in both cardiovascular and non-cardiovascu-lar tissues. Similarly, the vasorelaxant actions of the ET isopeptides are also sensitive to cyclooxygenase inhibitors (indomethacin, aspirin; Armstead et al. 1989). Since many cardiovascular disorders (atherosclerosis, hypertension, subarachnoid hemorrhage, restenosis) are associated with endothelial dys-function, as discussed in Sect. L, an abrogation in nitric oxide/PGI_2 function may augment the vasoconstrictor/mitogenic actions of ET-1 at the site of such lesions.

C. Multiple Receptor Subtypes Mediating the Vascular Actions of Endothelin-1

Prior to the molecular cloning of ET receptor subtypes (see Chap. 4), struc-ture-activity studies demonstrated that the rank order of ET isopeptide vasorelaxant potency (ET-3 = ET-1; endothelial ET_B-receptor, linked to nitric oxide release) differed from that for vasoconstriction (ET-1 > ET-3; smooth muscle ET_A-receptor subtype; De Nucci et al. 1988; Hiley et al. 1989; Randall et al. 1989; Warner et al. 1989a, b; Douglas and Hiley 1990). In accord, high affinity binding sites for ET-3 were identified on aortic endothelial cells and ET_B-selective ligands (devoid of contractile activity in isolated aortae) were shown to induce endothelium/cGMP-dependent vasodilation via a methylene blue/L-NMMA-sensitive mechanism (Emori et al. 1990; Takayanagi et al. 1991; James et al. 1993). Studies performed using piebald (s′) mice, a rodent strain possessing dysfunctional, mutant ET_B receptors, support the hypothesis that the predominant role of the ET_B receptor was to mediate the vasoactive actions of ET in vivo (see Sect. F).

The hypothesis that vascular smooth muscle contraction and dilation are mediated by ET_A and ET_B receptors, respectively, is however recog-nized as an oversimplification. ET_B-selective agonists (sarafotoxin S6c,

[Ala^{1,3,11,15}]ET-1, IRL 1620) induce vasoconstriction in numerous isolated arterial and venous tissues via a mechanism resistant to ET$_A$-selective antagonists (FR 139317, BQ-123; HARRISON et al. 1992; MORELAND et al. 1992; PANEK et al. 1992; SUMNER et al. 1992; AUGUET et al. 1993; JAMES et al. 1993; LA DOUCEUR et al. 1993; LODGE and HALAKA 1993; SUDJARWO et al. 1993; WARNER et al. 1993b; WHITE et al. 1993; OHLSTEIN et al. 1994a, b; WELLINGS et al. 1994). Indeed, the contractile activities of BQ-3020 and IRL-1620 are antagonized by the ET$_B$ selective antagonist, BQ-788 (ISHIKAWA et al. 1994; SATO et al. 1995). Thus, it has been proposed that multiple subtypes of the ET$_B$ receptor exist (see DOUGLAS et al. 1994b). For example, WARNER et al. (1993a, b) demonstrated that sarafotoxin S6c caused vasodilation in the isolated rat mesentery in a manner which is sensitive to BQ-123. However, this ET$_B$ receptor differs from that which mediates contraction of the rat stomach strip and rabbit pulmonary artery since the ET$_B$-mediated response in the mesentery (tentatively termed ET$_{B1}$) is sensitive to PD 142893, whereas that in the rat stomach strip and rabbit pulmonary artery (tentatively termed ET$_{B2}$) is insensitive to both BQ-123 and PD 142893. However, despite mounting functional evidence, ET$_B$ receptor subclassification has not, as of yet, been adopted by the IUPHAR Committee on Receptor Nomenclature and Drug Classification (MASAKI et al. 1994) since the existence of multiple ET$_B$ gene products is unclear (see Chap. 4). Functional differences could merely reflect differences in post-translational modification of the ET$_B$ receptor, i.e., differences in the extent of ET$_B$ receptor palmitoylation and phosphorylation etc. (OKAMOTO et al. 1998).

Irrespectively, however, functional studies clearly show that numerous experimental (SUDJARWO et al. 1993, 1994; DOUGLAS et al. 1995a) and human (GODFRAIND et al. 1994; SEO et al. 1994; MACLEAN et al. 1998) vascular tissues contain heterogeneous subpopulations of "non-ET$_A$" coupled to vasoconstriction (either ET$_B$ and/or "atypical ETC-like" receptors; ET-3 > ET-1). Indeed, ET$_B$ receptor-mediated contraction may be of particular importance in low pressure sections of the vasculature (venous and pulmonary circulation; MORELAND et al. 1994). Clearly, ET$_B$-selective agonists are systemic pressor agents in vivo, inducing increases in renal, mesenteric, and coronary vascular resistances in the rat and dog (DOUGLAS and HILEY 1991; BIGAUD and PELTON 1992; CLOZEL et al. 1992; JAMES et al. 1993; TEERLINK et al. 1994a). Nevertheless, although ET$_B$ receptors make a significant contribution to the pressor actions of ET-1 in specific vascular beds (i.e., rat renal and mesenteric, ET$_B$ >> ET$_A$; hindquarters, ET$_A$ > ET$_B$; GARDINER et al. 1994a, b), such responses are species-dependent (renal ET$_B$ receptors are less important in mediating the constrictor actions of ET-1 in the dog; BROOKS et al. 1994). Indeed, specific roles for functional ET receptor subtypes within the human vasculature are contentious (DAVENPORT and MAGUIRE 1994; GODFRAIND 1994; BACON and DAVENPORT 1996; PIERRE and DAVENPORT 1998), and require further detailed investigation, both at a pharmacological and molecular biological level (see Sect. F).

D. Cardiac Actions of Endothelin-1

Although low doses of ET-1 produce an initial coronary vasodilatation in isolated hearts and in vivo, the primary function of this peptide is as a potent and efficacious vasoconstrictor in isolated coronary arteries/Langendorff hearts and in vivo (BAYDOUN et al. 1989; FOLTA et al. 1989; LEMBECK et al. 1989; NEUBAUER et al. 1990; NICHOLS et al. 1990), leading to impaired ventricular function, delayed filling of distal coronary branches, and, in some cases, total occlusion of epicardial vessels (and, ultimately, ventricular fibrillation; CLOZEL and CLOZEL 1989a; EZRA et al. 1989; KURIHARA et al. 1989; LARKIN et al. 1989; NICHOLS et al. 1990; DOMENECH et al. 1991). In addition, ET-1 contracts isolated papillary muscle and is a potent positive inotropic agent (ISHIKAWA et al. 1988; EGLEN et al. 1989; SHAH et al. 1989) in paced left atria. The inotropic effect of ET-1 has been proposed to be an ET_B-mediated effect, and is most evident in vivo when afterload is reduced in the presence of the ET_A-selective antagonist, BQ-610 (BEYER et al. 1996, 1998). However, whereas LEMBECK et al. (1989) noted a positive chronotropic action in spontaneously beating right atria, this response is variable (HU et al. 1988; VIGNE et al. 1989).

Although both the atrioventricular conducting tissue, the atrial and ventricular myocardium and the coronary vasculature possess $ET_{A/B}$ receptors (MOLENAAR et al. 1993; GODFRAIND 1994), there remains controversy regarding the receptor subtype(s) responsible for mediating the cardiohemodynamic actions of ET-1. BQ-123, but not BQ-788, was reported to be able to attenuate the coronary contractile actions of ET-1 in the anesthetized goat (GARCIA et al. 1996). In contrast, however, the ET_B-selective agonists, IRL-1620 and [Ala[1,3,11,15]]ET-1, induced coronary vasoconstriction in the pig (WANG et al. 1995) and rat (FILEP et al. 1996). Indeed, in the rat, the coronary constrictor actions of ET-1 are blocked by both FR139317 and BQ-788 (MIKI et al. 1998).

Although ET-1 is released by the endocardial endothelium, it is distinct from the putative cardiotonic molecule endocardin (SHAH and LEWIS 1993; SHAH 1994; TØNNESSEN et al. 1994). ET-1 increases the amplitude and duration of the plateau phase of the cardiac action potential in guinea-pig left atria. The predominant current flowing during this phase is the slow inward current (I_{si}) and, since potentiation still occurs in tissues which have been depolarized with high K^+, ISHIKAWA et al. (1988) proposed that ET-1 augmented Ca^{2+} influx through verapamil-sensitive L-type channels. Obviously, the regulation of cardiac electrophysiology is complex and not solely driven by ET-1; rather, it is the result of a complex interplay between several distinct neurohumoral factors. JAMES et al. (1994) and ONO et al. (1994) proposed that ET-1-induced regulation of cardiac ion channels (Ca^{2+}, Cl^- conductances) was mediated by the ET_A receptor (a G-protein-coupled effect associated with adenylate cyclase inhibition). However, ET-1- and ET-3-induced alterations in L-type Ca^{2+} current (ICa) exhibited varying degrees of sensitivity to ET_A- (PD 145065 and PD 155080) and ET_B- (BQ-788 and RES-701-1) selective antagonists in rabbit cardiomyocytes (KELSO et al. 1998). Indeed, it has even been proposed that ET-1 exerts a cardioprotective mechanism, opposing the proarrhythmogenic activity associated with catecholamine-induced β-adrenoceptor activation. This would be supported by the finding that BQ-123 enhanced the incidence of irreversible ventricular fibrillation observed in an anesthetized rat ischemia model (GARJANI et al. 1995). However, numerous in vitro and in vivo studies demonstrate that exposure to both endogenous and exogenous ET-1 is associated with disturbances in cardiac rhythm, an effect blocked by ET$_A$-selective (BQ-485 and LU135252; ERCAN et al. 1996; RASCHACK et al. 1998) and mixed ET$_{A/B}$ (PD 142893 and SB 209670; BRUNNER et al. 1996; DOUGLAS et al. 1998) antagonists. Thus, it has been proposed that ET-1 itself is pro-arrhythmogenic via an indirect mechanism, i.e., disturbances in cardiac conduction are the indirect result of intense focal ischemia. However, TOOTH et al. (1998) have observed the development of (exogenous) ET-1-induced cardiac arrhythmias in the absence of any signs of myocardial ischemia in open chest dogs (see Sect. E).

In addition to the direct effects of ET-1 on cardiac and coronary contractility and myocardial electrophysiology, this peptide has also been proposed to influence directly/indirectly the endocrine functions of the heart regulating the synthesis and release of factors such as BNP and ANP (BRUNEAU et al. 1997). For example, bosentan attenuates mechanical stretch-induced BNP gene expression in isolated rat hearts (an effect not inhibited by the AT1-receptor antagonist, CV-11974, but one limited to the atrium; MAGGA et al. 1997), and ET-1 is also able to influence erythropoietin-induced ANP secretion in rat atria and myocytes (an ET$_A$-mediated effect; PORAT et al. 1996).

E. Integrated Cardiovascular Actions of Endothelin-1

Bolus i.v. ET-1 produces a biphasic change in arterial blood pressure consisting of an initial, transient depressor response rapidly followed by a sustained pressor response (YANAGISAWA et al. 1988), a profile observed in anesthetized rats, rabbits, cats, dogs, sheep, and squirrel monkeys (although the transient depressor response is variable in the anesthetized pigs and, in the rat, is more sensitive than the secondary pressor response to tachyphylaxis; LE MONNIER DE GOUVILLE et al. 1989, 1990). Neither component of this biphasic response is reflex in origin, and occurs in pithed rats and following bilateral vagotomy, autonomic/ganglionic blockade etc. (LE MONNIER DE GOUVILLE et al. 1989).

Regional hemodynamic studies demonstrate that the systemic depressor response to ET-1 is associated with an initial vasoactive response in the hindquarters and carotid vascular beds, while the conductances through the renal and mesenteric arterial beds are reduced (WRIGHT and FOZARD 1988, 1990). During the secondary pressor phase, however, blood flow through all four vascular beds is reduced due to regional vasoconstriction. Since cardiac output either remains unaltered or is reduced, this systemic pressor response is the result of an increase in total peripheral resistance (Le Monnier de Gouville et al. 1989). Radiolabeled microsphere techniques show that the increase in total peripheral resistance results from vasoconstriction mainly in the gastrointestinal tract, kidneys, and skeletal muscle (Ais et al. 1989; MacLean et al. 1989; Walder et al. 1989).

Whereas the pressor response is the direct result of an interaction between ET-1 and specific receptors present on the vascular smooth muscle (see below), the mechanism underlying the initial hypotensive response is less clearly defined. This initial, transient reduction in total peripheral resistance is resistant to a variety of G-protein-coupled receptor antagonists, ion channel blockers, and protease inhibitors (Le Monnier de Gouville et al. 1989). Although ET-1 stimulates the release of ANP (see Sect. E), this vasoactive peptide does not account for the hypotensive actions of ET-1. Fozard and Part (1990) were able to record a hypotensive response to ET-1 in rats rendered insensitive to the relaxant actions of exogenous ANP and, unlike ET-1, ANP cannot produce hindquarter vasodilatation in Brattleboro rats (Gardiner et al. 1990b).

As detailed above (Sect. B), ET-isopeptides induce endothelial-prostanoid/NO release. Initially, cyclooxygenase (meclofenamate or indomethacin) or lipooxygenase (with BW 755c) inhibition was reported to have no effect on the systemic and regional depressor effects of ET-1 (Le Monnier de Gouville et al. 1989). Paradoxically, however, both indomethacin and piroxicam have been reported to augment significantly the secondary systemic pressor actions of ET-1 (an effect that was abolished by adrenalectomy; De Nucci et al. 1988; Walder et al. 1989). Thus, as detailed by Gardiner et al. (1990a), vasoactive eicosanoids possess complex modulatory effects on the actions of the ET isopeptide family, and these show subtle regional variations. Similarly, Whittle et al. (1989) initially reported that the initial dilator response to ET-1 was inhibited by L-NMMA. Although initial reports (using L-NMMA and methylene blue) appeared equivocal (Gardiner et al. 1989, 1990c,d, 1991a, b; Hoffman et al. 1989a), the development of more potent NO synthase inhibitors (L-NAME) indicated that a significant component of this systemic response to ET-1 was also mediated by NO (although regional variations existed; Gardiner et al. 1990b; Fozard and Part 1992).

In vivo, ET-1 is more potent than ET-3 as a systemic pressor agonist, whereas these two isopeptides are equipotent as vasodepressor agonists leading to the hypothesis that the pressor and dilator actions of ET-1 are mediated by ET_A and ET_B receptor subtypes, respectively (Spokes et al. 1989; Douglas and Hiley 1991; Douglas and Ohlstein 1993). This proposition is supported by the observation that the systemic pressor actions of ET-1 are inhibited selectively by BQ-123 (Douglas et al. 1992). However, in accord with more recent findings described in vitro (Sect. B), such a classification is oversimplified, e.g., ET-1-induced mesenteric constriction is not abolished by BQ-123 in the intact rat (DOUGLAS et al. 1992). Indeed, several ET_B-selective agonists (e.g., BQ-3020, sarafotoxin S6c, [Ala^{1,3,11,15}]ET-1) increased vascular tone in vivo via mechanism which is insensitive to the ET_A-selective antagonist, FR 139317, and, thus, it is now appreciated that ET_B-receptors make a significant contribution to the pressor actions observed to ET-isopeptides (HILEY et al. 1989; DOUGLAS and HILEY 1991; BIGAUD and PELTON 1992; GARDINER et al. 1994).

However, the precise receptor subtype(s) mediating the regional dilator actions of ET-1 are less clearly defined. Several chemically distinct ET receptor antagonists such as SB 209670, Ro 46-2005, bosentan, PD 145065, and PD 142893 (CLOZEL et al. 1993a, b; DOHERTY et al. 1993; GARDINER et al. 1994) with equal or greater affinity for the ET_A receptor subtype over the ET_B receptor subtype selectively inhibit the systemic and regional vasodepressor actions of ET-1 (although SB 209670 and PD 142893 exhibit preferential selectivity for the ET_B receptor present in endothelial cells relative to those found on smooth muscle cell membrane, such values do not exceed the affinity reported at the ET_A receptor; SCHROEDER et al. 1998). Alternatively, it has been postulated that preferential inhibition of the depressor actions of ET-1 result from a pharmacokinetic phenomenon (preferential access to an endothelial receptor over those located on the smooth muscle deeper within the tunica media). However, such an explanation remains unsatisfactory since: (a) several chemically diverse antagonists exhibit this preferential effect; and (b) it was observed following sustained i.v. infusion of antagonist (DOUGLAS et al. 1993).

As detailed in Sect. D, in addition to any peripheral contractile actions, the ET isopeptides also have direct and indirect actions upon the heart itself. The systemic hemodynamic actions of ET-1 in vivo are associated with bradycardia and disturbances in cardiac rhythm, i.e., ET-induced ST segment elevation/elongation progresses into AV block and, ultimately, ventricular fibrillation (TERASHITA et al. 1989; HAN et al. 1990a; NICHOLS et al. 1990; OTSUKA et al. 1990a; DOUGLAS and HILEY 1991). Similar disturbances have been reported following administration of sarafotoxin isopeptides (WEISER et al. 1984; LEE et al. 1986; WOLLBERG et al. 1988; BDOLAH et al. 1989a). It has been suggested that the EKG changes observed are the result of coronary insufficiency/myocardial ischemia (EZRA et al. 1989; MIR et al. 1989). Microsphere techniques have shown that ET-1 causes a hyperemic response which is most profound in the subepicardial layer (CLOZEL and CLOZEL 1989a, b), an effect associated with myocardial ischemia as assessed by an increase in local ventricular lactate levels (EZRA et al. 1989; IGARASHI et al. 1989; LARKIN et al. 1989; ADACHI et al. 1991; DOMENECH et al. 1991). However, since ST segment elevation and ventricular fibrillation develop after detectable ischemic changes have subsided (YORIKANE and KOIKE 1990), vasoconstriction and arrhythmogenicity appear to be two distinct events.

F. Functional Genomics

Since exogenous ET-1 has profound actions on the cardiovascular system, investigators have been tempted to postulate that, like norepinephrine and angiotensin II, it plays an important role in the control of systemic hemodynamics. One approach employed to address this question has been application of functional genomics to the field of ET research. Conditional gene expression/deletion are important experimental approaches for examining the functions of particular gene products. To this end, investigators have used a variety of approaches to perturb the function of the ET system (ET isopeptides, receptors and converting enzymes; see Chap. 6). The identification of spontaneous mutations in rodents has been supplemented by alternate approaches designed to both abolish the activity of the ET system (i.e., knockout mice) and those designed to increase the activity of the ET system both on a transient (adenovirus-mediated transfer in rats) and chronic (transgenic mice) basis. Although the pleiotropic effects of ET-related gene deletions (the variable requirement of several ET-related genes) result in complex phenotypes in knockout mice (some fatal to the embryo/fetus), data generated in this fashion has given significant insight into the role of the ET system in the control of systemic hemodynamics.

I. Endothelin Isopeptides

Kurihara et al. (1994) have succeeded in producing targeted disruption of the ET-1 gene in mice (see Chap. 6). Although deletion of the ET-1 gene is fatal (homozygous ET-1$^{(-/-)}$ knockout mice die perinatally, usually within a few minutes post partum), ET-1$^{(+/-)}$ mice are reported to be mildly hypertensive relative to wild type litter mates (there was no differences in the systemic pressor response obtained to exogenous ET-1). Although somewhat paradoxical, the pleiotropic nature of this isopeptide means that such observations must be interpreted with some caution since several different compensatory mechanisms could be responsible for the reported elevation in blood pressure, e.g., altered receptor expression, altered regional vascular reactivity, neuroendocrine activation. Although Morita et al. (1998) reported that ET-1$^{(+/-)}$ mice are not salt-sensitive (this may be the case with ET$_{B(+/-)}$ mice; see below), a plethora of mechanisms may underlie the alterations in the contractile responses observed in these mice to ET-1. For example, in addition to the well characterized role in the development of neural crest-derived craniofacial structures, targeted deletion of the ET-1 gene results in profound cardiorespiratory deficits (impaired respiratory reflex), sympathetic overactivity, etc. (Kuwaki et al. 1997). Indeed, the transgenic overexpression of ET-1 reported by Hocher et al. (1997), predominantly in the brain, lung, and kidney, resulted in profound renal dysfunction (renal cysts, interstitial fibrosis, glomerulosclerosis), a phenotype which obviously complicates any observed alterations in cardiovascular function. Interestingly, the mice studied by Hocher et al. (1997)

were reported to be normotensive. Similarly, LIEFELDT et al. (1995) has previously reported that overexpression of ET-2 was not associated with an alteration in systemic blood pressure.

In contrast to the observation that ET-1$^{(+/-)}$ mice were hypertensive, adenovirus-mediated hepatic overexpression of prepro-ET-1 results in an elevation of systemic arterial blood pressure in rats. A ca. sixfold elevation in circulating plasma ET-1 levels was associated with a 30 mmHg increase in mean arterial blood pressure, a response selectively attenuated by FR 139317 (but not BQ-788; NIRANJAN et al. 1996).

II. ET$_A$ and ET$_B$ Receptor

Administration of ET-1 results in renal and mesenteric vasoconstriction in wild type ET$_{A(+/+)}$ and ET$_{B(+/+)}$ mice; contractile actions of ET-1 are sensitive to both BQ-123 and BQ-788 (BERTHIAUME et al. 1998). This implies that both the ET$_A$ and the ET$_B$ receptor contribute to the vasoconstrictor activities of ET-1 in these particular vascular beds. Accordingly, it has been reported that the contractile actions of ET-1, but not the ET$_B$-selective agonist, IRL-1620, are attenuated in ET$_{A(+/-)}$ mice (whereas, the response to both spasmogens is attenuated in ET$_{B(+/-)}$ mice). However, the observation that piebald (s') mice, a rodent strain possessing dysfunctional, mutant ET$_B$ receptors, are hypertensive supports the hypothesis that the predominant role of the ET$_B$ receptor is to mediate the vasoactive actions of ET in vivo. Nevertheless, such data has to be interpreted with caution since the observed elevation in blood pressure could be secondary to alterations in, for example, renal function, renin-angiotensin activity, etc. (indeed, ET$_B$ knockout mice are salt-sensitive, refractory to captopril, etc.). Although sarafotoxin S6c is devoid of dilator activity in both piebald (s') and ET$_B$ knockout mice (both in isolated aortae and in vivo; GILLER et al. 1997; MIZUGUCHI et al. 1997), such observations remain equivocal (see DOUGLAS 1997).

III. Endothelin Converting Enzyme

Simultaneous overexpression of ECE-1 with prepro-ET-1 elevates blood pressure in rats by ~50% more than when ET-1 is overexpressed on its own, suggesting the proET-1 conversion by ECE-1 is rate limiting (TELEMAQUE et al. 1998). This hypertensive response is selectively attenuated by administration of an ECE inhibitor or an ET$_A$ receptor antagonist (CGS 26303 or FR 139317).

G. Hemodynamic Actions of ET-1 in Man

As discussed in Sect. C, the demonstration that isolated human arteries and veins contract in response to ET$_B$ agonists suggests that functional ET$_B$ receptors make a significant contribution to the contractile actions of ET-1 in man. However, such findings remain controversial and more recent clinical investigations have centered on studying the role of ET_A and ET_B receptors in mediating the vascular actions of ET-1 in vivo. Although the clinical effects of the ET isopeptides and related antagonists are described in detail elsewhere (Chap. 19), a brief overview of their cardiovascular actions in humans is provided below.

I. Endothelin Isopeptide Administration

The hemodynamic actions of ET isopeptides are typically studied by measuring changes in forearm blood flow or skin microvascular responses (strain-gauge venous occlusion plethysmography, ultrasound, laser Doppler flowmetry, etc.) following brachial artery infusion or i.d. injection of peptide. Low dose infusion of ET-1 (Hughes et al. 1989) or ET-3 (Clarke et al. 1989) induces transient forearm hyperemia, a response attenuated by aspirin (Haynes et al. 1995b). At higher doses, however, this is followed by a sustained vasoconstriction. Indeed, since sarafotoxin S6c causes progressive reductions in flow under similar conditions, the predominant role for $ET_{A/B}$ receptors is to mediate the contractile actions of the ET isopeptides (at least within forearm resistance and capacitance vessels; Haynes et al. 1995b). In accord, big ET-1 also induces profound systemic hemodynamic changes consisting of a decrease in cardiac output (stroke volume and heart rate) accompanied by a concomitant increase in mean arterial pressure (elevated pulmonary and peripheral vascular resistance; Ahlborg et al. 1996).

II. Endothelin Receptor Antagonism

Systemic i.v. administration of the $ET_{A/B}$ antagonist, TAK-044 (10–1000 mg), blocks ET-1-induced forearm vasoconstriction (and decreases SVR and, to a lesser extent, MAP, increasing circulating ET-1; Haynes et al. 1996; Ferro et al. 1997). Similarly, the forearm hemodynamic response to big ET-1 (but not ET-1) is blocked by phosphoramidon (Haynes et al. 1995a; Webb 1995; Plumpton et al. 1995). To support further a predominant pressor role for the ET isopeptides in vivo, inhibition of endogenous ET-1 with BQ-123 induces vasodilation of human forearm resistance vessels (Haynes and Webb 1994; Haynes et al. 1995a; Berrazueta et al. 1997; Verhaar et al. 1998).

In contrast to selective ET_A-blockade with BQ-123, however, ET_B receptor blockade with BQ-788 (administered either alone or on a background of ET_A receptor antagonism) was reported to cause local vasoconstriction in the forearm (Verhaar et al. 1998), indicative of a predominantly vasoactive role of the ET_B receptor. Such a finding may appear paradoxical (since sarafotoxin S6c is a vasoconstrictor), but such observations should be interpreted with caution: first, only one ET_B-selective antagonist has been studied under these conditions and, second, agonist-induced ET_B receptor activation using exogenous peptide may not mimic the actions of endogenous ET-1. Similarly, the existence of functional ET_{B1}/ET_{B2} subtypes within the human vasculature has not been addressed in vivo; Sect. C3). Nevertheless, in support of these findings, STRACHAN et al. (1999) demonstrated that selective ET_B receptor blockade is associated with an increase (24%) in total peripheral vascular resistance (along with bradycardia and attenuated cardiac index) suggesting that the predominant role of the ET_B receptor is to mediate vasodilation.

The hemodynamic response to exogenous ET-1 administration is clearly altered under pathophysiological conditions. In CHF patients, both BQ-123 (and ECE inhibition with phosphoramidon) increase forearm blood flow (LOVE et al. 1996). However, responses to exogenous ET-1, either at rest or during exercise, are severely abrogated in this patient population (in contrast, vasoconstriction to sarafotoxin S6c was augmented; LOVE et al. 1996; KRUM and KATZ 1998). Altered microvascular reactivity to ET-1 and BQ-123 has also been examined in additional alternate patient populations including subjects with syndrome X (NEWBY et al. 1998), glaucoma (GASS et al. 1997), type-II insulin-resistant diabetes (NUGENT et al. 1996), and chronic renal failure (HAND et al. 1995), and in several studies substantial differences have been reported between the responses observed in patients compared to those initially reported using healthy subjects.

H. Endothelin-1 and the Central Control of Cardiovascular Function

In addition to modulating peripheral vascular function directly, ET-1 may also control systemic hemodynamics via central mechanisms, modulating neuronal function within the specific cardiovascular centers of the brain, altering sympathetic outflow and systemic neuroendocrine activity. ET-1 and ET-3 isopeptides and $ET_{A/B}$ receptors are present within regions of the brain involved in the central control of cardiovascular function including the hypothalamus (supraoptical and paraventricular nuclei), nucleus of the solitary tract (NTS), rostroventrolateral medulla (RVLM), and several circumventricular loci (KOSEKI et al. 1989; MACCUMBER et al. 1989; LEE et al. 1990; MASAKI 1993).

The peripheral hemodynamic responses to centrally administered ET-1 are dependent on the site of administration: intraventricular or intracisternal administration increases heart rate and mean arterial blood pressure (although the latter is preceded by a transient hypotensive response and bradycardia), whereas administration into the NTS and RVLM produces hypotension and bradycardia (although the latter is preceded by an increased blood pressure; OUCHI et al. 1989; MOSQUEDA-GARCIA et al. 1992). Intraventricular ET-1 redistributes cardiac output to the skin, primarily at the expense of the brain and kidneys (Rebello et al. 1995). Van den Busse and Itoh (1993) demonstrated that intracisternal administration of ET-1 sensitized the baroreceptor heart rate reflex, although direct exposure of the carotid sinus to ET-1 was also shown to suppress baroreceptor discharge in the anesthetized dog (Chapleau et al. 1992).

Clearly, the role of ET-1 in the control of systemic hemodynamics may not be limited to a direct action on the vascular smooth muscle. ET-1 modulates peripheral neurotransmission, exhibiting both prejunctional inhibitory and postjunctional stimulatory actions on both cholinergic and sympathetic neurotransmission (see Ohlstein et al. 1995). Furthermore, ET-1 increases the release of epinephrine, norepinephrine, ANP, AVP, substance P, aldosterone, ACTH, cortisol, and gonadotrophin, but inhibits the release of renin and prolactin, all of which may have important acute and long-term consequences for systemic cardiovascular function.

I. Endothelin-1 and "Endothelium-Derived Contracting Factor (EDCF)"

One of the first indications that the endothelium produced vasoconstrictor factors, originally termed EDCFs, was the observation that hypoxia induced the endothelium-dependent contraction of isolated vascular tissue (Rubanyi 1992). It was suggested that ET-1 might mediate such a response since hypoxia (i) stimulated ET-1 release, (ii) altered [^{125}I]ET-1 binding site density and affinity, and (iii) altered vascular actions of ET-1 both in vitro and in vivo (Liu 1989, 1990a, b; MacLean and Hiley 1989; MacLean et al. 1989; Hieda and Gomez-Sanchez 1990; Rakugi et al. 1990; Douglas et al. 1991; Hiley et al. 1993). Nevertheless, although the role of this peptide in mediating hypoxic vasoconstriction is controversial (duration of action, susceptibility to wash out, insensitivity to BQ-123; Vanhoutte et al. 1989; Douglas et al. 1993; Wong et al. 1993; Fukue et al. 1994), ET-1 may still have a physiological role to play in the control of regional hemodynamics in response to hypoxia, e.g., the closing of the ductus arteriosus following parturition (Coceani et al. 1989, 1991).

J. Endothelin-1 and the Control of Smooth Muscle and Myocyte Structure

Angiogenesis/vasculogenesis are usually processes associated with embryonic development. However, geometric and structural remodeling are dynamic processes evident throughout life. The intrinsic capacity of a blood vessel to sense changes in its physical and chemical environment (both normal and injurious stimuli) endows the circulatory system with the ability to adapt to the long-term metabolic requirements of the organ it perfuses and to respond appropriately when such a system is perturbed. Normally, such adaptation is under strict control, but a disturbance in the release of factors able to influence the remodeling process (i.e., vascular cell hypertrophy, hyperplasia, migration, matrix elaboration, etc.) can be maladaptive and, therefore, of pathological consequences.

As with angiotensin II and norepinephrine, ET-1 is not only a vasoconstrictor but is also a potent mitogen in several cultured cell lines of cardiovascular origin, stimulating smooth muscle cell hypertrophy and protein synthesis (DUBIN et al. 1989; ITO et al. 1991; CHUA et al. 1992; OHLSTEIN and DOUGLAS 1993; see Chap. 12). ET-1 stimulates [³H]thymidine incorporation and increases cell number in cultured rat and human vascular smooth muscle cells. Similarly, transfection of cultured aortic smooth muscle cells with a prepro-ET-1 expression plasmid elevates the growth rate of these cells six-fold via an autocrine action (ALBERTS et al. 1994). Consistent with data generated using exogenous ET-1, smooth muscle mitogenesis results from ET_A receptor activation (OHLSTEIN et al. 1992; ZAMORA et al. 1993; ALBERTS et al. 1994). However, such a response is subject to phenotypic modulation (SERRADEIL-LE GAL et al. 1991; EGUCHI et al. 1994; ORLANDI et al. 1994). Early passaged rat aortic smooth muscle cells ("contractile" phenotype) express a single population of ET_A receptors, whereas in later passage cells ("synthetic" phenotypes) ET_B receptor expression predominates (such that BQ-123 can no longer inhibit ET-1-induced mitogenesis).

In addition to the peripheral vasculature, ET-1 also influences the growth properties of cardiomyocytes. Cellular hypertrophy (e.g., elevated [³H]phenylalanine incorporation) results from ET_A receptor activation (BQ-123-sensitive) and is the consequence of both (a) the direct autocrine/paracrine ET-1 release (from perivascular cardiac fibroblasts and myocytes) and (b) an indirect response resulting from angiotensin-II-mediated release (HARADA et al. 1997; SUZUKI et al. 1997; PONICKE et al. 1997). Similarly, ET_A-receptor activation influences the release/actions of additional extracellular hypertrophic factors associated with cardiac fibroblast and myocyte remodeling including ANP, osteopontin (SKVORAK et al. 1995; GRAF et al. 1997; LESKINEN et al. 1997) and norepinephrine (upregulation of α_{1C}-adrenoceptor expression; ROKOSH et al. 1996).

In addition to a mitogenic/hypertrophic action in smooth muscle/endothelial cells and myocytes, ET-1 also functions as an autocrine/paracrine survival factor in rat aortic endothelial cell (and fibroblast) cultures where apoptosis is induced by serum-deprivation (assessed by nucleosomal laddering, flow cytometry/FACS, and TdT-mediated dUTP biotin nick-end labeling; SHICHIRI et al. 1998a–c). This ET_B-mediated cytoprotective response, blocked by BQ-788 and PD 142893 but not by BQ-123, is not directly mediated by the generation of IP3/DAG second messengers since the protective response is refractory to inhibition of phospholipase C (U73122). Similarly, neither inhibition of tyrosine kinase (ST638), MEK (PD98059), nor PI-3-kinase (wortmannin, LY294002) abrogates this response (the effect in fibroblasts was attributed to a c-myc/MAPK-dependent pathway). A similar role as a survival factor has been proposed based on studies performed in human pericardial and prostatic smooth muscle cells (Wu-Wong et al. 1997). To date, however, the role of this peptide in myocyte apoptosis is unknown (a terminally differentiated cell type where ET-1 is hypertrophic rather than mitogenic). To this end, it of interest to note that Okazawa et al. (1998) have observed ET-1-induced apoptosis in a human melanoma cell line (evidenced by DNA laddering, morphological changes, chromatin condensation, nuclear staining, p53 elevation), an effect attributed to ET_B-mediated activation of a pertussis toxin-sensitive G-protein.

Typically, inositol phosphate second messenger generation results in the rapid signal transduction of ET-receptor interactions. However, as detailed in Chap. 5, ET-1 also regulates the expression of several target genes involved in long-term signaling, e.g., phosphorylation of cytosolic and membrane-bound proteins including $p42^{MAPK}$, S6 kinase, and $pp60^{c-src}$ (Wang et al. 1992, 1993; Simonson et al. 1992; Simonson and Herman 1993). ET-1 induces the expression of several protooncogenes (c-fos, c-jun, c-myc) and growth factors (PDGF, EGF, TGF-β, bFGF, insulin) and may function as a co-mitogen (Brown and Littlewood 1989; Kusuhara et al. 1989; Resink et al. 1990; Weissberg et al. 1990; Chua et al. 1992; Weber et al. 1994). In primary cultures, ET-1 induces neonatal and adult rat myocyte hypertrophy as a result of modulation (phosphorylation and activation) of the p38/JNK MAPK pathway (Clerk et al. 1998; Nemoto et al. 1998).

In addition to cellular hypertrophy and hyperplasia, cellular migration and matrix elaboration are also important steps in the remodeling of the vasculature. ET-1 has been shown to stimulate vascular smooth muscle and endothelial cell migration (Wren et al. 1993; Yue et al. 1994; Lauder et al. 1997), a response linked to ET_B-receptor activation in human endothelial cells (Morbidelli et al. 1995; Tao et al. 1995; Noiri et al. 1997). Accordingly, it has been reported that ET-1 super-induces the expression of the early response gene Ets-1 in vascular smooth muscle cells, a transcription factor known to activate the expression of matrix-degrading proteinases including collagenase I and stromelysin (Naito et al. 1998). Indeed, ET-1 is also able to influence cellular adhesion molecule (ICAM-1, VCAM-1, E-selectin) expression in endothelial cells and fibroblasts and to induce type I and III collagen and to reduce collagenase activity (Guarda et al. 1993; López Farré et al. 1993; McCarron et al. 1993) allowing the requisite cells to infiltrate regions of the vasculature and, once present, to begin to synthesize extracellular matrix, thereby facilitating the "wound healing" process. Further, ET-1 increases the release of numerous chemotactic factors (TNFα, MCP-1, and interleukins-1β, -6, and -8) from monocytes and, therefore, may promote leukocyte migration both indirectly (Helset et al. 1993, 1994) and directly through ET_A (monocytes) and ET_B (neutrophils receptor stimulation (Elferink and de Koster 1996a, b; Achmad et al. 1997).

K. Hemostatic Actions of Endothelin-1

As with angiotensin II, vasopressin, and norepinephrine, ET-1 stimulates the release of numerous vasoactive factors which influence hemostasis by indirectly modulating platelet aggregation (DE NUCCI et al. 1988; THIEMERMANN et al. 1988; OHLSTEIN et al. 1990; FILEP et al. 1991; LEADLEY et al. 1993), the result of ET_B-mediated PGI_2/nitric oxide production (an effect sensitive to PD 145065, but not FR 139317; McMURDO et al. 1993; see Chap. 11). In addition, ET-1 modulates vascular permeability/fluid extravasation via direct vasoactive/constrictor actions on resistance vessels and indirectly through the release of numerous vasoactive factors (e.g., PAF, PGI_2, nitric oxide; FILEP et al. 1991, 1992; KUROSE et al. 1991). [^{51}Cr]erythrocyte and [^{125}I]albumin labeling studies demonstrate significant total body albumin escape (liver, lung, and heart) in splenectomized rats in response to ET-1, an $ET_{A/B}$ effect (FILEP et al. 1992; ZIMMERMAN et al. 1992; McMURDO et al. 1993). Interestingly, the F-actin stabilizer, phalloidin (but not BQ-123), inhibits the ET-3-induced increases in microvascular permeability, suggestive that ET_B receptor-mediated endothelial cell contraction also contributes to this phenomenon (KUROSE et al. 1993).

L. Endothelin and the Pathogenesis of Cardiovascular Dysfunction

Indirect evidence (e.g., changes in circulating ET-1 levels) has led to the proposition that ET-1 is involved in the etiology of numerous cardiovascular disorders characterized by abnormal smooth muscle/cardiomyocyte contractile function and growth. However, these initial observations alone did not establish a cause-effect relationship (i.e., ET-1 levels could be elevated in response to the "stress effects" resulting from critical illness rather than a specific disease mechanism(s); HAAK et al. 1994). Nevertheless, the rapid preclinical development of diverse ET inhibitors (a testimony to the perceived pathological significance of this endothelium-derived peptide within the medical community), and their successful progression into Phase II and III clinical development has established a definitive clinical utility for these therapeutic modalities.

I. Congestive Heart Failure

The prevalence of congestive heart failure (CHF) is estimated to be as high as 10% in individuals >65 years. Worldwide, CHF affects nearly 15 million people with 4.7 million diagnosed cases in the United States alone in 1995. Although significant advances have been made in the treatment of CHF in the past decade (manipulation of the neurohumoral axis with ACE inhibitors and β-blockers such as carvedilol), CHF remains the only major cardiovascular disorder that is increasing in prevalence. Furthermore, the prognosis of CHF is extremely poor: half of men diagnosed with heart failure die within 1.7 years, and half of women die within 3 years. Indeed, the number of deaths ascribed to CHF has risen steadily over the last 20 years (from 130000 in 1970 to 267000 in 1988 in the United States alone). Except for heart transplantation and surgical repair of underlying conditions, there are currently no treatments available that definitively halt/reverse disease progression.

Although the etiology of CHF is diverse (coronary artery disease and hypertension account for the majority of cases), the condition is usually associated with the activation of distinct "adaptive mechanisms" which attempt to compensate for disturbances in myocardial contractility or excessive hemodynamic burden. Generally this is achieved by increasing preload (to optimize the length-tension, Frank-Starling mechanism), myocardial hypertrophy (to increase the contractile mass), and activation of neurohumoral mechanism (to augment myocardial contractility and redistribute blood flow to vital organs). In the short term these responses maintain myocardial performance and organ perfusion; however, the long-term effects of such compensatory responses are maladaptive. Based on the in vitro and in vivo profile determined for ET-1 in vascular preparations, this peptide has the capacity to influence both cardiac and systemic hemodynamic function. Indeed, evidence has accumulated over the last decade suggesting that ET-1 may serve as a third arm to the neurohumoral axis and, as such, ET-1 clearly has the potential to influence etiology and/or progression of CHF.

Circulating ET-1-like immunoreactivity is elevated in patients suffering from CHF (CAVERO et al. 1990), a pathology clearly associated with altered ET-1 binding site density and vascular reactivity in rats (CAVERO et al. 1990; FU et al. 1993). While it is possible that, at least acutely, an elevation in circulating ET-1 may be essential in maintaining end-organ perfusion by elevating preload (CLAVELL et al. 1993), a sustained activation of the ET system may be detrimental in view of the ability of this peptide to act as a vasoconstrictor and hypertrophic/hyperplastic factor in smooth muscle and myocardial tissue. Indeed, chronic systemic elevation of ET-1 has been correlated with an increase in ventricular mass in rat (coronary artery ligation), rabbit (aortic valvular insufficiency and aortic stenosis), and dog (chronic, rapid ventricular pacing) models of CHF (CAVERO et al. 1990; MARGULIES et al. 1990; LÖFFLER et al. 1993; TEERLINK et al. 1994b). Similarly, chronic pressure overloading resulting in left ventricular hypertrophy enhances preproET-1 mRNA expression and ET-1 immunoreactivity in rats, supporting a pathogenic role for ET-1 in the chronic myocardial remodeling (YORIKANE et al. 1993). Indeed, BQ-123 reduces left ventricular hypertrophy, at least acutely, in an aortic banding hemodynamic overload rat model (ITO et al. 1994). Since there is a close association between established hypertension and left ventricular hypertrophy (a major risk factor in cardiovascular mortality), the growth-promoting activities of this vasoactive peptide might not be limited to the peripheral vasculature. Indeed, the antihypertensive effects of Ro 47-0203 are also associated with a reduction in left ventricular hypertrophy in DOCA-salt rats (DOUGLAS et al. 1994b).

SAKAI et al. (1996) reported an upregulated myocardial ET system in a rat coronary artery ligation model of CHF. However, in addition to demonstrating beneficial effects on left ventricular performance (dP/dt, LVEDP etc.), long-term treatment with BQ-123 greatly improved the survival of rats with chronic heart failure by preventing the ventricular remodeling (increase in the ventricular mass and cavity enlargement) associated with this disorder. Similar observations have been made in this model using bosentan (100 mg/kg per day) in a 9-month mortality study (MULDER et al. 1997).

Plasma levels of ET-1 are elevated in experimental models of CHF and in patients with chronic heart failure. However, such observations do not necessarily reflect disease progression or clinical response to therapy. In patients with moderate to severe chronic heart failure, KRUM et al. (1996) demonstrated that the improvement in clinical outcome observed in those receiving carvedilol (e.g., symptom severity, NYHA class, 6-min walk distance, ejection fraction, plasma noradrenaline levels) was correlated strongly with changes in plasma ET-1 levels (an independent, noninvasive predictor of functional and hemodynamic responses to therapy by stepwise regression analysis). Such findings suggest that measurement of plasma ET-1 may therefore be a useful, non-invasive approach to the prognostic evaluation of clinical response to drug therapy and disease progression (PACHER et al. 1996; POUSSET et al. 1997; HAUG et al. 1998). Furthermore, they are suggestive of a role for ET-1 in the etiology of CHF (indeed, a recent study has demonstrated that carvedilol, a therapeutic agent used to reduce morbidity and mortality in this patient population, attenuates ET-1 production in human coronary artery endothelial cells via a mechanism not shared by other β-blockers such as propranolol and celiprolol; OHLSTEIN et al. 1998). Consequently, clinical studies have assessed the utility of ET receptor antagonists in man using both acute hemodynamic and chronic morbidity/mortality endpoints.

Administration of either phosphoramidon or BQ-123 increase forearm blood flow in CHF patients, evidence that ET-1 contributes to the maintenance of hemodynamic tone in such individuals (LOVE et al. 1996). Furthermore, although hemodynamic responses to ET_A receptor activation appear to be attenuated relative to those seen in normal subjects, selective ET_B-mediated vasoconstriction is enhanced. In a random, double blind placebo-controlled trial, KIOWSKI et al. (1995) demonstrated that bosentan (100 mg i.v. followed 60 min later by 200 mg) reduced mean arterial pressure by 8%, pulmonary artery pressure by 14%, right atrial pressure by 18%, and pulmonary artery wedged pressure by 9%. Acute therapy was also associated with a 14% increase in cardiac index (without a change in heart rate) concomitant with a decrease in both systemic (16%) and pulmonary (33%) vascular resistance. Thus, in accord with experimental findings, it appears that, at least acutely, ET-1 contributes to maintenance of vascular tone in humans with CHF.

In a double-blind, randomized placebo-controlled study, Sutsch et al. (1998) recently reported the therapeutic efficacy of 2 week oral therapy with bosentan (1g, b.i.d.) in 36 males with symptomatic heart failure (NYHA class III; left ventricular ejection fraction, 22.4 ± 4.5%). Acute (1 day) administration of bosentan significantly decreased mean arterial pressure (14%), pulmonary artery mean (13%) and capillary wedge (14%) pressures and right atrial pressure (20%). Cardiac output increased (15%) without a concomitant change in heart rate. Both systemic (24%) and pulmonary (20%) vascular resistance were reduced (therapy was discontinued in one patient due to symptomatic hypotension). After 2 weeks of therapy, cardiac output had further increased (by 15%) and both systemic (9%) and pulmonary (10%) resistances further decreased. There was no evidence of neurohumoral (renin-angiotensin or sympathetic) activation. Thus, short-term oral ET-receptor antagonist therapy improved systemic and pulmonary hemodynamics in symptomatic heart failure patients. However, the REACH-1 trial, an investigation designed to characterize the effects of chronic bosentan therapy (500mg, b.i.d. for 6 months), was terminated early (Packer, et al. 1998). Bosentan therapy in this population of CHF patients (NYHA class IIIB-IV patients receiving concomitant diuretic and ACE inhibitor therapy) was associated with abnormal liver function tests (elevated hepatic transaminase levels). Nevertheless, those subjects who had completed bosentan therapy prior to the termination of this trial were shown to have received significant clinical benefit (composite clinical endpoint, symptomatic and event-related). Since the liver abnormalities were both asymptomatic and reversible, these encouraging findings have stimulated enrollment for a larger Phase III trial aimed at examining the effects of chronic ET receptor blockade on morbidity and mortality in this patient population.

II. Hypertension

According to statistics released by the American Heart Association, essential hypertension is believed to have contributed, directly and indirectly, to the death of 250000 Americans in 1996 (a per capita rate which has risen 7% in the last decade). Indeed, currently, it is estimated that in excess of 20% of the adult population suffers from hypertension in the developed world. Since this disease is, by and large, asymptomatic, almost one-third of such individuals are unaware of their elevated blood pressure. Despite decades of pharmacological intervention using a diverse panoply of agents, over a quarter of patients remain refractory to their current medication (treatment which often constitutes a polypharmaceutical approach). This problem is even more evident in particular ethnic groups such as non-Hispanic blacks and Mexican-Americans. As such, tremendous therapeutic opportunities exist in the management of this disease.

Several groups have observed changes in receptor kinetics and vascular reactivity to ET-1 in tissues isolated from different hypertensive animal models (Tomobe et al. 1988, 1991; Miyauchi et al. 1989a; Catelli de Carvalho et al. 1990; Criscione et al. 1990; Gu et al. 1990; Kamata et al. 1990; Suzuki et al. 1990b; Wright and Fozard 1990). ET-1 synthesis is augmented in numerous preclinical animal models of hypertension such as spontaneous hypertensive rats (SHR), salt-sensitive rats (DOCA salt and Dahl salt-sensitive hypertensive rats), stroke-prone SHRs, angiotensin II-infused rats, fructose-fed rats, and 1-kidney 1-clip Goldblatt rats (low renin 2-kidney, 1 clip 1C Goldblatt hypertensive rats and nitric oxide-deficient L-NAME-treated do not exhibit an ET-1 component; Schiffrin 1998). As such, the data suggest that ET-1 is involved in the development and/or maintenance of hypertension. Furthermore, plasma levels of ET-1 are raised in hypertensive humans although, clinically, this may depend on the renal status of the patient (Davenport et al. 1990; Randall 1991; Khraibi et al. 1993; Larivière et al. 1993; Lüscher 1993). However, since such changes may be a consequence of an unrelated primary hypertensive mechanism, they do not necessarily establish a cause-effect relationship.

Infusion of phosphoramidon reduces systemic arterial blood pressure in SHR (although this is a hypotensive, rather than antihypertensive, effect since it is also observed in WKY; McMahon et al. 1991). In contrast, however, systemic administration of specific neutralizing antibodies to ET-1 or sustained i.v. infusion of BQ-123 and SB 209670 (but not, interestingly, FR 139317) selectively decrease mean arterial pressure in SHR and renin-dependent hypertensive rats (no comparable drop in blood pressure is observed in normotensive WKY; Ohno et al. 1992; Sogabe et al. 1992; Nishikibe et al. 1993; Ohlstein et al. 1993, 1994b). The non-peptide receptor antagonists, bosentan (Ro 47-0203), BMS-182874, and SB 209670, also lower blood pressure in hypertensive rats following enteric administration (Stein et al. 1993; Douglas et al. 1994a, 1995b). The drop in blood pressure observed following antagonist administration is the result of a selective decrease in total peripheral resistance, principally in the renal and mesenteric beds, whereas cardiac output is unaltered (since bradycardia is accompanied by an enhanced stroke volume; Douglas et al. 1994a). The antihypertensive effects of these receptor antagonists are generally slow in onset (possibly reflecting the slow peripheral and/or central distribution kinetics), and may explain why Bazil et al. (1992) failed to observe an antihypertensive effect following bolus i.v. administration of BQ-123 in high renin models of hypertension (however, it is of interest to note that the ET_A-selective antagonists LU 135252 and A-127722.5 only attenuated elevated blood pressures in DOCA-salt hypertensive rats and not in those strains which lacked a pronounced overexpression of endothelin-1 gene, e.g., SHRs and 1-kidney 1-clip hypertensive rats; Schiffrin et al. 1997). In view of the neuroendocrine actions of ET-1 (see Ohlstein et al. 1995) and effects on hematocrit, it is also possible that ET-1 promotes hypertension indirectly, for example via (i) activation/synergism with the renin-angiotensin system or (ii) by increasing blood viscosity, e.g., low dose infusions of exogenous ET-1 elevate systemic arterial blood pressure in normotensive rats in a fashion which can be reversed by co-administration of an ACE inhibitor (Dohi et al. 1992; Mortensen et al. 1992; Lüscher et al. 1993).

In addition to essential hypertension, ET-1 has been implicated in the development of primary pulmonary hypertension, a disease characterized by increased pulmonary vascular resistance and pronounced medial thickening and intimal fibrosis. Circulating ET-1 levels are elevated in humans with pulmonary hypertension, and the pulmonary vasculature may be responsible for an enhanced production of ET-1 (Stewart et al. 1991; Cacoub et al. 1993; Giaid et al. 1993). Interestingly, continuous administration of BQ-123 attenuates pulmonary arterial pressure and vascular resistance, right ventricular hypertrophy, and pulmonary artery medial thickening in both monocrotaline and hypobaric rat models of pulmonary hypertension (Miyauchi et al. 1993; Bonvallet et al. 1994).

ET-1 has also been implicated in the etiology of pre-eclampsia (gestational proteinuric hypertension), a potentially life-threatening hypertensive disorder that occurs during the third trimester of pregnancy. In addition to proteinuria and abnormalities in the coagulation system, pre-eclampsia is characterized by a marked hypertension which results from elevated total peripheral resistance (particularly in the renal bed), reduced cardiac output, and endothelial dysfunction within the umbilicus. ET-1 is synthesized by the placental amnion and the uteroplacental vasculature. Relative to normotensive pregnancies, maternal plasma levels of ET-1 are elevated during pre-eclampsia, especially in the umbilical vein (Kamoi et al. 1989; Nova et al. 1991; Florijn et al. 1991; Nisell et al. 1991; Schiff et al. 1992). Interestingly, the elevated systemic ET-1 levels usually return to normal within 1 week post partum (Tsunoda et al. 1992). Although the correlation between elevated ET-1 levels and the degree of hypertension is poor, ET-1 has been implicated as a pathogenic mediator of this disorder since it is a potent contractor of the placental vasculature (Haegerstrand et al. 1989; Sunnergren et al. 1990; Wada et al. 1990). However, pre-eclampsia is an extremely difficult disorder to study both preclinically and in humans; therefore, the available data implicating ET-1 as a pathogenic mediator is, to a great extent, circumstantial. Indeed, Benigni et al. (1992) have argued against such a role for ET-1 based on the observation that, relative to control tissue, preproET-1 mRNA expression and big ET-1 immunoreactivity are not enhanced during pre-eclampsia.

The contribution of ET-1 to aberrant blood pressure regulation in humans has recently been assessed in patients with mild to moderate essential hypertension (Krum et al. 1998). Compared to placebo-treated subjects, 4 week bosentan administration (100 mg, 500 mg, or 1000 mg u.i.d. or 1000 mg b.i.d.) resulted in a significant reduction in diastolic pressure with a daily dose of 500 mg or 2000 mg (an absolute reduction of 5.7 mmHg with no significant change in heart rate). This anti-hypertensive effect was similar to that observed with 20 mg enalapril, u.i.d. (5.8 mmHg). The fall in blood pressure observed did not result in a reflexive activation of the sympathetic nervous system (plasma norepinephrine levels) or the renin-angiotensin system (plasma renin activity). Thus, ongoing clinical trials indicate that ET-1 contributes to elevated blood pressure in hypertensive patients and, as such, ET receptor antagonists may constitute novel therapeutic entities in the clinical management of elevated blood pressure.

III. Atherosclerosis and Vascular Restenosis

Advanced coronary and peripheral (i.e., cerebral) atherosclerosis is the primary cause of death in the United States, Europe, and Japan (angina/myocardial infarction and stroke). According to 1996 estimates from the American Heart Association, 17 million Americans suffer from coronary artery disease. Although the incidence has declined over the last decade, some 500000 deaths resulted from advanced coronary artery disease in 1996, making it the single leading cause of death today. Consequently, the pathogenesis and treatment of atherosclerosis and its sequelae constitute a major focus of research interest both within academia and the pharmaceutical industry.

Vasospasm and abnormal vascular smooth muscle proliferation are important complications of both atherosclerosis and vascular wall trauma such as seen following percutaneous transluminal balloon angioplasty (PTCA). Interestingly, oxidized-LDL, a well-established atherogenic risk factor, stimulates ET-1 synthesis in human and porcine macrophage and endothelial cell cultures (MARTIN-NIZARD et al. 1991; BOULANGER et al. 1992; OHLSTEIN and DOUGLAS 1993). Since both atherosclerosis and angioplasty are associated with (i) reduced endothelial-nitric oxide release and (ii) enhanced ET-1/ECE immunoreactivity (LERMAN et al. 1991; BATH and MARTIN 1991; INO et al. 1992; TAHARA et al. 1992; ZEIHER et al. 1995; GRANTHAM et al. 1998), this peptide may be involved in the etiology of this disease; an imbalance between nitric oxide (spasmolytic/cytostatic) and ET-1 (spasmogenic/mitogenic) release may promote acute vasospasm and chronic vascular remodeling.

Both atherosclerosis and balloon angioplasty augment the contractile actions of ET-1 in vessels isolated from primates, rabbits and rats (LOPEZ et al. 1990; DAVIES et al. 1993; DOUGLAS et al. 1994d). Systemic administration of exogenous [^{125}I]ET-1 accumulates within the atherosclerotic plaques of hypercholesterolemic rabbits (PRAT et al. 1993) and enhanced [^{125}I]ET-1 binding has been detected in hyperplastic lesions induced in pig femoral arteries and in atheromatous regions of human saphenous veins and coronary arteries (DASHWOOD et al. 1993, 1994). Based on their sensitivity to the ET_A-selective antagonist, FR 139317, it appears that atherosclerotic human coronary artery specimens express both ET_A- and non-ET_A receptors (DASHWOOD et al. 1994).

Radioligand binding studies, RT-PCR, and immunohistochemical assays clearly demonstrate induction of ET-1, ECE, and ET_A/ET_B mRNA/protein expression in association with neointima formation following experimental balloon injury (DOUGLAS et al. 1994c; WANG et al. 1995; LOESCH et al. 1997;

Minamino et al. 1997; Viswanathan et al. 1997). Moreover, both acute (Douglas and Ohlstein 1993; Douglas et al. 1994c) and chronic (Trachtenberg et al. 1993) administration of exogenous ET-1 to rats dose-dependently augments the degree of neointima formation associated with carotid artery balloon angioplasty. Indeed, administration of SB 209670 and SB 217242, but not BQ-123 (at doses shown to block the hemodynamic actions of exogenous ET-1), results in a reduction in neointima formation in this model and, therefore, suggests a putative role for ET$_B$ receptors in mediating this phenomenon (Douglas et al. 1994c; Chandra et al. 1998). Similarly, bosentan has shown efficacy in a rat aorta angioplasty model (using surrogate biochemical endpoints; Hele et al. 1995). Interestingly, chronic ECE inhibition with phosphoramidon has also been reported to be protective in the rat carotid artery (Minamino et al. 1997). However, the relative role of the ET$_A$ and ET$_B$ receptors in the pathogenesis of restenosis remains controversial (Kirchengast and Münter 1998). In addition the ET$_{A/B}$ antagonists SB 209670 (Douglas et al. 1994c), SB 217242 (Chandra et al. 1998), and TAK-044 (Tsujino et al. 1995) and the ET$_A$-selective antagonists BMS 182874 (Ferrer et al. 1995), FR 139317 (Takiguchi and Sogabe 1996), and LU 135252 (Münter et al. 1996) have been reported to be vasculoprotective in this model. Similarly, the ET$_A$-selective antagonist A-122722 reduces neointima formation in the pig iliac artery following balloon injury (Burke et al. 1997; McKenna et al. 1998). Thus, the optimal pharmacological profile for an efficacious ET receptor antagonist designed to prevent restenosis and similar cardiovascular disorders characterized by abnormal smooth muscle proliferation remains to be elucidated.

Nevertheless, it is of interest to note that, following rat aortic catheter-induced denudation, the smooth muscle cells which migrate from the tunica media to form the neointima undergo acute, yet profound, phenotype changes resulting in upregulation of the ET$_B$ receptor (see Sect. J). Furthermore, whereas both the mixed ET$_{A/B}$ antagonist bosentan and the ET$_B$-selective antagonist BQ-788 reduce neointima formation in human vein organ cultures (saphenous vein grafts), this activity is not shared by the ET$_A$-selective antagonist BQ-123 (Porter et al. 1998). In contrast, however, it has been suggested that in vivo ET$_A$ receptors may be involved in smooth muscle proliferation (tunica media and intima) whereas the ET$_B$ receptor is involved primarily in adventitial microangiogenesis in porcine saphenous vein grafts (Dashwood et al. 1998).

Furthermore, in addition to catheter-induced vascular trauma, activation of the ET system is also associated with hyperplastic neointima formation such as is observed in both experimental (heterotopic cardiac allografts; Watschinger et al. 1995; Okada et al. 1998) and clinical (human chronic renal allograft rejection, transplant coronary artery disease; Ravalli et al. 1996; Simonson et al. 1998) accelerated transplant atherosclerosis. Indeed, chronic bosentan administration attenuates neointima formation in a rat heterotopic heart transplantation model (Okada et al. 1998).

IV. Myocardial Ischemia

Since reductions in pO_2 promote ET-1 synthesis, alter ET-1 binding site density and affinity, and augment the contractile actions of ET-1, this potent coronary artery vasoconstrictor has been implicated in the pathogenesis of angina pectoris, coronary artery vasospasm and myocardial infarction (clinical conditions which are all associated with elevated plasma ET-1-like immunoreactivity; MIYAUCHI et al. 1989b; SALMINEN et al. 1989; YASUDA et al. 1990; MATSUYAMA et al. 1991). Recently, selective upregulation of ECE-1 mRNA expression has been observed in patients who have experienced a prior myocardial infarction (a response abrogated in subjects receiving β-blockade; BOHNEMEIER et al. 1998). By prolonging the ischemic period experienced by the myocardium, enhanced ET-1-induced coronary vasospasm may exacerbate the degree of subsequent myocardial infarction. Interestingly, both a polyclonal Ab to ET-1 and phosphoramidon have been shown to reduce rat and rabbit left ventricular infarction resulting from coronary artery ligation-reperfusion (a process which produces a five-fold increase in plasma ET-1-like immunoreactivity; GROVER et al. 1992; WATANABE et al. 1990; KUSUMOTO et al. 1993). In contrast, however, the ET_A-selective antagonist, FR 139317, has produced conflicting results in a rabbit model of myocardial ischemia-reperfusion. Whereas McMURDO et al. (1994) failed to observe a protective effect with FR 139317, NELSON et al. (1994) and LEE et al. (1994) found that FR 139317 attenuated infarct size both in the rabbit and in the rat (in view of these paradoxical reports, it is interesting to note that the antiarrhythmic effect of BQ-123 is lost in a rat coronary artery occlusion model when high doses are used; GARAJANI et al. 1994).

V. Cerebral Vasospasm, Stroke, and Subarachnoid Hemorrhage

The role of ET-1 in the pathogenesis of cerebral vasospasm is discussed in detail elsewhere (see Chap. 13). Briefly, ET-1 is a potent contractor of isolated canine basilar arteries, and angiographic studies have shown that local administration of ET-1 causes long-lasting vasoconstriction within the cerebrovasculature (ASANO et al. 1989, 1990). Since plasma levels of ET-1-like immunoreactivity are elevated in patients with subarachnoid hemorrhage, ET-1 may be involved in the pathogenesis of the cerebral vasospasm associated with this disorder (MASAOKA et al. 1989). Whereas COSENTINO et al. (1993) reported that BQ-123 and phosphoramidon did not ameliorate the degree of subarachnoid hemorrhage-induced cerebral vasospasm in a two-bleed canine model, ITOH et al. (1993) and FOLEY et al. (1994) were able to demonstrate a protective effect with BQ-485 and BQ-123. CLOZEL and WATANABE (1993) suggest that central administration of BQ-123 is required in order to see a protective effect since this antagonist does not penetrate the blood brain barrier in the rat. In addition, Ro 46-2005 also reverses the cerebral vasospasm observed in a rat model of subarachnoid hemorrhage (CLOZEL and WATANABE

1993). Ro 47-0203 and SB 209670 attenuate the acute (day 2) and chronic vasospasm (day 7) observed in rabbit and dog models of subarachnoid hemorrhage (ROUX et al. 1993; WILLETTE et al. 1994). Furthermore, BQ-123 and SB 209670 are neuroprotective in a gerbil stroke model, inhibiting the degree of CA_1 neurodegeneration associated with focal cerebral ischemia resulting from cerebral artery ligation (FEUERSTEIN et al. 1994; OHLSTEIN et al. 1994a,b). One of the consequences of subarachnoid hemorrhage is direct exposure of cerebral tissue to blood constituents such as oxyhemoglobin. In addition to being a cerebral vasoconstrictor, oxyhemoglobin stimulates [^{3}H]thymidine incorporation in rat vascular smooth muscle in a manner which is inhibited by BQ-123 (OHLSTEIN and DOUGLAS 1993). Since oxyhemoglobin also increases ET-1 biosynthesis in cultured endothelial cells over a concentration range similar to that observed following subarachnoid hemorrhage, ET-1 may also be involved in the cerebral vessel hypertrophy and fibrosis associated with the chronic stages of subarachnoid hemorrhage.

VI. Diabetes

Elevated levels of insulin and glucose modulate ET-1 synthesis and receptor expression in cultured endothelial and vascular smooth muscle cells (YAMAUCHI et al. 1990; OLIVER et al. 1991; KWOK et al. 1993). Levels of immunoreactive-ET-1 detected in plasma, urine, and peritoneal fluid are elevated in patients suffering from non insulin-dependent diabetes mellitus (LAM et al. 1991; NAKAYAMA et al. 1991; TOTSUNE et al. 1991; KIRILOV et al. 1993), a condition associated with endothelial dysfunction, abnormal vascular tone, and vasculopathy (although no such elevations were detected by KANNO et al. 1991). Thus, it has been postulated that ET-1 is involved in the etiology of the peripheral and coronary artery macrovascular disease associated with hyperglycemia. ET-1 is a potent vasoconstrictor in the porcine ophthalmic microcirculation and, thus, may play an important role in the physiological control of ocular perfusion (via ET_A-receptor activation; MEYER et al. 1993). However, since ET-1 receptor density and vascular reactivity are altered in rats with streptozotocin-induced diabetes, it has been suggested that this vasoactive peptide is involved in the microvascular complications (i.e., nephropathy and retinopathy) associated with hyperglycemia (AWAZU et al. 1991; LAWRENCE and BRAIN 1992).

VII. Endotoxic Shock

Systemic administration of *E. coli* endotoxin elevates circulating levels of ET-1-like immunoreactivity in rats, pigs, and sheep (MOREL et al. 1989; SUGIURA et al. 1989b; PERNOW et al. 1990). Such changes are accompanied by increased renal and splenic vascular resistance leading to the proposition that, by causing vasoconstriction, ET-1 may contribute to the end-organ failure associated with septic shock. However, it is equally plausible that the endotoxemia-induced release of ET-1 is acting in a protective hemostatic role. This explanation has been forwarded in hemorrhagic shock (where circulating levels are also enhanced) since the increase in systemic vascular resistance associated with this phenomenon is antagonized by BQ-123 (ZIMMERMAN et al. 1994).

VIII. Migraine

Relatively little attention has been focused on the role of ET-1 in the etiology of migraine, a surprising fact since (i) this disorder is characterized by endothelial cell dysfunction, (ii) ET-1 is known to synergize with 5-hydroxytryptamine; and (iii) as previously discussed, ET-1 has been implicated in the control of cerebrovascular blood flow following subarachnoid hemorrhage and stroke. Interestingly, however, GALLAI et al. (1994) have found elevated plasma levels of ET-1 in migraine patients proposing that ET-1 could mediate the hemodynamic changes associated with migraine, especially during the acute phases of the attack. Nevertheless, no data is available currently to establish a cause-effect relationship for ET-1 in this disorder.

IX. Reynaud's Phenomenon

A pathophysiological role for ET-1 in Reynaud's phenomenon is controversial (RANDALL 1991) since, relative to control patients, some investigators have seen elevated levels of ET-1-like immunoreactivity, whereas others have failed to document any such difference (ZAMORA et al. 1990; SMIT et al. 1991; FERRI et al. 1993).

M. Pharmacological Profile for the Treatment of Cardiovascular Disease

The rapid preclinical development of pharmacologically distinct ET receptor antagonists (see Chap. 9) bears evidence to the perceived pathological significance of ET-1 within both the medical and pharmaceutical communities. The development of antagonists with divergent pharmacodynamic profiles has been meteoric. Initially the trend was to develop "mixed/dual" nonpeptide $ET_{A/B}$ receptor antagonists. However, prudence has dictated that a diverse portfolio of antagonists be established, a reflection of the current uncertainty regarding the specific pathobiological roles of the ET_A and ET_B receptor in man. As such, the optimal therapeutic profile of an ET antagonist remains somewhat controversial and will, ultimately, depend upon the specific indications that are targeted. Regardless, the hypothetical disadvantages of untoward ET_B-blockade have stimulated the development ET_A-selective candidates (SB 234551, SB 247083, TBC-11251, PD 156707, LU 135252, A-127722). Typically, such antagonists exhibit ≥500-fold selectivity for the ET_A receptor.

As discussed, the putative deleterious effects associated with ET_B receptor blockade include (a) inhibition of the vasorelaxant actions of ET-1 and (b) elevation of circulating ET-1 levels. Data exist supporting a predominant vasoactive role for the ET_B receptor blockade, (Sect. G) antagonism of which might counteract any antagonist-induced reduction in blood pressure in, for example, a hypertensive patient subjected to concomitant ET_A receptor antagonism. Furthermore, ET_B receptor antagonism is associated with an elevation in circulating ET-1 levels due to displacement of ET-1 from "ET_B-like" clearance receptors in the lung (Willette et al. 1998). It has been proposed that a "rebound" elevation in systemic ET-1 levels could precipitate "secondary" responses. However, to date this issue remains speculative and open to debate. For instance, it is argued that, since a "mixed" antagonist would block both receptors, any "clearance receptor" phenomenon would be negated. In addition, there is a growing body of evidence indicating that antagonism of ET_B receptors is required to influence the growth properties of myocytes and smooth muscle cells. Similarly, since both ET_A and ET_B receptors are involved in mediating the contractile actions of ET-1, it is argued that mixed antagonists would be more efficacious than an ET_A-selective antagonist as an anti-hypertensive therapeutic. Clearly, the optimal profile required for the efficacious treatment of cardiovascular disorders is dependent upon the specific indication in question. Such complex clinical issues are only likely to be resolved following the lengthy clinical evaluation of antagonists (preferably in head-to-head comparative studies) in man.

N. Summary

ET-1 is a potent vasoconstrictor and smooth muscle mitogen in a wide variety of species, including man. These actions are modulated by the vascular endothelium through the concomitant, peptide-induced release of vasoactive factors such as nitric oxide. The direct effects of ET-1 on the cardiovascular system suggest an important role for ET-1 in the physiological control of vascular tone and structure. Furthermore, the recent availability of ET receptor antagonists has established a causal role for ET-1 in several preclinical cardiovascular disease models. The development and subsequent progression of such molecules into the clinic will help define a role for this peptide as pathogenic mediator of human cardiovascular disease.

Acknowledgements. The Authors wish to express their gratitude to Ms Susan Tirri for her expert assistance during the preparation of this manuscript.

References

Achmad TH, Winterscheidt A, Lindemann C, Rao GS (1997) Oxidized low density lipoprotein acts on endothelial cells in culture to enhance endothelin secretion and monocyte migration. Meth Find Expl Clin Pharmacol 19:153–159

Adachi H, Shoji T, Goto K (1991) Comparison of the effects of endothelin and BAY k 8644 on cardiohemodynamics in anesthetized pigs. Eur J Pharmacol 193:57–65

Ahlborg G, Ottosson-Seeberger A, Hemsen A, Lundberg JM (1996) Central and regional hemodynamic effects during infusion of big endothelin-1 in healthy humans. J Appl Physiol 80:1921–1927

Ais G, Novo C, López-Farré A, Romeo JM, López-Novoa JM (1989) Effect of endothelin on systemic and regional hemodynamics in rats. Eur J Pharmacol 170:113–116

Alberts GF, Peifley KA, Johns A, Kleha JF, Winkles JA (1994) Constitutive endothelin-1 over expression promotes smooth muscle proliferation via an external autocrine loop. J Biol Chem 269:10112–10118

Armstead WM, Mirro R, Leffler CW, Busija DW (1989) Influence of endothelin on piglet cerebral microcirculation. Am J Physiol 257:H707–H710

Asano T, Ikegaki I, Suzuki Y, Satoh SI, Shibuya M (1989) Endothelin and the production of cerebral vasospasm in dogs. Biochem Biophys Res Commun 159:1345–1351

Asano T, Ikegaki I, Satoh SI, Suzuki Y, Shibuya M, Sugita K, Hidaka H (1990) Endothelin: a potential modulator of cerebral vasospasm. Eur J Pharmacol 190:365–372

Auguet M, Delaflotte S, Chabrier PE, Braquet PC (1989) Comparative effects of endothelin and phorbol 12-13 dibutyrate in rat aorta. Life Sci 45:2051–2059

Auguet M, Delaflotte S, Chabrier PE, Braquet PC (1993) Characterization of endothelin receptors mediating contraction and relaxation in rabbit saphenous artery and vein. Can J Physiol Pharmacol 71:818–823

Awazu M, Parker RE, Harvie BR, Ichikawa I, Kon V (1991) Down-regulation of endothelin-1 receptors by protein kinase C in streptozotocin diabetic rats. J Cardiovasc Pharmacol 17 [Suppl 7]:S500–S502

Bacon CR, Davenport AP (1996) Endothelin receptors in human coronary artery and aorta. Brit J Pharmacol 117:986–992

Bath PMW, Martin JF (1991) Serum platelet-derived growth factor and endothelin concentrations in human hypercholesterolaemia. J Inter Med 230:313–317

Baydoun AR, Peers SH, Cirino G, Woodward B (1989) Effects of endothelin-1 on the rat isolated heart. J Cardiovasc Pharmacol 13 [Suppl 5]:S193–S196

Baydoun AR, Peers SH, Cirino G, Woodward B (1990) Vasodilator action of endothelin-1 in the perfused rat heart. J Cardiovasc Pharmacol 15:759–763

Bazil MK, Lappe RW, Webb RL (1992) Pharmacologic characterization of an endothelin$_A$ (ET$_A$) receptor antagonist in conscious rats. J Cardiovasc Pharmacol 20:940–948

Bdolah A, Wollberg Z, Ambar I, Kloog Y, Sokolovsky M, Kochva E (1989a) Disturbances in the cardiovascular system caused by endothelin and sarafotoxin. Biochem Pharmacol 38:3145–3146

Benigni A, Orisio S, Gaspari F, Frusca T, Amuso G, Remuzzi G (1992) evidence against a pathogenic role for endothelin in pre-eclampsia. Br J Obstet Gynaecol 99:798–802

Berrazueta JR, Bhagat K, Vallance P, MacAllister RJ (1997) Dose- and time-dependency of the dilator effects of the endothelin antagonist, BQ-123, in the human. Brit J Clin Pharmacol 44:569–571

Berthiaume N, Yanagisawa M, Yanagisawa H, deWit D, D'Orleans-Juste P (1998) Pharmacology of endothelins in vascular circuits of normal heterozygous endothelin-A or endothelin-B knockout transgenic mice. J Cradiovasc Pharmacol 31 [Suppl 1]:S561–S564

Beyer ME, Slesak G, Hoffmeister HM (1996) Significance of endothelinB receptors for myocardial contractility and myocardial energy metabolism. J Pharmacol Exp Ther 278:1228–1234

Beyer ME, Slesak G, Brehm BR, Hoffmeister HM (1998) Hemodynamic and inotropic effects of the endothelin A antagonist BQ-610 in vivo. J Cardiovasc Pharmacol 31 [Suppl 1]:S258–S261

Bialecki RA, Izzo NJ, Colucci WS (1989) endothelin-1 increases intracellular calcium mobilization but not calcium uptake in rabbit vascular smooth muscle cells. Biochem Biophys Res Commun 164:474–479

Bigaud M, Pelton JT (1992) discrimination between ET_A- and ET_B-receptor-mediated effects of endothelin-1 and [Ala ^{1,3,11,15}]endothelin-1 by BQ-123 in the anaesthetized rat. Br J Pharmacol 107:912–918

Bohnemeier H, Pinto YM, Horkay F, Toth M, Juhasz-Nagy A, Orzechowski HD, Bohm M, Paul M (1998) Endothelin converting enzyme-1 mRNA expression in human cardiovascular disease. J Cardiovasc Pharmacol 31 (Suppl 1):S52–S54

Bonvallet ST, Zamora MR, Hasunuma K, Sato K, Hanasato N, Anderson D, Sato K, Stelzner TJ (1994) BQ-123, an ET_A-receptor antagonist, attenuates hypoxic pulmonary hypertension in rats. Am J Physiol 266:H1327–H1331

Borges R, Carter DV, Von Grafenstein H, Knight DE (1989a) Ionic requirements of the endothelin response in aorta and portal vein. Circ Res 65:265–271

Borges R, Von Grafenstein H, Knight DE (1989b) Tissue selectivity of endothelin. Eur J Pharmacol 165:223–230

Boulanger CM, Tanner FC, Beau ML, Han AWA, Werner A, Lüscher TF (1992) Oxidized low density lipoproteins induce mRNA expression and release of endothelin from human and porcine endothelium. Circ Res 70:1191–1197

Brain SD, Tippins JR, Williams TJ (1988) endothelin induces potent microvascular constriction. Br J Pharmacol 95:1005–1007

Brain SD, Crossman DC, Buckley TL, Williams TJ (1989) endothelin-1: demonstration of potent effects on the microcirculation of humans and other species. J Cardiovasc Pharmacol 13 [Suppl 5]:S147–S149

Brooks DP, DePalma PD, Pullen M, Nambi P (1994) characterization of canine endothelin receptor subtypes and their function. J Pharmacol Exp Ther 268:1091–1097

Brown KD, Littlewood CJ (1989) endothelin stimulates DNA synthesis in Swiss 3T3 cells. Synergy with polypeptide growth factors. Biochem J 263:977–980

Bruneau BG, Piazza LA, de Bold AJ (1997) BNP gene expression is specifically modulated by stretch and ET-1 in a new model of isolated rat atria. Am J Physiol 273:H2678–H2686

Brunner F, Kukovetz WR (1996) Postischemic antiarrhythmic effects of angiotensin-converting enzyme inhibitors. Role of suppression of endogenous endothelin secretion. Circulation 94:1752–1761

Burke SE, Lubbers NL and Gagne GD (1997) Selective antagonism of the ET_A receptor reduces neointimal hyperplasia after balloon-induced injury in pigs. J Cardiovasc Pharmacol 30:33–41

Cacoub P, Dorent R, Maistre G, Nataf P, Carayon A, Piette JC, Godeau P, Cabrol C, Gandjbakhch I (1993) Endothelin-1 in primary pulmonary hypertension and the eisenmenger syndrome. Am J Cardiol 71:448–450

Catelli De Carvalho MH, Nigro D, Scivoletto R, Barbiero HV, De Oliveira MA, De Nucci G, Fortes ZB (1990) comparison of the effect of endothelin on microvessels and macrovessels in Golblatt ii and deoxycorticosterone acetate-salt hypertensive rats. Hypertension 15 [Suppl I]:I68–I71

Cavero PG, Miller WL, Heublein DM, Margulies KB, Burnett JC (1990) Endothelin in experimental congestive heart failure in the anesthetized dog. Am J Physiol 259:F312–F317

Chandra S, Vickery-Clark L, Coatney RW, Phan L, Sarkar SK, Ohlstein EH (1998) Application of serial *in vivo* magnetic resonance imaging to evaluate the efficacy of the endothelin receptor antagonist SB 217242 in the rat carotid artery model of neointima formation. Circulation 97:2252–2258

Chapleau MW, Hajduczok G, Abboud FM (1992) suppression of baroreceptor discharge by endothelin at high carotid sinus pressure. Am J Physiol 263:R103–R108

Chua BHL, Krebs CJ, Chua CC, Diglio CA (1992) Endothelin stimulates protein synthesis in smooth muscle cells. Am J Physiol 262:E412–E416

Clarke JG, Benjamin N, Larkin SW, Webb DJ, Davies GJ, Maseri A (1989) Endothelin is a potent long-lasting vasoconstrictor in men. Am J Physiol 257:H2033–H2035

Clavell A, Stingo A, Margulies K, Lerman A, Underwood D, Burnett JC Jr (1993) Physiological significance of endothelin: its role in congestive heart failure. Circulation 87 ([Suppl V]:V45–V50

Clerk A, Michael A, Sugden PH (1998) Stimulation of the p38 mitogen-activated protein kinase pathway in neonatal rat ventricular myocytes by the G-protein-coupled receptor agonists, endothelin-1 and phenylephrine: a role for cardiac myocyte hypertrophy. J Cell Biol 142:523–535

Clozel J-P, Clozel M (1989a) Effects of endothelin on the coronary vascular bed in open-chest dogs. Circ Res 65:1193–1200

Clozel M, Clozel J-P (1989b) Effects of endothelin on regional blood flows in squirrel monkeys. J Pharmacol Exp Ther 250:1125–1131

Clozel M, Gray GA, Breu V, Löffler B-M, Osterwalder R (1992) the endothelin ET_B receptor mediates both vasodilation and vasoconstriction in vivo. Biophys Res Commun 186:867–873

Clozel M, Breu V, Burri K, Cassal J-M, Fischli W, Gray GA, Hirth G, Löffler B-M, Müller M, Neidhart W, Ramutz H (1993a) Pathophysiological role of endothelin revealed by the first orally active endothelin receptor antagonist. Nature 365:759–761

Clozel M, Breu V, Gray GA, Löffler BM (1993b) In vivo pharmacology of Ro 46–2005, the first synthetic nonpeptide endothelin receptor antagonist: implications for endothelin physiology. J Cardiovasc Pharmacol 22 [Suppl 8]:S377–S379

Clozel M, Watanabe H (1993) BQ-123, a peptidic endothelin ET_A receptor antagonist, prevents the early cerebral vasospasm following subarachnoid hemorrhage after intracisternal but not intravenous injection. Life Sci 52:825–834

Coceani F, Armstrong C, Kelsey L (1989) Endothelin is a potent constrictor of the lamb ductus arteriosus. Can J Physiol Pharmacol 67:902–904

Coceani F, Kelsey L (1991) Endothelin-1 release from lamb ductus arteriosus: relevance to postnatal closure of the vessel. Can J Physiol Pharmacol 69:218–223

Cocks TM, Faulkner NL, Sudhir K, Angus J (1989) reactivity of endothelin-1 on human and canine large veins compared with large arteries in vitro. Eur J Pharmacol 171:17–24

Cosentino F, McMahon EG, Carter JS, Katusic ZS (1993) effect of the endothelin$_A$-receptor antagonist BQ-123 and phosphoramidon on cerebral vasospasm. J Cardiovasc Pharmacol 22 [Suppl 8]:S332–S335

Criscione L, Nellis P, Riniker B, Thomann H, Burdet R (1990) Reactivity and sensitivity of mesenteric vascular beds and aortic rings of spontaneously hypertensive rats to endothelin: effects of calcium entry blockers. Br J Pharmacol 99:31–36

Dashwood M, Turner M, Jacobs M (1989) Endothelin-1: contractile responses and autoradiographic localization of receptors in rabbit blood vessels. J Cardiovasc Pharmacol 13 [Suppl 5]:S183–S185

Dashwood M, Barker SGE, Muddle JR, Yacoub MH, Martin JF (1993) [^{125}I]-endothelin-1 binding to vasa vasorum and regions of neovascularization in human and porcine blood vessels: a possible role for endothelin in intimal hyperplasia and atherosclerosis. J Cardiovasc Pharmacol 22 [Suppl 8]:S343–S347

Dashwood MR, Allen SP, Luu TN, Muddle JR (1994) The effect of the ET_A receptor antagonist, FR 139317, on [^{125}I]-ET-1 binding to the atherosclerotic human coronary artery. Br J Pharmacol 112:386–389

Dashwood MR, Mehta D, Izzat MB, Timm M, Bryan AJ, Angelini GD, Jeremy JY (1998) Distribution of endothelin-1 (ET) receptors (ET_A and ET_B) and immunoreactive ET-1 in porcine saphenous vein-carotid artery interposition grafts. Atherosclerosis 137:233–242

Davenport AP, Ashby MJ, Easton P, Ella S, Bedorford J, Dickerson C, Nunez DJ, Capper SJ, Brown MJ (1990) A sensitive radioimmunoassay measuring endothelin-like immunoreactivity in human plasma: a comparison of levels in patients with essential hypertension and normotensive control subjects. Clin Sci 78:261–264

Davenport AP, Maguire JJ (1994) Is endothelin-induced vasoconstriction mediated only by ET_A receptors in humans? Trends Pharmacol Sci 15:9–11

Davies MG, Klyachkin ML, Kim JH, Hagen P-O (1993) Endothelin and vein bypass grafts in experimental atherosclerosis. J Cardiovasc Pharmacol 22 [Suppl 8]:S348–S351

De Nucci G, Thomas R, D'Orleans-Juste P, Antunes E, Walder C, Warner TD, Vane JR (1988) Pressor effects of circulating endothelin are limited by its removal from the pulmonary circulation and by the release of prostacyclin and endothelium-derived relaxing factor. Proc Natl Acad Sci (USA) 85:9797–9800

Doherty AM, Cody WL, He JX, Depue PL, Chang XM, Welch KM, Flynn MA, Reynolds EE, Ladouceur D, Davis LS, Keiser JA, Haleen SJ (1993) In vitro and in vivo studies with a series of hexapeptide endothelin antagonists. J Cardiovasc Pharmacol 22 [Suppl 8]:S98–S102

Dohi Y, Hahn AWA, Boulanger C, Bühler FR, Lüscher TF (1992) Endothelin stimulated by angiotensin II augments contractility of SHR resistance arteries. Hypertension 19:131–137

Domenech R, Macho P, González R, Huidobro-Toro JP (1991) Effect of endothelin on total and regional coronary resistance and on myocardial contractility. Eur J Pharmacol 192:409–416

D'Orleans-Juste P, De Nucci G, Vane JR (1989) Endothelin-1 contracts isolated vessels independently of dihydropyridine-sensitive Ca^{2+} channel activation. Eur J Pharmacol 165:289–295

Douglas SA, Hiley CR (1990) Endothelium-dependent vascular activities of endothelin-like peptides in the isolated superior mesenteric arterial bed of the rat. Br J Pharmacol 101:81–88

Douglas SA, Hiley CR (1991) endothelium-dependent mesenteric vasorelaxant effects and systemic actions of endothelin (16-21) and other endothelin-related peptides in the rat. Br J Pharmacol 104:311–320

Douglas SA, James S, Hiley CR (1991) Endothelial modulation and changes in endothelin pressor activity during hypoxia in the rat isolated perfused superior mesenteric arterial bed. Br J Pharmacol 103:1441–1448

Douglas SA, Elliott JD, Ohlstein EH (1992) Regional vasodilation to endothelin-1 is mediated by a non-ET_A receptor subtype in the anaesthetized rat: effect of BQ-123 on systemic haemodynamic responses. Eur J Pharmacol 221:315–324

Douglas SA, Vickery-Clark LM, Ohlstein EH (1993) Endothelin-1 does not mediate hypoxic vasoconstriction in canine isolated blood vessels: effect of BQ-123. Br J Pharmacol 108:418–421

Douglas SA, Gellai M, Ezekiel M, Ohlstein EH (1994a) BQ-123, a selective endothelin (ET_A) receptor antagonist, lowers blood pressure in different models of hypertension. J Hypertension 12:561–567

Douglas SA, Meek TD, Ohlstein EH (1994b) Novel receptor antagonists welcome a new era in endothelin biology. Trends Pharmacol Sci 15(9):313–316

Douglas SA, Vickery-Clark LM, Ohlstein EH (1994c) Functional evidence that angioplasty results in transient nitric oxide synthase induction. Eur J Pharmacol 255:81–89

Douglas SA, Vickery-Clark LM, Storer BL, Hart T, Louden C, Elliott JD, Ohlstein EH (1994d) A role for endogenous endothelin-1 in neointima formation following rat carotid artery balloon angioplasty: antiproliferative effects of the non-peptide endothelin receptor antagonist, SB 209670. Circ Res 75:190–197

Douglas SA, Beck GR Jr, Elliott JD, Ohlstein EH (1995a) Pharmacological evidence for the presence of three distinct functional endothelin receptor subtypes in the rabbit lateral saphenous vein. Brit J Pharmacol 114:1529–1540

Douglas SA, Gellai M, Ezekiel M, Feuerstein GZ, Elliott JD, Ohlstein EH (1995b) Antihypertensive actions of the novel nonpeptide endothelin receptor antagonist SB 209670. Hypertension 25:818–822

Douglas SA (1997) Clinical development of endothelin receptor antagonists. Trends Pharmacol Sci 18:408–412

Douglas SA, Ohlstein EH (1997) Signal transduction mechanisms mediating the vascular actions of endothelin. J Vasc Res 34:152–164

Douglas SA, Nichols AJ, Feuerstein GZ, Elliott JD, Ohlstein EH (1998) SB 209670 inhibits the arrhythmogenic actions of endothelin-1 in the anesthetized dog. J Cardiovasc Phramcol 31 [Suppl 1]:S99–S102

Dubin D, Pratt RE, Cooke JP, Dzau VJ (1989) endothelin, a potent vasoconstrictor, is a vascular smooth muscle mitogen. J Vasc Med Biol 1:150–153

Eglen RM, Michel AD, Sharif NA, Swank SR, Whiting RL (1989) the pharmacological properties of the peptide, endothelin. Br J Pharmacol 97:1297–1307

Eguchi S, Hirata Y, Imai T, Kanno K, Marumo F (1994) phenotypic change of endothelin receptor subtype in cultured rat vascular smooth muscle cells. Endocrinology 134:222–228

Elferink JG, de Koster BM (1996a) Modulation of human neutrophil chemotaxis by the endothelin-B receptor agonist sarafotoxin S6c. Chem Bio Int 101:165–174

Elferink JG, de Koster BM (1996b) The effect of endothelin-2 (ET-2) on migration and changes in cytosolic free calcium of neutrophils. Naunyn-Schmied Arch Pharmacol 353:130–135

Emori T, Hirata Y, Marumo F (1990) Specific receptors for endothelin-3 in cultured bovine endothelial cells and its cellular mechanism of action. FEBS Lett 263:261–264

Ercan ZS, Ilhan M, Kilinc M, Turker RK (1996) Arrhythmogenic action of endothelin peptides in isolated perfused whole hearts from guinea pigs and rats. Pharmacology 53:234–240

Ezra D, Goldstein RE, Czaja JF, Feuerstein GZ (1989) Lethal ischemia due to coronary endothelin in pigs. Am J Physiol 257:H339–H343

Ferrer P, Valentine M, Jenkins-West T (1995). Orally active endothelin receptor antagonist BMS-182874 suppresses neointimal development in balloon-injured rat carotid arteries. J Cardiovasc Pharmacol 26:908–915

Ferri C, Latorraca A, Catapano G, Greco F, Mazzoni A, Clerico A, Pedrinella R (1993) Increased plasma endothelin-1 immunoreactive levels in vasculitis: a clue to the use of endothelin-1 as a marker of vascular damage. J Hypertension 11 [Suppl 5]:S142–S143

Ferro CJ, Haynes WG, Johnston NR, Lomax CC, Newby DE, Webb DJ (1997) Brit J Clin Pharmacol 44:377–383

Feuerstein GZ, Gu J-L, Ohlstein EH, Barone FC, Yue T-L (1994) Peptidic endothelin-1 receptor antagonist, BQ-123, and neuroprotection. Peptides 15:467–469

Filep JG, Sirois MG, Rousseau A, Fournier A, Sirois P (1991) Effect of endothelin-1 on vascular permeability in the conscious rat: interactions with platelet-activating factor. Br J Pharmacol 104:797–804

Filep JG, Foldes-Filep E, Rousseau A, Fournier A, Sirois P, Yano M (1992) Endothelin-1 enhances vascular permeability in the rat heart through the ET_A receptor. Eur J Pharmacol 219:343–344

Filep JG, Skrobik Y, Fournier A, Foldes-Filep E (1996) Effects of calcium antagonists on endothelin-1-induced myocardial ischaemia and oedema in the rat. Br J Pharmacol 118:893–900

Florijn KW, Derkx FHM, Visser W, Hofman JA, Rosmalen FMA, Wallenburg HCS, Schalekamp MADH (1991) Elevated plasma levels of endothelin in pre-eclampsia. J Hypertension 9 [Suppl 6]:S166–S167

Foley PL, Caner HH, Kassell NF, Lee KS (1994) Reversal of subarachnoid hemorrhage-induced vasoconstriction with an endothelin receptor antagonist. Neurosurg 34:108–113

Folta A, Joshua IG, Webb RC (1989) Dilator actions of endothelin in coronary resistance vessels and the abdominal aorta of the guinea-pig. Life Sci 45:2627–2635

Fozard JR, Part ML (1990) No major role for atrial natriuretic peptide in the vasoactive response to endothelin-1 in the spontaneously hypertensive rat. Eur J Pharmacol 180:153–159

Fozard JR, Part ML (1992) The role of nitric oxide in the regional effects of endothelin-1 in the rat. Br J Pharmacol 105:744–750

Fu L-X, Sun X-Y, Hedner T, Feng Q-P, Liang Q-M, Hoebeke J, Hjalmarson A (1993) Decreased density of mesenteric arteries but not of myocardial endothelin receptors and function in rats with chronic ischemic heart failure. J Cardiovasc Pharmacol 22:177–182

Fukuda Y, Hirata Y, Yoshimi H, Kojima T, Kobayashi Y, Yanagisawa M, Masaki T (1988) Endothelin is a potent secretagonue for atrial natriuretic peptide in clutured rat atrial myocytes. Biochem Biophys Res Commun 155:167–172

Fukue M, Ishikawa S, Miyauchi T, Onizuka M, Jerome EH (1994) No attenuation of hypoxic vasoconstriction by endothelin receptor blocker in perfused sheep lungs. FASEB J 8:A330

Gallai V, Sarchielli P, Firenze C, Trequattrini A, Paciaroni M, Usai F, Palumbo R (1994) Endothelin-1 in migraine and tension-type headache. Acta Neurol Scand 89:47–55

Garcia JL, Fernandez N, Garcia-Villalon AL, Monge L, Gomez B, Dieguez G (1996) Coronary vasoconstriction by endothelin-1 in anesthetized goats: role of endothelin receptors, nitric oxide and prostanoids. Eur J Pharmacol 315:179–186

Gardiner SM, Compton AM, Bennett T, Palmer RMJ, Moncada S (1989) N^G-monomethyl-L-arginine does not inhibit the hindquarters vasoactive action of endothelin-1 in conscious rats. Eur J Pharmacol 171:237–240

Gardiner SM, Compton AM, Bennett T (1990a) Effects of indomethacin on the regional haemodynamic responses to low doses of endothelins and sarafotoxin. Br J Pharmacol 100:158–162

Gardiner SM, Compton AM, Bennett T (1990b) Regional haemodynamic effects of endothelin-1 and endothelin-3 in conscious Long Evans and Brattleboro rats. Br J Pharmacol 99:107–112

Gardiner SM, Compton AM, Bennett T, Palmer RMJ, Moncada S (1990c) Regional haemodynamic changes during oral ingestion of N^G-nitro-L-arginine methyl ester in conscious Brattleboro rats. Br J Pharmacol 101:10–12

Gardiner SM, Compton AM, Kemp PA, Bennett T (1990d) Regional and cardiac haemodynamic responses to glyceryl trinitrate, acetylcholine, bradykinin and endothelin-1 in conscious rats: effects of N^G-nitro-L-arginine methyl ester. Br J Pharmacol 101:632–639

Gardiner SM, Compton AM, Kemp PA, Bennett T (1991a) effects of n^G-nitro-L-arginine methyl ester or indomethacin on differential regional and cardiac haemodynamic actions of arginine vasopressin and lysine vasopressin in conscious rats. Br J Pharmacol 102:65–72

Gardiner SM, Compton AM, Kemp PA, Bennett T, Foulkes R, Hughes B (1991b) Haemodynamic effects of human a-calcitonin gene-related peptide following administration of endothelin-1 or N^ω-nitro-L-arginine methyl ester in conscious rats. Br J Pharmacol 103:1256–1262

Gardiner SM, Kemp PA, March JE, Bennett T (1994) Effects of bosentan (Ro 47-0203), an et_A-/ET_B-receptor antagonist, on regional haemodynamic responses to endothelins in conscious rats. Br J Pharmacol 112:823–830

Gardiner SM, Kemp PA, March JE, Bennett T, Davenport AP, Edvinsson L (1994) Effects of an ET_A-receptor antagonist, FR 139317, on regional haemodynamic responses to endothelin-1 and [Ala^{11,15}]Ac-endothelin-1 (6-21) in conscious rats. Br J Pharmacol 112:477–486

Garjani A, Wainwright CL, Zeitlin IJ, Wilson C, Slee S-J (1995) Effects of endothelin-1 and the ET_A-receptor antagonist, BQ123, on ischemic arrhythmias in anesthetized rats. J Cardiovasc Pharmacol 25:634–642

Gass A, Flammer J, Linder L, Romerio SC, Gasser P, Haefeli WE (1997) Inverse correlation between endothelin-induced peripheral microvascular vasoconstriction and blood pressure in glaucoma patients. Graefes Arch Clin Exp Ophthalmol 235:634–638

Giaid A, Yanagisawa M, Langleben D, Michel RP, Levy R, Shennib H, Kimura S, Masaki T, Duguid WP, Stewart DJ (1993) expression of endothelin-1 in the lungs of patients with pulmonary hypertension. New Eng J Med 328:1732–1739

Giller T, Breu V, Valdenaire O, Clozel M (1997) Absence of ET(B)-mediated contraction in Piebald-lethal mice. Life Sci 61:255–263

Godfraind T (1994) Endothelin receptors in human coronary arteries. Trends Pharmacol Sci 15:136–137

Graf K, Do YS, Ashizawa N, Meehan WP, Giachelli CM, Marboe CC, Fleck E, Hsueh WA (1997) Myocardial osteopontin expression is associated with left ventricular hypertrophy. Circulation 96:3063–3071

Grantham JA, Schirger JA, Williamson EE, Heublein DM, Wennberg PW, Kirchengast M, Münter K, Subkowski T, Burnett JC (1998) Enhanced endothelin-converting enzyme immunoreactivity in early atherosclerosis. J Cardiovasc Pharmacol 31 [Suppl 1]:S22–S26

Grover GJ, Sleph PG, Fox M, Trippodo NC (1992) Role of endothelin-1 and big endothelin-1 in modulating coronary vascular tone, contractile function and severity of ischemia in rat hearts. J Pharmacol Exp Ther 263:1074–1082

Gu X-H, Casley DJ, Cincotta M, Nayler WG (1990) ^{125}I-Endothelin-1 binding to brain and cardiac membranes from normotensive and spontaneously hypertensive rats. Eur J Pharmacol 177:205–209

Guarda E, Katwa LC, Myers PR, Tyagi SC, Weber KT (1993) effects of endothelins on collagen turnover in cardiac fibroblasts. Cardiovasc Res 27:2130–2134

Haak T, Jungmann E, Kasper-Dahm G, Ehrlich S, Usadel KH (1994) Elevated endothelin-1 levels in critical illness. Clin Invest 72:580–584

Haegerstrand A, Hemsén A, Gillis C, Larsson O, Lundberg JM (1989) Endothelin: presence in human umbilical vessels, high levels in fetal blood and potent constrictor effect. Acta Physiol Scand 137:541–542

Han S-P, Knuepfer MM, Trapani AJ, Fok KF, Westfall TC (1990a) cardiac and vascular actions of sarafotoxin S6b and endothelin-1. Life Sci 46:767–775

Hand MF, Haynes WG, Johnstone HA, Anderton JL, Webb DJ (1995) Erythropoietin enhances vascular responsiveness to norepinephrine in renal failure. Kidney Int 48:806–813

Harada M, Saito Y, Nakagawa O, Miyamoto Y, Ishikawa M, Kuwahara K, Ogawa E, Nakayama M, Kamitani S, Hamanaka I, Kajiyama N, Masuda I, Itoh H, Tanaka I, Nakao K (1997) Role of cardiac nonmyocytes in cyclical mechanical stretch-induced myocyte hypertrophy. Heart and Vessels 12:198–200

Harrison VJ, Randriantsoa A, Schoeffer P (1992) Heterogeneity of endothelin-sarafotoxin receptors mediating contraction of pig coronary artery. Br J Pharmacol 105:511–513

Haug C, Koenig W, Hoeher M, Kochs M, Hombach V, Gruenert A, Osterhues, H. (1998) Direct enzyme immunometric measurement of plasma big endothelin-I concentrations and correlation with indicators of left ventricular function. Clin Chem 44:239–243.

Haynes WG, Webb DJ (1994) Contribution of endogenous generation of endothelin-1 to basal vascular tone. Lancet 344:832–833

Haynes WG, Ferro CE, Webb DJ (1995a) Physiologic role of endothelin in maintenance of vascular tone in humans. J Cardiovasc Pharmacol 26 [Suppl 3]:S183–S185

Haynes WG, Strachan FE, Gray GA, Webb DJ (1995b) Forearm vasoconstriction to endothelin-1 is mediated by ET_A and ET_B receptors in vivo in humans. J Cardiovasc Pharmacol 26 [Suppl 3]:S40–S43

Haynes WG, Ferro, CJ, O'Kane KP, Somerville D, Lomax CC, Webb DJ (1996) Systemic endothelin receptor blockade decreases peripheral vascular resistance and blood pressure in humans. Circulation 93:1860–1870

Hele DJ, Birell M, Bush RC (1995) Selective ET_A and mixed ET_A/ET_B receptor antagonists inhibit balloon catheter-induced smooth muscle cell proliferation in the rat thoracic aorta. Br J Pharmacol 114:P199

Helset E, Sildnes T, Seljelid R, Konopski ZS (1993) endothelin-1 stimulates human monocytes in vitro to release TNFα, IL-1? and IL-6. Mediat Inflamm 2:417–422

Helset E, Sildnes T, Konopski ZS (1994) endothelin-1 stimulates monocytes in vitro to release chemotactic activity identified as interleukin-8 and monocyte chemotactic protein-1. Mediat Inflamm 3:155–160

Hieda HS, Gomez-Sanchez CE (1990) hypoxia increases endothelin release in bovine endothelial cells in culture, but epinephrine, norepinephrine, serotonin, histamine and angiotensin II do not. Life Sci 47:247–251

Hiley CR, Douglas SA, Randall MD (1989) pressor effects of endothelin-1 and some analogs in the perfused superior mesenteric arterial bed of the rat. J Cardiovasc Pharmacol 13 [Suppl 5]:S197–S199

Hiley CR, Patel KCR, Douglas SA (1993) effects of hypoxic perfusion on relaxation to endothelin peptides in the isolated perfused superior mesenteric arterial bed of the rat. Br J Pharmacol 108:155P

Hocher B, Thone-Reineke C, Rohmeiss P, Schmager F, Slowinski T, Burst V, Siegmund F, Quertermous T, Bauer C, Neumayer HH, Schleuning WD, Theuring F (1997) Endothelin-1 transgenic mice develop glomerulosclerosis, interstitial fibrosis and renal cysts but not hypertension. J Clin Invest 99:1380–1389

Hoffman A, Grossman E, Öhman P, Marks E, Keiser HR (1989a) endothelin induces an initial increase in cardiac output associated with selective vasodilation in rats. Life Sci 45:249–255

Hu JR, Berninger UG, Lang RE (1988) endothelin stimulates atrial natriuretic peptide (ANP) release from rat atria. Eur J Pharmacol 158:177–178

Hughes AD, Thom SA, Woodall N, Schachter M, Hair WM, Martin GN, Sever PS (1989) Human vascular responses to endothelin-1: observations in vivo and in vitro. J Cardiovasc Pharmacol 13 [Suppl 5]:S225–S228

Igarashi Y, Aizawa Y, Tamura M, Ebe K, Yamaguchi T, Shibata A (1989) vasoconstrictor effect of endothelin on the canine coronary artery: is a novel endogenous peptide involved in regulating myocardial blood flow and coronary spasm? Am Heart J 118:674–678

Ino T, Ohkubo M, Shimazaki S, Akimoto K, Nishimoto K, Iwahara M, Yabuta K, Hosoda Y (1992) plasma endothelin concentration: relation with vascular resistance and comparison before and after balloon dilatation procedures. Eur J Pediatr 151:416–419

Ishikawa K, Ihara M, Noguchi K, Mase T, Mino N, Saeki T, Fukuroda T, Fukami T, Ozaki S, Nagase T (1994) Biochemical and pharmacological profile of a potent and selective endothelin B-receptor antagonist, BQ-788. Proc Natl Acad Sci USA 91:4892–4896

Ishikawa T, Yanagisawa M, Kimura S, Goto K, Masaki T (1988) Positive inotropic action of novel vasoconstrictor peptide endothelin on guinea-pig atria. Am J Physiol 255:H970–H973

Ito H, Hirata Y, Hiroe M, Tsujino M, Adachi S, Takamoto T, Nitta M, Taniguchi K, Marumo F (1991) Endothelin-1 induces hypertrophy with enhanced expression of muscle specific genes in cultured neonatal rat cardiomyocytes. Circ Res 69:209–215

Ito H, Hiroe M, Hirata Y, Fujisaki H, Adachi S, Akimoto H, Ohta Y, Marumo F (1994) Endothelin ET_A receptor antagonist blocks cardiac hypertrophy provoked by hemodynamic overload. Circulation 89:2198–2203

Itoh S, Sasaki T, Ide K, Ishikawa K, Nishikibe M, Yano M (1993) A novel endothelin ET_A receptor antagonist, BQ-485, and its preventive effect on experimental vasospasm in dogs. Biochem Biophys Res Commun 195:969–975

James AF, Urade Y, Webb RL, Karaki H, Umemura I, Fujitani Y, Oda K, Okada T, Lappe RW, Takai M (1993) IRL 1620, succinyl-[glu^{9}Ala^{11,15}]-endothelin (8-21), a highly specific agonist of the ET$_B$ receptor. Cardiovasc Drug Rev 11:253–270

James AF, Xie LH, Fujitani Y, Hatashi S, Horie M (1994) Inhibition of the cardiac protein kinase A-dependent chloride conductance by endothelin-1. Nature 370:252–253

Kamata K, Miyata N, Kasuya Y (1990) Effects of endothelin on the portal vein from spontaneously hypertensive and Wistar Kyoto rats. Gen Pharmacol 21:127–129

Kamoi K, Sudo N, Ishibashi M, Yamaji T (1990) Plasma endothelin-1 levels in patients with pregnancy-induced hypertension. New Engl J Med 323:1486–1487

Kanno K, Hirata Y, Shichiri M, Marumo F (1991) Plasma endothelin-1 levels in patients with diabetes mellitus with or without vascular complication. J Cardiovasc Pharmacol 17 [Suppl 7]:S475–S476

Kelso EJ, Spiers JP, McDermott BJ, Schofield CN, Silke B (1998) Receptor-mediated effects of endothelin on the L-type Ca++ current in ventricular myocytes. J Pharmacol Exp Ther 286:662–669

Khraibi AA, Heublein DM, Knox FG, Burnett JC Jr (1993) Increased plasma level of endothelin-1 in the okamoto spontaneously hypertensive rat. Mayo Clin Proc 68:42–46

Kim S, Morimoto S, Koh E, Miyashita Y, Ogihara T (1989) Comparison of effects of a potassium channel opener, BRL34915, a specific potassium ionophore, valinomycin and calcium channel blockers on endothelin-induced vascular contraction. Biochem Biophys Res Commun 164:1003–1008

Kiowski W, Sutsch G, Hunziker P, Muller P, Kim J, Oechslin E, Schmitt R, Jones R, Bertel O (1995) Evidence for endothelin-1-mediated vasoconstriction in severe chronic heart failure. Lancet 346:732–736

Kirchengast M, Münter K (1998) Endothelin and restenosis. Cardiovasc Res 39:550–555

Kirilov G, Dakovska L, Borisova AM, Krivoshiev S, Nentchev N (1994) Increased plasma endothelin levels in patients with insulin-dependent diabetes mellitus and end-stage vascular complications. Horm Metab Res 26:119–120

Kodama M, Kanaide H, Abe S, Hirano K, Kai H, Nakamura M (1989) Endothelin-induced Ca-independent contraction of the porcine coronary artery. Biochem Biophys Res Commun 160:1302–1308

Koseki C, Imai Y, Hirata Y, Yanagisawa M, Masaki T (1989a) Autoradiographic distribution in rat tissues of binding sites for endothelin: a neuropeptide? Am J Physiol 256:R858–R866

Krum H, Gu A, Wilshire-Clement M, Sackner-Bernstein J, Goldsmith R, Medina N, Yushak M, Miller M, Packer M (1996) Changes in plasma endothelin-1 levels reflect clinical response to beta-blockade in chronic heart failure. Am Heart J 131:337–341

Krum H, Katz SD (1998) Effect of endothelin-1 on exercise-induced vasodilation in normal subjects and in patients with heart failure. Am J Cardiol 81:355–358

Krum H, Viskoper RJ, Lacourciere Y, Budde M, Charlon V (1998) The effect of an endothelin-receptor antagonist, bosentan, on blood pressure in patients with essential hypertension: bosentan Hypertension Investigators. N Engl J Med 338:784–790

Kurihara H, Yoshizumi M, Sugiyama T, Yamaoki K, Nagai R, Takaku F, Satoh H, Inui J, Yanagisawa M, Masaki T, Yazaki Y (1989) The possible role of endothelin-1 in the pathogenesis of coronary vasospasm. J Cardiovasc Pharmacol 13 [Suppl 5]:S132–S137

Kurihara Y, Kurihara H, Suzuki H, Kodama T, Maemura K, Nagai R, Oda H, Kuwaki T, Cao W-H, Kamada N, Jishage K, Ouchi Y, Azuma S, Toyoda Y, Ishikawa T, Kumada M, Yazaki Y (1994) Elevated blood pressure and craniofacial abnormalities in mice deficient in endothelin-1. Nature 368:703–710

Kurose I, Miura S, Suematsu M, Fukumura D, Nagata H, Sekizuka E, Tsuchiya M (1991) Involvement of platelet-activating factor in endothelin-induced vascular smooth muscle cell contraction. J Cardiovasc Pharmacol 17 [Suppl 7]:S279–S283

Kurose I, Miura S, Suematsu M, Fukumura, Tsuchiya M (1993) Mechanisms of endothelin-induced macromolecular leakage in microvascular beds of rat mesentery. Eur J Pharmacol 250:85–94

Kusuhara M, Yamaguchi K, Ohnishi A, Abe K, Kimura S, Oono H, Hori S, Nakamura Y (1989) Endothelin potentiates growth factor-stimulated dna synthesis in swiss 3T3 cells. Jpn J Cancer Res 80:302–305

Kusumoto K, Fujiwara S, Awane Y, Watanabe T (1993) The role of endogenous endothelin in extension of rabbit myocardial infarction. J Cardiovasc Pharmacol 22 (Suppl. 8):S339–S342

Kuwaki T, Kurihara H, Cao WH, Kurihara Y, Unekawa M, Yazaki Y, Kumada M (1997) Physiological role for brain endothelin in the central and autonomic control: from neuron to knockout mouse. Prog Neurobiol 51:545–579

Kwok CF, Chen ML, Ho LT (1993) Insulin increases the endothelin-1 receptor in cultured rat aortic smooth muscle cells. Diabetelog 36 (Suppl. 1):A203

Ladouceur DM, Flynn MA, Keiser JA, Reynolds E, Haleen SJ (1993) ET_A and ET_B receptors coexist on rabbit pulmonary artery vascular smooth muscle mediating contraction. Biochem Biophys Res Commun 196:209–215

Lam H-C, Takahashi K, Ghatei MA, Warrens AN, Rees AJ, Bloom SR (1991) Immunoreactive endothelin in human plasma, urine, milk and saliva. J Cardiovasc Pharmacol 17 [Suppl 7]:S390–S393

Larivière R, Thibault G, Schiffrin E (1993) Increase endothelin-1 content in blood vessels of deoxycorticosterone acetate-salt hypertension but not in spontaneously hypertensive rats. Hypertension 21:294–300

Larkin SW, Clarke JG, Keogh BE, Araujo L, Rhodes C, Davies GJ, Taylor KM, Maseri A (1989) Intracoronary endothelin induces myocardial ischemia by small vessel constriction in the dog. Am J Cardiol 64:956–958

Lauder H, Sellers LA, Fan TP, Feniuk W, Humphrey PP (1997) Somatostatin sst5 inhibition of receptor-mediated regeneration of rat aortic vascular smooth muscle cells. Brit J Pharmacol 122:663–670

Lawrence E, Brain SD (1992) Altered microvascular reactivity to endothelin-1, endothelin-3 and N^{ω}-nitro-L-arginine methyl ester in streptozotocin-induced diabetes mellitus. Br J Pharmacol 106:1035–1040

Leadley RJ, Lee P, Erickson LA, Shebuski RJ (1993) The snake venom peptide sarafotoxin S6b inhibits repetitive platelet thrombus formation in the stenosed canine coronary artery. J Cardiovasc Pharmacol 22 [Suppl 8]:S199–S203

Lee JY, Warner RB, Adler AL, Opgenorth TJ (1994) Endothelin ET_A receptor antagonist reduces myocardial infarction induced by coronary artery occlusion and reperfusion in the rat. Pharmacology 49:319–324

Lee ME, DeLa Monte SM, Ng SC, Bloch KD, Quertermous T (1990) Expression of the potent vasoconstrictor endothelin in the human central nervous system. J Clin Invest 86:141–147

Lee S-Y, Lee CY, Chen YM, Kochva E (1986) Coronary vasospasm as the primary cause of death due to the venom of the burrowing asp, *Atractaspis engaddensis*. Toxicon 24:285–291

Lee TS, Chao T, Hu K-Q, King GL (1989) Endothelin stimulates a sustained 1,2-diacylglycerol increase and protein kinase C activation in bovine aortic smooth muscle cells. Biochem Biophys Res Commun 162:381–386

Lembeck F, Decrinis M, Pertl C, Amann R, Donnerer J (1989) Effects of endothelin on the cardiovascular system and on smooth muscle preparations in different species. Naunyn-Schmiedeberg's Arch Pharmacol 340:744–751

Lerman A, Edwards BS, Hallett JW, Heublein DM, Sandberg SM, Burnett JC Jr (1991) Circulating and tissue endothelin immunoreactivity in advanced atherosclerosis. N Engl J Med 325:997–1001

Leskinen H, Vuolteenaho O, Ruskoaho H (1997) Combined inhibition of endothelin and angiotensin II receptors blocks volume load-induced cardiac hormone release. Circ Res 80:114–123

Le Monnier De Gouville A-C, Lippton HL, Cavero I, Summer WR, Hyman AL (1989) Endothelin – a new family of endothelium-derived peptides with widespread biological properties. Life Sci 45:1499–1513

Liefeldt L, Bocker W, Schonfelder G, Zintz M, Paul M (1995) Regulation of the endothelin system in transgenic rats expressing the human endothelin-2 gene. J Cardiovasc Pharmacol 26 [Suppl 3:S32–S33

Liu J, Casley DJ, Nayler WG (1989) Ischaemia causes externalization of endothelin-1 binding sites in rat cardiac membranes. Biochem Biophys Res Commun 164:1220–1225

Liu J, Chen R, Casley DJ, Nayler WG (1990a) Ischaemia and reperfusion increase ^{125}I-labelled endothelin-1 binding in rat cardiac membranes. Am J Physiol 258:H829–H835

Liu J, Gu X-H, Casley DJ, Nayler WG (1990b) Reoxygenation, but neither hypoxia nor intermittent ischaemia, increases [^{125}I]endothelin-1 binding to rat cardiac membranes. J Cardiovasc Pharmacol 15:436–443

Lodge NJ, Halaka NN (1993) Endothelin receptor subtypes(s) in rabbit jugular vein smooth muscle. J Cardiovasc Pharmacol 22 [Suppl 8]:S140–S143

Loesch A, Milner P, Anglin SC, Crowe R, Miah S, McEwan JR, Burnstock G (1997) Ultrastructural localisation of nitric oxide sythase, endothelin and binding sites for lectin (from Bandeirea simplicifolia) in the rat carotid artery after balloon catheter injury. J Anatomy 190:93–104

Löffler B-M, Roux S, Kalina B, Clozel M, Clozel J-P (1993) Influence of congestive heart failure on endothelin levels and receptors in rabbits. J Mol Cell Cardiol 25:407–416

Lopez JAG, Armstrong ML, Piegors DJ, Heistad DD (1990) Vascular responses to endothelin-1 in atherosclerotic primates. Artherosclerosis 10:1113–1118

López Farré A, Riesco A, Espinosa G, Digiuni E, Cernadas MR, Alvarez V, Montón M, Rivas F, Gallego MJ, Egido J, Casado S, Caramelo C (1993) Effect of endothelin on neutrophil adhesion to endothelial cells and perfused heart. Circulation 88:1166–1171

Love MP, Haynes WG, Gray GA, Webb DJ, McMurray JJ (1996) Vasodilator effects of endothelin-converting enzyme inhibition and endothelin ET_A receptor blockade in chronic heart failure patients treated with ACE inhibitors. Circulation 94:2131–2137

Lüscher TF, Seo B-G, Bühler FR (1993) Potential role of endothelin in hypertension: controversy on endothelin in hypertension. Hypertension 21:752–757

Maccumber MW, Ross CA, Glaser BM, Snyder SM (1989) Endothelin: visualization of mRNAs by in situ hybridization provides evidence for local action. Proc Natl Acad Sci USA 86:7285–7289

MacLean MR, Hiley CR (1989) Blood pressure changes induced by [Ala^{1,3,11,15}]endothelin in pithed rats: effects of lowering ventilation volume. Br J Pharmacol 97:529P

MacLean MR, Randall MD, Hiley CR (1989) Effects of moderate hypoxia, hypercapnia and acidosis on haemodynamic changes induced by endothelin-1 in the pithed rat. Br J Pharmacol 98:1055–1065

MacLean MR, McGrath JC (1990) Effects of pre-contraction with endothelin-1 on α_2-adrenoceptor- and (endothelium-dependent) neuropeptide Y-mediated contractions in the isolated vascular bed of the rat tail. Br J Pharmacol 101:205–211

MacLean MR, MacKenzie JF, Docherty CC (1998) Heterogeneity of endothelin-B receptors in rabbit pulmonary resistance arteries. J Cardiovasc Pharmacol 31 [Suppl 1]:S115–S118

Magga J, Vuolteenaho O, Marttila M, Ruskoaho H (1997) Endothelin-1 is involved in stretch-induced early activation of B-type natriuretic peptide gene expression in atrial but not in ventricular myocytes: acute effects of the mixed ET_A/ET_B and AT1 receptor antagonists *in vivo* and *in vitro*. Circulation 96:3053–3062

Margulies KB, Hildebrand FL, Lerman A, Perrella MA, Burnett JC (1990) Increased endothelin in experimental heart failure. Circulation 82:2226–2230

Martin-Nizard F, Houssaini HS, Lestavel-Delattre S, Duriez P, Fruchart J-C (1991) Modified low density lipoproteins activate human macrophages to secrete immunoreactive endothelin. FEBS Lett 293:127–130

Masaki T (1993) Endothelins: homeostatic and compensatory actions in the circulatory and endocrine systems. Endocrine Rev 14:256–268

Masaki T, Vane JR, Vanhoutte PM (1994) International union of pharmacology nomenclature of endothelin receptors. Pharmacol Rev 46:137–142

Masaoka H, Suzuki R, Hirata Y, Emori T, Marumo F, Hirakawa K (1989) Raised plasma endothelin in aneurysmal subarachnoid haemorrhage. Lancet 2:1402

Matsuyama K, Saito Y, Nakao K, Jougasaki M, Okumura K, Yasue H, Imura H (1991) Increased plasma level of endothelin-1-like immunoreactivity during coronary spasm in patients with coronary spastic angina. Am J Cardiol 68:991–995

McCarron RM, Wang L, Stanimirovic DB, Spatz M (1993) Endothelin induction of adhesion molecule expression on human brain microvascular endothelial cells. Neurosci Lett 156:31–34

McKenna CJ, Burke SE, Opgenorth TJ, Padley RJ, Camrud LJ, Camrud AR, Johnson J, Carlson PJ, Lerman A, Holmes DR, Schwartz RS (1998) Selective ET_A receptor antagonism reduces neointimal hyperplasia in a porcine coronary stent model. Circulation 97:2551–2556

McMahon EG, Palomo MA, Moore WM (1991) Phosphoramidon blocks the pressor activity of big endothelin (1-39) and lowers blood pressure in spontaneously hypertensive rats. J Cardiovasc Pharmacol 17 [Suppl 7]:S29–S33

McMurdo L, Lidbury PS, Corder R, Thiemermann C, Vane JR (1993) Heterogenous receptors mediate endothelin-1-induced changes in blood pressure, hematocrit, and platelet aggregation. J Cardiovasc Pharmacol 22 [Suppl 8]:S185–S188

McMurdo L, Thiemermann C, Vane JR (1994) The effects of the endothelin ET_A receptor antagonist, FR 139317, on infarct size in a rabbit model of acute myocardial ischemia and reperfusion. Br J Pharmacol 112:75–80

Meyer P, Flammer J, Lüscher TF (1993) Endothelium-dependent regulation of the ophthalmic microcirculation in the perfused porcine eye: role of nitric oxide and endothelins. Invest Ophthalmol Vis Sci 34:3614–3621

Miasiro N, Yamamoto H, Kanaide H, Nakamura M (1988) Does endothelin mobilize calcium from intracellular store sites in rat aortic vascular smooth muscle cells in primary culture? Biochem Biophys Res Commun 156:312–317

Miki S, Takeda K, Kiyama M, Hatta T, Morimoto S, Kawa T, Itoh H, Nakata T, Sasaki S, Nakagawa M (1998) Augmented response of endothelin-A and endothelin-B receptor stimulation in coronary arteries of hypertensive hearts. J Cardiovasc Pharmacol 31 [Suppl 1]:S94–S98

Minamino T, Kurihara H, Takahashi M, Shimada K, Maemura K, Oda H, Ishikawa T, Uchiyama T, Tanzawa K, Yazaki Y (1997) Endothelin coverting enzyme expression in the rat vascular injury model and human coronary atherosclerosis. Circulation 95:221–230

Mir AK, Berthold H, Scholtysik GE, Fozard JR (1989) Cardiovascular effects of endothelin 1 in pithed spontaneously hypertensive rats: evaluation of its mechanism(s) of action. Naunyn-Schmiedeberg's Arch Pharmacol 340:424–430

Miyauchi T, Ishikawa T, Tomobe Y, Yanagisawa M, Kimura S, Sugishita Y, Ito I, Goto K, Masaki T (1989a) Characteristics of pressor response to endothelin in spontaneously hypertensive and Wistar-Kyoto rats. Hypertension 14:427–434

Miyauchi T, Yanagisawa M, Tomizawa T, Sugishita Y, Suzuki M, Fuzino M, Ajisaka R, Goto K, Masaki T (1989b) Increased plasma concentrations of endothelin-1 and big endothelin-1 in acute myocardial infarction. Lancet 2:53–54

Miyauchi T, Yorikane R, Sakai S, Sakurai T, Okada M, Nishikibe M, Yano M, Yamaguchi I, Sugishita Y, Goto K (1993) Contribution of endogenous endothelin-1 to the progression of cardiopulmonary alterations in rats with monocrotaline-induced pulmonary hypertension. Circ Res 73:887–897

Mizuguchi T, Nishiyama M, Moroi K, Tanaka H, Saito T, Masuda Y, Masaki T, deWit D, Yanagisawa M, Kimura S (1997) Analysis of two pharmacologically predicted endothelin B receptor subtypes by using the endothelin B receptor gene knockout mouse. Brit J Pharmacol 120:1427–1430

Molenaar P, O'Reilly G, Sharkey A, Kuc RE, Harding DP, Plumpton C, Gresham GA, Davenport AP (1993) Characterization and localization of endothelin receptor subtypes in the human atrioventricular conducting system and myocardium. Circ Res 72:526–538

Morbidelli L, Orlandi C, Maggi CA, Ledda F, Ziche M (1995) Proliferation and migration of endothelial cells is promoted by endothelins via activation of ET_B receptors. Am J Physiol 269:H686–H695

Morel DR, Lacroix JS, Hemsen A, Steinig DA, Pittet J-F, Lundberg JM (1989) Increased plasma and pulmonary lymph levels of endothelin during endotoxin shock. Eur J Pharmacol 167:427–428

Moreland S, McMullen DM, Delaney CL, Lee VG, Hunt JT (1992) Venous smooth muscle contains vasoconstrictor ET_B-like receptors. Biochem Biophys Res Commun 184:100–106

Moreland S, McMullen D, Abboa-Offei B, Seymour A (1994) Evidence for a differential location of vasoconstrictor endothelin receptors in the vasculature. Br J Pharmacol 112:704–708

Morita H, Kurihara H, Kurihara Y, Shindo T, Kuwaki T, Kumada M, Yazaki Y (1998) Systemic and renal response to salt loading in endothelin-1 knockout mice. J Cardiovasc Pharmacol 31 [Suppl 1]:S557–S560

Mortensen LH, Fink GD (1992) Captopril prevents chronic hypertension produced by infusions of endothelin-1 in rats. Hypertension 19:676–680

Mosqueda-Garcia R, Inagami T, Appalsamy M, Sugiura M, Robertson RM (1992) Endothelin as a neuropeptide. Cardiovascular effects in the brainstem of normotensive rats. Circ Res 72:20–35

Mulder P, Richard V, Derumeaux G, Hogie M, Henry JP, Lallemand F, Compagnon P, Mace B, Comoy E, Letac B, Thuillez C (1997) The role of endogenous endothelin in chronic heart failure: effect of long-term treatment with an endothelin antagonist on survival, hemodynamics, and cardiac remodeling. Circulation 96:1976–1982

Münter K, Hergenröder S, Unger L, Kirchengast M (1996) Oral treatment with an ET_A-receptor antagonist inhibits neointima formation induced by endothelial injury. Pharm Pharmacol Lett 6:90–92

Nakayama K, Ishigai Y, Uchida H, Tanaka Y (1991) Potentiation by endothelin-1 of 5-hydroxytryptamine-induced contraction in coronary artery of the pig. Br J Pharmacol 104:978–986

Nakayama T, Numata Y, Kawada M, Kurokawa H, Nakamura T, Hasimoto K (1993) Hyperpermeability of abdominal capillary vessels to endothelin-1 in patients with diabetes mellitus. J Cardiovasc Pharmacol 22 [Suppl 8]:S360–S363

Naito S, Shimizu S, Maeda S, Wang J, Paul R, Fagin JA (1998) ET-1 is an early response gene activated by ET-1 and PDGF-BB in vascular smooth muscle cells. Am J Physiol 274:C472–C480

Nelson RA, Burke SE, Opgenorth T (1994) Endothelin receptor antagonist FR 139317 reduces infarct size in a rabbit coronary artery occlusion model. FASEB J 8:A854

Nemoto S, Sheng Z, Lin A (1998) Opposing effects of Jun kinase and p38 mitogenactivated protein kinases on cardiomyocyte hypertrophy. Mol Cell Biol 18:3518–3526

Neubauer S, Ertl G, Haas U, Pulzer F, Kochsiek K (1990) Effects of endothelin-1 in isolated perfused rat heart. J Cardiovasc Pharmacol 16:1–8

Newby DE, Flint LL, Fox KA, Boon NA, Webb DJ (1998) Reduced responsiveness to endothelin-1 in peripheral resistance vessels of patients with syndrome X. J Am Coll Cardiol 31:1585–1590

Nichols AJ, Koster PF, Ohlstein EH (1990) The effect of diltiazem on the coronary haemodynamic and cardiac functional effects produced by intracoronary administration of endothelin-1 in the anaesthetized dog. Br J Pharmacol 99:597–601

Niranjan V, Telemaque S, deWit D, Gerard RD, Yanagisawa M (1996) Systemic hypertension induced by hepatic overexpression of human preproendothelin-1 in rats. J Clin Invest 98:2364–2372

Nisell H, Wolff K, Hemsen A, Lindblom B, Lunell NO, Lundberg JM (1991) Endothelin, a vasoconstrictor important to the uteroplacental circulation in pre-eclampsia. J Hypertens 9(Suppl 6):S168–S169

Nishikibe M, Tsuchida S, Okada M, Fukuroda T, Shimamoto K, Yano M (1993) Antihypertensive effect of a newly synthesized endothelin antagonist, BQ-123, in a genetic hypertensive model. Life Sci 52:717–724

Noiri E, Hu Y, Bahou WF, Keese CR, Giaever I, Goligorsky MS (1997) Permissive role of nitric oxide in endothelin-induced migration of endothelial cells. J Biol Chem 272:1747–1752

Nova A, Sibai BM, Barton JR, Mercer BM, Mitchell MD (1991) Maternal plasma level of endothelin is increased in preeclampsia. Am J Obstet Gynecol 165:724–727

Nugent AC, McGurk C, Hayes JR, Johnston GD (1996) Impaired vasoconstriction to endothelin-1 in patients with NIDDM. Diabetes 45:105–107

Ohlstein EH, Horohonich S, Hay DWP (1989) Cellular mechanisms of endothelin in rabbit aorta. J Pharmacol Exp Ther 250:548–550

Ohlstein EH, Vickery L, Sauermelch C, Willette RN (1990) Vasodilation induced by endothelin: role of EDRF and prostanoids in rat hindquarters. Am J Physiol 259:H1835–H1841

Ohlstein EH, Arleth A, Bryan H, Elliott JD, Sung C-P (1992) The selective endothelin ET_A receptor antagonist BQ-123 antagonizes endothelin-1-mediated mitogenesis. Eur J Pharmacol 225:347–350

Ohlstein EH, Douglas SA (1993) Endothelin-1 modulates vascular smooth muscle structure and vasomotion: implications in cardiovascular pathology. Drug Develop Res 29:108–128

Ohlstein EH, Douglas SA, Ezekiel M, Gellai M (1993) Antihypertensive effects of the endothelin antagonist BQ-123 in conscious spontaneously hypertensive rats. J Cardiovasc Pharmacol 22 [Suppl 8]:S321–S324

Ohlstein EH, Beck G, Douglas SA, Nambi P, Gleason JG, Ruffolo RR Jr, Feuerstein GZ, Elliott JD (1994a) Non-peptide endothelin receptor antagonists. ii: pharmacological characterization of SB 209670. J Pharmacol Exp Ther. 271:762–768

Ohlstein EH, Nambi P, Douglas SA, Edwards RM, Gellai M, Lago A, Leber JD, Cousins RD, Gao A, Frazee JS, Peishoff CE, Bean JW, Eggleston DS, Elshourbagy NA, Kumar C, Lee JA, Brooks DP, Ruffolo RR Jr, Feuerstein GZ, Weinstock J, Gleason JG, Elliott JD (1994b) SB 209670, rationally designed potent nonpeptide endothelin receptor antagonist. Proc Natl Acad Sci USA 91:8052–8056

Ohlstein EH, Douglas SA, Brooks DP, Hay DWP, Feuerstein GZ, Ruffolo RR Jr (1995) Functions mediated by peripheral endothelin receptors. In Endothelin receptors: from the gene to the human. Ed Ruffolo RR Jr. pp109–186. CRC Press, Boca Raton, FL.

Ohlstein EH, Arleth AJ, Storer B, Romanic AM (1998) Carvedilol inhibits endothelin-1 biosynthesis in cultured human coronary artery endothelial cells. J Mol Cell Cardiol 30:167–173

Ohno A, Naruse M, Kato S, Hosaka M, Karuse K, Demura H (1992) Endothelin-specific antibodies decrease blood pressure and increase glomerular filtration rate and renal plasma flow in spontaneously hypertensive rats. J Hypertension 10:781–785

Okada K, Nishida Y, Murakami H, Sugimoto I, Kosaka H, Morita H, Yamashita C, Okada M (1998) Role of endogenous endothelin in the development of graft arteriosclerosis in rat cardiac allografts: antiproliferative effects of bosentan, a non selective endothelin receptor antagonist. Circulation 97:2346–2351

Ono K, Tsujimoto G, Sakamoto A, Eto K, Masaki T, Ozaki Y, Satake M (1994) Endothelin-A receptor mediates cardiac inhibition by regulating calcium and potassium currents. Nature 370:252–253

Oliver FJ, De La Rubia G, Feener E (1991) Stimulation of endothelin-1 gene expression by insulin in endothelial cells. J Biol Chem 266:23251–23256

Okamoto Y, Ninomiya H, Masaki T (1998) Posttranslational modifications of endothelin receptor type B. Trends Cardiovasc Med 8:327–329

Okazawa M, Shiraki T, Ninomiya H, Kobayashi S, Masaki T (1998) Endothelin-induced apoptosis of A375 human melanoma cells. J Biol Chem 273:12584–12592

Orlandi A, Ehrlich HP, Ropraz P, Spagnoli LG, Gabbiani G (1994) Rat aortic smooth muscle cells isolated from different layers at different times after endothelial denudation shows distinct biological features in vitro. Arterioscler Thromb 14:982–989

Otsuka A, Mikami H, Katahira K, Tsunetoshi T, Kohara K, Minamitani K, Moriguchi A, Ogihara T (1990a) Haemodynamic effect of endothelin, a novel potent vasoconstrictor in dogs. Clin Exp Pharmacol Physiol 17:351–360

Ouchi Y, Kim S, Souza AC, Iijima S, Hattori A, Orimo H, Yoshizumi M, Kurihara H, Yazaki Y (1989) Central effect of endothelin on blood pressure in conscious rats. Am J Physiol 256:H1747–H1751

Pacher R, Stanek B, Hulsmann M, Koller-Strametz J, Berger R, Schuller M, Hartter E, Ogris E, Frey B, Heinz G, Maurer G (1996) Prognostic impact of big endothelin-1 plasma concentrations compared with invasive hemodynamic evaluation in severe heart failure. J Am Coll Cardiol 27:633–641

Packer M, Caspi A, Charlton V, Cohen-Solal A, Kiowski W, Kostuk W, Krum H, Levine B, Massie B, McMurray J, Rizzon P, Swedberg K (1998) Multicenter, double-blind, placebo-controlled study of long-term endothelin blockade with bosentan in chronic heart failure: results of the REACH-1 trial. Circulation 98:I3

Panek RL, Major TC, Hingorani GP, Doherty AM, Taylor DG, Rapundalo ST (1992) Endothelin and structurally related analogs distinguish between endothelin receptor subtypes. Biochem Biophys Res Commun 184:566–571

Pernow J, Hemsén A, Hallén A, Lundberg JM (1990) Release of endothelin-like immunoreactivity in relation to neuropeptide Y and catecholamines during endotoxin shock and asphyxia in the pig. Acta Physiol Scand 140:311–312

Pierre LN, Davenport AP (1998) Endothelin receptor subtypes and their functional relevance in human small coronary arteries. Brit J Pharmacol 124:499–506

Plumpton C, Haynes WG, Webb DJ, Davenport AP (1995) Measurement of C-terminal fragment of big endothelin-1: a novel method for assessing the generation of endothelin-1 in humans. J Cardiovasc Pharmacol 26 [Suppl 3]:S34–S36

Ponicke K, Heinroth-Hoffmann I, Becker K, Brodde OE (1997) Trophic effect of angiotensin II in neonatal rat cardiomyocytes: role of endothelin and non-myocyte cells. Brit J Pharmacol 121:118–124

Porat O, Neumann D, Zamir O, Nachshon S, Feigin E, Cohen J, Zamir N (1996) Erythropoietin stimulates atrial natriuretic peptide secretion from adult rat cardiac atrium. J Pharmacol Exp Ther 276:1162–1168

Porter KE, Olojugba DH, Masood I, Pemberton M, Bell PR, London NJ (1998) Endothelin B receptors mediate intimal hyperplasia in an organ culture of human saphenous vein. J Vasc Surg 28:695–701

Pousset F, Isnard R, Lechat P, Kalotka H, Carayon A, Maistre G, Escolano S, Thomas D, Komajda M (1997) Prognostic value of plasma endothelin-1 in patients with chronic heart failure. Eur Heart J 18:254–258

Prat L, Carrió I, Roca M, Riambau V, Berné L, Estorch M, Ferrer I, Garcia C (1993) Polyclonal ^{111}In-IgG, ^{125}I-LDL and ^{125}I-endothelin-1 accumulation in experimental arterial wall injury. Eur J Nucl Med 20:1141–1145

Rakugi H, Tabuchi Y, Nakamaru M, Nagano M, Higashimori K, Mikami H, Ogihara T, Suzuki N (1990) Evidence for endothelin-1 release from resistance vessels of rats in response to hypoxia. Biochem Biophys Res Commun 169:973–977

Randall MD, Douglas SA, Hiley CR (1989) Vascular activities of endothelin-1 and some alanyl substituted analogues in resistance beds of the rat. Br J Pharmacol 98:685–699

Randall MD (1991) Vascular activities of the endothelins. Pharmacol Ther 50:73–93

Raschack M, Juchelka F, Rozek-Schaefer G (1998) The endothelin-A antagonist, LU 135252, suppresses ischemic ventricular extrasystoles and fibrillation in pigs and prevents hypoxic cellular decoupling. J Cardiovasc Pharmacol 31 [Suppl 1]: S145–S148

Ravalli S, Szabolcs M, Albala A, Michler RE, Cannon PJ (1996) Increased immunoreactive endothelin-1 in human transplant coronary artery disease. Circulation 94:2096–2102

Rebello S, Roy S, Saxena PR, Gulati A (1995) Systemic hemodynamic and regional circulatory effects of centrally administered endothelin-1 are mediated through ET_A receptors. Brain Res 676:141–150

Reiser G (1990) Endothelin and a Ca^{2+} ionophore raise cyclic GMP levels in a neuronal cell line via formation of nitric oxide. Br J Pharmacol 101:722–726

Resink TJ, Scott-Burden T, Bühler FR (1990b) Activation of multiple signal transduction pathways by endothelin in cultured human vascular smooth muscle cells. Eur J Biochem 189:415–421

Rokosh DG, Stewart AF, Chang KC, Bailey BA, Karliner JS, Camacho SA, Long CS, Simpson PC (1996) Alpha1-adrenergic receptor subtype mRNAs are differentially regulated by alpha1-adrenergic and other hypertrophic stimuli in cardiac myocytes in culture and in vivo. Repression of alpha1B and alpha1D but induction of alpha 1C. J Biol Chem 271:5839–5843

Roux S, Loffler BM, Gray GA, Sprecher U, Clozel M, Clozel JP (1995) The role of endothelin in experimental cerebral vasospasm. Neurosurgery 37:78–85

Rubanyi GM (1992) Endothelium-derived vasoconstrictor factors: an overview. In: Ryan US, Rubanyi GM (eds) Endothelial regulation of vascular tone, Marcel Dekker Inc, New York, pp. 375–386

Salminen K, Tikkanen I, Saijonmaa O, Nieminen M, Fyhrquist F, Frick MH (1989) Modulation of coronary tone in acute myocardial infarction by endothelin. Lancet 2:747

Sakai S, Miyauchi T, Kobayashi M, Yamaguchi I, Goto K, Sugishita Y (1996) Inhibition of myocardial endothelin pathway improves long-term survival in heart failure. Nature 384:353–355

Sato K, Oka M, Hasunuma K, Ohnishi M, Sato K, Kira S (1995) Effects of separate and combined ET_A and ET_B blockade on ET-1-induced constriction in perfused rat lungs. Am J Physiol 269:L668–L672

Schiff E, Ben-Baruch G, Peleg E, Rosenthal T, Alcalay M, Devir M, Mashiach S (1992) Immunoreactive circulating endothelin-1 in normal and hypertensive pregnancies. Am J Obstet Gynecol 166:624–628

Schiffrin EL, Turgeon A, Deng LY (1997) Effect of chronic ET_A-selective endothelin receptor antagonism on blood pressure in experimental and genetic hypertension in rats. Br J Pharmacol 121:935–940

Schiffrin EL (1998) Endothelin and endothelin antagonists in hypertension. J Hypertension 16:1891–1895

Schroeder RL, Jeiser JA, Cheng XM, Haleen SJ (1998) PD 142893, SB 209670 and BQ 788 selectively antagonize vascular endothelial versus vascular smooth muscle ET_B-receptor activity in the rat. J Cardiovasc Pharmacol 32:935–943

Seo B, Oemar BS, Siebenmann R, Von Segesser L, Lüscher TF (1994) Both ET_A and ET_B receptors mediate contraction to endothelin-1 in human blood vessels. Circulation 89:1203–1208

Serradeil-Le Gal C, Herbert JM, Garcia C, Boutin M, Maffrand JP (1991) Importance of the phenotypic state of vascular smooth muscle cells on the binding and the mitogenic activity of endothelin. Peptides 12:575–579

Shah AM (1994) Endothelin secretion by endocardial endothelial cells. Cardiovasc Res 28:722–1994

Shah AM, Lewis MJ, Henderson AH (1989) Inotropic effects of endothelin in ferret ventricular myocardium. Eur J Pharmacol 163:365–367

Shah AM, Lewis MJ (1993) Modulation of myocardial contraction by endocardial and coronary vascular endothelium. Trends Cardiovasc Med 3:98–103

Shichiri M, Kato H, Marumo F, Hirata Y (1998a) Endothelin-1 as an autocrine/paracrine apoptosis survival factor for endothelial cells. Hypertension 30:1198–1203

Shichiri M, Marumo F, Hirata Y (1998b) Endothelin-B receptor-mediated suppression of endothelial apoptosis. J Cardiovasc Pharmacol 31 [Suppl 1]:S138–S141

Shichiri M, Sadivy JM, Marumo F, Hirata Y (1998c) Endothelin-1 is a potent survival factor for c-Myc-dependent. Mol Endocrinol 12:172–180

Shubeita HE, McDonough PM, Harris AN, Knowlton KU, Glembotski CC, Brown JH, Chien KR (1990) Endothelin induction of inositol phospholipid hydrolysis, sarcomere assembly, and cardiac gene expression in ventricular myocytes. A paracrine mechanism for myocardial cell hypertrophy. J Biol Chem 265:20555–20562

Skvorak JP, Nazian SJ, Dietz JR (1995) Endothelin acts as a paracrine regulator of stretch-induced atrial natriuretic peptide release. Am J Physiol 269:R1093–R1098

Simonson MS, Jones JM, Dunn MJ (1992) Cytosolic and nuclear signaling by endothelin peptides: mesangial response to glomerular injury. Kidney Int 41:542–545

Simonson MS, Herman WH (1993) Protein kinase C and protein tyrosine kinase activity contribute to mitogenic signaling by endothelin-1. J Biol Chem 268:9347–9357

Simonson MS, Emancipator SN, Knauss T, Hricik DE (1998) Elevated neointimal endothelin-1 in transplantation-associated arteriosclerosis of renal graft recipients. Kidney Int 54:960–971

Smits P, Hofman H, Rosmalen F, Wollersheim H, Thein T (1991) Endothelin-1 in patients with Reynaud's phenomenon. Lancet: 337:236

Sogabe K, Nirei H, Shoubo M, Nomoto A, Ao S, Notsu Y, Ono T (1992) Pharmacological profile of FR 139317, a novel, potent endothelin ET_A receptor antagonist. J Pharmacol Exp Ther 264:1040–1144

Spokes RA, Ghatei MA, Bloom SR (1989) Studies with endothelin-3 and endothelin-1 on rat blood pressure and isolated tissues: evidence for multiple endothelin receptor subtypes. J Cardiovasc Pharmacol 13 [Suppl 5]:S191–S192

Stein PD, Hunt JT, Floyd DM, Moreland S, Dickinson KEJ, Mitchell C, Liu EC-K, Webb ML, Murugesan N, Dickey J, McMullen D, Zhang R, Lee VG, Serafino R, Delaney C, Schaeffer TR, Kozlowski M (1994) The discovery of sulfonamide endothelin antagonists and the development of the orally active ET_A antagonist 5-(Dimethylamino)-N-(3,4,-dimethyl-5-isoxazolyl)-1-naphthalene-sulfonamide. J Med Chem 37:329–331

Stewart DJ, Levy RD, Cernacek P, Langleben D (1991) Increased plasma endothelin-1 in pulmonary hypertension: marker or mediator of disease? Ann Int Med 114:464–469

Strachan FE, Spratt JC, Wilkinson IB, Johnston NR, Gray GA, Webb DJ (1999) Systemic blockade of the endothelin-B receptor increases peripheral vascular resistance in healthy men. Hypertension 33:581–585

Sudjarwo SA, Hori M, Takai M, Urade Y, Okada T, Karaki H (1993) A novel subtype of endothelin B receptor-mediating contraction in swine pulmonary vein. Life Sci 53:431–437

Sudjarwo SA, Hori M, Tanaka T, Matsuda Y, Okada T, Karaki H (1994) Subtypes of endothelin ET_A and ET_B receptors mediating venous smooth muscle contraction. Biochem Biophys Res Commun 200:627–633

Sugiura M, Inagami T, Hare GMT, Johns JA (1989a) Endothelin action: inhibition by a protein kinase C inhibitor and involvement of phosphoinositols. Biochem Biophys Res Commun 158:170–176

Sugiura M, Inagami T, Kon V (1989b) Endotoxin stimulates endothelin-release in vivo and in vitro as determined by radioimmunoassay. Biochem Biophys Res Commun 161:1220–1227

Sumner MJ, Cannon TR, Mundin JW, White DG, Watts IS (1992) Endothelin ET_A and ET_B receptors mediate vascular smooth muscle contraction. Br J Pharmacol 107:858–860

Sunnergren KP, Word RA, Sambrook JF, Macdonald PC, Casey ML (1990) Expression and regulation of endothelin precursor mRNA in a vascular human amnion. Mol Cell Endocrinol 68:R7–R14

Suzuki N, Miyauchi T, Tomobe Y, Matsumoto H, Goto K, Masaki T, Fujino M (1990b) Plasma concentrations of endothelin-1 in spontaneously hypertensive rats and DOCA-salt hypertensive rats. Biochem Biophys Res Commun 167:941–947

Suzuki T, Tsuruda A, Katoh S, Kubodera A, Mitsui Y (1997) Purification of endothelin from a conditioned medium of cardiac fibroblastic cells using beating rate assay of myocytes cultured in a serum-free medium. J Mol Cell Cardiol 29:2087–2093

Tahara A, Kohno M, Yanagi S, Itagane H, Toda I, Akioka K, Teragaki M, Yasuda M, Takeuchi K, Takeda T (1992) Circulating immunoreactive endothelin in patients undergoing percutaneous transluminal coronary angioplasty. Metabolism 40:1235–1237

Takayanagi R, Kitazumi K, Takasaki C, Ohnaka K, Aimoto S, Tasaka K, Ohashi M, Nawata H (1991) Presence of non-selective type of endothelin receptor on vascular endothelium and its linkage to vasodilation. FEBS Lett 282:103–106

Takiguchi Y, Sogabe K (1996) The selective endothelin ET_A receptor antagonist FR139317 inhibits neointimal thickening in the rat. Eur J Pharmacol 309:59–62

Tao W, Liou GI, Wu X, Abney TO, Reinach PS (1995) ET_B and epidermal growth factor receptor stimulation of wound closure in bovine corneal epithelial cells. Invest Ophthal Vis Sci 36:2614–2622

Teerlink JR, Breu V, Sprecher U, Clozel M, Clozel J-P (1994a) Potent vasoconstriction mediated by ET_B receptors in canine coronary arteries. Circ Res 74:105–114

Teerlink JR, Loffler BM, Hess P, Maire JP, Clozel M, Clozel JP (1994b) Role of endothelin in the maintenance of blood pressure in conscious rats with chronic heart failure. Acute effects of the endothelin receptor antagonist Ro 47–0203 (bosentan). Circulation 90:2510–2518

Telemaque S, Emoto N, deWit D, Yanagisawa M (1998) In vivo role of endothelin-converting enzyme-1 as examined by adenovirus-mediated overexpression in rats. J Cardiovasc Pharmacol 31 [Suppl 1]:S548–S550

Terashita Z-I, Shibouta Y, Imura Y, Iwasaki K, Nishikawa K (1989) Endothelin-induced sudden death and the possible involvement of platelet activating factor (PAF). Life Sci 45 1911–1918

Thiemermann C, Lidbury PS, Thomas GR, Vane JR (1988) Endothelin inhibits ex vivo platelet aggregation in the rabbit. Eur J Pharmacol 158:181–182

Tomobe Y, Miyauchi T, Saito A, Yanagisawa M, Kimura S, Goto K, Masaki T (1988) Effects of endothelin on the renal artery from spontaneously hypertensive and Wistar Kyoto rats. Eur J Pharmacol 152:373–374

Tomobe Y, Ishikawa T, Yanagisawa M, Kimura S, Masaki T, Goto K (1991) Mechanisms of altered sensitivity to endothelin-1 between aortic smooth muscles of spontaneously hypertensive and Wistar-Kyoto rats. J Pharmacol Exp Ther 257:555–561

Tønnessen T, Naess PA, Kirkebøen KA, Ilebekk A, Christensen G (1994) The authors respond. Cardiovasc Res 28:722

Topouzis S, Huggins JP, Pelton JT, Miller RC (1991) Modulation by endothelium of the responses induced by endothelin-1 and by some of its analogues in rat isolated aorta. Br J Pharmacol 102:545–549

Toth M, Solti F, Merkely B, Kekesi V, Szokodi I, Horkay F, Juhasz-Nagy A (1998) Bradycardia increases the arrhythmogenic effect of endothelin. J Cardiovasc Pharmacol 31 [Suppl 1]:S431–S433

Totsune K, Sone M, Takahashi K, Ohneda M, Itoi K, Murakami O, Saito T, Mouri T, Yoshinaga K (1991) Immunoreactive endothelin in urine of patients with and without diabetes mellitus. J Cardiovasc Pharmacol 17 [Suppl 7]:S423–S424

Trachtenberg JD, Sun S, Choi ET, Callow AD, Ryan US (1993) Effect of endothelin-1 infusion on the development of intimal hyperplasia after balloon catheter injury. J Cardiovasc Pharmacol 22 [Suppl 8]:S355–359

Tsujino M, Hirata Y, Eguchi S (1995) Nonselective ET_A/ET_B receptor antagonist blocks proliferation of rat vascular smooth muscle cells after balloon angioplasty. Life Sci 56:PL449–PL454

Tsunoda K, Abe K, Yoshinaga K, Furuhashi N, Kimura H, Tsujei M, Yajima A (1992) Maternal and umbilical venous levels of endothelin in women with pre-eclampsia. J Hum Hypertension 6:61–64

Van Den Busse M, Itoh S (1993) Central effects of endothelin on baroreflex of spontaneously hypertensive rats. J Hypertension 11:379–387

Van Renterghem C, Vigne P, Barhanin J, Schmid-Alliana A, Frelin C, Lazdunski M (1988) Molecular mechanism of action of the vasoconstrictor peptide endothelin. Biochem Biophys Res Commun 157:977–985

Vanhoutte PM, Auch-Schwelk W, Boulanger C, Janssen PA, Katusic ZS, Komori K, Miller VM, Schini VB, Vidal M (1989) Does endothelin-1 mediate endothelium-dependent contractions during anoxia? J Cardiovasc Pharmacol 13 [Suppl 5]: S124–S128

Verhaar MC, Strachan FE, Newby DE, Cruden NL, Koomans HA, Rabelink TJ, Webb DJ (1998) Endothelin-A receptor antagonist-mediated vasodilation is attenuated by inhibition of nitric oxide synthesis and by endothelin-B blockade. Circulation 97:752–756

Vigne P, Lazdunski M, Frelin C (1989) Inotropic effect of endothelin-1 on rat atria involves hydrolysis of phosphatidylinositol. FEBS Lett 249:143–146

Viswanathan M, De Oliveira AM, Johren O, Saavedra JM (1997) Increased endothelin ET(A) receptor expression in rat carotid arteries after balloon injury. Peptides 18:247–255

Wada K, Tabuchi H, Ohba R (1990) Purification of an endothelin receptor from human placenta. Biochem Biophys Res Commun 167:251–257

Walder CE, Thomas GR, Thiemermann C, Vane JR (1989) The hemodynamic effects of endothelin-1 in the pithed rat. J Cardiovasc Pharmacol 13 [Suppl 5]:S93–S97

Wang QD, Hemsen A, Li XS, Lundberg JM, Uriuda Y, Pernow J (1995) Local overflow and enhanced tissue content of endothelin following myocardial ischaemia and reperfusion in the pig: modulation by L-arginine. Cardiovasc Res 29:44–49

Wang X, Douglas SA, Feuerstein GZ, Ohlstein EH (1995) Temporal expression of ECE-1, ET-1, ET-3, ET_A and ET_B receptor mRNAs after balloon angioplasty in the rat. J Cardiovasc Pharmacol 26 [Suppl 3]:S22–S25

Wang Y, Simonson MS, Pousségur J, Dunn MJ (1992) Endothelin rapidly stimulates mitogen-activated protein kinase in rat mesangial cells. Biochem J 287:589–594

Wang Y, Pousségur J, Dunn MJ (1993) Endothelin stimulates mitogen-activated protein kinase p42 activity through the phosphorylation of the kinase in rat mesangial cells. J Cardiovasc Pharmacol 22 [Suppl 8]:S164–S167

Warner TD, De Nucci G, Vane JR (1989a) Rat endothelin is a vasoactive in the isolated perfused mesentery of the rat. Eur J Pharmacol 159:325–326

Warner TD, Mitchell JA, De Nucci G, Vane JR (1989b) Endothelin-1 and endothelin-3 release EDRF from isolated perfused arterial vessels of the rat and rabbit. J Cardiovasc Pharmacol 13 [Suppl 5]:S85–S88

Warner TD, Allcock GH, Mickley EJ, Corder R, Vane JR (1993a) Comparative studies with endothelin receptor antagonists BQ-123 and PD 142893 indicate at least three endothelin receptors. J Cardiovasc Pharmacol 22 [Suppl 8]:S117–S120

Warner TD, Battistini B, Allcock GH, Vane JR (1993b) Endothelin ET_A and ET_B receptors mediate vasoconstriction and prostanoid release in the isolated kidney of the rat. Eur J Pharmacol 250:447–453

Watanabe T, Suzuki N, Shimamoto N, Fujino M, Imada A (1990) Endothelin in myocardial infarction. Nature 344:114

Watschinger B, Sayegh MH, Hancock WW, Russell ME (1995) Upregulation of endothelin-1 mRNA and peptide expression in rat cardiac allografts with rejection and arteriosclerosis. Am J Pathol 146:1065–1072

Webb DJ (1995) Endogenous endothelin generation maintains vascular tone in humans. J Hum Hypertens 9:459–463

Weber H, Webb ML, Serafino R, Taylor DS, Moreland S, Norman J, Molloy CJ (1994) Endothelin-1 and angiotensin II stimulate delayed mitogenesis in cultured rat aortic smooth muscle cells: evidence for common signaling mechanisms. Mol Endocrinol 8:148–158

Weiser E, Wollberg Z, Kochva E, Lee S (1984) Cardiotoxic effects of the venom of the burrowing asp, Atractaspis engaddensis (Atractasididae, Ophidia). Toxicon 22:767–774

Weissberg PL, Witchell C, Davenport AP, Hesketh TR, Metcalfe JC (1990) The endothelin peptides ET-1, ET-2, ET-3 and sarafotoxin S6b are co-mitogenic with platelet-derived growth factor for vascular smooth muscle cells. Atherosclerosis 85:257–262

Wellings RP, Corder R, Warner TD, Cristol J-P, Thiemermann C, Vane JR (1994) Evidence from receptor antagonists of an important role for ET_B receptor-mediated vasoconstrictor effects of endothelin-1 in the rat kidney. Br J Pharmacol 111: 515–520

White DG, Cannon TR, Garratt H, Mundin JW, Sumner MJ, Watts IS (1992) Endothelin ET_A and ET_B receptors mediate vascular smooth muscle contraction. J Cardiovasc Pharmacol 22 [Suppl 8]:S144–S148

Whittle BJR, Lopez-Belmonte J, Rees DD (1989) Modulation of the vasodepressor actions of acetylcholine, bradykinin, substance P and endothelin in the rat by a specific inhibitor of nitric oxide formation. Br J Pharmacol 98:646–652

Willette RN, Zhang H, Mitchell MP, Sauermelch CF, Ohlstein EH, Sulpizio AC (1994) Nonpeptide endothelin antagonist: cerebrovascular characterization and effects on delayed cerebral vasospasm. Stroke 25:2450–2456

Willette RN, Sauermelch CF, Storer BL, Guiney S, Luengo JI, Xiang J-N, Elliott JD, Ohlstein EH (1998) Plasma- and cerebrospinal fluid-immunoreactive endothelin-1: effects of nonpeptide endothelin receptor antagonists with diverse affinity profiles for endothelin-A and endothelin-B receptors. J Cardiovasc Pharmacol 31 [Suppl 1]:S149–S157

Wollberg Z, Shabo-Shina R, Intrator N, Bdolah A, Kochva E, Shavit G, Oron Y, Vidne BA, Gitter S (1988) A novel cardiotoxic polypeptide from the venom of atractaspis engaddensis (burrowing asp): cardiac effects in mice and isolated rat and human heart preparations. Toxicon 26:525–534

Wong J, Vanderford PA, Winters JW, Chang R, Soifer SJ, Fineman JR (1993) Endothelin-1 does not mediate acute hypoxic pulmonary vasoconstriction in the intact newborn lamb. J Cardiovasc Pharmacol 22 [Suppl 8]:S262–S266

Wren A, Hiley CR, Fan T-P (1993) Endothelin-3-mediated proliferation in wounded human umbilical vein endothelial cells. Biochem Biophys Res Commun 196:369–375

Wright CE, Fozard JR (1988) Regional vasodilation is a prominent feature of the haemodynamic response to endothelin in anaesthetized, spontaneously hypertensive rats. Eur J Pharmacol 155:201–203

Wright CE, Fozard JR (1990) Differences in regional vascular sensitivity to endothelin-1 between spontaneously hypertensive and normotensive Wistar-Kyoto rats. Br J Pharmacol 100:107–113

Wu-Wong JR, Chiou WJ, Dickinson R, Opgenorth TJ (1997) Endothelin attenuates apoptosis in human smooth muscle cells. Biochem J 328:733–737

Yamauchi T, Ohnaka K, Takakyanagi R (1990) Enhanced secretion of endothelin-1 by elevated glucose levels from cultured bovine aortic endothelial cells. FEBS Lett 267:16–18

Yanagisawa M, Kurihara H, Kimura S, Tomobe Y, Kobayashi M, Mitsui Y, Yazaki Y, Goto K, Masaki T (1988) A novel potent vasoconstrictor peptide produced by vascular endothelial cells. Nature 332:411–415

Yasuda M, Kohno M, Tahara A, Itagane H, Toda I, Akioka K, Teragaki M, Oku H, Takeuchi K, Takeda T (1990) Circulating immunoreactive endothelin in ischemic heart disease. Am Heart J 119:801–806

Yorikane R, Koike H (1990) The arrhythmogenic action of endothelin in rats. Jpn J Pharmacol 53:259–263

Yorikane R, Sakai S, Miyauchi T, Sakurai T, Sugishita Y, Goto K (1993) Increased production of endothelin-1 in the hypertrophied rat heart due to pressure overload. FEBS Lett 332:31–34

Yue T-L, Wang X, Sung C-P, Olson B, McKenna PJ, Gu J-L, Feuerstein GZ (1994) Interleukin-8: a mitogen and chemoattractant for vascular smooth muscle cells. Circ Res 75:1–7

Zamora MA, Dempsey EC, Walchak SJ, Stelzner TJ (1993) BQ-123, an ET_A receptor antagonist, inhibits endothelin-1-mediated proliferation of human pulmonary artery smooth muscle cells. Am J Respir Cell Mol 9:429–433

Zamora MR, O'Brien RF, Rutherford RB, Weil JV (1990) Werum endothelin-1 concentrations and cold provocation in primary Reynaud's phenomenon. Lancet 336:1144–1147

Zeiher AM, Goebel H, Schachinger V, Ihling C (1995) Tissue endothelin-1 immunoreactivity in the active coronary atherosclerotic plaque. A clue to the mechanism of increased vasoreactivity of the culprit lesion in unstable angina. Circulation 91:941–947

Zimmerman RS, Martinez AJ, Maymind M, Barbee RW (1992) Effect of endothelin on plasma volume and albumin escape. Circ Res 70:1027–1034

Zimmerman RS, Maymind M, Barbee RW (1994) Endothelin blockade lowers total peripheral resistance in hemorrhagic shock recovery. Hypertension 23:205–210

CHAPTER 17

Endothelins Within the Liver and the Gastrointestinal System

B. BATTISTINI

A. Introduction: The ET System in the Liver and the GI Tract

This chapter contains description of the pharmacological responses induced by ET-1 and related isopeptides in the hepatic system and gastrointestinal (GI) smooth muscle preparations, and the relationships of these responses to the expression (mRNA), synthesis, production, and degradation of the endothelins and the distribution of ET receptors. Attention is also given to the possible physiological and pathophysiological roles of ETs in the gastrointestinal tract and related organs.

I. ETs and ET Receptors

Together with ET-1, -2 and -3 (YANAGISAWA et al. 1988; INOUE et al. 1989) and the four associated sarafotoxins (KLOOG et al. 1988), there is also VIC, vasoactive intestinal contractor, another isopeptide of the ET family, which is only present in the mouse small intestine (SAIDA et al. 1989; ISHIDA et al. 1989). VIC is identical to human ET-2 except for amino acid position 4, where Ser is replace by Asn (SAIDA et al. 1989).

ET-1 mRNA is present in numerous rat tissues, including the rat GI system (from stomach to colon; TAKAHASHI et al. 1990). Northern blot analysis revealed the presence of ET-1 mRNA in the rat colon and the mucosa of both the ileum and the colon, but not in the jejunum. The signal found from the whole colon was twice as intense as that from the colonic mucosa (TAKAHASHI et al. 1990; GHATEI et al. 1994). ET-1 mRNA was also found in HCA-7-Col-26 cells, a human colorectal adenocarcinoma cell line established from a primary tumor from the GI epithelium (GHATEI et al. 1994). Both ET-1 and ET-3 genes are expressed in the human pancreas (BLOCH et al. 1989, 1991).

High densities of ET-binding sites (see Chap. 4) have been reported in numerous tissues, more especially in smooth muscle cells, pulmonary and renal organs. Quantitative autoradiographic studies of ET-1 binding sites have also shown high densities of high-affinity binding sites in the human liver and intestine (Hoyer et al. 1989). In the GI tract, using membrane preparations from rat, specific binding was found mostly in the fundus of stomach, the jejunum, the ileum, the longitudinal and circular muscle layer of the small intestine, and the colon (Ghatei et al. 1994). Cross-linking of ^{125}I-ET-1 to ET receptors located in the guinea-pig ileum revealed a receptor with a molecular weight of 70 KDa (Galron et al. 1991). Noticeably, the distribution of ET-1 binding sites did not correspond to the content of ir-ET in various tissues along the GI tract. For instance, the duodenum has a high ir-ET content but no specific binding, whereas the longitudinal muscle layer has the highest specific binding but the lowest content of ir-ET. When assessing possible roles for ET-1 in the GI tract, based on both the content of ET-1 and the presence of adjacent binding sites, it is likely that ET-1 has a function regulating the motility of the stomach, small (jejunum, ileum) and large (colon) intestine (see below).

II. In Situ and/or Immunocytochemical Distribution

Nerve fibers and cell bodies of the myenteric plexus of the opossum esophagus showed VIC immunoreactivity (Fang et al. 1994). ET-like immunoreactivity was also found in nerve bundles throughout the human colon (Inagaki et al. 1991a, b). The distribution of ET-1 and big ET-1 was also studied in adult and developing human gut. ET-1 was detected in extracts of adult GI tract and localized by immunocytochemistry to the submucous plexus only. Conversely, ET-1 immunoreactivity was not detected by immunocytochemistry in the fetal human gut until the 32nd week of gestation, whereas big ET-1 was found as early as 11 weeks in developing neural structures and epithelial cells (Escrig et al. 1992).

Autoradiographic localization of [^{125}I]ET-1 was also reported in various GI tissues of the guinea pig (Yoshinaga et al. 1992) and rat (Takahashi et al. 1990), including the liver (Neuser et al. 1989), as well as the human liver (Hoyer et al. 1989).

Determination of ET_A receptor distribution by *in situ* hybridization was reported for the rat intestine (Lin et al. 1991) and the rat liver (Hori et al. 1992) while ET_B receptors were found in the porcine liver (Elshourbagy et al. 1992), the bovine intestine (Mizuno et al. 1992), and the rat colon (Hori et al. 1992). Nevertheless, the highest densities of ET-1 binding sites have been observed in the heart, the lung, and the kidney.

III. Production, Release and Tissue Levels of ETs

ET-1 is produced by numerous vascular and non-vascular cells including endothelial cells, epithelial cells, leukocytes, macrophages, cancer cells, and neurons (see Battistini et al. 1993). Its production or release is also regulated by various vasoactive substances, growth factors, cytokines, procoagulants, etc. In the GI system, ETs have been reported to be secreted from human liver cells (Hep G2) (SUZUKI et al. 1989; TOKITO et al. 1991), human colon cells (SUZUKI et al. 1989), rat intestinal cells (SAIDA et al. 1989; MASTUMOTO et al. 1989) and HCA-7-Col-26 cells in culture (GHATEI et al. 1994).

Interestingly, ET are also secreted from a number of tissues in body fluids where, in some cases, their concentrations are greater than those found in plasma (see BATTISTINI et al. 1993). ETs are normal constituents of human milk (LAM et al. 1990, 1991; KEN-DROR et al. 1997) and saliva (LAM et al. 1991). Levels of ir-ET in the milk of post-colostrum lactating mothers are 11 times higher than those found in plasma. It was suggested that such ET levels would imply an active concentrating mechanism within the mammary gland and/or an extravascular origin for the ET peptides. It was also suggested that ETs may be involved in the development of gastrointestinal motility of the suckling neonate (LAM et al. 1990). Indeed, ET-1 causes contraction in rat gut (see below) and ET receptors are present in the gastrointestinal tract (TAKAHASHI et al. 1990). Ir-ET present in saliva could also have a physiological role in modulating oral and gastric mucosal integrity (LAM et al. 1991).

The content of ir-ET in the rat GI tract measured by RIA revealed that ETs were present in all GI tissues, mostly in the duodenum (48 fmol/g wet weight tissue), the jejunum (35 fmol/g), the ileum (37 fmol/g), and the colon (33 fmol/g) (TAKAHASHI et al. 1990; GHATEI et al. 1994). The highest concentration of ir-ET was also found in the circular muscle layer. Mucosal layers of the ileum or the colon and the stomach showed half or less of the amount present in other portion of the GI tract. When ir-ET was characterized by FPLC, the ir-ET detected in the rat jejunum, ileum or colon was revealed to be mostly ET-1 (TAKAHASHI et al. 1990; GHATEI et al. 1994). High concentrations of ir-ET-3 were also reported in the rat intestine and the brain (greater than 100 pg/g wet tissue) while ir-ET-1 showed widespread distribution but with very large amount in the rat colon (1000 pg/g wet tissue) (MATSUMOTO et al. 1989). The abundance of ET-3 in the intestine, but also in the brain, suggests a role for ETs as new brain-gut peptides as suggested by MATSUMOTO et al. (1989), similar to cholecystokinin, for which receptors are present in both tissues. A significant amount of ir-ET was also measured in the rat liver, just like in the heart, the lung, and the kidney (YOSHIMI et al. 1989).

IV. Metabolism

Injection of radioiodinated ET-1 or ET-3 into the left ventricle revealed that both peptides are rapidly removed form the rat circulation (ANGGARD et al. 1989; SHIBA et al. 1989). One of the organs with the highest uptake of radioactivity was the liver, together with the lung and the kidney.

B. The Pharmacology of ETs in the Hepatopancreatic System

I. ET-Induced Responses in the Liver and Pancreas

ET-1 caused sustained concentration-dependent vasoconstriction of the portal vasculature in the perfused rat liver with an EC_{50} of 1 nmol/l (Gandhi et al. 1990, 1992a, b, 1993). The ET-1-induced responses in the dog liver were not altered by prior infusion of phenylephrine, isoproterenol, AII, glucagon, PAF, or BN52021, a PAF antagonist (Gandhi et al. 1990). In open-chest rabbits, ET-1 infusion caused vasodilatation of the hepatic artery, through biphasic (spleen) to pure vasoconstrictor (pancreas, stomach, colon) effects, estimated with tracer microspheres (Hof et al. 1989). In the blood perfused liver of the dog (via the hepatic portal circuit), ET-1 caused a biphasic response characterized by an initial transient increase in flow (vasodilation) of short duration followed by a prolonged vasoconstriction (Withrington et al. 1989). ET-1 injected intra-portally also increased the portal perfusion pressure and inflow resistance.

I.v. administration of ET-1, -2, or -3 induced dose-dependent decreases in pancreatic tissue blood flow by 45.4%, 19.6%, and 51.9%, respectively, as measured by a laser Doppler flow meter in anesthetized dogs, whereas systemic arterial blood pressure was not significantly affected, suggesting a possible role of these agents in regulating the pancreatic microcirculation (Takaori et al. 1992).

I.v. bolus injection of ET-1 increased dose-dependently the vascular permeability of the stomach and duodenum, as measured by the extravasation of Evans blue dye, but not of the liver and pancreas. Pretreatment with WEB 2086 or BN 52021, both PAF receptor antagonists, reduced the protein extravasation in the stomach and duodenum (Filep et al. 1991).

II. ET-Induced Responses in Hepatic and Pancreatic Cells

The vasoactive effect of ET-1 in the rat perfused liver was accompanied by dose-dependent increased glycogenolysis (increase in hepatic glucose output) and alterations in hepatic oxygen consumption (Gandhi et al. 1990; Roden et al. 1992). ET-1 also stimulated the metabolism of inositol phospholipids in isolated hepatocytes and Kupffer cells in primary culture (Gandhi et al. 1990).

In rat pancreatic acinar cells, ET-1 increased $[Ca^{2+}]_i$ through release from the same intracellular stores as cholecystokinin, but with less potency. ET-1 also increased the accumulation of IP_3, but failed (up to 1 μmol/l) to increase amylase secretion (Yule et al. 1992). A single class of specific binding sites for $[^{125}I]$-ET-1 was demonstrated on rat pancreatic acini, with a relative potency order for displacing $[^{125}I]ET$ that was greater for ET-1 than that for ET-2, which was greater than that for ET-3. Conversely, HILDEBRAND et al. (1993) demonstrated two classes of receptors, an ET_A receptor with a high affinity for ET-1 but a low affinity for ET-3, and an ET_B receptor with equally high affinities for ET-1 and -3. Pancreatic secretagogues that activate phospholipase C (PLC) inhibited binding of ^{125}I-ET-1or ^{125}I-ET-3, whereas agents that act through cAMP did not. ETs neither stimulate nor alter changes in enzyme secretion, intracellular calcium, cAMP, or IP_3. In mice, ET-1 dose-dependently stimulated insulin secretion by direct action on isolated mouse islets of Langerhans (GREGERSEN et al. 1996). The effect of ET-1 was glucose-dependent, and critically dependent on influx via Ca^{2+}-channels, without inhibition of the potassium permeability. GREGERSEN et al. (2000) further showed that the insulinotropic action of ET-1 was via the activation of the ET_A receptor (blocked by BQ-123) and the protein kinase C pathway.

C. Responses of Isolated Smooth Muscle Preparations Related to the GI System

I. The Esophagus

ET-1 elicits contractile responses of the guinea-pig *oesophageal muscularis mucosae*. Amongst many non-vascular smooth muscle preparations, this one was the most sensitive to ET-1, exhibiting a sensitivity similar to that observed in the endothelium-denuded aorta of the rat (EGLEN et al. 1989; see Chap. 4). In the *esophageal muscularis mucosae* isolated from guinea-pigs, ET-1, -2, -3, and SX6c produced contractions in a concentration-dependent manner, that were modulated by FR139317, BQ-123, and RES-701-1 as endothelin receptor antagonists (UCHIDA et al. 1998a, b; see Chap. 9). The order of potency ($-\log EC_{50}$) was ET-1 (8.61) = SX6c (8.65) > ET-2 (8.40) > ET-3 (8.18). Both ET_A receptor antagonists caused parallel rightward shifts of the concentration-response curve to ET-1. Since SX6c-densensitized preparations still contracted to ET-1, in a manner that was blocked by FR139317, both ET_A and ET_B receptors are present. In addition, the contractile responses to ET-1 and SX6c were abolished in a Ca^{2+}-free EGTA-containing medium, weakly inhibited by nicardipine, and markedly inhibited by SK&F 96365. In addition, both H-7 and U-73122 strongly inhibited the ET-induced contractions, whereas U-73343 only weakly inhibited these responses, suggesting that ET-1/SC6c activate receptors that are coupled mainly to receptor-operated Ca^{2+} influx and linked with the phospholipase C-protein kinase C pathway (UCHIDA et al. 1998a, b). VIC, which is localized in the opossum esophagus (see above), caused an atropine-resistant increase in the amplitude of nerve-induced contractions of the circular muscle. VIC also contracted longitudinal muscle, an effect that was nearly eliminated by atropine (FANG et al. 1994).

II. The Stomach and Gastric Cells

1. ET-Induced Vascular Response of the Stomach

In anesthetized pithed rats, ET-1 caused an increase in gastric vascular resistance that was potentiated by indomethacin (WALDER et al. 1989; see Chap. 11). In anesthetized rats, ET-1 was 5–10 times more potent as a vasoconstrictor in the stomach than ET-3 (WALLACE et al. 1989b). In open-chest rabbits, ET-1 infusion caused vasoconstriction and decreased blood flow to the stomach, as measured by tracer microspheres (HOF et al. 1989). The vasoconstriction apparently involved L-channel activation (HOF et al. 1990). Conversely, in anesthetized squirrel monkeys, ET-1 infusion did not decrease blood flow to the stomach (CLOZEL and CLOZEL 1989). In chloralose-anesthetized dogs, ET analogues (ET-2 > [Ala3,11]ET-1 (Ala-ET) >> ET-3) infused intra-arterially to the stomach, produced dose-related sustained increases in vascular resistance or no effect (COOH-terminal hexapeptide (ET-C) (WOOD et al. 1992). When comparing the effects of ET-1 and big ET-1 on vascular resistance of a blood-perfused *ex vivo* stomach segment of chloralose-anesthetized dogs, big ET-1 caused a small phosphoramidon-sensitive, but statistically significant, vasoconstriction only at the highest concentration (WOOD et al. 1994). Thus, ETs act in the stomach in a similar way to most vascular beds, promoting vasoconstriction and reducing blood flow.

2. ET-Induced Contraction of the Fundus

ET-1 was reported to contract rat superfused stomach strips (D'ORLÉANS-JUSTE et al. 1988), with ET-1 and ET-3 being apparently equipotent (SPOKES et al. 1989). Furthermore, SX6c and IRL 1620 also induced equipotent concentration-dependent contractions (ALLCOCK et al. 1995). In this preparation the E_{max}s of ET-1, ET-3, SX6c, and IRL 1620 have been reported not to be different (ALLCOCK et al. 1995) or even higher for SX6c (+32%) compared to ET-1 (GRAY and CLOZEL 1994). The reasons for this discrepancy are unclear at present but might be attributed to differences in tissue preparations and/or differences in the experimental protocols (initial load and exposure to KCl). Contractions were unaffected by the selective ET_A receptor antagonist, BQ-123, but were attenuated by the non-selective ET_A/ET_B receptor antagonist, PD 145065, and even more by the selective ET_B receptor antagonist, BQ-788, thus providing further evidence that ET/SX-induced contractions of rat stomach strips are mediated by ET_B receptors (ALLCOCK et al. 1995). These data partly differ from those of GRAY and CLOZEL (1994) who found that BQ-123 alone or in the presence of the non-selective ET_A/ET_B receptor antagonist Ro 47–0203 (Bosentan) enhanced contractions of rat stomach strips to ET-1 or SX6c. GRAY and CLOZEL (1994) attributed their findings to the presence of inhibitory ET_A or ET_B receptors in the rat stomach that would attenuate the response to ET-1 or SX6c. We believed that such a receptor may be found on the surface of gastric epithelial cells releasing dilators. Part of this observation is supported by the presence of ET_A receptors in the stomach (TAKAHASHI et al. 1990). In the rat stomach preparation, selective receptors to ET-1 are also present, activation of which reduced phasic activity (FULGINITI et al. 1993) or gastric motility (OLSEN and WEIS 1992) and caused transient relaxation of the smooth muscle in precontracted stomach strips (OLSEN and WEIS 1992). In the rat isolated stomach strip precontracted with PGE_2, ET-1 produced a transient relaxation (40% of the induced tone). This relaxation was blocked by BQ-123 thus confirming an effect mediated by ET_A receptors. In the presence of BQ-788, the transient relaxation was changed to a sustained effect (ALLCOCK et al. 1995).

Human and porcine big ET-1, inactive precursors, contracted rat stomach strips, suggesting the presence of an ECE activity (ALLCOCK et al. 1995). These responses were not significantly different from those induced by ET-1, although it took about ten times longer to develop. The ECE appears to be selective for big ET-1 over big ET-3, since big ET-3 was inactive in this preparation. The rat stomach ECE activity was sensitive to phosphoramidon (ALLCOCK et al. 1995).

In summary, the fundal part of the rat stomach contains a population of ET_B receptors that mediate contraction, a population of ET_A receptors that mediate relaxation and a phosphoramidon-sensitive ECE which is selective for big ET–1 over big ET-3.

3. ET-Induced Effects in Gastric Cells

In isolated gastric smooth muscle cells of the guinea pig, [^{125}I]ET-1, -2, and -3 bound in a time- and temperature-dependent manner, which was specific and saturable (KITSUKAWA et al. 1994). Inhibition by both ET-1 and ET-3 revealed the presence of two ET receptors, ET_A and ET_B, since the dose-inhibition curve for ET-3 against [^{125}I]ET-1 was biphasic. ET-3 was 1400-fold less potent than ET-1 for binding to the ET_A receptors. Thus, both receptors are present in the guinea-pig stomach (KITSUKAWA et al. 1994). ET-1 also stimulated contraction of isolated gastric smooth muscle cells while ET-3 had no effect on contraction or relaxation. Thus, there is a species-related difference between rat and guinea-pig stomach.

III. The Gallbladder

ET-1, -2, and -3 genes are expressed, and ET-1 released (HOUSSET et al. 1993a), by human gallbladder-derived biliary epithelial cells in primary culture (FOUASSIER et al. 1998). These cells also displayed ET receptor mRNAs and high-affinity binding sites for ET-1, mostly of the ET_B type.

ET receptors mediating contractions induced by ET-1 (MOUMMI et al. 1992a), -2, and -3, and the ET_B-selective receptor agonists SX6c, IRL 1620, BQ-3020, [Ala^{1,3,11,15}] ET-1 and ET-1(16–21) were identified in the isolated gallbladder of the guinea-pig (BATTISTINI et al. 1993; CARDOZO et al. 1997), with the use of ET_A (BQ-123) and ET_A/ET_B (PD 145065) receptor antagonists. Results showed that ET_B receptors were involved in the contractions induced by ETs in this preparation. However, since SX6c and other selective ET_B agonists produced only half or less than half of the contractile response induced by non-selective agonists, and as responses to ET-1, but not to ET-3, were insensitive to the antagonist action of BQ-123, whereas BQ-123 or PD 145065 strongly antagonized contractions induced by ET-3, the presence of an additional ET receptor, not conforming to the established ET_A/ET_B receptor subtype classification, was suggested (BATTISTINI et al. 1993). ET-1 was 5 times less potent than cholecystokinin, but 20 and 40 times more potent than, respectively, carbachol and histamine (MOUMMI et al. 1992a).

In addition, phosphoramidon-sensitive conversion and biological activation of big ET-1 (into ET-1) and big ET-2 but not big ET-3, were also observed in the isolated gallbladder of the guinea-pig. Thiorphan, a selective inhibitor of NEP, had no effect on contractions induced by exogenous big ET-1 in this preparation (BATTISTINI et al. 1994).

Thus ET-1 is locally expressed and produced in the biliary tract and by a paracrine route, and could play a role in choledochal motility and gallbladder contraction.

IV. The Duodenum

At the level of smooth muscle cells, rat isolated duodenal strips contracted in response to ET-1 (LE MONNIER DE GOUVILLE et al. 1989). ET-1, -3 and IRL 1620, a selective ET_B agonist, (1-100nmol/l) also elicited sustained isotonic contraction of neonatal (1-week-old) duodenum, in a concentration-dependent manner, with a potency order of ET-1 = ET-3 > IRL 1620, unaffected by FR139317, a selective ET_A receptor antagonist. The response to ET-1 and -3 (10–1000nmol/l) of adult duodenum was biphasic, i.e., transient relaxation followed by contraction, with a potency order of ET-1 > ET-3, and significantly antagonized with FR139317, thus suggesting that the duodenal response to ETs changes from a sustained contraction in neonates to a biphasic response in adults, via ET_B and ET_A receptors, respectively (IRIE et al. 1995).

At the level of the rat duodenal mucosa, administration of ET-1 (0.6nmol/kg and 1nmol/kg) to a perfused proximal duodenal loop caused an ET_A-mediated (blocked by BQ-123) increase of duodenal HCO_3^- secretion, whereas ET-1 caused a significant decrease in histamine-stimulated acid secretion, also antagonized by BQ-123, suggest an effect on the duodenal mucosal integrity by modifying both gastric acid and duodenal HCO_3^- secretions (TAKEUCHI et al. 1999). Bolus injection of ET-1 (1nmol kg-1, i.v.) enhanced the vascular permeability of the rat duodenum as measured by the extravasation of Evans blue dye, that was completely inhibited by BQ-123 (FILEP et al. 1993).

At the cellular level, subepithelial fibroblasts of rat duodenal villi were cultured and their physiological characteristics were studied using fura-2 fluorescence. ET-1 and ET-3 (>0.1–1 nmol/l) induced a transient response that consisted of an initial Ca^{2+} release from the intracellular store and a sustained Ca^{2+} influx, that was not blocked by BQ-123 (FURUYA et al. 1994)

V. The Jejunum

In the rat, ET-1 mRNA was not demonstrated in the whole jejunum, whereas ET-like immunoreactivity in jejunum extracts revealed the presence of ET-1 and ET-3. Furthermore, analysis of specific ET-1 binding in the rat gastrointestinal tract showed it to be particularly high in the jejunum (TAKAHASHI et al. 1990).

ET-1 caused a long-lasting contraction of rabbit isolated jejunum, but did not interfere with the spontaneous phasic activity of the rabbit jejunum or with contractions induced by histamine or carbachol (LEMBECK et al. 1989).

ET-3, but not ET-1, decreased fluid and NaCl absorption across the jejunum in anesthetized dogs, via an effect that was not mediated by nitric oxide or soluble guanylate cyclase (CHOWDHURY et al. 1993a,b). In addition, ET-1 potently stimulated chloride secretion (reduced by tetrodotoxin) and inhibited Na^+-glucose absorption in human jejunal mucosa in vitro, by reducing the activity of the Na^+-glucose, SGLT1 (KUHN et al. 1997).

VI. The Ileum

VIC induced a prolonged contraction in mouse ileum (ISHIDA et al. 1989). The maximum ileum contraction induced by VIC was much higher than induced by ET-1 in both guinea pig and mouse systems. ET-1 also elicited contractions of the guinea-pig isolated ileum but not at concentrations that caused vasoconstriction of rat portal vein (BORGES et al. 1989). Together with the esophagus, the ileum represents an ET-sensitive non-vascular smooth muscle preparation when compared to the guinea-pig urinary bladder, trachea, or rat vas deferens (EGLEN et al. 1989). The rate or the amplitude of the spontaneous rhythmic contractions of the ileal smooth muscle were essentially not affected by any of the peptides (WOLLBERG et al. 1991). ET-2 or SX6b caused equipotent concentration-dependent contractions of the guinea-pig ileum (MAGGI et al. 1989). ET-3 was 10–12 times less potent than ET-1 at contracting the guinea pig ileum (SPOKES et al. 1989). ET (16–21) was inactive (MAGGI et al. 1989). Furthermore, since VIC = ET-1 = SX6a,b > ET-3 >> SX6C, and such contractile effects were antagonized by BQ-123 or PD 142893, the contractile response observed in the guinea-pig ileum was associated to ET_A receptors (SPOKES et al. 1989; TAKAHASHI et al. 1990; WOLLBERG et al. 1991; WARNER et al. 1993; GHATEI et al. 1994). In other parts of the guinea-pig small intestine, ET-1 and ET-3 were equipotent, suggesting that muscle contraction is mediated by ET_B receptors (YOSHINAGA et al. 1992).

Later on it was noticed that ET-1 caused a biphasic effect on spontaneous smooth muscle tone, an initial relaxation followed by a late contraction in the guinea-pig ileum (LIN and LEE 1990; MIASIRO et al. 1990). Both the relaxing and the contractile phases were not affected by pretreatment with tetrodotoxin, phentolamine, tolazoline, propranolol, guanethidine, 8-phenyl-theophylline, naloxone, methylsergide, [D-Pro^4,D-Try^{7,9}]substance P-(4–11), diphenylhydralamine, indomethacin, or atropine, but inhibited by verapamil (Ca^{2+} influx through voltage-dependent Ca^{2+} channels) and H7 (protein kinase C activation). Thus, ET-1-induced relaxation was due neither to the indirectly evoked release of inhibitory neurotransmitters nor to the direct activation of adrenoceptors, purine or opiate receptors. Cross-tachyphylaxis studies between ET-1 and ET-3 suggested the existence of at least two ET receptor subtypes in the guinea pig ileum (MIASIRO and PAIVA 1992; HORI et al. 1994). Additional results suggested that two ET_B receptors with distinct signal trans-duction mechanism mediate the biphasic response: (1) an ET_{B1} and (2) an ET_{B2} receptor (MIASIRO et al. 1993, 1995, 1999).

ET-1 also affected cholinergic neurotransmission in the guinea-pig ileum (WIKLUND et al. 1989a, b). Nerve-induced contractions (i.e., transmural nerve stimulation) were inhibited by ET-1 while basal muscle tone was increased as described above. ET-1 inhibited the nerve-induced release of Ach, whereas it potentiated contractions induced by exogenous ACh. Thus, it is suggested that ET-1 is a modulator of cholinergic neuroeffector transmission in the guinea-pig ileum acting via both inhibitory pre- and stimulatory post-junctional mechanisms (WIKLUND et al. 1989). These effects of ET-1, -2, and -3 (caused graded inhibitions of nerve-mediated responses followed by sus-tained contractions) on the guinea pig field-stimulated ileum, were shown to be mediated by two receptors coupled to pertussis toxin-insensitive mecha-nisms (GUIMARAES and RAE 1992). Further studies comparing the relative activities of the ET/SX6 peptides and the effectiveness of the ET receptor antagonists were consistent with postjunctional ET_A receptors mediating the effect (WARNER et al. 1993).

VII. The Colon

ET-1 causes a long-lasting contraction of the rat isolated colon (LEMBECK et al. 1989). Furthermore, ET-1 is twice as potent as ET-3 and produces a much higher maximal response (ET-3: 45% of ET-1) (TAKAHASHI et al. 1990; GHATEI et al. 1994). Interestingly, this preparation was five times more sensitive to ET-1 than the guinea-pig ileum and ten times more sensitive than the rat stomach (TAKAHASHI et al. 1990; GHATEI et al. 1994). The high concentrations of ET binding sites in the rat colon might explain these potent effects. ET-1 and its binding sites are present in the human enteric nervous system, where they are displayed in nerve bundles and most of the ganglion cells in both the myenteric and submucous plexuses, providing evidence that ET-1 is a neuropeptide in the human with a possible role in the modulation of motility and secretion in the human intestine (INAGAKI et al. 1991a, b).

In open-chest rabbits, ET-1 infusion caused vasoconstriction of the colonic vascular bed as measured with tracer microspheres (HOF et al. 1989). Similarly, in anesthetized pithed rats, ET-1 increased vascular resistance in the large intestine, as measured by radiolabeled microspheres. The response was attenuated by indomethacin, which potentiated the vasoconstriction seen in the gastric vasculature (see above; WALDER et al. 1989).

D. Possible Roles of ETs in Gastrointestinal Diseases and Clinical Relevance

Plasma levels of ir-ET in healthy humans are low but measurable and are not sex-related, age-related, or increased with physical exercise. However, in patients suffering from diabetes mellitus, respiratory diseases, uremia, sepsis, and other shock states, post-surgery, or with various renal or cardiovascular diseases, ET-like immunoreactivity within the circulation is elevated (see BATTISTINI et al. 1993).

I. Animal Models of GI Disorders: Modulation of the ET System and Treatment with ET Antibodies, ECE Inhibitors, or ET Receptor Antagonists

1. Portal Hypertension

The mRNA expression of ET-1, ET_A, and ET_B receptors was significantly increased by 2.2-, 2.5-, and 1.5-fold, respectively, in esophageal specimens from rats with portal hypertension compared to controls operated rats (OHTA et al. 2000). The ET-1 peptide content was also significantly increased by 2.2-fold vs controls. It was suggested that since ET-1 and its receptors could promote vascular proliferation and induce mucosal damage, the over expressed ET-1 may play an important role in the development and rupture of esophageal varices in portal hypertension (OHTA et al. 2000). Similarly, over expression of ET-1 mRNA and protein was observed in the gastric mucosa of portal hypertensive rats (OHTA et al. 1997). The extent of ethanol-induced gastric mucosal necrosis in PHT was reduced after administering FR 139317, a selective ET_A receptor antagonist (OHTA et al. 1997). The role of ET-1 in congestive gastropathy in portal hypertensive rats was further studied (MIGOH et al. 2000). The mixed $ET_{A/B}$ receptor antagonist Bosentan attenuated the vascular hyperpermeability of the gastric mucosal in portal hypertensive rats.

2. Gastric/Duodenal Mucosal Injury and Ulceration

Local i.a. infusion of ET-1 caused hemorrhagic and necrotic damage to the mucosa suggesting that ETs might be potent ulcerogenic agents in rats

(WHITTLE and ESPLUGUES 1988). ET-1-induced injury was not attenuated by atropine, cimetidine, adrenoceptor antagonists or the 5-lipoxygenase inhibitor BW A4C (WHITTLE and ESPLUGUES 1988). Capsaicin-pretreatment, two weeks earlier to deplete sensory neuropeptides from primary afferent neurones, augmented the mucosal damage induced by ET-1, as assessed by both macroscopic and histological examination. The damage induced by threshold doses of ET-1 alone or in capsaicin-pretreated rats was further enhanced by administration of indomethacin, indicating a modulatory influence of endogenous prostanoids (WHITTLE and LOPEZ-BELMONTE 1991). L-Arginine had a gastroprotective effect, while Bosentan antagonized the effects of ET-1 on rat gastric blood flow and mucosal integrity (LARAZATOS et al. 1994, 1995).

Modulation of the gastric mucosal integrity by ET-1 was also examined using an *ex vivo* chamber preparation of the rat stomach (MACNAUGHTON et al. 1989; WALLACE et al. 1989a, b). Under these conditions, ET-1 augmented gastric hemorrhagic damage induced by topical application of ethanol (20%) or hydrochloric acid. ET-3 was 5–10 times less potent than ET-1 to increase the susceptibility to ethanol-induced damage but produced similar gastric mucosal hemorrhage to ET-1 in the absence of any exogenous irritant (WALLACE et al. 1989a, b). Pretreatment with indomethacin potentiated the effect of ET-1, probably by inhibiting the formation of prostacyclin, a cytoprotective agent (MACNAUGHTON et al. 1989; WALLACE et al. 1989a). PAF or LTD_4 antagonists did not affect the ulcerogenic actions of ET-1 but sodium nitroprusside, a NO donor, reduced the damage induced by i.v. ET-1 and topically applied ethanol (WALLACE et al. 1989a).

Gastric mucosal injury induced via various protocols in rats also suggested a role for ETs. *In vivo*, i.v. administration of ET-1 greatly increased acid-induced gastric damage caused by oral administration of HCl in rats (MACNAUGHTON et al. 1989). In indomethacin-, hemorrhagic-, or ethanol-increased (vulnerability of the stomach to HCl) or -induced gastric mucosal damage in rats, treatment with an ET antiserum almost completely abolished lesions (KITAJIMA et al. 1992, 1993, 1994, 1995; MORALES et al. 1992; MASUDA et al. 1991, 1993; MICHIDA et al. 1994). ET antiserum also decreased ethanol-induced gastric (MASUDA et al. 1992) or hepatic (OSHITA et al. 1993) vasoconstriction in rats. Burn-induced gastric mucosal injury (Curling's ulcer) in rats was reduced by TAK-044, a mixed ET_A/ET_B receptor antagonist (BATTAL et al. 1997). In addition, the severity of cysteamine-induced solitary duodenal ulcers (involving only the duodenum and not the stomach) in rats was dose-dependently decreased by pretreatment with ET-1 antibodies or the mixed ET_A/ET_B receptor antagonist Bosentan (SZABO et al. 1998). In rabbits, i.p. infusion of ethanol to raise blood ethanol increased the vulnerability of the stomach to HCl gastric mucosal injury. Ethanol infusion caused a dose-dependent increase in gastric vascular resistance in perfused rabbit stomach, which was accompanied by an increased production of ET-1, suggesting that ET may be involved in the pathogenesis (MASUDA et al. 1991). LPS-induced septic shock in rats is also associated to gastric mucosal injury attributed to microcirculatory disturbances (hypoperfusion and arteriolar vasoconstriction). In this case, ET antiserum restored blood flow and decreased vasoconstriction (WILSON et al. 1993).

In summary, gastroduodenal mucosal injury is a complex process, attributed to (1) the heterogenous structure and (2) multiple functions of the gut. The action of exogenous ET-1 and related inhibitors in animal models did suggest that mucosal injury is mediated in part – or amplified by – ET-1, causing direct vascular damage (ischemic injury), together with the release of other proinflammatory and vasoactive mediators such as leukotrienes, thromboxanes, and or PAF.

3. Chronic or Acute Pancreatitis

Since microcirculatory disturbance may play an important role in the development of severe pancreatitis, ET-1 was considered as a good candidate for the development of severe pancreatitis in rats. To that effect, ET-1 induced pancreatitis-like microvascular deterioration and acinar cell injury which was similar to the microcirculatory failure found in sodium taurocholate-induced experimental pancreatitis using *in vivo* microscopy, red blood cell (RBC) velocities, functional capillary density (FCD), and capillary diameters (PLUSCZYK et al. 1999). Thus, ET released by injured endothelial cells during acute biliary pancreatitis may promote microcirculatory failure and ischemia in acute pancreatitis, eventually leading to acinar cell necrosis. In acute pancreatitis induced by i.p. injections of cerulein, bolus i.a. injection of ET-1 caused a dose-dependent increase in pancreatic microcirculatory disturbance (local pancreatic blood flow decreased significantly) that aggravated acute pancreatitis (LIU et al. 1995). In addition, ET receptor blockade by the ET receptor antagonist LU-135252, 12h after the induction of necrotizing pancreatitis induced by intraductal bile acid infusion and cerulein hyperstimulation, was proven effective. LU-135252 significantly improved fluid sequestration, improved urinary output, decreased ascites, increased pancreatic capillary blood flow in the pancreas and colon, reduced leukocyte rolling, stabilized capillary permeability, decreased fluid loss into the third space, and improved renal function and survival (FOITZIK et al. 1998, 1999, 2000). Overall, treatment reduced the mortality rate, from 50% in untreated animals to 8%. A more detailed study showed that ET-1 increased pancreatic and colonic capillary permeability in both healthy animals and animals with mild acute pancreatitis but had no additional adverse effect in severe acute pancreatitis. ET receptor blockade decreased pancreatic capillary permeability in the pancreas of sham operated rats and stabilized increased capillary permeability in mild and severe acute pancreatitis (EIBL et al. 2000). Conversely, the mixed ET receptor antagonist Bosentan, administered intravenously in 12-h intervals for 3 days starting 1h after induction of bile acid pancreatitis, did not improve survival of rats with severe experimental pancreatitis induced by intraductal infusion into the pancreatic duct of sodium taurocholate (Fiedler et al. 1999). The involvement of ET-1 in acute pancreatitis was further examined by another group, using ET-1 and its receptor antagonist BQ-123 in cerulein-induced pancreatitis in rats. They showed that ET-1, when infused i.v. with cerulein, decreased the extent of pancreatic edema whereas BQ-123 further augmented pancreatic edema caused by cerulein (Kogire et al. 1995). Thus, endogenous ET-1 had a protective role.

In a rat model of chronic pancreatitis (injection of oleic acid into the common bile/pancreatic duct), characterized by the presence of an inflammatory infiltrate with progressive destruction of acinar cells and fibrosis, ET-1 staining in the pancreas was diffused in the cytoplasm of vascular endothelial, acinar, and ductal cells, compared to control pancreas where focal staining was found only in acinar cells, at 3 weeks. The elevation in the local production of ET-1 may be associated with morphological and hemodynamic changes during chronic pancreatitis (Kakugawa et al. 1996).

There are a number of models, in addition to a multitude of approaches (way of administration, dose, time) that complicate the interpretation of the results. A more methodical testing of ET receptor antagonists and/or ECE inhibitors would be optimal. Nevertheless, I believe that extrapolating from experimental data to the clinical situation, blockade of the local vasoconstriction by ET inhibitors may prove beneficial in pancreatitis.

4. Inflammatory Bowel Disease

In colitis induced by intra-rectal administration of trinitrobenzene sulphonic acid (TNBS) in rats, treatment [24h and 2h prior to (pre-dose) or 1h after the induction (post-induction) of colitis and every 24h thereafter for 5 days] with the orally active mixed ET_A/ET_B receptor antagonist Bosentan, markedly reduced colonic damage and myeloperoxidase (MPO) activity (Hogaboam et al. 1996). In another established model of colitis induced by intra-rectal administration of two haptens, TNBS or dinitrobenzene sulphonic acid (DNBS) in rats, oral treatment with Ro 48–5695, daily for 5 days, almost completely prevented TNBS-induced damage, as measured by MPO and the incidence of diarrhea and adhesions (Padol et al. 2000). Such results suggest that ETs could play an active role in gut inflammation.

5. Intestinal Mucosal Acidosis

Intestinal mucosal acidosis is associated to septic shock. In a porcine model, Bosentan, a mixed ET_A/ET_B receptor antagonist, was reported to restore gut oxygen delivery and significantly improved the notably deteriorated intestinal mucosal pH, as measured by a tonometer in the ileum, and mucosal-arterial PCO_2 gap (Oldner et al. 1998). This effect may be attributed to inhibiting the vasocontractile effect of ET-1 in the splanchnic circulation, and thus gut ischemia in septic shock.

II. Human Subjects and Patients

1. Peptic Ulcer

Peptic ulcer in the human stomach is associated with localized destruction of the gastric wall. Plasma and gastric mucosal ET-1 concentrations in patients with peptic ulcer are significantly elevated compared to normal subjects, correlating with the severity and area of ulcer. This further supports ET-1's participation as a modulator of regional blood flow in the stomach, as described above (MASUDA et al. 1997).

2. Chronic Pancreatitis

Semi-quantitative analyses of immunostaining showed that ET-1-like immunoreactivity in islet cells of patients with chronic pancreatitis was greater than in normal subjects. ET-1 mRNA was expressed in sites similar to those of the immunostaining, as well as in vascular endothelial cells. Northern blot analysis showed an increase in the expression of ET-1 mRNA in the patient population. There was a significant correlation between the intensity of ET-1 immunostaining and the severity of fibrosis in the patients with chronic pancreatitis, suggesting that an elevation in local expression of ET-1 may be associated with the morphological and hemodynamic changes of chronic pancreatitis (KAKUGAWA et al. 1996).

3. Inflammatory Bowel Diseases: Crohn's Disease and Ulcerative Colitis

Immunological hypersensitivity, vascular abnormalities, and neovascularization have been implicated in the pathogenesis of inflammatory bowel disease (IBD). Interactions between macrophages and endothelial cells are also important. Monocytes are also believed to play an important role in IBD (SALH et al. 1998). All these cells express, produce, and release ET-1. Interestingly, plasmatic (LETIZIA et al. 1998) and local production (MURCH et al. 1992) of ET-1 is increased in patients with Crohn's disease and ulcerative colitis. The percentage of ET-immunoreactive cells in the lamina propria and submucosa was also increased in the two disease groups (MURCH et al. 1992). Similarly, ET concentrations in supernatants of homogenized colonic samples were increased in one study (MURCH et al. 1992) while they did not differ from controls in another, much smaller study (RACHMILEWITZ et al. 1992). In this latter study, however, the protocol used for the extraction of ir-ET from the tissue was quite different. *In vitro* autoradiography has also identified ET-1 binding sites in normal human GI tissue and an increase in ET receptor density in patients with Crohn's disease and ulcerative colitis (HUSDON et al. 1994). The density was increased in both groups in submucosal vessels, while it was increased in villi and muscle only in Crohn's disease. It was suggested by MURCH et al. (1992) that local ET production by inflammatory cells may contribute to vasculitis in chronic inflammatory bowel disease by inducing intestinal ischemia through vasoconstriction. Furthermore, ET-1 can be produced by human endothelial cells, polymorphonuclear leukocytes (SESSA et al. 1991), macrophages (EHRENREICH et al. 1990), and monocytes (SALH et al. 1998), which migrate and aggregate in inflamed bowel mucosa and around submucosal blood vessels in IBD. The cytokine IL-1, which is associated with chronic IBD, also increases the production of ET-1 by cultured endothelial cells (YOSHIZUMI et al. 1990).

E. Discussion and Conclusions

Much research has been oriented towards elucidating the roles of ETs in the vascular system and related pathophysiologies. It is certain that ETs are potent vasoconstrictor peptides with growth-promoting properties influencing vascular remodeling. However, accumulating evidence, such as that reported above, strongly supports roles for ETs in hepatopancreatic function and throughout the GI system. Gene and protein expression of ETs and their receptors are readily demonstrable in these systems, and they are modulated in both animal models and human diseases. Like in the vasculature (endothelium) and the pulmonary system (epithelium and other cells releasing pro-inflammatory mediators), there is both balance and interactions between ETs and other contracting and relaxing factors that interrelate in the pathogenesis of GI disease such as gastric mucosal damage. Thus, ETs are important regulators of GI physiology and may emerge as key mediators in GI pathologies such as ulcers, colitis, etc. Thus, ET related antagonists and/or inhibitors are candidates for further clinical testing.

Acknowledgements. BB is a recipient of a Junior I Research Scholarship from the Fonds de Recherche en Santé du Québec (FRSQ). Research in his lab is supported by grants from the FRSQ, les Fonds pour les chercheurs et l'aide à la recherche (FCAR), the Medical Research Council of Canada (MRCC)/PMAC, the Heart and Stroke Foundation of Quebec (HSFQ), the British Columbia Lung Foundation (BCLF), la Fondation de l'Institut de cardiologie de Québec (FICQ), and la Fondation Bégin.

References

Allcock Gh, Battistini B, Fournier A, Warner TD, Vane JR (1995) Characterization of endothelin receptors mediating mechanical responses to the endothelins in the isolated stomach strip of the rat. J Pharmacol Exp Ther 275:120–126

Anggard E, Galton S, Rae G, Thomas R, McLoughlin L, De Nucci G, Vane JR (1989) The fate of radioiodinated endothelin-1 and endothelin-3 in the rat. J Cardiovasc Pharmacol 13:S46–S49

Battal MN, Hata Y, Matsuka K, Ito O, Matsuda H, Yoshida Y, Kawazoe T (1997) Reduction of progressive burn injury by using a new nonselective endothelin-A and endothelin-B receptor antagonist, TAK-044: an experimental study in rats. Plast Reconstr Surg 99:1610–1619

Battistini B, D'Orleans-Juste P, Sirois P (1993) Endothelins: circulating plasma levels and presence in other biologic fluids. Lab Invest 68(6):600–628

Battistini B, O'Donnell LDJ, Warner TD, Fournier A, Farthing MJG, Vane JR (1995) Characterisation of an endothelin-converting enzyme activity in the isolated gall-bladder of the guinea-pig. Br J Pharmacol 112:1244–1250

Battistini B, Warner TD, O'Donnell LCD, Fournier A, Vane JR (1994) Endothelin-1 induces the release of prostaglandin E2 from the isolated gallbladder of the guinea-pig

Bloch KD, Eddy RL, Shows TB, Quertermous T (1989) cDNA cloning and chromosomal assignment of the gene encoding endothelin 3. J Biol Chem 264:18156–18161

Bloch KD, Hong CC, Eddy RL, Shows TB, Quertermous T (1991) Chromosomal assignment of the endothelin 2 gene: vasoactive intestinal contractor peptide is rat endothelin 2. Genomics 10:236–242

Borges R, Von-Grafenstein H, Knight DE (1989) Tissue selectivity of endothelin. Eur J Pharmacol 165:223–230

Cardozo AM, D'Orleans-Juste P, Yano M, Frank PA Jr, Rae GA (1997) Influence of endothelin ET(A) and ET(B) receptor antagonists on endothelin-induced contractions of the guinea pig isolated gall bladder. Regul Pept 69:15–23

Chowdhury MRK, Uemura N, Suzuki S, Nishida Y, Morita H, Hosomi H (1993a) Effects of endothelin on fluid and NaCl absorption across the jejunum in anesthetized dogs. J Cardiovasc Pharmacol 22:S189–S191

Chowdhury MR, Uemura N, Nishida Y, Morita H, Hosomi H (1993b) Effects of endothelins on fluid and NaCl absorption across the jejunum anesthetized dogs. Jpn J Physiol 43(5):709–26

Clozel M, Clozel JP (1989) Effects of endothelin on regional blood flows in squirrel monkeys. J Pharmacol Exp Ther 250:1125–1131

D'Orléans-Juste P, Antunes E, de Nucci G, Vane JR (1988) A profile of endothelin on isolated vascular and others smooth muscle preparations. Br J Pharmacol 95:809P

Eglen RM, Michel AD, Sharif NA, Swank SR, Whiting RL (1989) The pharmacological properties of the peptide, endothelin. Br J Pharmacol 97:1297–1307

Ehrenreich H, Anderson RW, Fox CH, Rieckman P, Hoffman GS, Travis WD, Coligan GE, Kehrl JS, Fauci AS (1990) Endothelins, peptides with potent vasoactive properties, are produced by human macrophages. J Exp Med 172:1741–1748

Eibl G, Hotz HG, Faulhaber J, Kirchengast M, Buhr HJ, Foitzik T (2000) Effect of endothelin and endothelin receptor blockade on capillary permeability in experimental pancreatitis. Gut 46:390–394

Elshourbagy NA, Lee JA, Korman DR, Nuthalaganti P, Sylvester DR, Dilella AG, Sutiphong JA, Kumar CS (1992) Molecular cloning and characterization of the major endothelin receptor subtype in porcine cerebellum. Mol Pharmacol 41:465–473

Escrig C, Bishop AE, Inagaki H, Moscoso G, Takahashi K, Varndell IM, Ghatei MA, Bloom SR, Polak JM (1992) Localisation of endothelin like immunoreactivity in adult and developing human gut. Gut 33:212–217

Fang S, Ledlow A, Murray JA, Christensen J, Conklin JL (1994) Vasoactive intestinal contractor: localization in the opossum esophagus and effects on motor functions. Gastroenterology 107:1621–1626

Fiedler F, Ayasse D, Rohmeiss P, Gretz N, Rehbein C, Keim V (1999) The endothelin antagonist bosentan does not improve survival in severe experimental pancreatitis in rats. Int J Pancreatol 26:147–154

Filep JG, Sirois MG, Rousseau A, Fournier A, Sirois P (1991) Effects of endothelin-1 on vascular permeability in the conscious rat: interactions with platelet-activating factor. Br J Pharmacol 104:797–804

Filep JG, Sirois MG, Foldes-Filep E, Rousseau A, Plante GE, Fournier A, Yano M, Sirois P (1993) Enhancement by endothelin-1 of microvascular permeability via the activation of ETA receptors. Br J Pharmacol 109(3):880–886

Foitzik T, Hotz HG, Hot B, Kirchengast M, Buhr HJ (1998) Endothelin-1 mediates the alcohol-induced reduction of pancreatic capillary blood flow. J Gastrointest Surg 2:379–384

Foitzik T, Hotz HG, Eibl G, Hotz B, Kirchengast M, Buhr HJ (1999) Therapy for microcirculatory disorders in severe acute pancreatitis: effectiveness of platelet-activating factor receptor blockade vs. endothelin receptor blockade. J Gastrointest Surg 3:244–251

Foitzik T, Eibl G, Buhr HJ (2000) Therapy for microcirculatory disorders in severe acute pancreatitis: comparison of delayed therapy with ICAM-1 and a specific endothelin A receptor antagonist. J Gastrointest Surg 4:240–247

Fouassier L, Chinet T, Robert B, Carayon A, Balladur P, Mergey M, Paul A, Poupon R, Capeau J, Barbu V, Housset C (1998) Endothelin-1 is synthesized and inhibits cyclic adenosine monophosphate- dependent anion secretion by an autocrine/paracrine mechanism in gallbladder epithelial cells. J Clin Invest 101:2881–2888

Fulginiti J, Cohen MM, Moreland RS (1993) Endothelin differentially affects rat gastric longitudinal and circular smooth muscle. J Pharmacol Exp Ther 265:1413–1420

Furuya K, Furuya S, Yamagishi S (1994) Intracellular calcium responses and shape conversions induced by endothelin in cultured subepithelial fibroblasts of rat duodenal villi. Pflugers Arch 428:97–104

Galron R, Bdolah A, Kochva E, Wollberg Z, Kloog Y, Sokolovsky M (1991) Kinetic and cross-linking studies indicate different receptors for endothelins and sarafotoxins in the ileum and cerebellum. Fed Eur Bioch Soc 283:11–14

Gandhi CR, Behal RH, Harvey SA, Nouchi TA, Olson MS (1992a) Hepatic effects of endothelin. Receptor characterization and endothelin-induced signal transduction in hepatocytes. Biochem J 28:897–904

Gandhi CR, Harvey SA, Olson MS (1993) Hepatic effects of endothelin: metabolism of [125I]endothelin-1 by liver-derived cells. Arch Biochem Biophys 305:38–46

Gandhi CR, Stephenson K, Olson MS (1990) Endothelin, a potent peptide agonist in the liver. J of Biol Chem 265:17432–17435

Gandhi CR, Stephenson K, Olson MS (1992b) A comparative study of endothelin- and platelet-activating-factor-mediated signal transduction and prostaglandin synthesis in rat Kupffer cells. Biochem J 281:485–492

Ghatei MA, Takahashi K, Kirkland SC, Jones PM, Perera T, Wright NA, Bloom SR (1994) Endothelin. In: Gut peptides: biochemistry and physiology, pp 389–400

Gray GA, Clozel M (1994) ET-1 activates multiple receptor subtypes, including an inhibitory ET_A receptor, in the rat fundic strip. Br J Pharmacol 112:121P

Gregersen S, Thomsen JL, Brock B, Hermansen K (1996) Endothelin-1 stimulates insulin secretion by direct action on the islets of Langerhans in mice. Diabetologia 39:1030–1035

Gregersen S, Thomsen JL, Hermansen K (2000) Endothelin-1 (ET-1)-potentiated insulin secretion: involvement of protein kinase C and the ET(A) receptor subtype. Metabolism 49:264–269

Guimaraes CL, Rae GA (1992) Dual effects of endothelins –1 -2 and -3 on guinea pig field-stimulated ileum: possible mediation by two receptors coupled to pertussis toxin-insensitive mechanisms. J Pharmacol Exp Ther 261:1253–1259

Hildebrand P, Mrozinski JE Jr, Mantey SA, Patto RJ, Jensen RT (1993) Pancreatic acini possess endothelin receptors whose internalization is regulated by PLC-activating agents. Am J Physiol 264:G984–G993

Hof RP, Hof A, Takiguchi Y (1989) Massive regional differences in the vascular effects of endothelin. J Hypertens 7:S274–S275

Hof RP, Hof A, Takiguchi Y (1990) Attenuation of endothelin-induced regional vasoconstriction by isradipine: a nonspecific antivasoconstrictor effect. J Cardiovasc Pharmacol 15:S48–54

Hogaboam CM, Muller MJ, Collins SM, Hunt RH (1996) An orally active non-selective endothelin receptor antagonist, bosentan, markedly reduces injury in a rat model of colitis. Eur J Pharmacol 309: 261–269

Hori M, Sudjarwo SA, Oda K, Urade Y, Karaki H (1994) Two types of endothelin B receptors mediating relaxation in the guinea pig ileum. Life Sci 54:645–652

Hori S, Komatsu Y, Shigemoto R, Mizuno N, Nakanishi S (1992) Distinct tissue distribution and cellular localization of two messenger ribonucleic acids encoding different subtypes of rat endothelin receptors. Endocrinology 130:1885–1895

Housset C, Carayon A, Housset B, Legendre C, Hannoun L, Poupon R (1993a) Endothelin-1 secretion by human gallbladder epithelial cells in primary culture. Lab invest 69:750–755

Housset C, Rockey DC, Bissell DM (1993b) Endothelin receptors in rat liver: lipocytes as a contractile target for endothelin-1. Proc Natl Acad Sci USA 90:9266–9270

Hoyer D, Waeber C, Palacios JM (1989) [125I]endothelin-1 binding sites: autoradiographic studies in the brain and periphery of various species including humans. J Cardiovasc Pharmacol 13:S162–S165

Husdon M, Dashwood MR, Pounder RE, Wakefield AJ (1994) [125I] Endothelin-1 in normal human intestine and inflammatory bowel disease. Gastro (Suppl) 106:A239

Inagaki H, Bishop AE, Escrig C, Wharton J, Allen-Mersh TG, Polak JM (1991a) Localization of endothelin like immunoreactivity and endothelin binding sites in human colon. Gastroenterology 101:47–54

Inagaki H, Bishop AE, Yura J, Polak JM (1991b) Localization of endothelin-1 and its binding sites to the nervous system of the human colon. J Cardiovasc Pharmacol 17:S455–S457

Inoue A, Yanagisawa M, Kimura S, Kasuya Y, Miyauchi T, Goto K, Masaki T (1989) The human endothelin family: three structurally and pharmacologically distinct isopeptides predicted by three separate genes. Proc Natl Acad Sci USA 86:2863–2867

Irie K, Uchida Y, Fujii E, Muraki T (1995) Developmental changes in response to endothelins and receptor subtypes of isolated rat duodenum. Eur J Pharmacol 1995 Feb 24;275(1):45–51

Ishida N, Tsujioka K, Tomoi M, Saida K, Mitsi Y (1989) Differential activities of two distinct endothelin family peptides on ileum and coronary artery. Fed Eur Bioch Soc 247:337–340

Kakugawa Y, Paraskevas S, Metrakos P, Giaid A, Qi SJ, Duguid WP, Rosenberg L (1996) Alterations in pancreatic microcirculation and expression of endothelin-1 in a model of chronic pancreatitis. Pancreas 13:89–95

Ken-Dror S, Weintraub Z, Yechiely H, Kahana L (1997) Atrial natriuretic peptide and endothelin concentrations in human milk during postpartum lactation. Acta Paediatr 86:793–795

Kitajima T, Tani K, Kubota Y, Yamaguchi T, Okuhira M, Mizuno T, Inoue K (1994) Role of endogenous endothelin in gastric mucosal injury induced by endotoxin shock in rats. Digestion 54:156–159

Kitajima T, Yamaguchi T, Tani K, Fujimura K, Okuhira M, Kubota Y, Hiramatsu A, Mizuno T, Inoue K, Yamada H (1992) Role of vasoactive substance on indomethacin induced gastric mucosal lesions-evaluation of endothelin and platelet activating factor. Gastroenterology 120:A97

Kitajima T, Yamaguchi T, Tani K, Kubota Y, Okuhira M, Inoue K, Yamada H (1993) Role of endothelin and platelet-activating factor in indomethacin-induced gastric mucosal injury in rats. Digestion 54:156–159

Kitajima T, Tani K, Yamaguchi T, Kubota Y, Okuhira M, Mizuno T, Inoue K (1995) Role of endogenous endothelin in gastric mucosal injury induced by hemorrhagic shock in rats. Digestion 56:111–116

Kitsukawa Y, Gu ZF, Hildebrand P, Jensen RT (1994) Gastric smooth muscle cells possess two classes of endothelin receptors but only one alters contraction. Am J Physiol 266:G713–G721

Kloog Y, Ambar I, Sokolowsky M, Kochva E, Wollberg Z, Bdolah A (1988) Sarafotoxin, a novel vasoconstrictor peptide: phosphoinositide hydrolysis in rat heart and brain. Science 242:268–270

Kogire M, Inoue K, Higashide S, Takaori K, Echigo Y, Gu YJ, Sumi S, Uchida K, Imamura M (1995) Protective effects of endothelin-1 on acute pancreatitis in rats. Dig Dis Sci 40:1207–1212

Kuhn M, Fuchs M, Beck FX, Martin S, Jahne J, Klempnauer J, Kaever V, Rechkemmer G, Forssmann WG (1997) Endothelin-1 potently stimulates chloride secretion and inhibits Na(+)-glucose absorption in human intestine in vitro. J Physiol (Lond) 499:391–402

Lam HC, Takahashi K, Ghatei MA, Bloom SR (1990) Presence of immunoreactive endothelin in human milk. FEBS 261:184–186

Lam HC, Takahashi K, Ghatei MA, Katsuyuki S, Kanse SM, Bloom SR (1991) Presence of immunoreactive endothelin in human saliva and rat parotid gland. Peptides 12:883–885

Lazaratos S, Kashimura H, Nakahara A, Fukutomi H, Osuga T, Goto K (1994) L-Arginine has gastroprotective and vasoactive actions against the endothelin-1-induced gastric ulcer in rats. Gastro [Supp 1] 106:A123

Larazatos S, Nakahara A, Goto K, Fukutomi H (1995) Bosentan antagonizes the effects of endothelin-1 on rat gastric blood flow and mucosal integrity. Life Sci 56:195–200

Le Monnier de Gouville AC, Lippton HL, Cavero I, Summer WR, Hyman AL (1989) Endothelin: a new family of endothelium-derived peptides with widespread biological properties. Life Sci 45(17):1499–1513

Lembeck F, Decrinis M, Pertl C, Amann R, Donnerer J (1989) Effects of endothelin on the cardiovascular system and on smooth muscle preparations in different species. Naunyn-Schmiedeberg's Arch Pharmacol 340:744–751

Letizia C, Boirivant M, De Toma G, Cerci S, Subioli S, Scuro L, Ferrari P, Pallone F (1998) Plasma levels of endothelin-1 in patients with Crohn's disease and ulcerative colitis. Ital J Gastroenterol Hepatol 30:266–269

Lin WW, Lee CY (1990) Biphasic effects of endothelin in the guinea-pig ileum. Eur J Pharmacol 176:57–62

Lin HY, Kaji EH, Winkel GK, Ives HE, Lodish HF (1991) Cloning and functional expression of a vascular smooth muscle endothelin 1 receptor. Proc Natl Acad Sci U S A 88:3185–3189

Liu XH, Nakano I, Ito T, Yamaguchi H, Migita Y, Miyahara T, Koyanagi S, Ohgoshi K, Nawata H (1995) Role of endothelin in the development of hemorrhagic pancreatitis in rats. J Gastroenterol 30:275–277

MacNaughton WK, Keenan CM, McKnight GW, Wallace JL (1989) The modulation of gastric mucosal integrity by endothelin-1 and prostacyclin. J Cardiovasc Pharmacol 13:S118–S122

Maggi CA, Giuliani S, Patacchini R, Rovero P, Giachetti A, Meli A (1989) The activity of peptides of the endothelin family in various mammalian smooth muscle preparations. Eur J Pharmacol 174:23–31

Masuda E, Kawano S, Nagano K, Ogihara T, Tsuji S, Tanimura H, Ishigami Y, Tsujii M, Hayashi N, Sasayama-Y, et al (1991) Role of blood ethanol on gastric mucosal injury and gastric hemodynamics. Alcohol 1:335–338

Masuda E, Kawano S, Michida T, Tsuji S, Nagano K, Fusamoto H, Kamada T (1997) Plasma and gastric mucosal endothelin-1 concentrations in patients with peptic ulcer. Dig Dis Sci 42:314–318

Masuda E, Kawano S, Nagano K, Tsuji S, Ishigami Y, Hayashi N, Tsujii M, Sasayama Y, Michida T, Fusamoto H, Kamada T (1991b) Effect of ethanol on endothelin-1 release from gastric vasculature. Gastroenterol Jpn 26:81–82

Masuda E, Kawano S, Nagano K, Tsuji S, Takei Y, Hayashi N, Tsujii M, Oshita M, Michida T, Kobayashi I, Peng H-B, Fusamoto H, Kamada T (1993) Role of endogenous endothelin in pathogenesis of ethanol-induced gastric mucosal injury in rats. Am J Physiol 265:G474–G481

Matsumoto H, Suzuki N, Onda H, Fujino M (1989) Abundance of endothelin-3 in rat intestine, pituitary gland and brain. Biochem Biophys Res Comm 164:74–80

Miasiro N, Karaki H, Matsuda Y, Paiva AC, Rae GA (1999) Effects of endothelin ET(B) receptor agonists and antagonists on the biphasic response in the ileum. Eur J Pharmacol 1999 Mar 19;369(2):205–213

Miasiro N, Karaki H, Paiva AC (1995) Heterogeneous endothelin receptors mediate relaxation and contraction in the guinea-pig ileum. Eur J Pharmacol 285(3): 247–254

Miasiro N, Paiva AC (1990) Effects of endothelin-1 on the isolated guinea-pig ileum: role of Na$^+$ ions. Naunyn-Schmiedeberg's Arch Pharmacol 342:706–712

Miasiro N, Paiva ACM (1992) Effects of endothelin-1 and endothelin-3 on the isolated guinea pig ileum : role of Na$^+$ ions and endothelin receptor subtypes. J Cardiovasc Pharmacol 20 [Suppl 12]:S37–S40

Miasiro N, Paiva ACM, Ihara M, Yano M (1993) Two receptor subtypes are involved in the contractile component of the guinea pig ileum responses to endothelins. J Cardiovasc Pharmacol 22:S211-S213

Michida T, Kawano S, Masuda E, Kobayashi I, Nishimura Y, Tsujii M, Hayashi N, Takei Y, Tsuji S, Nagano K, Fusamoto H, Kamada T (1994) Role of endothelin 1 in hemorrhagic shock-induced gastric mucosal injury in rats. Gastroenterology 106:988–993

Migoh S, Hashizume M, Tsugawa K, Tanoue K, Sugimachi K (2000) Role of endothelin-1 in congestive gastropathy in portal hypertensive rats. J Gastroenterol Hepatol 15:142–147

Mizuno T, Imai T, Itakura M, Hirose S, Hirata Y, Marumo F, Hagiwara H (1992) Structure of the bovine endothelin-B receptor gene and its tissue-specific expression revealed by northern analysis. J Cardiovasc Pharmacol 20:S8–10

Morales RE, Johnson BR, Szabo S (1992) Endothelin induces vascular and mucosal lesions, enhances the injury by HCl/ethanol, and the antibody exerts gastroprotection. FASEB 6:2354–2360

Moummi C, Gullikson GW, Gaginella TS (1992a) Effect of endothelin-1 on guinea pig gallbladder smooth muscle in vitro. J Pharmacol Exp Ther 260:549–553

Moummi C, Xie Y, Kachur JF, Gaginella TS (1992b) Endothelin-1 stimulates contraction and ion transport in the rat colon: different mechanisms of action. J Pharmacol Exp Ther 262:409–414

Murch SH, Braegger CP, Sessa WC, MacDonald TT (1992) High endothelin-1 immunoreactivity in Crohn's disease and ulcerative colitis. Lancet 339:381–385

Neuser D, Steinke W, Theiss G, Stasch JP (1989) Autoradiographic localization of [125I]endothelin-1 and [125I]atrial natriuretic peptide in rat tissue: a comparative study. J Cardiovasc Pharmacol 13:S67–S73

Ohta M, Nguyen TH, Tarnawski AS, Pai R, Kratzberg YP, Sugimachi K, Sarfeh IJ (1997) Overexpression of endothelin-1 mRNA and protein in portal hypertensive gastric mucosa of rats: a key to increased susceptibility to damage? Surgery 122:936–942

Ohta M, Pai R, Kawanaka H, Ma T, Sugimachi K, Sarfeh IJ, Tarnawski AS (2000) Expression of endothelin-1, and endothelin A and B receptors in portal hypertensive esophagus of rats. J Physiol Pharmacol 51:57–67

Oldner A, Wanecek M, Goiny M, Weitzberg E, Rudehill A, Alving K, Sollevi A (1998) The endothelin receptor antagonist bosentan restores gut oxygen delivery and reverses intestinal mucosal acidosis in porcine endotoxin shock. Gut 42:696–702

Olsen UB, Weis J (1992) Rat gastric relaxation induced by stimulation of endothelin-1 selective receptors. Regulatory Peptides 39:113–119

Oshita M, Takei Y, Kawano S, Yoshihara H, Hijioka K, Fukui H, Goto M, Masuda E, Nishimura Y, Fusamoto H, Kamada T (1993) Roles of endothelin-1 and nitric oxide in the mechanism for ethanol-induced vasoconstriction in rat liver. J Clin Invest 91:1337–1342

Padol I, Huang JQ, Hogaboam CM, Hunt RH (2000) Therapeutic effects of the endothelin receptor antagonist Ro 48–5695 in the TNBS/DNBS rat model of colitis. Eur J Gastroenterol Hepatol 12:257–265

Plusczyk T, Bersal B, Westermann S, Menger M, Feifel G (1999) ET-1 induces pancreatitis-like microvascular deterioration and acinar cell injury. J Surg Res 85:301–310

Rachmilewitz D, Eliakim R, Ackerman Z, Karmeli F (1992) Colonic endothelin-1 immunoreactivity in active ulcerative colitis. Lancet 339:1062

Roden M, Vierhapper H, Liener K, Waldhausl W (1992) Endothelin-1 stimulated glucose production in vitro in the isolated perfused rat liver. Metabolism 41: 290–295

Saida K, Mitsui Y, Ishida N (1989) A novel peptide, vasoactive intestinal contractor, of a new (endothelin) peptide family. Molecular cloning, expression, and biological activity. J Biol Chem 264:14613–14616

Salh B, Hoeflick K, Kwan W, Pelech S (1998) Granulocyte-macrophage colony-stimulating factor and interleukin-3 potentiate interferon-gamma-mediated endothelin production by human monocytes: role of protein kinase C. Immunology 95: 473–479

Sessa WC, Kaw S, Hecker M, Vane JR (1991) The biosynthesis of endothelin-1 by human polymorphonuclear leukocytes. Biochem Biophys Res Commun 174:613–618

Shiba R, Yanagisawa M, Miyauchi T, Ishii Y, Kimura S, Uchiyama Y, Masaki T, Goto K (1989) Elimination of intravenously injected endothelin-1 from the circulation of the rat. J Cardiovasc Pharmacol 13:S98–101

Spokes RA, Ghatei MA, Bloom SR (1989) Studies with endothelin-3 and endothelin-1 on rat blood pressure and isolated tissues: evidence for multiple endothelin receptor subtypes. J Cardiovasc Pharmacol 13:S191–S192

Suzuki N, Matsumoto H, Kitada C, Kimura S, Fujino M (1989) Production of endothelin-1 and big-endothelin-1 by tumor cells with epithelial-like morphology. J Biochem (Tokyo) 106:736–741

Szabo S, Vincze A, Sandor Z, Jadus M, Gombos Z, Pedram A, Levin E, Hagar J, Iaquinto G (1998) Vascular approach to gastroduodenal ulceration: new studies with endothelins and VEGF. Dig Dis Sci 43:40S–45S

Takahashi K, Jones PM, Kanse SM, Lam H-C, Spokes RA, Ghatei MA, Bloom SR (1990) Endothelin in the gastrointestinal tract. Presence of endothelin like immunoreactivity, endothelin-1 messenger RNA, endothelin receptors, and pharmacological effect. Gastroenterology 99:1660–1667

Takaori K, Inoue K, Kogire M, Higashide S, Tun T, Aung T, Doi R, Fujii N, Tobe T (1992) Effects of endothelin on microcirculation of the pancreas. Life Sci 51:615–622

Takeuchi K, Sugamoto S, Suzuki K, Kawauchi S, Furukawa O (1999) Effects of endothelin-1 on duodenal bicarbonate secretion and mucosal integrity in rats. Chin J Physiol 199 30;42(3):129–35

Tokito F, Suzuki N, Hosoya M, Matsumoto H, Ohkubo S, Fujino M (1991) Epidermal growth factor (EGF) decreased endothelin-2 (ET-2) production in human renal adenocarcinoma cells. FEBS Lett 295:17–21

Uchida K, Yuzuki R, Kamikawa Y (1998a) Pharmacological characterization of endothelin-induced contraction in the guinea-pig oesophageal muscularis mucosae. J Cardiovasc Pharmacol 31:S504–506

Uchida K, Yuzuki R, Kamikawa Y (1998b) Pharmacological characterization of endothelin-induced contraction in the guinea-pig oesophageal muscularis mucosae. Br J Pharmacol 125:849–857

Walder CE, Thomas GR, Thiemermann C, Vane JR (1989) The hemodynamic effects of endothelin-1 in the pithed rat. J Cardiovasc Pharmacol 13:S93–S97

Wallace JL, Cirino G, De Nucci G, McKnight W, MacNaughton WK (1989a) Endothelin a potent ulcerogenic and vasoconstrictor actions in the stomach. Am J Physiol 256:G661–G666

Wallace JL, Keenan CM, MacNaughton WK, McKnight GW (1989b) Comparison of the effects of endothelin-1 and endothelin-3 on the rat stomach. Eur J Pharmacol 167:41–47

Warner TD, Allcock GH, Mickley EJ, Vane JR (1993) Characterization of endothelin receptors mediating the effects of the endothelin/sarafotoxin peptides on autonomic neurotransmission in the rat vas deferens and guinea-pig ileum. Br J Pharmacol 002–007

Whittle BJR, Esplugues JV (1988) Induction of rat gastric damage by the endothelium-derived peptide, endothelin. Br J Pharmacol 95:1011–1013

Whittle BJR, Lopez-Belmonte J (1991) Interactions between the vascular peptide endothelin-1 and sensory neuropeptides gastric mucosal injury. Br J Pharmacol 102:950–954

Wiklund NP, Ohlen A, Wiklund CU, Cederqvist B, Hedqvist P, Gustafsson LE (1989a) Neuromuscular actions of endothelin on smooth, cardiac and skeletal muscle from guinea-pig, rat and rabbit. Acta Physiol Scand 137:399–407

Wiklund NP, Wiklund CU, Ohlen A, Gustafsson LE (1989b) Cholinergic neuromodulation by endothelin in guinea pig ileum. Neurosci Lett 101:342–346

Wilson MA, Steeb GD, Garrison RN (1993) Endothelins mediate intestinal hypoperfusion during bacteremia. J Surg Res 55:168–175

Withrington PG, de Nucci G, Vane JR (1989) Endothelin-1 causes vasoconstriction and vasodilation in the blood perfused liver of the dog. J of Cardiovasc Pharmacol 13:S209–S210

Wollberg Z, Bdolah A, Galron R, Sokolovsky M, Kochva E (1991) Contractile effects and binding properties of endothelins/sarafotoxins in the guinea pig ileum. Eur J Pharmacol 198:31–36

Wood JG, Yan ZY, Cheung LY (1992) Relative potency of endothelin analogues on changes in gastric vascular resistance. Am J Physiol 262:G977–G982

Wood JG, Yan ZY, Davis JM, Cheung LY (1994) Phosporamidon attenuates big endothelin-1 induced vasoconstriction in canine stomach. Am J Physiol 266:G311–G317

Yanagisawa M, Kurihara H, Kimura S, Tomobe Y, Kobayashi M, Mitsui Y, Yazaki Y, Goto K, Masaki T (1988) A novel potent vasoconstrictor peptide produced by vascular endothelial cells. Nature 332:411–415

Yoshimi H, Hirata Y, Fukuda Y, Kawano Y, Emori T, Kuramochi M, Omae T, Marumo F (1989) Regional distribution of immunoreactive endothelin in rats. Peptides 10:805–808

Yoshinaga M, Chijiiwa Y, Misawa T, Harada N, Nawata H (1992) EndothelinB receptor on guinea pig small intestinal smooth muscle cells. Am Physiol Soc G308–G311

Yoshizumi M, Kurihara H, Morita T, Yamashita T, Oh-hashi Y, Sugiyama T, Takaku F, Yanagisawa M, Masaki T, Yazaki Y (1990) Interleukin 1 increases the production of endothelin-1 by cultured endothelial cells. Biochem Biophys Res Commun 166:324–329

Yule DI, Blevins GT Jr, Wagner AC, Williams JA (1992) Endothelin increases [Ca2+] in rat pancreatic acinar cells by intracellular release but fails to increase amylase secretion. Biochim Biophys Act 1136:175–180

CHAPTER 18
Endothelin and the Kidney

D.M. POLLOCK

A. Introduction

The kidney contains an abundant quantity of the machinery necessary for the paracrine and autocrine actions of endothelin-1 (ET-1). ET-1 is synthesized by a variety of cell types within the kidney including vascular and glomerular endothelium, mesangial cells and tubular epithelial cells. Sites of synthesis are in close proximity to receptors. ET-1 can produce a variety of responses through both ET_A and ET_B receptor subtypes located throughout the kidney. These cells include vascular smooth muscle, mesangial cells, pericytes within the renal medullary circulation and tubular epithelium, most predominantly, inner medullary collecting duct cells. The paracrine and autocrine functions are important to consider since some of the actions of ET-1 are in apparent opposition (e.g., vasoconstriction and inhibition of tubular reabsorption). Despite the local actions, ET-1 appears to participate in several mechanisms that are important in fluid volume homeostasis and arterial pressure regulation. Furthermore, the role of ET-1 in several forms of renal failure suggest potential therapeutic targets for endothelin receptor antagonists. This chapter reviews the localization of the endothelin system (ET-1 synthesis, endothelin converting enzyme and ET_A and ET_B receptor expression). More importantly, the actions of ET-1 in the kidney will be reviewed including the physiological and pathophysiological roles that may be ascribed to ET-1 in the kidney.

B. Localization of ET-1, ECE, and Endothelin Receptors in the Kidney

The renal vascular endothelium is a source of ET-1 production although a variety of other cell types also have been shown to synthesize and release ET-1. In culture, these include cells originating from renal tubules (SCHNER-MANN et al. 1996; TODD-TURLA et al. 1996), mesangium (SAKAMOTO et al. 1992; KOHAN 1992), glomerular endothelium (MARSDEN et al. 1991), and medullary interstitium (WILKES et al. 1991a). In terms of intact tissue, immunohisto-chemical analysis has consistently demonstrated higher amounts of ET-1-like immunoreactivity in the renal medulla compared to the cortex in a variety of species including human (KARET et al. 1993; MORITA et al. 1991; WILKES et al. 1991b; KITAMURA et al. 1989). In the rat kidney, Wilkes et al. reported that immunostaining density was greatest in vasa recta while lesser staining was observed within the cytosol of collecting duct cells in the papilla (WILKES et al. 1991b). Within the renal cortex, ET-like immunoreactivity was predominantly localized to endothelial cells of arcuate arteries, arterioles, peritubular capillaries, and veins. Lesser amounts of staining appeared in both glomerular capillaries and mesangial cells along with the early proximal tubule.

Studies examining ET-1 gene expression confirm that higher amounts of the peptide are produced within the medullary regions of the kidney (FIRTH and RATCLIFFE 1992; NUNEZ et al. 1991). Specific cellular localization using microdissection and RT-PCR or RNAse protection reveal that in both rat and human, glomeruli, proximal tubules, thick ascending limbs and outer and inner medullary collecting ducts express ET-1 mRNA (CHEN et al. 1993; PUPILLI et al. 1994). In situ hybridization studies also have identified ET-1 mRNA expression within the vasa recta (NUNEZ et al. 1991). Kohan detected immunoreactive ET-1 in primary cell cultures of rabbit proximal tubules thick ascending limbs, and cortical and inner medullary collecting ducts (KOHAN 1991). Glomerular expression of ET-1 mRNA has been reported in the rat, but not the human, although technical differences may not have allowed detection of a low expression level in human glomeruli (CHEN et al. 1993; PUPILLI et al. 1994).

Synthesis of biologically active ET-1 requires conversion of the inactive precursor, Big ET-1, by endothelin converting enzyme-1 (ECE-1) (see Chaps. 3 and 7). In rats, RT-PCR and Western blot analysis was used to identify ECE-1 mRNA and protein, respectively, in glomeruli, proximal straight tubule, cortical and medullary thick ascending limb, and inner medullary collecting duct (DISASHI et al. 1997). Tubular expression of ECE-1 was higher in spontaneously hypertensive rats compared to normotensive controls. In humans, Pupilli et al. reported ECE-1 immunoreactivity in vascular and glomerular endothelial cells as well as vasa recta capillaries (PUPILLI et al. 1997). In tubular epithelium, ECE-1 immunostaining was detected in outer and inner medullary collecting ducts and thin limbs of Henle's loops. Similar results were obtained for mRNA using in situ hybridization. Localization of ECE-1 was also observed in proximal and distal tubules although the distribution pattern was inconsistent within specific tubular segments.

Autoradiographic binding studies have shown a pattern of endothelin receptor localization consistent with its role as a paracrine factor, that is, the receptors are present at sites where endothelin synthesis occurs. A particularly high density of ET-1 binding sites have been identified in the inner medulla, inner stripe of the outer medulla, and glomeruli, with moderate to low levels of binding distributed throughout the outer cortex (JONES et al. 1989; KOHZUKI et al. 1989). In vitro radioligand binding indicated the presence of receptors in the renal vascular smooth muscle, mesangial cells, and cultured tubular cells

(JONES et al. 1989; KOHZUKI et al. 1989; NEUSER et al. 1990). Intravenous injection of ^{125}I-labeled ET-1 in the rat indicated specific binding localized to endothelial cells of glomeruli and peritubular capillaries of the cortex (DEAN et al. 1996). In terms of receptor subtypes (see Chap. 4), kidneys from all species including human contain both ET_A and ET_B receptors although the role of each of these receptors in producing vasoconstriction are species dependent. Binding studies suggest a mixed population of both ET_A and ET_B receptors in kidneys of all species tested thus far although the relative quantities of each subtype are not correlated to functional responses (POLLOCK and OPGENORTH 1993; POLLOCK and OPGENORTH 1994; TÉLÉMAQUE et al. 1993; YAMASHITA et al. 1991; NAMBI et al. 1994b; FUKURODA et al. 1994b). For example, the vascular response to ET-1 appears to be ET_A-dependent in the human kidney even though the tissue contains a majority of ET_B receptors (KARET et al. 1993). In the rat, both ET_A and ET_B receptor activation produce vasoconstriction in the kidney (POLLOCK and OPGENORTH 1993, 1994). In the dog, ET_B agonists or low doses of ET-3 will produce renal vasodilation while having no effect in the rabbit kidney (TÉLÉMAQUE et al. 1993; YAMASHITA et al. 1991). Specific cellular localization of receptor subtypes has not been completely elucidated. Russell et al. have recently shown that Big ET-1 binds to glomeruli, distal tubule, collecting duct and endothelial cells in human kidneys and that binding was abolished by functional inhibition of endothelin converting enzyme or ET_B, but not ET_A, receptor blockade (RUSSELL et al. 1998). These findings suggest an important functional role for ET_B receptors in the kidney due to their presumed co-localization with the endothelin converting enzyme.

C. Hemodynamic Effects of Exogenous ET-1 in the Kidney

Originally it was thought that the ET_A receptor mediated all of the vasoconstrictor properties of ET-1 and that ET-1-induced vasodilation was due to nitric oxide release stimulated via the ET_B receptors located on the vascular endothelium. Intravenous infusion of ET-1 in the dog or rat produces a prolonged reduction in renal blood flow and glomerular filtration rate (MILLER et al. 1989; CHOU et al. 1990). Direct infusion into the renal artery of dogs produced a transient vasodilation followed by prolonged constriction consistent with the notion of both ET_A and ET_B receptors in the renal circulation (BANKS 1990). Consistent with these observations is that ET-3 also produces vasodilation at low doses but vasoconstriction at higher doses in the dog (YAMASHITA et al. 1991). Rats appear to have fewer ET_B receptors responsible for vasodilation since ET_B agonists reduce renal blood flow without any transient vasodilator actions (BIRD and GIANCARLI 1996; BANKS et al. 1997). Using implanted single-fiber optical probes, Gurbanov and colleagues demonstrated that ET-1 vasodilates the renal medullary circulation while simultaneously reducing renal cortical blood flow (GURBANOV et al. 1996). These observations suggest ET_B-dependent vasodilation in the renal medulla.

Not all responses to endothelin peptides conform to the initial classification of ET_A-mediated vasoconstriction and ET_B-mediated vasodilation. Soon after specific ET_A receptor antagonists such as BQ-123 and FR139317 became available, it was observed that non-ET_A receptors mediate at least some of the vasoconstrictor actions of ET-1 (POLLOCK and OPGENORTH 1993; WELLINGS et al. 1994). Of particular importance, there is pharmacological evidence that the ET_B receptor responsible for non-ET_A-mediated vasoconstriction represents a subtype of the ET_B receptor. Several studies have shown that the receptor present on vascular endothelium responsible for the release of nitric oxide (termed ET_{B1}) has a unique pharmacological profile compared to the ET_B receptor that mediates vasoconstriction, termed ET_{B2} (WARNER et al. 1993). In isolated smooth muscle preparations, Warner et al. observed that the non-selective antagonist, PD 142893, blocks ET_A-induced contractions and ET_B-induced relaxations, yet has no effect on non-ET_A contractions (WARNER et al. 1993). Infusion of BQ-123 intravenously into the rat will completely prevent the rise in mean arterial pressure and hematocrit produced by a simultaneous infusion of relatively high doses of ET-1 (POLLOCK and OPGENORTH 1993). However, ET-1-induced decreases in renal blood flow and glomerular filtration rate were totally unaffected even when the doses of BQ-123 were higher than that required to block the systemic actions of ET-1. Binding studies and other in vitro experiments revealed that the renal vasculature and several other arterial and venous circulations contain ET_B receptors that mediate vasoconstriction (KARET 1996; KARET et al. 1993; MAGUIRE et al. 1994). Several binding and mRNA expression studies have confirmed the existence of ET_B receptors in vascular smooth muscle. In fact, in most species, infusion of ET_B agonists will elicit a hypertensive response.

The ET_{B2} receptor on vascular smooth muscle may also have a lower affinity for ET-1 compared to the ET_{B1} or ET_A receptor. ET_A receptor blockade completely prevents ET-1-induced decreases in renal blood flow at low doses but has no effect on higher doses of ET-1 (POLLOCK and OPGENORTH 1993, 1994). These results cannot be explained by insufficient blockade of ET_A receptors since the pressor response was completely prevented at both high and low doses of ET-1 and higher doses of ET_A antagonist yielded the same results.

At present, there is no clear biochemical or molecular evidence for ET_{B1} and ET_{B2} receptor subtypes. In fact, RT-PCR and in situ hybridization experiments have revealed the presence of mRNA encoding both ET_A and ET_B receptors (the latter using probes from the endothelial ET_B receptor) in human vascular smooth muscle (KARET 1996). These data suggest that differences between the endothelium ET_B receptor, ET_{B1}, and the smooth muscle ET_B receptor, ET_{B2}, are either extremely minor or may be post-translational in nature. There is some limited molecular evidence for variants of the ET_B receptor in human tissues. Shyamala et al. have described two distinct human

ET_B receptors generated by alternative splicing from a single gene located within human brain, placenta, lung, and heart (SHYAMALA et al. 1994). Similarly, Cheng et al. discovered a novel cDNA encoding a non-selective endothelin receptor found in rat brain and possibly other tissues as well (CHENG et al. 1993). It has yet to be determined whether these findings could help explain some of the pharmacological characterization studies.

In arterioles isolated from the renal circulation, ET-1 directly constricts both afferent and efferent arterioles. Edwards et al. reported similar potency in these vessels (EDWARDS et al. 1990) while Lanese et al. reported that ET-1 was more potent in efferent compared to afferent arterioles (LANESE et al. 1992). This vasoconstriction appears to be predominantly via the ET_A receptor since ET-3 had the effect of producing vasoconstriction but at a much lower potency compared to ET-1. Using the hydronephrotic kidney to examine segmental vascular resistance, ET-1 constricts throughout the renal circulation (FRETSCHNER et al. 1991; CAVARAPE and BARTOLI 1998) although Loutzenhiser et al. have reported that the decrease in vessel diameter was larger in afferent compared to efferent arterioles (LOUTZENHISER et al. 1990). More recently, Endlich et al. used this model to characterize receptor subtype-specific responses (ENDLICH et al. 1996). These investigators reported that the preglomerular vasoconstriction was predominantly ET_A receptor mediated while ET_B-dependent vasoconstriction was more evident in postglomerular vessels.

D. Tubular Actions of Endothelin

There is now considerable evidence that, in addition to hemodynamic actions, endothelin peptides can directly influence renal tubular function. As discussed above, renal tubular epithelium express endothelin receptors, primarily the ET_B subtype, as well as synthesize ET-1. In isolated proximal tubule, endothelin has been reported to have a biphasic effect on transport by stimulating flux at low concentrations and inhibiting at high concentrations (GARCIA and GARVIN 1994). Endothelin also inhibits transport in the thick ascending limb, the cortical, outer medullary and inner medullary collecting ducts.

ET-1 may stimulate transport in proximal tubules by influencing several transport mechanisms. In brush border membrane preparations from rats and rabbits, ET-1 has been shown to stimulate Na^+/P_i co-transport, Na^+/H^+ exchange, and Na^+/HCO_3^- transporters (GARVIN and SANDERS 1991; EIAM-ONG et al. 1992; GUNTUPALLI and DUBOSE 1994). Membrane binding studies have identified both ET_A and ET_B receptors in renal tubule preparations (KNOTEK et al. 1996; KOHAN et al. 1992). Several studies from Alpern's laboratory have also demonstrated that ET-1 activates and phosphorylates a specific isoform of the Na^+/H^+ antiporter (NHE3) via a tyrosine kinase and Ca^{2+} dependent pathway in OKP cells expressing the ET_B receptor (CHU et al. 1996a, b; PENG et al. 1999).

In thick ascending limb, Plato and Garvin have demonstrated that ET_B receptor activation inhibits chloride flux via an NO-dependent mechanism. In inner medullary collecting duct (IMCD) cells, ET_B receptors stimulate NO and cGMP production (EDWARDS et al. 1992; POLLOCK et al. 1998). ET-1 also reduces Na^+ transport in freshly isolated IMCD cells (ZEIDEL et al. 1989). This effect appears to involve inhibition of Na^+/K^+-ATPase by stimulation of PGE_2 synthesis although the specific endothelin receptor subtype responsible for this effect is not known. Both ET_A and ET_B receptors have been identified in IMCD cells using membrane binding techniques (KOHAN et al. 1992); however, autoradiographic studies have localized ET_A receptors only to vascular structures and not tubules. ET-1 synthesis and receptor expression appears to be highest in the inner medullary collecting duct suggesting a more important role for ET-1 in the IMCD compared to other nephron segments (KOHAN et al. 1992; KOHAN 1991; KOHAN and PADILLA 1992).

ET-1 also has a well-characterized effect on the IMCD to inhibit ADH-induced changes in water permeability (OISHI et al. 1991; SCHNERMANN et al. 1992; EDWARDS et al. 1993). Several laboratories have demonstrated that ET-1 inhibits ADH-induced increases in cAMP accumulation (KOHAN et al. 1993; MIGAS et al. 1993; EDWARDS et al. 1993). ET-1 also stimulates PGE_2 production in IMCD cells although this effect does not mediate inhibition of ADH actions (KOHAN et al. 1993). This effect occurs primarily via ET_B receptors located on inner medullary collecting duct cells (KOHAN et al. 1993; EDWARDS et al. 1993). While ET_A receptors are present in the IMCD, their function remains speculative. In terms of regulating tubular function, it is possible that ET_A receptors present on medullary interstitial cells may also influence water excretion by control of medullary hemodynamics (WILKES et al. 1991a).

E. Role of Endothelin in Control of Excretory Function

Due to its potent vasoconstrictor actions, ET-1 has been considered to be important in regulating vascular tone. However, it would appear that this putative role may not be a simple one. Acute administration of ET_A receptor antagonists have little effect on renal and cardiovascular function in healthy normotensive animals suggesting that ET-1 does not exert a tonic influence on vascular resistance via the ET_A receptor (QIU et al. 1995b; MATSUURA et al. 1997). However, the non-selective ET_A/ET_B receptor antagonist, bosentan, had no effect on arterial pressure or renal blood flow but did produce a small, but significant, decrease in GFR (QIU et al. 1995b). Using glomerular micropuncture techniques, these same investigators have also reported that bosentan produced a fall in glomerular blood flow as the result of a significant increase in preglomerular arteriolar resistance (QIU and BAYLIS 1999). These results suggest that endothelin may exert a tonic vasodilator influence on GFR by release of either NO or prostacyclin (HIRATA and EMORI 1993; WARNER et al. 1989; QIU and BAYLIS 1999). Nonetheless, it is possible that endothelin also plays a role in a chronic setting by modulating the activity of other vasoactive systems. Due to its profound effects on kidney function, ET-1 may influence arterial pressure through chronic regulation of renal sodium and water excretion.

ET-1 has been reported to have potent diuretic and natriuretic effects when given at doses that do not produce significant decreases in GFR (HARRIS et al. 1991; SCHNERMANN et al. 1992). Big ET-1 also has been reported to produce a natriuretic and diuretic response which is blocked by an ET_B receptor antagonist and is independent of changes in renal perfusion pressure or GFR (POLLOCK and OPGENORTH 1994; ABASSI et al. 1998). These observations are consistent with an overall inhibitory effect on tubular transport as discussed in the previous section. It would appear that most of the diuretic and natriuretic responses to endothelin peptides are mediated by tubular ET_B receptors due to their relative abundance. However, increases in renal perfusion pressure as well as changes in intrarenal hemodynamics which may be ET_A-mediated can also contribute to the excretory effects. Much of the diuretic and natriuretic actions of exogenous ET-1 can be prevented by holding renal perfusion pressure constant (KING et al. 1989; UZUNER and BANKS 1993). Furthermore, ET-1 may increase medullary blood flow and interstitial pressure as a means of reducing sodium reabsorption since the medullary vasculature also contains ET_B receptors. Gurbanov et al. showed that ET-1 increases medullary blood flow while simultaneously decreasing cortical blood flow (GURBANOV et al. 1996). Although the presumption is that the vasodilatory effect of ET-1 on medullary blood flow is via the ET_B receptor, experimental evidence has not been reported. In addition, it is not known whether changes in medullary hemodynamics produced by ET-1 play a physiological role in determining the excretion of sodium and water. It is well-established that alterations in medullary hemodynamics may play an important role in the control of sodium excretion (COWLEY and ROMAN 1996) and so the extent to which ET-1 hemodynamic factors vs tubular actions may control excretion in a physiological setting needs further investigation.

New information regarding the role of ET_B receptors in regulating fluid volume balance has been obtained from studies using rats and mice deficient in the ET_B receptor gene (see Chap. 6). In addition to the vascular endothelium and renal tubules, ET_B receptors are located on prostatic stroma, melanocytes, astrocytes, and enteric neurons. Pertinent to the latter location, the spotting lethal rat carries a naturally occurring deletion of the ET_B receptor gene that prevents expression of functional ET_B receptor and results in aganglionic megacolon (GARIEPY et al. 1996). Gariepy and colleagues have reported that the dopamine-β-hydroxylase promoter can be used to direct tissue-specific ET_B transgene expression to support normal enteric nervous system development (GARIEPY et al. 1998). Studies using these rats and a similar mouse model have shown that they also represent a model of salt-sensitive hypertension. The ET_B-deficient rat is normotensive until challenged with a high salt diet which significantly increases arterial pressure. Interestingly, the hypertension can be completely ameliorated by the epithelial sodium channel (ENaC) inhibitor, amiloride. These findings support the hypothesis that the predominant role of ET_B receptors in pressure regulation is through the regulation of sodium excretion and that ET_B receptor activation may function to down-regulate ENaC. Recent studies have also demonstrated that ET_B receptors in the renal medulla are up-regulated in response to salt loading consistent with a role of the ET_B receptor in chronic regulation of sodium balance (POLLOCK and POLLOCK 1999; POLLOCK et al. 2000). Despite these important findings, the ET_B receptor may also serve as a "clearance" receptor by removing ET-1 from the circulation since ET-1 binding appears to be irreversible and blockade of ET_B receptors increases circulating levels of ET-1 (LÖFFLER et al. 1993; FUKURODA et al. 1994a). This clearance function operates mostly in the lungs and so the extent to which renal ET_B receptors serve in this capacity would appear to be minimal. ET_B receptor blockade potentiates the renal vasoconstrictor actions of ET-1 which supports a potential role for ET_B receptors in removing ET-1 from the circulation although these results cannot exclude an important contribution of ET_B-dependent vasodilation (ONO et al. 1998).

F. Interaction Between Endothelial Factors in the Kidney

There are several possible links between the ET-1 and NO systems within the kidney. As mentioned above, ET_B receptors can stimulate NO-dependent vasodilation although the degree to which this occurs in the cortical circulation is different among species (CERNACEK et al. 1998; BROOKS et al. 1995; NAMBI et al. 1994b). In the medullary circulation, the predominant action of ET-1 is vasodilation with little vasoconstrictor activity (GURBANOV et al. 1996). Nitric oxide may also serve to tonically regulate the vasoconstrictor activity of ET-1 either by inhibition of ET-1 release from endothelial cells or by regulating ET-1-dependent tone at the level of the vascular smooth muscle. There is evidence for both of these mechanisms. First, it has been demonstrated that NOS inhibitors stimulate ET-1 release and NO donors inhibit ET-1 release from cultured endothelial cells (BOULANGER and LÜSCHER 1990; SAIJONMAA et al. 1990). This mechanism has been demonstrated by the in vivo observation that inhibition of NOS increased plasma levels of ET-1 (RICHARD et al. 1995). It has also been well documented that NO can directly antagonize the constrictor actions of ET-1 in vascular smooth muscle (HIRATA et al. 1991; LERMAN et al. 1992).

Studies examining the effects of endothelin antagonists on the renal hemodynamic response to NO synthase inhibition reveal complex changes in glomerular hemodynamics that involve numerous factors. Several groups have reported that ET_A or non-selective ET_A/ET_B receptor antagonists can attenuate the hypertension and increased renal vascular resistance produced by acute NOS inhibition (QIU et al. 1995a; RICHARD et al. 1995). However, angiotensin II blockade can inhibit most of the vasoconstrictor effects of NOS inhibition (SIGMON and BEIERWALTES 1993a, b; BAYLIS et al. 1994). Qui and Baylis have more carefully examined the role of endothelin and angiotensin II in mediating the glomerular microcirculatory changes during acute systemic NOS inhibition using glomerular micropuncture techniques (QIU and BAYLIS 1999). These investigators reported that both ET-1 and angiotensin II mediate the increases in renal vascular resistance (primarily efferent arteriolar resistance) in an additive manner. Complete blockade of the vasoconstrictor responses to NOS inhibition were achieved by combined ET-1 and angiotensin II receptor blockade.

During chronic NO depletion, acute administration of ET_A receptor antagonists have no significant effect on renal or systemic hemodynamics (FUJITA et al. 1995). Recent evidence suggests that ET-1 may play a significant role in hypertension produced by NOS inhibition only in the early phase of L-NAME treatment (BAYLIS and SCHMIDT 1996; POLLOCK et al. 1998; POLLOCK 1999). Bouriquet et al. have gone on to propose that ET-1 may be a mediator in the development of renal vascular lesions associated with chronic L-NAME treatment (BOURIQUET et al. 1996).

G. Role of Renal ET-1 in Hypertension

Over-expression of ET-1 is present in several models of hypertension (SCHIFFRIN 1999). It is not clear in every model whether the involvement of the endothelin system is a cause of the hypertension or a response to other factors that elevate arterial pressure. A role for ET-1 in regulating arterial pressure through fluid-volume control is supported by the observations that ET-1 has been consistently shown to contribute to the hypertension associated with salt-dependent models such as the DOCA-salt or Dahl salt-sensitive rat (ALLCOCK et al. 1998; BARTON et al. 1998; KASSAB et al. 1997, 1998; LI et al. 1994; SCHIFFRIN et al. 1995, 1997b). These forms of hypertension are uniquely sensitive to blood pressure-lowering effects of ET_A receptor blockade and relatively insensitive to inhibitors of the renin-angiotensin system. Chronic treatment with deoxycorticosterone acetate (DOCA) and high salt in rats produces a severe, malignant form of hypertension. Blockade of ET_A receptors significantly attenuates the elevation of blood pressure in the DOCA-salt model of hypertension (ALLCOCK et al. 1998; LI et al. 1994; SCHIFFRIN et al. 1995). However, ET_A antagonism does not improve GFR, as measured by creatinine clearance, in the DOCA-salt rat suggesting that the renal vasoconstriction associated with this model is not due to ET_A-induced constriction. More direct evidence that endothelin "activity" is increased comes from Northern blot analysis and in situ hybridization studies demonstrating that prepro-ET-1 mRNA expression is increased in blood vessels of DOCA-salt hypertensive rats (DENG et al. 1996; SCHIFFRIN et al. 1997a). Furthermore, plasma levels of ET-1 have been reported to be elevated in DOCA-salt hypertensive rats

(ALLCOCK et al. 1998). Acute administration of an ET_B receptor antagonist reportedly has little effect or even lowers arterial pressure in normal rats (CLOZEL and BREU 1996). In contrast, chronic ET_B receptor blockade produces a further significant increase in arterial pressure in DOCA-salt hypertensive rats (POLLOCK et al. 2000). In normotensive rats, long-term treatment with an ET_B receptor antagonist raises arterial pressure in a manner that is directly dependent upon salt intake (POLLOCK 2000). These observations are consistent with the hypothesis that the ET_B receptor serves to maintain a lower arterial pressure by enhancing sodium excretion.

The mechanism for elevating ET-1 expression in the DOCA-salt rat has not been completely identified. However, Intengan et al. recently provided evidence that vasopressin may be an important stimulus in this model (INTENGAN et al. 1998, 1999). These authors demonstrated that pharmacological blockade of vasopressin V_1 receptors abolished the increase in vascular ET-1 gene expression and attenuated the development of hypertension in the DOCA-salt model (INTENGAN et al. 1998). Similar findings were also observed in rats genetically deficient in vasopressin (INTENGAN et al. 1999). Homozygous vasopressin-deficient Brattleboro rats had a significantly attenuated hypertensive response to DOCA-salt treatment compared to Long-Evans control rats. DOCA-salt treatment increased ET-1 mRNA expression in both strains of rat although Brattleboro rats maintained significantly lower ET-1 expression whether treated with DOCA-salt or not. These studies demonstrate an important role for vasopressin in the DOCA-salt model which are mediated, at least in part, by stimulating ET-1 expression.

In genetically salt-sensitive rats, ET_A receptor blockade has been shown to attenuate the hypertensive response to salt-loading (BARTON et al. 1998; KASSAB et al. 1997, 1998; D'USCIO et al. 1997). Infusion of an ET_A receptor antagonist, either systemically or directly into the renal interstitium of the Dahl salt-sensitive rat, significantly increased GFR and renal blood flow. This finding in the Dahl rat is different from the lack of effect on GFR in the DOCA-salt rat suggesting unique aspects of endothelin involvement in these models. It is important to note that both the Dahl salt-sensitive and DOCA-salt models of hypertension are not renin-dependent and that models which respond well to angiotensin antagonism, such as SHR and 2-kidney, 1-clip Goldblatt hypertensive rats do not exhibit generalized over-expression of ET-1. A somewhat paradoxical finding is that hypertension and renal vasoconstriction associated with chronic angiotensin II infusion can be significantly attenuated by endothelin antagonists (HERIZI et al. 1998; RAJAGOPALAN et al. 1997). Therefore, additional studies are needed to investigate the relationship between ET-1 and the renin-angiotensin system in models of hypertension.

The participation of ET-1 in the various models of hypertension is complicated by the apparent contrasting effects of reducing GFR and inhibiting sodium excretion. The functional role of ET-1 will be sorted out when we are better able to distinguish these effects. Presuming that the endothelin system functions in an autocrine or paracrine manner, we can hypothesize that ET-1 is produced in response to salt loading as a means of eliminating the salt load. ET-1 then enhances sodium excretion by inhibiting reabsorption in the collecting duct and possibly other tubular segments along with medullary vasodilation. Under extreme conditions such as during DOCA-salt treatment, ET-1 production results in systemic increases in ET-1 expression to produce elevations in blood pressure. Despite the dependence of arterial pressure on ET-1, ET_A receptor expression is down-regulated both within the kidney and systemically while ET_B expression is elevated (GIULUMIAN et al. 1998; POLLOCK et al. 2000). In genetic salt-sensitivity, it is possible that there is a defect in several mechanisms. Salt loading may result in an over-production of ET-1 which results in hypertension. However, it seems more likely that there is a decreased sensitivity of the mechanisms that respond to ET-1 such as a deficiency of ET_B receptors. Results demonstrating salt-sensitivity in the ET_B deficient rat and during chronic pharmacological blockade of ET_B receptors support this possibility.

H. Role of ET-1 in Acute Renal Failure

In addition to a putative physiological role in regulating renal excretory function, there is considerable evidence that ET-1 contributes to the pathophysiology associated with several forms of renal failure. The following discussion focuses on some areas in which pharmacological intervention appears to be most promising. It is suggested that the reader consult other recent reviews for more specific information concerning this subject (IVIC and STEFANOVIC 1998; RABELINK et al. 1996, 1998; HUNLEY and KON 1997; CLAVELL and BURNETT 1994). As with other therapeutic targets, there remains considerable controversy as to which endothelin receptor antagonist, if any, might be considered for treating kidney disease. It is not clear whether selective ET_A antagonists or non-selective ET_A/ET_B antagonists (see Chap. 9) would produce the most benefit because the functional significance of ET_A and ET_B receptors in the human kidney are poorly defined.

In most studies examining the role of ET-1 in disease, either plasma concentration or urinary excretion of immunoreactive ET-1 has been measured. Such measurements are limited in their utility because of the unique, complex, and paracrine nature of the endothelin system. Urinary endothelin is derived primarily from renal tubular production since the brush border of proximal tubules contain neutral endopeptidases which degrade most of the filtered ET-1. Also, the renal tubules, and in particular, collecting ducts have the capacity to synthesized large amounts of ET-1. The utility of plasma measurements has not been well established since circulating immunoreactive ET-1 levels reflect the end result of synthesis, metabolism, and clearance by ET_A and ET_B receptors. The release of ET-1 is polarized in endothelial cells with most of the ET-1 being released towards the vascular smooth muscle. Since receptor binding is characterized as being irreversible and ET-1 is synthesized in close proximity to its receptors, changes in plasma levels probably do not always reflect changes in production. A number of studies have used receptor block-ade in animal models of renal disease as a means of defining a role for endothelin. However, not all studies have been consistent in their findings.

Due to the potency and long-lasting nature of endothelin-induced renal vasoconstriction, it was speculated that ET-1 may participate in the patho-genesis of acute renal failure (ARF). ARF is defined as a sudden decline in glomerular filtration rate (GFR) which is caused by preglomerular vasocon-striction. The factors responsible for the vasoconstriction have yet to be clearly identified but can be generally classified as being produced by either pro-longed periods of ischemia or exposure to nephrotoxic agents. Hypoxia can be a potent stimulus for ET-1 production in various organs including the kidney. Ischemia-induced ARF is produced in animal models by renal artery occlusion and has been reported to increase plasma ET-1 concentrations, renal tissue ET-1 content, and endothelin receptor binding (SHIBOUTA et al. 1990; NAMBI et al. 1993). In addition, mRNA expression of prepro-ET-1, ET_A, and ET_B receptors are elevated following renal ischemia (WILHELM et al. 1999; RUSCHITZKA et al. 1999). Studies on both rats and dogs have indicated that endothelin receptor antagonists can attenuate the decrease in GFR produced by renal artery occlusion or aortic cross-clamping (MINO et al. 1992; GELLAI et al. 1995; STINGO et al. 1993; KRAUSE et al. 1997). However, while the effect of endothelin receptor blockers may be statistically significant, they are relatively weak in their ability to return GFR and renal blood flow to normal indicating that other factors contribute to ischemia-induced renal vasoconstriction.

There are many studies indicating that ET-1 plays an important role in toxin-related nephropathies including that produced by radiocontrast dye, cyclosporine, and endotoxin. As with ischemic injury, investigators have not come to a clear consensus on what particular role, if any, ET-1 actually plays in producing ARF as opposed to ET-1 production being stimulated as a result of ARF. It is also not clear what mechanisms these toxic agents might have in common although there is evidence that they all may directly stimulate ET-1 release from vascular endothelial cells. It should also be considered that, in addition to the vascular endothelium, the renal source of ET-1 could be tubular epithelium since these cells have the capacity to make large amounts of ET-1.

Several lines of evidence suggest that endothelin may be an important mediator of ARF produced by infusion of radiocontrast media. Circulating levels of endothelin have been reported in rats and dogs following intravenous infusion of several different radiocontrast agents (HEYMAN et al. 1992, 1993; MARGULIES et al. 1991). This is probably due to a direct stimulation of endothe-lin release from endothelial cells, as has been demonstrated in vitro, and does not appear to be related to the high osmolality of the radiocontrast media (HEYMAN et al. 1993). ARF produced by radiocontrast media occurs primar-ily in patients with compromised vascular function, such as in diabetes and hypertension (RICH and CRECELIUS 1990; WEISBERG et al. 1992). Animal models have typically utilized maneuvers designed to increase the probability of producing acute renal failure such as a low sodium diet and pre-treatment with prostaglandin synthesis inhibitors since radiocontrast media alone rarely have any detrimental effect on renal function in otherwise healthy animals. Margulies et al. have reported that ET_A receptor blockade in the dog prevented the acute renal vasoconstriction produced by radiocontrast agents in the kidney (MARGULIES et al. 1991). In rats, Cantley et al. observed that the ET_A receptor-selective antagonist, CP170687, had no effect on the transient decrease in renal blood flow produced by the contrast agent, iothalamate (CANTLEY et al. 1993). However, the antagonist significantly attenuated the more prolonged decreases in renal blood flow produced by iothalamate in rats on a low sodium diet and pre-treated with indomethacin. Both ET_A-selective and non-selective antagonists have long-term beneficial effects in a different rat model in which endothelial function was first compromised by inhibiting NO and prostaglandin production followed by contrast dye administration (POLLOCK et al. 1997; BIRD et al. 1996). In the dog kidney, Brooks and DePalma also observed that non-selective endothelin receptor antagonist, SB 209670, abolished renal vasoconstriction produced by the contrast dye, Hypaque (BROOKS and DEPALMA 1996). Investigation in humans is clearly warranted given the consistent findings in animal models, although it is not clear what dosing regimen or antagonist selectivity should be chosen.

ET-1 appears to contribute to the renal vasoconstrictor actions of cyclosporine probably due to the ability of cyclosporine, and other immunosuppressants, to directly stimulate ET-1 release from endothelial, mesangial, and epithelial cells. Some investigators have observed that the renal vasoconstrictor effects of cyclosporine can be attenuated with either anti-ET-1 antibodies or receptor antagonists although this has not been uniformly observed (FOGO et al. 1992; HUNLEY et al. 1995; MEYER-LEHNERT et al. 1997). In terms of chronic cyclosporine treatment in salt-depleted rats, Kon et al. have observed that a non-selective ET_A/ET_B receptor antagonist can attenuate the renal vasoconstriction but does not prevent the development of interstitial fibrosis (KON et al. 1995). These authors provided evidence that angiotensin II was a contributor to the development of fibrosis in this model.

Bacterial endotoxin produces profound renal vasoconstriction and increases the concentration of ET-1 in renal tissue (SUGIURA et al. 1989). Similar to other toxins, lipopolysaccharide directly stimulates ET-1 release from vascular endothelial cells (KADDOURA et al. 1996; NAMBI et al. 1994a). As with other forms of ARF, blockade of ET-1 action either with antibodies or receptor antagonists attenuate the renal vasoconstriction associated with endotoxin (MORISE et al. 1994; ÖZSAN et al. 1994). However, blockade of ET-1 could potentially exacerbate the hypotension associated with endotoxin suggesting that ET-1 may actually play a protective role in septic shock. A recent study by Mikata and co-workers demonstrated in dogs that a non-selective antagonist, TAK-044, had no effect on lipopolysaccharide-induced hypotension but prevented metabolic acidosis and improved associated decreases in urine volume, creatinine clearance and renal blood flow (MIKATA et al. 1999).

I. Role of ET-1 in Chronic Renal Failure

In addition to being a vasoconstrictor, it is well-known that ET-1 can also function as a mitogen and growth promoter. This has led to speculation that, in addition to producing long-lasting decreases in renal perfusion and GFR, ET-1 may also play a role in chronic renal failure associated with glomerular injury, progressive increases in proteinuria, and interstitial inflammation and fibrosis that occurs in chronic renal disease (REMUZZI and BENIGNI 1997; BENIGNI and REMUZZI 1995; BROOKS 1996; RABELINK et al. 1998).

In experimental studies, many investigators have used surgical reduction of roughly five-sixths of the total renal mass in rats as a model for chronic renal failure. The model is characterized by an increase in the activity of the renin-angiotensin system and high blood pressure which contributes to progressive glomerular injury as a consequence of hyperperfusion and hyperfiltration. Functional indices of the severity of renal failure include development of proteinuria, a substantial decrease in urinary concentrating ability, and inability to excrete adequately fluid and electrolytes. Angiotensin II appears to play a role in the systemic hypertension, renal vasoconstrictor mechanisms, and possibly malfunction of mesangial cells and the filtration barrier, but cannot account for all of the vasoconstrictor and growth-related changes observed in progressive renal disease.

Benigni and colleagues reported that the rate of urinary endothelin excretion was increased in rats with reduced renal mass although tissue concentrations of the peptide were similar between sham and renal ablated rats (BENIGNI et al. 1991). In a study from the same laboratory, ORISIO et al. (1993) reported that expression of endothelin mRNA was increased in the remnant kidney. To provide further evidence for endothelin involvement in the pathophysiology of this model, Benigni et al. reported that treatment with the peptidic ET_A receptor antagonist, FR139317, normalizes systemic blood pressure and attenuates the development of proteinuria (BENIGNI et al. 1993). Also, Brooks et al. reported that urinary excretion of ET-1 is increased in rats with renal mass reduction when calculated as a fraction of creatinine clearance (BROOKS et al. 1991). Several other studies using animal models have suggested a therapeutic utility for endothelin antagonists in treating the progression of chronic renal failure suggesting increased renal ET-1 synthesis (BROCHU et al. 1999; WOLF et al. 1999; LARIVIERE et al. 1997; POTTER et al. 1997). Although there is a fairly large amount of additional evidence for endothelin involvement in other forms of renal failure, there have been a number of conflicting reports related to the use of endothelin antagonists as therapeutic agents. Torralbo et al. reported that urinary excretion of endothelin is actually decreased in conjunction with the decrease in creatinine clearance and the severity of tubu-lointerstitial damage (TORRALBO et al. 1995). Pollock and Polakowski observed that a non-peptide, orally active antagonist had no effect on the development of hypertension or proteinuria in rats with reduced renal mass (POLLOCK and POLAKOWSKI 1997). It is not clear why all the animal data are not consistent, but there are differences between laboratories in the method of renal abla-tion and the genetic background of the rats used in these studies.

Studies in patients with chronic renal failure have suggested that end-othelin production is elevated (PECO-ANTIC et al. 1996; EBIHARA et al. 1997; VLACHOJANNIS et al. 1997) although, similar to animal studies, these findings are not always observed (NAKAMURA et al. 1995a; SHICHIRI et al. 1990). Ohta et al. also reported that the excretion rate of immunoreactive ET-1 was ele-vated in patients with renal disease (OHTA et al. 1991). Furthermore, these investigators determined that the renal clearance of ET-1 was elevated in these patients and that the clearance of ET-1 was greater than the clearance of cre-atinine. These findings indicate that urinary ET-1 is most likely derived from renal tubular sources and not an indication of filtered ET-1. Once again, the therapeutic utility of endothelin receptor antagonists needs to be studied. As with all clinical trials using endothelin blockers, it is not clear whether a non-selective or ET_A-selective antagonist would provide the most benefit. Fur-thermore, the time and duration of intervention is needed will have to be established.

There is also limited evidence that ET-1 production is enhanced in exper-imental and human glomerulonephritis (YOSHIMURA et al. 1995; MURER et al. 1994; NAKAMURA et al. 1995a; ROCCATELLO et al. 1994). ET-1 mediates some of the proliferative effects of several cytokines (BAKRIS and RE 1993; KOHNO et al. 1994; NITTA et al. 1995). Also, endothelin receptor antagonists reduce mesangial cell proliferation in experimental mesangial proliferative glomeru-lonephritis and decrease renal injury in murine lupus nephritis (NAKAMURA et al. 1995b; FUKUDA et al. 1996). It is unknown whether chronic treatment with endothelin antagonists would provide any therapeutic benefit in human glomerulonephritis.

TGFβ1 mediates the overproduction of extracellular matrix proteins in a variety of fibrotic diseases and contributes to glomerular and interstitial fibro-sis that occurs in most forms of chronic renal failure (BORDER and NOBLE 1997; OISHI et al. 1996). Understanding the relationship of ET-1 to growth factors and cytokines, such as TGFβ1, may provide the key to understanding the role of ET-1 and the potential benefit of inhibiting the endothelin system in chronic renal failure. TGFβ1 also has been shown to stimulate ET-1 release from a variety of cell types including vascular endothelium and collecting duct cells (SCHNERMANN et al. 1996; UJIIE et al. 1992; BROWN et al. 1991). ET-1 also has TGFβ1-like properties in terms of promoting cell growth and synthesis of extracellular matrix proteins suggesting that these two mediators may work synergistically (PÖNICKE et al. 1997; KIRCHENGAST and MUNTER 1998; BENIGNI and REMUZZI 1998; HUTCHINSON 1998; GOMEZ-GARRE et al. 1996).

Table 1. Summary of endothelin actions in the kidney

Pre-glomerular vasculature
  Predominantly ET_A-mediated vasoconstriction, ET_B-mediated vasodilation
Post-glomerular vasculature
  ET_A and ET_B-mediated vasoconstriction, ET_B-mediated vasodilation
Vasa recta capillaries
  ET_B-mediated vasodilation
Proximal tubule
  ET_B-mediated urinary acidification (increased Na^+/H^+ antiport activity)
Thick ascending limb
  ET_B-mediated inhibition of Na^+ reabsorption (NO dependent)
Inner medullary collecting duct
  ET_B-mediated inhibition of ADH-induced cAMP production
  ET_B-mediated inhibition of Na^+ reabsorption (NO and prostacyclin dependent)

J. Conclusion

There is now considerable evidence that endothelin modulates a variety of tubular and hemodynamic functions within the kidney (see Table 1). It is also apparent that this role is highly localized within very specific regions of both the renal cortex and medulla. Endothelin modulates both pre- and post-glomerular resistance as a means of influencing renal blood flow and GFR and plays a role in the delicate balance between vascular regulatory agents through both ET_A-receptor mediated vasoconstriction and ET_B-mediated vasodilation. However, the most important physiological role for endothelin in the kidney is probably a tubular action to promote sodium and water excretion. This pathway appears to be extremely important in salt-dependent forms of hypertension. In renal failure, endothelin not only reduces renal perfusion and filtration but appears to promote in vascular growth, proliferation and fibrosis. There is a clear lack of carefully designed clinical trials examining the therapeutic utility of endothelin receptor antagonists for the treatment of human renal disease.

References

Abassi Z, Brodsky S, Rubinstein I, Ramadan R, Winaver J, Hoffman A (1998) Endothelin B receptor (ETB) mediates the diuretic and natriuretic effects of big endothelin (BET) in the rat (Abstract). J Am Soc Nephrol 9:396A

Allcock GH, Venema RC, Pollock DM (1998) ET_A receptor blockade attenuates the hypertension but not renal dysfunction in DOCA-salt rats. Am J Physiol 275:R245–R252

Bakris GL, Re RN (1993) Endothelin modulates angiotensin II-induced mitogenesis of human mesangial cells. Am J Physiol 264:F937–F942

Banks RO (1990) Effects of endothelin on renal function in dogs and rats. Am J Physiol 258:F775–F780

Banks RO, Pollock DM, Novak J (1997) The renal and systemic hemodynamic actions of endothelin. In: Highsmith RF (ed) Endothelin: molecular biology, physiology, and pathology. Humana Press, Totowa, NJ, p 167

Barton M, d'Uscio LV, Shaw S, Meyer P, Moreau P, Luscher TF (1998) ET(A) receptor blockade prevents increased tissue endothelin-1, vascular hypertrophy, and endothelial dysfunction in salt-sensitive hypertension. Hypertension 31:499–504

Baylis C, Harvey J, Engels K (1994) Acute nitric oxide blockade amplifies the renal vasoconstrictor actions of angiotensin II. J Am Soc Nephrol 5:211–214

Baylis C, Schmidt R (1996) The aging glomerulus. Sem Nephrol 16:265–276

Benigni A, Perico N, Gaspari F, Zoja C, Bellizzi L, Babanelli M, Remuzzi G (1991) Increased renal endothelin production in rats with reduced renal mass. Am J Physiol 260:F331–F339

Benigni A, Zoja C, Corna D, Orisio S, Longaretti L, Bertani T, Remuzzi G (1993) A specific endothelin subtype A receptor antagonist protects against injury in renal disease progression. Kidney Int 44:440–444

Benigni A, Remuzzi G (1995) Endothelin in the progressive renal disease of glomerulopathies. Min Elect Metab 21:283–291

Benigni A, Remuzzi G (1998) Mechanism of progression of renal disease: growth factors and related mechanisms. [Review] [50 refs]. J Hypertension 16:S9–12

Bird JE, Giancarli MR (1996) Cardiovascular and renal effects of endothelin B receptor selective agonists in spontaneously hypertensive rats. J Cardiovasc Pharmacol 28:381–384

Bird JE, Giancarli MR, Megill JR, Durham SK (1996) Effects of endothelin in radiocontrast-induced nephropathy in rats are mediated through endothelin-A receptors. J Am Soc Nephrol 7:1153–1157

Border WA, Noble NA (1997) TGF-beta in kidney fibrosis: a target for gene therapy. Kidney Int 51:1388–1396

Boulanger C, Lüscher TF (1990) Release of endothelin from the porcine aorta: inhibition by endothelium-derived nitric oxide. J Clin Invest 85:587–590

Bouriquet N, Dupont M, Herizi A, Mimran A, Casellas D (1996) Preglomerular sudanophilia in L-NAME hypertensive rats: involvement of endothelin. Hypertension 27:382–391

Brochu E, Lacasse S, Moreau C, Lebel M, Kingma I, Grose JH, Lariviere R (1999) Endothelin ET(A) receptor blockade prevents the progression of renal failure and hypertension in uraemic rats. Nephrol Dialysis Transplant 14:1881–1888

Brooks DP, Contino LC, Storer B, Ohlstein EH (1991) Increased endothelin excretion in rats with renal failure induced by partial nephrectomy. Br J Pharmacol 104:987–989

Brooks DP, DePalma PD, Pullen M, Gellai M, Nambi P (1995) Identification and function of putative ET_B receptor subtypes in the dog kidney. J Cardiovasc Pharmacol 26 [Suppl 3]:S322–S325

Brooks DP (1996) Role of endothelin in renal function and dysfunction. Clin Exptl Pharmacol Physiol 23:345–348

Brooks DP, DePalma PD (1996) Blockade of radiocontrast-induced nephrotoxicity by the endothelin receptor antagonist, SB 209670. Nephron 72:629–636

Brown MR, Vaughan J, Jimenez LL, Vale W, Baird A (1991) Transforming growth factor-beta: role in mediating serum-induced endothelin production by vascular endothelial cells. Endocrinol 129:2355–2360

Cantley LG, Spokes K, Clark B, McMahon EG, Carter J, Epstein FH (1993) Role of endothelin and prostaglandins in radiocontrast-induced renal artery constriction. Kidney Int 44:1217–1223

Cavarape A, Bartoli E (1998) Effects of BQ-123 on systemic and renal hemodynamic responses to endothelin-1 in the rat split hydronephrotic kidney. J Hypertension 16:1449–1458

Cernacek P, Strmen J, Levy M (1998) Acute renal effects of endothelin-A blockade: interspecies differences. J Cardiovasc Pharmacol 31 [Suppl 1]:S269–S272

Chen M, Todd-Turla K, Wang WH, Cao X, Smart A, Brosius FC, Killen PD, Keiser JA, Briggs JP, Schnermann J (1993) Endothelin-1 mRNA in glomerular and epithelial cells of kidney. Am J Physiol 265:F542–F550

Cheng HF, Su YM, Yeh JR, Chang KJ (1993) Alternative transcript of the non-selective-type endothelin receptor from rat brain. Mol Pharmacol 44:533–538

Chou S-Y, Dahhan A, Porush JG (1990) Renal actions of endothelin: interaction with prostacyclin. Am J Physiol 259:F645–F652

Chu TS, Peng Y, Cano A, Yanagisawa M, Alpern RJ (1996a) Endothelin$_B$ receptor activates NHE-3 by a Ca^{2+}-dependent pathway in OKP cells. J Clin Invest 97:1454–1462

Chu TS, Tsuganezawa H, Peng Y, Cano A, Yanagisawa M, Alpern RJ (1996b) Role of tyrosine kinase pathways in ET$_B$ receptor activation of NHE3. Am J Physiol 271:C763–C771

Clavell AL, Burnett JC,Jr. (1994) Physiologic and pathophysiologic roles of endothelin in the kidney. Curr Op Nephrol Hyperten 3:66–72

Clozel M, Breu V (1996) The role of ET$_B$ receptors in normotensive and hypertensive rats as revealed by the non-peptide selective ET$_B$ receptor antagonist Ro 46–8443. FEBS Lett 383:42–45

Cowley AW Jr, Roman RJ (1996) The role of the kidney in hypertension. J Am Med Assoc 275:1581–1589

d'Uscio LV, Barton M, Shaw S, Moreau P, Lüscher TF (1997) Structure and function of small arteries in salt-induced hypertension – Effects of chronic endothelin-subtype-A-receptor blockade. Hypertension 30:905–911

Dean R, Zhuo J, Alcorn D, Casley D, Mendelsohn FAO (1996) Cellular localization of endothelin receptor subtypes in the rat kidney following in vitro labelling. Clin Exptl Pharmacol Physiol 23:524–531

Deng LY, Day R, Schiffrin EL (1996) Localization of sites of enhanced expression of endothelin-1 in the kidney of DOCA-salt hypertensive rats. J Am Soc Nephrol 7:1158–1164

Disashi T, Nonoguchi H, Iwaoka T, Naomi S, Nakayama Y, Shimada K, Tanzawa K, Tomita K (1997) Endothelin converting enzyme-1 gene expression in the kidney of spontaneously hypertensive rats. Hypertension 30:1591–1597

Ebihara I, Nakamura T, Takahashi T, Tomino Y, Shimada N, Koide H (1997) Increased endothelin-1 mRNA expression in peripheral blood monocytes of dialysis patients. Peritoneal Dial Int 17:595–601

Edwards RM, Trizna W, Ohlstein EH (1990) Renal microvascular effects of endothelin. Am J Physiol 259:F217–F221

Edwards RM, Pullen M, Nambi P (1992) Activation of endothelin ET$_B$ receptors increases glomerular cGMP via an L-arginine-dependent pathway. Am J Physiol 263:F1020–F1025

Edwards RM, Stack EJ, Pullen M, Nambi P (1993) Endothelin inhibits vasopressin action in rat inner medullary collecting duct via the ET$_B$ receptor. J Pharmacol Exptl Therap 267:1028–1033

Eiam-Ong S, Hilden SA, King AJ, Johns CA, Madias NE (1992) Endothelin-1 stimulates the Na^+/H^+ and Na^+/HCO_3^{-} transporters in rabbit renal cortex. Kidney Int 42:18–24

Endlich K, Hoffend J, Steinhausen M (1996) Localization of endothelin ET$_A$ and ET$_B$ receptor-mediated constriction in the renal microcirculation of rats. J Physiol (London) 497:211–218

Firth JD, Ratcliffe PJ (1992) Organ distribution of the three rat endothelin messenger RNAs and the effects of ischemia on renal gene expression. J Clin Invest 90:1023–1031

Fogo A, Hellings SE, Inagami T, Kon V (1992) Endothelin receptor antagonism is protective in in vivo acute cyclosporine toxicity. Kidney Int 42:770–774

Fretschner M, Endlich K, Gulbins E, Lang RE, Schlottmann K, Steinhausen M (1991) Effects of endothelin on the renal microcirculation of the split hydronephrotic kidney. Renal Physiol Biochem 14:112–127

Fujita K, Matsumura Y, Miyazaki Y, Takaoka M, Morimoto S (1995) Role of endothelin-1 in hypertension induced by long-term inhibition of nitric oxide synthase. Eur J Pharmacol 280:311–316

Fukuda K, Yanagida T, Okuda S, Tamaki K, Ando T, Fujishima M (1996) Role of endothelin as a mitogen in experimental glomerulonephritis in rats. Kidney Int 49:1320–1329

Fukuroda T, Fukikawa T, Ozaki S, Ishikawa K, Yano M, Nishikibe M (1994a) Clearance of circulating endothelin-1 by ET_B receptors in rats. Biochem Biophys Res Commun 199:1461–1465

Fukuroda T, Kobayashi M, Ozaki S, Yano M, Miyauchi T, Onizuka M, Sugishita Y, Goto K, Nishikibe M (1994b) Endothelin receptor subtypes in human versus rabbit pulmonary arteries. J Appl Physiol 76:1976–1982

Garcia NH, Garvin JL (1994) Endothelin's biphasic effect on fluid absorption in the proximal straight tubule and its inhibitory cascade. J Clin Invest 93:2572–2577

Gariepy CE, Cass DT, Yanagisawa M (1996) Null mutation of endothelin receptor type B gene in spotting lethal rats causes aganglionic megacolon and white coat color. Proc Natl Acad Sci USA 93:867–872

Gariepy CE, Williams SC, Richardson JA, Hammer RE, Yanagisawa M (1998) Transgenic expression of the endothelin-B receptor prevents congenital intestinal aganglionosis in a rat model of Hirschsprung disease. J Clin Invest 102:1092–1101

Garvin JL, Sanders K (1991) Endothelin inhibits fluid and bicarbonate transport in part by reducing the Na^+/K^+-ATPase activity in the rat proximal straight tubule. J Am Soc Nephrol 2:976–982

Gellai M, Jugus M, Fletcher T, Nambi P, Ohlstein EH, Elliott JD, Brooks DP (1995) Nonpeptide endothelin receptor antagonists. V: Prevention and reversal of acute renal failure in the rat by SB 209670. J Pharmacol Exp Ther 275:200–206

Giulumian AD, Pollock DM, Clarke N, Fuchs LC (1998) Coronary vascular reactivity is improved by endothelin A receptor blockade in DOCA-salt hypertensive rats. Am J Physiol 274:R1613–R1618

Gomez-Garre D, Ruiz-Ortega M, Ortego M, Largo R, Lopez-Armada MJ, Plaza JJ, Gonzalez E, Egido J (1996) Effects and interactions of endothelin-1 and angiotensin II on matrix protein expression and synthesis and mesangial cell growth. Hypertension 27:885–892

Guntupalli J, DuBose TD Jr (1994) Effects of endothelin on rat renal proximal tubule Na^+-P_i cotransport and Na^+/H^+ exchange. Am J Physiol 266:F658–F666

Gurbanov K, Rubinstein I, Hoffman A, Abassi Z, Better OS, Winaver J (1996) Differential regulation of renal regional blood flow by endothelin-1. Am J Physiol 271:F1166–F1172

Harris PJ, Zhuo J, Mendelsohn FA, Skinner SL (1991) Haemodynamic and renal tubular effects of low doses of endothelin in anaesthetized rats. J Physiol (London) 433:25–39

Herizi A, Jover B, Bouriquet N, Mimran A (1998) Prevention of the cardiovascular and renal effects of angiotensin II by endothelin blockade. Hypertension 31:10–14

Heyman SN, Clark BA, Kaiser N, Spokes K, Rosen S, Brezis M, Epstein FH (1992) Radiocontrast agents induce endothelin release in vivo and in vitro. J Am Soc Nephrol 3:58–65

Heyman SN, Clark BA, Cantley L, Spokes K, Rosen S, Brezis M, Epstein FH (1993) Effects of ioversol versus iothalamate on endothelin release and radiocontrast nephropathy. Invest Radiol 28:313–318

Hirata Y, Matsuoka H, Kimura K, Sugimoto T, Hayakawa H, Suzuki E (1991) Role of endothelium-derived relaxing factor in endothelin-induced renal vasoconstriction. J Cardiovasc Pharmacol 17 [Suppl 7]:S169–S171

Hirata Y, Emori T (1993) Cellular mechanism of endothelin-induced nitric oxide synthesis by cultured bovine endothelial cells. J Cardiovasc Pharmacol 22 [Suppl 8]:S225–S228

Hunley TE, Fogo A, Iwasaki S, Kon V (1995) Endothelin A receptor mediates functional but not structural damage in chronic cyclosporine nephrotoxicity. J Am Soc Nephrol 5:1718–1723

Hunley TE, Kon V (1997) Endothelin in ischemic acute renal failure. Curr Op Nephrol Hyperten 6:394–400

Hutchinson IV (1998) An endothelin-transforming growth factor beta pathway in the nephrotoxicity of immunosuppressive drugs. Curr Op Nephrol Hyperten 7:665–671

Intengan HD, He G, Schiffrin EL (1998) Effect of vasopressin antagonism on structure and mechanics of small arteries and vascular expression of endothelin-1 in deoxycorticosterone acetate-salt hypertensive rats. Hypertension 32:770–777

Intengan HD, Park JB, Schiffrin EL (1999) Blood pressure and small arteries in DOCA salt-treated genetically AVP-deficient rats. Hypertension 34:907–913

Ivic MA, Stefanovic V (1998) Endothelins in hypertension and kidney diseases. Pathol Biol 46:723–730

Jones CR, Hiley CR, Pelton JT, Miller RC (1989) Autoradiographic localisation of endothelin binding sites in kidney. Eur J Pharmacol 163:379–382

Kaddoura S, Curzen NP, Evans TW, Firth JD, Poole-Wilson PA (1996) Tissue expression of endothelin-1 mRNA in endotoxaemia. Biochem Biophys Res Commun 218:641–647

Karet FE, Kuc RE, Davenport AP (1993) Novel ligands BQ123 and BQ3020 characterize endothelin receptor subtypes ET_A and ET_B in human kidney. Kidney Int 44:36–42

Karet FE (1996) Endothelin peptides and receptors in human kidney. Clin Sci 91:267–273

Kassab S, Novak J, Miller T, Kirchner K, Granger J (1997) Role of endothelin in mediating the attenuated renal hemodynamics in Dahl salt-sensitive hypertension. Hypertension 30:682–686

Kassab S, Miller MT, Novak J, Reckelhoff J, Clower B, Granger JP (1998) Endothelin-A receptor antagonism attenuates the hypertension and renal injury in Dahl salt-sensitive rats. Hypertension 31:397–402

King AJ, Brenner BM, Anderson S (1989) Endothelin: a potent renal and systemic vasoconstrictor peptide. Am J Physiol 256:F1051–F1058

Kirchengast M, Munter K (1998) Endothelin and restenosis. Cardiovasc Res 39:550–555

Kitamura K, Tanaka T, Kato J, Ogawa T, Eto T, Tanaka K (1989) Immunoreactive endothelin in rat kidney inner medulla: marked decrease in spontaneously hypertensive rats. Biochem Biophys Res Commun 162:38–44

Knotek M, Jaksic O, Selmani R, Skoric B, Banfic H (1996) Different endothelin receptor subtypes are involved in phospholipid signalling in the proximal tubule of rat kidney. Pflügers Arch 432:165–173

Kohan DE (1991) Endothelin synthesis by rabbit renal tubule cells. Am J Physiol 261:F221–F226

Kohan DE (1992) Production of endothelin-1 by rat mesangial cells: regulation by tumor necrosis factor. J Lab Clin Med 119:477–484

Kohan DE, Hughes AK, Perkins SL (1992) Characterization of endothelin receptors in the inner medullary collecting duct of the rat. J Biol Chem 267:12336–12340

Kohan DE, Padilla E (1992) Endothelin-1 is an autocrine factor in rat inner medullary collecting ducts. Am J Physiol 263:F607–F612

Kohan DE, Padilla E, Hughes AK (1993) Endothelin B receptor mediates ET-1 effects on cAMP and PGE_2 accumulation in rat IMCD. Am J Physiol 265:F670–F676

Kohno M, Horio T, Yokokawa K, Yasunari K, Kurihara N, Takeda T (1994) Endothelin modulates the mitogenic effect of PDGF on glomerular mesangial cells. Am J Physiol 266:F894–F900

Kohzuki M, Johnston CI, Chai SY, Casley DJ, Mendelsohn FAO (1989) Localization of endothelin receptors in the rat kidney. Eur J Pharmacol 160:193–194

Kon V, Hunley TE, Fogo A (1995) Combined antagonism of endothelin A/B receptors links endothelin to vasoconstriction whereas angiotensin II effects fibrosis. Studies in chronic cyclosporine nephrotoxicity in rats. Transplantation 60:89–95

Krause SM, Walsh TF, Greenlee WJ, Ranaei R, Williams DL,Jr., Kivlighn SD (1997) Renal protection by a dual ETA/ETB endothelin antagonist, L-754,142, after aortic cross-clamping in the dog. J Am Soc Nephrol 8:1061–1071

Lanese DM, Yuan BH, McMurtry IF, Conger JD (1992) Comparative sensitivities of isolated rat renal arterioles to endothelin. Am J Physiol 263:F894–F899

Lariviere R, D'Amours M, Lebel M, Kingma I, Grose JH, Caron L (1997) Increased immunoreactive endothelin-1 levels in blood vessels and glomeruli of rats with reduced renal mass. Kidney Blood Press Res 20:372–380

Lerman A, Sandok EK, Hildebrand FL Jr, Burnett JC,Jr. (1992) Inhibition of endothelium-derived relaxing factor enhances endothelin-mediated vasoconstriction. Circulation 85:1894–1898

Li JS, Lariviere R, Schiffrin EL (1994) Effect of a nonselective endothelin antagonist on vascular remodeling in deoxycorticosterone acetate-salt hypertensive rats. Evidence for a role of endothelin in vascular hypertrophy. Hypertension 24:183–188

Loutzenhiser R, Epstein M, Hayashi K, Horton C (1990) Direct visualization of effects of endothelin on the renal microvasculature. Am J Physiol 258:F61–F68

Löffler B-M, Breu V, Clozel M (1993) Effect of different endothelin receptor antagonists and of the novel non-peptide antagonist Ro 46–2005 on endothelin levels in rat plasma. FEBS Lett 333:108–110

Maguire JJ, Kuc RE, O'Reilly G, Davenport AP (1994) Vasoconstrictor endothelin receptors characterized in human renal artery and vein in vitro. Br J Pharmacol 113:49–54

Margulies KB, Hildebrand FL, Heublein DM, Burnett JC, Jr. (1991) Radiocontrast increases plasma and urinary endothelin. J Am Soc Nephrol 2:1041–1045

Marsden PA, Dorfman DM, Collins T, Brenner BM, Orkin SH, Ballermann BJ (1991) Regulated expression of endothelin 1 in glomerular capillary endothelial cells. Am J Physiol 261:F117–F125

Matsuura T, Miura K, Ebara T, Yukimura T, Yamanaka S, Kim S, Iwao H (1997) Renal vascular effects of the selective endothelin receptor antagonists in anaesthetized rats. Br J Pharmacol 122:81–86

Meyer-Lehnert H, Bokemeyer D, Friedrichs U, Bäcker A, Kramer HJ (1997) Cellular mechanisms of cyclosporine A-associated side-effects: Role of endothelin. Kidney Int 52 [Suppl 61]:S27–S31

Migas I, Bäcker A, Meyer-Lehnert H, Michel H, Wulfhekel U, Kramer HJ (1993) Characteristics of endothelin receptors and intracellular signalling in porcine inner medullary collecting duct cells. Am J Hypertens 6:611–618

Mikata C, Hirata Y, Yokoyama K, Nagura T, Tsunoda Y, Amaha K (1999) Improvement of renal dysfunction in dogs with endotoxemia by a nonselective endothelin receptor antagonist. Crit Care Med 27:146–153

Miller WL, Redfield MM, Burnett JC, Jr. (1989) Integrated cardiac, renal, and endocrine actions of endothelin. J Clin Invest 83:317–320

Mino N, Kobayashi M, Nakajima A, Amano H, Shimamoto K, Ishikawa K, Watanabe K, Nishikibe M, Yano M, Ikemoto F (1992) Protective effect of a selective endothelin receptor antagonist, BQ-123, in ischemic acute renal failure in rats. Eur J Pharmacol 221:77–83

Morise Z, Ueda M, Aiura K, Endo M, Kitajima M (1994) Pathophysiologic role of endothelin-1 in renal function in rats with endotoxin shock. Surgery 115:199–204

Morita S, Kitamura K, Yamamoto Y, Eto T, Osada Y, Sumiyoshi A, Koono M, Tanaka K (1991) Immunoreactive endothelin in human kidney. Annal Clin Biochem 28:267–271

Murer L, Zacchello G, Basso G, Scarpa A, Montini G, Chiozza ML, Zacchello F (1994) Immunohistochemical distribution of endothelin in biopsies of pediatric nephrotic syndrome. Am J Nephrol 14:157–161

Nakamura T, Ebihara I, Fukui M, Osada S, Tomino Y, Masaki T, Goto K, Furuichi Y, Koide H (1995a) Modulation of glomerular endothelin and endothelin receptor gene expression in aminonucleoside-induced nephrosis. J Am Soc Nephrol 5:1585–1590

Nakamura T, Ebihara I, Tomino Y, Koide H (1995b) Effect of a specific endothelin A receptor antagonist on murine lupus nephritis. Kidney Int 47:481–489

Nambi P, Pullen M, Jugus M, Gellai M (1993) Rat kidney endothelin receptors in ischemia-induced acute renal failure. J Pharmacol Exp Ther 264:345–348

Nambi P, Pullen M, Slivjak MJ, Ohlstein EH, Storer B, Smith EF (1994a) Endotoxin-mediated changes in plasma endothelin concentrations, renal endothelin receptor and renal function. Pharmacology 48:147–156

Nambi P, Pullen M, Spielman W (1994b) Species differences in the binding characteristics of [^{125}I]IRL-1620, a potent agonist specific for endothelin-B receptors. J Pharmacol Exp Ther 268:202–207

Neuser D, Zaiss S, Stasch JP (1990) Endothelin receptors in cultured renal epithelial cells. Eur J Pharmacol 176:241–243

Nitta K, Uchida K, Kimata N, Kawashima A, Yumura W, Nihei H (1995) Endothelin-1 mediates erythropoietin-stimulated glomerular endothelial cell-dependent proliferation of mesangial cells. Eur J Pharmacol 293:491–494

Nunez DJR, Taylor EA, Oh VMS, Schofield JP, Brown MJ (1991) Endothelin-1 mRNA expression in the rat kidney. Biochem J 275:817–819

Ohta K, Hirata Y, Shichiri M, Kanno K, Emori T, Tomita K, Marumo F (1991) Urinary excretion of endothelin-1 in normal subjects and patients with renal disease. Kidney Int 39:307–311

Oishi R, Nonoguchi H, Tomita K, Marumo F (1991) Endothelin-1 inhibits AVP-stimulated osmotic water permeability in rat inner medullary collecting duct. Am J Physiol 261:F951–F956

Oishi T, Ogura T, Yamauchi T, Harada K, Ota Z (1996) Effect of renin-angiotensin inhibition on glomerular injuries in DOCA-salt hypertensive rats. Regul Pept 62:89–95

Ono N, Matsui T, Yoshida M, Suzuki-Kusaba M, Hisa H, Satoh S (1998) Renal effects of endothelin in anesthetized rabbits. Eur J Pharmacol 359:177–184

Orisio S, Benigni A, Bruzzi I, Corna D, Perico N, Zoja C, Benatti L, Remuzzi G (1993) Renal endothelin gene expression is increased in remnant kidney and correlates with disease progression. Kidney Int 43:354–358

Özsan K, Türker RK, Ercan ZS (1994) Possible involvement of endothelin peptides and L-arginine-nitric oxide pathway on the effect of endotoxin in the rabbit isolated perfused kidney. Mediators Inflamm 3:211–214

Peco-Antic A, Nastic-Miric D, Popovic-Rolovic M, Adanja G, Marsenic O, Kostic M, Parezanovic V (1996) Acute changes of endothelin 1 in children on hemodialysis. Nephron 73:482–483

Peng Y, Moe OW, Chu TS, Preisig P, Yanagisawa M, Alpern RJ (1999) ET$_B$ receptor activation leads to activation and phosphorylation of NHE3. Am J Physiol 276:C938–C945

Pollock DM, Opgenorth TJ (1993) Evidence for endothelin-induced renal vasoconstriction independent of ET$_A$ receptor activation. Am J Physiol 264:R222–R226

Pollock DM, Opgenorth TJ (1994) ET$_A$ receptor-mediated responses to endothelin-1 and big endothelin-1 in the rat kidney. Br J Pharmacol 111:729–732

Pollock DM, Polakowski JS (1997) ET$_A$ receptor blockade prevents hypertension associated with exogenous endothelin-1 but not renal mass reduction in the rat. J Am Soc Nephrol 8:1054–1060

Pollock DM, Polakowski JS, Wegner CD, Opgenorth TJ (1997) Beneficial effect of ET$_A$ receptor blockade in a rat model of radiocontrast-induced nephropathy. Renal Fail 19:753–761

Pollock DM, Polakowski JS, Opgenorth TJ, Pollock JS (1998) Role of endothelin ET$_A$ receptors in the hypertension produced by 4-day L-nitroarginine methyl ester and cyclosporine treatment. Eur J Pharmacol 346:43–50

Pollock DM (1999) Chronic studies on the interaction between nitric oxide and endothelin in cardiovascular and renal function. Clin Exp Pharmacol Physiol 26:258–261

Pollock DM, Pollock JS (1999) Role for ET_B receptors in the regulation of arterial pressure (Abstract). FASEB J 13:A782

Pollock DM (2000) High salt intake increases the hypertension produced by chronic ET_B receptor blockade in Sprague-Dawley rats (Abstract). FASEB J 14:A132

Pollock DM, Allcock GH, Pollock JS (2000) Up-regulation of endothelin B receptors in kidneys of DOCA-salt hypertensive rats. Am J Physiol 278:F279–F286

Pollock JS, Allcock GH, Thakkar J, Xin J, Pollock DM (1998) Evidence for endothelin-1-mediated release of nitric oxide from inner medullary collecting duct cells (Abstract). FASEB J 12:A53

Potter GS, Johnson RJ, Fink GD (1997) Role of endothelin in hypertension of experimental chronic renal failure. Hypertension 30:1578–1584

Pupilli C, Brunori M, Misciglia N, Selli C, Ianni L, Yanagisawa M, Mannelli M, Serio M (1994) Presence and distribution of endothelin-1 gene expression in human kidney. Am J Physiol 267:F679–F687

Pupilli C, Romagnani P, Lasagni L, Bellini F, Misciglia N, Emoto N, Yanagisawa M, Rizzo M, Mannelli M, Serio M (1997) Localization of endothelin-converting enzyme-1 in human kidney. Am J Physiol 273:F749–F756

Pönicke K, Heinroth-Hoffmann I, Becker K, Brodde OE (1997) Trophic effect of angiotensin II in neonatal rat cardiomyocytes: Role of endothelin-1 and non-myocyte cells. Br J Pharmacol 121:118–124

Qiu C, Engels K, Baylis C (1995a) Endothelin modulates the pressor actions of acute systemic nitric oxide blockade. J Am Soc Nephrol 6:1476–1481

Qiu C, Samsell L, Baylis C (1995b) Actions of endogenous endothelin on glomerular hemodynamics in the rat. Am J Physiol 269:R469–R473

Qiu C, Baylis C (1999) Endothelin and angiotensin mediate most glomerular responses to nitric oxide inhibition. Kidney Int 55:2390–2396

Rabelink TJ, Kaasjager KAH, Stroes ESG, Koomans HA (1996) Endothelin in renal pathophysiology: From experimental to therapeutic application. Kidney Int 50:1827–1833

Rabelink TJ, Stroes ES, Bouter KP, Morrison P (1998) Endothelin blockers and renal protection: a new strategy to prevent end-organ damage in cardiovascular disease? Cardiovasc Res 39:543–549

Rajagopalan S, Laursen JB, Borthayre A, Kurz S, Keiser J, Haleen S, Giaid A, Harrison DG (1997) Role for endothelin-1 in angiotensin II-mediated hypertension. Hypertension 30:29–34

Remuzzi G, Benigni A (1997) Progression of proteinuric diabetic and nondiabetic diseases: A possible role for renal endothelin. Kidney Int 51 [Suppl 58]:S66–S68

Rich MW, Crecelius CA (1990) Incidence, risk factors and clinical course of acute renal insufficiency after cardiac catheterization in patients 70 years of age or older. Arch Intern Med 150:1237–1242

Richard V, Hogie M, Clozel M, Löffler B-M, Thuillez C (1995) In vivo evidence of an endothelin-induced vasopressor tone after inhibition of nitric oxide synthesis in rats. Circulation 91:771–775

Roccatello D, Mosso R, Ferro M, Polloni R, De Filippi PG, Quattrocchio G, Bancale E, Cesano G, Sena LM, Piccoli G (1994) Urinary endothelin in glomerulonephritis patients with normal renal function. Clin Nephrol 41:323–330

Ruschitzka F, Shaw S, Gygi D, Noll G, Barton M, Luscher TF (1999) Endothelial dysfunction in acute renal failure: role of circulating and tissue endothelin-1. J Am Soc Nephrol 10:953–962

Russell FD, Coppell AL, Davenport AP (1998) In vitro enzymatic processing of radiolabelled big ET-1 in human kidney. Biochem Pharmacol 55:697–701

Saijonmaa O, Ristimaki A, Fyhrquist F (1990) Atrial natriuretic peptide, nitroglycerine, and nitroprusside reduce basal and stimulated endothelin production from cultured endothelial cells. Biochem Biophys Res Commun 173:514–520

Sakamoto H, Sasaki S, Nakamura Y, Fushimi K, Marumo F (1992) Regulation of endothelin-1 production in cultured rat mesangial cells. Kidney Int 41:350–355

Schiffrin EL, Lariviere R, Li JS, Sventek P, Touyz RM (1995) Deoxycorticosterone acetate plus salt induces overexpression of vascular endothelin-1 and severe vascular hypertrophy in spontaneously hypertensive rats. Hypertension 25:769–773

Schiffrin EL, Deng LY, Sventek P, Day R (1997a) Enhanced expression of endothelin-1 gene in resistance arteries in severe human essential hypertension. J Hypertension 15:57–63

Schiffrin EL, Turgeon A, Deng LY (1997b) Effect of chronic ET_A-selective endothelin receptor antagonism on blood pressure in experimental and genetic hypertension in rats. Br J Pharmacol 121:935–940

Schiffrin EL (1999) Role of endothelin-1 in hypertension. Hypertension 34: 876–881

Schnermann J, Lorenz JN, Briggs JP, Keiser JA (1992) Induction of water diuresis by endothelin in rats. Am J Physiol 263:F516–F526

Schnermann JB, Zhu XL, Shu XQ, Yang TX, Huang YG, Kretzler M, Briggs JP (1996) Regulation of endothelin production and secretion in cultured collecting duct cells by endogenous transforming growth factor-β. Endocrinol 137:5000–5008

Shibouta Y, Suzuki N, Shino A, Matsumoto H, Terashita Z-I, Kondo K, Nishikawa K (1990) Pathophysiological role of endothelin in acute renal failure. Life Sci 46:1611–1618

Shichiri M, Hirata Y, Ando K, Emori T, Ohta K, Kimoto S, Ogura M, Inoue A, Marumo F (1990) Plasma endothelin levels in hypertension and chronic renal failure. Hypertension 15:493–496

Shyamala V, Moulthrop TH, Stratton-Thomas J, Tekamp-Olson P (1994) Two distinct human endothelin B receptors generated by alternative splicing from a single gene. Cell Mol Biol Res 40:285–296

Sigmon DH, Beierwaltes WH (1993a) Renal nitric oxide and angiotensin II interaction in renovascular hypertension. Hypertension 22:237–242

Sigmon DH, Beierwaltes WH (1993b) Angiotensin II: nitric oxide interaction and the distribution of blood flow. Am J Physiol 265:R1276–R1283

Stingo AJ, Clavell AL, Aarhus LL, Burnett JC Jr (1993) Biological role for the endothelin-A receptor in aortic cross-clamping. Hypertension 22:62–66

Sugiura M, Inagami T, Kon V (1989) Endotoxin stimulates endothelin-release in vivo and in vitro as determined by radioimmunoassay. Biochem Biophys Res Commun 161:1220–1227

Todd-Turla KM, Zhu XLL, Shu XQ, Chen M, Yu TX, Smart A, Killen PD, Fejes-Toth G, Briggs JP, Schnermann JB (1996) Synthesis and secretion of endothelin in a cortical collecting duct cell line. Am J Physiol 271:F330–F339

Torralbo A, Trobo JI, Borque M, Herrero JA, Velasco E, Marcello M, González-Mate A, Barrientos A (1995) Alterations in renal endothelin production in rats with reduced renal mass. Am J Kid Dis 25:918–923

Télémaque S, Gratton J-P, Claing A, D'Orléans-Juste P (1993) Endothelin-1 induces vasoconstriction and prostacyclin release via the activation of endothelin ET_A receptors in the perfused rabbit kidney. Eur J Pharmacol 237:275–281

Ujiie K, Terada Y, Nonoguchi H, Shinohara M, Tomita K, Marumo F (1992) Messenger RNA expression and synthesis of endothelin-1 along rat nephron segments. J Clin Invest 90:1043–1048

Uzuner K, Banks RO (1993) Endothelin-induced natriuresis and diuresis are pressure-dependent events in the rat. Am J Physiol 265:R90–R96

Vlachojannis J, Tsakas S, Chinari E, Orphanos V, Zoumbos N, Kurz P (1997) Increased endothelin-1 content in the platelets of hemodialysed patients. Clin Nephrol 48:185–190

Warner TD, Mitchell JA, De Nucci G, Vane JR (1989) Endothelin-1 and endothelin-3 release EDRF from isolated perfused arterial vessels of the rat and rabbit. J Cardiovasc Pharmacol 13 [Suppl 5]:S85–88

Warner TD, Allcock GH, Mickley EJ, Corder R, Vane JR (1993) Comparative studies with the endothelin receptor antagonists BQ-123 and PD 142893 indicate at least three endothelin receptors. J Cardiovasc Pharmacol 22 Suppl 8:S117–S120

Weisberg LS, Kurnik PB, Kurnik BRC (1992) Radiocontrast-induced nephropathy in humans: role of renal vasoconstriction. Kidney Int 41:1408–1415

Wellings RP, Corder R, Warner TD, Cristol JP, Thiemermann C, Vane JR (1994) Evidence from receptor antagonists of an important role for ET_B receptor-mediated vasoconstrictor effects of endothelin-1 in the rat kidney. Br J Pharmacol 111:515–520

Wilhelm SM, Simonson MS, Robinson AV, Stowe NT, Schulak JA (1999) Endothelin up-regulation and localization following renal ischemia and reperfusion. Kidney Int 55:1011–1018

Wilkes BM, Ruston AS, Mento P, Girardi E, Hart D, Vander Molen M, Barnett R, Nord EP (1991a) Characterization of endothelin 1 receptor and signal transduction mechanisms in rat medullary interstitial cells. Am J Physiol 260:F579–F589

Wilkes BM, Susin M, Mento PF, Macica CM, Girardi EP, Boss E, Nord EP (1991b) Localization of endothelin-like immunoreactivity in rat kidneys. Am J Physiol 260:F913–F920

Wolf SC, Brehm BR, Gaschler F, Brehm S, Klaussner M, Smykowski J, Amann K, Osswald H, Erley CM, Risler T (1999) Protective effects of endothelin antagonists in chronic renal failure. Nephrol Dialysis Transplant 14 [Suppl 4]:29–30

Yamashita Y, Yukimura T, Miura K, Okumura M, Yamamoto K (1991) Effect of endothelin-3 on renal functions. J Pharmacol Exp Ther 259:1256–1260

Yoshimura A, Iwasaki S, Inui K, Ideura T, Koshikawa S, Yanagisawa M, Masaki T (1995) Endothelin-1 and endothelin B type receptor are induced in mesangial proliferative nephritis in the rat. Kidney Int 48:1290–1297

Zeidel ML, Brady HR, Kone BC, Gullans SR, Brenner BM (1989) Endothelin, a peptide inhibitor of Na^+-K^+-ATPase in intact renal tubular epithelial cells. Am J Physiol 257:C1101–C1107

CHAPTER 19

Endothelin Ligands and their Experimental Effects Within the Human Circulation

S.J. Leslie and D.J. Webb

A. Introduction

A number of endothelin (ET) receptor antagonists have been developed as research tools and as potential medicines (see Chap. 9). Some of these are selective for the ET_A or ET_B receptor while others are mixed $ET_{A/B}$ receptor antagonists. Peptide antagonists are inactive orally and, therefore, of limited use as long term treatments. However, these compounds have been useful as research tools and are currently being developed as short term therapeutic agents. Due to ease of administration, non-peptide, orally active antagonists are more promising candidates for the treatment of chronic conditions and several are at various stages of clinical evaluation.

Current interest in endothelin antagonists (ETAs) as possible therapeutic agents is predominantly focused on cardiovascular diseases such as chronic heart failure (CHF), essential hypertension (see Chap. 16) and chronic renal failure (CRF) (see Chap. 18). However, ETAs may also be of value in the treatment of a number of other, unrelated conditions such as bronchospasm (FUKURODA et al. 1996) (see Chap. 14), prostatic hypertrophy and prostate cancer (NELSON et al. 1995), portal hypertension (GUNAL et al. 1996) (see Chap. 17), pulmonary hypertension (McCULLOCH et al. 1995, 1998) (see Chap. 15), acute renal failure (NORD 1997) (see Chap. 18), stroke (LAMPL et al. 1997) (see Chap. 13) and gram-positive sepsis (UOSAKI et al. 1996). It is still unclear whether selective ET_A receptor blockade or non-selective $ET_{A/B}$ receptor blockade will confer greater clinical benefit; indeed, this may differ between different conditions (FERRO et al. 1996a).

B. Research Tools Used in Human Subjects

I. Local Study Techniques

Endothelin-1 causes a marked and sustained vasoconstriction in many vascular beds, the coronary, renal and cerebral circulations being particularly sensitive. There are, therefore, potential risks with administration of systemic doses of endothelin agonists. The use of local techniques, such as local brachial artery infusion coupled with forearm plethysmography, local intra-venous drug administration in hand vein studies, and intra-dermal administration with laser Doppler microcirculatory flow measurement has allowed the relatively safe observation of the in vivo effects of the endothelin system in individual vascular beds. Because forearm muscle blood flow constitutes only ~1% of the cardiac output, locally effective doses of drug can be administered that are 10–100 times lower than the systemically active dose (BENJAMIN et al. 1995; WEBB 1995). These small doses also minimise the confounding systemic effects of these agents on the heart, kidney and other organs, and the activation of neurohumoral reflexes. This has facilitated a systematic pharmacological approach to the investigation of the endothelin system, in vivo, in humans.

Therefore, forearm plethysmography coupled with unilateral brachial artery drug infusion represents a powerful tool in the investigation of the vascular effects of study compounds, the contralateral arm acting as a contemporaneous control. It is now widely used in many areas of vascular research and is accurate and reproducible with little inter- or intra-observer variation. Vascular changes demonstrated in forearm studies have been shown to correlate well with responses in other, more critical, vascular beds (WEBB 1995).

The effects of compounds on capacitance vessel tone can be assessed by using local intra-venous administration of study compounds while measuring changes in hand vein diameter during venous occlusion (NACHEV et al. 1971; AELLIG 1981). This technique involves the displacement of a light-weight magnetised rod resting on the summit of a dorsal hand vein kept distended by an upper arm cuff, inflated to above venous but below arterial pressure. Deflation of the cuff results in collapse of the vein and downward displacement of the rod, which passes through the core of a linear variable differential transformer, causing a linear change in the generated voltage, thus allowing determination of the internal diameter of the vein.

Intra-dermal studies are performed using a 0.4-mm needle and skin blood flow can be estimated using a laser Doppler flowmeter (NILSSON et al. 1980). There is current interest in the differential changes in skin blood flow at the site of injection and the surrounding area of skin following local injection of endothelin.

Taken together, these are powerful techniques that facilitate the investigation of the endothelin system in vivo. However, care should be taken when extrapolating results from local techniques to the whole subject as endothelin and its antagonists, at systemic doses, may affect the heart and other vascular beds in a complex fashion.

II. Systemic Study Techniques

Systemic studies have been conducted using endothelin ligands in both healthy volunteers and patients. The sensitive, accurate and reproducible measurement of cardiovascular parameters is central to this research. A variety of invasive and non-invasive methods have been used for measuring and calculating cardiac parameters. Invasive monitoring techniques include central venous, pulmonary artery and arterial catheterisation. Non-invasive methods have been especially useful in healthy volunteer studies where the insertion of pulmonary artery catheters should be avoided where possible. Impedance cardiography was first developed for clinical use over 30 years ago and measures changes in electrical impedance of injected current through skin surface electrodes (KUBICEK et al. 1966; THOMAS et al. 1992). This technique has been criticised when used in the clinical setting for its inability to provide accurate absolute values of cardiac parameters (JENSEN et al. 1995). However, it is a useful technique for short term studies, in an experimental setting, because is easy to use, non-invasive, and gives a reliable measure of changes in cardiac parameters.

Renal plasma flow and glomerular filtration rate, can be measured using the p-aminohippuric acid (PAH) and inulin plasma clearance methods respectively. These techniques are widely used and described elsewhere in detail (BRADLEY 1987).

C. Endothelin Ligands Used in Human Studies

Synthetic ET-1, ET-2, ET-3 and sarafotoxin 6c (SX6c), an ET_B agonist, are available commercially. The ET_A receptors have a high affinity for ET-1 and ET-2 and lower affinity for ET-3 whereas the ET_B receptor have equal affinity for ET-1, ET-2 and ET-3. ET-3 and especially SX6c have been used as tools to investigate the ET_B receptor as ET-3 is 2000-fold selective for the ET_B receptor and SX6c is 30,000-fold selective for the ET_B receptor (see Chap. 4).

Agonist studies have been useful in defining the target organs and the receptor subtypes involved in physiological responses to the endothelins. However, as the endothelins act predominantly via autocrine and paracrine mechanisms (HOCHER et al. 1997), administration of exogenous agonist is unlikely to reproduce physiological responses and results from such studies may be misleading. Therefore, it has been the study of the effects of ET antagonists which has been central to unravelling the complexities of the pathophysiology of the endothelin system in man.

Several selective and non-selective ET antagonists have been developed. Some, such as BQ-123, 2000-fold selectivity for the ET_A receptor (IHARA et al. 1992); BQ-788, 200-fold selectivity for the ET_B receptor (ISHIKAWA et al. 1994) and TAK-044, a mixed $ET_{A/B}$ antagonist (KIKUCHI et al. 1994) are peptides and therefore inactive when given orally. Other, orally active, non-peptide endothelin antagonists such as bosentan, a mixed $ET_{A/B}$ receptor antagonist, have been developed (PETER et al. 1996). Human studies using these agents will be discussed below.

With the current optimism regarding the endothelin system as a target in a number of pathological conditions, there are many other ET antagonists undergoing pre-clinical and clinical studies. The results of these studies are awaited with interest.

D. Endothelin Ligands and their Effects in Healthy Volunteers

I. Local Activation of the Endothelin System in Healthy Volunteers

1. Resistance Vessels

Infusion of ET-1 (5 pmol/min) into the brachial artery caused a slow onset, dose-dependent sustained reduction in blood flow of 40% at 60 min in the forearm of healthy volunteers. This persisted for up to 2 h following discontinuation of the infusion (CLARK et al. 1989). Similar vasoconstriction has been demonstrated by other groups and in addition ET-1 caused a non-significant trend towards causing early vasodilatation (HAYNES et al. 1995b). In one study, sustained vasodilatation has been described with ET-1 at low concentrations (0.2 pmol/min/100 ml forearm tissue) (KIOWSKI et al. 1990). The low doses used in this study may have been insufficient to cause the sustained vasoconstriction response seen in other studies but at the same time sufficient to cause the initial vasodilatation. The mechanisms behind this are not clearly understood.

SX6c and ET-3, both ET_B agonists, also caused constriction of forearm resistance vessels in vivo but to a lesser degree than ET-1 (HAYNES et al. 1995a), thus implicating the ET_B receptor in vasoconstriction. However, ET-1, ET-3 and SX6c caused a transient vasodilatation prior to a sustained vasoconstriction. The transient vasodilatation is more marked with ET-3 and SX6c suggesting that this is mediated via endothelial ET_B receptors (HAYNES et al. 1995b). Thus, it seems that ET_B receptors can mediate both vasodilatation and vasoconstriction.

It is postulated that the ET_B receptors on the vascular smooth muscle cells cause vasoconstriction (WILLIAMS et al. 1991; MORELAND et al. 1992; SUMNER et al. 1992) while the endothelial cell receptors mediate vasodilatation (TAKAYANAGI et al. 1991) possibly also acting as clearance receptors for ET-1.

The exact mechanism of ET-3 and SX6c vasoconstriction is unclear. It may be mediated by direct binding to the ET_B receptors on the vascular smooth muscle cells resulting in vasoconstriction. However, it may be that ET_B receptor ligands may decrease the clearance or displace ET-1 from the receptor, thus increasing binding at the ET_A receptor, resulting in further vasoconstriction. In addition, there is animal evidence that SX6c may act through non-endothelin dependent mechanisms, and the pressor responses may be independent of the endothelin system (FLYNN et al. 1995).

Local infusions of big ET-1 (50 pmol/min) produce vasoconstriction which can be blocked by phosphoramidon, a combined endothelin converting enzyme (ECE) and neutral endopeptidase (NEP) inhibitor (HAYNES et al. 1994b) (see Chap. 7). There is limited plasma ECE activity (WATANABE et al. 1991) and this result, therefore, suggests that forearm resistance vessels have local ECE activity.

2. Capacitance Vessels

Constriction of dorsal hand veins is seen with ET-1 infusions of 5 pmol/min (CLARK et al. 1989; HAYNES et al. 1993a, 1995a). There is no venoconstriction to local infusion of big endothelin-1 (50 pmol/min) suggesting no local ECE activity in hand veins (HAYNES et al. 1995b). The mechanism of action of ET-1 in hand veins is thought to be mediated via both the opening of voltage operated Ca^{2+} channels and the closure of ATP sensitive K^+ channels, thus offering other targets for therapeutic intervention (HAYNES et al. 1993b). Sarafotoxin (SX6c) and ET-3 (ET_B agonists) also caused constriction of hand capacitance vessels in vivo but to a lesser degree than ET-1 (HAYNES et al. 1995a), thus implicating the ET_B receptor in venoconstriction.

Dorsal hand veins have no intrinsic tone, however, in preconstricted human hand veins, as in arteries, ET-1, ET-3 and SX6c also caused a transient vasodilatation prior to a sustained vasoconstriction. The transient vasodilatation was more marked with ET-3 and SX6c suggesting that it is mediated via endothelial ET_B receptors (HAYNES et al. 1995b). This effect is blocked by aspirin and therefore it is postulated that it is prostanoid dependent (HAYNES et al. 1993a).

3. Microcirculation

Intra-dermal injection of ET-1 results in vasoconstriction of the microcirculation (WENZEL et al. 1994). Intra-dermal injection of ET-3 does not cause vasoconstriction, suggesting that vasoconstriction in the skin microcirculation is an ET_A mediated response. More recently, studies have demonstrated vasodilatation 1–2 cm from the site of injection (WENZEL et al. 1998a). It appears that this is an ET_A receptor mediated effect through stimulation of polimodal nociceptor fibres leading to nitric oxide release because this effect is blocked by BQ-123 and pre-treatment with L-NMMA, lignocaine or capsaicin. This potentially implicates endothelins in the process of neurogenic inflammation, suggesting ETAs may possibly have a role to play in the treatment of inflammatory conditions.

The overall effect of ET-1 at the ET_B receptor will therefore be determined by the balance of effects between its actions at the endothelial and vascular smooth muscle ET_B receptors. As discussed, the results of agonist studies should be interpreted with caution as the endothelins act in a paracrine and autocrine fashion and the administration of supra-physiological concentrations of exogenous agonists may not mimic in vivo physiology. The results of antagonist studies are likely to be much more illuminating.

II. Local Inhibition of the Endothelin System in Healthy Volunteers

1. Resistance Vessels

Local infusion of BQ-123 (HAYNES et al. 1994b; BERRAZUETA et al. 1997; VERHAAR et al. 1998) or TAK-044 (HAYNES et al. 1996) results in vasodilatation in the forearm arteries of healthy volunteers. The larger effect seen with BQ-123 suggests that vasoconstriction to ET-1 is mediated predominantly through the ET_A receptor located on vascular smooth muscle cells. Thus, the net effect of inhibition of the ET_B receptor is vasoconstriction. The role of the ET_B receptor is further clarified by studies with the ET_B receptor selective antagonist, BQ-788 (ISHIKAWA et al. 1994). BQ-788 causes vasoconstriction in the forearm vessels of healthy volunteers. This vasoconstriction persists on a background of ET_A antagonism (VERHAAR et al. 1998), reinforcing the hypothesis that ET-1 stimulation of the endothelial ET_B receptor causes dilatation, and is likely to be a direct effect, whereas the vascular smooth muscle ET_B receptor causes vasoconstriction.

2. Microcirculation

In the skin microcirculation of healthy volunteers, intradermal injection of a mixed $ET_{A/B}$ antagonist caused a vasodilatation which was no greater than that seen with a selective ET_A receptor antagonist suggesting vasoconstriction to endothelin is solely ET_A receptor mediated (WENZEL et al. 1994). However, in patients with coronary artery disease there was increased vasodilatation with mixed $ET_{A/B}$ antagonism compared to ET_A antagonism, suggesting there may be ET_B receptor mediated vasoconstriction in these patients (WENZEL et al. 1996). In addition, intravenous administration of Bosentan reversed the vasoconstrictor effect of ET-1 measured in the skin microcirculation (WEBER et al. 1996). This study did not report a distal flare following intra-dermal ET-1 administration. However, in a more recent study, pretreatment with intradermal ET_A receptor antagonist prevented the ET-1 induced vasoconstriction and also the 'flare reaction' caused by vasodilatation in the surrounding area. There was no additional effect with a mixed $ET_{A/B}$ antagonist suggesting that the flare reaction is an ET_A receptor mediated response (WENZEL et al. 1998a). Thus, in the skin microcirculation in healthy volunteers, ET_A receptors appear to be involved in endothelin mediated vasoconstriction and to mediate the distal vasodilatation.

III. Systemic Activation of the Endothelin System in Healthy Volunteers

1. Cardiovascular Effects

The administration of ET-1(5 ng/kg/min for 15 min) to healthy volunteers results in an increase in mean blood pressure of 5–10 mmHg and a reduction in heart rate, which is probably reflex in nature. This dose of ET-1 increased plasma concentrations by 50-fold (VIERHAPPER et al. 1990). In more recent studies, both ET-1 (8 pmol/kg/min for 10 min) and big ET (8 pmol/kg/min) infusions, sufficient to cause increases in plasma ET-1 of 30- and 2.4-fold respectively, caused similar increases in blood pressure (BP) and a reduction in heart rate which persisted following cessation of the infusion for 30 min and 90 min, respectively. These doses of ET-1 and big ET-1 also reduced coronary sinus blood flow, by a maximum of ~25%, and increased coronary vascular resistance by ~50 and 100% respectively (PERNOW et al. 1996). In other studies, similar haemodynamic changes following big ET-1 infusion (8 pmol/min for 20 min; twofold increase in plasma ET-1) persisted for up to 2 h (AHLBORG et al. 1994). Further studies have shown that doses of ET-1 insufficient to cause systemic or pulmonary pressor effects (0.75 pmol/kg/min) can impair left and right diastolic dysfunction and are negatively inotropic (KIELY et al. 1997). Thus ET may have subtle effects in the cardiovascular system at concentrations lower than those required to cause changes in BP or systemic vascular resistance (SVR).

2. Renovascular Effects

A reduction in splanchnic and renal blood flow by 34% and 26% respectively is seen following ET-1 infusion (4 pmol/kg/min for 20 min; 12-fold increase in plasma ET-1) which persisted for 1 h and 3 h respectively (WEITZBERG et al. 1991). More recently, changes in renal parameters have been demonstrated following administration of lower doses of ET-1 (1 pmol/kg/min for 60 min: 11-fold increase in plasma ET-1) to healthy volunteers causing an increase in diastolic BP of 8% and decrease in heart rate of 14% but no significant change in systolic BP. Renal plasma flow and glomerular filtration rate (GFR) were both reduced, by 35% and 16% respectively. Urine flow was reduced by 40% and urinary sodium excretion by 58%. Lithium clearance in these subjects suggested that the reduction in sodium excretion occurred at the distal rather than proximal tubule (SORENSEN et al. 1994). Infusion of lower doses of ET-1 (0.4 pmol/kg/min) for a longer period (6 h) results in decreased renal blood flow by 43%, associated with a 10% decrease in heart rate. This dose of ET-1 increased plasma endothelin concentration by ~300% (JILMA et al. 1997). In a similar study the administration of exogenous ET-1 (2.5 ng/kg/min) to healthy volunteers sufficient to cause a threefold increase in plasma ET concentrations caused renal vasoconstriction and sodium retention. In this study, administration of a comparable dose of ET-3 produced no effect on blood pressure, renal blood flow or electrolyte excretion, suggesting that these responses are predominantly ET_A mediated (KAASJAGER et al. 1997). Interestingly, ET-1 produced sodium retention in humans even at doses insufficient to reduce renal plasma flow or glomerular filtration rate (RABELINK et al. 1994) suggesting that endothelin may have subtle intra-renal effects on sodium handling.

IV. Systemic Inhibition of the Endothelin System in Healthy Volunteers

1. Cardiovascular Effects

The results of systemic studies with ET receptor antagonists have, in general, confirmed the predictions made from the results of local vascular studies.

a) Non-Selective Inhibition

The administration of the $ET_{A/B}$ antagonist TAK-044 (1000mg over 15min) reduced systolic blood pressure, diastolic blood pressure and systemic vascular resistance by 4%, 18% and 26% respectively and resulted in an increase in both heart rate and cardiac index. Plasma ET concentrations increased by ninefold following administration of TAK-044 (1000mg). These results suggest an effect predominantly in resistance vessels (HAYNES et al. 1996). The increase in plasma ET is thought to be the result of displacement of bound ET-1 and reduced ET-1 clearance as there is no associated increase in big ET-1 or its C-terminal fragment (PLUMPTON et al. 1996). Similar results are found in healthy volunteers given oral or intra-venous bosentan. Bosentan, 2400mg administered orally, resulted in a maximal decrease in systolic blood pressure of 9 mmHg 2h after dosing (WEBER et al. 1996). This study also demonstrated a dose-dependent increase in plasma ET-1 after administration of bosentan; twofold following 2400mg (oral) and threefold following 750mg (IV).

b) Selective Inhibition

BQ-123 (3000nmol/min for 15min) administered to healthy volunteers resulted in a reduction in mean arterial blood pressure and systemic vascular resistance index of 12% and 23% respectively with an associated increase in cardiac index and a non significant increase in heart rate. Similar effects were seen with lower doses (1000nmol/min for 15min) (SPRATT et al. 1999). Forearm vasoconstriction to local infusions of ET-1 were blocked with the higher systemic doses of BQ-123 (3000nmol/min for 15min), in this study, thus confirming the dominant role of the ET_A receptor as a vasoconstrictor in peripheral resistance vessels. Interestingly, there was no consistent increase in plasma big ET-1 or plasma ET-1 with systemic blockade of the ET_A receptor.

The administration of BQ-788, a selective ET_B receptor antagonist, at systemic doses (300nmol/min for 15min) to healthy men causes systemic vasoconstriction with a reduction of heart rate and cardiac index but no effect on mean arterial blood pressure, again suggesting that endogenous activity at the vascular ET_B receptors mediates a predominantly vasodilator tone (STRACHAN et al. 1999). In common with previous studies plasma ET-1 increased, twofold, with ET_B receptor blockade.

These results would suggest that ET_A receptor blockade may have beneficial effects in conditions characterised by systemic vasoconstriction, e.g. CHF.

2. Renovascular Effects

BQ-123 (~100nmol/min IV for 60min) administered to healthy men had no renal or systemic effects but did prevent ET-1 (1pmol/kg/min for 120min) induced reductions in renal plasma flow and GFR (SCHMETTERER et al. 1998). These results suggest that the vasoconstriction of renal vessels caused by ET-1 are mediated through the ET_A receptor and are more sensitive to ET receptor antagonists than the peripheral vasculature. The authors claim that the lack of change in renal parameters following administration of BQ-123 suggests there is no endothelin mediated resting tone in the renal vasculature. However, despite reversing the effects of exogenous ET-1, the lack of a systemic effect with this dose of BQ-123 would suggest that this dose may be insufficient to cause renal effects.

More recently, administration of the mixed $ET_{A/B}$ receptor antagonist, TAK-044 (750mg) resulted in a decrease in effective renal vascular resistance, suggesting there may be endothelin mediated resting vascular tone in the kidney. However there was a decrease in GFR and a significant decrease in the filtration fraction with both doses of TAK-044 (FERRO 1998).

Thus, the question as to whether there is basal endothelin mediated tone and the role of the ET_A and ET_B receptor in the renal vasculature remain to be clarified.

E. Endothelin Ligands and their Effects in Patients

I. Heart Disease

1. Congestive Heart Failure

The first clinical trial of an endothelin antagonist in CHF involved the systemic administration of bosentan (100mg IV followed by 200mg IV after 60min) and resulted in a reduction of mean arterial pressure, systemic and pulmonary vascular resistance and an increase in cardiac output without reflex increases in heart rate and importantly no increase in plasma angiotensin II or noradrenaline (KIOWSKI et al. 1995). Pulmonary vascular resistance was decreased to a greater degree than systemic vascular resistance, in contrast to the effect usually seen with other vasodilators. These patients, however, do not reflect the general patient population as they were not being treated with angiotensin converting enzyme (ACE) inhibitors.

Subsequently, local studies were performed in CHF patients on standard treatments (including ACE inhibitors). Local brachial artery infusion of ET-1 (5 pmol/min for 60 min) caused less vasoconstriction than in matched controls in resistance (Love et al. 1996b) and capacitance vessels (Love et al. 1996c). In these patients potentially beneficial haemodynamic effects are seen with local brachial artery infusion of BQ-123 (100 nmol/min for 60 min) which causes vasodilatation. Although there is increased vasoconstriction to the ET_B receptor agonist, S6Xc (5 pmol/min for 60 min), in CHF patients in comparison to matched controls, BQ-788 also causes vasoconstriction, suggesting the net effect of stimulation of the ET_B receptor is vasodilatation. The effects of S6Xc have been discussed previously.

Systemic studies in patients on treatment including ACE inhibitors, digitalis glycosides and diuretics demonstrated similar haemodynamic improvements in the short term with a small increase in heart rate when bosentan was given orally (0.5 g twice a day for 14 days) (Sutsh et al. 1997). A further study demonstrated improved haemodynamic parameters in CHF patients treated with oral bosentan (1 g twice daily) with an increase in cardiac output of 15% and a decrease in systemic and pulmonary vascular resistance, each of ~10% respectively. This is important, as these beneficial effects were seen in patients already treated with ACE inhibitors, diuretics and digoxin (Sutsh et al. 1998).

More recently the results from a larger study, REACH-1, were reported. A total of 370 patients with NYHA class IIIB-IV on standard treatments, including ACE inhibitors, received either bosentan 500 mg bd or placebo for 6 months. The bosentan group had a reduction in the numbers of hospital admissions for any reason by 41%. The trial was stopped early due to concerns over raised hepatic transaminases. However, in patients followed for the intended duration of the study (6 months) bosentan significantly increased the likelihood of clinical improvement and decreased the likelihood of CHF deterioration (Packer et al. 1998).

These studies have demonstrated the beneficial haemodynamics of short term non-selective ET receptor blockade in CHF patients. In order to investigate whether selective antagonism will confer clinical benefit, BQ-123 (200 nmol/min for 60 min) was administered by intravenous infusion to 10 CHF patients. This caused a systemic vasodilatation and rise in cardiac index (CI), with no change in heart rate (HR). There was a non-significant fall in pulmonary vascular resistance (PVR) (Cowburn et al. 1998a).

In further studies, consistent with predictions from local studies in CHF patients (Love 1996a,b), the effects of a selective ET_B receptor antagonist in 8 CHF patients were investigated. BQ-788 (50–100 nmol/min for 45 min) caused systemic vasoconstriction, a rise in MAP and SVR, with a reduction in cardiac index. The subsequent infusion of the selective ET_A receptor antagonist BQ-123 resulted in a reversal of these effects and an increase in cardiac index (Cowburn et al. 1998b).

The selective ET_A antagonist TBC11251 (3 mg/kg/min for 15 min) administered to 24 CHF patients resulted in a significant fall in mean pulmonary artery pressure of 12% and PVR of 39% following systemic administration of the study drug for 15 min. An effect was seen at 15 min, becoming significant at 1 h and maximal at 2–3 h. Conversely, there was no significant change in HR, MAP, SV, PCWP or SVR (GIVERTZ et al. 1998).

Thus, current evidence suggests that selective ET_B receptor blockade has deleterious effects on the haemodynamics of patients with CHF. However, it is still unclear, when the ET_A receptor is blocked, whether ET_B blockade will improve systemic haemodynamics in these patients.

2. Coronary Artery Disease (CAD)

There have been few studies performed specifically in patients with coronary artery disease. In healthy volunteers, intradermal injection of a mixed $ET_{A/B}$ receptor antagonist caused a similar vasodilatation compared with selective ET_A receptor antagonism in the skin microcirculation. However, in patients with CAD, mixed inhibition caused a greater vasodilatation (WENZEL et al. 1996). This suggests that the ET_B receptor may have increased functional significance in patients with CAD.

In a recent systemic study, 28 patients with angiographically documented coronary artery disease were given bosantan 200 mg IV. As may have been predicted from previous studies, this resulted in a decrease in systolic blood pressure and a small increase in heart rate. In addition, there was an increase in coronary artery diameter which appeared to be maximal as no further increase was noted after treatment with glycerol trinitrate. This suggests there is a basal coronary artery vasoconstrictor tone, in vivo, mediated by endogenous ET (WENZEL et al. 1998b).

This has been confirmed by another study demonstrating endogenous ET_A receptor mediated coronary artery tone in patients undergoing coronary arteriography (KYRIAKIDES et al. 1999).

3. Hypertension

In patients with essential hypertension there is an increased venoconstrictor response to local ET-1(5 pmol/min) and sympathetically mediated venoconstriction of capacitance vessels is potentiated by ET-1 (HAYNES et al. 1994). Vasodilatation in patients with essential hypertension following intra-arterial administration of BQ-123 was shown to be no different to that in normal healthy volunteers, suggesting that there may be no major dysfunction of endothelium dependent vasodilatation (FERRO et al. 1996b).

Several animal and in vitro experiments on human tissue have demonstrated close interactions between the endothelin, renin-angiotensin and sympathetic nervous systems in several disease states although there is limited human in vivo data.

Results of systemic studies performed with hypertensive patients have recently been reported. Bosentan treatment for 4 weeks in 293 hypertensive patients caused significant lowering of blood pressure without reflex neurohormonal stimulation of the sympathetic nervous system or the renin-angiotensin system. Observation of the effects of 4 doses of bosentan demonstrated a plateau reached at 500 mg of bosentan which was similar to the reduction in blood pressure seen with 20 mg of the ACE inhibitor enalapril (KRUM et al. 1998).

II. Renal Disease

1. Chronic Renal Failure

Compared with the vast animal literature concerning endothelin ligands and their effects on the renal system (see Chap. 18) there are relatively little human data. Plasma concentrations of ET-1 are raised in chronic haemodialysis patients (HAND et al. 1999). Reduced sensitivity of hand veins in CRF patients to exogenous ET has been demonstrated (HAND et al. 1994) which may be the result of receptor down regulation following increased occupancy secondary to increased ET concentration in patients CRF. In a recent study of preconstricted hand veins of CRF patients, infusion of BQ-123 (3 mg/kg/min for 45 min) caused an increase in venodilation of 74% compared with 28% in controls, suggesting that in these subjects responses to endogenous ET may be increased (BUSSEMAKER et al. 1998). In contrast, vasodilatation in the forearm resistance vessels in response to BQ-123 (3 mg/h) has been shown to be reduced in patients with CRF (HAND et al. 1999). From the current research, changes in the ET system in CRF are only poorly understood and further studies are required.

Administration of the mixed $ET_{A/B}$ receptor antagonist, TAK-044 (100 mg and 750 mg IV over 15 min) resulted in a reduction of MAP and SVRI of 11% and 24% respectively and an increase in cardiac output. TAK-044 at both doses had no significant effect on GFR or effective renal plasma flow. However, effective renal vascular resistance was lowered by ~10% by both doses. Neither dose had any effect on sodium or lithium clearance (FERRO et al. 1998). This study suggests that there is ET mediated renovascular tone and that ET receptor antagonists reduce blood pressure in CRF and may have a potential beneficial role as vasodilators in these patients.

2. Hepatorenal Failure

The hepatorenal syndrome (HRS) is characterised by renal vasospasm in the face of systemic vasodilatation, resulting in sodium retention which persists despite adequate correction of plasma volume and cardiac output. Plasma ET-1 concentrations are increased in this condition even when compared with patients with hepatic and renal failure of different aetiology. Recently, in a small study, 3 patients with HRS received BQ-123 (10 nmol/min, 25 nmol/min and 100 nmol/min for 60 min). There was a dose-dependent increase in renal plasma flow and glomerular filtration rate as measured by PAH and inulin clearance but no significant changes in HR, MAP or SVR (SOPER et al. 1996).

These results correspond well with the results of studies in healthy volunteers (RABELINK et al. 1994).

3. Contrast Nephropathy

Despite previous suggestions that contrast-stimulated, endothelin mediated intrarenal vasoconstriction may contribute to contrast nephrotoxicity, a recent study of 158 high risk patients undergoing coronary angiography demonstrated that the non-selective endothelin antagonist SB 209670 did not protect against contrast nephropathy. Indeed, more patients in the treatment group compared with the control group developed contrast nephropathy (56% vs 29%) (WANG et al. 1998).

F. Conclusion

The endothelin system has been extensively studied over the last 10 years. Much of our understanding of the role of endothelin in normal physiology and pathophysiology comes as a result of carefully designed preclinical and clinical studies with ET receptor antagonists. In addition to their use in endothelin research, ET receptor antagonists are now being developed as potential therapeutic agents in various cardiovascular conditions. It appears that they may be of most use in clinical conditions that involve chronic vasoconstriction, such as heart failure and hypertension, as well as in conditions resulting from vasospasm such as subarachnoid haemorrhage. However, the best target for these potential medicines is not yet known, and it is unclear whether mixed $ET_{A/B}$ antagonism or selective ET_A antagonism will provide the best clinical treatment; indeed, this may differ between conditions. In addition, it remains unclear whether it will be possible to target selectively the endothelial or vascular smooth muscle ET_B receptor.

In conclusion, there is a need for continuing research to define further the roles of the ET_A and ET_B receptors in health and disease and for well designed clinical trials to confirm whether ET receptor antagonists, or even synthesis inhibitors, will provide benefits in terms of morbidity and mortality in different target populations.

Acknowledgements. Professor DJ Webb is supported by a Research Leave Fellowship from the Wellcome Trust (WT 0526330). Dr SJ Leslie is a British Heart Foundation Junior Research Fellow (FS/98040).

References

Aellig WH (1981) A new technique for measuring compliance of human hand veins. Br J Clin Pharmacol 11:237–243
Ahlborg G, Ottosson-Seeberger A, Hemsen A, Lundberg JM (1994) Big ET-1 infusion in man causes renal ET-1 release, renal and splanchnic vasoconstriction, and increased mean arterial blood pressure. Cardiovasc Res 28:1559–1563

Benjamin N, Calver A, Collier J, Robinson B, Vallance P, Webb D (1995) Measuring forearm blood flow and interpreting the responses to drugs and mediators. Hypertension 25:918–923

Berrazueta JR, Bhagat K, Vallance P, MacAllister RJ (1997) Dose- and time-dependency of the dilator effects of the endothelin antagonist, BQ-123, in the human forearm. Br J Clin Pharmacol 44:569–571

Bradley SE (1987) Clearance concepts in renal physiology. In: Gottschalk CW, Berliner RW, Giebisch GH (eds) Renal Physiology – People and Ideas. American Physiological Society, Bethesda Maryland USA, pp 63–100

Bussemaker E, Wichmann G, Passauer J, Gross P (1998) Role of endothelin and nitric oxide in the regulation of vascular tone in normotensive patients on chronic hemodialysis (Abstract). Circulation 98;Suppl 1:A1793

Clark JG, Benjamin N, Larkin SW, Webb DJ, Keogh BE, Davies GJ, Maseri A (1989) Endothelin is a potent long-lasting vasoconstrictor in men. Am J Physiol 257: H2033–2035

Cowburn PJ, Cleland JGF, MacArthur JD, MacLean MR, McMurray JJV, Dargie HJ (1998a) Short-term haemodynamic effects of BQ-123, a selective endothelin ETA-receptor antagonist, in chronic heart failure. Lancet 352:201–202

Cowburn PJ, Cleland JGF, McDonagh A, MaArthur JD, MacLean MR, Dargie HJ (1998b) Adverse haemodynamic effects of a selective endothelin ETB receptor antagonist in patients with chronic heart failure: reversal with a selective endothelin ETA receptor antagonist (abstract). Circulation 98;Suppl 1:718

Davenport AP, O'Reilly G, Molenaar P, Maguire JJ, Kuc RE, Sharkey A, Bacon CR, Ferro A (1993) Human endothelin receptors characterised using reverse transcriptase polymerase chain reaction, in-situ hybridisation and subtype selective ligands. BQ123 and BQ3020: evidence for expression of ET-B receptors in human vascular smooth muscle. J Cardiovasc Pharmacol 22 Suppl 8:S22–S25

Douglas SA (1997) Meeting Report – Clinical development of endothelin receptor antagonists. TiPS 18:408–412

Ferro CJ, Webb DJ (1996a) The clinical potential of endothelin receptor antagonists in cardiovascular medicine. Drugs 51:12–27

Ferro CJ, Haynes WG, Hand MF, Webb DJ (1996b) The vascular endothelin and nitric oxide systems in essential hypertension. J Hypertens 14:S50

Ferro CJ, Strachan FE, Plumpton C, Davenport AP, Cumming AD, Haynes WG, Webb DJ (1998) Actions of the Endothelin A/B Receptor Antagonist TAK-044 in Chronic Renal Failure. Nephrol Dial Trans 13:A66

Flynn DA, Sargent CA, Brazdil R, Brown TJ, Roach A (1995) Sarafotoxin S6c elicits a non-ETA or non-ETB mediated pressor response in the pithed rat. J Cardiovasc Pharmacol 26;Suppl 3:S219–221

Fukuroda T, Ozaki M, Ihara K, Ishikawa K, Yano M, Miyauchi T, Ishikawa S, Onizuka M, Goto K, Nishikibe M (1996) Necessity of dual blockade of endothelin ETA and ETB receptor subtypes for antagonism of endothelin-1-induced contraction in human bronchi. Br J Pharmacol 117:995–999

Givertz MM, Colucci WS, Gottlieb SS, Hare JM, LeJemtel T, Slawsky MT, Leier CV, Loh E, Nicklas JM, Lewis BE (1998) Acute ETA receptor blockade reduces pulmonary vascular resistance in patients with chronic heart failure (abstract). Circulation 98;Suppl 1:578

Gunal O, Yegen C, Aktan O, Yalin R, Yegen B (1996) Gastric function in portal hypertension: role of endothelin. Dig Dis Sci 41:585–590

Hand MF, Haynes WG, Anderton JL, Winney R, Webb DJ (1994) Responsiveness of veins to endothelin-1 in chronic renal failure. Nephrol Dial Trans 9:1349–1350

Hand MF, Haynes WG, Webb DJ (1999) Reduced endogenous endothelin-1-mediated vascular tone in chronic renal failure. Kidney Int 55:613–620

Haynes WG, Webb DJ (1993a) Endothelium-dependent modulation of responses to endothelin-1 in human veins. Clin Sci 84:427–433

Haynes WG, Webb DJ (1993b) Venoconstriction to endothelin-1 in humans: role of calcium and potassium channels. Am J Physiol 265:H1676–1681

Haynes WG, Hand MF, Johnstone HA, Padfield PL, Webb DJ (1994a) Direct and sympathetically mediated venoconstriction in essential hypertension: enhanced response to endothelin. J Clin Invest 94:1359–1364

Haynes WG, Webb DJ (1994b) Contribution of endogenous generation of endothelin-1 to basal vascular tone. Lancet 344:852–854

Haynes WG, Moffat S, Webb DJ (1995a) An investigation into the direct and indirect venoconstrictor effects of endothelin-1 and big endothelin-1 in man. Br J Clin Pharmacol 40:307–311

Haynes WG, Strachan FE, Webb DJ (1995b) Endothelin ETA and ETB receptors cause vasoconstriction of human resistance and capacitance vessels in vivo. Circulation 92:357–363

Haynes WG, Ferro CJ, O'Kane KPJ, Somerville D, Lomax CC, Webb DJ (1996) Systemic endothelin receptor blockade decreases peripheral vascular resistance and blood pressure in man. Circulation 93:1860–1870

Hocher B, Thone-Reineke C, Bauer C, Raschack M, Neumayer HH (1997) The paracrine endothelin system: pathophysiology and implications in clinical medicine. Eur J Clin Chem Clin Biochem 35:175–189

Ihara M, Noguchi K Saeki T, Fukuroda T, Tsuchida S, Kimura S, Fukami T, Ishikawa K, Nishikibe M, Yano M (1992) Biological profiles of highly potent novel endothelin antagonists selective for the ETA receptor. Life Sci 50:247–255

Ishikawa K, Ihara M, Noguchi K, Mase T, Mino N, Saeki T, Fukuroda T, Fukami T, Ozaki S, Nagase T, Nishikibe M, Yano M (1994) Biochemical and pharmacological profile of a potent and selective endothelin B-receptor antagonist, BQ-788. Proc Natl Acad Sci USA 91:4892–4896

Jensen L, Yakimets J, Teo KK (1995) A review of impedance cardiography. Heart Lung 24:183–93

Jilma B, Szalay T, Dirnberger E, Eichler HG, Stohlawetz P, Schwarzinger I, Kapiotis S, Wagner OF (1997) Effects of endothelin-1 on circulating adhesion molecules in man. Eur J Clin Invest 27:850–856

Kaasjager KA, Shaw S, Koomans HA, Rabelink TJ (1997) Role of endothelin receptor subtypes in the systemic and renal responses to endothelin-1 in humans. J Am Soc Neph 8:32–39

Kiely DG, Cargill RI, Struthers AD, Lipworth BJ (1997) Cardiopulmonary effects of endothelin-1 in man. Cardiovasc Res 33:378–386

Kikuchi T, Ohtaki T, Kawata A, Imada T, Asami T, Masuda Y, Sugo T, Kusumoto K, Kubo K, Watanabe T, Wakimasu M, Fujino M (1994) Cyclic hexapeptide endothelin receptor antagonists highly potent for both receptor subtypes ETA and ETB. Biochem Biophys Res Commun 200:1708–1712

Kiowski W, Luscher TF, Linder L, Buhler FR (1990) Endothelin –1-induced vasoconstriction in humans, reversal by calcium channel blockade but not by nitrovasodilators or endothelium-derived relaxing factor. Circulation 83:469–475

Kiowski W, Sutsch G, Hunziker P, Muller P, Kim J, Oechslin E, Schmitt R, Jones R, Bertel O (1995) Evidence for endothelin-1 mediated vasoconstriction in severe chronic heart failure. Lancet 346:732–736

Krum H, Viskoper RJ, Lacourciere Y, Budde M, Charlon V (1998) The effect of an endothelin-receptor antagonist, bosentan, on blood pressure in patients with essential hypertension. N Engl J Med 338:784–790

Kubicek WG, Karnegis JN, Patterson RP, Witsoe DA, Mattson RH (1966) Development and evaluation of an impedance cardiac output system. Aerospace Med 37:1208–1212

Kyriakides ZS, Bofilis E, Tousoulis D, Antoniadis A, Kremastinos DTh, Webb DJ (1998) Endothelin-A receptor antagonist BQ123 provokes coronary vasodilatation and blood flow increases in humans (abstract) J Am Coll Cardiol 31

Lampl Y, Fleminger G, Gilad R, Galron R, Sarova-Pinhas I, Sokolovsky M (1997) Endothelin in cerebrospinal fluid and plasma of patients in the early stage of ischemic stroke. Stoke 28:1951–1955

Love MP, Ferro CJ, Haynes WG, Webb DJ, McMurray JJ (1996a) Selective or non-selective endothelin receptor blockade in chronic heart failure? Circulation 94(Suppl I):2899–2900

Love MP, Haynes WG, Gray GA, Webb DJ, McMurray JJ (1996b) Vasodilator effects of endothelin-converting enzyme inhibition and endothelin ETA receptor blockade in chronic heart failure patients treated with ACE inhibitors. Circulation 94:2131–2137

Love MP, Haynes WG, Webb DJ, McMurray JJ (1996c) Endothelin receptor function in the capacitance vessels of patients with chronic heart failure. Circulation 94(Suppl I):2899–2900

Moreland S, McMullen DM, Delaney CL, Lee VG, Hunt JT (1992) Venous smooth muscle contains vasoconstrictor ET-B like receptors. Biochem Biophys Res Commun 184:100–106

McCulloch KM, MacLean MR (1995) Endothelin B receptor-mediated contraction of human and rat pulmonary resistance arteries and the effect of pulmonary hypertension on endothelin responses in the rat. J Cardiovasc Pharmacol 26 [Suppl 3]:S169–S176

McCulloch KM, Docherty C, MacLean MR (1998) Endothelin receptors mediating contraction of rat and human pulmonary resistance arteries: effect of chronic hypoxia in the rat. Brit J Pharmacol 123:1621–1630

Nachev C, Collier J, Robinson B (1971) Simplified method for measuring compliance of superficial veins. Cardiovasc Res 5:147–156

Nelson JB, Hedican SP, George DJ, Reddi AH, Piantadosi S, Eisenberger MA, Simon JW (1995) Identification of endothelin-1 in the pathophysiology of metastatic adenocarcinoma of the prostate. Nature Med 9:944–949

Nilsson GE, Tenland T, Oberg PA (1980) A new instrument for continuous measurement of tissue blood flow by light beating spectroscopy. IEEE Trans Biomed Eng BME-27:12–19

Nord EP (1997) Role of endothelin in acute renal failure. Blood Purification 15:273–285

Packer M, Caspi A, Charlon V, Cohen-Solal A, Kiowski W, Kostuk W, Krum H, Levine B, Massie B, McMurray J, Rizzon P, Swedberg K (1998) Multicenter, double-blind, placebo controlled study of long-term endothelin blockade with bosentan in chronic heart failure – results of the REACH-1 trial (abstract). Circulation 98 [Suppl 1]:3

Pernow J, Kaijser L, Lundberg JM, Ahlborg G (1996) Comparable potent coronary constrictor effects of endothelin-1 and big endothelin-1 in humans. Circulation 94:2077–2082

Peter MG, Davenport AP (1996) Characterization of the endothelin receptor selective agonist, BQ3020 and antagonists BQ123, FR139317, BQ788, 50235, Ro462005 and bosentan in the heart. Br J Pharmacol 117:455–462

Plumpton C, Ferro CJ, Haynes WG, Webb DJ, Davenport AP (1996) The increase in human plasma immunoreactive endothelin but not endothelin-1 or its C-terminal fragment induced by systemic administration of the endothelin antagonist TAK-044. Br J Pharmacol 119:311–314

Rabelink TJ, Kaasjager KAH, Boer P, Stroes EG, Braam B, Koomans HA (1994) Effects of endothelin-1 on renal function in humans: implications for physiology and pathophysiology. Kidney Int 46:376–381

Schmetterer L, Dallinger S, Bobr B, Selenko N, Eichler HG, Wolzt M (1998) Systemic and renal effects of an ETA receptor subtype-specific antagonist in healthy subjects. Br J Pharmacol 124:930–934

Soper CPR, Latif AB, Bending MR (1996) Amelioration of hepatorenal syndrome with selective endothelin-A antagonist. Lancet 347:1842–1843

Sorensen SS, Madsen JK, Pedersen EB (1994) Systemic and renal effects of intravenous infusion of endothelin-1 in healthy volunteers. Am J Physiol 266:F411–418

Spratt JCS, Goddard J, Labinjoh C, Strachan F, Patel N, Webb DJ (1999) The haemodynamic effects of systemic endothelin A receptor antagonism in healthy humans in vivo. Br J Clin Pharmacology 47:575P-620P

Strachan FE, Spratt JC, Wilkinson IB, Johnston NR, Gray GA, Webb DJ (1998) Systemic blockade of the ETB receptor increases peripheral vascular resistance in healthy men. Hypertension 33:581–585

Sumner MJ, Cannon TR, Mundin JW, White DG, Watts IS (1992) Endothelin ET-A and ET-B receptors mediated vascular smooth muscle contraction. Br J Pharmacol 107:858–860

Sutsch G, Bertel O, Kiowski W (1997) Acute and short-term effects of the nonpeptide endothelin-1 receptor antagonist bosentan in humans. Cardiovasc Drugs Ther 10:171–725

Sutsch G, Kiowski W, Yan XW, Hunziker P, Christen S, Strobel W, Kim JH, Rickenbacher P, Bertel O (1998) Short-term oral endothelin-receptor antagonist therapy in conventionally treated patients with symptomatic severe chronic heart failure. Circulation 98:2262–2268

Takayanagi R, Kitazumi K, Takasaki C, Ohnaka K, Aimoto S, Tasaka K, Ohashi M, Nawata H (1991) Presence of a non-selective type of endothelin receptor on vascular endothelium and its linkage to vasodilation. FEBS Letters 282:103–106

Thomas SHL (1992) Impedance cardiography using the Sramek-Bernstein method: accuracy and reproducibility at rest and during exercise. Br J Clin Pharmacol 34:467–476

Uosaki Y, Yoshida M, Ogawa T, Saitoh YJ (1996) RES 1149–1 and –2, novel nonpeptidic endothelin B receptor antagonist produced by Aspergillus sp. II. Structure derivation. J Antibiotics 49:6–12

Verhaar MC, Strachan FE, Newby DE, Cruden NL, Koomans HA, Rabelink TJ, Webb DJ (1998) Endothelin-A receptor antagonist mediated vasodilation is attenuated by inhibition of nitric oxide synthesis and by endothelin-B receptor blockade. Circulation 97:752–756

Vierhapper H, Wagner O, Nowotny P, Waldhausl W (1990) Effects of endothelin-1 in man. Circulation 81:1415–1418

Vierhapper H, Wagner OF, Nowotny P, Waldhausl W (1991) Effect of endothelin-1 in man: pretreatment with nifedipine, with indomethacin and with cyclosporine A. Eur J Clin Invest 22:55–59

Vierhapper H, Nowotny P, Waldhausl W (1995) Effects of endothelin-1 in man: impact on basal and adrenocorticotropin-stimulated concentrations of aldosterone. J Clin Endocrinol Metab 80:948–951

Wang A, Bashore T, Holcslaw T, Freed M, Shusterman N (1998) Randomized prospective double blind multicenter trial of an endothelin receptor antagonist in the prevention of contrast nephrotoxicity. J Am Soc Nephrol 7:574A

Watanabe Y, Naruse M, Monzen C, Naruse K, Ohsumi K, Horiuchi J et al. (1991) Is big endothelin converted to endothelin-1 in circulating blood? J Cardiovasc Pharmacol 17 Suppl 7:S503-S505

Webb DJ (1995) The pharmacology of human blood vessels in vivo. J Vasc Res 32:2–15

Weber C, Schmitt R, Birnboeck H, Hopfgartner G, van Marle SP, Peeters PAM, Jonkman JHG, Jones CR (1996) Pharmacokinetics and drug disposition: pharmacokinetics and pharmacodynamics of the endothelin-receptor antagonist bosentan in healthy human subjects. Clin Pharmacol Ther 60:124–137

Weitzberg E, Ahlborg G, Lundberg JM (1991) Long-lasting vasoconstriction and efficient regional extraction of endothelin-1 in human splanchnic and renal tissues. Biochem Biophys Res Commun 180:1298–1303

Wenzel RR, Noll G, Luscher TF (1994) Endothelin receptor antagonists inhibit endothelin in human skin microcirculation. Hypertension 23:581–586

Wenzel RR, Duthiers N, Noll G, Bucher J, Kaufmann U, Luscher TF (1996) Endothelin and calcium antagonists in the skin microcirculation of patients with coronary artery disease. Circulation 94:316–322

Wenzel RR, Zbinden S, Noll G, Meier B, Luscher TF (1998a) Endothelin-1 induces vasodilation in human skin by nociceptor fibres and release of nitric oxide. Br J Clin Pharmacol 45:441–446

Wenzel RR, Fleisch M, Shaw S, Noll G, Kaufmann U, Schmitt R, Jones R, Clozel M, Meier B, Luscher TF (1998b) Hemodynamic and coronary effects of the endothelin antagonist bosentan in patients with coronary artery disease. Circulation 98:2235–2240

Williams DL, Jones KL, Pettibone DJ, Lis EV, Clineschmidt BV (1991) Sarafotoxin S6c: an agonist which distinguishes between endothelin receptor subtypes. Biochem Biophys Res Commun 175:556–561

Yanagisawa M, Kurihara H, Kimura S, Tomobe Y, Kobayashi M, Mitsui Y, Yazaki Y, Goto K, Masaki T (1988) A novel potent vasoconstrictor peptide produced by vascular endothelial cells. Nature 332:411–415

Subject Index